Plant Phenolics in Sustainable Agriculture

Rafiq Lone • Razia Shuab • Azra N. Kamili
Editors

Plant Phenolics in Sustainable Agriculture

Volume 1

Springer

Editors
Rafiq Lone
Department of Botany
Sant Baba Bhag Singh University
Jalandhar, Punjab, India

Razia Shuab
Department of Botany
University of Kashmir
Srinagar, Jammu and Kashmir, India

Azra N. Kamili
Center of Research for Development
University of Kashmir
Srinagar, Jammu and Kashmir, India

ISBN 978-981-15-4889-5 ISBN 978-981-15-4890-1 (eBook)
https://doi.org/10.1007/978-981-15-4890-1

This Springer imprint is published by the registered company Springer Nature Singapore Pte Ltd.
The registered company address is: 152 Beach Road, #21-01/04 Gateway East, Singapore 189721, Singapore

Preface

Ever since the evolution of mankind on Earth, their interaction with plants has never ceased. In fact it will be fit to say that man co-evolved with plants. If the only living planet draws its sustenance from the plants, then mankind cannot escape their absolute dependence on anything and everything that plants produce and provide them with – be it food, clothing, shelter, health, entertainment, luxury, wealth, etc.

In natural ecosystems, plants face a plethora of antagonists that necessitate the possession of many types of defenses, and as a result plants have evolved multiple defense mechanisms by which they are able to survive various kinds of biotic and abiotic stresses by adaptation. This displays a tremendous metabolic plasticity that is well illustrated by their ability to synthesize masses of so-called secondary organic compounds ('natural products') that seems to be crucial for growth and development

One of the major secondary plant metabolites are the phenolics. Approximately 8000 naturally occurring compounds belong to the different categories of phenolics. The ability of plant phenolics to provide resistance against various abiotic and biotic stresses and their role in sustainable agriculture has been explored in this edition.

Phenolics being ubiquitous have been extracted from every part of plants. In spite of their wide-spread presence in plant kingdom, understanding of their physiological role in plants is very limited. Some investigators believe that phenolics function as inhibitory while some believe they are promoters to growth and developmental phenomena. However, some researchers claim against any basic role of phenolic constituents in plant growth regulation.

Every plant's biological activity is influenced by more than one hormone, thus the biological phenomenon often reproduces the combined interaction of several different hormones. Though salicylates have now almost been established for having growth and developmental functions in plants and being recognized as plant growth regulators, other phenolics like cinnamic acid despite their ubiquitous presence are yet to be identified as one of any importance in plant growth regulation.

The first edition of the book *Plant Phenolics* gives exemplary insight into the advancements in plant phenolics research. This edition comprehensively sheds light on recent studies in the plant phenolics field. Efficient management of plant phenolics has the potential to support sustainable agriculture ensuring at the same time environmental quality for future generations. It is hoped that this first edition will interest readers in the latest outcomes of plant phenolics research and also encourage young researchers to prove the challenging field of these studies. The book demonstrates different evolutionary developments with modern concepts from different academic and research perspectives to achieve sustainable use of plant phenolics. This volume consists of 25 chapters covering the diverse and different role of plant phenolics in sustainable agriculture by 86 eminent academicians, learned researchers and subject specialists. We are grateful to the scholarly and learned personalities who have lent unconditional help to bring this volume to light. Finally, special thanks are due to our families for their valuable and unconditional support.

Jalandhar, Punjab, India — Rafiq Lone
Srinagar, Jammu and Kashmir, India — Razia Shuab
Srinagar, Jammu and Kashmir, India — Azra N. Kamili

Contents

Editors and Contributors

About the Editors

Dr. Rafiq Lone is currently an Assistant Professor of Botany at SBBS University, India. He completed his M.Sc and Ph.D. in Botany at Jiwaji University, India and has worked as a lecturer (academic arrangement) in higher education in Jammu and Kashmir, as a teaching assistant at the School of Studies in Botany, Jiwaji University, and as a fellow in a project funded by MPCST-Bhopal-(M.P)-India. Dr Lone has 5 years of teaching and 10 years of research experience, and has published 25 research papers in national and international journals, as well as 14 book chapters. Dr Lone has attended 25 national/international conferences and has presented papers on plant microbe interaction, plant nutrition and plant phenolics. He has received numerous awards. His research interests include on plant-microbe interaction, plant nutrition, plant phenolics and invasion biology.

Dr. Razia Shuab completed her M.Sc., M.Phil. and Ph.D. in Botany at Jiwaji University, India, and received an N-PDF (SERB-DST, GOI) from the University of Kashmir, Jammu and Kashmir. Dr Razia has 1 year of teaching and 7 years of research experience. She has published 18 research papers in national and international journals, as well as 5 book chapters. She has attended 12 national/international conferences and has presented papers on plant phenolics and invasion biology.

Prof. Azra N. Kamili is currently a Professor of Botany at the Center of Research for Development, University of Kashmir Srinagar, Jammu and Kashmir, India. She completed her M.Phil. and Ph.D. in Botany at the University of Kashmir and has significantly contributed to various fields of botany, and her current research focuses on plant tissue culture, microbial ecology, stress biology, plant microbe interaction and molecular biology. She has 30 years of teaching and 35 years of research experience. She has attended 30 national and international conferences and is /has been the principal investigator /co-principal investigator/mentor of 16 major research projects supported by various funding agencies, like DST, ICAR, DBT and UGC. She has also published 8 books and over 170 research papers in respected journals. She was a member of the Syndicate of the University of Kashmir, Jammu and Kashmir, of the J&K Biodiversity board and of the Academic council, SKUAST-K.

Contributors

Md. Minhajul Abedin Department of Microbiology, School of Life Sciences, Sikkim University, Gangtok, Sikkim, India

Ambreen Ahmed Department of Botany, University of the Punjab, Lahore, Pakistan

Dulce L. Ambriz-Pérez Escuela de Ingeniería y Ciencias, Tecnologico de Monterrey, Monterrey, NL, Mexico

M. Aravind Department of Plant Sciences, Central University of Kerala, Kasaragod, Kerala, India

Dalia Aseel Plant Protection and Biomolecular Diagnosis Department, Arid Lands Cultivation Research Institute, City of Scientific Research and Technological Applications, Alexandria, Egypt

Sara Aviles-Gaxiola Department of Food Science and Technology, Centro de Investigación en Alimentación y Desarrollo, A.C., Culiacán, Sinaloa, México

Abdelfattah Badr Botany and Microbiology Department, Faculty of Science, Helwan University, Cairo, Egypt

Mudasir Bashir Department of School Education, Jammu and Kashmir, India

Sheikh Ummar Bashir Department of Botany, SGB Amravati University, Amravati, Maharashtra, India

Monika Bhadauria Toxicology and Pharmacology Laboratory, Department Of Zoology, Guru Ghasidas Vishwavidyalaya, Bilaspur, Chhattisgarh, India

Bharat Bhushan ICAR- Indian Institute of Maize Research, Ludhiana, Punjab, India

David J. Burritt Department of Botany, University of Otago, Dunedin, New Zealand

Luis A. Cabanillas-Bojórquez Department of Food Science and Technology, Centro de Investigación en Alimentación y Desarrollo, A.C., Culiacán, Sinaloa, México

Prince Choyal Directorate of Medicinal and Aromatic Plants Research, Anand, Gujarat, India

Laura A. Contreras-Angulo Department of Food Science and Technology, Centro de Investigación en Alimentación y Desarrollo, A.C., Culiacán, Sinaloa, México

Saurav Das Department of Agronomy and Horticulture, University of Nebraska, Lincoln, NE, USA

Sophia Dominic Department of Biochemistry, Jyoti Nivas Degree College, Bangalore, Karnataka, India

C. Alicia Elizalde-Romero Department of Food Science and Technology, Centro de Investigación en Alimentación y Desarrollo, A.C., Culiacán, Sinaloa, México

Hanaa H. El-Shazly Biological and Geological sciences Department, Faculty of Education, Ain Shams University, Cairo, Egypt

Zeeshan Fatima Amity Institute of Biotechnology, Amity University Haryana, Gurugram, Haryana, India

Ajay Kumar Gautam Department of Plant Sciences, Central University of Punjab, Bathinda, Punjab, India

Ranjana Gautum Department of Plant Sciences, University of Hyderabad, Hyderabad, Telangana, India

Vinoth Prasanna Gunasekaran Department of Biomedical Science, Bharathidasan University, Tiruchirappalli, Tamil Nadu, India

O. P. Gupta ICAR-Indian Institute of Wheat and Barley Research, Karnal, Haryana, India

Erick P. Gutiérrez-Grijalva Cátedras CONACYT – Centro de Investigación en Alimentación y Desarrollo, AC, Culiacán, Sinaloa, México

Sabiha Habib Department of Botany, University of the Punjab, Lahore, Pakistan

Younis Ahmad Hajam Department of Biosciences, Division Zoology, Career Point University, Hamirpur, Himachal Pradesh, India

Saif Hameed Amity Institute of Biotechnology, Amity University Haryana, Gurugram, Haryana, India

Saad Hammad Plant Protection and Biomolecular Diagnosis Department, Arid Lands Cultivation Research Institute, City of Scientific Research and Technological Applications, Alexandria, Egypt

J. Basilio Heredia Department of Food Science and Technology, Centro de Investigación en Alimentación y Desarrollo, A.C, Culiacán, Sinaloa, México

Tahsina Sharmin Hoque Department of Soil Science, Bangladesh Agricultural University, Mymensingh, Bangladesh

Mohammad Anwar Hossain Department of Genetics and Plant Breeding, Bangladesh Agricultural University, Mymensingh, Bangladesh

Charanjit Kaur ICAR- Indian Agricultural Research Institute, New Delhi, India

Mahesh Kumar ICAR- National Institute of Abiotic Stress Management, Baramati, Maharashtra, India

Manoj Kumar ICAR-Central Institute for Research on Cotton Technology, Mumbai, Maharashtra, India

Pavan Kumar Department of Livestock Products Technology, College of Veterinary Science, Guru Angad Dev Veterinary and Animal Sciences University, Ludhiana, Punjab, India

Rajesh Kumar Department of Zoology, School of Basic and Applied Sciences, Career Point University, Hamirpur, Himachal Pradesh, India

Santosh Kumar Department of Microbiology, School of Life Sciences, Sikkim University, Gangtok, Sikkim, India

Satish Kumar ICAR- National Institute of Abiotic Stress Management, Baramati, Maharashtra, India

Rafiq Lone Department of Botany, Sant Baba Bhag Singh University, Jalandhar, Punjab, India

Leticia X. López-Martínez Cátedras CONACYT – Centro de Investigación en Alimentación y Desarrollo, AC, Hermosillo, Sonora, México

Om Prakash Malav Department of Livestock Products Technology, College of Veterinary Science, Guru Angad Dev Veterinary and Animal Sciences University, Ludhiana, Punjab, India

Javid Ahmad Malik Department of Zoology, Guru Ghasidas Vishwavidyalaya, Bilaspur, Chhattisgarh, India

Ofelia Marquez-Molina Centro Universitario UAEM Amecameca, Universidad Autónoma del Estado de México, Toluca, Estado de México, México

Kamlesh Kumar Meena ICAR- National Institute of Abiotic Stress Management, Baramati, Maharashtra, India

Nand Lal Meena ICAR-National Bureau of Plant Genetic Resources, New Delhi, India

Nitin Mehta Department of Livestock Products Technology, College of Veterinary Science, Guru Angad Dev Veterinary and Animal Sciences University, Ludhiana, Punjab, India

Mohammad Yaseen Mir Centre of Research for Development, University of Kashmir, Srinagar, Jammu and Kashmir, India

Heba I. Mohamed Biological and Geological sciences Department, Faculty of Education, Ain Shams University, Cairo, Egypt

Gabriela Olivo-Vázquez Department of Food Science and Technology, Centro de Investigación en Alimentación y Desarrollo, A.C., Culiacán, Sinaloa, México

Chella Perumal Palanisamy Department of Plant Sciences, University of the Free State, Phuthaditjhaba, QwaQwa, South Africa

Rachna Pande ICAR-Central Institute for Cotton Research, Nagpur, Maharashtra, India

Jayashree Potkule ICAR- Central Institute for Research on Cotton Technology, Mumbai, Maharashtra, India

Shah Rafiq Plant Tissue Culture Laboratory, Department of Botany, University of Kashmir, Srinagar, Jammu and Kashmir, India

Seema Rai Department of Zoology, Guru Ghasidas Vishwavidayalaya, Bilaspur, Chhattisgarh, India

Jagadish Rane ICAR- National Institute of Abiotic Stress Management, Baramati, Maharashtra, India

Younes Rashad Plant Protection and Biomolecular Diagnosis Department, Arid Lands Cultivation Research Institute, City of Scientific Research and Technological Applications, Alexandria, Egypt

Gulab Khan Rohela Biotechnology Section, Moriculture Division, Central Sericultural Research & Training Institute, Central Silk Board, Ministry of Textiles, Government of India, Pampore, Jammu and Kashmir, India

Liliana Santos-Zea Maestría en Ciencias Aplicadas, Universidad Politécnica de Sinaloa, Mazatlán, Sinaloa, México

Aftab Ahmad Shabnam Mulberry Breeding & Genetics Section, Moriculture Division, Central Sericultural Research & Training Institute, Central Silk Board, Ministry of Textiles, Government of India, Pampore, Jammu and Kashmir, India

Gaurav Sharma ICAR- Accredited School of Agriculture, Lovely Professional University, Phagwara, Punjab, India

Pawan Shukla Biotechnology Section, Moriculture Division, Central Sericultural Research & Training Institute, Central Silk Board, Ministry of Textiles, Government of India, Pampore, Jammu and Kashmir, India

Anil Kumar Singh Department of Botany, Sant Baba Bhag Singh University, Khiala, Punjab, India

Ashish Kumar Singh Department of Microbiology, School of Life Sciences, Sikkim University, Gangtok, Sikkim, India

Pramod Kumar Singh Department of Biosciences, Christian Eminent College, Indore, Madhya Pradesh, India

Shweta Singh Amity Institute of Biotechnology, Amity University Haryana, Gurugram, Haryana, India

Poonam Singla Department of Chemistry and Centre for Advance Studies in Chemistry, Panjab University, Chandigarh, India

Abdullah Al Manum Sohag Department of Biochemistry and Molecular Biology, Bangladesh Agricultural University, Mymensingh, Bangladesh

Sonia School of Pharmaceutical and Healthcare Sciences, CT University, Ludhiana, Punjab, India

Yamini Tak Agriculture University, Kota, Rajasthan, India

Aqsa Tariq Department of Botany, University of the Punjab, Lahore, Pakistan

Maharishi Tomar Indian Grassland and Fodder Research Institute, Jhansi, Uttar Pradesh, India

Kamal Tyagi Department of Viticulture and Enology, University of California, Davis, Davis, CA, USA

Pramila Umaraw Department of Livestock Products Technology, College of Veterinary & Animal Sciences, Sardar Vallabhbhai Patel University of Agriculture & Technology, Meerut, Uttar Pradesh, India

Atul Kumar Upadhyay Department of Biotechnology, Thapar Institute of Engineering and Technology, Patiala, Punjab, India

Department of Biotechnology, Thapar institute of engineering and technology, Patiala, Punjab, India

Gabriela Vázquez-Olivo Department of Food Science and Technology, Centro de Investigación en Alimentación y Desarrollo, A.C., Culiacán, Sinaloa, México

Akhilesh Kumar Verma Department of Livestock Products Technology, College of Veterinary & Animal Sciences, Sardar Vallabhbhai Patel University of Agriculture & Technology, Meerut, Uttar Pradesh, India

Pooja Verma ICAR-Central Institute for Cotton Research, Nagpur, Maharashtra, India

Nasir Aziz Wagay Botany Research Laboratory, Department of Botany, Vidyabharati Mahavidyalya College, Amravati, Maharashtra, India

Goraksha Chimaji Wakchaure ICAR- National Institute of Abiotic Stress Management, Baramati, Maharashtra, India

Anshul Watts ICAR- Indian Institute of Plant Biotechnology, New Delhi, India

Tran Dang Xuan Graduate School for Advanced Science and Engineering, Transdisciplinary Science and Engineering Program, Hiroshima University, Hiroshima, Japan

Salicylic Acid-Mediated Salt Stress Tolerance in Plants

1

Tahsina Sharmin Hoque, Abdullah Al Manum Sohag,
David J. Burritt, and Mohammad Anwar Hossain

Abstract

Salt stress is one of the most important abiotic stresses threatening agricultural production worldwide. Salt stress affects vital physiological, biochemical and molecular processes in crop plants, leading to reduced plant growth and yields or even plant death. To cope with salt stress, plants have evolved many adaptive mechanisms, including the development of salt-associated signal transduction cascades that contain a wide range of second messengers. During stress, regulatory molecules, including plant hormones, play key roles in controlling developmental processes and signalling networks, and these molecules have been recognized as having the potential to be used to develop stress-tolerant plants. Salicylic acid (SA) is a phenolic compound involved in the regulation of plant growth, development and defence responses. SA is a critical signalling molecule that is known to participate in the responses of plants to salinity stress, through extensive signalling crosstalk with other hormones that results in physiological and biochemical responses in plants and changes in gene expression. SA is an important regulator of Na^+ exclusion and sequestration, through the modulation of sodium and potassium transporters, and is associated with the control of photosynthesis and nutrient metabolism, proline and glycinebetaine synthesis,

T. S. Hoque
Department of Soil Science, Bangladesh Agricultural University, Mymensingh, Bangladesh

A. A. M. Sohag
Department of Biochemistry and Molecular Biology, Bangladesh Agricultural University, Mymensingh, Bangladesh

D. J. Burritt
Department of Botany, University of Otago, Dunedin, New Zealand

M. A. Hossain (✉)
Department of Genetics and Plant Breeding, Bangladesh Agricultural University, Mymensingh, Bangladesh
e-mail: anwargpb@bau.edu.bd

R. Lone et al. (eds.), *Plant Phenolics in Sustainable Agriculture*,
https://doi.org/10.1007/978-981-15-4890-1_1

reactive oxygen species metabolism, and plant–water relations, in plants under salt stress. Furthermore, applying SA has been shown to improve plant tolerance to salinity by regulating multiple stress-responsive pathways and processes. Recent studies with transgenic and mutant plants have shown the diverse roles SA in plant stress biology. This chapter summarizes the current knowledge of the roles of SA in salinity tolerance, the responses of plants to salt, and the potential mechanisms underlying SA-mediated salinity tolerance in plants.

Keywords

Salicylic acid · Salt stress · Signalling · Stress responses · Stress tolerance · Hormones · Transgenic plants

1.1 Introduction

Global agricultural crop production is severely affected by a variety of abiotic and biotic stressors. Salinity is one of the most serious abiotic stressors that restricts crop yields in many regions of the world, affecting approximately 20% of the world's total land area and negatively impacting more than 45 million ha of irrigated area (Pitman and Lauchli 2002; Munns and Tester 2008). Salinization of arable land is an ongoing process that is occurring as a result of climate change and also due to poor agricultural practices associated with irrigation and/or land drainage. Scientists have already predicted the possible devastating effects that salt stress could have on food production by the middle of the twenty-first century, which could cause a 50% loss of cultivable land (Mahajan and Tuteja 2005) and cost the agricultural sectors billions of dollars (Qadir et al. 2014). In order to meet the demand for food from an ever-increasing world population, it is necessary that salt-tolerant cultivars of our most important crop plants be developed by plant breeding. However, to do this an understanding plant salinity stress and tolerance mechanisms is required (Jayakannan et al. 2015a).

Plants are sessile organisms and salt stress can be lethal as it disturbs their vital physiological processes (Tavakkoli et al. 2010). High levels of salinity induce primary effects, such as hyperionic and hyperosmotic stress, that lead to plant death (Roy et al. 2014; Nimir et al. 2017). Salt stress can also cause many secondary effects such as reduced cell growth, membrane damage, nutrient imbalances, altered production of reactive oxygen species (ROS), reactive nitrogen species (RNS), reactive carbonyl species (RCS), enzyme inhibition and metabolic dysfunction, and lower photosynthetic rates, etc. (Sudhir and Murthy 2004; Mahajan and Tuteja 2005; Hossain et al. 2009, 2011; El-Shabrawi et al. 2010; Upadhyaya et al. 2011; Hasanuzzaman et al. 2012; Kumar 2013; Mostofa et al. 2015; Acosta-Motos et al. 2017; Tahjib-Ul-Arif et al. 2018a; Gupta et al. 2018; Hoque et al. 2018). To cope with salinity, plants employ a wide range of adaptations and mitigation strategies involving the initiation and modulation of signal transduction pathways (Zhu 2002; Tuteja and Sopory 2008; Hoque et al. 2018). Phytohormones and associated signalling networks play vital roles in regulating plant growth and developmental processes and, thus, they have direct and indirect involvement in plant stress responses and stress tolerance (Khan et al. 2012a, b, c; Asgher et al. 2015). Salicylic acid (SA; 2-hydroxybenzoic acid) is a phenolic phytohormone and an important signalling compound that regulates different stress responses in plants often by interacting with other phytohormones (Achard et al. 2006; Spoel and Dong 2008; Tuteja and Sopory 2008; Vlot et al. 2009; Wolters and Jürgens 2009; Horváth et al. 2007; Asensi-Fabado and Munné-Bosch 2011). SA has been shown to play a significant role not only in the regulation of plant physiological processes like photosynthesis, transpiration, ion uptake and transport, nutrient metabolism, osmolyte production and antioxidant defence (Syeed and Khan 2010; Khan et al. 2010, 2012a, b, c, 2013, 2014; Nazar et al. 2011; Miura and Tada 2014), but also in the modulation of plant adaptive and defence responses to abiotic stressors, including salinity and osmotic stress (Borsani et al. 2001; Singh and Usha 2003; Khodary 2004; Naser Alavi et al. 2014).

The importance of SA in plant stress tolerance has been well established, and many recent studies have identified mutant plants with altered SA synthesis and accumulation, which show altered salinity tolerance (Borsani et al. 2001; Cao et al. 2009; Asensi-Fabado and Munné-Bosch 2011; Miura et al. 2011; Hao et al. 2012). Also, exogenous application of SA to various species of plants under salinity stress has shown to mitigate the toxicity effects of salt by upregulation of salinity tolerance mechanisms (Horváth et al. 2007; Ashraf et al. 2010; Hayat et al. 2010; Palma et al. 2013; Khan et al. 2014; Lee et al. 2014; Ardebili et al. 2014). Although progress has been made to explore the novel roles of SA in salt-stressed plants, the fundamental mechanisms underlying SA-mediated salt stress tolerance still need to be elucidated. In this chapter, we discuss the synthesis of SA in plants under salt stress and focus on the involvement of SA in salt stress responses, the signalling networks of SA and the potential mechanisms of SA-mediated salt stress tolerance in plants. We evaluate the recent findings of exogenous SA-mediated salt stress tolerance in connection to multiple stress tolerance mechanisms and stress responsive pathways.

1.2 Biosynthesis of Salicylic Acid in Plants in Response to Salt Stress

Salicylic acid is a lipophilic beta-hydroxy phenolic compound that is ubiquitous in plants and can function as a phytohormone. Plant phenolics are mainly synthesized via the shikimic acid pathway (Khan et al. 2015) and though this pathway, carbohydrate precursors (obtained from glucose and pentose phosphate pathway) are converted into aromatic amino acids like phenylalanine, which is the precursor of SA (Herrmann and Weaver 1999). In plants, two major pathways are known to be involved in the synthesis of SA. The phenylalanine ammonia-lyase (PAL) pathway, which is present in cytoplasm and the isochorismate synthase (ICS) pathway, which is present in chloroplasts (Jayakannan et al. 2015a). Both the PAL and ICS pathways require chorismic acid, which is the end product of shikimic acid pathway (Fig. 1.1).

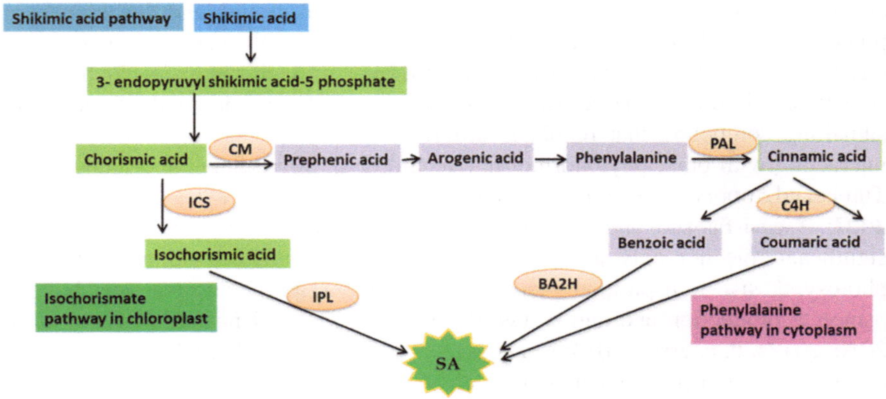

Fig. 1.1 Schematic representation of biosynthesis of salicylic acid in plants. Salicylic acid is synthesised by the shikimic acid pathway in the cytoplasm or the isochorismate pathway in chloroplasts. In the cytoplasm, chorismic acid is transformed into prephenic acid by the enzyme CM, which is then transformed into phenylalanine. Via the phenylalanine pathway, phenylalanine is then converted to cinnamic acid by the enzyme PAL. Cinnamic acid is then converted to benzoic acid, by a process linked to the β-oxidation of fatty acids, or transformed into coumaric acid by the action of the enzyme C4H. SA is then produced from benzoic acid by the enzyme BA2H or from hydroxylation of coumaric acid. In the isochorismate pathway, chorismic acid is converted to isochorismic acid by the enzyme ICS. Isochorismic acid is then converted into SA by the enzyme IPL. *BA2H* benzoic acid 2-hydroxylase, *C4H* cinnamate 4-hydroxylase, *CM* chorismate mutase, *ICS* isochorismate synthase, *IPL* isochorismate pyruvate lyase, *PAL* phenylalanine ammonia-lyase, *SA* salicylic acid

In the phenylalanine pathway, phenylalanine is used to produce cinnamic acid, via a non-oxidative deamination reaction, and this conversion is catalysed by the enzyme of PAL. Cinnamic acid conversion to benzoic acid can then occur by a non-oxidative route or the conversion can be linked to the β-oxidation of fatty acids (Verberne et al. 1999; Horváth et al. 2007; Mustafa et al. 2009). SA is then produced from benzoic acid by the action of benzoic acid 2-hydroxylase (BA2H). Alternatively cinnamic acid can be converted into coumaric acid by cinnamate 4-hydroxylase (C4H) and the coumaric acid then subsequently into SA. In the isochorismate pathway, SA is synthesized from chorismic acid in a two-step enzymatic (ICS and IPL) process with isochorismic acid as an intermediate (Verberne et al. 1999; Strawn et al. 2007).

Basal tissue SA levels vary between various plant species, for example in tomato (*Lycopersicum esculentum*), soybean (*Glycine max*) and rice (*Oryza sativa*) basal SA levels far exceed the higher levels associated with systemic acquired resistance in tobacco (*Nicotiana tabacum*) and Arabidopsis without apparent deleterious biological effects (Chen et al. 1997; Hammond-Kosack et al. 1996; Silverman et al. 1995; Raskin et al. 1990). It has been shown that endogenous SA levels increase dramatically in response to both biotic and abiotic stressors (Yalpain et al. 1994; Shim et al. 2003).

During salinity and osmotic stress, SA is involved in the oxidative responses of plant cells (Borsani et al. 2001) and endogenous SA levels can increase manifold, depending on plant species (Sawada et al. 2006). Although many recent studies have highlighted the function of SA as a major signalling molecule in plant responses to salt stress, the relationship between SA and the changes in metabolism in plants responding to salt stress remains obscure. Importantly, in different SA–biosynthetic pathways (PAL and ICS pathways), many enzymes play key roles in regulating a series of chemical reactions and their functions can be modulated by abiotic stressor including salinity. Endogenous SA levels are closely related to the activities of enzymes known to be necessary for salinity tolerance in plants. For example, overproduction of endogenous SA via enhanced biosynthesis of BA2H in rice (*Oryza sativa*) plants results in increased tolerance to salinity (Sawada et al. 2006). Recently, Su et al. (2018) reported that salinity stress resulted in an increase in PAL and ICS activities accompanied by an increase in SA levels in *Scutellaria baicalensis* plants. Higher activities of the enzymes ICS, C4H and PAL were also found to be associated with increased SA levels in *Carthamus tinctorius* plants exposed to salt stress, with a positive relationship between activities of these enzymes and with SA concentrations also observed (Sadeghi et al. 2013; Dehghan et al. 2014).

1.3 Mechanisms of Salt Stress Tolerance in Plants Related to Salicylic Acid

1.3.1 Regulation of Plant Developmental Processes

During salt stress, SA regulates seed germination and plant growth in a concentration-dependent manner, with the degree of regulation varying according to plant species. An application of SA reduced seed germination in maize (*Zea mays*) (Guan and Scandalios 1995), barley (*Hordeum vulgare*) (Xie et al. 2007) and Arabidopsis (*Arabidopsis thaliana*) (Nishimura et al. 2005; Lee et al. 2010) wild-type plants under normal condition, whereas in SA-deficient mutant plants SA application gave different results with respect to seed germination. The characterization study of Arabidopsis mutant plants with different SA levels showed its role in plant developmental processes during salinity. Rajjou et al. (2006) reported that exogenous SA application to SA-deficient *NahG* mutants increased seed germination under salinity stress. In SA-deficient *sid2* mutant plants, SA application improved seed germination under salinity stress when applied at concentrations up to 50 μM, but at 100 μM SA no improvement was observed (Lee et al. 2010). Increased seed germination due to SA application under salinity stress has also been reported for wheat (*Triticum aestivum*), chickpea (*Cicer arietinum*) and alfalfa (*Medicago sativa*) (Kaydan et al. 2007; Deef 2007; Asadi et al. 2013; Torabian 2010). The mechanism of SA improvement in seed germination under salinity stress might be attributed to the ROS homeostasis by SA which decreased H_2O_2 level and reduced oxidative damage (Lee et al. 2010). The role of SA in regulating salinity stress responses suggests that SA acts as a powerful tool in enhancing growth, development and productivity of plants under stress (Rajeshwari and Bhuvaneshwari 2017). In the case of Arabidopsis wild-type plants, SA-pretreatment improved shoot growth and water contents under salinity stress (Jayakannan et al. 2013), although the effect of altered endogenous SA levels in mutant plants is still not clear. Some researchers (Rivas-San Vicente and Plasencia 2011, Borsani et al. 2001; Cao et al. 2009; Hao et al. 2012) have demonstrated higher growth in SA-deficient mutant plants such as *sid2, eds5/sid1 and NahG* compared to SA-hyperaccumulating mutants such as *cpr1/5/6, acd1/5/6/11, dnd1/2, isd1, nudt7, agd2, snc1 and siz1* during salinity stress. These studies confirmed that plant growth is enhanced at low SA concentrations (less than 0.1 mM), while it is reduced at higher SA concentration (more than 1 mM) under salinity (Rivas-San Vicente and Plasencia 2011). The aforementioned phenomena could be due to the negative regulatory role of SA has on cell division and cell enlargement (Xia et al. 2009; Hao et al. 2012). However, in contrast, other studies have shown enhanced growth in SA-hyperaccumulating *siz1* and *NahG siz1* double mutants and reduced growth in SA-deficient *NahG*, *sid2* and *eds* mutants (Miura et al. 2011; Hao et al. 2012; Asensi-Fabado and Munné-Bosch 2011). According to Hao et al. (2012) SA-deficient *NahG* mutants exhibited higher chlorophyll contents and fluorescence ratios compared to SA-hyperaccumulating *snc1* mutants whereas SA-deficient mutants (*sid2* and *eds5*) and the SA-hyperaccumulating mutant *aba3* showed no variation in chlorophyll contents and fluorescence ratios under salinity stress (Asensi-Fabado and Munné-Bosch 2011). Thus, the mechanism by which SA regulates plant growth during salinity stress still need to be elucidated.

1.3.2 Regulation of Stomatal Conductance, Photosynthesis and Transpiration

Stomatal movement regulates gas exchange, transpiration and photosynthesis, which are all vital for plant adaptation to stressors, including salinity. Both endogenous and exogenous SA promote stomatal closure via production of ROS (Dong et al. 2001; Melotto et al. 2006; He et al. 2007) and stomatal closure can affect gas exchange. For example, the application of 0.4 mM SA induces stomatal closure within 2 h, which reduces gas exchange by several fold (Mateo et al. 2004; Rivas-San Vicente and Plasencia 2011). SA-induced stomatal closure was observed in *Arabidopsis thaliana*, *Vicia faba* and *Cucumis sativus* (Khokon et al. 2011; Miura et al. 2013; Miura and Tada 2014; Mori et al. 2001; Hao et al. 2011), and this closure is accompanied with extracellular ROS production mediated by salicylhydroxamic acid (SHAM)-sensitive guaiacol peroxidases, intracellular ROS accumulation in guard cell and K^+_{in} channel inactivation (Mori et al. 2001; Khokon et al. 2011). Recently, Khokon et al. (2017) demonstrated that two mitogen-activated protein kinases (MAPKs), MPK9 and MPK12, are involved in SA-induced stomatal closure in Arabidopsis. Prodhan et al. (2018) suggested that guard cell SA signalling is associated with abscisic acid (ABA) signalling via the Ca^{2+}/CPK (Ca^{2+}-dependent protein kinase)-dependent pathway whereas SA requires *CPK3* and *CPK6* protein kinases for induction of stomatal closure. Under salt stress, the SA-hyperaccumulating Arabidopsis mutant *siz1* showed increased stomatal closure, demonstrating the importance of SA-mediated stomatal closure for salinity responses (Miura et al. 2013). In addition, an Arabidopsis SA-hyperaccumulating double mutant *wrky54wrky70* showed osmotic stress tolerance, which was associated with increased stomatal closure and improved water retention (Li et al. 2013a).

1.3.3 Interaction with Nutrients

For growth, development and survival under salinity stress, maintaining good mineral nutrition is a fundamental requirement and the mineral nutrient status of plants plays a vital role in ameliorating the adverse effects of salt stress (Nazar et al. 2011; Nazar et al. 2015). The role of SA on nutrient acquisition depends on the SA concentration, the plant species, and the duration and intensity of salinity stress (Horváth et al. 2007). Notably, SA can modulate nutrient uptake and metabolism and thereby regulate membrane integrity and maintain ion homeostasis under salt stress (Alpaslan and Gunes 2001; Gunes et al. 2007; Tufail et al. 2013; Nazar et al. 2015). Maintenance of higher K^+/Na^+ and Ca^{2+}/Na^+ ratios by SA was considered a key factor for improved growth and yield, gas exchange and salt tolerance in plants and both can be influenced by SA (Tufail et al. 2013). Moreover, SA and Ca treatment can improve salinity tolerance in *Triticum aestivum* via increasing proline (Pro) levels (Al-Whaibi et al. 2012). According to a previous report by Kawano et al. (1998), SA can induce a rapid and transient generation of ROS, namely the superoxide anion, followed by an increase in cytosolic-free Ca^{2+}concentrations in *Nicotiana tabacum*. Salt stress can also induce the expression of calmodulin-binding protein (a Ca-containing protein), which was also reported to be induced by SA in Arabidopsis under multiple abiotic stresses, including salinity (Yang and Poovaiah 2002). Besides, calcium-dependent protein kinases (CDPKs) are also involved in abiotic stress responses, which might also be induced by SA (Chung et al. 2004).

1.3.4 Regulation of Membrane Transporters

During salt stress, activation of several transporters, ion channels and proton pumps is strongly associated with salt tolerance in plants (Kerkeb et al. 2001). Under salinity, the entry of Na^+ and its redistribution through the plasma membrane transporter high-affinity potassium transporter (HKT) and nonselective cation channels (NSCC) cause a strong membrane depolarization that favours K^+ leakage from the cytosol via depolarization-activated K^+ outward-rectifying (KOR) channels (Shabala and Cuin 2008; Miller et al. 2008, 2009). Reduction of Na^+ influx and prevention of K^+ loss through these channels during salt stress are critical for salt tolerance in plants. Importantly pretreatment of Arabidopsis roots with less than

0.5 mM SA reduced K^+ leak through guard cell outward-rectifying K^+ (GORK) channels, which suggests that prevention of K^+ loss through GORKs is the major mode of action of SA for salt tolerance in Arabidopsis (Jayakannan et al. 2013). In Arabidopsis WT and SA-hyperaccumulating *nudt7* mutant plants, SA pretreatment reduced shoot Na^+ concentrations during prolong salinity, but in the NPR-1 (nonexpresser of pathogenesis-related gene 1) signalling blockage mutant *npr1–5*, Na^+ influx was high. Thus, reduced K^+ leak through GORK channels is nonexpresser of pathogenesis-related gene 1 (NPR1)-mediated, because the *npr1–5* mutant is unable to limit K^+ loss through depolarization-activated KOR channels. This shows the significance of NPR-1-dependent SA signalling for plant salt tolerance (Jayakannan et al. 2015b). A subgroup of NSCC named glutamate receptor-like channels (GLRs) has been suggested to be a downstream target of SA, and importantly SA may modulate the activity of GLRs and control K^+ leakage, which is essential for salt tolerance in plants (Jayakannan et al. 2015a).

The enhanced activity of proton pumps, such as H^+-ATPases, regulates voltage-dependent KOR channels preventing K^+ leakage (Chen et al. 2007). H^+ pumping provides a driving force for the plasma membrane Na^+/H^+ exchanger (SOS1-salt overly sensitive1) to remove Na^+ from the cytoplasm to the apoplast (Shi et al. 2000; Apse and Blumwald 2007), reducing the Na^+ concentration in the cytoplasm. Interestingly, during salt stress, SA pretreatment in Arabidopsis increased H^+-ATPase activity in a time and dose-dependent manner, which minimized the extent of plasma-membrane depolarization and reduced NaCl-induced K^+ efflux via KOR channels (Jayakannan et al. 2013), with plants showing SA-mediated salt stress tolerance. Again, a stress-inducible plasma membrane localized plasma membrane protein 3 (PMP3) has been shown to participate in the Na^+ efflux, which depends on a Na^+/H^+exchanger or Na^+-ATPase during salt stress (Inada et al. 2005; Mitsuya et al. 2005). A study by Inada et al. (2005) showed that in the halophyte sheep grass *Aneurolepidium chinense*, AcPMP3 expression was upregulated within 15 min of H_2O_2, and 30 min of SA treatments, showing that SA may control AcPMP3 operation during salt stress. In *Nicotiana tabacum* cells, SA can induce the activation of a MAPK, namely salicylic acid–induced protein kinase (SIPK), during osmotic stress (Mikolajczyk et al. 2000). In summary, SA-mediated control of ion homeostasis under salinity stress is mainly NPR1-dependent and involves minimizing Na^+ entry into roots, increasing H^+-ATPase activity in the roots, preventing stress-induced K^+ leakage from the roots, via KOR and NSCC, and enhancing K^+ concentrations in shoots (Jayakannan et al. 2015a).

1.3.5 Interaction with Osmoprotectants

Plants have evolved protective osmoregulation mechanisms, mediated by various osmolytes such as glycinebetaine (GB), Pro, soluble sugars, amines, etc., which contribute to turgor maintenance when plants are under salinity and osmotic stress (Misra and Saxena 2009). For example, accumulation of GB and Pro in plants during osmotic and salinity stress adjusts the cellular osmotic balance, stabilizes proteins, protects membrane integrity, prevents polypeptide dissociation from the PSII complex and detoxifies ions (Ashraf and Foolad 2007; Iqbal et al. 2014). Importantly, under high salt stress, SA and aspirin (an analogue of SA) can stimulate GB accumulation in the concentration range of 0.5–2.5 mM (Jagendorf and Takabe 2001). The induction of GB by SA can also activate protein kinases during hyperosmotic stress (Hoyos and Zhang 2000). The SA-mediated higher GB levels in plants can reduce the adverse effects of salt stress and facilitate plant adaptations to stress that include improvements in growth and biomass production in *Rauwolfia serpentina* (Misra and Misra 2012) and increased photosynthesis in *Vigna radiata* (Khan et al. 2014). These studies also showed that application of either SA (0.5 mM) or SA analogue 2, 6, dichloroisonicotinic acid can induce GB accumulation, increase the methionine content and inhibit ethylene formation in plants during salinity stress. SA can also increase Pro metabolism in plants under salt stress, for example 0.5 mM SA significantly enhanced the activity of the Pro-biosynthesis enzymes such as pyrroline-5-carboxylate reductase and γ-glutamyl kinase and also increased the Pro contents, which led to salt tolerance, of *Lens esculenta* plants (Misra and Saxena 2009). According to Misra and Misra (2012), SA also improves the salt tolerance of *R. serpentina* by protecting the protein turnover machinery against salt-induced damage, upregulating stress protective proteins through Pro-metabolizing enzymes and ultimately maintaining cell turgor with higher proline level. Also, soluble sugars and sugar alcohols are considered as important osmoprotectants and their accumulation in plants provides protection against salt stress.

1.4 Salicylic Acid Signalling in Regulating Salt Stress Tolerance in Plants

1.4.1 Interaction with Other Phytohormones

Phytohormones regulate plant growth, developmental processes and signalling plexus and they are directly or indirectly associated with responses to a wide spectrum of abiotic and biotic stressors (Khan et al. 2012a; Asgher et al. 2015). During salinity and other abiotic stresses, SA can modulate plant stress responses through signalling crosstalk with other phytohormones (Asensi-Fabado and Munné-Bosch 2011; Khan et al. 2012b, 2014) and alter phytohormone biosynthesis (Pieterse et al. 2009). Under normal, as well as salt stress conditions, SA can interact in a diverse manner with phytohormones, including namely auxin (Iglesias et al. 2011; Fahad and Bano 2012; Esan et al. 2017), gibberellins (Alonso-Ramírez et al. 2009; Lee et al. 2010; Lee and Park 2010; Hamayun et al. 2010; Dai et al. 2012), abscisic acid (Yasuda et al. 2008; Szepesi et al. 2009; Miura et al. 2011; Asensi-Fabado and Munné-Bosch 2011), ethylene (Khan et al. 2014), and brassinosteroids (Divi et al. 2010; Hayat et al. 2012) (Table 1.1). These interactions (either synergistic or antagonistic) play a crucial roles in plant defence signalling pathways (Bali et al. 2017). Being a signalling molecule SA stimulates stress tolerance against salinity and other abiotic stresses (Guzmán-Téllez et al. 2014; Wildermuth et al. 2001), and

Table 1.1 Crosstalk of SA with other hormones during salt stress

Plant hormone	Plant species	Mode of interaction with SA	Response phenotype	References
GAs	Zoysia grass (*Zoysia japonica*)	Synergistic	Increased endogenous SA levels and mitigation of growth inhibition and alleviation of plant death	Dai et al. (2012)
GAs	Arabidopsis (*Arabidopsis thaliana*)	Synergistic	Modulation of GA–SA signalling and improvement in seed germination	Lee et al. (2010)
GAs	Arabidopsis (*Arabidopsis thaliana*)	Synergistic	Stimulation of SA biosynthesis by inducing *sid2* gene and genes encoding GA biosynthetic enzymes and modulation of antioxidant activities	Lee and Park (2010)
GAs	Arabidopsis (*Arabidopsis thaliana*)	Synergistic	Improved germination of SA-deficient *sid2* mutant via SA–GA signalling	Alonso-Ramirez et al. (2009)
GAs	Soybean (*Glycine max*)	Antagonistic	Decreased SA content and improved growth and development	Hamayun et al. (2010)
IAA	Okra (*Abelmoschus esculentus*)	Synergistic	Stimulation of antioxidant activities via IAA–SA interaction	Esan et al. (2017)
IAA	Maize (*Zea mays*)	Synergistic	Increased chlorophyll, carotenoids, sugars, proline, protein and antioxidant activities through SA–IAA communication	Fahad and Bano (2012)
IAA	Arabidopsis (*Arabidopsis thaliana*)	Antagonistic	SA increased transcript levels of PR1in tir1/afb2 mutant and repressed auxin signalling	Iglesias et al. (2011)
IAA and ABA	Wheat (*Triticum aestivum*)	Synergistic	SA increased accumulation of both IAA and ABA and improved plant growth under salinity	Shakirova et al. (2003)
ABA	Tomato (*Solanum lycopersicum*)	Synergistic	SA-activated ABA accumulation, enhanced osmotic adjustment, increased growth and photosynthetic pigment	Szepesi et al. (2009)
ABA	Arabidopsis (*Arabidopsis thaliana*)	Antagonistic	Exogenous ABA hindered SA-mediated stress defence mechanism, inhibited SAR induction, and SA suppressed ABA signalling	Yasuda et al. (2008)
ABA	Arabidopsis (*Arabidopsis thaliana*)	Antagonistic	Enhanced SA content in ABA-sensitive mutant *aba3*	Asensi-Fabado and Munné-Bosch (2011)

(continued)

Table 1.1 (continued)

Plant hormone	Plant species	Mode of interaction with SA	Response phenotype	References
ABA	Arabidopsis (*Arabidopsis thaliana*)	Antagonistic	Enhanced SA content in ABA-sensitive mutant *siz1*	Miura et al. (2011)
ET	*Mung bean* (*Vigna radiata*)	Antagonistic	SA repressed ET synthesis, enhanced plant growth and photosynthesis	Khan et al. (2014)
BR	Arabidopsis (*Arabidopsis thaliana*)	Synergistic	BR upregulated SA-mediated defence gene	Divi et al. (2010)
BR	Mustard (*Brassica juncea*)	Synergistic	Reduced toxic effects of salinity through SA–BR communication	Hayat et al. (2012)

Abbreviations: *ABA* abscisic acid, *BS* brassinosteroid, *ET* ethylene, *GA* gibberellic acid, *IAA* indole acetic acid, *SA* salicylic acid

the ability of the signalling crosstalk between SA with and other phytohormones to regulate stress tolerance depends on the plant species as well as the intensity, nature and time of exposure of the plant to stress (Khan et al. 2015). As plant salt stress responses and signal molecules are very complex, and they are species- or genotype-dependent, the studies on salinity tolerance of different species of plants, under similar conditions, may result in different conclusions.

1.4.2 Interaction with Reactive Oxygen Species and Modulation of Antioxidants Metabolism

Salt stress induces the generation of various ROS such as superoxide ($O_2^{\bullet -}$), singlet oxygen ($^1O_2$), hydrogen peroxide (H_2O_2) and the hydroxyl radical ($OH^{\cdot}$) in plants, which leads to oxidative stress (reviewed in Parida and Das 2005). Plant salt tolerance is usually positively correlated with increased activities of antioxidant enzymes (Horváth et al. 2007; Munns and Tester 2008; Ashraf et al. 2010). SA performing multiple functions has been found very effective in detoxifying ROS in plant cells (Hasanuzzaman et al. 2017). Recently some researchers (Kang et al. 2013; Khan et al. 2014) have shown that SA plays significant roles in the modulation of antioxidant metabolism and controlling cellular ROS levels, and the involvement of SA on antioxidant metabolism has been suggested to regulate plant tolerance not only to salt stress, but also to other abiotic stressors (Khan et al. 2010, 2012a, b, c; Nazar et al. 2011, 2015). Generally, SA at low concentrations facilitates salt tolerance by acting as a signalling molecule through the production of low ROS levels and the induction of antioxidant systems. However, SA at high concentrations triggers oxidative stress due to greatly increased production of ROS (Love et al. 2008; Quan et al. 2008; Lee et al. 2010; Poór et al. 2011b; Miura and Tada 2014;

Belkadhi et al. 2014). ROS such as H_2O_2 can cause increases in endogenous SA concentration levels by increasing the activity of the BA2H enzyme, which catalyses the production of SA from benzoic acid (Dempsey and Klessig 1995), while at high concentrations SA can increase H_2O_2 concentrations by inhibiting antioxidant enzymes like catalase (CAT) and ascorbate peroxidase (APX), through SA binding (Durner Darner and Klessig 1995; Durner and Klessig 1996; Horváth et al. 2002). Moreover, enhanced salt tolerance was observed in SA-hyperaccumulating mutants, namely *siz1* and *aba3–1*, suggesting that higher SA levels may be necessary to reduce salt-induced oxidative stress (Asensi-Fabado and Munné-Bosch 2011; Miura et al. 2011). In particular, study comparing two SA-hyperaccumulating Arabidopsis mutants, namely *nudt7* and *npr1–5*, under salinity and oxidative stresses. Jayakannan et al. (2015b) revealed that an active NPR1-mediated SA signalling pathway is crucial for salt-induced H_2O_2 production and also for tolerance to salt-induced oxidative stress in plants.

1.4.3 Interactions with Nitric Oxide

In plant systems, nitric oxide (NO) is a major ubiquitous signalling molecule having antioxidant properties, which plays significant protective roles during salinity and oxidative stress by quenching ROS and maintaining ion homeostasis (Fatma et al. 2016; Sohag et al. 2020a). SA is one of the main phytohormones with which NO interacts and it functions as a second messenger and regulator of stomatal movement (Hao et al. 2010). In Arabidopsis, SA can also stimulate the synthesis of NO in a dose-dependent manner, by enhancing the activity of NO-synthesizing enzymes. This suggests the involvement of a regulatory loop to amplify the signal between SA and NO (Zottini Bottini et al. 2007). Interestingly, an interaction between SA and NO has been shown to be effective in ameliorating the adverse effects of salinity and osmotic stress in plants. For example, in osmotic stressed *T. aestivum* seedlings NO functioned as downstream regulator of SA-signalling and improved plant growth by mitigating oxidative damages by increasing antioxidant enzyme activities (Naser Alavi et al. 2014). The protective action of SA and NO was also observed in salt-stressed *Gossypium hirsutum*, where the synergistic effects of combined SA and SNP (sodium nitroprusside) treatments, NO donors, lowered lipid peroxidation, improved ion absorption, activated the metabolism of osmotic-regulated substances and facilitated the membrane transport and ROS detoxification (Dong et al. 2015). Moreover, Yadu et al. (2017) studied the imperative interactive roles of SA and NO for improving salinity tolerance in *Pisum sativum* and reported that combining SA and SNP treatments had significantly increased antioxidant enzyme activities, for example superoxide dismutase (SOD), glutathione peroxidase (GPX), APX and the levels of osmoprotectants such as Pro, GB and sugars. In another study, Simaei et al. (2011) showed enhanced salt tolerance in *Glycine max* by exogenous application of SA plus SNP that led to the activation of CAT, APX, GPX activities and a decline in the oxidative damage. In most of the above studies, the protective action of SA plus

SNP against salt-induced oxidative damage was more efficient than the effects of SA or SNP alone.

1.4.4 Induction of Stress-Responsive Genes

Salicylic acid has been shown as to be a potent inducer of defence gene expression in plants (Innes 2018). SA signalling is mediated by two mechanisms: (i) *NPR1*-dependent and (ii) *NPR1*-independent (Shah 2003; Yuan and Lin 2008). It should be mentioned that the *NPR1* gene is a master regulator of SA-mediated defence genes, and that *NPR1*-dependent SA signalling controls multiple physiological traits, which are pivotal for improving salt tolerance in plants (Jayakannan et al. 2015a, b). SA can also induce abiotic-stress-protective genes through the *NPR1*-independent mechanism (Yuan and Lin 2008). Wu et al. (2012) showed that the transcriptional co-regulator NPR1 is critical for the activation of SA-dependent defence genes as SA binds to the C-terminal domain of NPR1, which is required for activation of defence genes in response to SA. According to Innes (2018), SA binding to NPR1 activates transcription, whereas SA binding to NPR3 and NPR4 blocks the transcriptional repression activity of the C-terminal domain. Thus, a robust activation of SA-responsive promoters occurs in the presence of high SA levels, and finetuning of transcriptional responses occurs at medium SA levels. In fact, SA-mediated activation of defence gene transcription is not fully understood. Therefore, a thorough study of post-transcriptional regulation of all NPR proteins is necessary, because all three NPR proteins (NPR1, NPR3 and NPR4) significantly influence responses to SA (Innes 2018). Again, genes involved in signal transduction, such as protein kinases like MAPKs, and transcription factors also could be promoted by NPR1 (Blanco et al. 2005).

Transmission of the SA signal takes place via at least three MAPK signalling cascades (MPK3, MPK4 and MPK6), and gene targets of SA signalling through MAPK should be considered when investigating salt stress tolerance in plants, because MAPKs are activated by various abiotic stresses including salinity and are regarded as key factors for SA–ROS signal transduction (Baier et al. 2005; Kangasjarvi et al. 2005; Fujita et al. 2006). During salinity and osmotic stress in tobacco cells, SA can induce SIPK, which is a member of the MAPK family and a homolog of protein kinase Arabidopsis serine/threonine kinase 1 (ASK1), which is a member of sucrose nonfermenting 1 (SNF1) kinase family, as demonstrated by Mikolajczyk and his co-workers (Mikolajczyk et al. 2000). Besides, in response to SA, the transcript levels of dehydrin-like proteins, heat shock protein (HSP) and alternative oxidase (AOX) can also be much elevated (Salzman et al. 2005; Rajjou et al. 2006). Some reports suggest that SA regulates gene expression, which is involved in stress responses via a Ca^{2+}-dependent pathway (Du et al. 2009; Wang et al. 2009; Coca and San Segunda 2010).

1.5 Exogenous Salicylic Acid–Mediated Salt Stress Tolerance in Plants

In recent years, a series of experiments have shown the ameliorative role of SA in salinity-affected plants through the modulation of morphological, physiological and biochemical traits and also by regulating multiple stress responsive pathways in different plant varieties (Table 1.2). Exogenous application of SA improved germination indexes (Anaya et al. 2015), growth parameters (Lee et al. 2014) and biomass production (Jayakannan et al. 2013); promoted the uptake of beneficial minerals (Ca^{2+}, K^+, P, Mg^{2+} etc.) (Syeed et al. 2011); and arrested the uptake of excess detrimental ions (Na^+, Cl^-, etc.) (Nazar et al. 2011). In addition, exogenous SA played a vital role in improving photosynthetic attributes (photosynthetic rate, transpiration rate, intercellular CO_2 concentration, water-use efficiency, etc.) (Li et al. 2014), in increasing pigment contents (chlorophyll and carotenoids) (Ma et al. 2017), minimizing excess ROS generation ($O_2^{\bullet-}$, H_2O_2, OH^-, etc.) (Nazar et al. 2011), in reducing membrane degradation products (malondialdehyde, MDA; thiobarbituric acid reactive substance; TABRS) (Ahmad et al. 2018), in enhancing antioxidant enzyme activities (SOD, CAT, APX, POX, GR, etc.) (Ardebili et al. 2014) and non-enzymatic antioxidants (AsA, GSH, phenolic compounds, etc.) (Csiszár et al. 2018), in regulating osmoprotectants (GB, Pro, sugar, protein etc.) (Shaki et al. 2018; Fayez and Bazaid 2014), and in upregulating and downregulating the expression of some genes (*MYB* and *P5CS*) (Zheng et al. 2018). Apart from SA-induced salinity stress tolerance in the seedling, exogenous SA improved salt stress tolerance in plants at the reproductive stage, as indicated by increased number of panicles, number of grains per panicle, number of filled grains per panicle, 1000 grain weight and yield (Jini and Joseph 2017; Pirasteh-Anosheh et al. 2017; Tahjib-Ul-Arif 2018b). Some plant species which are salt tolerant showed much, much greater increases in protection against salinity tolerance compared to salt-susceptible plant species by displaying increased activity in defence system. However, both species showed a considerable tolerance after exogenous supply of SA in growth media (Jini and Joseph 2017). At the time of exogenous treatment, various concentrations of SA have been applied in salt-stressed plants to evaluate its effective concentration, and among them low concentration of SA (0.1 and 0.5 mM) showed better protection against salt stress in most of the cases whereas high concentration of SA displayed inhibitory role (Anaya et al. 2015; Syeed et al. 2011). In addition, the duration of exposure and mode of application of SA also play noteworthy roles in enhancing tolerance against salt stress. In Table 1.2, we summarized some of the most successful recent studies related to using exogenous application of SA to increase salinity stress tolerance in various plant species.

Table 1.2 Salicylic acid–mediated salt stress tolerance in various plant species, with the suggested mode of protection

Name of the crop	Salinity dose and duration	SA dose, mode of application and duration	Mode of action of exogenous SA	References
Maize (*Zea mays* L.)	75 mM NaCl; 40 days	1000 μM; foliar spray; 40 days	Enhancement of photosynthetic rate	Tahjib-Ul-Arif et al. (2018b)
			Increased carboxylation efficiency and water-use efficiency	
			Increased chlorophyll content (SPAD value)	
			Decreased lipid peroxidation (MDA content)	
			Enhanced the activity of CAT and APX	
			Increased RWC	
			Improved the yield-contributing traits including 100 kernel weight	
Mung bean (*Vigna radiata* L.)	100 mM NaCl; 20 days	500 μM; foliar spray; 7 days	Enhanced GB and Met production	Khan et al. (2014)
			Suppressed ET formation	
			Reduce oxidative stress	
			Decrease Na^+ and Cl^- ion accumulation	
			Inhibited ACS activity	
			Increased GSH content	
			Improved photosynthesis and growth parameters	
Mung bean (*Vigna radiata* L.)	50 mM NaCl; 30 days	500 and 1000 μM; foliar spray; 15 days	Decreased Na^+ and Cl^- uptake	Nazar et al. (2011)
			Increased ATPase and S content	
			Decreased H_2O_2, TBARS and EL	
			Decreased GSSG	
			Increased GSH	
			Decreased water and osmotic potential	
			Increased SOD, APX and GR activity	
			Increased F_v/F_m, net photosynthesis	

(continued)

Table 1.2 (continued)

Name of the crop	Salinity dose and duration	SA dose, mode of application and duration	Mode of action of exogenous SA	References
Tobacco (*Nicotiana tabacum* L.)	50, 100, 150 mM NaCl; 35 days	100 μM; growth media; 35 days	Improved growth parameters Enhanced the activities of Rubisco	Lee et al. (2014)
Lentil (*Lens esculenta*)	100 mM NaCl; 10 days	500 μM; growth media; 10 days	Improved germination percentage Increased SL, SFW, SDW, RL, RFW and RDW Increased Pro and GB contents Increased activities of P5CR and γ-glutamyl kinase	Misra and Saxena (2009)
Chinese nutmeg (*Torreya grandis*)	0.2 and 0.4% NaCl; 60 days	500 μM; foliar spray; 30 days	Increased the biomass Enhanced the Chl content Enhanced activity of SOD, CAT and POX Increased the photosynthetic parameters (Pn, Ci, Tr, Gs) Alleviated membrane injury by decreasing MDA content and EL	Li et al. (2014)
Barley (*Hordeum vulgare* L.)	50, 100 and 150 mM NaCl; 14 days	50 μM; foliar spray; 14 days	Increase the growth parameters Enhanced Chl and Car content Enhance antioxidant activity Decreased Pro content Decreased soluble protein and sugar content Decreased Na^+ and increased K^+ uptake Decreased MDA content	Fayez and Bazaid (2014)
Barley (*Hordeum vulgare* L.)	12 dSm^{-1} NaCl; crop growth stages	500, 1000, 1500, and 2000 μM; foliar spray	Decreased Na^+ uptake Increased K^+ uptake Reduced Na^+/K^+ ratio Lowered Cl^- content Increased Ca^{2+} uptake Increased grain yield and biomass	Pirasteh-Anosheh et al. (2017)

(continued)

Table 1.2 (continued)

Name of the crop	Salinity dose and duration	SA dose, mode of application and duration	Mode of action of exogenous SA	References
Wheat (*Triticum aestivum*)	250 mM NaCl; 3 days	0.5 mM; Growth media; 3 days	Increased the contents of AsA and GSH	Li et al. (2013b)
Wheat (*Triticum aestivum*)	150 mM NaCl; 30 days	0.25, 0.50, 0.75 and 1 mM; growth media, 30 days	Increased fresh and dry masses of root and shoots	Arfan et al. (2007)
			Reduced salt-induced damages in grain yield, 100-grain weight and number of grains	
Arabidopsis (*Arabidopsis thaliana*)	100 mM NaCl; 7 days	0.01–10 μM; pretreatment; 14 days	Elevated AsA and GSH levels	Csiszár et al. (2018)
			Facilitated preservation of the redox potential	
			Increased GSH/GSSG ratio	
Arabidopsis (*Arabidopsis thaliana*)	100 mM NaCl; 14 days	50 μM; pretreatment; 1 h	Improved biomass and water content	Jayakannan et al. (2013)
			Lowered H^+ influx	
			Upregulated H^+-ATPase	
			Reduced Na^+ accumulation	
			Decreased the extent of plasma membrane depolarization	
Safflower (*Carthamus tinctorius* L.)	100, and 200 mM NaCl; 21 days	1000 μM; foliar spray; 21 days	Increased GB, total soluble protein, carbohydrates, Chl, Car, flavonoid and anthocyanin contents	Shaki et al. (2018)
			Decreased Pro content	
			Increased the activity of PAL gene	
Rice (*Oryza sativa*)	150 mM NaCl; 30 days	2000 μM; priming; 1 day	Enhanced seed germination and seedling growth	Sheteiwy et al. (2018)
			Enhanced Pn, Gs, Ci, and Tr	
			Decreased EL	
			Improved SOD, CAT, APX and POX activity	
			Increased K^+,Ca^{2+} and Mg^{2+}	
			Decreased Na^+ accumulation	

(continued)

Table 1.2 (continued)

Name of the crop	Salinity dose and duration	SA dose, mode of application and duration	Mode of action of exogenous SA	References
			Decreased H_2O_2, MDA content and EL	
			Increased PAL activity	
			Increased total Chl and Car	
Rice (*Oryza sativa* L.)	100 mM NaCl; 30 days	1000 μM; pretreatment; 7 days; treatment in soil; 30 days	Improved germination, growth, yield and nutrient value	Jini and Joseph (2017)
			Increased K^+ and Ca^{2+} uptake	
			Decreased accumulation of Na^+ and Cl^-	
Rice (*Oryza sativa* L.)	100 mM NaCl; 5 days	500 and 1000 μM growth medium; 2 days	Decreased Na^+ uptake	Kim et al. (2018)
			Increased growth and biomass	
			Increased endogenous SA content	
			Decreased H_2O_2 and MDA content	
Maize (*Zea mays* L.)	40 mM NaCl; 56 days	0.5 mM; growth media; 56 days	Reduced MDA content	Gunes et al. (2007)
			Ameliorated membrane deterioration	
			Lowered accumulation of Na^+ and Cl^- ions	
			Increased N concentration	
Tomato (*Lycopersicum esculentum*)	200 mM NaCl; 15 days	10 μM; foliar spray; 3 days	Restored growth parameters	Fariduddin et al. (2017)
			Improved CAT, POX and SOD activity	
			Increased SPAD chlorophyll	
			Improved photosynthetic parameters	
			Increased CA and NR	
Tomato (*Solanum lycopersicum*)	200 mM NaCl; 7 days	0.01 and 100 μM; pretreatment 21 days	Lowered accumulation of ABA	Horváth et al. (2015)
			Prevented higher production of ET	
			Enhanced net CO_2 fixation rate	

(continued)

Table 1.2 (continued)

Name of the crop	Salinity dose and duration	SA dose, mode of application and duration	Mode of action of exogenous SA	References
Tomato (*Solanum lycopersicum*)	100 mM NaCl; 10 days	0.1 and 10 μM Pretreatment 21 days	Increased water potential	Szepesi et al. (2009)
			Improved Chl content	
			Reduced MDA content	
Tomato (*Solanum lycopersicum*)	100 mM NaCl; 7 days	0.1 and 100 μM; Pretreatment 21 days	Increased the activities of SOD, CAT, APX and GR	Tari et al. (2015)
			Increased total AsA and GSH	
Ethiopian mustard (*Brassica carinata*)	50,100 and 150 mM NaCl; 28 days	500 μM; Foliar Spray; 28 days	Enhanced growth and biomass	Husen et al. (2018)
			Enhanced photosynthetic efficiency by increasing Chl synthesis	
			Modulated cells redox balance	
			Enhanced Chl fluorescence, Gs, Pn, Tr and WUE	
			Increased total Chl and Car	
			Decreased TBARS content	
			Decreased Pro content	
			Increased NR	
			Increased SOD, CAT and POD activity	
Indigenous berry (*Nitraria tangutorum*)	200–400 mM NaCl; 9 days	500–1500 μM nutrient solution; 9 days	Improved plants growth and biomass	Yan et al. (2018)
			Decreased superoxide and H_2O_2 content	
			Promoted the ratios of AsA/DHA and GSH/GSSG	
			Decreased content of TBARS	
			Increased APX, DHAR, MDHAR and GR	
Faba bean (*Vicia faba* L.)	50 and 100 mM NaCl; 60 days	1000 μM; foliar spray 50 days	Enhanced growth, biomass and yield	Ahmad et al. (2018)
			Increased total Chl and Car content	
			Increase RWC	
			Increased Pro and GB content	

(continued)

Table 1.2 (continued)

Name of the crop	Salinity dose and duration	SA dose, mode of application and duration	Mode of action of exogenous SA	References
			Increased uptake of Ca^{2+} and K^+	
			Decreased NA^+ accumulation	
			Enhanced the IAA and IBA	
			Decreased ABA content	
			Increased SOD, CAT, APX and GR activity	
			Declined H_2O_2, MDA content and EL	
Broad bean (*Vicia faba* L.)	90, 120, 150 and 200 mM NaCl	250, 500 and 1000 μM growth medium; 7 days	Increased seed germination percentage, precocity of germination, TG, MGT, GRI, germinated seeds fresh and dry weight	Anaya et al. (2015)
Soybean (*Glycine max*)	4, 7, and 10 dSm^{-1} NaCl; 109 days	1000 μM; foliar spray; 23 days	Increased biomass and yield	Farhangi-Abriz and Ghassemi-Golezani (2018)
			Increased total Chl content	
			Increased soluble sugar, soluble protein, GB, Pro and RWC	
			Decreased Na^+	
			Increased K^+ and Ca^{2+}	
			Increased SOD, CAT, APX and POX activity	
			Decreased MDA	
			Increased MSI	
Soybean (*Glycine max*)	100 mM NaCl; 7 days	500 μM; Foliar Spray; 7 days	Increased the activity of CAT, APX and POX	Ardebili et al. (2014)
			Reduced oxidative stress	
			Increased Pro content	
			Enhanced photosynthetic pigments (Car content)	
			Promotion of AsA content	
			Decrease in Na^+/K^+ ratio	
			Increased Mg^{2+} uptake	

(continued)

Table 1.2 (continued)

Name of the crop	Salinity dose and duration	SA dose, mode of application and duration	Mode of action of exogenous SA	References
Cucumber (*Cucumis sativus* L.)	25, 50, and 100 mM NaCl; 7 days	400 μM; Pretreatment; 12 h	Increased plant growth, biomass and yield	Gurmani et al. (2018)
			Decreased salt injury index	
			Increased Pn and WUE	
			Enhanced SOD, CAT and POX	
			Reduced Na^+ uptake	
			Increased K^+ uptake	
			Increased total Chl content	
Gerbera (*Gerbera jamesonii* L.)	100 mM NaCl; 15 days	500 μM; pretreatment; 2 days	Decreased MDA and EL	Kumara et al. (2010)
			Increased the activities of SOD, POX, CAT and APX	
Chickpea (*Cicer arietinum*)	0.50, 4.0, 6.2, and 8.3 dSm^{-1} NaCl; 35 days	500 μM; priming; overnight	Increased growth and biomass	Garg and Bharti (2018)
			Increased number of arbuscules, number of vesicles and vesicle to arbuscule ratio	
			Reduced flower abortion	
			Improved pod, seed formation	
			ION-homeostasis, nutrient uptake and sugar metabolism	
			Strengthen the plasma membrane stability	
			Increased Chl and anthocyanin	
			Reduced Na^+ uptake	
			Increased K^+ uptake	
Strawberry (*Fragaria* × *ananassa*)	30 and 60 mM NaCl; 25 days	100, 500, and 750 μM; foliar spray; 7 days	Increased the activities of APX, POX and SOD	Faghih et al. (2017)
			Increased Pn and Gs	
			Decreased Ci	
			Increased biomass	
			Decreased Na^+/K^+ ratio	
Dianthus (*Dianthus superbus*)	0.3, 0.6, and 0.9% NaCl; 45 days	500 μM; foliar spray; 45 days	Enhanced plant growth and development	Ma et al. (2017)
			Increased Chl and Car content	

(continued)

Table 1.2 (continued)

Name of the crop	Salinity dose and duration	SA dose, mode of application and duration	Mode of action of exogenous SA	References
			Decreased MDA content	
			Increased REC and Pro content	
			Decreased superoxide anion and H_2O_2 content	
			Increase SOD, CAT and POX activity	
			Increased stomatal density	
			Compensated salt-induced damage in chloroplast ultra-structure	
Dianthus (*Dianthus superbus*)	0.3, 0.6, and 0.9% NaCl; 45 days	500 μM; foliar spray; 45 days	Increased leaf biomass	Zheng et al. (2018)
			Increased soluble protein and sugar content	
			Upregulated the expression of *MYB* and *P5CS*	
			Increased the thickness of mesophyll, palisade tissue, spongy parenchyma and upper epidermis	
Olive (*Olea europaea* L.)	200 mM NaCl; 75 days	500 and 1000 μM; pretreatment with foliar spray; 45 days	Enhanced growth and biomass	Methenni et al. (2018)
			Decreased Na^+ uptake	
			Increased K^+ uptake	
			Increased K^+ and Na^+ ratio	
			Decreased MDA and H_2O_2 content	
			Increased phenolic compound	
			Increased flavonoid content	
			Decreased DPPH value	
			Increased CO_2 assimilation rate, Tr and Gs	

(continued)

Table 1.2 (continued)

Name of the crop	Salinity dose and duration	SA dose, mode of application and duration	Mode of action of exogenous SA	References
Alfalfa *(Medicago sativa)*	200 mM NaCl; 12 days	0.1 and 0.5 mM; pretreatment; 2 days	Increased F_v/F_m ratio	Palma et al. (2013)
			Prevented the decrease of nodule mass	
			Increased the activities of SOD, POX, DHAR, APX and GR	
Potato (*Solanum tuberosum*)	50 mM NaCl; 22 days	500 μM; foliar spray; 15 days	Improved activities of SOD, CAT and POX	Faried et al. (2017)
			Regulating osmotic adjustment (Pro contents)	
			Increased RWC	
			Increased K^+ uptake	
			Reduced Na^+ uptake	
			Increased Gs	
			Decreased Ci	
			Increased protein	
			Increased total phenolic compound content	
			Decreased MDA content	
Mustard (*Brassica juncea* L.)	100 mM NaCl; 30 days	500 μM; foliar spray; 15 days	Increased APX, GR, DHAR activity	Nazar et al. (2015)
			Increased N and S assimilation	
			Increased Cys, AsA and GSH content	
			Decreased Na^+ and Cl^- uptake	
			Decreased H_2O_2 and MDA content	
			Increased ATPase, serine acetyl transferase and rubisco activity	
			Increased net photosynthesis	
Mustard (*Brassica juncea* L.)	50 mM NaCl; 15 days	100, 500, and 1000 μM; foliar spray; once at 15 days	Increased the nutrients content	Syeed et al. (2011)
			Increased biomass	
			Increased GSH	
			Decreased the content of H_2O_2, TBARS and EL	
			Increased SOD, APX and GR activity	
			Increase K^+, Ca^{2+}, P and N content	

(continued)

Table 1.2 (continued)

Name of the crop	Salinity dose and duration	SA dose, mode of application and duration	Mode of action of exogenous SA	References
			Improved photosynthetic parameters	
			Reduced the content of Na^+ and Cl^-	

Abbreviations: *ABA* abscisic acid, *ACS* aminocyclopropane carboxylic acid synthase, *AsA* ascorbic acid, *APX* ascorbate peroxidase, *CA* carbonic anhydrase, *Car* carotenoid, *CAT* catalase, *Chl* chlorophyll, *CVG* coefficient of velocity of germination, *Cys* cysteine, *DHA* dehydroascorbate, *DHAR* dehydroascorbate reductase, *DPPH* 2,2-diphenyl-1-picrylhydrazyl, *EL* electrolyte leakage, *ET* ethylene, *GRI* germination rate index, *GSH* glutathione (reduced), *GSSG* glutathione (oxidized), *GR* glutathione reductase, *GB* glycinebetaine, *IAA* indole acetic acid, *IBA* indole butyric acid, *Ci* intercellular CO_2 concentration, *LA* leaf area, *MDA* malondialdehyde, F_v/F_m maximal PS II photochemical efficiency, *MGR* mean germination rate, *MGT* mean germination time, *MSI* membrane stability index, *Met* methionine, *MDHAR* monodehydroascorbate reductase, *NR* nitrate reductase, *POX* peroxidase, *PAL* phenylalanine ammonia-lyase, *Pn* photosynthetic rate, *Pro* proline, *REC* relative electric conductivity, *RWC* relative water content, *RDW* root dry weight, *RFW* root fresh weight, *RL* root length, *SDW* shoot dry weight, *SFW* shoot fresh weight, *SL* shoot length, *SOD* superoxide dismutase, *Gs* stomatal conductance, *TG* total germination, *Tr* transpiration rate, *TABRS* thiobarbituric acid reactive substance, *WUE* water-use efficiency

1.5.1 Exogenous SA: Influences on Ingrowth, Photosynthesis, Pigment Production and Sugar Contents in Plants Grown Under Saline Conditions

During stress, exogenous SA has been found very potential in enhancing the germination percentage, shoot and root length, fresh and dry weight of both the roots and shoots of plants. In general, SA treatment during salinity improved plant growth and increased photosynthetic pigment, such as chlorophyll a, b and carotenoids levels in various plant species including *Lycopersicum esculentum*, *Triticum aestivum*, *Zea mays*, *Helianthus annuus*, *Torreya grandis*, *Cicer arietinum*, *Phaseolus vulgaris*, *Fragaria* x *ananasa*, *Hordeum vulgaris*, *Vigna radiata* and *Glycine max*, etc. (Kaydan et al. 2007; Wasti et al. 2012; Agami 2013; Agami et al. 2013; Tufail et al. 2013; Noreen and Ashraf 2010; Li et al. 2014; Asadi et al. 2013; Hadi et al. 2014; Karlidag et al. 2009; El-Tayeb 2005a, b, Khan et al. 2014; Jaiswal et al. 2014; Sohag et al. 2020b). Recently, Jini and Joseph (2017) showed increased growth of salt-stressed *Oryza sativa* plants when they were treated with 1 mM SA at the germination and at vegetative growth stages. In salt-stressed *Hordeum vulgare*, SA treatment was found to stimulate photosynthetic performance and carbohydrate metabolism leading to improved salt tolerance (Khodary 2004). The effect of SA on photosynthesis is also concentration-dependent although the actual role of SA on photosynthetic parameters is still unclear. In various plant species, SA at low concentrations (less than 10 μM) mitigated salt-induced reductions in photosynthetic rates (Stevens et al. 2006; Nazar et al. 2011), carbon fixation, transpiration, stomatal movement (Stevens et al. 2006; Poór et al. 2011a) and antioxidant enzyme activities (Szepesi et al. 2008). In contrast, the reverse effects were observed with higher SA concentrations (above 1 mM), with reduced photosynthetic rates (Nemeth et al. 2002), disruption of antioxidant activity (Pancheva and Popova 1997), reduction in chlorophyll content (Moharekar et al. 2003), increased chloroplast volumes, swelling of the thylakoid grana and coagulation of the stroma (Uzunova and Popova 2000). Previously, Khodary (2004) reported increased sugar levels after application of SA to salt-stressed *Hordeum vulgare* plants and suggested that SA treatment disrupts the enzymatic systems involved with polysaccharide hydrolysis and activates metabolism associated with the consumption of soluble sugars by increasing osmotic pressure. Improved plant growth was observed with increased contents of polysaccharides and soluble sugars, with less than 1 mM SA treatment in *Dendrobium officinale* (Yuan et al. 2014). In addition, Sahar et al. (2011) and Jaiswal et al. (2014) also showed increased sugar contents in *Salvia officinalis* and in *Glycine max* with SA application to plants under salinity stress.

1.5.2 Exogenous SA-Mediated Ion Homeostasis During Salinity

SA-mediated inhibition of cellular concentration of Na^+ and Cl^- ions and stimulation of the contents of nutrients like N, P, K, S, Ca, Mg, Fe, Mn and Cu were reported in *Brassica juncea*, *Zea mays*, *Hordeum vulgaris*, *Cucumis sativus* and *Vigna radiata* plants that provided tolerance to salt stress (Syeed et al. 2011; Nazar et al. 2015; Fayez and Bazaid 2014; El-Tayeb 2005a, b; Gunes et al. 2005, 2007; Yildirim et al. 2008; Khan et al. 2010). Apart from exogenous application or pretreatment, addition of SA to soil also showed salt-ameliorative effects in maize and mustard by reducing the accumulation of toxic ions (Gunes et al. 2007). However, some studies have shown a contradictory role for SA with respect to ion homeostasis under salt stress. Gunes et al. (2007) and El-Tayeb (2005a, b) showed that exogenous application of SA reduced the concentration of K^+ and P in root and shoot tissues of *Z. mays* and *H. vulgaris* plants subjected to salinity. In salt-affected *Spinacia oleracea*, the concentrations of Na^+ and Cl^- were not influenced by SA (Eraslan et al. 2008). SA-mediated inhibition of K^+ uptake and increase of Na^+ were reported in *Lycopersicum esculentum* plants under salt stress (Szepesi et al. 2009). Further study is required to clarify the interaction between SA with various nutrient elements in plants under salinity to help develop salt-tolerant cultivars.

1.5.3 Exogenous SA as a Regulator of Antioxidant Metabolism

During acclimation to salinity, exogenous SA application at physiological concentrations caused moderate oxidative stress as it caused H_2O_2 production and induced both enzymatic [SOD, CAT, APX and GPX] and non-enzymatic antioxidant defence systems (glutathione, ascorbic acid, carotenoids and tocopherols) in plants (Durner and Klessig 1995, 1996; Gill and Tuteja 2010). Increased growth and higher antioxidant enzyme activities namely SOD, CAT and POD were observed in salt-stressed *Helianthus annuus* with exogenous application of SA to plants (Noreen et al. 2009). Arfan (2009) also showed increased SOD activity caused by exogenous SA application with an increase in shoot Ca^{2+} resulting in a transient increase in H_2O_2 to induce antioxidant enzymes, which reduced cellular ROS in spring wheat under salinity. Palma et al. (2013) reported that SA can act to protect against salinity by increasing the activities of antioxidant enzymes including POX, SOD, APX and dehydroascorbate reductase (DHAR) in *Medicago sativa.* It was observed that SA-pretreatment could alleviate the toxic effects of salt stress on photosynthesis and growth in *V. radiata* plants by increasing the activities of SOD, CAT, GPX, APX and GR (Khan et al. 2014). In addition, increased salt tolerance, as a result of increased activities of AsA–GSH pathway enzymes, was found in *T. aestivum* as well as in *B. Juncea* plants with exogenous SA application at 0.5 mM(Li et al. 2013b; Nazar et al. 2015). The fundamental mechanisms underlying this SA-mediated salt tolerance via the AsA–GSH cycle was attributed to the differential regulation of the levels of transcript of genes encoding enzymes such as GPX (*GPX1*), phospholipid hydroperoxide (*GPX2*), and DHAR (*DHAR*), GR (*GR*), GST (*GST1* and *GST2*), MDHAR (*MDHAR*), and GSH synthetase (*GS*) as suggested by Li et al.(2013b). Furthermore, in *S. lycopersicum*, SA-priming reduced the effects of salt stress by upregulating the expression of GST-supergene family (*SIGSTs*) in the leaves and roots of plants, in a concentration-dependent manner (Csiszár et al. 2014). Thus, multiple antioxidant metabolic enzymes and peroxidase activities are mediated by SA signals which act to reduce oxidative damage and provide increased tolerance to salinity stress.

1.6 Conclusions and Future Perspectives

Genetic engineering of plant hormones manifests a dominant platform for stress tolerance and provides a novel approach for improving crop performance under stress. SA signalling is highly potential and exploitation of this phytohormone as a salt stress management tool has received immense importance in developing salt-tolerant crop varieties. Recently, many studies have shown exogenous SA application to be beneficial for plants, not only in normal condition, but also under salt-stressed conditions. In under salinity stress, SA regulates plant developmental and metabolic processes, controls stomatal movement, maintains the plant nutrient status, modulates the production of various osmolytes, reacts with various antioxidants, interacts with different hormones and ROS, and thereby can mediate salt stress tolerance (Fig. 1.2). Further in-depth investigation of the physiological, biochemical as well as metabolic changes observed in SA-overexpressing and SA-deficient mutant plants is also essential to better explore the regulatory roles of SA in plants. Therefore, much works still needs to be performed to unravel the exact mode of action of SA interactions with osmoprotectants, hormones and other signalling compounds, as well as SA biosynthesis upon salinity considering plant species and physiological SA concentrations. A pragmatic and integrated approach

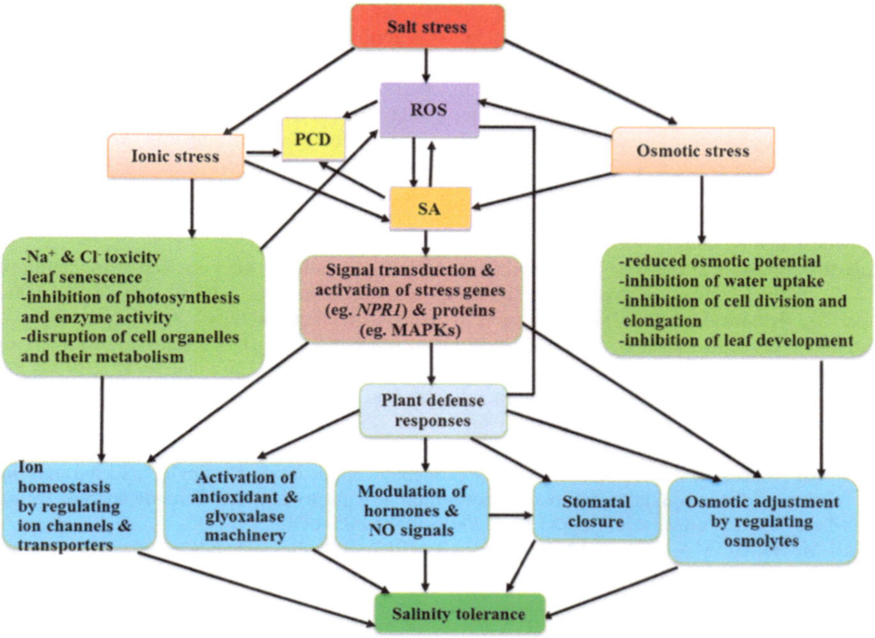

Fig. 1.2 A schematic representation of SA-mediated salt stress tolerance in plants. Salt stress causes both ionic stress and osmotic stress in plants, resulting in ion toxicity, inhibition of growth and water uptake, programmed cell death and excess production of ROS. ROS can induce SA production and correspondingly SA can also stimulate ROS generation, where SA and ROS comprise a positive feedback. As a part of defence responses, both SA and/or ROS, are involved in signal transduction and activate the expression of stress-associated genes such as *NPR1* and protein kinases like MAPKs, which subsequently regulate various cellular processes, such as ion homeostasis by ion channels and transporters, osmotic adjustment by osmolytes, activation of antioxidant enzymes and the glyoxalase pathway, and modulation of hormone and NO levels that control stomatal conductance, leading to salt tolerance

should be adopted incorporating genetics, molecular biology, biochemistry, genomics, proteomic and bioinformatics techniques to unravel the SA-mediated plant defence networks and SA-mediated control of other hormone signalling pathways, in order to develop robust salt-tolerant crop varieties.

References

Achard P, Cheng H, De Grauwe L, Alioua M, Hedden P, Genschik P (2006) Integration of plant responses to environmentally activated phytohormonal signals. Science 311:91–94

Acosta-Motos JR, Ortuño MF, Bernal-Vicente A, Diaz-Vivancos P, Sanchez-Blanco MJ, Hernandez JA (2017) Plant responses to salt stress: adaptive mechanisms. *Agronomy 7*:18

Agami RA (2013) Alleviating the adverse effects of NaCl stress in maize seedlings by pretreating seeds with salicylic acid and 24-epibrassinolide. South Afri J of Bot 88:171–177

Ahmad P, Alyemeni MN, Ahanger MA, Egamberdieva D, Wijaya L, Alam P (2018) Salicylic acid (SA) induced alterations in growth, biochemical attributes and antioxidant enzyme activity in faba bean (*Vicia faba* L.) seedlings under NaCl toxicity. Russ J Plant Physl 65(1):104–114

Alonso-Ramirez A, Rodriguez D, Reyes D, Jimenez JA, Nicolas G, Lopez-Climent M, Gomez-Cadenas A, Nicolas C (2009) Evidence for a role of gibberellins in salicylic acid-modulated early plant responses to abiotic stress in Arabidopsis seeds. Plant Physiol 150:1335–1344

Alpaslan M, Gunes A (2001) Interactive effect of boron and salinity stress on the growth, membrane permeability and mineral composition of tomato and cucumber plants. Plant Soil 236:123–128

Al-Whaibi MH, Siddiqui MH, Basalah MO (2012) Salicylic acid and calcium-induced protection of wheat against salinity. Protoplasma 249:769–778

Anaya F, Fghire R, Wahbi S, Loutfi K (2015) Influence of salicylic acid on seed germination of *Vicia faba* L. under salt stress. J Saudi Soc Agr Sci. https://doi.org/10.1016/j.jssas.2015.10.002

Apse MP, Blumwald E (2007) Na^+ transport in plants. FEBS Lett 581:2247–2254

Ardebili NO, Saadatmand S, Niknam V, Khavari-Nejad RA (2014) The alleviating effects of selenium and salicylic acid in salinity exposed soybean. Acta Physiol Plant 36(12):3199–3205

Arfan M (2009) Exogenous application of salicylic acid through rooting medium modulates ion accumulation and antioxidant activity in spring wheat under salt stress. Intl J Agri Biol 11:437–442

Arfan M, Athar HR, Ashraf M (2007) Does exogenous application of salicylic acid through the rooting medium modulate growth and photosynthetic capacity in two differently adapted spring wheat cultivars under salt stress? J Plant Physiol 164:685–694

Asadi M, Heidari MA, Kazemi M, Filinejad AR (2013) Salicylic acid induced changes in some physiological parameters in chickpea (*Cicer arietinum* L.) under salt stress. J Agric Technol 9:243–248

Asensi-Fabado M, Munné-Bosch S (2011) The *aba3-1* mutant of *Arabidopsis thaliana* withstands moderate doses of salt stress by modulating leaf growth and salicylic acid levels. J Plant Growth Regul 30:456–466

Asgher M, Khan MIR, Anjum NA, Khan NA (2015) Minimizing toxicity of cadmium in plants–role of plant growth regulators. Protoplasma 252:399–413

Ashraf M, Foolad M (2007) Roles of glycinebetaine and proline in improving plant abiotic stress resistance. Environ Exp Bot 59:206–216

Ashraf M, Akram NA, Arteca RN, Foolad MR (2010) The physiological, biochemical and molecular roles of brassinosteroids and salicylic acid in plant processes and salt tolerance. Crit Rev Plant Sci 29:162–190

Agami RA, Hafez EE, Taha TH (2013) Acquired resistant motivated by salicylic acid applications on salt stressed tomato (*Lycopersicon esculentum* Mill.). Am Eurasian J Agric Environ Sci 13 (1):50–57

Baier M, Kandlbinder A, Golldack D, Dietz KJ (2005) Oxidative stress and ozone: perception, signalling and response. Plant Cell Environ 28:1012–1020

Bali S, Poonam GV, Kaur P, Khanna K, Kaur R, Vig AP, Ohri P, Bhardwaj R (2017) Interaction of salicylic acid with plant hormones in plants under abiotic stress. In: Nazar R, Iqbal N, Khan NA (eds) Salicylic acid: a multifaceted hormone. Springer, Singapore, pp 201–219

Belkadhi A, DeHaro A, Obregon S, Chaïbi W, Djebali W (2014) Positive effects of salicylic acid pretreatment on the composition of flax plastidial membrane lipids under cadmium stress. Environ Sci Pollu Res 22:1457–1467

Blanco F, Garreton V, Frey N, Dominguez C, Pérez-Acle T, Van der Straeten D, Jordana X, Holuigue L (2005) Identification of NPR1-dependent and independent genes early induced by salicylic acid treatment in *Arabidopsis*. Plant Mol Biol 59:927–944

Borsani O, Valpuesta V, Botella MA (2001) Evidence for a role of salicylic acid in the oxidative damage generated by NaCl and osmotic stress in Arabidopsis seedlings. Plant Physiol 126:1024–1030

Cao Y, Zhang ZW, XueL W, Du JB, Shang J, Xu F, Yuan S, Lin HH (2009) Lack of salicylic acid in *Arabidopsis* protects plants against moderate salt stress. Z Naturforsch C64:231–238

Chen Z, Iyer S, Caplan A, Klessig DF, Fan B (1997) Differential accumulation of salicylic acid and salicylic acid – sensitive catalase in different rice tissue. Plant Physiol 114:193–201

Chen Z, Pottosin II, Cuin TA, Fuglsang AT, Tester M, Jha D, Zepeda-Jazo I, Zhou M, Palmgren MG, Newman IA, Shabala S (2007) Root plasma membrane transporters controlling K^+/Na^+ homeostasis in salt-stressed barley. Plant Physiol 145:1714–1725

Chung E, Park JM, Oh SK, Joung YH, Lee S, Choi D (2004) Molecular and biochemical characterization of the *Capsicum annuum* calcium- dependent protein kinase3 (CaCDPK3) gene induced by abiotic and biotic stresses. Planta 220:286–295

Coca M, San Segundo B (2010) AtCPK1 calcium-dependent protein kinase mediates pathogen resistance in Arabidopsis. Plant J 63:526–540

Csiszár J, Horváth E, VáryZ GÁ, Bela K, Brunner S, Tari I (2014) Glutathione transferase supergene family in tomato:salt stress- regulated expression of representative genes from distinct GST classes in plants primed with salicylic acid. Plant Physiol Biochem 78:15–26

Csiszár J, Brunner S, Horváth E, Bela K, Ködmön P, Riyazuddin R, Gallé Á, Hurton Á, Papdi C, Szabados L, Tari I (2018) Exogenously applied salicylic acid maintains redox homeostasis in salt-stressed *Arabidopsis gr1* mutants expressing cytosolic roGFP1. Plant Growth Regul:1–4. https://doi.org/10.1007/s10725-018-0420-6

Dai X, Cheng X, Li Y, Tang W, Han L (2012) Differential expression of gibberellin 20 oxidase gene induced by abiotic stresses in Zoysia grass (*Zoysia japonica*). Biologia 67:681–688

Deef HE (2007) Influence of salicylic acid on salt stress tolerance during seed germination of *Triticum aestivum* and *Hordeum vulgare*. Adv Biol Res 1:40–48

Dehghan S, Sadeghi M, Poppel A, Fischer R (2014) Differential inductions of phenylalanine ammonia-lyase and chalcone synthase during wounding, salicylic acid treatment, and salinity stress in safflower, *Carthamus tinctorious*. Biosci Rep 34:3

Dempsey DA, Klessig DF (1995) Signals in plant disease resistance. Bulletin de Inst Pasteur 93:167–186

Divi UK, Rahman T, Krishna P (2010) Brassinosteroid-mediated stress tolerance in *Arabidopsis* shows interactions with abscisic acid, ethylene and salicylic acid pathways. BMC Plant Biol 10:151

Dong FC, Wang PT, Song CP (2001) The role of hydrogen peroxide in salicylic acid-induced stomatal closure in *Vicia faba* guard cells. Acta Phytophysiol Sin 27:296–302

Dong YJ, Wang ZL, Zhang JW, Liu S, He ZL, He MR (2015) Interaction effects of nitric oxide and salicylic acid in alleviating salt stress of *Gossypium hirsutum* L. J Soil Sci Plant Nutr 15:561–573

Du L, Ali GS, Simons KA, Hou J, Yang T, Reddy AS, Poovaiah BW (2009) Ca^{2+}/calmodulin regulates salicylic-acid-mediated plant immunity. Nature 457:1156–1158

Durner J, Klessig DF (1995) Inhibition of ascorbate peroxidase by salicylic acid and 2, 6-dichloroisonicotinic acid, two inducers ofplant defense responses. Proc Natl Acad Sci 92:11312–11316

Durner JR, Klessig DF (1996) Salicylic acid is a modulator of tobaccoand mammalian catalases. J Biol Chem 271:28492–28501

El-Shabrawi H, Kumar B, Kaul T, Reddy MK, Singla-Pareek SL, Sopory SK (2010) Redox homeostasis, antioxidant defense, and methylglyoxal detoxification as markers for salt tolerance in Pokkali rice. Protoplasma 24:85–96

El-Tayeb MA (2005a) Response of barley grains to the interactive effect of salinity and salicylic acid. Plant Growth Regul 45:215–224

El-Tayeb MA (2005b) Response of barley grains to the interactive effect of salinity and salicylic acid. Plant Growth Regul 45:215–224

Eraslan F, Inal A, Pilbeam DJ, Gunes A (2008) Interactive effects of salicylic acid and silicon on oxidative damage and antioxidant activity in spinach (*Spinacia oleracea* L. cv. Matador) grown under boron toxicity and salinity. Plant Growth Regul 55:207–219

Esan AM, Masisi K, Dada FA, Olaiya CO (2017) Comparative effects of indole acetic acid and salicylic acid on oxidative stress marker and antioxidant potential of okra (*Abelmoschus esculentus*) fruit under salinity stress. Sci Hort 216:278–283

Faghih S, Ghobadi C, Zarei A (2017) Response of strawberry plant cv. Camarosato salicylic acid and methyl jasmonate application under salt stress condition. J Plant Growth Regul 36 (3):651–659

Fahad S, Bano A (2012) Effect of salicylic acid on physiological and biochemical characterization of maize grown in saline area. Pak J Bot 44:1433–1438

Farhangi-Abriz S, Ghassemi-Golezani K (2018) How can salicylic acid and jasmonic acid mitigate salt toxicity in soybean plants? Ecotoxicol Environ Saf 147:1010–1016

Fariduddin Q, Khan TA, Yusuf M, Aafaqee ST, Khalil RR (2017) Ameliorative role of salicylic acid and spermidine in the presence of excess salt in *Lycopersicon esculentum*. Photosynthetica 1:1–3

Faried HN, Ayyub CM, Amjad M, Ahmed R, Wattoo FM, Butt M, Bashir M, Shaheen MR, Waqas MA (2017) Salicylic acid confers salt tolerance in potato plants by improving water relations, gaseous exchange, antioxidant activities and osmoregulation. J Sci Food Agric 97 (6):1868–1875

Fatma M, Masood A, Per TS, Khan NA (2016) Nitric oxide alleviates salt stress inhibited photosynthetic performance by interacting with sulfur assimilation in mustard. Front Plant Sci 7:521

Fayez KA, Bazaid SA (2014) Improving drought and salinity tolerance in barley by application of salicylic acid and potassium nitrate. J Saudi Soc Agric Sci 13:45–55

Fujita M, Fujita Y, Noutoshi Y, Takahashi F, Narusaka Y, Yamaguchi-Shinozaki K, Shinozaki K (2006) Crosstalk between abiotic and biotic stress responses: a current view from the points of convergence in the stress signalling networks. Curr Opin Plant Biol 9:436–442

Garg N, Bharti A (2018) Salicylic acid improves arbuscular mycorrhizal symbiosis, and chickpea growth and yield by modulating carbohydrate metabolism under salt stress. Mycorrhiza 24:1–20

Gill SS, Tuteja N (2010) Reactive oxygen species and antioxidant machinery in abiotic stress tolerance in crop plants. Plant Physiol Biochem 48:909–930

Guan L, Scandalios JG (1995) Developmentally related responses of maize catalase genes to salicylic acid. Proc Natl Acad Sci 92:5930–5934

Gunes A, Inal A, Alpaslan M, Cicek N, Guneri E, Eraslan F, Guzelordu T (2005) Effects of exogenously applied salicylic acid on the induction of multiple stress tolerance and mineral nutrition in maize (*Zea mays* L.). Arch Agron Soil Sci 51:687–695

Gunes A, Inal A, Alpaslan M, Eraslan F, Bagci EG, Cicek N (2007) Salicylic acid induced changes on some physiological parameters symptomatic for oxidative stress and mineral nutrition in maize (*Zea mays* L.) grown under salinity. J Plant Physiol 164:728–736

Gupta BK, Sahoo KK, Ghosh A, Tripathi AK, Anwar K, Das P, Singh AK, Pareek A, Sopory SK, Singla-Pareek SL (2018) Manipulation of glyoxalase pathway confers tolerance to multiple stresses in rice. Plant Cell Environ 41(5):1186–1200

Gurmani AR, Khan SU, Ali A, Rubab T, Schwinghamer T, Jilani G, Farid A, Zhang J (2018) Salicylic acid and kinetin mediated stimulation of salt tolerance in cucumber (*Cucumis sativus* L.) genotypes varying in salinity tolerance. Hortic Environ Biotechnol 59(4):461–471

Guzmán-Téllez E, Montenegro DD, Benavides-Mendoza A (2014) Concentration of salicylic acid in tomato leaves after foliar aspersions of this compound. Am J Plant Sci 5:2048–2056

Hadi H, Najafabadi A, Amirnia R (2014) Comparison of different treatment methods of salicylic acid on some physiological traits of white bean under salinity stress. Cerceti Agro Moldova 3:97–105

Hamayun M, Shin JH, Khan SA, Ahmad B, Shin DH, Khan AL, Lee IJ (2010) Exogenous gibberellic acid reprograms soybean to higher growth and salt stress tolerance. J Agric Food Chem 70:7226–7232

Hammond-Kosack KE, Silverman P, Raskin I, Jones JDG (1996) Race-specific elicitors of *Cladosporium fulvum* induce changes in cell morphology and the synthesis of ethylene and salicylic acid in tomato plants carrying the corresponding Cf disease resistance gene. Plant Physiol 10:1381–1394

Hao F, Zhao S, Dong H, Zhang H, Sun L, Miao C (2010) Nia1and Nia2 are involved in exogenous salicylic acid-induced nitric oxide generation and stomatal closure in *Arabidopsis*. J Integ Plant Biol 52:298–307

Hao L, Zhao Y, Jin D, Zhang L, Bi X, Chen H, Xu Q, Ma C, Li G (2012) Salicylic acid-altering Arabidopsis mutants response to salt stress. Plant Soil 354:81–95

Hasanuzzaman M, Hossain MA, da Silva JAT, Fujita M (2012) Plant responses and tolerance to abiotic oxidative stress: antioxidant defenses is a key factors. In: Bandi V, Shanker AK, Shanker C, Mandapaka M (eds) Crop stress and its management: perspectives and strategies. Springer, Berlin, pp 261–316

Hasanuzzaman M, Nahar K, Bhuiyan TF, Anee TI, Inafuku M, Oku H, Fujita M (2017) Salicylic acid: an all rounder in regulating abiotic stress responses in plant. In: El-Esawi M (ed) Phytohormones signalling mechanisms and crosstalk in plant development and stress responses. IntechOpen, pp 31–72

Hayat Q, Hayat S, Irfan M, Ahmad A (2010) Effect of exogenous salicylic acid under changing environment: a review. Environ Exp Bot 68:14–25

Hayat S, Maheshwari P, Wani AS, Irfan M, Alyemeni MN, Ahmad A (2012) Comparative effect of 28 homobrassinolide and salicylic acid in the amelioration of NaCl stress in *Brassica juncea* L. Plant Physiol Biochem 53:61–68

He JH, Wang PT, Wang WJ, Song CP (2007) Effect of salicylic acid on stomatal movement and K^+ channel on plasma membrane of guard cells in *Vicia faba* L. Acta Agro Sin 33:1779–1787

Herrmann KM, Weaver LM (1999) The shikmate pathway. Annu Rev Plant Physiol Plant Mol Biol 50:473–503

Hoque TS, Burritt DJ, Hossain MA (2018) Glyoxalase system and salt stress tolerance in plants. In: Kumar V, Wani SH, Suprasanna P, LSP T (eds) Novel perspectives for salinity responses and tolerance in plants: targeting sensory, ion-transport and signalling mechanisms. Springer, Cham, pp 257–281

Horváth E, Pál M, Szalai G, Páldi E, Janda T (2007) Exogenous 4-hydroxy benzoic acid and salicylic acid modulate the effect of short-term drought and freezing stress on wheat plants. Biol Plant 51:480–487

Horváth E, Csiszár J, Gallé Á, Poór P, Szepesi Á, Tari I (2015) Hardening with salicylic acid induces concentration-dependent changes in abscisic acid biosynthesis of tomato under salt stress. J Plant Physiol 183:54–63

Hossain MA, Hossain MZ, Fujita M (2009) Stress-induced changes of methylglyoxal level and glyoxalase I activity in pumpkin seedlings and cDNA cloning of glyoxalase I gene. Aust J Crop Sci 3:53–64

Hossain MA, Hasanuzzaman M, Fujita M (2011) Coordinate induction of antioxidant defense and glyoxalase system by exogenous proline and glycinebetaine is correlated with salt tolerance in mung bean. Front Agric China 5(1):1–14

Hoyos ME, Zhang S (2000) Calcium-independent activation of salicylic acid-induced protein kinase and a 40-kilodalton protein kinase by hyperosmotic stress. Plant Physiol 122:1355–1364

Husen A, Iqbal M, Sohrab SS, Ansari MK (2018) Salicylic acid alleviates salinity-caused damage to foliar functions, plant growth and antioxidant system in Ethiopian mustard (*Brassica carinata*A. Br.). Agric Food Secur 7(1):44

Hao JH, Wang XL, Dong CJ, Zhang ZG, Shang QM (2011) Salicylic acid induces stomatal closure by modulating endogenous hormone levels in cucumber cotyledons. Russ J Plant Physiol 58 (5):906–913

Horváth E, Janda T, Szalai G, Páldi E (2002) In vitro salicylic acid inhibition of catalase activity in maize: differences between the isozymes and a possible role in the induction of chilling tolerance. Plant Sci 163:112–135

Iglesias MJ, Terrile MC, Casalongué CA (2011) Auxin and salicylic acid signalling counteract during the adaptive response to stress. Plant Signal Behav 6:452–454

Inada M, Ueda A, Shi W, Takabe T (2005) A stress-inducible plasmamembrane protein 3 (AcPMP3) in a monocotyledonous halophyte, *Aneurolepidium chinense*, regulates cellular Na^+ and K^+ accumulation under salt stress. Planta 220:395–402

Innes R (2018) The positives and negatives of NPR: a unifying model for salicylic acid signalling in plants. Cell 173:1314–1315

Iqbal N, Umar S, Khan NA, Khan MIR (2014) A new perspective of phytohormones in salinity tolerance: regulation of proline metabolism. Environ Exp Bot 100:34–42

Jagendorf AT, Takabe T (2001) Inducers of glycinebetaine synthesis in barley. Plant Physiol 127:1827–1835

Jaiswal A, Pandurangam V, Sharma SK (2014) Effect of salicylic acid in soybean (*Glycine max* L. meril) under salinity stress. Bioscan 9:671–676

Jayakannan M, Bose J, Babourina O, Rengel Z, Shabala S (2013) Salicylic acid improves salinity tolerance in *Arabidopsis* by restoring membrane potential and preventing salt-induced K^+ loss via a GORK channel. J Exp Bot 64(8):2255–2268

Jayakannan M, Bose J, Babourina O, Rengel Z, Shabala S (2015a) Salicylic acid in plant salinity stress signalling and tolerance. Plant Growth Regul 76:25–40

Jayakannan M, Bose J, Babourina O, Shabala S, Massart A, Poschenrieder C, Rengel Z (2015b) NPR1-dependent salicylic acid signalling pathway is pivotal for enhanced salt and oxidative stress tolerance in Arabidopsis. J Exp Bot. https://doi.org/10.1093/jxb/eru528

Jini D, Joseph B (2017) Physiological mechanism of salicylic acid for alleviation of salt stress in rice. Rice Sci 24:97–108

Kang GZ, Li GZ, Liu GQ, Xu W, Peng XQ, Wang CY, Zhu YJ, Guo TC (2013) Exogenous salicylic acid enhances wheat drought tolerance by influence on the expression of genes related to ascorbate-glutathione cycle. Biol Plant 57:718–724

Kangasjarvi J, Jaspers P, Kollist H (2005) Signalling and cell death in ozone-exposed plants. Plant Cell Environ 28:1021D1036

Karlidag H, Yildirim E, Turan M (2009) Salicylic acid ameliorates the adverse effect of salt stress on strawberry. Sci Agric 66:180–187

Kawano T, Sahashi N, Takahashi K, Uozumi N, Muto S (1998) Salicylic acid induces extracellular superoxide generation followed by an increase in cytosolic calcium ion into bacco suspension culture: the earliest events in salicylic acid signal transduction. Plant Cell Physiol 39:721–730

Kaydan D, Yagmur M, Okut N (2007) Effects of salicylic acid on the growth and some physiological characters in salt stressed wheat (*Triticum aestivum* L.). Tarim Bilimleri Dergisi 13:114–119

Kerkeb L, Donaire JP, Rodríguez-Rosales MP (2001) Plasmamembrane H+-ATPase activity is involved in adaptation of tomato calli to NaCl. Physiol Plant 111:483–490

Khan NA, Syeed S, Masood A, Nazar R, Iqbal N (2010) Application of salicylic acid increases contents of nutrients and antioxidative metabolism in mungbean and alleviates adverse effects of salinity stress. Intl J Plant Biol 1:e1

Khan NA, Nazar R, Iqbal N, Anjum NA (2012a) Phytohormones and abiotic stress tolerance in plants. Springer, Berlin. https://doi.org/10.1007/978-3-642-25829-9

Khan MIR, Syeed S, Nazar R (2012b) An insight into the role of salicylic acid and jasmonic acid in salt stress tolerance. In: Khan NA, Nazar R, Iqbal N, Anjum NA (eds) Phytohormones and abiotic stress tolerance in plants. Springer, Berlin, pp 277–300

Khan MIR, Iqbal N, Masood A, Khan NA (2012c) Variation in salt tolerance of wheat cultivars: role of glycinebetaine and ethylene. Pedosphere 22:746–754

Khan MIR, Iqbal N, Masood A, Per TS, Khan NA (2013) Salicylic acid alleviates adverse effects of heat stress on photosynthesis through changes in proline production and ethylene formation. Plant Signal Behav 8:e26374

Khan MIR, Asgher M, Khan NA (2014) Alleviation of salt-induced photosynthesis and growth inhibition by salicylic acid involves glycinebetaine and ethylene in mungbean (*Vigna radiata* L.). Plant Physiol Biochem 80:67–74

Khan MIR, Fatma M, Per TS, Anjum NA, Khan NA (2015) Salicylic acid-induced abiotic stress tolerance and underlying mechanisms in plants. Front Plant Sci 6:462

Khodary SEA (2004) Effect of salicylic acid on growth, photosynthesis and carbohydrate metabolism in salt stressed maize plants. Int J Agric Biol 6:5–8

Khokon MAR, Okuma E, Hossain MA, Munemasa S, Uraji M, Nakamura Y, Murata Y (2011) Involvement of extracellular oxidative burst in salicylic acid-induced stomatal closure in *Arabidopsis*. Plant Cell Environ 34:434–443

Khokon MAR, Salam MA, Jammes F, Ye W, Hossain MA, Okuma E, Nakamura Y, Mori IC, Kwak JM, Murata Y (2017) MPK9 and MPK12 function in SA-induced stomatal closure in *Arabidopsis thaliana*. Biosci Biotechnol Biochem 81:1394–1400

Kim Y, Kim S, Shim IS (2017) Exogenous salicylic acid alleviates salt-stress damage in cucumber under moderate nitrogen conditions by controlling endogenous salicylic acid levels. Hort Environ Biotechnol 58(3):247–253

Kim Y, Mun BG, Khan AL, Waqas M, Kim HH, Shahzad R, Imran M, Yun BW, Lee IJ (2018) Regulation of reactive oxygen and nitrogen species by salicylic acid in rice plants under salinity stress conditions. PloS one 13(3):e0192650

Kumar M (2013) Crop plants and abiotic stresses. J Biomol Res Ther 3:e125

Kumara GDK, Xia Y, Zhu ZJ, Basnayake BMVS, Beneragama CK (2010) Effects of exogenous salicylic acid on antioxidative enzyme activity and physiological characteristics in gerbera (*Gerbera jamesonii* L.) grown under NaCl stress. J Zhejiang Univ Agric Life Sci 36:591–601

Lee S, Park CM (2010) Modulation of reactive oxygen species by salicylic acid in *Arabidopsis* seed germination under high salinity. Plant Signal Behav 5:1534–1536

Lee S, Kim SG, Park CM (2010) Salicylic acid promotes seed germination under high salinity by modulating antioxidant activity in Arabidopsis. New Phytol 188:626–637

Lee SY, Damodaran PN, Roh KS (2014) Influence of salicylic acid on rubisco and rubisco activase in tobacco plant grown under sodium chloride in vitro. Saudi J Biol Sci 21(5):417–426

Li J, Besseau S, Törönen P, Sipari N, Kollist H, Holm L, Palva ET (2013a) Defense-related transcription factors WRKY70 and WRKY54 modulate osmotic stress tolerance by regulating stomatal aperture in Arabidopsis. New Phytol 200:457–472

Li G, Peng X, Wei L, Kang G (2013b) Salicylic acid increases the contents of glutathione and ascorbate and temporally regulates the related gene expression in salt-stressed wheat seedlings. Gene 529:321–325

Li T, Hu Y, Du X, Tang H, Shen C, Wu J (2014) Salicylic acid alleviates the adverse effects of salt stress in *Torreya grandis* cv. Merrillii seedlings by activating photosynthesis and enhancing antioxidant systems. PLoS One 9(10):e109492

Love AJ, Milner JJ, Sadanandom A (2008) Timing is everything: regulatory overlap in plant cell death. Trends Plant Sci 13:589–595

Ma X, Zheng J, Zhang X, Hu Q, Qian R (2017) Salicylic acid alleviates the adverse effects of salt stress on *Dianthus superbus* (Caryophyllaceae) by activating photosynthesis, protecting morphological structure, and enhancing the antioxidant system. Front Plant Sci 8:600

Mahajan S, Tuteja N (2005) Cold, salinity and drought stresses: an overview. Arch Biochem Biophys 444:139–158

Mateo A, Mühlenbock P, Rustérucci C, Chang CC, Miszalski Z, Karpinska B, Parker JE, Mullineaux PM, Karpinski S (2004) Lesion Simulating disease 1 is required for acclimation to conditions that promote excess excitation energy. Plant Physiol 136:2818–2830

Melotto M, Underwood W, Koczan J, Nomura K, He SY (2006) Plant stomata function in innate immunity against bacterial invasion. Cell 126:969–980

Methenni K, Abdallah MB, Nouairi I, Smaoui A, Zarrouk M, Youssef NB (2018) Salicylic acid and calcium pretreatments alleviate the toxic effect of salinity in the Oueslati olive variety. Sci Hort 233:349–358

Mikolajczyk M, Awotunde OS, Muszynska G, Klessig DF, Dobrowolska G (2000) Osmotic stress induces rapid activation of a salicylic acid-induced protein kinase and a homolog of protein kinase ASK1 in tobacco cell. Plant Cell 12:165–178

Miller G, Suzuki N, Ciftci-Yilmaz S, Mittler R (2009) Reactive oxygen species homeostasis and signalling during drought and salinity stresses. Plant Cell Environ 33:453–467

Misra N, Misra R (2012) Salicylic acid changes plant growth parameters and proline metabolism in *Rauwolfia serpentina* leaves grown under salinity stress. Am Eurasian J Agri Environ Sci 12:1601–1609
Misra N, Saxena P (2009) Effect of salicylic acid on proline metabolism in lentil grown under salinity stress. Plant Sci 177:181–189
Miura K, Tada Y (2014) Regulation of water, salinity, and cold stress responses by salicylic acid. Front Plant Sci 5:4
Miura K, Sato A, Ohta M, Furukawa J (2011) Increased tolerance to salt stress in the phosphate-accumulating Arabidopsis mutants *siz1* and *pho2*. Planta 234:1191–1199
Miura K, Okamoto H, Okuma E, Shiba H, Kamada H, Hasegawa PM, Murata Y (2013) SIZ1 deficiency causes reduced stomatal aperture and enhanced drought tolerance via controlling salicylic acid-induced accumulation of reactive oxygen species in *Arabidopsis*. Plant J 73:91–104
Moharekar S, Lokhande S, Hara T, Tanaka R, Tanaka A, Chavan P (2003) Effect of salicylic acid on chlorophyll and carotenoid contents of wheat and moong seedlings. Photosynthetica 147:315–317
Mori IC, Pinontoan R, Kawano T, Muto S (2001) Involvement of superoxide generation in salicylic acid-induced stomatal closure in *Vicia faba*. Plant Cell Physiol 42:1383–1388
Mostofa MG, Hossain MA, Fujita M (2015) Trehalose pretreatment induces salt tolerance in rice seedlings: oxidative damage and co-induction of antioxidant defense and glyoxalase systems. Protoplsasma 252:461–475
Munns R, Tester M (2008) Mechanisms of salinity tolerance. Annu Rev Plant Biol 59:651–681
Mustafa NR, Kim HK, Choi YH, Erkelens C, Lefeber AW, Spijksma G, van der Haijen R, Verpoorte R (2009) Biosynthesis of salicylic acid in fungus elicited *Catharanthus roseus* cells. Phytochemistry 70:532–539
Miller G, Shulaev V, Miller R (2008) Reactive oxygen species homeostasis and abiotic stress. Physiol Plant 133:481–489
Mitsuya S, Taniguchi M, Miyake H, Takabe T (2005) Disruption of RCI2A leads to over-accumulation of Na+ and increased salt sensitivity in *Arabidopsis thaliana* plants. Planta 222:1001–1009
Naser Alavi SM, Arvin MJ, Kalantari KM (2014) Salicylic acid and nitric oxide alleviate osmotic stress in wheat (*Triticum aestivum* L.) seedlings. J Plant Interact 9:683–688
Nazar R, Iqbal N, Syeed S, Khan NA (2011) Salicylic acid alleviates decreases in photosynthesis under salt stress by enhancing nitrogen and sulfur assimilation and antioxidant metabolism differentially in two mungbean cultivars. J Plant Physiol 168(8):807–815
Nazar R, Umar S, Khan NA (2015) Exogenous salicylic acid improves photosynthesis and growth through increase in ascorbate-glutathione metabolism and S assimilation in mustard under salt stress. Plant Signal Behav 10(3):e1003751
Nemeth M, Janda T, Horváth E, Paldi E, Szalai G (2002) Exogenous salicylic acid increases polyamine content but may decrease drought tolerance in maize. Plant Sci 162:569–574
Nimir NEA, Zhou G, Guo W, Ma B, Lu S, Wang Y (2017) Effect of foliar application of GA3, kinetin, and salicylic acid on ions content, membrane permeability, and photosynthesis under salt stress of sweet sorghum [*Sorghum bicolor* (L.) Moench]. Can J Plant Sci 97:525–535
Nishimura N, Kitahata N, Seki M, Narusaka Y, Narusaka M, Kuromori T, Asami T, Shinozaki K, Hirayama T (2005) Analysis of ABA hypersensitive germination 2 revealed the pivotal functions of PARN in stress response in Arabidopsis. Plant J 44:972–984
Noreen S, Ashraf M (2010) Modulation of salt (NaCl)-induced effects on oil composition and fatty acid profile of sunflower (*Helianthus annuus* L.) by exogenous application of salicylic acid. J Sci Food Agric 90:2608–2616
Noreen S, Ashraf MH, Jamil A (2009) Exogenous application of salicylic acid enhances antioxidative capacity in salt stressed sunflower (*Helianthus annuus* L.) plants. Pak J Bot 41:473–479

Palma F, López-Gómez M, Tejera NA, Lluch C (2013) Salicylic acid improves the salinity tolerance of *Medicago sativa* in symbiosis with *Sinorhizobium meliloti* by preventing nitrogen fixation inhibition. Plant Sci 208:75–82

Pancheva TV, Popova LV (1997) Effect of salicylic acid on the synthesis of ribulose-1, 5-bisphosphate carboxylase/oxygenase inbarley leaves. J Plant Physiol 220:381–386

Parida AK, Das AB (2005) Salt tolerance and salinity effects on plants: a review. Ecotoxicol Environ Saf 60:324–349

Pieterse CM, Leon-Reyes A, Van Der Ent S, Van Wees SC (2009) Networking by small molecule hormones in plant immunity. Nat Chem Biol 5:308–316

Pirasteh-Anosheh H, Emam Y, Rousta MJ, Ashraf M (2017) Salicylic acid induced salinity tolerance through manipulation of ion distribution rather than ion accumulation. J Plant Growth Regul 36(1):227–239

Pitman MG, Läuchli A (2002) Global impact of salinity and agricultural ecosystems. In: Läuchli A, Lüttge U (eds) Salinity: environment – plants- molecules. Kluwer, Dordrecht, pp 3–20

Poór P, Gémes K, Horváth F, Szepesi Á, Simon ML, Tari I (2011a) Salicylic acid treatment via the rooting medium interferes with stomatal response, CO_2 fixation rate and carbohydrate metabolism in tomato, and decreases harmful effects of subsequent salt stress. Plant Biol 13:105–114

Poór P, Szopkó D, Tari I (2011b) Ionic homeostasis disturbance is involved in tomato cell death induced by NaCl and salicylic acid. In Vitro Cell Dev Biol Plant 48:377–382

Prodhan MY, Munemasa S, Nahar MNEN, Nakamura Y, Murata Y (2018) Guard cell salicylic acid signalling is integrated into abscisic acid signalling via the Ca/CPK-dependent pathway. Plant Physiol 178:441–450

Qadir M, Nangia V, Singh M, Murtaza G, Singh M, Thomas RJ, Drechsel P, Noble AD (2014) Economics of salt-induced land degradation and restoration. Nat Resour Forum 38:282–295

Quan LJ, Zhang B, Shi WW, Li HY (2008) Hydrogen peroxide in plants: a versatile molecule of the reactive oxygen species network. J Integr Plant Biol 50:2–18

Rajeshwari V, Bhuvaneshwari V (2017) Salicylic acid induced salt stress tolerance in plants. Int J Plant Biol Res 5:1067

Rajjou L, Belghazi M, Huguet R, Robin C, Moreau A, Job C, Job D (2006) Proteomic investigation of the effect of salicylic acid on Arabidopsis seed germination and establishment of early defense mechanisms. Plant Physiol 141:910–923

Raskin I, Skubatz H, Tang W, Meeuse BJD (1990) Salicylic acid levels in thermogenic and non-thermogenic plants. Ann Bot 66:369–373

Rivas-San Vicente M, Plasencia J (2011) Salicylic acid beyond defence: its role in plant growth and development. J Exp Bot 62:3321–3338

Roy SJ, Negrão S, Tester M (2014) Salt resistant crop plants. Curr Opin Biotechnol 26:115–124

Sadeghi M, Dehghan S, Fischer R, Wenzel U, Vilcinskas A, Kavousi HR, Rahnamaeian M (2013) Isolation and characterization of isochorismate synthase and cinnamate 4-hydroxylase during salinity stress, wounding, and salicylic acid treatment in *Carthamus tinctorius*. Plant Signal Behav 8:11

Sahar K, Baghizadeh A, Taher NM (2011) The salicylic acid effect on the *Salvia officianlis* sugar, protein and proline contents under salinity (NaCl) stress. J Stress Physiol Biochem 7:80–87

Salzman RA, Brady JA, Finlayson SA, Buchanan CD, Summer EJ, Sun F, Klein PE, Klein RR, Pratt LH, Cordonnier-Pratt MM, Mullet JE (2005) Transcriptional profiling of sorghum induced by methyl jasmonate, salicylic acid, and amino cyclopropane carboxylic acid reveals cooperative regulation and novel gene responses. Plant Physiol 138:352–368

Sawada H, Shim I-S, Usui K (2006) Induction of benzoic acid 2-hydroxylase and salicylic acid biosynthesis – modulation by salt stress in rice seedlings. Plant Sci 171:263–270

Shabala S, Cuin TA (2008) Potassium transport and plant salt tolerance. Physiol Plant 133:651–669

Shah J (2003) The salicylic acid loop in plant defense. Curr Opin Plant Biol 6:365D371

Shaki F, Maboud HE, Niknam V (2018) Growth enhancement and salt tolerance of safflower (*Carthamus tinctorius* L.), by salicylic acid. Curr Plant Biol. https://doi.org/10.1016/j.cpb.2018.04.001

Shakirova FM, Sakhabutdinova AR, Bezrukova MV, Fatkhutdinova RA, Fatkhutdinova DR (2003) Changes in the hormonal status of wheat seedlings induced by salicylic acid and salinity. Plant Sci 164:317–322

Sheteiwy MS, An J, Yin M, Jia X, Guan Y, He F, Hu J (2018) Cold plasma treatment and exogenous salicylic acid priming enhances salinity tolerance of *Oryza sativa* seedlings. Protoplasma 8:1–20

Shi HZ, Ishitani M, Kim CS (2000) The *Arabidopsis thaliana* salt tolerance gene SOS1 encodes a putative Na+/H+antiporter. Proc Natl Acad Sci USA 97:6896–6901

Shim LS, Momose Y, Yamamoto A, Kim DW, Usui K (2003) Inhibition of catalase activity by oxidative stress and its relationship to salicylic acid accumulation in plants. Plant Growth Regul 39:285–292

Silverman P, Seskar M, Kanter D, Schweizer P, Métraux JP, Raskin I (1995) Salicylic acid in rice: biosynthesis, conjugation, and possible role. Plant Physiol 108:633–639

Simaei M, Khavarinejad RA, Saadatmand S, Bernard F, Fahimi H (2011) Interactive effects of salicylic acid and nitric oxide on soybean plants under NaCl salinity. Russ J Plant Physiol 58:783

Singh B, Usha K (2003) Salicylic acid induced physiological and biochemical changes in wheat seedlings under water stress. Plant Growth Regul 39:137–141

Spoel SH, Dong X (2008) Making sense of hormone crosstalk during plant immune responses. Cell Host Microbe 3:348–351

Stevens J, Senaratna T, Sivasithamparam K (2006) Salicylic acid induces salinity tolerance in tomato (*Lycopersicon esculentum* cv. Roma): associated changes in gas exchange, water relations and membrane stabilisation. Plant Growth Regul 49:77–83

Strawn MA, Marr SK, Inoue K, Inada N, Zubieta C, Wildermuth MC (2007) Arabidopsis isochorismate synthase functional in pathogen-induced salicylate biosynthesis exhibits properties consistent with a role in diverse stress responses. J Biol Chem 282:5919–5933

Su H, Song S, Yan X, Fang L, Zeng B, Zhu Y (2018) Endogenous salicylic acid shows different correlation with baicalin and baicalein in the medicinal plant *Scutellaria baicalensis* Georgi subjected to stress and exogenous salicylic acid. PLoS ONE 13:e0192114

Sudhir P, Murthy SDS (2004) Effects of salt stress on basic processes of photosynthesis. Photosynthetica 42:481–486

Syeed S, Khan NA (2010) Physiological aspects of salicylic acid-mediated salinity tolerance in plants. Plant Stress 4:39–46

Syeed S, Anjum NA, Nazar R, Iqbal N, Masood A, Khan NA (2011) Salicylic acid-mediated changes in photosynthesis, nutrients content and antioxidant metabolism in two mustard (*Brassicajuncea* L.) cultivars differing in salt tolerance. Acta Physiol Plant 33(3):877–886

Szepesi A, Poór P, Gémes K, Horváth E, Tari I (2008) Influence of exogenous salicylic acid on antioxidant enzyme activities in the roots of salt stressed tomato plants. Acta Biol Szeged 52:199–200

Szepesi A, Csiszára J, Gémesa K, Horvátha E, Horvátha F, Simonb ML, Tari I (2009) Salicylic acid improves acclimation to salt stress by stimulating abscisic aldehyde oxidase activity and abscisic acid accumulation, and increases Na^+ content in leaves without toxicity symptoms in *Solanum lycopersicum* L. J Plant Physiol 166:914–925

Sohag AAA, Tahjib-Ul-Arif M, Afrin S, Khan MK, Hannan MA, Skalicky M, Mortuza MG, Brestic M, Hossain MA, Murata Y (2020a) Insights into nitric oxide-mediated water balance, antioxidant defence and mineral homeostasis in rice (*Oryza sativa* L.) under chilling stress. Nitric Oxide 100:7–16. https://doi.org/10.1016/j.niox.2020.04.001

Sohag AAA, Tahjib-Ul-Arif M, Brestič M, Afrin S, Sakil MA, Hossain MT, Hossain MA, Hossain MA (2020b) Exogenous salicylic acid and hydrogen peroxide attenuates drought stress in rice. Plant Soil Environ 66:7–13. https://doi.org/10.17221/472/2019-PSE

Tahjib-Ul-Arif M, Siddiqui MN, Sohag AA, Sakil MA, Rahman MM, Polash MA, Mostofa MG, Tran LS (2018a b) Salicylic acid-mediated enhancement of photosynthesis attributes and antioxidant capacity contributes to yield improvement of maize plants under salt stress. J Plant Growth Regul 37:1318-1330

Tahjib-Ul-Arif M, Roy PR, Sohag AA, Afrin S, Rady MM, Hossain MA (2018b a) Exogenous calcium supplementation improves salinity tolerance in BRRI Dhan28; a salt-susceptible high-yielding *Oryza sativa* cultivar. J Crop Sci and Biotech 21:383-394.

Tari I, Csiszár J, Horváth E, Poór P, Takács Z, Szepesi A (2015) The alleviation of the adverse effects of salt stress in the tomato plant by salicylic acid shows a time and organ-specific antioxidant response. Acta Biol Cracov Ser Bot 57:21–30

Tavakkoli E, Rengasamy P, McDonald GK (2010) High concentrations of Na^+ and Cl^- ions in soil solution have simultaneous detrimental effects on growth of faba bean under salinity stress. J Exp Bot 61:4449–4459

Torabian AR (2010) Effect of salicylic acid on germination and growth of alfalfa (*Medicago sativa* L.) seedlings under water potential loss at salinity stress. Plant Ecophysiol 2:151–155

Tufail A, Arfan M, Gurmani AR, Khan A, Bano A (2013) Salicylic acid induced salinity tolerance in maize (*Zea mays*). Pak J Bot 45:75–82

Tuteja N, Sopory SK (2008) Chemical signaling under abiotic stress environment in plants. Plant Signal Behav 3:525–536

Upadhyaya CP, Venkatesh J, Gururani MA, Asnin L, Sharma K, Ajappala H, Park SW (2011) Transgenic potato overproducing L-ascorbic acid resisted an increase in methylglyoxal under salinity stress via maintaining higher reduced glutathione level and glyoxalase enzyme activity. Biotechnol Lett 33:2297–2307

Uzunova A, Popova L (2000) Effect of salicylic acid on leaf anatomy and chloroplast ultra structure of barley plants. Photosynthetica 38:243–250

Verberne MC, BudiMuljono RA, Verpoorte R (1999) Salicylic acid biosynthesis. In: Hooykaas PJJ, Hall MA, Libbenga KR (eds) Biochemistry and molecular biology of plant hormones. Elsevier Science BV, Amsterdam, pp 295–314

Vlot AC, Dempsey DA, Klessig DF (2009) Salicylic acid, a multifaceted hormone to combat disease. Annu Rev Phytopathol 47:177–206

Wang L, Tsuda K, Sato M, Cohen JD, Katagiri F, Glazebrook J (2009) Arabidopsis CaM binding protein CBP60g contributes to MAMP-induced SA accumulation and is involved in disease resistance against *Pseudomonas syringae*. PLoS Pathog 5:e1000301

Wasti S, Mimouni H, Smiti S, Zid E, Ben Ahmed H (2012) Enhanced salt tolerance of tomatoes by exogenous salicylic acid applied through rooting medium. OMICS 16:200–207

Wildermuth MC, Dewdney J, Wu G, Ausubel FM (2001) Isochorismate synthase is required to synthesize salicylic acid for plant defence. Nature 414:562–565

Wolters H, Jürgens G (2009) Survival of the flexible: hormonal growth control and adaptation in plant development. Nat Rev Genet 10:305–317

Wu Y, Zhang D, Chu JY, Boyle P, Wang Y, Brindle ID, De Luca V, Després C (2012) The Arabidopsis NPR1 protein is a receptor for the plant defense hormone salicylic acid. Cell Rep 1:639–647

Xia J, Zhao H, Liu W, Li L, He Y (2009) Role of cytokinin and salicylic acid in plant growth at low temperatures. Plant Growth Regul 57:211–221

Xie Z, Zhang ZL, Hanzlik S, Cook E, Shen QJ (2007) Salicylic acid inhibits gibberellin-induced alpha-amylase expression and seed germination via a pathway involving an abscisic-acid-inducible WRKY gene. Plant Mol Biol 64:293–303

Yadu S, Dewangan TL, Chandrakar V, Keshavkant S (2017) Imperative roles of salicylic acid and nitric oxide in improving salinity tolerance in *Pisum sativum* L. Physiol Mol Biol Plant 23:43–58

Yalpani N, Enyedi AJ, Leon R, Raskin I (1994) Ultraviolet light and ozone stimulate accumulation of salicylic acid, pathogen-related proteins and virus resistance in tobacco. Planta 193:273–376

Yan Y, Pan C, Du Y, Li D, Liu W (2018) Exogenous salicylic acid regulates reactive oxygen species metabolism and ascorbate–glutathione cycle in *Nitraria tangutorum* Bobr. under salinity stress. Physiol Mol Biol Plants 24(4):577–589

Yang T, Poovaiah BW (2002) A calmodulin-binding / CGCG box DNA-binding protein family involved in multiple signaling pathways in plants. J Biol Chem 277:45049–45058

Yasuda M, Ishikawa A, Jikumaru Y, Seki M, Umezawa T, Asami T, Maruyama-Nakashita A, Kudo T, Shinozaki K, Yoshida S (2008) Antagonistic interaction between systemic acquired resistance and the abscisic acid-mediated abiotic stress response in Arabidopsis. Plant Cell 20:1678–1692

Yildirim E, Turan M, Guvenc I (2008) Effect of foliar salicylic acid applications on growth, chlorophyll, and mineral content of cucumber grown under salt stress. J Plant Nutr 31:593–612

Yuan S, Lin HH (2008) Role of salicylic acid in plant abiotic stress. Z Naturforsch 63c:313–320

Yuan Z, Cong G, Zhang J (2014) Effects of exogenous salicylic acid on polysaccharides production of *Dendrobium officinale*. South Afr J Bot 95:78–84

Zheng J, Ma X, Zhang X, Hu Q, Qian R (2018) Salicylic acid promotes plant growth and salt-related gene expression in *Dianthus superbus* L. (Caryophyllaceae) grown under different salt stress conditions. Physiol Mol Plant 24(2):231–238

Zhu JK (2002) Salt and drought stress signal transduction in plants. Annu Rev Plant Biol 53:247–273

Zottini M, Costa A, DeMichele R, Ruzzene M, Carimi F, Lo Schiavo F (2007) Salicylic acid activates nitric oxide synthesis in *Arabidopsis*. J Exp Bot 58:1397–1405

Biotechnology for Extraction of Plant Phenolics

2

Luis A. Cabanillas-Bojórquez, Erick P. Gutiérrez-Grijalva, Laura A. Contreras-Angulo, Sara Aviles-Gaxiola, and J. Basilio Heredia

Abstract

Several biotechnological techniques are increasingly being employed in plants to enhance their nutritional value and phenolic compound content. As a result, the number of studies regarding the effects of different biotechnological techniques has risen in the last years. Current studies are specially focused on the high phenolic extraction conditions, and recently, the use of industrial by-products as a potential source of bioactive compounds has been studied. To fully comprehend the potential benefits of biotechnological techniques on plant phenolic improvement, research should focus on the advantages, characteristics, and difference of biotechnological techniques in contrast to conventional and emerging process extraction in plants. This chapter aims to describe the state of the art on plant phenolics enhanced by biotechnological techniques.

Keywords

Plant phenolics · Extraction · Bioactive compounds · Biotechnological techniques

L. A. Cabanillas-Bojórquez · L. A. Contreras-Angulo · S. Aviles-Gaxiola · J. B. Heredia (✉)
Department of Food Science and Technology, Centro de Investigación en Alimentación y Desarrollo, A.C., Culiacán, Sinaloa, Mexico
e-mail: jbheredia@ciad.mx

E. P. Gutiérrez-Grijalva
Cátedras CONACYT – Centro de Investigación en Alimentación y Desarrollo, AC, Culiacán, Sinaloa, México

R. Lone et al. (eds.), *Plant Phenolics in Sustainable Agriculture*,
https://doi.org/10.1007/978-981-15-4890-1_2

2.1 Introduction

Phenolic compounds are a vastly diverse group of phytochemicals; they are produced by the secondary metabolism of plants. Structurally, phenolic compounds constitute a basic aromatic ring with a hydroxyl group (-OH) attached. Plant phenolics are biosynthesized as defense metabolites aimed to protect the plant against biotic and abiotic stresses. Phenolic compounds can be classified depending on their basic structure of phenolic acids and flavonoids. Flavonoids are characterized by the presence of three rings (C6-C3-C6) named A, B, and C. Flavonoids can be subclassified depending on their distribution of -OH radicals and prenylation and glycosylation into flavonols, flavones, flavan-3-ols, anthocyanidins, flavanones, isoflavones, etc. Similarly, phenolic acids can be classified depending on whether they are derived from hydroxycinnamic or hydroxybenzoic acid (Vermerris and Nicholson 2006; Kutchan et al. 2015).

2.1.1 Bioactive Properties of Phenolic Compounds

Phenolic compounds are compounds that have been associated in the prevention of various diseases; also, the beneficial effect of phenolic compounds has been attributed mainly to their antioxidant activity. Phenolic compounds have been largely reported because of their anticancer, anti-inflammatory, cardioprotective, and antimicrobial properties (Fig. 2.1).

2.1.1.1 Antioxidant Activity

It has been reported that phenolic compounds possess antioxidant activity due to their defined structures that confer the ability to accept or donate electrons or hydrogen atoms, which in turn stabilizes free radicals as well as chelates metal cations and prevents oxidative stress (Balasundram et al. 2006). Despite the high activity evaluated in in vitro studies, the human body cannot absorb 100% of the phenolic compounds; in addition, the absorption mechanism is not yet completely elucidated (Balasundram et al. 2006).

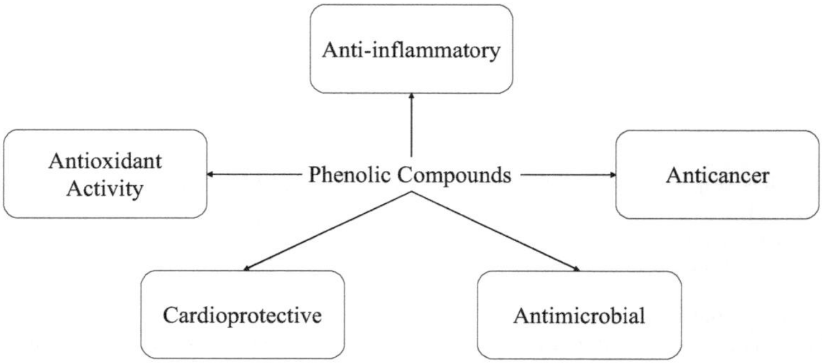

Fig. 2.1 Properties of phenolic compounds

2.1.1.2 Anticancer

Cancer is a disease of great worldwide interest; it develops due to various factors such as environment, physical, chemical, genetic, or metabolic factors, which trigger various reactions in the body such as abnormal cell division and spread to adjacent tissues. A strategy for the prevention of this disease is a diet with high levels of phytochemicals such as fruits and vegetables, because these compounds have been attributed high antioxidant activity; also they are linked to reducing the incidence of cancer. Among these phytochemicals are the phenolic compounds. It has been reported that these are widely distributed and are found mainly in vegetables; on the other hand, their antioxidant activity has been related to carcinogenesis modulation through in vitro and in vivo studies (Dai and Mumper 2010).

The cancer cell lines that have been studied in vitro to test the anticarcinogenic effect of phenolic compounds are found in the breast (MCF-7), prostate (LNCaP, DU-145), colon (HT-29, HTC-116), and human oral cavity (KB, CAL-27), reporting a tumor growth reduction after treatment with phenolic compounds of different sources (Dai and Mumper 2010; Seeram et al. 2006; Zhang et al. 2008). The effect of phenolic compounds has been reported in in vivo models, and it was found that these compounds when added to the diet of rats induced to different types of tumors (colon carcinoma, skin papilloma, human lung carcinoma, among others) have a beneficial effect; therefore, the consumption of phenolic compounds could help prevent various conditions in humans (Dai and Mumper 2010; Ding et al. 2006; Lala et al. 2006).

2.1.1.3 Anti-inflammatory

Polyphenols have been studied to elucidate their effects in human health. In this sense, it has been reported that these inhibited enzymes related to the inflammatory process; among them, mitogen-activated protein kinase (MAPK), protein kinase-C, nitric oxide synthase (iNOS), nuclear factor-kappa B (NF-kB), and phase II antioxidant detoxifying and activating protein-1 enzymes have been inhibited by polyphenols, and thus, polyphenols possess an anti-inflammatory effect (Dziadek et al. 2019; Santangelo et al. 2007).

2.1.1.4 Cardioprotective

It has been reported that polyphenols can prevent LDL oxidation. In fact, flavonoids have been shown to have a cardioprotective effect in different in vivo models (Heim et al. 2002). In this sense, it has been demonstrated that the intake of foods with high phenolic compound content such as flavonoids have a positive effect on human health, since a decrease in the incidence of heart attacks and mortality due to coronary and ischemic heart disease has been observed in adults (Folsom et al. 1999; Heim et al. 2002; Hertog et al. 1993). Likewise, the cardioprotective effect of flavonoids has been demonstrated in murine models with a high-flavonoid diet, due to reduction in myocardial postischemic damage (Facino et al. 1999).

2.1.1.5 Antimicrobial

According to several authors, phenolic compounds have a high antimicrobial activity; this could be due to the fact that most of these compounds are capable of inhibiting the formation of biofilms. Likewise, they neutralize bacterial toxins and reduce the host ligands' adhesion; therefore, a study of a great variety of phenolic compounds on several microorganisms has been carried out. It has been reported that among the phenolic compounds, flavonols and tannins have the highest antimicrobial activity. On the other hand, due to the latest advances in studies of antimicrobial activity, phenolic compounds have been proposed for the development of new preservatives in food (Daglia 2012; Davies and Deroles 2014).

2.2 Plant Phenolic Extraction

Plants are rich in phenolic compounds, so to know all its characteristics, it is necessary to separate, extract, and purify them; however, there is no specific method that allows extraction of all phenolic compounds due to the complexity and diversity found in a single plant. Phenolic compounds have different polarities (hydrophilic and lipophilic), molecular weights (from simple molecules to very complex ones), and molecular structure (isomers). In addition, depending on the matrix where they are found, they can be free or associated with weak or strong bonds, making their extraction complicated and extensive. Other factors such as solvent type, solid-liquid ratio, temperature, pressure, and extraction times are involved in the efficient extraction of phenolic compounds (Jones and Kinghorn 2006; Khoddami et al. 2013; Sáyago-Ayerdi et al. 2016; Wang and Weller 2006).

The first step in the extraction of plant phenolic compounds is the sample preparation, whether it is fresh, dried (air-drying or freeze-drying), or frozen; in addition, the sample must be processed by milling, grinding, and homogenization (Dai and Mumper 2010). These processes must be carried out avoiding high temperatures that can decompose the plants thermolabile compounds, as well as protecting themselves from sunlight due to changes that may arise from exposure to ultraviolet radiation. Exposure to temperatures greater than 100 °C may influence the recovery of some phenolic compounds. Likewise, in the milling, the sizes of particles allow a more efficient extraction, due to the fact that the surface area increases; however if it is too fine, it is difficult for the solvent to pass through the particle (Rajha et al. 2014; Velavan 2015).

The solvent selection is based on the compounds that are to be extracted, in addition to considerations such as low toxicity, evaporation with low heating, easy to concentrate, preservation capacity without alterations in the compounds, and should not cause interference. For the extraction of hydrophilic compounds, polar solvents such as water, methanol, ethanol, or ethyl acetate are generally used, while nonpolar lipophilic agents such as dichloromethane, acetone, besides, chloroform, or hexane are used to eliminate interferences in the process such as chlorophyll (Sasidharan et al. 2011; Velavan 2015). In the extraction of phenolic acids and flavonoids, methanol, ethanol, or water or a mixture of them is used; in the case of extraction of tannins and anthocyanins, because they are of high molecular weight, alkaline or acidified water is commonly required. In this sense, for stilbene (resveratrol) solvents, such as acetone, ethyl acetate and alcohols are very effective (Sandoval and Villanueva-Rodriguez 2018). The different techniques or extraction methods include conventional methods or very sophisticated emergent technologies (Fig. 2.2) in order to extract largest amounts of bioactive compounds.

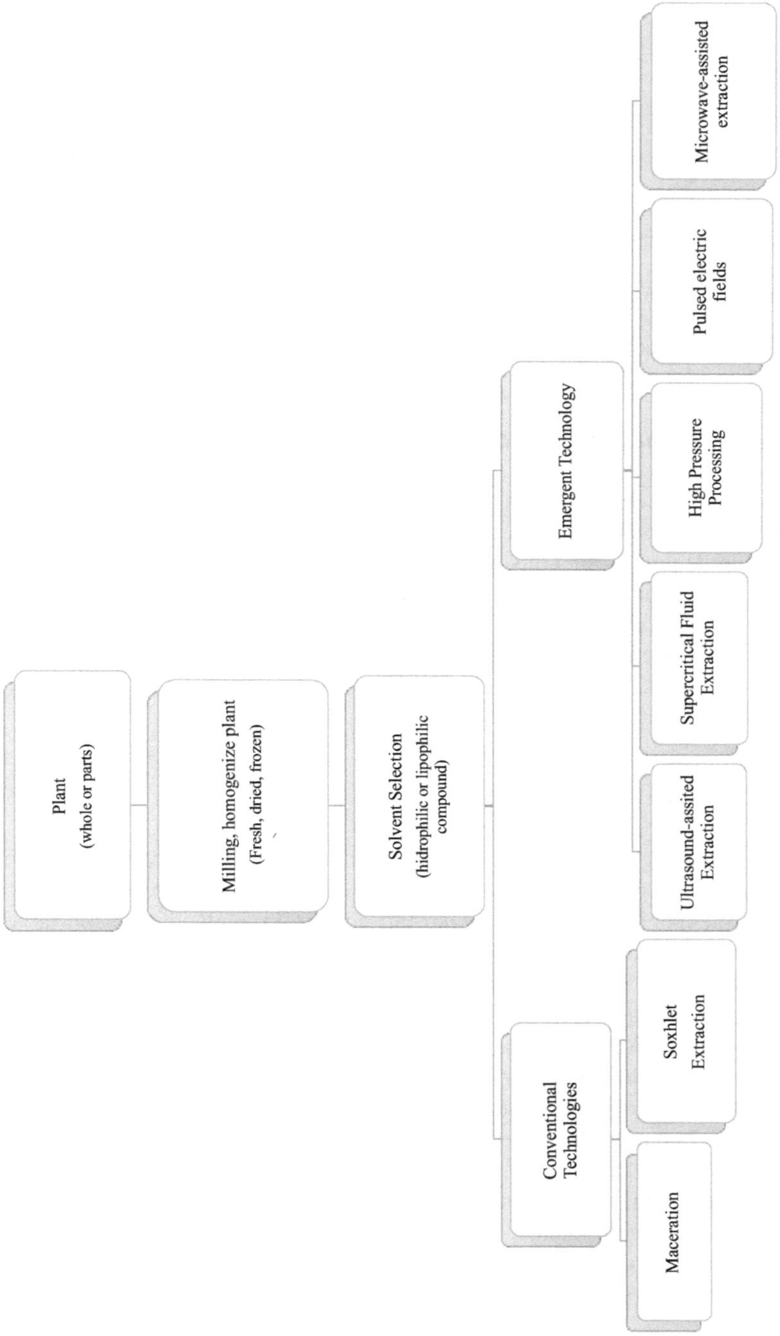

Fig. 2.2 Scheme process of phenolic compound extraction in plants

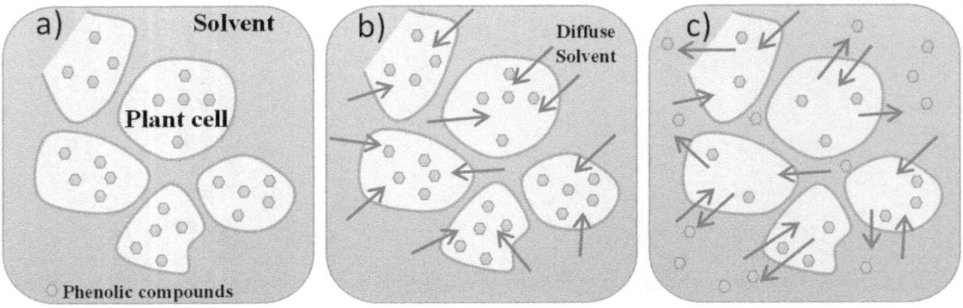

Fig. 2.3 General representation of a conventional extraction process: (**a**) solvent-solvent contact, (**b**) break down of cell barriers and diffusion of the solvent, (**c**) transfer of phenolic compounds to the medium (outside the cell)

2.2.1 Conventional Technologies

Changes in the cell structure occur in the extraction process to free the compounds of interest. First, it is necessary to break down barriers so that the transfer of compounds is made to the medium. In the conventional method, mass transfer occurs from the inside out (Fig. 2.3) and heat transfer from the outside to the inside (Sadeghi et al. 2017). There are several methods for the extraction of phenolic compounds; however, conventional methods are mostly used such as maceration and Soxhlet (Annegowda et al. 2012).

2.2.2 Maceration

This method was designed for a long time to produce tonics; however, it became very useful in obtaining essential oils and bioactive compounds such as phenolics, because it is inexpensive and it can be carried out in vials, flasks, or large containers (Jones and Kinghorn 2006). In maceration, the sample is kept in contact with the solvent for a period (at least 72 h) with constant agitation at room temperature (possible application of heating) until all soluble compounds are dissolved. Later, filtration methods, or decanting, are used to obtain the extract. However, this method has disadvantages such as the following: it requires large volumes of solvents and long extraction periods and can be affected by external factors such as light and high temperatures (Jones and Kinghorn 2006; Khoddami et al. 2013; Velavan 2015).

The extraction conditions such as temperature and solvents are determinants for the yield of bioactive compounds; in this sense, previous studies have reported efficient extraction of phenolic compounds through the maceration method in different plants such as Cretan barberry herb (Kukula-Koch et al. 2013), *Moringa*

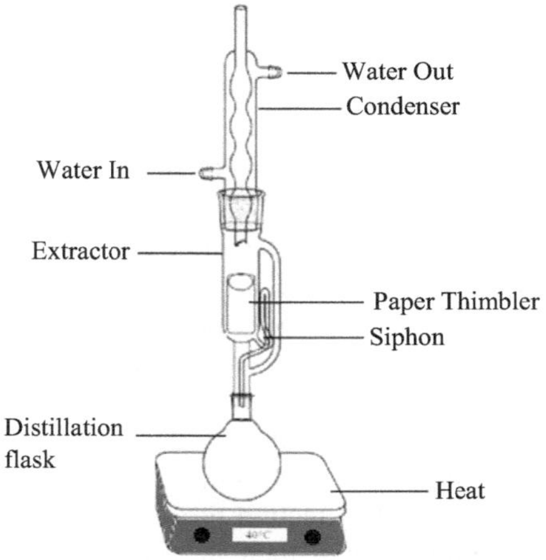

Fig. 2.4 Soxhlet extraction equipment

oleifera L. leaves (Vongsak et al. 2013), aerial parts of *Salvia veneris* (Gulsoy Toplan et al. 2017), *Cassia fistula* L., *Punica granatum*, *Juglans regia* (Khan et al. 2012), *Thymus serpyllum* (Jovanović et al. 2017), *Alpinia zerumbet* leaves (Victório et al. 2009), olive leaves (Rafiee et al. 2011), *Pinus radiata* bark (Aspé and Fernández 2011), and *Opuntia ficus-indica* flowers (Benayad et al. 2014), among others.

2.2.3 Soxhlet Extraction

Soxhlet is a continuous extraction process of solid material with organic solvent; it is an extraction method proposed by the German chemist Franz Ritter Von Soxhlet in the year 1879. It was initially designed for the extraction of lipids; however, it has been widely used for the extraction of phenolic compounds (Azmir et al. 2013). This method uses a Soxhlet extraction equipment (Fig. 2.4), which continuously supplies fresh solvent to the sample contained in the thimble, this through a series of cycles of solvent evaporation and condensation (Theegala 2015). This process has some advantages as follows: the sample is repeatedly in contact with fresh solvent, which facilitates the diffusion and extraction; the temperature is relatively constant; and it does not need filtration, is a very simple method, and is easy to use. Nevertheless, the disadvantages are the amount of sample per extraction, the long-time process, and the high solvent consumption and it cannot be used with thermo-labile compounds (De Castro and Priego-Capote 2010; Velavan 2015). Like maceration, several reports of plant phenolic compound extraction by the Soxhlet process are available; in this sense, it has been studied on different plants such as *Bauhinia purpurea* (Annegowda et al. 2012), different parts of *Salvia halophila* and *Salvia virgata* (Akkol et al. 2008), *Euphorbia neriifolia* leaves (Sharma and Janmeda 2017), henna leaves (Mahkam et al. 2014), *Opuntia ficus-indica* (Ammar et al. 2015), and different salvia species (*S. officinalis, S. aegyptiaca,* and *S. argentea*) (Farhat et al. 2013).

2.2.4 Emerging Technologies

In recent years, new technologies have been studied to obtain bioactive compounds, which minimize the problems caused by conventional extractions. In this sense, emerging technologies (also called unconventional technologies) in comparison with conventional technologies have some advantages such as energy saving, sensitive extraction process, environment-friendly, higher yield, and quality extracts. However, the principal disadvantage of these technologies is the moderate extraction costs. In addition, it has been reported that they are more selective and efficient than conventional extraction processes and offer superior yields as well as short extraction times. Among them ultrasound-assisted extraction, supercritical fluid extraction, high-pressure processing, pulsed electric fields, and microwave-assisted extractions have been studied (Bursać Kovačević et al. 2018; Moreira et al. 2019; Rocchetti et al. 2019).

2.2.5 Ultrasound-Assisted Extraction

Ultrasound is an emerging technology that in recent years has been widely studied in various areas such as the food industry, because it has been reported that ultrasound has the ability to penetrate through the medium by a phenomenon called cavitation, which creates a great vibration between the molecules damaging their structure and facilitating mass transfer. Therefore, compound extraction can be improved by combining this technology and the use of solvents; in recent years, it has been proven that the rate and yield extraction of bioactive compounds have increased when using ultrasound-assisted extraction in plants (Gayathri et al. 2018; Moreira et al. 2019). Likewise, it has been reported that ultrasound-assisted extraction is a promising technique for the extraction of bioactive compounds from plants (Azmir et al. 2013). However, extraction at high ultrasound waves could induce an elevated production of free radicals and decrease the content of bioactive compounds in plants, so specific studies on the effect of ultrasound on plants for the extraction of bioactive compounds such as phenolic compounds are necessary (Ameer et al. 2017; Bursać Kovačević et al. 2018).

2.2.6 Supercritical Fluid Extraction

Supercritical fluid is any substance that is above its point pressure and critical temperature; also, this supercritical fluid behaves like a gas and maintains the solvent properties of a liquid. In addition, it has high diffusivity and low viscosity which allows it to more easily cross porous media increasing the transfer of mass compared to a liquid. The properties of a supercritical fluid depend on many factors such as pressure, temperature, and the supercritical fluid substance (da Silva et al. 2016; Moreira et al. 2019). Carbon dioxide is the most used supercritical fluid due to its pressure and temperature supercritical moderate properties; it is also considered as safe for the environment as well as human health. Another advantage of supercritical fluid extraction is that it is considered a generally recognized as safe (GRAS) solvent (mainly ethanol, due to its food grade properties and its ease of separation of the final product) in which, depending on the interest product, the addition of the cosolvents favors a better extraction. Likewise, it has been reported that higher yields, lower extraction times, and solvent consumption are obtained with this technology (Galanakis 2012); on the other hand, the supercritical extraction equipment can be adapted with different chromatographic techniques (da Silva et al. 2016). Therefore, its application in bioactive compound extraction has been growing in recent years (Moreira et al. 2019; Rocchetti et al. 2019; Uddin et al. 2015). In this sense, this technology is mainly used in the extraction of solid matrix compounds. Likewise, in recent years, it has been studied for bioactive compound extraction from by-products to extract specific phenolic compounds and give them added value (Ameer et al. 2017; Herrero et al. 2010).

2.2.7 High-Pressure Processing

High-pressure processing is a technology that consists of exposing the material to pressures ranging from 100 to 600 MPa. It has been reported that during the extraction, the covalent bonds are not damaged; however, the extraction temperature remains with a variation of 3 °C per 100 MPa. Therefore, it has been used for the extraction of compounds that are thermolabile. In addition, the high pressures allow the extraction of compounds without damaging or denaturation; it is also recognized by the Food and Drug Administration (FDA) as environmentally ecofriendly (Moreira et al. 2019; Sevenich and Mathys 2018). In this sense, phenolic compounds have been extracted from different plants through high-pressure processing.

2.2.8 Pulsed Electric Fields

Pulsed electric field is a promising technology for food processing and conservation that has attracted strong interest in food engineering; extraction by electrical pulses is based on cell membrane destruction by an electric field, because the molecules of the membrane are separated by their charge which results in a repulsion between the molecules, forming pores and increasing the permeability (Azmir et al. 2013). Moreover, mass transfer is increased by the destruction of the plant membrane to improve compound extraction and reduce extraction times (Azmir et al. 2013; Toepfl et al. 2006). On this sense, different studies have been reported its advantage in comparison with the traditional extraction process such as selective molecule extraction and induction of structural modification of the cell membrane to improve bioactive compound extraction (Comuzzo et al. 2018; Moreira et al. 2019; Vorobiev and Lebovka 2016).

2.2.9 Microwave-Assisted Extraction

Microwave-assisted extraction involves the heating of solvents due to electromagnetic waves; in the case of plants, indoor water heats up and evaporates causing an increase in pressure, so the cell wall ruptures and bioactive compounds become available for extraction. This technology reduces the extraction time, as well as increases the yield (Bursać Kovačević et al. 2018; Tatke and Jaiswal 2011). According to Alupului et al. (2012), the extraction of plant compounds through microwave-assisted extraction is carried out in three steps: first, the solutes are separated from the matrix due to the increase in pressure and temperature caused by the collision of molecules while under microwave frequencies, then solvent diffusion occurs through the matrix, and finally, the solutes are separated from the matrix to the solvent (Azmir et al. 2013).

However, compound extracting conditions should be monitored because long periods of extraction and high microwave power could greatly increase the temperature and cause compound degradation; therefore, optimization studies are necessary for the extraction of bioactive compounds in plants (Bursać Kovačević et al. 2018; Périno-Issartier et al. 2011).

2.3 Improving Plant Phenolic Extraction by Biotechnological Processes

Biotechnological processes have been studied to counteract the problems caused by conventional processes (high consumption of chemical agents, prolonged extraction times, and high temperatures), which are also used in in vitro (controlled) conditions to produce natural products from diverse sources such as plants, as well as for bioprocessing applications. In the literature, there are diverse studies of biotechnological processes for the improvement of bioactive compounds such as phenolics; among these, enzymatic, fermentation, germination, and plant cell cultures have been reported (Table 2.1).

2.3.1 Enzymatic Process

Enzymatic process is an emerged technology to overcome conventional phenolic compound extraction. Enzymes are proteins responsible for accelerating the metabolic reactions of all organisms and are characterized for having high specificity. Enzymatic reactions take place at the active side with the interaction between the enzyme's active side and the substrate for product transformation (Martínez Cuesta et al. 2015; Nam et al. 2012). Despite enzymatic processes being expensive, in recent

Table 2.1 Advantages and disadvantages of the biotechnological processes used to enhance the phenolic content in plants

Biotechnology process	Advantage	Disadvantage	Reference
Enzymatic process	Environment-friendly Release bound phenolic compounds	Expensive process Stages of extraction and purification of phenolic compounds are needed	Bhanja Dey et al. (2016)
Fermentation process	Rapid growth rates Cost-effectiveness Simple processing and cultivation Environmentally friendly Low temperature and water requirements	Long fermentation times Multiple parameters to control It requires specialized infrastructure	Bhanja Dey et al. (2016), Fowler and Koffas (2009), and Martins et al. (2011)
Germination process	Simple processing Low temperature and water requirements It does not require specialized infrastructure Inexpensive Environmentally friendly process	Long relative processing periods Multiple parameters to control	Cevallos-Casals and Cisneros-Zevallos (2010), and Gan et al. (2017)
Plant cell cultures	High yields It does not need toxic chemicals It can get specific compounds No product variability Easy growth control	It requires specialized infrastructure Multiple parameters to control Expensive process It needs genetically modified plants to increase yield and purity	Davies and Deroles (2014), James and Lee (2001), and Wilson and Roberts (2012)

years biotechnological processes advanced and the costs of enzymes have decreased, and thus, it can currently be found in different industrial processes (Laroze et al. 2010). Nevertheless, the enzymatic process has some advantages in comparison with conventional processes such as it is cost-effective, eco-friendly, and easy to obtain and maintain, as well as degrade the cell wall components. On the other hand, enzymes have then been isolated from different organisms such as bacteria, fungi, animal organs, and vegetable/fruit extracts and are being used for a wide range of applications such as food industry and medicine (Raveendran et al. 2018).

In the last years, enzymes have been used to improve phytochemical extraction from food matrices. In this sense, it has been reported that phenolic compounds are increased by enzymatic process, which could be due to the breakdown of protein structures that are linked phenolics, and also by cellulosic composite destruction of the cell wall by catalyzing hydrolytic reactions. Likewise, some enzymes also improve phenolic solubility, another important factor for their extraction (Mushtaq et al. 2015). In the literature, fruits, vegetables, and by-product industry have been studied for increasing phenolic compound recovery by enzymatic process. In this sense, cellulase, hemicellulase, β-glucanase, xylanase, and pectinase have been studied to digest vegetable cell walls and improve phenolic compounds (Table 2.2). In fact, cellulase catalyzes the hydrolysis of endo-1,4-β-D-glycosidic linkages in cellulose, lichenin, barley glucan, and the cello-oligosaccharide cellotriose to cellohexaose. Hemicellulase, on the other hand, hydrolyzes hemicellulose and pectinase separates polygalacturonic acid into mono-galacturonic acid by opening glycosidic linkages (Peng et al. 2015).

Laroze et al. (2010) reported a total phenolic compound increase of 35% by enzymatic process from raspberry solid waste compared with hydroalcoholic extraction. They used enzymes extracted from different organisms such as *Aspergillus niger*. The best enzyme concentration to increase phenolic compounds varies according to raw material composition. Some authors suggest that the enzymatic process must be used together with an extraction technology (such as microwave-assisted extraction, supercritical extraction, ultrasound-assisted extraction, among others) to increase phenolic compound extraction.

Likewise, cell-wall-degrading enzymes, such as cellulase, have also been used to improve phenolic compounds from *Gingko biloba* leaves. The extraction yield was twofold higher than hydroalcoholic extraction. This could be achieved by transglycosylating flavonol aglycones into more polar glucosides (Chen et al. 2011). Wang et al. (2017) reported an increase in quercetin and kaempferol by cellulase and β-glucosidase-assisted extraction from guava (*Psidium guajava* L.) leaves. They found that the enzymatic process could release insoluble-bound and soluble-conjugated phenolic compounds.

Proteases such as alcalase and neutrase-hydrolyzing internal peptide bonds have been used in high-protein sources to enhance phenolic compounds. Further, tannase is used to improve phenolic extraction, which is extracted from filamentous fungi and catalyze and hydrolyze the ester present in hydrolysable tannins and gallic esters such as epigallocatechin O-gallate (EGCG) or epicatechin O-gallate (ECG), releasing gallic acid (GA). In this sense, tannase in combination with cellulase has increased total phenolic extraction from grape pomace after 24 h. It is important to understand that gallic acid content increased by fivefold compared to hydroalcoholic extracts (Chamorro et al. 2012). Similarly, green tea (*Camellia sinensis*) has been studied to increase phenolic compounds through the enzymatic process. In this sense, it has been reported that tannase combined with Viscozyme increased total

Table 2.2 Enzymes used to increase phenolic compound extraction

Enzyme	Sources	Activity	Optimal pH	Optimal temperature (°C)	Reference
Cellulase	*Aspergillus niger*, *Trichoderma reesei*, *Clostridium thermocellum*, and *Dictyoglomus turgidum*	Catalyzes the hydrolysis of endo-1,4-β-D-glycosidic linkages in cellulose and cellohexaose	5.0	37	Sohail et al. (2009)
Hemicellulase	*Aspergillus niger*, *Aureobasidium pullulans*, *Thermomyces lanuginosus*, and *Trichoderma viride*	Catalyzes the endohydrolysis of 1,4-β-D-xylosidic linkages in xylans	4.5	40	Fan et al. (2009)
Pectinase	*Aspergillus niger*, *Aspergillus aculeatus*, and *Rhizopus species*	Catalyzes the random hydrolysis of 1,4-α-D-galactosiduronic linkages in pectin and other galacturonans	4.0	25	Sharma et al. (2013)
β-Glucanase	*Aspergillus niger* and *Trichoderma longibrachiatum*	Degrade β-1,4-glucans of cellulose, xyloglucan, and β-1,4-xylan	5.0	55	Lafond et al. (2012)
Xylanase	*Aureobasidium pullulans*, *Thermomyces lanuginosus*, and *Trichoderma viride*	Catalyzes the endohydrolysis of 1,4-β-D-xylosidic linkages in xylans	4.5	30	Qiu et al. (2010)
Pronase	*Streptomyces griseus*	Hydrolyzes peptide bonds	7.5	37	Walker and Sweeney (2009)
Viscozyme (multienzyme complex)	*Aspergillus aculeatus*	Catalyzes different types of breakdown over carbohydrates	3.3–5.5	25–55	Charoensiddhi et al. (2016)
β-Glucosidase	*Almonds*	Hydrolytic removal of aglycone moiety from flavonoid and isoflavonoid glycosides	5.0	37	Hamzah et al. (2011)
Neutrase	*Bacillus amyloliquefaciens*	Hydrolyzes peptide bonds	5.5–7.5	30–55	Lee et al. (2011)
Alcalase	*Bacillus licheniformis*	Hydrolyzes peptide bonds	6.5–8.5	60	Ahn et al. (2012)
Umamizyme (multienzyme complex)	*Aspergillus oryzae*	Hydrolyzes peptide bonds	7.0	45	Boschin et al. (2014)

(continued)

Table 2.2 (continued)

Enzyme	Sources	Activity	Optimal pH	Optimal temperature (°C)	Reference
β-Glucosidase	*Almonds*	Hydrolytic removal of aglycone moiety from flavonoid and isoflavonoid glycosides	5.0	37	Hamzah et al. (2011)
Neutrase	*Bacillus amyloliquefaciens*	Hydrolyzes peptide bonds	5.5–7.5	30–55	Lee et al. (2011)
Alcalase	*Bacillus licheniformis*	Hydrolyzes peptide bonds	6.5–8.5	60	Ahn et al. (2012)
Flavorzyme (multi-enzyme complex)	*Aspergillus oryzae*	Hydrolyses peptides bonds	8.0	50	Betancur-Ancona et al. (2014)
Papain	*Carica papaya*	Cysteine endopeptidase	5.0–7.0	65	Homaei et al. (2010)
Pepsin	Pig gastric mucosa	Hydrolyses peptides bonds	2.0	37	Esmaeilipour et al. (2012)
Tannase	*Aspergillus niger*, *Aspergillus oryzae*, and *Aspergillus flavus*	Catalyzes the chemical reaction from digallate to gallate	5.0	37	Marco et al. (2009)

phenolic content by around twofold in comparison with cold water extraction (Hong et al. 2013).

Cellulase has been used in combination with microwaving for the recovery of phenolic compounds from raw materials. Yang et al. (2010) reported an increase in corilagin and geranin from geranium (*Geranium sibiricum*) through the enzymatic process coupled with microwave-assisted extraction. Corilagin and geranin increased 64.01% and 72.95%, respectively, in comparison with water extraction. Besides, the enzymatic process coupled with microwave-assisted extraction did not affect the structure and functionality of phenolic compounds and was better than solvent extraction.

Similarly, the enzymatic process coupled with microwave-assisted extraction has been studied for polyphenol extraction from other raw materials such as peanut shells. Zhang et al. (2013) found a total phenolic increase of 62.73% by the enzymatic process coupled with microwave-assisted extraction. Furthermore, these combined technologies were more effective in extracting phenolic compounds than others such as ultrasound-assisted extraction. On the other hand, supercritical extraction coupled with the enzymatic process (using pectinase, protease, and cellulase) doubled the total phenolic extraction from food sources such as pomegranate waste (Mushtaq et al. 2015).

Li et al. (2017) reported a total phenolic extraction of 30% by cellulase enzymatic process coupled with ultrasound-assisted extraction from sijiaoling pericarps (*Trapa quadrispinosa Roxb.*). Enzymes are biological units with high efficiency and are used to improved bioactive phenolic compounds in plants. The use of enzymes coupled with extraction methods has been reported to increase the antioxidant and biological activities of extracts. Also, these extracts are useful for preserving food and are used to produce cosmetic and pharmaceutical products. Enzyme-assisted extraction should be further utilized for the development of functional food ingredients and nutraceuticals with high market value.

2.3.2 Fermentation Process

The fermentation process has been studied for the recovery of bioactive compounds such as phenolic acids (Table 2.3). This process is divided into two systems, submerged fermentation and solid-state fermentation. Fermentation processes are based on the growth of microorganisms in liquid or solid medium in the absence or near absence of free liquid. These microorganisms consume the substrates and produce extracellular enzymes that weaken the matrix and release bound phenolic compounds; in addition, the phenolic compound increase in the fermentative processes may be due to the fact that these microorganisms can produce new phenolic compounds through their secondary metabolism (Bhanja Dey et al. 2016).

Table 2.3 Summary of studies using fermentation as a strategy to enhance the extraction of plant phenolic compounds

Source	Fermentation process	Effect on phenolic compounds	Reference
Corn stover	Submerged fermentation with *Inonotus obliquus* for 216 h	Increased the content of both extracellular (118.9–135.7 gallic acid equivalent/L) and intracellular (21.2–23.7 gallic acid equivalent/g) phenolic compounds.	Xu and Zhu (2011)
Apple pomace sludge	Submerged fermentation with *Phanerochaete chrysosporium* ATCC 24275 by 67 h	They reported a 1.5-fold increase in polyphenol content.	Gassara et al. (2012)
Soy whey	Fermented by *Lactobacillus plantarum* (B1–6) for 10 h	Total phenolic compound increase (22.48%, approximate), and the highest total phenolic compound content was 64.66 ± 2.79 mg gallic acid equivalent/g at fermentation time of 10 h.	Xiao et al. (2015)
Cowpeas	Fermented by *Lactobacillus plantarum* (ATCC 14917)	They reported a higher identification of phenolic compounds in the fermented than raw cowpea.	Dueñas et al. (2005)
Pate olive cake	Fermented with *Saccharomyces cerevisiae* and *Leuconostoc mesenteroides*	A phenolic compound decrease in fermented sample in comparison with unfermented was reported; nevertheless, fermented sample had higher antioxidant activity.	Tufariello et al. (2019)
Wheat and corn straw	Fermentation with 15% *Lactobacillus plantarum* (DSMZ 20174)	They reported that *Lactobacillus plantarum* fermentation increased the phenolic compounds of wheat (39%) and corn straw (51%).	Đorđević et al. (2019)
Quinoa seeds	Solid-state fermentation with different molds	*Rhizopus oligosporus* was the most efficient mold for enhanced phenolic compounds; they reported an increase of 129% and 220% for black and red quinoa seeds, respectively.	Starzyńska-Janiszewska et al. (2019)
Pineapple-soy flour mixture	Solid-state fermentation by *Rhizopus oligosporus*	Fermentation process increased the total phenolic compounds of pineapple-soy flour by 79.4% at 12-day fermentation with 5 g of pineapple and 5 g of soy flour.	Correia et al. (2004)

(continued)

Table 2.3 (continued)

Source	Fermentation process	Effect on phenolic compounds	Reference
Berry pomaces	Solid-state fermentation by *Aspergillus niger*	They reported an increase of total phenolic and flavonoid content of 18.82% and 11.11% for *Sambucus ebulus* L. and *Sambucus nigra* L., respectively.	Dulf et al. (2015)
Apricot (*Prunus armeniaca* L.) pomace	Solid-state fermentation by two fungal strains (*Aspergillus niger* [ATCC-6275] and *Rhizopus oligosporus* [ATCC-22959])	They found a total phenolic (78% and 36%) and flavonoid (34% and 12%) increase after solid-state fermentation with both *Rhizopus oligosporus* and *Aspergillus niger* fermentation, respectively.	Dulf et al. (2017)

2.3.3 Submerged Fermentation

It has been reported that the submerged fermentation process increases phenolic compounds from various materials such as agricultural wastes, food, and plants.

In this sense, Xu and Zhu (2011) found an extracellular and intracellular phenolic compound increase by submerged fermentation with *Inonotus obliquus* of a corn stover medium. The highest extracellular (118.9–135.7 gallic acid equivalent/L) and intracellular (21.2–23.7 gallic acid equivalent/g) phenolic compounds were at 216 h and 144 h, respectively. Similarly, Gassara et al. (2012) reported a 1.5-fold increase in polyphenol content of apple pomace sludge by submerged fermentation with *Phanerochaete chrysosporium* ATCC 24275 at fermentation time of 67 h.

Xiao et al. (2015) reported a total phenolic compound increase (22.48%, approximate) of fermented soy whey by *Lactobacillus plantarum* (B1–6). The highest total phenolic compounds of fermented soy whey were 64.66 ± 2.79 mg gallic acid equivalent/g at fermentation time of 10 h. Similarly, it has been reported that *Lactobacillus plantarum* (ATCC 14917) fermentation could modify the phenolic compound content of legumes. Dueñas et al. (2005) reported an enhanced phenolic compound content from fermented cowpeas by *Lactobacillus plantarum* (ATCC 14917); in fact, they reported a higher identification of phenolic compounds in the fermented rather than raw cowpea, which could be synthesized during *Lactobacillus plantarum* (ATCC 14917) fermentation.

In the literature, the mechanism of some bioactive phenolics has been hypothesized. Moreover, it has been established that the increase of phenolics could be attributed to bioactive compounds released by enzymatic degradation of lignocelluloses produced by the microorganisms present in the fermentation (Bhanja Dey et al. 2016; Xu and Zhu 2011).

Recently, Tufariello et al. (2019) reported a small decrease in polyphenol content (in a pate olive cake fermented with *Saccharomyces cerevisiae* and *Leuconostoc mesenteroides*) compared to the unfermented sample. It is important to mention that the fermented samples had an increase in antioxidant activity.

Đorđević et al. (2019) found a total phenolic and flavonoid increase after wheat and corn straw fermentation with 15% *Lactobacillus plantarum* (DSMZ 20174). They reported that wheat straw before and after fermentation has higher total phenolic and flavonoid content than corn straw. Likewise, total phenolic content of fermented wheat and corn was 77.9 ± 5.1 and 68.0 ± 4.9 mg gallic acid equivalent/g, respectively. Besides, total flavonoid content of fermented wheat and corn was 56.0 ± 4.8 and 44.8 ± 4.2 mg quercetin equivalent, respectively. On the other hand, total phenolic and flavonoid increase could be due to the bound phenolic acids released by the cell wall breakdown induced during fermentation as well as enzyme hydrolysis (Martins et al. 2013).

2.3.4 Solid-State Fermentation

Solid-state fermentation is a promising method to enhance phenolic compounds from plants, cereals, agriculture, and industrial by-products. It is a microbial process that is grown on solid substrate in the absence of or lower free liquid. Likewise, the substrates are moist so that the microorganism grows and metabolizes them (Martins et al. 2011). It has been reported that a key process in solid-state fermentation is the selection of the microorganism as well as the substrate, because the adaptability and metabolism of the microorganism can give an efficient release of phenolic compounds (Dulf et al. 2015, 2017; Ruiz et al. 2005). The most appropriate microorganisms for solid-state fermentation are filamentous fungi, followed by yeasts, and later some bacteria species; some microorganism characteristics for the success of the fermentation process are their high susceptibility to grow in environments with low water activity as well as their ability to synthesize high amounts of enzymes and other metabolites (Soccol et al. 2017).

Likewise, Starzyńska-Janiszewska et al. (2019) studied the effect of solid-state fermentation by different molds on the enhancement of free phenolics on quinoa seeds. They reported that *Rhizopus oligosporus* was the most efficient mold to enhance bioactive compounds such as free phenolic of quinoa seeds. In comparison with raw material, fermented quinoa seeds have higher free phenolic content (quinoa black seeds increase by 220% and quinoa red seeds by 129%). It has been reported that mold enzymes may be responsible for the bound phenolic compound increase, because these enzymes breakdown the cell wall, making phenolic compounds available (Huynh et al. 2014).

Several reports have informed that industrial by-products possess a high phenolic compound content. Therefore, researches have carried out studies to improve its content. In this sense, solid-state fermentation has been highlighted as an effective method. Correia et al. (2004) studied the effect of solid-state fermentation by *Rhizopus oligosporus* on total phenolic content of pineapple-soy flour mixture. They reported that the highest total phenolic content (increase of 79.4%) was at 12-day fermentation with 5 g of pineapple and 5 g of soy flour.

Besides, Dulf et al. (2015) reported the increase of total phenolic and flavonoid content of *Sambucus nigra* L. and *Sambucus ebulus* L. berry pomaces by solid-state fermentation of *Aspergillus niger*. In the case of total phenolic compounds, the highest yield was found on the third day of fermentation for both berry pomaces with an increase of 18.82% and 11.11% for *Sambucus ebulus* L. and *Sambucus nigra* L., respectively. Similarly, Dulf et al. (2017) studied the effect of solid-state fermentation by two fungal strains (*Aspergillus niger* [ATCC-6275] and *Rhizopus oligosporus* [ATCC-22959]) on phenolic and flavonoid content of apricot (*Prunus armeniaca* L.) pomace. They found a total phenolic and flavonoid increase after solid-state fermentation with both fungal strains. In fact, an increase of 78% of total phenolic content was in 9 days for *Rhizopus oligosporus* fermentation, while an increase of 34% of total phenolic content was in 6 days for *Aspergillus niger* fermentation. On the other hand, the total flavonoid content of solid-state-processed apricot increase 36% and 12% with 9-day *Rhizopus oligosporus* fermentation and 6-day *Aspergillus niger* fermentation, respectively.

2.3.5 Germination

The germination process transformed macronutrients to micronutrients and other compounds; sprouts are synthetized during this process and it has been reported that it may have various biological activities. Likewise, the germination process has been reported to increase the nutritive value and health characteristics of plants and improves the digestibility, palatability, and availability of certain nutrients like bioactive compounds such as γ-aminobutyric acid, vitamins, and polyphenols, which are accumulated and increased. On the other hand, it is an inexpensive, simple, and environment-friendly method (Gan et al. 2017; Kumar et al. 2011; Liu et al. 2019). According to Danisova et al. (1994), during the germination process, diverse reactions occur that change the characteristics of the materials; in this sense, the increase of compounds may be due to the accumulation of soluble bioactive compounds due to the release of bound compounds by the generation of new cells that separate them from the cell wall. It is important to note that the increase of these compounds depends on many factors, and there are several works where the effect of germination on different materials such as cereals, edible seeds, sprouts, and legumes has been studied. On the other hand, the germinated products have a similar total phenolic content as fruits and vegetables (Azmir et al. 2013; Gan et al. 2017). Also decreases in the concentration of antinutritional factors such as phytate, trypsin inhibitor, α-galactosidase, and alkaloids are diminished by germination which could enhance phenolic compounds. Different seeds, plants, and other foods have been germinated for human consumption; however, the germination strategies depend on the type of material and can be carried out in several simple procedures such as sterilization, soaking, and sprouting (Gan et al. 2017).

Table 2.4 Plant cell culture to increase phenolic compounds

Source	Plant cell culture conditions	Effect on phenolic compounds	Reference
Hypericum perforatum	The best bioreactor cell growth conditions were at adventitious root inoculums of 6 g/l in ½ Murashige and Skoog medium supplemented with a 1 mg/l and 0.1 mg/l of indole-3-butyric acid and kinetin concentrations	They reported a total flavonoid and phenolic content of 28.31 ± 0.56 and 35.01 ± 1.26 mg catechin equivalent/g dry roots and 56.47 ± 0.22 mg gallic acid equivalent/g dry roots, respectively.	Cui et al. (2010)
Angelica sinensis Diels (female ginseng)	The best cell suspension culture was at 16 days after cultivation cycle in ½ B5 basal medium, a sucrose concentration of 30 g/L, initial pH of 5, and inoculum density of 10.8275 g/L	The cell suspension culture under illumination increased the ferulic acid content.	Liu et al. (2019)
Dracocephalum polychaeta Bornm.	They reported that 30 mT of static magnetic field and 100 ppm of Fe_2O_3 magnetic nanoparticles on cell suspension cultures have a significant effect on phenolic compounds	They found a 1.73-fold increase of phenolic acid content; besides, the highest total phenol content was at Fe_2O_3 magnetic nanoparticle treatments (405.65 μg g^{-1} fresh weight).	Taghizadeh et al. (2019)
Gymnosporia buxifolia	They found that the highest phenolic and flavonoid content in callus biomass from solid culture was at 5 μM phloroglucinol, 1 μM picloram, and 1.5 μM benzyladenine in a Murashige and Skoog basal nutrient medium	They reported that cell suspension increases the phenolic compounds of *Gymnosporia buxifolia* with growth regulators.	Kumari et al. (2018)

2.3.6 Plant Cell Culture

Plant cell culture is a promising method to produce secondary plant metabolites such as phenolic compounds (Table 2.4), due to the complex chemical synthesis of bioactive compounds as well as the environmental problems to natural harvest (Wilson and Roberts 2012). In recent years, the increase of secondary metabolites by plant cell culture has been studied. In this sense, the following strategy has been used: firstly, the plant secondary metabolite of interest is identified, and then it is sought to increase its genomic expression by means of metabolic engineering. Once the modified cell with high yield in the metabolite is obtained, the culture conditions are optimized (culture medium, elicitors, among other factors), and subsequently, the cell culture is carried out in bioreactors to increase the production of metabolites through the cultivation of cells as well as organs in plants, and finally extraction is carried out (Cui et al. 2010; Yue et al. 2016). Plant cell culture has some advantage over full chemical synthesis, since it avoids the problems related to the synthesis of specific compounds as well as the use of toxic chemicals. Likewise, this technology coupled with metabolic engineering offers a great field of study for the improvement of target bioactive compounds through genetically modified plants. Further, through the cell culture, similar yields can be obtained, because it is easier to control the growth conditions compared to the whole plant growth systems (Davies and Deroles 2014; James and Lee 2001; Wilson and Roberts 2012).

It is important to highlight that to carry out a cell culture process, we must know how to improve the interest metabolite content before selecting the reactor, because it has been reported that continuous reactors are used when the metabolites are excreted in the medium. On the contrary, discontinuous reactors are used when the metabolite is inside the cells; therefore in both cases, to increase their expression, it is necessary to provide them with the best conditions (Alfermann 2010; Davies and Deroles 2014; Steingroewer et al. 2013).

In this sense, Cui et al. (2010) studied the best culture conditions to enhance the phenolic compounds of *Hypericum perforatum*. They reported that the total phenolic and flavonoids from the adventitious root of *Hypericum perforatum* were affected by the inoculum sizes, auxin and auxin/cytokinin concentration, and Murashige and Skoog medium dilutions during culture growth. Besides, the best bioreactor cell growth conditions were at adventitious root inoculums of 6 g/l in ½ Murashige and Skoog medium supplemented with a 1 mg/l and 0.1 mg/l of indole-3-butyric acid and kinetin concentrations, respectively. Moreover, at these conditions, they reported a growth ratio and total flavonoid and phenolic content of 28.31 ± 0.56 and 35.01 ± 1.26 mg catechin equivalent/g dry roots and 56.47 ± 0.22 mg gallic acid equivalent/g dry roots, respectively. On the other hand, the certain phenolic compound improvement by plant cell culture has been studied. In this sense, Liu et al. (2019) reported that the best conditions for ferulic acid content of *Angelica sinensis* Diels (namely, "female ginseng") by cell suspension culture were at 16 days after cultivation cycle in ½ B5 basal medium, a sucrose concentration of 30 g/L, initial pH of 5, and inoculum density of 10.8275 g/L. In addition, the cell suspension culture under illumination had higher ferulic acid than in the dark.

Recently, different strategies to increase bioactive compound content in the cell culture of plants have been studied. Additionally, the use of static magnetic fields and addition of magnetic nanoparticles have a significant effect on the phenolic acid content of suspension-cultured plant cell. Taghizadeh et al. (2019) reported that 30 mT of static magnetic field and 100 ppm of Fe_2O_3 magnetic nanoparticles on the cell suspension culture of *Dracocephalum polychaeta* Bornm. increased the phenolic acid content by 1.73-fold. Besides, the highest total phenol content was at Fe_2O_3 magnetic nanoparticle treatments (405.65 $\mu g\ g^{-1}$ fresh weight). Therefore, a static magnetic field coupled with Fe_2O_3 magnetic nanoparticles could be considered elicitors of phenolic compounds.

In order to increase the bioactive compounds of plant cell cultures, the use of plant growth regulators has been studied, and it has been reported that these growth regulators increase the production of bioactive compounds (Baskaran et al. 2015; Kumari et al. 2018). In this sense, Kumari et al. (2018) studied the effect of different growth regulators on the phenolic compounds of *Gymnosporia buxifolia* cell culture. They found that the highest phenolic and flavonoid content in callus biomass from solid culture was at 5 μM phloroglucinol, 1 μM picloram and 1.5 μM benzyladenine in a Murashige and Skoog basal nutrient medium, while the cell biomass from liquid culture with the highest phenolic and flavonoid content was at 5 μM phloroglucinol and 1 μM benzyladenine in a Murashige and Skoog basal nutrient medium. Therefore, growth regulators are an efficient strategy to enhance the phenolic compound content of plant cell cultures.

Despite the different reports for obtaining bioactive compounds such as phenolics by cell culture, commercial success has not yet been achieved, and this could be due to various factors such as the difference in prices between production and product market price, as well as the discrepancy between the amount of metabolites found in cell cultures and the progenitor plant organs (Davies and Deroles 2014).

2.4 Conclusions

In the last years, different biotechnological methods are emerging to improve the extraction of bioactive compounds, due to environmental problems caused by conventional extraction methods. The most studied methods are enzymatic, fermentation, germination, and plant cell cultures. These strategies must be coupled to extraction methods to enhance the extraction of bioactive compounds. Despite the success of biotechnical methods to improve the extraction of bioactive compounds in different materials, there are few processes that are being carried out at the industrial level; therefore, studies must be carried out to ensure that conventional methods are displaced by biotechnological processes.

References

Ahn C-B, Jeon Y-J, Kim Y-T, Je J-Y (2012) Angiotensin I converting enzyme (ACE) inhibitory peptides from salmon byproduct protein hydrolysate by Alcalase hydrolysis. Process Biochem 47(12):2240–2245

Akkol EK, Göger F, Koşar M, Başer KHC (2008) Phenolic composition and biological activities of Salvia halophila and Salvia virgata from Turkey. Food Chem 108(3):942–949

Alfermann AW (2010) Production of natural products by plant cell and organ cultures. In: Wink M (ed) Annual plant reviews. Wiley-Blackwell, Chichester

Alupului A, Calinescu I, Lavric VJUSB, Series B (2012) Microwave extraction of active principles from medicinal plants. UPB Sci Bull Ser B Chem Mater Sci 74(2):1454–2331

Ameer K, Shahbaz HM, Kwon J-H (2017) Green extraction methods for polyphenols from plant matrices and their byproducts: a review. Compr Rev Food Sci Food Saf 16(2):295–315. https://doi.org/10.1111/1541-4337.12253

Ammar I, Ennouri M, Attia H (2015) Phenolic content and antioxidant activity of cactus (Opuntia ficus-indica L.) flowers are modified according to the extraction method. Ind Crop Prod 64:97–104

Annegowda H, Mordi M, Ramanathan S, Hamdan M, Mansor S (2012) Effect of extraction techniques on phenolic content, antioxidant and antimicrobial activity of Bauhinia purpurea: HPTLC determination of antioxidants. Food Anal Methods 5(2):226–233

Aspé E, Fernández K (2011) The effect of different extraction techniques on extraction yield, total phenolic, and anti-radical capacity of extracts from Pinus radiata bark. Ind Crop Prod 34 (1):838–844

Azmir J, Zaidul ISM, Rahman MM, Sharif KM, Mohamed A, Sahena F et al (2013) Techniques for extraction of bioactive compounds from plant materials: a review. J Food Eng 117(4):426–436. https://doi.org/10.1016/j.jfoodeng.2013.01.014

Balasundram N, Sundram K, Samman S (2006) Phenolic compounds in plants and agri-industrial by-products: antioxidant activity, occurrence, and potential uses. Food Chem 99(1):191–203. https://doi.org/10.1016/j.foodchem.2005.07.042

Baskaran P, Kumari A, Naidoo D, Staden JV (2015) In vitro propagation and biochemical changes in Aloe pruinosa. Ind Crop Prod 77:51–58. https://doi.org/10.1016/j.indcrop.2015.08.036

Benayad Z, Martinez-Villaluenga C, Frias J, Gomez-Cordoves C, Es-Safi NE (2014) Phenolic composition, antioxidant and anti-inflammatory activities of extracts from Moroccan Opuntia ficus-indica flowers obtained by different extraction methods. Ind Crop Prod 62:412–420

Betancur-Ancona D, Sosa-Espinoza T, Ruiz-Ruiz J, Segura-Campos M, Chel-Guerrero L (2014) Enzymatic hydrolysis of hard-to-cook bean (Phaseolus vulgaris L.) protein concentrates and its effects on biological and functional properties. Int J Food Sci Technol 49(1):2–8. https://doi.org/10.1111/ijfs.12267

Bhanja Dey T, Chakraborty S, Jain KK, Sharma A, Kuhad RC (2016) Antioxidant phenolics and their microbial production by submerged and solid state fermentation process: a review. Trends Food Sci Technol 53:60–74. https://doi.org/10.1016/j.tifs.2016.04.007

Boschin G, Scigliuolo GM, Resta D, Arnoldi A (2014) Optimization of the enzymatic hydrolysis of Lupin (Lupinus) proteins for producing ACE-inhibitory peptides. J Agric Food Chem 62 (8):1846–1851. https://doi.org/10.1021/jf4039056

Bursać Kovačević D, Maras M, Barba FJ, Granato D, Roohinejad S, Mallikarjunan K et al (2018) Innovative technologies for the recovery of phytochemicals from Stevia rebaudiana Bertoni leaves: a review. Food Chem 268:513–521. https://doi.org/10.1016/j.foodchem.2018.06.091

Cevallos-Casals BA, Cisneros-Zevallos L (2010) Impact of germination on phenolic content and antioxidant activity of 13 edible seed species. Food Chem 119(4):1485–1490. https://doi.org/10.1016/j.foodchem.2009.09.030

Chamorro S, Viveros A, Alvarez I, Vega E, Brenes A (2012) Changes in polyphenol and polysaccharide content of grape seed extract and grape pomace after enzymatic treatment. Food Chem 133(2):308–314. https://doi.org/10.1016/j.foodchem.2012.01.031

Charoensiddhi S, Lorbeer AJ, Lahnstein J, Bulone V, Franco CMM, Zhang W (2016) Enzyme-assisted extraction of carbohydrates from the brown alga Ecklonia radiata: effect of enzyme type, pH and buffer on sugar yield and molecular weight profiles. Process Biochem 51 (10):1503–1510. https://doi.org/10.1016/j.procbio.2016.07.014

Chen S, Xing X-H, Huang J-J, Xu M-S (2011) Enzyme-assisted extraction of flavonoids from Ginkgo biloba leaves: improvement effect of flavonol transglycosylation catalyzed by Penicillium decumbens cellulase. Enzym Microb Technol 48(1):100–105. https://doi.org/10.1016/j.enzmictec.2010.09.017

Comuzzo P, Marconi M, Zanella G, Querzè M (2018) Pulsed electric field processing of white grapes (cv. Garganega): effects on wine composition and volatile compounds. Food Chem 264:16–23. https://doi.org/10.1016/j.foodchem.2018.04.116

Correia RTP, McCue P, Magalhães MMA, Macêdo GR, Shetty K (2004) Production of phenolic antioxidants by the solid-state bioconversion of pineapple waste mixed with soy flour using Rhizopus oligosporus. Process Biochem 39(12):2167–2172. https://doi.org/10.1016/j.procbio.2003.11.034

Cui X-H, Chakrabarty D, Lee E-J, Paek K-Y (2010) Production of adventitious roots and secondary metabolites by Hypericum perforatum L. in a bioreactor. Bioresour Technol 101 (12):4708–4716. https://doi.org/10.1016/j.biortech.2010.01.115

da Silva RPFF, Rocha-Santos TAP, Duarte AC (2016) Supercritical fluid extraction of bioactive compounds. TrAC Trends Anal Chem 76:40–51. https://doi.org/10.1016/j.trac.2015.11.013

Daglia M (2012) Polyphenols as antimicrobial agents. Curr Opin Biotechnol 23(2):174–181. https://doi.org/10.1016/j.copbio.2011.08.007

Dai J, Mumper RJ (2010) Plant phenolics: extraction, analysis and their antioxidant and anticancer properties. Molecules 15(10):7313–7352. https://doi.org/10.3390/molecules15107313

Danisova C, Holotnakova E, Hozova B, Buchtova V (1994) Effect of germination on a range of nutrients of selected grains and legumes. Acta Aliment 23(3):287–298

Davies KM, Deroles SC (2014) Prospects for the use of plant cell cultures in food biotechnology. Curr Opin Biotechnol 26:133–140. https://doi.org/10.1016/j.copbio.2013.12.010

De Castro ML, Priego-Capote F (2010) Soxhlet extraction: past and present panacea. J Chromatogr A 1217(16):2383–2389

Ding M, Feng R, Wang SY, Bowman L, Lu Y, Qian Y et al (2006) Cyanidin-3-glucoside, a natural product derived from blackberry, exhibits chemopreventive and chemotherapeutic activity. J Biol Chem 281(25):17359–17368. https://doi.org/10.1074/jbc.M600861200

Đorđević T, Sarić-Krsmanović M, Gajić UJ (2019) Phenolic compounds and allelopathic potential of fermented and unfermented wheat and corn straw extracts. Chem Biodivers 16(2):e1800420. https://doi.org/10.1002/cbdv.201800420

Dueñas M, Fernández D, Hernández T, Estrella I, Muñoz R (2005) Bioactive phenolic compounds of cowpeas (Vigna sinensis L). Modifications by fermentation with natural microflora and with Lactobacillus plantarum ATCC 14917. J Sci Food Agric 85(2):297–304. https://doi.org/10.1002/jsfa.1924

Dulf FV, Vodnar DC, Dulf E-H, Toşa MI (2015) Total phenolic contents, antioxidant activities, and lipid fractions from berry pomaces obtained by solid-state fermentation of two Sambucus species with Aspergillus niger. J Agric Food Chem 63(13):3489–3500. https://doi.org/10.1021/acs.jafc.5b00520

Dulf FV, Vodnar DC, Dulf EH, Pintea A (2017) Phenolic compounds, flavonoids, lipids and antioxidant potential of apricot (Prunus armeniaca L.) pomace fermented by two filamentous fungal strains in solid state system. Chem Cent J 11:10. https://doi.org/10.1186/s13065-017-0323-z

Dziadek K, Kopeć A, Tabaszewska M (2019) Potential of sweet cherry (Prunus avium L.) by-products: bioactive compounds and antioxidant activity of leaves and petioles. J Eur Food Res Technol 245(3):763–772. https://doi.org/10.1007/s00217-018-3198-x

Esmaeilipour O, Van Krimpen MM, Jongbloed AW, De Jonge LH, Bikker P (2012) Effects of temperature, pH, incubation time and pepsin concentration on the in vitro stability of intrinsic phytase of wheat, barley and rye. Anim Feed Sci Technol 175(3):168–174. https://doi.org/10.1016/j.anifeedsci.2012.05.007

Facino RM, Carini M, Aldini G, Berti F, Rossoni G, Bombardelli E et al (1999) Diet enriched with procyanidins enhances antioxidant activity and reduces myocardial post-ischaemic damage in rats. Life Sci 64(8):627–642. https://doi.org/10.1016/S0024-3205(98)00605-5

Fan Z, Werkman JR, Yuan L (2009) Engineering of a multifunctional hemicellulase. Biotechnol Lett 31(5):751–757. https://doi.org/10.1007/s10529-009-9926-3

Farhat MB, Landoulsi A, Chaouch-Hamada R, Sotomayor JA, Jordán MJ (2013) Characterization and quantification of phenolic compounds and antioxidant properties of Salvia species growing in different habitats. Ind Crop Prod 49:904–914

Folsom AR, Yochum L, Kushi LH, Meyer K (1999) Dietary flavonoid intake and risk of cardiovascular disease in postmenopausal women. Am J Epidemiol 149(10):943–949. https://doi.org/10.1093/oxfordjournals.aje.a009738%J

Fowler ZL, Koffas MAG (2009) Biosynthesis and biotechnological production of flavanones: current state and perspectives. Appl Microbiol Biotechnol 83(5):799–808. https://doi.org/10.1007/s00253-009-2039-z

Galanakis CM (2012) Recovery of high added-value components from food wastes: conventional, emerging technologies and commercialized applications. Trends Food Sci Technol 26 (2):68–87. https://doi.org/10.1016/j.tifs.2012.03.003

Gan R-Y, Lui W-Y, Wu K, Chan C-L, Dai S-H, Sui Z-Q et al (2017) Bioactive compounds and bioactivities of germinated edible seeds and sprouts: an updated review. Trends Food Sci Technol 59:1–14. https://doi.org/10.1016/j.tifs.2016.11.010

Gassara F, Ajila CM, Brar SK, Verma M, Tyagi RD, Valero JR (2012) Liquid state fermentation of apple pomace sludge for the production of ligninolytic enzymes and liberation of polyphenolic compounds. Process Biochem 47(6):999–1004. https://doi.org/10.1016/j.procbio.2012.03.001

Gayathri S, Rajasree Radhika SR, Suman TY, Aranganathan L (2018) Ultrasound-assisted microextraction of β, ε-carotene-3, 3′-diol (lutein) from marine microalgae Chlorella salina: effect of different extraction parameters. Biomass Convers Biorefinery 8(4):791–797. https://doi.org/10.1007/s13399-018-0331-9

Gulsoy Toplan G, Kurkcuoglu M, Goger F, İşcan G, Ağalar HG, Mat A et al (2017) Composition and biological activities of Salvia veneris Hedge growing in Cyprus. Ind Crop Prod 97:41–48. https://doi.org/10.1016/j.indcrop.2016.11.055

Hamzah F, Idris A, Shuan TK (2011) Preliminary study on enzymatic hydrolysis of treated oil palm (Elaeis) empty fruit bunches fibre (EFB) by using combination of cellulase and β 1-4 glucosidase. Biomass Bioenergy 35(3):1055–1059. https://doi.org/10.1016/j.biombioe.2010.11.020

Heim KE, Tagliaferro AR, Bobilya DJ (2002) Flavonoid antioxidants: chemistry, metabolism and structure-activity relationships. J Nutr Biochem 13(10):572–584. https://doi.org/10.1016/S0955-2863(02)00208-5

Herrero M, Mendiola JA, Cifuentes A, Ibáñez E (2010) Supercritical fluid extraction: recent advances and applications. J Chromatogr A 1217(16):2495–2511. https://doi.org/10.1016/j.chroma.2009.12.019

Hertog MGL, Feskens EJM, Kromhout D, Hertog MGL, Hollman PCH, Hertog MGL et al (1993) Dietary antioxidant flavonoids and risk of coronary heart disease: the Zutphen Elderly Study. Lancet 342(8878):1007–1011. https://doi.org/10.1016/0140-6736(93)92876-U

Homaei AA, Sajedi RH, Sariri R, Seyfzadeh S, Stevanato R (2010) Cysteine enhances activity and stability of immobilized papain. Amino Acids 38(3):937–942. https://doi.org/10.1007/s00726-009-0302-3

Hong Y-H, Jung EY, Park Y, Shin K-S, Kim TY, Yu K-W et al (2013) Enzymatic improvement in the polyphenol extractability and antioxidant activity of green tea extracts. Biosci Biotechnol Biochem 77(1):22–29. https://doi.org/10.1271/bbb.120373

Huynh NT, Van Camp J, Smagghe G, Raes K (2014) Improved release and metabolism of flavonoids by steered fermentation processes: a review. Int J Mol Sci 15(11):19369–19388. https://doi.org/10.3390/ijms151119369

James E, Lee JM (2001) The production of foreign proteins from genetically modified plant cells. In: Zhong JJ, Byun SY, Cho GH, Choi JW, Haigh JR, Honda H et al (eds) Plant cells. Springer, Berlin/Heidelberg, pp 127–156

Jones WP, Kinghorn AD (2006) Extraction of plant secondary metabolites. Natural products isolation. Springer, pp 323–351

Jovanović AA, Đorđević VB, Zdunić GM, Pljevljakušić DS, Šavikin KP, Gođevac DM et al (2017) Optimization of the extraction process of polyphenols from Thymus serpyllum L. herb using maceration, heat-and ultrasound-assisted techniques. Sep Purif Technol 179:369–380

Khan BA, Akhtar N, Rasul A, Mahmood T, Khan HS, Iqbal M et al (2012) Investigation of the effects of extraction solvent/technique on the antioxidant activity of Cassia fistula L. J Med Plants Res 6(3):500–503

Khoddami A, Wilkes M, Roberts T (2013) Techniques for analysis of plant phenolic compounds. Molecules 18(2):2328–2375

Kukula-Koch W, Aligiannis N, Halabalaki M, Skaltsounis A-L, Glowniak K, Kalpoutzakis E (2013) Influence of extraction procedures on phenolic content and antioxidant activity of Cretan barberry herb. Food Chem 138(1):406–413

Kumar B, Verma SK, Singh HP (2011) Effect of temperature on seed germination parameters in Kalmegh (Andrographis paniculata Wall. ex Nees.). Ind Crop Prod 34(1):1241–1244. https://doi.org/10.1016/j.indcrop.2011.04.008

Kumari A, Naidoo D, Baskaran P, Doležal K, Nisler J, Van Staden J (2018) Phenolic and flavonoid production and antimicrobial activity of Gymnosporia buxifolia (L.) Szyszyl cell cultures. Plant Growth Regul 86(3):333–338. https://doi.org/10.1007/s10725-018-0432-2

Kutchan TM, Gershenzon J, Møller BL, Gang DR (2015) Natural products. In: Buchanan BB, Gruissem W, Jones RL (eds) Biochemistry & molecular biology of plants. American Society of Plants, Rockville, pp 1250–1318

Lafond M, Navarro D, Haon M, Couturier M, Berrin J-G (2012) Characterization of a broad-specificity β-glucanase acting on β-(1,3)-, β-(1,4)-, and β-(1,6)-glucans that defines a new glycoside hydrolase family. Appl Environ Microbiol 78(24):8540–8546. https://doi.org/10.1128/AEM.02572-12%J

Lala G, Malik M, Zhao C, He J, Kwon Y, Giusti MM et al (2006) Anthocyanin-rich extracts inhibit multiple biomarkers of colon cancer in rats. Nutr Cancer 54(1):84–93. https://doi.org/10.1207/s15327914nc5401_10

Laroze L, Soto C, Zúñiga ME (2010) Phenolic antioxidants extraction from raspberry wastes assisted by-enzymes. Electron J Biotechnol 13:11–12

Lee JK, Jeon J-K, Byun H-G (2011) Effect of angiotensin I converting enzyme inhibitory peptide purified from skate skin hydrolysate. Food Chem 125(2):495–499. https://doi.org/10.1016/j.foodchem.2010.09.039

Li F, Mao Y-D, Wang Y-F, Raza A, Qiu L-P, Xu X-Q (2017) Optimization of ultrasonic-assisted enzymatic extraction conditions for improving total phenolic content, antioxidant and antitumor activities in vitro from Trapa quadrispinosa Roxb. Residues. Molecules 22(3):–396

Liu H, Kang Y, Zhao X, Liu Y, Zhang X, Zhang S (2019) Effects of elicitation on bioactive compounds and biological activities of sprouts. J Funct Foods 53:136–145. https://doi.org/10.1016/j.jff.2018.12.019

Mahkam M, Nabati M, Rahbar KH (2014) Isolation, identification and characterization of lawsone from henna leaves powder with soxhlet technique. Iran Chem Commun 2:34–38

Marco M-G, Rodríguez LV, Ramos EL, Renovato J, Cruz-Hernández MA, Rodríguez R et al (2009) A novel tannase from the xerophilic fungus Aspergillus niger GH1. J Microbiol Biotechnol 1:1–10. https://doi.org/10.4014/jmb.0811.615

Martínez Cuesta S, Rahman Syed A, Furnham N, Thornton JM (2015) The classification and evolution of enzyme function. Biophys J 109(6):1082–1086. https://doi.org/10.1016/j.bpj.2015.04.020

Martins S, Mussatto SI, Martínez-Avila G, Montañez-Saenz J, Aguilar CN, Teixeira JA (2011) Bioactive phenolic compounds: production and extraction by solid-state fermentation. A review. Biotechnol Adv 29(3):365–373. https://doi.org/10.1016/j.biotechadv.2011.01.008

Martins S, Teixeira JA, Mussatto SI (2013) Solid-state fermentation as a strategy to improve the bioactive compounds recovery from Larrea tridentata leaves. Appl Biochem Biotechnol 171(5):1227–1239. https://doi.org/10.1007/s12010-013-0222-2

Moreira SA, Alexandre EMC, Pintado M, Saraiva JA (2019) Effect of emergent non-thermal extraction technologies on bioactive individual compounds profile from different plant materials. Food Res Int 115:177–190. https://doi.org/10.1016/j.foodres.2018.08.046

Mushtaq M, Sultana B, Anwar F, Adnan A, Rizvi SSH (2015) Enzyme-assisted supercritical fluid extraction of phenolic antioxidants from pomegranate peel. J Supercrit Fluids 104:122–131. https://doi.org/10.1016/j.supflu.2015.05.020

Nam H, Lewis NE, Lerman JA, Lee D-H, Chang RL, Kim D et al (2012) Network context and selection in the evolution to enzyme specificity. Science 337(6098):1101. https://doi.org/10.1126/science.1216861

Peng X, Qiao W, Mi S, Jia X, Su H, Han Y (2015) Characterization of hemicellulase and cellulase from the extremely thermophilic bacterium Caldicellulosiruptor owensensis and their potential application for bioconversion of lignocellulosic biomass without pretreatment. Biotechnol Biofuels 8(1):131. https://doi.org/10.1186/s13068-015-0313-0

Périno-Issartier S, Zill-e-Huma A-VM, Chemat F (2011) Solvent free microwave-assisted extraction of antioxidants from sea buckthorn (Hippophae rhamnoides) food by-products. Food Bioprocess Technol 4(6):1020–1028. https://doi.org/10.1007/s11947-010-0438-x

Qiu Z, Shi P, Luo H, Bai Y, Yuan T, Yang P et al (2010) A xylanase with broad pH and temperature adaptability from Streptomyces megasporus DSM 41476, and its potential application in brewing industry. Enzym Microb Technol 46(6):506–512. https://doi.org/10.1016/j.enzmictec.2010.02.003

Rafiee Z, Jafari S, Alami M, Khomeiri M (2011) Microwave-assisted extraction of phenolic compounds from olive leaves; a comparison with maceration. J Anim Plant Sci 21(4):738–745

Rajha HN, El Darra N, Hobaika Z, Boussetta N, Vorobiev E, Maroun RG et al (2014) Extraction of total phenolic compounds, flavonoids, anthocyanins and tannins from grape byproducts by response surface methodology. Influence of solid-liquid ratio, particle size, time, temperature and solvent mixtures on the optimization process. Food Nutr Sci 5(04):397

Raveendran S, Parameswaran B, Ummalyma SB, Abraham A, Mathew AK, Madhavan A et al (2018) Applications of microbial enzymes in food industry. Food Technol Biotechnol 56 (1):16–30. https://doi.org/10.17113/ftb.56.01.18.5491

Rocchetti G, Blasi F, Montesano D, Ghisoni S, Marcotullio MC, Sabatini S et al (2019) Impact of conventional/non-conventional extraction methods on the untargeted phenolic profile of Moringa oleifera leaves. Food Res Int 115:319–327. https://doi.org/10.1016/j.foodres.2018.11.046

Ruiz D, Egea J, Gil MI, Tomás-Barberán FA (2005) Characterization and quantitation of phenolic compounds in new apricot (Prunus armeniaca L.) varieties. J Agric Food Chem 53 (24):9544–9552. https://doi.org/10.1021/jf051539p

Sadeghi A, Hakimzadeh V, Karimifar B (2017) Microwave assisted extraction of bioactive compounds from food: a review. Int J Food Sci Nutr Eng 7:19–27

Sandoval G, Villanueva-Rodriguez SJ (2018) Extraction methods for phenolic compounds. In: Nollet LM, Gutierrez-Uribe JA (eds) Phenolic compounds in food: characterization and analysis, 1st edn. CRC Press, London, p 415

Santangelo C, Varì R, Scazzocchio B, Di Benedetto R, Filesi C, Masella R (2007) Polyphenols, intracellular signalling and inflammation. Annali-istituto superiore di sanita 43(4):394

Sasidharan S, Chen Y, Saravanan D, Sundram K, Latha LY (2011) Extraction, isolation and characterization of bioactive compounds from plants' extracts. Afr J Tradit Complement Altern Med 8:1

Sáyago-Ayerdi SG, Mercado-Mercado G, Ramos-Romero S, Torres JL, Pérez-Jiménez J (2016) Analysis and characterization of polyphenol extracts. Adv Technol Prod Food-relevant Polyphen 132

Seeram NP, Adams LS, Zhang Y, Lee R, Sand D, Scheuller HS et al (2006) Blackberry, black raspberry, blueberry, cranberry, red raspberry, and strawberry extracts inhibit growth and stimulate apoptosis of human cancer cells in vitro. J Agric Food Chem 54(25):9329–9339. https://doi.org/10.1021/jf061750g

Sevenich R, Mathys A (2018) Continuous versus discontinuous ultra-high-pressure Systems for Food Sterilization with focus on ultra-high-pressure homogenization and high-pressure thermal sterilization: a review. Compr Rev Food Sci Food Saf 17(3):646–662. https://doi.org/10.1111/1541-4337.12348

Sharma V, Janmeda P (2017) Extraction, isolation and identification of flavonoid from Euphorbia neriifolia leaves. Arab J Chem 10(4):509–514

Sharma N, Rathore M, Sharma M (2013) Microbial pectinase: sources, characterization and applications. Rev Environ Sci Bio/Technol 12(1):45–60. https://doi.org/10.1007/s11157-012-9276-9

Soccol CR, ESFD C, LAJ L, Karp SG, Woiciechowski AL, Vandenberghe LPDS (2017) Recent developments and innovations in solid state fermentation. Biotechnol Res Innov 1(1):52–71. https://doi.org/10.1016/j.biori.2017.01.002
Sohail M, Siddiqi R, Ahmad A, Khan SA (2009) Cellulase production from Aspergillus niger MS82: effect of temperature and pH. New Biotechnol 25(6):437–441. https://doi.org/10.1016/j.nbt.2009.02.002
Starzyńska-Janiszewska A, Stodolak B, Gómez-Caravaca AM, Mickowska B, Martin-Garcia B, Byczyński Ł (2019) Mould starter selection for extended solid-state fermentation of quinoa. LWT 99:231–237. https://doi.org/10.1016/j.lwt.2018.09.055
Steingroewer J, Bley T, Georgiev V, Ivanov I, Lenk F, Marchev A et al (2013) Bioprocessing of differentiated plant in vitro systems. Eng Life Sci 13(1):26–38. https://doi.org/10.1002/elsc.201100226
Taghizadeh M, Nasibi F, Kalantari KM, Ghanati F (2019) Evaluation of secondary metabolites and antioxidant activity in Dracocephalum polychaetum Bornm. Cell suspension culture under magnetite nanoparticles and static magnetic field elicitation. Plant Cell Tissue Organ Cult (PCTOC) 136(3):489–498. https://doi.org/10.1007/s11240-018-01530-1
Tatke P, Jaiswal Y (2011) An overview of microwave assisted extraction and its applications in herbal drug research. Res J Med Plant 5(1):21–31. https://doi.org/10.3923/rjmp.2011.21.31
Theegala CS (2015) Algal cell disruption and lipid extraction: a review on current technologies and limitations. In: Prokop A, Bajpai RK, Zappi ME (eds) Algal biorefineries: volume 2: products and refinery design. Springer, Cham, pp 419–441
Toepfl S, Mathys A, Heinz V, Knorr D (2006) Review: potential of high hydrostatic pressure and pulsed electric fields for energy efficient and environmentally friendly food processing. Food Rev Intl 22(4):405–423. https://doi.org/10.1080/87559120600865164
Tufariello M, Durante M, Veneziani G, Taticchi A, Servili M, Bleve G et al (2019) Patè olive cake: possible exploitation of a by-product for food applications. Front Nutr 6(3). https://doi.org/10.3389/fnut.2019.00003
Uddin MS, Sarker MZI, Ferdosh S, Akanda MJH, Easmin MS, Bt Shamsudin SH et al (2015) Phytosterols and their extraction from various plant matrices using supercritical carbon dioxide: a review. J Sci Food Agric 95(7):1385–1394. https://doi.org/10.1002/jsfa.6833
Velavan S (2015) Phytochemical techniques-a review. World J Sci Res 1(2):80–91
Vermerris W, Nicholson R (2006) Families of phenolic compounds and means of classification. Phenolic compound biochemistry. Springer, pp 1–34
Victório CP, Lage CLS, Kuster RM (2009) Flavonoid extraction from Alpinia zerumbet (Pers.) Burtt et Smith leaves using different techniques and solvents. Eclética Química 34(1):19–24
Vongsak B, Sithisarn P, Mangmool S, Thongpraditchote S, Wongkrajang Y, Gritsanapan W (2013) Maximizing total phenolics, total flavonoids contents and antioxidant activity of Moringa oleifera leaf extract by the appropriate extraction method. Ind Crop Prod 44:566–571
Vorobiev E, Lebovka N (2016) Selective extraction of molecules from biomaterials by pulsed electric field treatment. In: Miklavcic D (ed) Handbook of electroporation. Springer, Cham, pp 1–16
Walker JM, Sweeney PJ (2009) Production of protein hydrolysates using enzymes. In: Walker JM (ed) The protein protocols handbook. Humana Press, Totowa, pp 989–993
Wang L, Weller CL (2006) Recent advances in extraction of nutraceuticals from plants. Trends Food Sci Technol 17(6):300–312
Wang L, Wu Y, Liu Y, Wu Z (2017) Complex enzyme-assisted extraction releases antioxidative phenolic compositions from guava leaves. Molecules 22(10):1648
Wilson SA, Roberts SC (2012) Recent advances towards development and commercialization of plant cell culture processes for the synthesis of biomolecules. Plant Biotechnol J 10(3):249–268. https://doi.org/10.1111/j.1467-7652.2011.00664.x
Xiao Y, Wang L, Rui X, Li W, Chen X, Jiang M et al (2015) Enhancement of the antioxidant capacity of soy whey by fermentation with Lactobacillus plantarum B1–6. J Funct Foods 12:33–44. https://doi.org/10.1016/j.jff.2014.10.033

Xu X, Zhu J (2011) Enhanced phenolic antioxidants production in submerged cultures of Inonotus obliquus in a ground corn stover medium. Biochem Eng J 58–59:103–109. https://doi.org/10.1016/j.bej.2011.09.003

Yang Y-C, Li J, Zu Y-G, Fu Y-J, Luo M, Wu N et al (2010) Optimisation of microwave-assisted enzymatic extraction of corilagin and geraniin from Geranium sibiricum Linne and evaluation of antioxidant activity. Food Chem 122(1):373–380. https://doi.org/10.1016/j.foodchem.2010.02.061

Yue W, Q-l M, Lin B, Rahman K, Zheng C-J, Han T et al (2016) Medicinal plant cell suspension cultures: pharmaceutical applications and high-yielding strategies for the desired secondary metabolites. Crit Rev Biotechnol 36(2):215–232. https://doi.org/10.3109/07388551.2014.923986

Zhang Y, Seeram NP, Lee R, Feng L, Heber D (2008) Isolation and identification of strawberry phenolics with antioxidant and human cancer cell antiproliferative properties. J Agric Food Chem 56(3):670–675. https://doi.org/10.1021/jf071989c

Zhang G, Hu M, He L, Fu P, Wang L, Zhou J (2013) Optimization of microwave-assisted enzymatic extraction of polyphenols from waste peanut shells and evaluation of its antioxidant and antibacterial activities in vitro. Food Bioprod Process 91(2):158–168. https://doi.org/10.1016/j.fbp.2012.09.003

Exploitation of Plant Phenolics in Animal Farming

3

Javid Ahmad Malik, Monika Bhadauria, Rafiq Lone, and Younis Ahmad Hajam

Abstract

Secondary metabolites are produced by the plants besides the primary metabolites which are not needed for the daily functioning of the plant. However, they are responsible for a myriad number of potential roles in the living system, such as protection, attraction, or signaling. Herbs and spices have been used since ancient times as folk medicine and as preservatives in foods as they contain many biologically active compounds, especially polyphenols, which possess antimicrobial, antioxidant, antiparasitic, antifungal, and anti-inflammatory properties. The production of these polyphenols is affected by various factors including both internal and environmental factors. These polyphenolics are of many forms and beneficial to the animal world in each of this form. The demand for these plant derivatives has increased alarmingly because they are natural and eco-friendly. Polyphenols are secondary plant metabolites which contain bioactive components and deliver positive effects for humans and animals and also protect the plant from various diseases. Likewise, they act as allelochemicals and suppress weed growth markedly. For better yield and crop management, a few traditional practices like cover crops, green manures, and crop rotation are being utilized. Besides other useful roles, plant phenolics improve ruminant fermentation by acting as antioxidant agents. They also help to improve

J. A. Malik (✉)
Department of Zoology, Guru Ghasidas Vishwavidyalaya, Bilaspur, Chhattisgarh, India

M. Bhadauria
Toxicology and Pharmacology Laboratory, Department of Zoology, Guru Ghasidas Vishwavidyalaya, Bilaspur, Chhattisgarh, India

R. Lone
Department of Botany, Sant Baba Bhag Singh University, Jalandhar, Punjab, India

Y. A. Hajam
Department of Biosciences, Division Zoology, Career Point University, Hamirpur, Himachal Pradesh, India

R. Lone et al. (eds.), *Plant Phenolics in Sustainable Agriculture*,
https://doi.org/10.1007/978-981-15-4890-1_3

enzymatic activity and therefore favor high nutritive uptake by the particular animals. Condensed tannins, one of the groups of polyphenols, possess antiparasitic roles by repressing internal parasitism in animals. Their direct role in sustainable animal farming is with regard to improved lactation, trace element status, growth, resilience to internal parasites, and reductions in parasite larval development. Subsequently, the popularity of plant-based feed additives in animal farming has increased prominently in the last decade. Polyphenols have antioxidant properties, and they minimize the negative consequences of oxidative stress on meat quality and production. There is a proper need for managing administration of polyphenols in the animal farming to utilize all the benefits of these phenolic components.

Keywords

Secondary metabolites · Phenolics · Animal farming · Polyphenols · Herbs

3.1 Introduction

Plants synthesize a diverse number of organic compounds that serve the plant in several ways. The compounds are often portrayed as secondary, different from primary ones, which are involved in basic cell metabolic processes of the plant. Besides other functions, secondary metabolites function as anti-stress components against different forms of stress. Plants form secondary metabolites in some specific developmental stages or as a response to unfavorable environmental conditions. They are produced as a normal part of plant metabolism in specialized cells or tissues and accumulate in surprisingly high levels in some species. Many compounds can be characteristic for a particular plant family, genus, or even species and therefore might be used as taxonomic tools in classifying plants (Piasecka et al. 2015).

Their features in plants are now attracting attention as some appear to have a key role in mutualistic interactions with other animals as pollinators, seed dispersals, fungi as mycorrhiza, and bacteria by formation of nitrogen-fixing root nodules in legumes. They protect the plant against an ample number of stress forms as well as help as chemical plant defense mechanism against herbivores and pathogens (Hartmann 2008). In addition, they may also play a vital role in plant–plant interactions, in which they function as allelochemicals with either positive or negative effects on the neighboring plants. Secondary metabolites can also be of economic interest as dyes, fibers, oils, glues, waxes, drugs, flavoring agents, and perfumes and are viewed as potential sources of new natural drugs, antibiotics, insecticides, and herbicides (Akula and Ravishankar 2011). Plant phenolics are considered as one of the most important groups of secondary metabolites with above 10,000 molecules reported which are found throughout the plant kingdom. Phenolics range from simple, low-molecular-weight compounds with a single aromatic ring (phenolic acids) to large and complex molecules (tannins). Flavonoids are the most prominent group with two aromatic rings. They can be categorized into three groups as phenolics, terpenoids, and nitrogen-containing compounds on the basis of their biosynthetic origins; mostly, they are stored in vacuoles of the epidermal cell or specialized cells. Another important group of phenolics, which are chemically related to flavonoids, are stilbenes with resveratrol as the most common compound in this group (Bavaresco et al. 2012). Phenolics are attributed to numerous ecological functions; they are not only indicators of plant stress responses to various abiotic factors but also play a vital role in plant resistance and tolerance against pests and diseases. Phenolic-based polymers, such as lignin, suberin, or tannins, contribute substantially toward mechanical or environmental damage, such as drought or wounding with respect to the stability and robustness of plant tissues (Treutter 2010). Allelochemicals released from plants mainly affect the performance of neighboring plants in a negative way, with respect to gain an advantage in nutrient, water, or light acquisition.

3.2 Factors Influencing the Phenolics

Phenolic content differs among species of the same family, even between cultivars. Studies revealed that the phenolic content is genotype-dependent than influenced either by organic or integrated production (Veberic et al. 2005). Furthermore, various plant organs or tissues may be differentiated by different phenolic compositions. Ripening most importantly affects the phenolic content of crops, especially fruit. Environmental stress to the plant incites responses in its phenolic metabolism especially UV light which is often accompanied by high temperatures and results in photooxidative damage to the photosystem complex. The response to the stress events at the cellular level triggers repair processes in the cells by de novo synthesis of protective substances; most of these protective compounds belong to phenolics. Zupan et al. confirmed the excessive sun irradiation causes increased levels of phenolics (Zupan et al. 2014); part of the fruit damaged by sunburn was observed to possess highest levels of phenolics. This kind of study was supported by Jakopic et al. (2010) who observed the anthocyanin content in apple skin as strongly light-dependent.

Although high nitrogen doses improve plant growth but to some extent reduce the amount of phenolics, thus nutrient deficiency can cause accumulation of certain phenolics. Salt stress due to a high concentration of certain ions might be responsible for increasing the phenolic content as seen in various plant species (Akula and Ravishankar 2011). A proper water supply in cultivation technology is necessary for normal growth and yield as well as phenolic composition of crops.

3.3 Phenolics and Plant Disease

Plants are constantly in contact with other organisms including those that are pathogenic, and thus subsequently defense mechanisms in plants have developed to protect themselves, and pathogens have established countermeasures to overcome these mechanisms. Plants protect themselves from invading organisms by means of physical (e.g., cell wall, cuticle, trichomes) and chemical barriers (different secondary compounds). Each species, or closely related species, can form characteristic molecules that act as defense compounds against specific or nonspecific pathogens. The first line of the defense response to a pathogen attack is the recognition of the pathogen molecules. These elicitors may be oligosaccharides, peptides, or glycoconjugates of bacterial origin or are plant-derived molecules released at wounding.

These elicitors initiate early cellular responses, like alterations in plant cell walls, reactive oxygen species and reactive nitrogen species accumulation, changes in ion fluxes, and gene transcription in host organism (Newman et al. 2013). The second line of defense involves the induction of plant defense genes and biosynthesis of endogenous secondary metabolites, cell wall fortification, hypersensitive responses, and systemic acquired resistance (Newman et al. 2013; Amil-Ruiz et al. 2011). Moreover, not just the quantity but also the speed of the plant response is crucial for its increased resistance. When the induced responses are rapid during a plant–pathogen interaction, the plant is resistant to the disease. Contrarily, postinfection defense mechanisms are established at a slower rate in susceptible plants.

3.4 Phenolic Production

Synthetic chemicals such as pesticides, fertilizers, and fruit thinning agents are not used in the organic cultivation of fruit crops; only natural compounds are allowed, and their use is closely monitored. The lack of pesticides and insecticides in organic production permitted the incidence of injuries and pathogen infections due to higher disease pressure and potentially elevated the phenolic content in plant tissue. In organic farming, nutrients are often supplied through compost and manure and plant derived by products in which organic nitrogen is transformed into the inorganic form by soil microflora, so the level of nutrient availability to plants may be difficult to control. A number of studies have confirmed the impact of fertilization on the phenolic content. Toor et al. (2006) reported tomatoes fertilized with chicken manure were observed to have lower shoot biomass compared with those fertilized by mineral nutrient solution. Toor et al. (2006) revealed that surplus carbon, which was not utilized for growth, was utilized in the production of secondary metabolites, especially phenolics. A similar type of study was carried out by Bavec et al. (2010) who demonstrated that biodynamic red beet (*Beta vulgaris* L. ssp. *vulgaris*) increased higher levels of total phenolics than did red beets from a conventional farming system. Lower availability of nutrients in the biodynamic farming system among other factors was also listed by these authors. Nitrogen use efficiency and availability to plants is better in conventional production as inorganic nitrogen is applied.

It would be interesting to determine whether the different production types have any impact on storage life of fruits. Production technology had negotiable effect on the storage of cashew nuts, as up to 180 days of storage it did not influence the phenolic content. The soluble solids were higher in conventional farming (Soares et al. 2012). Roth et al. (2007) observed various quality parameters over 6 months of apple storage, and it was demonstrated that changes mainly occurred due to different storage conditions rather than possible effects of organic versus integrated production systems.

3.4.1 Allelochemicals

The common allelochemicals from crop plants are generally secondary metabolites. These include phenolics, terpenoids, alkaloids, flavonoids, coumarins, tannins, steroids, and quinines (Einhellig and Leather 1988). Flavonoids and phenolic acids show strong inhibition in bioassays but exhibit weak phytotoxicity in soil and less selectivity. However, phenolics may suppress weed growth markedly in field when they influence nutrient uptake (Booker et al. 1992). Terpenoids also show strong effects in crop–weed interactions. They exhibit marked phytotoxicity, and cinmethylin, a derivative of monoterpene 1,8-cineole, was marketed as bioherbicides (Dayan et al. 1999). *Artemisia annua* produces artemisinin which is highly phototoxic, strongly suppressing root growth and causing extreme chlorosis. It influences mitochondrial oxygen evolution and inhibits mitosis, which leads to appearance of aberrant mitotic phase (Dayan et al. 1999). In addition, terpenoids like Taxol and alkaloids (colchicines and vinblastine) suppress mitosis, which have a similar mode of action to certain synthetic herbicides (Vaughan and Vaughan 1988). Quinones such as juglone (Hejl et al. 1993) and sorgoleone (Einhellig et al. 1993) inhibit chloroplast oxygen evolution and strongly affect mitochondrial functions (Rasmussen et al. 1992). Allelopathic crops such as sorghum, alfalfa, barley, corn, wheat, asparagus, coffee, tea, tobacco, and sunflower which contain allelochemicals are considered as potent herbicides (Macias et al. 1996). However, it should be kept in mind that allelochemicals should be safe and technologically sustainable (from environmental, agronomical, and economical views), have new site of action and in sufficient number and quantity, and must be active at low concentrations and have a broad range of activities (Macias et al. 2001).

3.4.2 Weed Control by Allelopathic Crops

A crop is cultivated for food, fodder, fiber, and several other products. However, modern agroecosystems are characterized by the presence of synthetic agrochemicals (herbicides, pesticides, fungicides, and fertilizers), less diversity, and resistant pests, making it ecologically unsustainable polluting the soil and water. Sustainable agroecosystems are need of the hour which must be organic, regenerative, biodynamic, and resource conserving (Anaya 1999). Certain practices such as crop rotation, companion cropping, cover cropping, or polyculture cropping practices must be used in traditional cultivation in agricultural production (Singh et al. 2001). To achieve the goal of sustainable agriculture, soil fertility, plant breeding and tillage, crop protection, and cropping systems are being much focused.

Allelopathy is one of the chemical pathways of interaction among plants which may be both inhibitive and promotive. The inhibitive effects are exploited for pest and weed control in agricultural practice (Kohli et al. 1998). An allelopathic crop must affect the growth, productivity, and yield of other crops; may affect similar crop growing in monocultures or grown in succession; causes soil sickness and imbalance of nutrients and microbial population; and can be exploited to selectively suppress weeds through various manipulations (Einhellig 1985; Batish et al. 2001). Chemicals released from crop plants cause soil sickness which can be minimized by crop rotation (De Candolle 1832).

Akobundu in 1987 noted that weeds reduce crop yield by 5% in the most highly developed countries, 10% in the less developed countries, and 25% in the least developed countries. Weeds and cultivated crops compete with each other for growth factors like water, light, nutrients, and spaces and harbor plant pathogens and pests (Qasem and Foy 2001). A tremendous increase in usage of pesticides and herbicides has been witnessed in developing countries, as urbanization decreases labor force in the agricultural sector and farmers also tend to earn extra money by spending more time outside agricultural work to earn extra money (Xuan and Tsuzuki 2004). Various crops inhibit weeds by releasing phytotoxic chemicals and thus are promising for sustainable weed management. The magnitude of inhibition may decrease because of soil pH, organic carbon, organic matter, and available nitrogen which affect the allelopathic expression of a plant (Blum 1996). Thus, an allelopathic crop must be applied for weed control, under natural conditions, and the time and dosage required for maximum weed control should be examined (Xuan et al. 2005). The magnitude of weed suppression is dependent on the applied dose. The allelopathic impacts in fields may result in recurrence of weeds, which are then suppressed by crop shading (Xuan et al. 2005). The inhibition on weeds is selective; however, the effects differ among crop species (Xuan et al. 2005).

The sustainability of agriculture has its basis on the development of strategies that reduce the need for costly external inputs and minimize harmful effects on the environment. Agriculture integrated with allelopathy could lower the dependence on synthetic herbicides and other agrochemicals, and therefore environmental contamination, use of unsafe agricultural products, and effects on human health could be reduced. The allelopathy-integrated management of weeds includes various approaches such as the use of allelopathy in crop rotation, green manure, cover crops, mulch, and intercropping besides incorporation of allelopathic plants in soil. Many secondary metabolites involved in allelopathic activities have been identified; however, a lot remained unknown.

3.4.3 Crop Rotation, Green Manures, and Cover Crops

It is one of the traditional practices where some crops, particularly leguminous species, are grown in short rotation with the main crops. In crop rotation, an allelopathic crop is designed in rotational sequences which can repress weeds in the cultivated and next crops. Batish et al. (2001) reported that soil sickness or autotoxicity caused by allelopathy can be restricted by crop rotation. There are certain principles of selecting crops for rotational sequences, viz., alternating between annual and perennial crops, alternating between autumn and spring germinating crops, and alternating between closed crops which diminish weeds and open crops which encourage weeds and a variety of operations (Lampkin 1994).

Cover crops are defined as regular crops for soil and moisture conservation, enhancement of biomass production, nutrient recycling, temperature lowering, nuisance weed suppression, and forage supply (Swanton and Murphy 1996; Gallandt et al. 1999; Batish et al. 2001). Cover crops also can stagger weed growth by shading effects because of their dense population and fast growth (Foley 1999).

Alfalfa, foxtail millet (*Setaria italica*), buckwheat sorghum, and rye are the commonly used cover crops. Legume species and some cruciferous plants increase weed suppression and improve soil conditions by contributing organic matter and nitrogen to the soil. Buckwheat too is a good candidate for green manure to enhance soil fertility, owing to the rich nutrients contained in it. The nitrogen content in buckwheat tissue is about 1.2% (Valenzuela and Smith 2002). Kelling et al. (1981) reported that buckwheat when incorporated with rye as green manure was good for soil nutrient improvement. Plant diseases cause about 20% yield loss of major food and cash crops worldwide (Oerke et al. 1994). Allelopathic crops reduce the intensity of soilborne diseases through crop residues, which release inhibitory compounds (Yu 1999). Important substances discovered from cereals are DIMBOA and DIBOA, which repress predators such as insects, fungi, and even bacteria (Kutchan 1997). Saponins from alfalfa root showed potential suppression of both weed and numerous phytopathogenic fungi (Oleszek 1999; Oleszek et al. 1999).

Crucifers when incorporated into soil caused declining of severity of root rot of peas caused by *Aphanomyces euteiches* (Muehlchen et al. 1990). Emergence of potato dry rot (*Fusarium sambucinum*) was reduced by species of *Brassica*, which contain high concentration of allyl isothiocyanate (Mayton et al. 1996). Since the most allelopathic accessions are selected, they are important sources to examine at the molecular levels with modern highly sophisticated techniques like polymerase chain reaction, random amplified polymorphic DNA, restriction fragment length polymorphism, near-isogenic lines, or cloning genes, which are very useful to prepare the genetic maps of higher plants. Hoult and Lovett (1993) recognized that various wild accessions of modern-day crop plants possess allelopathic traits that can suppress weeds and pests. However, during the process of cultivation, aiming at selection of high-yielding varieties while ignoring weed and pest resistance characteristics has led to the loss of these traits (Singh et al. 2001). DNA recombinant technology or even conventional breeding methods can be used to transfer allelopathic-based gene from one locus to another in target crops (Singh et al. 2001).

3.4.4 Plant Phenolics and Ruminant Fermentation

Gram-positive rumen bacteria generate more ammonia, lactate, and hydrogen than Gram-negative species. The substances which selectively inhibit Gram-positive bacteria therefore improve animal productivity via increasing propionate production and decreasing methane production, proteolysis of dietary protein, and accumulation of lactate (Russell and Strobel 1989). However, the antibiotic use in animal nutrition has been prohibited by the European Union (Official Journal of European Union 2003) due to the risk of residues in animal products as well as to the concern about the manifestation of resistant strains of bacteria. So the focus was moved to natural feed additives. Plant extracts and bioactive plant secondary metabolites are the most common natural feed additives. Secondary metabolites are derived from the primary metabolisms which do not have any nutritive value and direct contribution to growth, reproduction, and development. However, they have a broad range of antimicrobial activity and serve to protect plants against pathogens, parasites, herbivores, predators, interplant competition, and abiotic stresses as desiccation and UV radiation (Greathead 2003). Plants usually localize secondary compounds in specialized vacuoles, glands, cell walls, or plant part surfaces to protect their own tissues. It has been revealed that defensive secondary metabolites are often synthesized in response to stress. So, secondary plant metabolites can be considered as survival and defense mechanisms of plants (Wink 1988).

Moreover, these substances give plant specific colors and odors (Greathead 2003). Phenolic compounds (carotenoids, alkaloids, phenolic acids, flavonoids, and tannins), saponins, and essential oils are the best known representatives of plant secondary metabolites. Recent research has been greatly focused to exploit these plant metabolites to mitigate methane emission, improve protein utilization, and improve animal productivity (Broudiscou et al. 2000; Kim et al. 2015; Patra and Yu 2014). Phenolic compounds, i.e., flavonoids, phenolic acids, and tannins, are the most common phytochemical groups found in plants which exhibit several bioactivities such as antimicrobial, antiviral, anti-inflammatory, and antioxidant. Other less common phenolic compounds include coumarins, lignans, quinones, and stilbenes (Kaçar D (Kaçar 2008).

3.4.5 Flavonoids and Phenolic Acids

The most studied group among plant phenolics is flavonoids, which consist of several subclasses: flavanols (quercetin, kaempferol, myricetin), flavons (luteolin and apigenin), flavanones (naringenin), anthocyanins, and isoflavonoids (genistein).

Phenolic acids comprise another important class of phenolic compounds, which mainly consist of two major groups: hydroxycinnamic acids (caffeic acid, ferulic acid, coumaric acid, chlorogenic acid, cinnamic acid, etc.) and hydroxybenzoic acids (gallic acid, protocatechuic acid, p-hydroxybenzoic acid, vanillic acid, etc.) (Kaçar 2008). Phenolic acids and flavonoids have been reported to possess antimicrobial effects (Mirzoeva et al. 1997). Kim et al. (2015) showed that flavonoid-rich plant extracts decreased methane accumulation, number of Gram-positive bacteria, and ciliate populations.

Flavonoids can also encourage the fermentative activity of rumen bacteria (Broudiscou et al. 2000). Flavonoid-rich *Lavandula officinalis* (lavender) and *Solidago virgaure* (goldenrod) have been reported to improve the ruminal fermentation via increasing production of total volatile fatty acids (VFA) (Broudiscou et al. 2000).

Increased propionate and total VFA production by cashew nut shell liquid (Watanabe et al. (2010) and *Olea europaea* (olive leaf) extract (Öztürk et al. 2012) containing phenolic compounds have been reported. In another study, propolis extract phenolic compounds increased acetate production (Paula et al. 2016).

3.4.6 Tannins

Tannins are hydrophilic polyphenolic compounds having molecular weight in the range of 500–5000 units which can be classified into two subtypes, condensed and hydrolysable types (Szumacher-Strabel and Cieślak 2010). Tannins are widely found in legumes, fruits, and cereals which limit nutritional value of plants significantly when present in amount higher than 5% (Atiku et al. 2016). Tannins have also been reported to indirectly inhibit fiber degradation (Bodas et al. 2012). Addition of tannic acid in the beef cattle ration has been observed to reduce digestibility of crude protein as well as methane production (Yang et al. 2017). These reducing effects of tannins on ruminal protein degradation and methane emission proved advantageous for ruminant nutrition.

3.4.7 Saponins

Saponins (*Gk.* "saponin" meaning "soap") constitute primarily sapogenins and glycosides which are found generally in angiosperms and are divided into two groups as steroidal and triterpenoid saponins. They have been observed to have antibacterial and antifungal protection properties for plants against various bacterial and fungal diseases (Szumacher-Strabel and Cieślak 2010). Soybeans are the main example of ruminant diet plants which are saponin-rich. Saponins act on rumen fermentation, chiefly by increasing the amino acid flow to the small intestine by decreasing protein degradation and consequently ammonia and urea concentrations in rumen. The saponin effects on nitrogen metabolism in the rumen were attributed mainly to their harmful effect on protozoa which are largely responsible from nitrogen retention in the rumen, because of their proteolytic activity on both dietary and microbial proteins (Patra and Saxena 2009). Patra (2012) reported that the reduction of methane production might be related to direct effect on archaea activity and/or indirect effect on protozoa abundance. However, in long-term studies, the activity of saponins seems to be inconsistent and even decreasing (Jouany and Morgavi (2007) probably due to microbial adaptation (Patra 2012). Besides this, saponins can raise propionate concentration at the expenditure of acetate and butyrate (Jouany and Morgavi 2007). Jayanegara et al. (2014) reported that saponins decrease methane emissions at both low and high levels. On the other hand, nutrient digestibility enhanced at low levels while diminished at high levels of saponins. Further, it was reported by Odenyo et al. that the effects of saponins are more prominent when they are directly introduced to the rumen than mixed with the diet (Odenyo et al. 1997).

3.4.8 Essential Oils

These are volatile components which are usually derived from leaves, barks, flowers, seeds, and roots of various plant species. They have been categorized into terpenoids and phenylpropanoids according to their chemical structures (Calsamiglia et al. 2007). Essential oils contain various bioactive substances such as thymol, eugenol, carvacrol, anethole, geraniol, capsaicin, limonene, etc., which protect the plant with their antimicrobial activity. Essential oils can penetrate and diffuse to cell walls of Gram-positive bacteria because of their hydrophobic and lipophilic structures. Thus, like antibiotics, they are responsible for disruption of the ionic balance between inside and outside of the bacterial cell (Griffin et al. 1999). Essential oils such as *Origanum*, clove, garlic, peppermint, and eucalyptus were reported to decrease methane production and reduce the abundance of archaea, protozoa, and major cellulolytic bacteria (Patra and Yu 2012). Furthermore, rosemary, cinnamon, vanillin oils (Patra and Yu 2014), and dill seeds (Cobellis et al. 2016) reduced the ammonia and methane production. Some molecules of essential oils have the potential to diffuse to membrane of Gram-negative bacteria (Jouany and Morgavi (2007).

It was reported that the effects of essential oils on ruminal fermentation were closely related with their received doses. Moreover, the effects of essential oils on ruminal fermentation can also vary depending on the chemical nature of essential oils which changes their antimicrobial spectrum.

Plant extracts and secondary plant metabolites potentially modify ruminal fermentation and improve animal productivity. The effects of plant extracts and secondary metabolites on rumen fermentation and feed digestibility vary greatly depending on their dose, antimicrobial spectrum, and the amount of active metabolites of the plant. Defining of minimal inhibitor concentrations of single active compounds independently on rumen bacteria cultures will be useful to determine antimicrobial spectrum and effective dose. Molecular profiling to detect microbial changes would provide more apparent and tangible information to the literature. Verification of results with in vivo trials is also essential to define the true value of plant metabolites for altering rumen fermentation.

3.5 Agroecology and Animal Farming

Agroecology was firstly defined as a scientific discipline that applies ecological theory to the design and management of agroecosystems in order to enhance their sustainability (Altieri 1987). However, its horizons reached beyond the agroecosystem scale toward a wider focus on the whole food system, encompassing food production, distribution, and consumption (Wezel et al. 2009). Animal production systems have negative impacts on the environment; thus, animal agriculture needs to be reconfigured to minimize its negative impacts and produce food and other ecosystem services.

Livestock farming has undergone a significant transformation in the past few decades. Production has shifted from smaller, family-owned farms to large farms that become much more efficient. Since 1960, milk, meat, and egg production has increased multiple folds (Pew Commission on Industrial Animal Farm Production 2009). This has been achieved through improvements of animal breeding, mechanical innovations, and the introduction of specially formulated feeds and animal pharmaceuticals. Livestock farming is under an elevating pressure to be more efficient (Garnett et al. 2013) to meet the needs of projected 9 (or even 11) billion people to be alive in 2050 (Godfray et al. 2010). Antibiotics are often used by farmers (World Health Organization 2014; O'Neill 2015), and so they are faced with the problem of how to rear more animals, more efficiently and with higher standards of food safety but without using antibiotics. Farmers needed to become more efficient and competitive while improving animal welfare and food safety and reducing medication and their environmental impact. Developments in agriculture during the twentieth century led to escalation and specialization of livestock production systems. However, crises may arise due to animal diseases and detrimental effects on farm income, animal welfare, and the environment (Ten Napel et al. 2011).

3.5.1 Polyphenols in Monogastric Nutrition

Polyphenols contain active ingredients; they exert nonspecific effects on living organisms and regulate the activity of enzymes and cell receptors (D'Archivioet al. 2007). Most importantly, polyphenols act as powerful antioxidants by preventing oxidative stress and reduce the risk of various oxidative pathologies like cancer and neurodegenerative and cardiovascular diseases (Scalbert et al. 2005; Petti and Scully 2009; Paszkiewicz et al. 2012). The structure and the presence of functional groups make them strong antioxidants. They also exert effects on cell metabolism and deliver a number of health benefits for humans and animals (Petti and Scully 2009; Kamboh et al. 2015). In recent research, much attention has also been given to the modification of animal products with biologically active compounds such as polyphenols. Modifications of the fatty acid profile, mostly by decreasing the concentrations of saturated fatty acids (SFAs) and increasing the levels of polyunsaturated fatty acids (PUFAs) in muscle tissue, can indirectly contribute to consumer health by improving the quality of food products and increasing their oxidative stability (Brenes et al. 2008; Kamboh and Zhu 2013).

3.5.2 Nutritional Role of Secondary Plant Compounds

Plant secondary metabolites, prevalent in the plant kingdom, have effects on ruminants ranging from improved nutrition and health to interference with feed intake, digestion, or metabolism of energy or nutrients and acute toxicity and death (Rosenthal and Janzen 1979). Secondary compounds are produced in response to pathogenic microorganisms, insects, and grazing by herbivores as natural defense mechanisms (Swain 1979). Some trees, shrubs, and forage legumes contain too high concentrations of CT that are only marginally acceptable for sheep and cattle. The condensed tannins, in contrast to most other phenolics, are beneficial to ruminant animal production only at low to medium concentrations. Condensed tannins bind to protein strongly in a pH-dependent manner (Asquith and Butler 1986). Condensed tannins are not absorbed from the digestive tract (Terrill et al. 1994). Medium CT concentrations reduce degradability of protein and increase amino acid absorption from the rumen small intestine (Bermingham et al.2001; Waghorn et al. 1994). Various plants contain varied CT with respect to their chemical structure and molecular mass. This appears to influence astringency (and hence feed intake) and the effectiveness of CT in precipitating protein and increasing absorption of essential amino acids from the small intestine in sheep (Aerts et al. 1999; Barry and McNabb 1999; Waghorn et al. 1994). Deer have evolved production of proline-rich salivary proteins which bind CT, as a means of counteracting plant defense mechanisms and the anti-nutritional effects of high CT concentrations. No such kind of proteins has been found in the saliva of domesticated sheep and cattle (Austin et al. 1989). Barry and McNabb (1999) suggest that the beneficial nutritional effects of CT could only be seen if the concerned ruminant species does not produce salivary CT-binding proteins. However, it is less likely for plant species with medium to high CT concentrations, where the concentration of CT in the diet surpasses the capacity of the salivary proteins for binding.

This relationship may also differ based on the CT-binding capacity of the saliva produced by the concerned species. Also, not all CT present in forages are likely to be bound to the same extent by salivary proteins. An important role of CT in diets of sheep and cattle is its ability in forages to bind to soluble protein in the rumen that reduces the incidence of frothy bloat. However, the high rumen outflow rate of liquid in deer (Domingue et al. 1991) means deer appear not to be susceptible to bloat. A recent study revealed that forages containing CT decreased methane emissions from dairy cows (Woodward et al. 2002) and sheep (Waghorn et al. 2002).

3.5.3 Secondary Plant Compounds and Internal Parasitism

A range of known plant secondary compounds are present in plant species currently fed, or suitable for feeding, to farmed animals. There are many compounds as yet unidentified in addition to knowing little about the majority of the known compounds, other than antioxidant (Duke 1992), antifungal, and antibacterial roles (Barry and Blaney 1987). Internal parasites, particularly abomasal nematodes, are significant contributors to production loss and risk of mortality in young farmed animals. There are direct and indirect mechanisms by which secondary compounds in forages may potentially reduce infection and thus ameliorate internal parasitism in ruminants. The direct effects may be due to direct binding to the parasites themselves in the digestive tract or feces of the animal. The indirect effects of CT are to improve protein status of the host which could increase the animal's tolerance of worm burdens.

Quebracho extract (*Schinopsis* sp.) containing CT fed to sheep directly repressed gastrointestinal parasitic growth (Athanasiadou et al. 2001). The effects of CT from seven plants on egg hatching and larval development of *Trichostrongylus colubriformis* from sheep (also found in deer) were determined in vitro (Molan et al. 2002). The most inhibitory effect on egg development was due to CT from dock (*Rumex obtusifolius*), followed by CT from *Dorycnium rectum*, sainfoin, *Dorycnium pentaphyllum*, lotus major, sulla, and bird's-foot trefoil. There is no difference in the hatching of eggs from deer grazing chicory compared with perennial ryegrass-based pasture (Schreurs et al. 2002), although in vitro study utilizing sesquiterpene lactones from chicory showed related inhibitory properties against deer parasite larvae to CT (Molan et al. 2000).

3.6 The Role of Plants Containing Secondary Compounds in Sustainable Farming

The health and productivity of farmed animals is improved by substituting perennial ryegrass with forages containing condensed tannins (CT) and/or sesquiterpene lactones. The benefits of CT usage are improved lactation, trace element status, growth, resilience to internal parasites, and reductions in parasite larval development.

The importance of secondary compounds with respect to other nutritive characteristics of alternative forage species, such as low fiber and highly digestible carbohydrate concentrations, has not been fully characterized, but they are likely to contribute to improved animal health. Unlike sheep and cattle, deer produce salivary proteins that bind CT, allowing them to consume plants containing higher CT concentrations than sheep and cattle.

A pervasive control of internal parasites using synthetic chemicals may be unsustainable in the long term, due to the increasing risk of anthelmintic resistance and the risk or perception of chemical residues in deer products. There is a growing consumer awareness of, and demand for, low chemical input or "naturally produced" deer products (Loza 2001). The focus on natural rather than manufactured dietary supplements to provide essential trace elements and vitamins for farmed animals is increasing.

3.6.1 Gut Health and Immunomodulatory Effects

Polyphenols are characterized by low bioavailability, and unabsorbed compounds exert a major influence on gut health (Etxeberria et al. 2013; Breneset al. 2016). Phenolic compounds have antibacterial (Malik et al. 2017) and bacteriostatic properties (Etxeberria et al. 2013); they minimize the adhesion of pathogenic bacteria (*E. coli, Clostridium*), inhibit the progression of infections in the digestive tract, and improve nutrient utilization and animal performance (Viveros et al. 2011; Dueñas et al. 2015; Brenes et al. 2016). Metal ions and oxygen promote the formation of phenoxyl radicals which induce cytotoxic effects and damage bacterial DNA. Anthocyanins present in cherries, raspberries, and berries demonstrate bacteriostatic and bactericidal effects (*Bacillus, Klebsiella, Helicobacter*). Due to the presence of hydroxyl groups, polyphenols like quercetin have the ability to incorporate into lipid membranes and increase their permeability, which makes pathogens more sensitive to antibacterial compounds (Chiva-Blanch and Visoli 2012; Paszkiewicz et al. 2012). By enhancing the proliferation of beneficial bacteria (*Bacillus* spp., *Lactobacillus* spp.) and stabilizing gut microflora, polyphenols indirectly enhance the host's immune system and overall health (Hashemi and Davoodi 2011; Paszkiewicz et al. 2012). Further, in monogastric animals, they can exert a positive influence on gut morphology and improve nutrient absorption (Kamboh et al. 2015). In a study by Hong et al. (2012), essential oils significantly amplified the height of intestinal villi in the duodenum of broiler chickens without inducing any changes in ileal microflora composition. Similarly, Kamboh and Zhu (2014) revealed that genistein, hesperidin, and flavonoids extracted from *Ginkgo biloba* leaves significantly increased the surface area of the small intestine and modified its structure in broilers exposed to lipopolysaccharide stress. Cranberry extract, a rich source of phenolic acids, anthocyanins, flavonols, and flavan-3-ols, significantly reduced the population size of *Enterococcus* spp. in broilers (Leusink et al. 2010). Hajati et al. (2015) observed that chickens whose diets were enriched with polyphenols or vitamin C had longer intestinal villi, and similar results were reported by Akbarian et al. (2013), in whose study *E. coli* counts decreased significantly in the ileum and cecum of chickens fed with lemon peel extract or *Curcuma xanthorrhiza* essential oil during chronic exposure to high temperatures. Viveros et al. (2011) observed an increase in *Lactobacillus* counts in the ileum of birds fed grape seed extract in comparison with birds receiving an antibiotic or grape pomace concentrate. Zhu et al. (2015) investigated the effect of soy isoflavones on weaned piglets intoxicated with LPS. Commercial polyphenol preparations (containing hydrolysable tannin or grape seed extract) were effective in decreasing the incidence of diarrhea in piglets infected with enterotoxigenic *Escherichia coli* (Verhelst et al. 2014). The use of polyphenols in monogastric nutrition can improve gut health owing to their bacteriostatic properties, and they reduce the incidence of diarrhea, in particular in piglets, while stimulating the growth of beneficial microflora. In animals exposed to oxidative stress, polyphenols can enhance immunity by activating immunoglobulins and inhibiting the secretion of proinflammatory cytokines.

3.6.2 Improving Meat Quality Through Natural Antioxidants

The demand of more natural foods has elevated nowadays, obliging the industries to include natural antioxidants in foods. Natural antioxidants are used to reduce lipid oxidation in foods to improve their quality and nutritional value. A diverse number of plants have been explored to possess natural antioxidants. The antioxidant activity of these plants is dependent on their phenolic compound content, which includes various volatile compounds also known as essential oils. The antioxidant activity of plant extracts is affected by various factors like type of solvent used during extraction, measurement method, and number of samples. Some studies have demonstrated that shelf life and meat quality can be improved by using natural antioxidants in some stages of meat production. The main effects of these compounds are reducing microbial growth and lipid oxidation during storage. Nevertheless, more research is needed to determine antimicrobial activity of natural antioxidants in meat during storage and identify the main metabolic pathway of these compounds and its effect on other meat quality parameters. Some plant extracts are an excellent source of natural antioxidants that can improve meat shelf life and quality mainly by retarding lipid oxidation and microbial growth. The effect of oregano essential oil on meat quality has been studied the most, whereas there is less information about other plants. Antimicrobial activity of spices and herbs has been extensively studied. However, most of the research was done in in vitro assays using microbiological media rather than actual foods. Further research is needed to determine their effect on bacterial growth in raw and cooked meats during storage. Active packaging containing natural antioxidants is a promising tool in this field. There is some evidence that dietary essential oils can improve meat quality. Since bioavailability of essential oils in meat cannot be directly demonstrated, more research is needed to identify the main metabolic pathway of these compounds and the key essential oil antioxidant compounds deposited in meat. Further research is needed to determine the effect of natural antioxidants on other meat quality characteristics.

The increasing preference for natural foods has obliged the food industry to include natural antioxidants in various products to delay oxidative degradation of lipids, improve quality and nutritional value of foods, and replace synthetic antioxidants (Fasseas et al. 2007; Wojdylo et al. 2007; Coma et al. 2008). Including antioxidants in the diet has beneficial effects on human health because they protect the biologically important cellular components, such as DNA, proteins, and membrane lipids, from reactive oxygen species (ROS) attack (Su et al. 2007). Synthetic antioxidants have been used to retard or minimize oxidative deterioration of foods, such as butylated hydroxyanisole (BHA), butylated hydroxytoluene (BHT), and tertiary butyl hydroquinone (TBHQ) (Fasseas et al. 2007). Recently, consumers have rejected synthetic antioxidants because of their carcinogenicity (Altmann et al. 1986; Van Esch 1986). The advantages of natural antioxidants in foods are high consumer acceptance and their safe use. The disadvantages are their higher cost and lower effectiveness (Fasseas et al. 2007). Many herbs, spices, and their extracts have been added in a variety of foods to improve their sensory characteristics and extend shelf life (Shahidi et al. 1992). Herbs of the Lamiaceae family, mainly oregano (*Origanum vulgare* L.), rosemary (*Rosmarinus officinalis* L.), and sage (*Salvia officinalis* L.), have been reported as having significant antioxidant capacity (Shan et al. 2005; Wojdylo et al. 2007). The short shelf life of refrigerated packed meat makes its commercialization more difficult. Some studies have demonstrated that meat shelf life and quality can be improved by natural antioxidants added in the preslaughter and post-slaughter stages, that is, incorporating natural antioxidants in animal diets, onto the meat surface, or in active packaging. Some authors have reported the effectiveness of rosemary and oregano extracts to reduce lipid oxidation (Djenane et al. 2002, 2003; Fasseas et al. 2007; Coma et al. 2008), color loss, and microbial growth (Djenane et al. 2002, 2003; Coma et al. 2008; Zinoviadou et al. 2009) in some types of meats.

3.7 Conclusions and Future Perspectives

Plants produce secondary metabolites besides primary metabolites in a different pathway. These compounds do not have a direct role in growth; however, these possess many other myriad functions. They help in protection, in attraction, and even in signaling aspects. They increase nutrient uptake by animals by acting as antiparasitic agents. Now there is a need for a proper management for utilizing their beneficial properties in sustainable animal farming and sustainable living. Many techniques can be manipulated to extract maximum useful aspects.

References

Aerts RJ, McNabb WC, Molan AL, Brand A, Peters JS, Barry TN (1999) Condensed tannins from Lotus corniculatus and Lotus pedunculatus effect the degradation of ribulose 1,5-bisphosphate carboxylase (Rubisco) protein in the rumen differently. J Sci Food Agric 79:79–85

Akbarian A, Golian A, Kermanshahi H, Farhoosh R, Raji AR, DeSmet S, Michiels J (2013) Growth performance and gut health parameters of finishing broilers supplemented with plant extracts and exposed to daily increased temperature. Span J Agric Res 11:109–119

Akobundu IO (1987) Weed science in the tropics: principles and practices. Wiley and Sons, Chichester

Akula R, Ravishankar GA (2011) Influence of abiotic stress signals on secondary metabolites in plants. Plant Signal Behav 6:1720–1731

Altmann HJ, Grunow W, Mohr U, Richter-Reichhelm HB, Wester PW (1986) Effects of BHA and related phenols on the forestomach of rats. Food Chem Toxicol 24:1183–1188

Altieri MA (1987) Agroecology: the scientific basis of alternative agriculture. Westview Press, Boulder

Amil-Ruiz F, Blanco-Portales R, Muñoz-Blanco J, Caballero JL (2011) The strawberry plant defense mechanism: a molecular review. Plant Cell Physiol 52:1873–1903

Anaya AL (1999) Allelopathy as a tool in the management of biotic resources. Crit Rev Plant Sci 18:697–739

Asquith TN, Butler LG (1986) Interactions of condensed tannins with selected proteins. Phytochemistry 25:1591–1593

Athanasiadou S, Kyriazakis I, Jackson F, Coop RL (2001) Direct anthelmintic effects of condensed tannins towards different gastrointestinal nematodes of sheep: *in vitro* and *in vivo* studies. Vet Parasitol 99:205–219

Atiku A, Oladipo OO, Forcados GE et al (2016) Anti-nutritional and phytochemical profile of some plants grazed upon by ruminants in North Central Nigeria during the dry season (January to April). Int J Livest Prod 7:19–23

Austin PJ, Suchar LA, Robbins CT, Hagerman AE (1989) Tannin-binding proteins in the saliva of deer and their absence in saliva of sheep and cattle. J Chem Ecol 15:1335–1347

Barry TN, Blaney BJ (1987) Secondary compounds of forages. In: Hacker JB, Ternouth JH (eds) Nutrition of herbivores. Academic Press, Sydney, pp 91–119

Barry TN, McNabb WC (1999) The implications of condensed tannins on the nutritive value of temperate forages fed to ruminants. Br J Nutr 81:263–272

Batish DR, Singh HP, Kohli RK, Kaur S (2001) Crop allelopathy and its role in ecological agriculture. J Crop Prod 4:121–162

Bavaresco L, Mattivi F, De Rosso M, Flamini R (2012) Effects of elicitors, viticultural factors, and enological practices on resveratrol and stilbenes in grapevine and wine. Mini Rev Med Chem 12:1366–1381

Bavec M, Turinek M, Grobelnik-Mlakar S, Slatnar A, Bavec F (2010) Influence of industrial and alternative farming systems on contents of sugars, organic acids, total phenolic content, and the antioxidant activity of red beet (Beta vulgaris L. ssp. vulgaris Rote Kugel). J Agric Food Chem 58:11825–11831

Blum U (1996) Allelopathic interactions involving phenolic acids. J Nematol 28:259–267

Bodas R, Prieto N, García-González R et al (2012) Manipulation of rumen fermentation and methane production with plant secondary metabolites. Anim Feed Sci Technol 176:78–93

Booker FL, Blum U, Fiscus EL (1992) Short-term effects of ferulic acid on ion uptake and water relations in cucumber seedling. J Exp Bot 93:649–655

Brenes A, Viveros A, Goñi I, Centeno C, Sayago-Ayerdi SG, Arija I, Saura-Calixto F (2008) Effect of grape pomace concentrate and vitamin E on digestibility of polyphenols and antioxidant activity in chickens. Poult Sci 87:307–316

Brenes A, Viveros A, Chamorro S, ArijaI (2016) Use of polyphenol-rich grape by-products in monogastric nutrition. A review. Anim Feed Sci Technol 211:1–17

Broudiscou LP, Papon Y, Broudiscou AF (2000) Effects of dry plant extracts on fermentation and methanogenesis in continuous culture of rumen microbes. Anim Feed Sci Technol 87:263–277

Calsamiglia S, Busquet M, Cardozo PW et al (2007) Invited review: essential oils as modifiers of rumen microbial fermentation. J Dairy Sci 90:2580–2595

Chiva-Blanch G, Visioli F (2012) Polyphenols and health: moving beyond antioxidants. J Berry Res 2:63–71

Cobellis G, Trabalza-Marinucci M, Marcotullioc MC et al (2016) Evaluation of different essential oils in modulating methane and ammonia production, rumen fermentation, and rumen bacteria in vitro. Anim Feed Sci Technol 215:25–36

Coma V (2008) Bioactive packaging technologies for extended shelf life of meat-based products. Meat Sci 78:90–103

Dayan F, Romagni J, Tellez M, Rimando A, Duke S (1999) Managing weeds with natural products. Pestic Outlook 10:185–188

De Candolle MAP (1832) Physiologie Vegetale.Tomme-III, pp 1474–1475. Bechet Jeune, Library, Faculty of Medicine, Paris

Djenane D, Sánchez-Escalante A, Beltrán JA, Roncalés P (2002) Ability of α-tocopherol, taurine and rosemary, in combination with vitamin C, to increase the oxidative stability of beef steaks packaged in modified atmosphere. Food Chem 76:407–415

Domingue BMF, Dellow DW, Wilson PR, Barry TN (1991) Comparative digestion in deer, goats and sheep. N Z J Agric Res 34:45–53

Dueñas M, Muñoz-González I, Cueva C, Jiménez-Girón A, Sánchez-Pa-tán F, Santos-Buelga C, Bartolomé B (2015) A survey of modulation of gut microbiota by dietary polyphenols. Bio Med Res Int, ID 850902, 15 pp

Duke JA (1992) Handbook of biologically active phytochemicals and their activities. CRC Press, Boca Raton

Einhellig FA (1985) Allelopathy – a natural protection, allelochemicals. In: Mandava NY (ed) Handbook of natural pesticides: methods, vol 1. Theory. CRC Press, Boca Raton

Einhellig FA, Leather GR (1988) Potentials for exploiting allelopathy to enhance crop production. J Chem Ecol 14:1829–1844

Einhellig FA, Rasmussen JA, Hejl AM, Souza IF (1993) Effects of root exudate sorgoleone on photosynthesis. J Chem Ecol 19:369–375

Etxeberria U, Fernández-Quintela A, Milagro FI, Aguirre L, Martínez JA, Portillo MP (2013) Impact of polyphenols and polyphenol-rich dietary sources on gut microbiota composition. J Agric Food Chem 61:9517–9533

Fasseas MK, Mountzouris KC, Tarantilis PA, Polissiou M, Zervas G (2007) Antioxidant activity in meat treated with oregano and sage essential oils. Food Chem 106:1188–1194

Foley ME (1999) Genetic approach to the development of cover crops for weed management. J Crop Prod 2:77–93

Gallandt ER, Liebman M, Huggins DR (1999) Improving soil quality: implications for weed management. J Crop Prod 2:95–121

Garnett T, Appleby MC, Balmford A, Bateman IJ, Benton TG, Bloomers P, Burlingame M, Dawkins M, Dolan L, Fraser D, Herrero M, Hoffman I, Smith P, Thornton PK, Toulmin C, Vermeulen SJ, Godfray HCJ (2013) Sustainable intensification in agriculture: premises and policies. Science 341:33–34

Godfray HCJ, Beddington JR, Crute IR, Haddad L, Lawrence D, Muir JF, Pretty J, Robinson S, Thomas SM, Tiulmin C (2010) Food security: the challenge of feeding 9 billion people. Science 327:812–818

Greathead H (2003) Plants and plant extracts for improving animal productivity. P Nutr Soc 62:279–290

Griffin SG, Wyllie SG, Markham JL et al (1999) The role of structure and molecular properties of terpenoids in determining their antimicrobial activity. Flavour Frag J 14:322–332

Hajati H, Hassanabadi A, Nassiri MH, Nassiri MR (2015) The effect of grape seed extract and vitamin C feed supplements carcass characteristics, gut morphology and ileal microflora in broiler chickens exposed to chronic heat stress. Iran J Appl Anim Sci 5:155–165

Hartmann T (2008) The lost origin of chemical ecology in the late 19th century. Proc Natl Acad Sci U S A 105:4541–4546

Hashemi SR, Davoodi H (2011) Herbal plants and their derivatives as growth and health promoters in animal nutrition. Vet Res Commun 35:169–180

Hejl AM, Einhellig FA, Rasmussen JA (1993) Effects of juglone on growth photosynthesis and respiration. J Chem Ecol 19:559–568

Hong JC, Steiner T, Aufy A, Lien TF (2012) Effects of supplemental essential oil on growth performance, lipid metabolites and immunity, intestinal characteristics, microbiota and carcass traits in broilers. Livest Sci 144:253–262

Hoult AHC, Lovett JV (1993) Biologically active secondary metabolites of barley III A method for identification and quantification of hordenine and gramine in barley by high-performance liquid chromatography. J Chem Ecol 19:2245–2254

Jakopic J, Stampar F, Veberic R (2010) Influence of hail net and reflective foil on cyanidin glycosides and quercetin glycosides in "Fuji" apple skin. HortScience 45:1447–1452

Jayanegara A, Ridla M, Laconi EB et al (2014) Can plant saponins lower methane emissions without hampering the nutrient digestibility of ruminants. In: Proceedings of the 16th AAAP Animal Science Congress, vol II, 10–14 November 2014, Gadjah Mada University, Yogyakarta, Indonesia

Jouany JP, Morgavi DP (2007) Use of 'natural' products as alternatives to antibiotic feed additives in ruminant production. Animal 1:1443–1466

Kaçar D (2008) Screening of some plant species for their total antioxidant and antimicrobial activities. Thesis of master degree.Graduate School of Engineering and Sciences of İzmir Institute of Technology, İzmir

Kamboh AA, Zhu WY (2013) Effect of increasing levels of bioflavonoids in broiler feed on plasma anti-oxidative potential, lipid metabolites, and fatty acid composition of meat. Poult Sci 92:454–461

Kamboh AA, Zhu WY (2014) Individual and combined effects of genistein and hesperidin on immunity and intestinal morphometry in lipopolysaccharidechallenged broiler chickens. Poult Sci 93:2175–2183

Kamboh AA, Arain MA, Mughal MJ, Zaman A, Arain ZM, Soomro AH (2015) Flavonoids: health promoting phytochemicals for animal production – a review. J Anim Health Prod 3:6–13

Kelling KA, Fixen PE, Schulte EE, Liegel EA, Simmon CR (1981) Soil test recommendation for field, vegetable and fruit crops. Bull. Wisconsin-Madison Univ. University of Wisconsin Cooperative Extension Publication (A20/809) (November issue)
Kim ET, Guan LL, Lee SJ et al (2015) Effects of flavonoid-rich plant extracts on in vitro ruminal methanogenesis, microbial populations and fermentation characteristics. Asian Australas J Anim Sci 28:530–537
Kohli RK, Batish D, Singh HP (1998) Allelopathy and its implications in agroecosystems. J Crop Prod 1:169–202
Kutchan TM (1997) Natural product munitions-new prospects for plant protection. Res News 2:449–450
Lampkin L (1994) Organic farming. Farming Press Ltd, Ipswich
Leusink G, Rempel H, Skura B, Berkyto M, White W, Yang Y, Fitzpatrick S (2010) Growth performance, meat quality, and gut microflora of broiler chickens fed with cranberry extract. Poult Sci 89:1514–1523
Loza MJ (2001) Sensitive issues for the deer industry. In: Proceedings of a Deer Course for Veterinarians, Deer Branch of the New Zealand Veterinarians Association, vol 18, pp 73–78
Macias FA, Torres A, Molinillo JMG, Varela RM, Castellano D (1996) Potential allelopathic sesquiterpene lactones from sunflower leaves. Phytochemistry 43:1205–1215
Macias FA, Molinillo JMG, Galindo JCG, Varela RM, Simonet AM, Castellano D (2001) The use of allelopathic studies in search for natural herbicides. J Crop Prod 4:237–256
Malik JA, Nirala SK, Bhadauria M (2017) Phytochemical and antibacterial evaluation of Tephrosia purpurea and its active constituent gallic acid. Flora Fauna 23:252–256
Mayton HS, Olivier C, Vaughn SF, Loria R (1996) Correlation of fungicidal activity of Brassica species with allyl isothiocyanate production in macerated leaf tissue. Phytopathology 86:267–271
Mirzoeva OK, Grishanin RN, Calder PC (1997) Antimicrobial action of propolis and some of its components: the effects on growth, membrane potential and motility of bacteria. Microbiol Res 152:239–246
Molan AL, Duncan A, Barry TN, WC MN (2000) Effect of condensed tannins and sesquiterpene lactones extracted from chicory on the viability of deer lungworm larvae. Proc N Z Soc Anim Prod 60:26–29
Molan AL, Waghorn GC, McNabb WC (2002) Effect of condensed tannins on egg hatching and larval development of *Trichostrongylus colubriformis* in vitro. Vet Rec 150:65–69
Muehlchen AM, Rand RE, Parke JL (1990) Evaluation of crucifer green manures for controlling aphanomyces root rot of peas. Plant Dis 74:651–654
Newman M-A, Sundelin T, Nielsen JT, Erbs G (2013) MAMP (microbe-associated molecular pattern) triggered immunity in plants. Front Plant Sci 4:139
Odenyo AA, Osuji PO, Karanfil O (1997) Effect of multipurpose tree (MPT) supplements on ruminal ciliate protozoa. Anim Feed Sci Technol 67:169–180
Oerke EC, Dehne HW, Schönbeck F, Weber A (1994) Crop protection and crop production. Elsevier, Amsterdam
OJEU (Official Journal of European Union) (2003) OJEU Regulation (EC) No 1831/2003 of the European Parliament and the Council of 22 September 2003 on additives for use in animal nutrition. Offic J Eur Union. Page L268/36 in OJEU of 18/10/2003 (2003)
Oleszek W (1999) Allelopathic significance of plant saponins. In: Macias FA, Galindo JCG, Molinillo JHG, Cutler HG (eds) Recent advances in allelopathy, vol 1. A science for the future. Servicio de Publicaciones, Universidad de Cadiz, Spain, pp 167—178
Oleszek W, Hoagland RE, Zablotowicz RM (1999) Ecological significance of plant saponins. In: Inderjit K, Dakshini MM, Foy CL (eds) Principles and practices in plant ecology: Allelochemical interactions. CRC Press, Boca Raton, pp 451–465
O'Neill J (2015) Chairman. Review on antimicrobial resistance. Wellcome trust. Available at http://amr-review.org/Publications [Verified 9 December 2015]
Öztürk H, Demirtaş A, Salgırlı Y et al (2012) Effects of olive leaf extract on rumen microbial fermentation in in vitro semi continuous culture system (RUSITEC). Vet J Ankara Univ 59:17–21

Paszkiewicz M, Budzynska A, Rozalska B, Sadowska B (2012) The immunomodulatory role of plant polyphenols (in Polish). Post Hig Med Dosw 66:637–646
Patra AK, Saxena J (2009) The effect and mode of action of saponins on the microbial populations and fermentation in the rumen and ruminant production. Nutr Res Rev 22:204–219
Patra AK, Yu Z (2012) Effects of essential oils on methane production and fermentation by, and abundance and diversity of, rumen microbial populations. Appl Environ Microb 78:4271–4280
Patra AK, Yu Z (2014) Effects of vanillin, quillaja saponin, and essential oils on in vitro fermentation and protein-degrading microorganisms of the rumen. Appl Microbiol Biotechnol 98:897–905
Paula EM, Samensari RB, Machado E et al (2016) Effects of phenolic compounds on ruminal protozoa population, ruminal fermentation, and digestion in water buffaloes. Livest Sci 185:136–141
Petti S, Scully C (2009) Polyphenols, oral health and disease: a review. J Dent 37:413–423
Piasecka A, Jedrzejczak-Rey N, Bednarek P (2015) Secondary metabolites in plant innate immunity: conserved function of divergent chemicals. New Phytol 206:948–964
Qasem JR, Foy CL (2001) Weed allelopathy, its ecological impacts and future prospect: a review. J Crop Prod 4:43–120
Rasmussen DA, Hejl AM, Einhellig FA, Thomas TA (1992) Sorgoleone from root exudate inhibits mitochondrial functions. J Chem Ecol 18:197–207
Rosenthal GE, Janzen DH (eds) (1979) Herbivores: their interaction with secondary plant metabolites. Academic Press, New York
Roth E, Berna A, Beullens K, Yarramraju S, Lammertyn J, Schenk A, Nicolai B (2007) Postharvest quality of integrated and organically produced apple fruit. Postharvest Biol Technol 45:11–19
Russell JB, Strobel HJ (1989) Mini-review: the effect of ionophores on ruminal fermentation. Appl Environ Microb 55:1–6
Schreurs NM, Molan AL, Lopez-Villalobos N, Barry TN, McNabb WC (2002) Effect of grazing undrenched weaner deer on chicory or perennial ryegrass/white clover pasture on gastrointestinal nematode and lungworm viability. Proc N Z Soc Anim Prod 62:143–144
Singh HP, Batish DR, Kohli RK (2001) Allelopathy in agroecosystem: an overview. J Crop Prod 4:1–42
Soares DJ, Cavalcante CEB, Cardoso TG, de Figueiredo EAT, Maia GA, de Sousa PHM, de Figueiredo RW (2012) Study of the stability of cashew nuts obtained from conventional and organic cultivation. Semin Cienc Agrar 33:1855–1867
Scalbert A, Manach C, Morand C, Remesy C, Jimenez L (2005) Dietary polyphenols and the prevention of diseases. CRC Crit Rev Food Sci Nutr 45:287–306
Shahidi F, Janitha PK, Wanasundara P (1992) Phenolic antioxidants. Crit Rev Food Sci Nutr 32:67–102
Shan B, Cai YZ, Sun M, Corke H (2005) Antioxidant capacity of 26 spice extracts and characterization of their phenolic constituents. J Agric Food Chem 53:7749–7759
Su L, Yin JJ, Charles D, Zhou K, Moore J, Yu L (2007) Total phenolic contents, chelating capacities, and radical-scavenging properties of black peppercorn, nutmeg, rosehip, cinnamon and oregano leaf. Food Chem 100:990–997
Swain A (1979) In: Rosenthal GA, Janzen DH (eds) Herbivores; their interaction with secondary plant metabolites. Academic Press, London, pp 657–682
Swanton CJ, Murphy SD (1996) Weed science beyond the weeds: the role of integrated weed management (IWM) in agroecosystem health. Weed Sci 44:437–445
Szumacher-Strabel M, Cieślak A (2010) Potential of phytofactors to mitigate rumen ammonia and methane production. J Anim Feed Sci 19:319–337
Ten Napel J, van der Veen AA, Oosting SJ, Koerkamp PWG (2011) A conceptual approach to design livestock production systems for robustness to enhance sustainability. Livest Sci 139:150–160
Terrill TH, Waghorn GC, Woolley DJ, McNabb WC, Barry TN (1994) Assay and digestion of 14Clabelled condensed tannins in the gastrointestinal tract of sheep. Br J Nutr 72:467–477

Toor RK, Savage GP, Heeb A (2006) Influence of different types of fertilisers on the major antioxidant components of tomatoes. J Food Compos Anal 19:20–27

Treutter D (2010) Managing phenol contents in crop plants by phytochemical farming and breeding-visions and constraints. Int J Mol Sci 11:807–857

Valenzuela H, Smith J (2002) Sustainable agriculture green manure crops: buckwheat. Cooperative extension service. College of Tropical Agriculture and Human Resource, Hawaii University, p 4

Van Esch GJ (1986) Toxicology of tert-butylhydroquinone (TBHQ). Food Chem Toxicol 24:1063–1065

Vaughan AD, Vaughan KC (1988) Mitotic disrupters from higher plants and their potential uses as herbicides. Weed Technol 2:533–539

Veberic R, Trobec M, Herbinger K, Hofer M, Grill D, Stampar F (2005) Phenolic compounds in some apple (Malus domestica Borkh.) cultivars of organic and integrated production. J Sci Food Agric 85:1687–1694

Verhelst R, Schroyen M, Buys N, Niewold T (2014) Dietary polyphenols reduce diarrhea in enterotoxigenic Escherichia coli (ETEC) infected post-weaning piglets. Livest Sci 160:138–140

Viveros A, Chamorro S, Pizarro M, Arija I, Centeno C, Brenes A (2011) Effects of dietary polyphenol-rich grape products on intestinal microflora and gut morphology in broiler chicks. Poult Sci 90:566–578

Waghorn GC, Shelton ID, McNabb WC, McCutcheon SH (1994) Effects of condensed tannins in Lotus pedunculatus on its nutritive value for sheep. 2. Nitrogenous aspects. J Agric Sci, Cambridge 123:109–119

Waghorn GC, Tavendale MH, Woodfield DR (2002) Methanogenesis from forages fed to sheep. Proc New Zealand Grassland Assoc 64:167–171

Watanabe Y, Suzuki R, Koike S et al (2010) In vitro evaluation of cashew nut shell liquid as a methane-inhibiting and propionate-enhancing agent for ruminants. J Dairy Sci 93:5258–5267

Wezel A, Bellon S, Doré T, Francis C, Vallod D, David C (2009) Agroecology as a science, a movement and a practice. A review. Agron Sustain Dev 29:503–515. Editorial 1203

Wink M (1988) Plant breeding: importance of plant secondary metabolites for protection against pathogens and herbivores. Theor Appl Genet 75:225–233

Woodward SL, Waghorn GC, Lassey KR, Laboyrie PG (2002) Does feeding sulla (Hedysarum coronarium) reduce methane emissions from dairy cows? Proc N Z Soc Anim Prod 62:227–230

Wojdylo A, Oszmiański J, Czemerys R (2007) Antioxidant activity and phenolic compounds in 32 selected herbs. Food Chem 105:940–949

World Health Organization (2014) Antimicrobial resistance. Global report on surveillance. Available at www.who.int/drugresistance/documents/surveillancereport/en/ [Verified 4 April 2016]

Xuan TD, Tsuzuki E (2004) Allelopathic plants: buckwheat. Allelopathy J 13:137–148

Xuan TD, Tawata S, Khanh TD, Chung IM (2005) Biological control of weeds and plant pathogens in paddy fields by exploiting plant allelopathy: an overview. Crop Prot 24:197–206

Yang K, Wei C, Zhao GY et al (2017) Effects of dietary supplementing tannic acid in the ration of beef cattle on rumen fermentation, methane emission, microbial flora and nutrient digestibility. J Anim Physiol Anim Nutr 101:302–310

Yu JQ (1999) Autotoxic potential of vegetable crops. In: Narwal SS (ed) Allelopathy update: basic and applied aspects. Science Publisher, Inc., New York, pp 149—162

Zhu C, Wu Y, Jiang Z, Zheng C, Wang L, Yang X, Hu Y (2015) Dietary soy isoflavone attenuated growth performance and intestinal barrier functions in weaned piglets challenged with lipopolysaccharide. Int Immunopharmacol 28:288–294

Zupan A, Mikulic-Petkovsek M, Slatnar A, Stampar F, Veberic R (2014) Individual phenolic response and peroxidase activity in peel of differently sun-exposed apples in the period favorable for sunburn occurrence. J Plant Physiol 171:1706–1712

Flavones and Flavonols: Bioactivities and Responses Under Light Stress in Herbs

4

Erick P. Gutiérrez-Grijalva, Liliana Santos-Zea, Dulce L. Ambriz-Pérez, Leticia X. López-Martínez, and J. Basilio Heredia

Abstract

Phenolic compounds are a widespread group of phytochemicals derived from the secondary metabolism of herbs and plants. Phytochemicals are mainly produced as a plant response to biotic and abiotic stresses to overcome/adapt these adverse conditions. The study of abiotic stress response of plants has led to numerous discoveries regarding flavonoids and phenolic acids biosynthetic pathways. Since phenolic compounds such as flavonoids and phenolic acids have been related to numerous bioactive health effects, it is of interest to enhance the content of these compounds in plants and herbs. One of the most studied abiotic stresses on herbs is manipulation of light, whether through use of different light colors, light intensity, and/or sunlight. One of the most active groups of compounds elicited in response of light stress is flavones, flavonols, and some phenolic acids. These groups of compounds have been related to health benefits to humans such as antioxidant, cardiovascular disease, and against metabolic syndrome. This chapter aims to comprehensive review recent works on this field.

E. P. Gutiérrez-Grijalva (✉)
Cátedras CONACYT – Centro de Investigación en Alimentación y Desarrollo, AC, Culiacán, Sinaloa, México
e-mail: erick.gutierrez@ciad.mx

L. Santos-Zea
Escuela de Ingeniería y Ciencias, Tecnologico de Monterrey, Campus Toluca, Toluca, Estado de México, México

D. L. Ambriz-Pérez
Maestría en Ciencias Aplicadas, Universidad Politécnica de Sinaloa, Mazatlán, Sinaloa, Mexico

L. X. López-Martínez
Cátedras CONACYT – Centro de Investigación en Alimentación y Desarrollo, AC, Hermosillo, Sonora, México

J. B. Heredia
Centro de Investigación en Alimentación y Desarrollo, AC, Culiacán, Sinaloa, Mexico

R. Lone et al. (eds.), *Plant Phenolics in Sustainable Agriculture*,
https://doi.org/10.1007/978-981-15-4890-1_4

Keywords

Phytochemicals · Phenolic compounds · Flavones · Stress · Herbs

4.1 Introduction

Flavonoids belong to a group of compounds named phytochemicals, which are derived from the secondary metabolism of plants. Flavonoids have many roles in plant defense and reproduction. For instance, flavonoids are defense molecules against biotic and abiotic stress conditions; moreover, flavonoids such as anthocyanins give plants different colors that may attract pollinators (Corradini et al. 2011; Crozier et al. 2009). In this sense, as a response to abiotic and biotic stresses, plant synthesizes flavonoids for their protection, which might be an advantage to human consumption by the increase/enhancement of flavonoids which have been related to numerous health-promoting properties (Kutchan et al. 2015; Torres-Contreras et al. 2018).

Flavonoids have gained major importance due to epidemiological and experimental studies reporting their health-promoting properties. Daily/regular intake of flavonoid-rich foodstuff has been inversely correlated with the onset of noncommunicable diseases such as Alzheimer, Parkinson's, several types of cancer, atherosclerosis, cardiovascular diseases, diabetes, and the metabolic syndrome (Feliciano et al. 2015; Kumar and Pandey 2013; Nijveldt et al. 2001).

4.2 Flavones and Flavonols

Within the secondary metabolite, flavonoid family is considered as highly diverse; it is formed by more than 9000 known structures based upon 15-carbon skeletons consisting of 2 benzene rings (A and B) linked via a heterocyclic pyrane ring (C) (Fig. 4.1).

Fig. 4.1 Graphical representation of the general structure of flavonoids

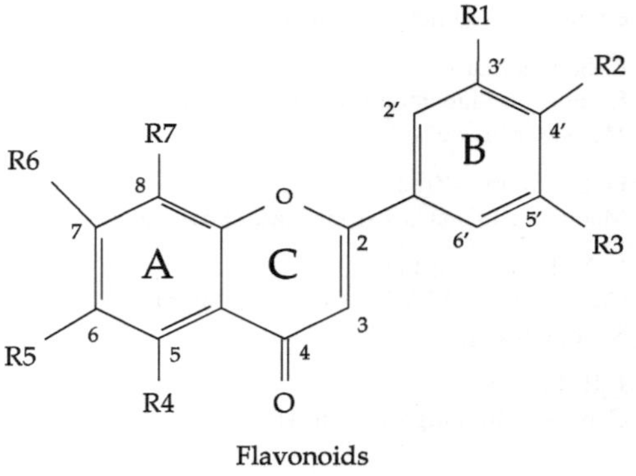

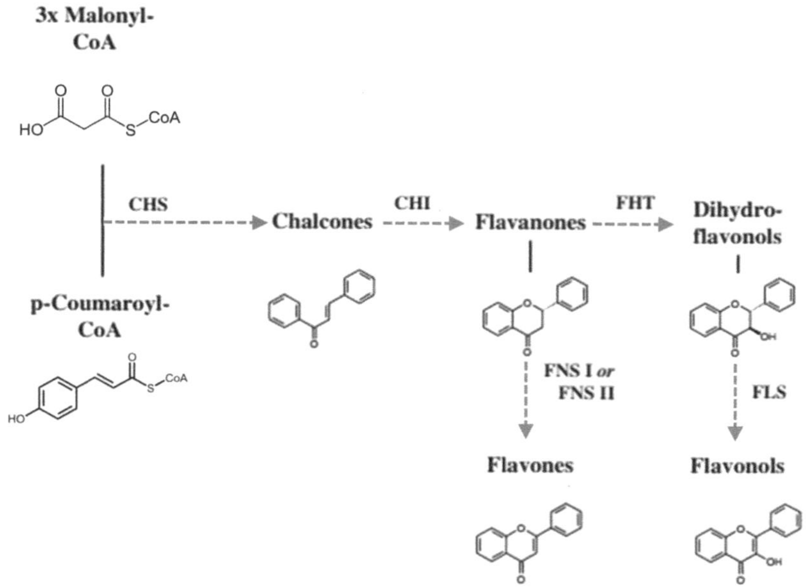

Fig. 4.2 General biosynthetic pathway of flavones and flavonols. Enzymes are abbreviated as follows: *CHS* chalcone synthase, *CHI* chalcone isomerase, *FHT* flavanone 3-b-hydroxylase, *FNS* flavone synthase, *FLS* flavonol synthase. (Adapted from Martens and Mithöfer 2005)

Flavonoids can be sub-classified into flavonols, flavones, flavanones, isoflavones, anthocyanins, and flavan-3-ols; among them, flavones and flavonols are the most abundant subgroups and almost ubiquitous. The biosynthetic pathway of flavonoids begins in the shikimate pathway, where the aromatic ring B and the chromane ring are considered to originate from the amino acid phenylalanine leading to p-coumaroyl CoA, whereas Ring A comes from the condensation of three units of malonyl-CoA, which is originated from the carboxylation of acetyl-CoA, a central intermediate in the Krebs tricarboxylic acid cycle. Flavonoids are derivatives of 1,3-diphenylpropan-1-one (C6–C3–C6); then occurs the condensation of three molecules malonyl-CoA with one molecule p-coumaroyl-CoA to a chalcone intermediate (two phenolic groups connected by an open three-carbon bridge); after chalcone is isomerized by chalcone flavanone isomerase (CHI) to a flavanone (structures containing three rings), so the three-carbon bridge is part of an additional heterocyclic six-membered ring that involves one of the phenolic groups on the adjacent ring; hence, flavanones lead to all other flavonoids, including flavones and flavonols (Fig. 4.2) (Martens and Mithöfer 2005; Brahmachari and Gorai 2006; Tsao 2010; Corradini et al. 2011; Kumar and Pandey 2013; Singh et al. 2014).

Flavones are colorless to yellow, depending on pH; they are characterized by the presence of a double bond between C2 and C3 in the heterocycle of the flavan skeleton. The B ring is attached to C2, and usually C3 do not have any substituent on

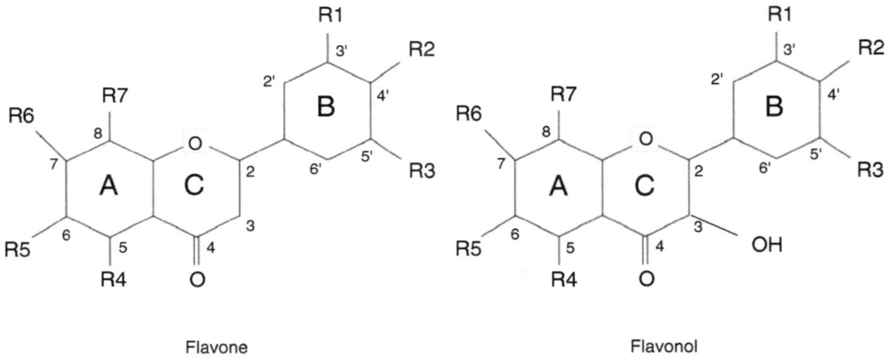

Fig. 4.3 Flavones and flavonols general structure

it. This is the difference between flavones and flavonols; the last ones have a hydroxyl group in C3 position (Fig. 4.3) (Martens and Mithöfer 2005; Singh et al. 2014).

Flavones have three functional groups, including hydroxy, carbonyl, and conjugated double bond; therefore they give typical reactions of all three functional groups such as reduction, degradation in the presence of base, oxidation, rearrangement, substitution, addition, condensation, and reaction with organometallic reagents (Singh et al. 2014).

Interestingly, higher plants can generate flavones using two different enzyme systems, flavone synthase I (FNSI) and flavone synthase II (FNSII), from the same substrates, but both enzymes never occur in the same organism. The function of FNS is the introduction of a double bond between C2 and C3 by the abstraction of two hydrogen atoms. Most of the plants use FNSII as catalytic converter, except family *Apiaceae* which use FNSI (Martens and Mithöfer 2005).

Flavones and flavonols, as all flavonoids, can exist both as free aglycones and as glycosidic conjugates; in fact, most of them exist naturally as glycosides; glycosylation is catalyzed by glucosyltransferase; glycosylation and hydroxylation degree makes the molecules less reactive, more polar therefore more water soluble, so that this modification is as an essential form of protection in plants to prevent cytoplasmic damage and to store flavones and flavonols in the cell vacuole (Corradini et al. 2011; Crozier et al. 2009; Tsao 2010). Glycosides can be linked to flavones and flavonols as O-glycosides or C-glycosides; in the first ones, the most frequent, one or more of the aglycone hydroxyl groups is bound to a sugar with formation of an O–C acid-labile acetal bond; in the second ones, glycosylation takes place by direct linkage of the sugar to the flavonoid basic nucleus, via an acid-resistant C–C bond (Corradini et al. 2011). Even each of the hydroxyl groups can be glycosylated, certain positions are more common; normally the sugar is located in positions 3 or 7, for example, the 7-hydroxyl group in flavones and the 3- and 7-hydroxyls in flavonols (Kumar and Pandey 2013). Glucose is the most commonly found sugar, followed by galactose, rhamnose, xylose, and arabinose, whereas glucuronic and

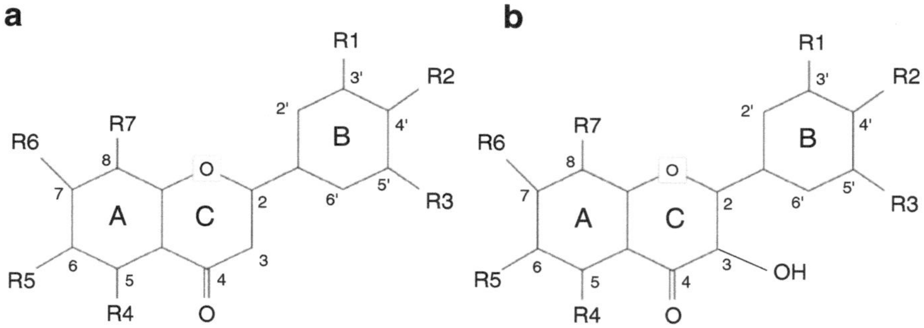

Flavones	
Apigenin	$R_1 = H, R_2 = OH$
Luteolin	$R_1 = OH, R_2 = OH$

Flavonols	
Kaempferol	$R_1 = H, R_2 = OH, R_3 = H$
Quercetin	$R_1 = OH, R_2 = OH, R_3 = H$
Isorhamnetin	$R_1 = OCH_3, R_2 = OH, R_3 = H$
Myricetin	$R_1 = OH, R_2 = OH, R_3 = OH$

Fig. 4.4 General structure of flavones (**a**) and flavonols (**b**). *R1*, R2, and R3 represent different substituent depending on the compound listed below structure

galacturonic acids are rare; also disaccharides or higher saccharides are also found; rutinose (6-*O*-L-rhamnosyl-D-glucose) and neohesperidose (2-*O*-L-rhamnosyl-D-glucose) are the most common (Corradini et al. 2011).

It should be said that studies by spectroscopy have revealed that most flavones and flavonols exhibit two major absorption bands: band I (320–385 nm) represents the B-ring absorption, while band II (250–285 nm) corresponds to the A-ring absorption, whereas functional groups, such as hydroxyl, attached to the flavonoid skeleton, may cause a shift in absorption, for example, from 367 nm in kaempferol (3,5,7,4′-hydroxyl groups) to 371 nm in quercetin (3,5,7,3′,4′-hydroxyl groups) and to 374 nm in myricetin (3,5,7,3′,4′,5′- hydroxyl groups), while flavones and flavonols may be distinguished by the absence of a 3-hydroxyl group in the first ones (Kumar and Pandey 2013).

4.2.1 Distribution of Flavones and Flavonols in Herbs

Flavones may have a wide range of substitutions including hydroxylation, methylation, alkylation, and glycosylation, being apigenin and luteolin the most representatives (Fig. 4.4a). Most flavones occur as 7-*O*-glycosides, but substitution can be found at any other position. There is a wide diversity of flavonoids, which is attributed to their affinity to other molecules like carbohydrates. For instance, quercetin alone have at least 279 different glycosidic combinations. Besides, these molecules can be found in all structural parts of the plants, above- and belowground, in vegetative and generative organs: stem, leaves, buds, bark, heartwood, thorns, roots, rhizomes, flowers, farina, fruit, seeds, and also in root and leaf exudates or resin. Flavones are present in relatively small quantities in grains, leafy vegetables, and herbs; however, plant species belonging to more than 70 different families are able to produce flavones, which means at least some members of these plant families synthesize them; for example the flavones chrysin, apigenin, rutin, luteolin, and its glucosides have been reported in fruit skins, red wine, buckwheat, red pepper, and tomato skin; nevertheless, these compounds seem to be absent in almost all of the about 3000 *Brassicaceae* species; there are only few reports of them, mostly 6-C-glycosides. Polymethoxylated flavones, such as tangeretin and nobiletin, have been found in citrus species (Crozier et al. 2009; Kumar and Pandey 2013; Martens and Mithöfer 2005; Sesso et al. 2009; Tsao 2010).

Flavonols are the most numerous of the phenolics, and they are found throughout the entire plant kingdom, mainly present in leafy vegetables, apples, onions, broccoli, and berries but absent in algae. Chemically, flavonols are simply 3-hydroxyflavones, but they are biosynthetically distinct flavonoid classes. The distribution and structural variations of flavonols are extensive, being the main dietary flavonols kaempferol, quercetin, isorhamnetin, and myricetin, mostly found as O-glycosides (Fig. 4.4b). Conjugation occurs most frequently at the C3 position, but substitutions can also occur at the C5, C7, C4′, C3′, and C5′. Although flavonol aglycones are limited, there are numerous flavonol glycosides, with more than 347 different sugar conjugates of kaempferol by itself. The amount of flavonols found in fruits, vegetables, and beverages varies depending on seasonal changes, varietal differences, and effects of processing (Crozier et al. 2009; Martens and Mithöfer 2005; Sesso et al. 2009; Tsao 2010).

The major sources identified of flavones and flavonols are tea (18–50 mg/L), onions (284–486 μg/g), and apple (21–72 μg/g); according to Stewart et al. (2000) also tomatoes can be considered as a good source with a flavonol content from 1.3 μg/g to 22.2 μg/g of fresh weight, depending on variety, size, country of origin, season, and climate characteristics; even tomato-based products, such as juice and puree, keep a good amount of flavonols containing 14–16 μg/mL and 70 μg/g fresh weight, respectively.

4.3 Flavonoids: Flavones and Flavonols in Human Health Promotion

As with flavonoids in general, flavones and flavonols are predominantly recognized for their antioxidant capacity (Kumar and Pandey 2013). Besides their antioxidant potential, flavones have demonstrated positive effects on cholesterol metabolism and anti-inflammatory effects (Ralston et al. 2017). With respect to flavonols, such as quercetin, a wide variety of health benefits have been documented on cardiovascular health, neurodegenerative disorders, as an anticancer and anti-inflammatory agent, antiviral, and antibacterial potential (Anand David et al. 2016). It is important to consider that chemical structure plays an important role on their bioactivity and bioavailability; therefore, the potential of each individual compound will differ according to hydroxylation and glycosylation patterns (Wang et al. 2018).

Additionally, the effect of other components of the matrix, including other flavonoids, phenolics, or essential oils, may act synergistically to generate the beneficial consequences in human health.

4.3.1 Antioxidant Capacity

Metabolic processes of living cells generate highly reactive molecules known as reactive oxygen species (ROS), including the superoxide radical ($O_2^{\cdot -}$) and hydroxyl radicals ($^{\cdot}OH$), as well as reactive nitrogen species (RNS), such as nitric oxide ($NO^{\cdot}$) (Pizzino et al. 2017). While these molecules are part of normal metabolism, a misbalance in the oxidative status in the body can be related to the development of chronic diseases, including metabolic syndrome, type 2 diabetes, cancer, and cardiovascular disorders (Marseglia et al. 2015). Both, ROS and RNS can target and damage biomolecules, such as membrane lipids, proteins, and DNA, compromising cellular function and integrity (Pizzino et al. 2017).

Since oxidative stress is one of the key processes underlying disease, antioxidants have become highly relevant in the prevention of many chronic ailments. An antioxidant is a molecule with the capacity to inhibit or delay oxidation reactions at a concentration lower than the substrate (Pisoschi and Pop 2015). Some of these molecules may be endogenous, comprised of a system of enzymes (superoxide dismutase (SOD), catalase (CAT), glutathione peroxidase (GPx), among others) and non-enzymatic molecules, including lipoic acid, glutathione, L-arginine, and coenzyme Q10, that are able to disable free radicals (Pizzino et al. 2017). On the other hand, there are exogenous antioxidants that can be obtained from the diet, particularly through the consumption of fruits, vegetables, and other foods from plant origin, including herbs (Pisoschi and Pop 2015). Among these compounds flavonoids stand out as antioxidants, functioning through mechanisms such as free radical scavenging, inhibition of oxidative enzymes, metallic ion chelation, and enhancement of endogenous antioxidant system (Kumar and Pandey 2013). The evaluation of antioxidant effect of flavones and flavonols in herbs has been widely studied, by chemical assays such as ORAC, ABTS, DPPH, or FRAP, in vitro using living cells and in vivo using animals models (Opara and Chohan 2014). Some select samples of these studies are presented in Table 4.1.

Peppermint (*Mentha piperita*) is commonly consumed as a herbal infusion; therefore Figueroa-Pérez et al. (2014) evaluated the radical scavenging effect of the infusion itself and identified the compounds in it. Their results showed that peppermint infusion can function as a free radical scavengers through the DPPH ($IC_{50} = 61$ μg/mL) and ABTS ($IC_{50} = 17$ μg/mL) methods (Table 4.1). However, these two methods have the disadvantage that both DPPH· and $ABTS\cdot^{+}$ radicals are not related to free radicals produced in living biological systems (López-Alarcón and Denicola 2013). In this same study, it was also seen that the infusion was able to inhibit nitric oxide radical ($IC_{50} = 48$ μg/mL), which is an important RNS in vivo related to the inflammatory pathway (Förstermann and Sessa 2011). Among the compounds related to these effects, the flavone luteolin was identified, along with the

Table 4.1 Summarization of the bioactive properties of selected herbs attributed to their flavone and flavonol content

Source	Flavones and flavonols identified	Biological effect	Reference
Antioxidant properties			
Peppermint (*Mentha piperita*) infusion	Luteolin, quercetin, rutin	Peppermint infusion showed DPPH ($IC_{50} = 61$ μg/mL), ABTS ($IC_{50} = 17$ μg/mL), and nitric oxide ($IC_{50} = 48$ μg/mL) radical scavenging capacity	Figueroa-Pérez et al. (2014)
Mexican oregano (*Hedeoma patens*, *Lippia graveolens*, *Lippia palmeri*) extracts	Derivatives from quercetin, scutellarein, luteolin, and methoxyflavones	Extracts inhibited over 75% nitric oxide radicals and ROS in vitro at concentrations from 50–400 μg/mL on LPS-activated macrophages	Leyva-López et al. (2016)
Mexican oregano (*Hedeoma patens*, *Lippia graveolens*, *Lippia palmeri*)	Derivatives from quercetin, scutellarein, luteolin, and apigenin	*L. graveolens* and *L. palmeri* extracts inhibited intracellular oxidation at 50 μg/mL from 96% to 98% on Caco-2 cells	Gutiérrez-Grijalva et al. (2019)
Parsley (*Petroselinum crispum*) supplemented diet	Quercetin and kaempferol	Hyperuremic rats fed a parsley supplemented diet (5 mg/kg) during 14 days observed a reduction in serum uric levels, xanthine oxidoreductase activity, and an increase in serum antioxidant capacity	Haidari et al. (2011)
Metabolic syndrome			
Mexican oregano (*Lippia graveolens*), rosemary (*Rosmarinus officinalis*)	Cirsimaritin, hispidulin	In vitro inhibition of DPP-IV (0.4–2.5 μM as pure compounds and 0.4–1.5 μM GAE as flavone-enriched fractions)	Bower et al. (2014)
Metabolic syndrome			
Mexican oregano (*Hedeoma patens*, *Lippia graveolens*, *Lippia palmeri*) submitted to simulated gastrointestinal digestion	Derivatives from quercetin, scutellarein, luteolin, and apigenin	In vitro inhibition of α-glucosidase (43–63%) by the undigested fraction and α-amylase (36–52%) and pancreatic lipase (47–55%) by the intestinal fraction at 400 μg/mL	Gutiérrez-Grijalva et al. (2019)

(continued)

Table 4.1 (continued)

Source	Flavones and flavonols identified	Biological effect	Reference
Peppermint (*Mentha piperita*) infusion	Luteolin, quercetin, ruti	At 10 mg/mL (equivalent to peppermint tea) inhibited pancreatic lipase by 45%	Figueroa-Pérez et al. (2014)
Salvia virgata ethanolic extract and purified compound	Chrysoeriol	Ethanolic extract and chrysoeriol inhibited 50% α-amylase activity at 19 mg/mL and 1.27 mM, respectively	Nickavar, Abolhasani (2013)
Rosemary (*Rosmarinus officinalis*) and thyme (*Thymus vulgaris*)	Rutin, quercetin, and other phenolics	Rosemary and thymus extract ameliorated the liver damage induced by gentamicin and had an hypolipidemic effect	Hegazy et al. (2017)
Cardiovascular			
Rooibos (*Aspalathus linearis*) fermented and unfermented infusion	Luteolin, apigenin, and quercetin derivatives	Wistar rats that consumed rooibos flavonoids experienced a 61% recovery rate after ischemia, possibly mediated by a decrease in apoptosis in heart cells	Pantsi et al. (2011)
Rooibos (*Aspalathus linearis*) extract	Luteolin, apigenin, and quercetin derivatives	Rooibos extract reduced ex vivo ischemic cardiomyocyte apoptosis and ROS generation to levels similar to non-ischemic controls	Dludla et al. (2014)
Parsley (*Petroselinum crispum*) flavonoid extract	Kaempferol, apigenin	Kaempferol and apigenin-rich extract (0.3 mg/mL) inhibited platelet aggregation in vitro by 77%	Gadi et al. (2012)
Antiviral/antibacterial			
Euphorbia humifusa Willd	Apigenin and luteolin glycosides and acylated derivatives	Apigenin monoglycosides inhibited hepatitis B virus antigens (IC_{50} 15–75 μg/mL) without cytotoxic effect (>80 μg/mL)	Tian et al. (2010)
Thyme (*Thymus vulgaris*) purified flavonoids	Chrysin, apigenin, kaempferol derivatives	Isolated flavonoids from *Thymus vulgaris* inhibited *E. coli* (15–18 mm) and *S. aureus* (14–15 mm)	Nadia, Rachid (2013)
Anticancer/chemopreventive			
Roman chamomile (*Chamaemelum nobile L.*)	Apigenin, luteolin, myricetin, kaempferol,	Methanolic extracts inhibited diverse cancer cell lines growth in vitro	Guimarães et al. (2013)

(continued)

Table 4.1 (continued)

Source	Flavones and flavonols identified	Biological effect	Reference
	quercetin, isorhamnetin derivatives	(82–168 μg/mL), while infusions were less effective (150–250 μg/mL)	
Mexican oregano (*Poliomintha glabrescens*) methanolic extract	Luteolin, quercetin, diosmetin, other flavones	Oregano extracts were able to inhibit colon cancer HT-29 cell line growth in vitro by 50% at concentrations 1.3–4.5 μg/mL. Apoptotic biomarkers were increased in treated cells	García-Pérez et al. (2013)
Peppermint (*Mentha piperita*) infusion	Diosmin, rutin, luteolin rutinoside, luteolin, myricetin, apigenin, diosmetin	Peppermint infusion (10 mg/mL) was consumed by healthy volunteers for 6 days, causing a significant decrease in NAT2 activity	Begas et al. (2017)
Mexican oregano (*Lippia graveolens*) methanolic extract	Galangin and flavanones	The methanolic extract presented photoprotective activity in vitro on *E. coli* cell cultures and inhibited photo-induced cancer in vivo on a mouse model	García-Bores et al. (2017)
Various herbal infusions	Chrysin, kaempferol, quercetin, luteolin, apigenin, and other flavone derivatives	Herbal infusions, particularly *Satureja thymbra* and *Mentha pulegium*, presented in vitro inhibitory effect on HT-29 and PC3 cell lines and a reduction in IL-8 expression	Kogiannou et al. (2013)

flavonol quercetin and its glycosylated derivative rutin (Figueroa-Pérez et al. 2014). While some information is obtained about the mechanism of action of these compounds, the complexity of biological reactions is not appropriately addressed in these assays.

Methanolic extracts from Mexican oregano, including the species *Hedeoma patens*, *Lippia graveolens*, and *Lippia palmeri*, have been phytochemically characterized as sources of derivatives of the flavonol quercetin, as well as the flavones scutellarein, luteolin, apigenin, and diverse methoxyflavones (Leyva-López et al. 2016; Gutiérrez-Grijalva et al. 2019). These extracts were evaluated on LPS-activated macrophages (RAW 264.7) and effectively inhibited over 75% of NO·, ROS, and superoxide at concentrations ranging from 50 μg/mL to 400 μg/mL (Table 4.1). In the study by Leyva-López et al. (2016), the extracts from *Lippia graveolens* had the strongest inhibitory effect, showing 80% inhibition at 50 μg/mL and almost complete inhibition at 100 μg/mL for NO·, complete inhibition of ROS at both concentrations and 83% inhibition of superoxide at 100 μg/mL (Leyva-López et al. 2016). By reducing the amount of these free radicals not only oxidative stress is relieved, but it also generates an anti-inflammatory effect.

These extracts from the same three oregano species were submitted to simulated gastrointestinal digestion and evaluated for cellular antioxidant activity (CAA), which measures the capacity of an antioxidant compound to prevent the oxidation of the fluorescent molecule dihydrodichlorofluorescein by the action of free radicals (López-Alarcón and Denicola 2013). In this study, *Lippia graveolens* and *Lippia palmeri* extracts were able to reduce Caco-2 cellular oxidation from 96 to 98% at a concentration of 50 μg/mL, and it was observed for undigested extract, as well as for gastric and intestinal digestion fractions (Table 4.1) (Gutiérrez-Grijalva et al. 2019). The authors also evaluated some of the individual polyphenols identified in the extracts, including apigenin, luteolin, and quercetin; however the effect was less potent (80–92%) at higher concentrations than evaluated for the extract (100–200 μg/mL). These results indicated that the action of the extract was likely a synergistic effect of the different polyphenols it contained, particularly since the antioxidant effect in CAA can occur through various mechanisms (Gutiérrez-Grijalva et al. 2019; López-Alarcón and Denicola 2013).

Antioxidant effect of herbs can also be observed in vivo, as was the case in hyperuremic rats fed a parsley (*Petroselinum crispum*) supplemented diet (5 g/kg weight) during 14 days (Table 4.1). In this work, the rats that received the supplemented diet presented a 40% reduction of serum uric acid levels. Additionally, xanthine oxidase and xanthine dehydrogenase enzymatic activities were inhibited 34% and 43%, respectively, a possible mechanism in the reduction of uric acid. Moreover, the parsley supplementation increased serum antioxidant capacity two-fold in comparison to non-treated, hyperuremic rats. These results were similar to those obtained by treating the rats with the main flavonols found in parsley, kaempferol, and quercetin (5 mg/kg weight) (Haidari et al. 2011).

More studies have been conducted on other culinary herbs, including Mediterranean oregano, thyme, mint, and dill, among others (Opara and Chohan 2014; Rubió et al. 2013). In these herbs, the recurrent flavonoids identified generally include derivatives of quercetin, luteolin, and apigenin, which also match the compounds mentioned in Table 4.1. Differences in effects seen in different herbs can be attributed to some structural changes in these derivatives, since the structure of the rings, including degree of hydroxylation and conjugation with sugars or acyl groups, can impact their antioxidant effect (Kumar and Pandey 2013). Additionally, the effect of the food matrix on the absorption and metabolization of these compounds has to be taken into consideration to determine their true potency as antioxidants in an in vivo system.

4.3.2 Metabolic Syndrome

Metabolic syndrome is a condition where risk factors for cardiovascular disease and type 2 diabetes prevail in an individual: imbalance in lipid and glucose metabolism, high blood pressure, chronic inflammation, and abnormalities in clotting process that increase risk for thrombosis (Grundy 2016). To identify dietary or medicinal herbs with potentially preventive effect against metabolic syndrome, carbohydrate and lipid digestive enzyme inhibition has been established as a target. Most often, experimental design includes α-amylase and α-glucosidase (starch and oligosaccharide digestion), as well as pancreatic lipase (50–70% triglyceride hydrolysis) to reduce carbohydrate and lipid assimilation (Figueroa-Pérez et al. 2014). Additionally, since a chronic inflammatory state is of the characteristics of metabolic syndrome and is related to oxidative stress, compounds with antioxidant potential, such as the ones mentioned in Sect. 4.3.1, are also important study subjects for this pathology (Grundy 2016; Marseglia et al. 2015; Pisoschi and Pop 2015).

Culinary herbs, including Mediterranean (*Origanum vulgare*) and Mexican (*Lippia graveolens*) oregano and rosemary (*Rosmarinus officinalis*), are a good source of flavonoids, including the flavones cirsimaritin and hispidulin (Table 4.1). Flavone-enriched fractions from Mexican oregano and rosemary were recognized as potent inhibitors of dipeptidyl peptidase (DPP-IV), a serine protease involved in the degradation of glucagon-like peptide 1, and protein tyrosine phosphatase 1B (PTP1B), both related to insulin sensitivity. The authors of this study determined that fractions from both plants containing the flavones cirsimaritin and hispidulin presented half-inhibitory concentrations (IC_{50}) from 0.4–2.5 μM GAE on DPP-IV in vitro. Hydroxylation in positions 4′ and 5′ of the B ring and *O*-methylation in position 6 of the A ring appear to be important for the binding affinity of these flavones to DPP-IV and therefore their inhibitory power (Bower et al. 2014).

Regarding Mexican oregano, various species, including *Hedeoma patens*, *Lippia graveolens*, and *Lippia palmeri*, were found to be inhibitors of various digestive enzymes (Table 4.1). Methanolic extracts from these species were submitted to a simulated gastrointestinal digestion, and the undigested, gastric, and intestinal fractions were evaluated against three digestive enzymes. The undigested fraction was an effective inhibitor of α-glucosidase (43–63%) at 400 μg/mL, while the intestinal fraction could inhibit α-amylase (36–52%) and pancreatic lipase (47–55%). These extracts contained glycosylated derivatives of the flavonol quercetin, as well as the flavones scutellarein, luteolin, and apigenin; however the results of this study did not allow to attribute the bioactivity to a particular compound (Gutiérrez-Grijalva et al. 2019).

Concerning pancreatic lipase inhibition, peppermint infusion was found to inhibit pancreatic lipase in vitro by 45% at a concentration equivalent to preparing peppermint tea (10 mg/mL) (Table 4.1). Analysis of this infusion indicated the presence of flavonoids, including luteolin, quercetin, rutin, and hesperidin, the latter showing the highest correlation to lipase inhibition. In this work, α-amylase and α-glucosidase were also evaluated; however, only poor inhibition was observed at a dose equivalent to the infusion (12% and 8%, respectively). Hesperidin was correlated to α-amylase inhibition, while luteolin was correlated to α-glucosidase inhibition (Figueroa-Pérez et al. 2014). These results indicate that while the enzymatic inhibition was not strong, it could be valuable to analyze fractions enriched in flavonols and flavonoids to determine whether the infusion contains antagonistic compounds that could hinder the effectiveness of these compounds.

Salvia spp. are a plant genus used in traditional Iranian medicine for diverse purposes, including as a hypoglycemic agent. Nickavar, Abolhasani (2013) determined that an ethanolic extract of the aerial parts of *Salvia virgata* was able to inhibit α-amylase with an IC_{50} of 19 mg/mL (Table 4.1). From this extract, the active component, the flavone chrysoeriol, was isolated and effectively inhibited the enzyme with an IC_{50} of 1.27 mM, equivalent to 0.38 mg/mL (Nickavar, Abolhasani 2013). In this case, the antagonistic effect of the matrix is clear, as the purified compound was 50 times more potent when compared to the extract.

In vivo studies are relevant to determine the effectiveness of herbal infusions or extracts on metabolism and organ damage associated with metabolic syndrome conditions. Rosemary and thyme extracts were evaluated in a gentamicin-treated rat model, which presented dyslipidemia and liver damage (Table 4.1). In this work, administration of both extracts helped rats maintain a weight similar to non-gentamicin-induced rats, and thyme extract ameliorated liver weight loss. Moreover, both extracts were able to reduce levels of alanine (ALT) and aspartate (AST) transaminases around 70% for ALT and 54% for AST, indicating a recovery of liver function with respect to gentamicin-treated animals. Lipids, including triglycerides and cholesterol levels, were also regularized to levels similar to non-induced rats by both extracts (Hegazy et al. 2017). While no association with a particular compound was established in this work, the role of polyphenols, including flavones and flavonols, was mentioned by the authors.

The results of these works give us an important indication that herbs can have help prevent conditions associated with the metabolic syndrome, particularly through effects in carbohydrate and lipid metabolism. However, as shown by the in vivo study, there can be multiple targets and effects when an extract with diverse components is administered.

4.3.3 Cardiovascular

The final outcome of metabolic syndrome is an elevated risk of cardiovascular disease. While the effects of several flavones and flavonols on prevention and amelioration of metabolic syndrome causes and symptoms were discussed in Sect. 4.3.2, these compounds can also have an effect once CVD has occurred.

Rooibos (*Aspalathus linearis*) is a South African plant consumed as a green (unfermented) or fermented infusion, rich in polyphenols with potential to prevent or alleviate CVD (Smith and Swart 2018). Such effects have been observed in diverse models, including ex vivo and in vivo (Table 4.1). A fermented rooibos aqueous extract rich in polyphenols, including luteolin, apigenin, and quercetin derivatives, presented a cardioprotective effect on cardiomyocytes cultured from hyperglycemic rats. Ischemia was induced ex vivo on the cells, which were treated with 1 and 10 μg/mL of rooibos extract. A significant decrease in cardiomyocyte apoptosis was observed, at 1 and 10 μg/mL of extract on cells exposed to ischemic induction, as well as normalization of ROS levels within the cells. Moreover, restoration of metabolic activity of cardiomyocytes exposed to ischemia occurred with 1 μg/mL treatment, as measured by ATP production, as well as preservation of glutathione, an endogenous antioxidant in cells (Dludla et al. 2014). It was interesting to note that the cardioprotective effect was higher at the lower extract concentration, and the authors hypothesized that this could be caused by a possible prooxidant effect at a higher concentration, which is documented in the literature (Procházková et al. 2011). However, for this case, it was not experimentally validated.

The cardioprotective of rooibos was also evaluated in vivo in a rat model (Table 4.1). The animals were fed fermented (RF) and unfermented rooibos (RU) infusion (2 g/100 mL) during 7 weeks, before sacrifice, heart excision, and ischemia induction. The animals consumed 69.00 ± 14.73 and 20.65 ± 3.41 and 13.61 ± 1.07 mg/day of flavonols in the fermented and unfermented rooibos infusion, respectively. A 15–18% decrease in the apoptotic proteins poly (ADP-ribose) polymerase (PARP) with RU and RF and 25% decrease in caspase-3 with RF were observed. Inhibition of these two proteins might indicate one of the mechanisms by which rooibos infusion could be cardioprotective, diminishing heart damage after ischemia. These effects were attributed to the flavonol content of the infusion, and therefore regular consumption of this infusion is likely to be beneficial to cardiovascular health (Pantsi et al. 2011).

Risk of thrombosis is also a prevalent factor of the metabolic syndrome, therefore agents that prevent this event from occurring. For instance, parsley contains flavonoids, such as kaempferol and apigenin, which may play a role in hindering thrombosis. In particular, an extract rich in these flavonoids in their aglycone form was able to inhibit platelet aggregation in vitro by 89% when induced by thrombin and collagen III at a 0.2 mg/mL concentration (Table 4.1). The extract was most effecting on aggregation induced by collagen, with an IC_{50} of 0.08 mg/mL. Moreover, under flow conditions similar to the circulatory system, collagen III-induced platelet aggregation was inhibited by 77% at 0.3 mg/mL of extract. This activity was attributed to kaempferol and apigenin present in the active fractions of the extract (Gadi et al. 2012).

The results from these studies indicate that flavones and flavonoids from herbs can actively be cardioprotective factors. On one hand, they can directly prevent damage that occurs during ischemic events, particularly when ROS are involved. Additionally, they can function as agents that can interfere in the process of thrombosis by delaying platelet aggregation and, therefore, preventing ischemia from happening.

4.3.4 Antibacterial and Antiviral

In recent year, bacterial resistance to antibiotics has become a concern as a public health threat. Antibiotic resistance is defined as the loss of an antibiotic's ability to inhibit bacterial growth at a therapeutic dose, and bacteria that can multiply in the presence of an antibiotic are classified as resistant (Zaman et al. 2017). On the other hand, viral diseases, such as human immunodeficiency virus, human papilloma, and hepatitis B virus, remain prevalent, as well as the latent risk of a widespread outbreak of Ebola virus and the recent outbreaks of formerly eradicated diseases, such as measles. Antivirals are drugs that play an important role in the prevention of viral epidemics, though viruses can also become resistance to these drugs (Zaman et al. 2017). Therefore, the development of new antiviral and antibiotics is an important research area in natural product chemistry and their applications.

Medicinal herbs are an interesting source of new antiviral drugs (Table 4.1), since they are often used in traditional medicine of different cultures for health purposes. In particular, the herb *Euphorbia humifusa* Willd is a herb used in traditional Chinese medicine to treat dysentery, enteritis, and hepatitis from viral origin (Tian et al. 2010). Tian et al. (2010) worked on validating this claim and determined the active principles of this plant against HBV. Using bioactivity-based screening, 13 flavone derivatives were isolated from *Euphorbia humifusa* Willd. Among these compounds, apigenin monoglycosides presented inhibitory activity (measured by the inhibition of antigens HBsAg and HBeAg in vitro) against this virus, with IC_{50}s between 15 and 75 μg/mL without causing cytotoxicity. While more studies are required to validate this activity, these compounds seem to be potential candidates for the development of new antivirals that can be used safely.

Cooking herbs can be a source of novel antibiotics, such as *Thymus vulgaris*, also known as thyme (Table 4.1). Several flavonoids, including derivatives from chrysin, apigenin, and kaempferol, were isolated from *Thymus vulgaris* and evaluated by the inhibition halo method on the bacteria *Escherichia coli* and *Streptococcus aureus*. Flavonoids with a substituent group on the hydroxyls of carbons 3′ of the B ring and 5 or 7 from the A ring presented the higher antibacterial activity. Chrysin 7-*O*-glycoside was the most effective antibiotic against *E. coli* (18 mm), followed by apigenin 5′*O*,7′*O*-diglycoside (16 mm) and kaempferol 3′*O*,7′*O*-diglycoside (15 mm). The same three compounds were also effective on *S. aureus* inhibition, with chrysin 7-*O*-glycoside and apigenin 5′*O*,7′*O*-diglycoside being similarly effective (15 mm), followed by kaempferol 3′*O*,7′*O*-diglycoside (14 mm) (García-Pérez et al. 2013). While this information is preliminary, it sets a background for future studies concerning the use of these compounds as new antibacterial agents.

Culinary herbs are an important source of new potential antibiotics. However, medicinal herbs from different traditions should not be overlooked, since their effect can be validated, including identification and isolation of active components, determination of their mechanism of action, and proposals for more effective formulations.

4.3.5 Anticancer and Chemopreventive

Cancer is one of the most important public health issues of the twenty-first century, but as a multifactorial disease, it requires a multifactorial approach in finding effective treatments (Gali-Muhtasib et al. 2015). Compounds derived from plant origin in the diet have been linked to a reduced risk in developing cancer (Manju et al. 2017), thus a valuable vault of new drugs that can prevent or treat this group of diseases. It is also essential to point out that oxidative stress has also been related to the onset and progression of cancer, due to ROS/RNS causing damage to DNA or by altering expression of oncogenes and gene suppressors (Pizzino et al. 2017). For this reason, the compounds mentioned in Sect. 4.3.1 could potentially be considered chemopreventive. In this section, other mechanisms for anticancer or chemopreventive activity are described (Table 4.1).

Roman chamomile is a medicinal plant antibacterial, antifungal, anti-inflammatory, antioxidant, and cytotoxic, among others (Kogiannou et al. 2013). It is also a source of flavonoids, and derivatives from apigenin, luteolin, myricetin, kaempferol, quercetin, and isorhamnetin have been identified (Guimarães et al. 2013). It was also determined that methanolic extracts were able to inhibit the growth of diverse cancer cell line in vitro, with IC_{50} between 82 and 168 μg/mL, while infusions were less effective (150–250 μg/mL). The extract was most effective of breast cancer line MCF-7 (82.52 ± 4.57 μg/mL) and lung cancer line NCI-H460 (82.75 ± 8.14 μg/mL) and least effective on hepatic cancer line HepG2 (168.40 ± 2.23 μg/mL). Meanwhile, the infusion was most effect on colon cancer cells HCT-15 (150.24 ± 5.47 μg/mL) and similarly effective on breast, lung, hepatic, and cervical (HeLa) cancer cell lines (226–250 μg/mL). No particular correlations were established with the identified compounds, but the authors hypothesize that the profile differences due to extraction methods are responsible for the effects seen in the different preparations (Guimarães et al. 2013).

Infusions are a common way in which herbs, including medicinal herbs, are consumed, as a way to concentrate the active components. These infusions can have health effects, including anticancer (Table 4.1). In a study where different herbal infusions were evaluated, determining that pink savory (*Satureja thymbra*) and pennyroyal (*Mentha pulegium*) infusions was the most effective against cancer (HT-29) and prostate (PC3) cell lines when tested in vitro. Both, pink savory and pennyroyal, showed 80 and 50% inhibition at 24 h on HT-29 and PC3 cells, respectively, and reached almost complete inhibition at 72 h on both lines in ranges from 0.2 to 0.6 μg/mL. It is also interesting to note that both infusions inhibited over 75% the expression of IL-8 on both cells lines, which is a marker of inflammation. No correlation with a particular compound was established, but derivatives of chrysin, luteolin, apigenin, and other flavones were identified, as well as kaempferol and quercetin (Kogiannou et al. 2013).

As mentioned in previous sections, diverse species of Mexican oregano constitute a source of bioactive compounds. Particularly the species *Polimintha glabrescens* and *Lippia graveolens* have presented in vitro anticancer and photoprotective effects (Table 4.1). In the case of *P. glabrescens*, methanolic extracts showed in vitro inhibitory potential (IC_{50} 1.3–4.5 mg/mL) on HT-29 cells, without harming healthy colon cell line CCD18-Co (>5.3 mg/mL). More importantly, apoptosis is the hypothesized mechanism of cell death, as seen by an increased expression of FAS, Bax, and Caspase-3 proteins in cells treated with the extracts. Identified compounds included luteolin, quercetin, diosmetin, and other flavones, luteolin derivatives being the most abundant (García-Pérez et al. 2013).

With respect to *L. graveolens*, it presented a photoprotective effect against UVB radiation (Table 4.1). Since UV damage contributes to the onset of cancer, an extract showing protection against it can constitute a chemopreventive agent. In this study, methanolic extract was evaluated in vitro on *E. coli* cultures, increasing their survival when exposed to UVB radiation. This protective effect was also observed in vivo on a mouse model, in which the number of lesions induced by UVB was inhibited, when compared to non-treated controls. In this case, the flavonol galangin was identified in the extracts, along with diverse flavanones (García-Bores et al. 2017). These results are promising, making this compound a potential component in sunscreen products to prevent the onset of skin cancer.

All these studies point out to the role of flavonols and flavones derived from different types of herbs in various important groups of diseases, ranging from infectious to chronic degenerative. It is interesting to note that some of these diseases are highly related to oxidative stress, and, therefore, finding strong antioxidants in this herb already contributes to the prevention of other diseases, including metabolic syndrome, cardiovascular disease, and cancer.

4.4 Flavone, Flavonol, and Phenolic Compounds Accumulation Under Light Stress Conditions

Herbs are regularly subjected to different type of stress that affects their growth, development, and yield. Stress can be classified in biotic and abiotic; biotic stress is inflicted by living organisms, while abiotic stress is imposed by environmental

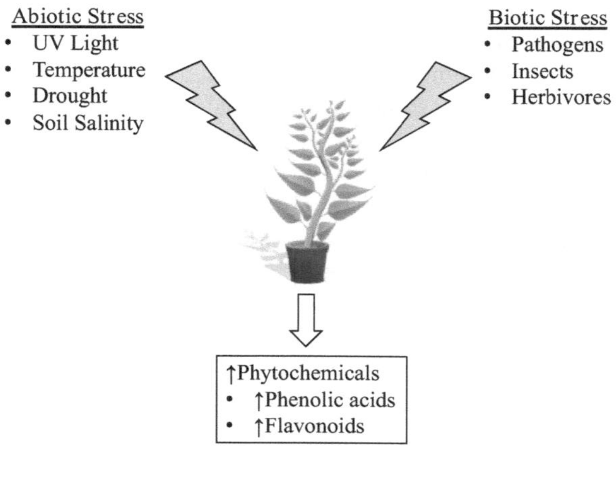

Fig. 4.5 Abiotic and biotic stresses increase the phytochemical content in plants

conditions such as flood, drought, high or low temperatures, inadequate soil salinity and nutrients, and UV light (Fig. 4.5) (Embuscado 2015; Kutchan et al. 2015; Bray et al. 2015).

Abiotic stress elicits a wide range of plant responses, such as alteration in gene expression and cellular metabolism, in order to adapt and overcome the adverse conditions, which in turn affect the plant's growth and yield. Alteration in phytochemical biosynthesis is one of the mechanisms of plants and herbs as result from stress-induced injury. Phytochemicals are a wide range of molecules from the secondary metabolism of plants (Bray et al. 2015; Torres-Contreras et al. 2018; Trivellini et al. 2016).

Herbs are rich sources of phytochemicals such as terpenes and polyphenols; the latter have been widely studied due to their health-promoting properties, such as delaying the onset of noncommunicable diseases such as Alzheimer, Parkinson's, atherosclerosis, cardiovascular diseases, metabolic syndrome, diabetes, and cancer. Most herbs are rich in polyphenolic compounds such as flavonoids, where the most abundant subclasses are flavones and flavonols, to which most of its bioactivity is attributed (Gutiérrez-Grijalva et al. 2019; Embuscado 2015). For that reason, strategies to enhance the flavone and flavonol content in herbs have been the subject of many research groups. This research arose from the studies focusing on the response of herbs to light stresses to understand the plant response on stress-induced conditions to enhance the content of phytochemicals. The most commonly applied abiotic stress conditions are water stress, temperature, and UV light (Szymańska et al. 2017). We will focus on light manipulation as a way to induce plant stress to enhance phytochemicals such as flavones and flavonols.

Furthermore, the ability of the photosynthetic apparatus and metabolism of plants to cope with light fluctuation and stress is highly related to their survival rate and reproductive success (Szymańska et al. 2017). The aforementioned is generally achieved by the accumulation of phenolic acids and flavonoids in the vacuoles of epidermal cells and chloroplast mesophilic cells of herbs and plants (Tattini et al. 2004; Conéjéro et al. 2014).

A recent work by Csepregi et al. (2017) evaluated the relationship between the developmental age of leaves from *Arabidopsis rosettes* and UV-B light manipulation; the authors exposed *Arabidopsis thaliana* to UV-B light with a PAR intensity of 60–80 μmol/m^{-2}s and supplemented by 1.6 W/m^2 UV-A; plants were exposed for 2 h for a total of 7 days which equals 0.6648 kJ m^{-2} day^{-1}. They reported that the developmental age of leaves influences the response of the plant on low UV-B light which enhances the content of pigments, flavonols, and the antioxidant capacity of these compounds. Low UV-B increased the concentration of the flavonols quercetin and kaempferol derivatives in leaves.

Flavonols have been implicated in the fertility of maize and other crops and also contribute to protecting plants from UV-B radiation. Falcone Ferreyra et al. (2010) used an *Arabidopsis thaliana* model to study and showed that a maize gene encoding a protein (ZmFLS1) is capable of converting the dihydrokaempferol (DHK) and dihydroquercetin (DHQ) and dihydroflavonols to the corresponding flavonols, kaempferol, and quercetin. The authors showed that ZmFLS1 is under control of the anthocyanin and 3-deoxy-flavonoid transcriptional regulators and is induced in maize seedlings by UV-B, and this production is in part mediated by the increased expression of P1, B, and PL1.

The geographical localization of crops can often have a negative influence on crop yield due to the combination of adverse factors such as low temperatures and ultraviolet radiation. On this sense, León-Chan et al. (2017) showed that low temperature and UV-B radiation are abiotic factors that cause stress in plants, and using a bell pepper model reported that the combination of low temperature and UV-B radiation caused degradation of chlorophyll in the bell pepper leaves and the higher accumulation of carotenoids, chlorogenic acid, and the flavonoids apigenin-7-*O*-glucoside and luteolin-7-*O*-glucoside. Moreover, UV-B irradiation alone induced a higher total flavonoid concentration than lower temperature alone. Furthermore, chlorogenic acid content significantly increased after combined treatment of lower temperature and UV-B radiation in comparison to lone treatments. Their results highly suggest that luteolin-7-*O*-glucoside is involved in reactive oxygen species quenching caused by low temperature and UV-B radiation.

In this regard, Radyukina et al. (2012) evaluated the effect of the combination of light and salinity stress on the phenolic content of *Artemisia lercheana*, *Ocimum basilicum*, and *Nigella sativa*; the authors reported that the accumulation of anthocyanins, soluble phenolics, and flavonoids provides protection against UV-B radiation and that anthocyanins are more involved in the salinity-resistance response.

Manukyan (2013) evaluated the effect of photosynthetically active radiation and UV-B radiation on the antioxidant capacity and polyphenols of *Nepeta cataria*, *Melissa officinalis*, and *Salvia officinalis* under controlled greenhouse cultivation. Low UV-B radiation at 1 kJ m^{-2} d^{-1} enhanced the biosynthesis as suggested by the increased total phenolic content in the evaluated herbs. Moreover, Peng et al. (2017) reported that flavone *O*-glycosides in rice are modulated by flavone 7-*O*-glucosyltransferase and flavone 5-*O*-glucosyltransferase and proved by UV-B exposure that allelic variation contributes to UV-B tolerance in nature in plants such as rice. It has also been reported that flavonol accumulation is upregulated by UV-B irradiation, however, as a result of microbe-associated molecular pattern-triggered immunity induced by the bacterial elicitor flg22 suppresses flavonoid production, which is suggested to be a mechanism of pathogen defense by directing phenylalanine from UV-B flavonol production toward production of phytoalexins and cell wall fortification by lignin incorporation (Zhou et al. 2017).

Another group of phenolic compounds that have been reported to be elicited in response to light stress are rosmarinic acid and salvianolic acids. For instance, Ma et al. (2013) showed that UV-B radiation enhances rosmarinic acid and lithospermic acid B by 6–17% and 3–5%, respectively, in *Salvia miltiorrhiza*. The authors suggested that the increased rosmarinic and lithospermic acid content was enhanced by the action of methyl jasmonate which in turn might coordinately induce gene transcripts of phenylalanine ammonia lyase, cinnamic acid 4-hydroxylase, tyrosine aminotransferase, and 4-hydroxyphenylpyruvate reductase of the rosmarinic acid and lithospermic acid biosynthetic pathway.

The quality and type of light source can also affect the flavonoid response of plants. For example, a study in *Salvia plebeian* under fluorescent and sunlight exposition during 4 months and 5–7 months, respectively, showed that phenolic compounds such as rosmarinic acid is reduced under sunlight, but levels of flavonoids such as homoplantaginin and luteolin 7-glucoside were enhanced (Jang et al. 2017). The intensity of the light irradiation can also influence the flavonoid response of herbs and plants. This was assessed by Ghasemzadeh et al. (2016) who evaluated the induction of phenolic compounds by UV-B irradiation of sweet basil leaves. For this, the authors tested three different irradiation intensities (2.30, 3.60, and 4.80 W/m^2) and found that the accumulation of phenolic compounds was dependent on the irradiation intensity. They report that the best UV-B intensity was 3.60 W/m^2 enhancing the content of compounds such as gallic acid, cinnamic acid, ferulic acid, quercetin, catechin, kaempferol, rutin, and luteolin. This has been reported to be a response against the reactive oxygen species generated during UV light damage, which also triggers the expression of enzymes of the phenylpropanoid metabolism from which phenolic compounds are biosynthesized, such as PAL and chalcone synthase. Interestingly, phenolic acids are at its highest content within the 6–8 h after treatment and then decreased, while flavonoids accumulation takes longer, from within the 8 to 10 h, which is consistent with the biosynthetic pathway of flavonoids where phenolic acids are precursors.

The wavelength of light can also affect the flavonoid response of herbs and plants. Recently, Taulavuori et al. (2018) reported the response of *Rumex sanguineus*, *Ocimum basilicum*, and *Eruca sativa* under irradiation with blue and blue-violet light on the phytochemical content. They showed that blue and blue-violet light induce the accumulation of phenolic compounds in basil and flavonoids in arugula. Specifically, an increase in chlorogenic acid derivative and chicoric acid content was detected in blue and blue-violet light treatments in basil leaves. In argula, light treatments enhanced the content of flavonoids such as isorhamnetin-diglycoside, luteolin-glycoside derivative 1, apigenin derivative 1, apigenin derivative 2, and luteolin-glycoside derivative 2. This study supports the hypothesis that the flavonoid response of herbs and plants is species-specific since not all the evaluated species reacted in the same way. However, overall blue light showed a significant enhancement of these compounds in all plants. Similarly, Stagnari et al. (2018) aimed to study the effect of light quantity and quality manipulation applying yellow, green, and blue cover films on basil plants and evaluating the response of their secondary metabolite content. Interestingly, in contrast with studies aforementioned, the application of colored films reduced the concentration of rosmarinic and caftaric acids in basil leaves by 29.8 and 33.2% (in average), respectively. On the other hand, caffeic acid was positively induced by the colored-light treatments.

Similarly, Nadeem et al. (2019) evaluated the effect of LED lights on the biosynthesis of phytochemicals in callus cultures of basil. Yellow light enhanced the biosynthesis of rosmarinic and chicoric acid; green light increased rosmarinic acid, eugenol, and chicoric acid content; rosmarinic acid content increased to 55.0 and 48.0 mg/g DW in blue light followed by yellow and white light, respectively, representing almost 2.46-, 1.4-, and 1.2-folds higher than control. It is suggested that this effect is caused by reactive oxygen species accumulated by the action of the metabolic enzyme CYP450 and the high antioxidant capacity of rosmarinic acid. The authors also found that the flavonoid response is specific to some compounds, for instance, they reported that chicoric acid is most increased (81.40 mg/g DW) by white light almost by 4.52-fold times higher than control basil. Another group of compounds elicited by LED lights are anthocyanins, which content particularly increased after treatment with red light which enhanced the quantity of cyanidin (0.121 mg/g DW) and peonidin (0.1271 mg/g DW), which is 3.5- and 4.53-fold greater than control.

Furthermore, the flavonoid response of herbs and plants is also cultivar-dependent. A differential response has been observed in cultivar green and red basil under combinations of different proportions of blue and red light. The phytochemical content in the green cultivar of basil was most enhanced with a higher proportion of red light, while red basil showed a better response to higher ratios of blue light. In this regard, the anthocyanin content of both basil cultivars was enhanced with either red or blue light in comparison to control white light; however caffeic acid, rosmarinic acid, and anthocyanins showed more accumulation in predominantly blue light, with an increase up to 4.15-, 15.12-, and 1.5-fold times higher content than in control white light (Lobiuc et al. 2017).

Also, some genes involved in the biosynthetic route of tanshinones in *Salvia miltiorrhiza* were evaluated through manipulation of light spectra. Chen et al. (2018) used five light spectra: red (R, 660 nm), green (G, 525 nm), blue (B, 450 nm), far-red (IR, 730 nm), and ultraviolet (UV, 380 nm). The content of the depside tanshinone IIA decreased up to 60% in all light treatments including blue light. Furthermore, ultraviolet light increased the level of rosmarinic acid. The aforementioned was explained as a result of the downregulation of the SmHMGR, SmDXS2, SmDXR, SmGGPPS, SmCPS, and CYP76AH1 genes. In contrast, combination of blue and red light treatment increased rosmarinic acid content up to 90% in comparison to white light control.

4.5 Conclusions

Flavonoids and phenolic acids play a physiological role in plant defense against abiotic stresses. Studies regarding the elicitation of plant flavonoids through abiotic stresses have allowed the elucidation of flavonoid and phenolic acid responses. Flavonoids and phenolic acids have been related to numerous health-promoting properties such as the prevention of the onset of noncommunicable diseases like cancer, cardiovascular diseases, metabolic syndrome, and diabetes. This has been an emphatic precedent for studies regarding the induction of flavonoids by abiotic stresses such as light manipulation, which aim to enhance the phytochemical content in herbs and plants for human use. Manipulation of light has proven to be an easy and cheap technology to potentiate phytochemical content in herbs and plants.

References

Anand David AV, Arulmoli R, Parasuraman S (2016) Overviews of biological importance of quercetin: a bioactive flavonoid. Pharmacogn Rev 10(20):84–89. https://doi.org/10.4103/0973-7847.194044

Begas E, Tsioutsiouliti A, Kouvaras E, Haroutounian SA, Kasiotis KM, Kouretas D et al (2017) Effects of peppermint tea consumption on the activities of CYP1A2, CYP2A6, Xanthine Oxidase, N-acetyltranferase-2 and UDP-glucuronosyltransferases-1A1/1A6 in healthy volunteers. Food Chem Toxicol 100:80–89

Bower AM, Real Hernandez LM, Berhow MA, de Mejia EG (2014) Bioactive compounds from culinary herbs inhibit a molecular target for type 2 diabetes management, dipeptidyl peptidase IV. J Agric Food Chem 62(26):6147–6158. https://doi.org/10.1021/jf500639f

Brahmachari G, Gorai D (2006) Progress in the research on naturally occurring flavones and flavonols: an overview. Curr Org Chem 10(8):873–898. https://doi.org/10.2174/138527206776894438

Bray EA, Bailey-Serres J, Weretilnyk E (2015) Responses to abiotic stresses. In: Buchanan B, Gruissem W, Jones R (eds) Biochemistry & molecular biology of plants, 2nd edn. American Society of Plants, Rockville, pp 1158–1203

Chen I-GJ, Lee M-S, Lin M-K, Ko C-Y, Chang W-T (2018) Blue light decreases tanshinone IIA content in Salvia miltiorrhiza hairy roots via genes regulation. J Photochem Photobiol B Biol 183:164–171

Conéjéro G, Noirot M, Talamond P, Verdeil J-L (2014) Spectral analysis combined with advanced linear unmixing allows for histolocalization of phenolics in leaves of coffee trees. Front Plant Sci 5(39). https://doi.org/10.3389/fpls.2014.00039

Corradini E, Foglia P, Giansanti P, Gubbiotti R, Samperi R, Laganà A (2011) Flavonoids: chemical properties and analytical methodologies of identification and quantitation in foods and plants. Nat Prod Res 25(5):469–495. https://doi.org/10.1080/14786419.2010.482054

Crozier A, Jaganath IB, Clifford MN (2009) Dietary phenolics: chemistry, bioavailability and effects on health. Nat Prod Rep 26(8):1001. https://doi.org/10.1039/b802662a

Csepregi K, Coffey A, Cunningham N, Prinsen E, Hideg É, Jansen MAK (2017) Developmental age and UV-B exposure co-determine antioxidant capacity and flavonol accumulation in Arabidopsis leaves. Environ Exp Bot 140:19–25. https://doi.org/10.1016/j.envexpbot.2017.05.009

Dludla PV, Muller CJF, Louw J, Joubert E, Salie R, Opoku AR et al (2014) The cardioprotective effect of an aqueous extract of fermented rooibos (Aspalathus linearis) on cultured cardiomyocytes derived from diabetic rats. Phytomedicine 21(5):595–601. https://doi.org/10.1016/j.phymed.2013.10.029

Embuscado ME (2015) Spices and herbs: natural sources of antioxidants – a mini review. J Funct Foods 18:811–819. https://doi.org/10.1016/j.jff.2015.03.005

Falcone Ferreyra ML, Rius S, Emiliani J, Pourcel L, Feller A, Morohashi K et al (2010) Cloning and characterization of a UV-B-inducible maize flavonol synthase. Plant J 62(1):77–91. https://doi.org/10.1111/j.1365-313X.2010.04133.x

Feliciano RP, Pritzel S, Heiss C, Rodriguez-Mateos A (2015) Flavonoid intake and cardiovascular disease risk. Curr Opin Food Sci 2:92–99

Figueroa-Pérez MG, Rocha-Guzmán NE, Pérez-Ramírez IF, Mercado-Silva E, Reynoso-Camacho R (2014) Metabolite profile, antioxidant capacity, and inhibition of digestive enzymes in infusions of peppermint (Mentha piperita) grown under drought stress. J Agric Food Chem 62 (49):12027–12033. https://doi.org/10.1021/jf503628c

Förstermann U, Sessa WC (2011) Nitric oxide synthases: regulation and function. Eur Heart J 33 (7):829–837. https://doi.org/10.1093/eurheartj/ehr304

Gadi D, Bnouham M, Aziz M, Ziyyat A, Legssyer A, Bruel A et al (2012) Flavonoids purified from parsley inhibit human blood platelet aggregation and adhesion to collagen under flow. J Complem Integr Med 9

Gali-Muhtasib H, Hmadi R, Kareh M, Tohme R, Darwiche N (2015) Cell death mechanisms of plant-derived anticancer drugs: beyond apoptosis. Apoptosis 20(12):1531–1562. https://doi.org/10.1007/s10495-015-1169-2

García-Bores AM, Espinosa-González AM, Reyna-Campos A, Cruz-Toscano S, Benítez-Flores JC, Hernández-Delgado CT et al (2017) Lippia graveolens photochemopreventive effect against UVB radiation-induced skin carcinogenesis. J Photochem Photobiol B Biol 167:72–81. https://doi.org/10.1016/j.jphotobiol.2016.12.014

García-Pérez E, Noratto GD, García-Lara S, Gutiérrez-Uribe JA, Mertens-Talcott SU (2013) Micropropagation effect on the anti-carcinogenic activity of polyphenolics from Mexican Oregano (Poliomintha glabrescens Gray) in human colon cancer cells HT-29. Plant Foods Hum Nutr 68(2):155–162. https://doi.org/10.1007/s11130-013-0344-2

Ghasemzadeh A, Ashkani S, Baghdadi A, Pazoki A, Jaafar HZE, Rahmat A (2016) Improvement in flavonoids and phenolic acids production and pharmaceutical quality of sweet basil (Ocimum basilicum L.) by ultraviolet-B irradiation. Molecules 21(9):15. https://doi.org/10.3390/molecules21091203

Grundy SM (2016) Metabolic syndrome update. Trends Cardiovasc Med 26(4):364–373. https://doi.org/10.1016/j.tcm.2015.10.004

Guimarães R, Barros L, Dueñas M, Calhelha RC, Carvalho AM, Santos-Buelga C et al (2013) Nutrients, phytochemicals and bioactivity of wild Roman chamomile: a comparison between the herb and its preparations. Food Chem 136(2):718–725. https://doi.org/10.1016/j.foodchem.2012.08.025

Gutiérrez-Grijalva EP, Antunes-Ricardo M, Acosta-Estrada BA, Gutiérrez-Uribe JA, Basilio HJ (2019) Cellular antioxidant activity and in vitro inhibition of α-glucosidase, α-amylase and pancreatic lipase of oregano polyphenols under simulated gastrointestinal digestion. Food Res Int 116:676. https://doi.org/10.1016/j.foodres.2018.08.096

Haidari F, Keshavarz SA, Mohammad Shahi M, Mahboob S-A, Rashidi M-R (2011) Effects of parsley (Petroselinum crispum) and its flavonol constituents, kaempferol and quercetin, on serum uric acid levels, biomarkers of oxidative stress and liver xanthine oxidoreductase activity inOxonate-induced hyperuricemic rats. Iran J Pharm Res IJPR 10(4):811–819

Hegazy AM, Abdel-Azeem AS, Zeidan HM, Ibrahim KS, Sayed EME (2017) Hypolipidemic and hepatoprotective activities of rosemary and thyme in gentamicin-treated rats. Hum Exp Toxicol 37(4):420–430. https://doi.org/10.1177/0960327117710534

Jang HJ, Lee SJ, Kim CY, Hwang JT, Choi JH, Park JH et al (2017) Effect of sunlight radiation on the growth and chemical constituents of Salvia plebeia R. Br. Molecules 22(8):16. https://doi.org/10.3390/molecules22081279

Kogiannou DAA, Kalogeropoulos N, Kefalas P, Polissiou MG, Kaliora AC (2013) Herbal infusions; their phenolic profile, antioxidant and anti-inflammatory effects in HT29 and PC3 cells. Food Chem Toxicol 61:152–159. https://doi.org/10.1016/j.fct.2013.05.027

Kumar S, Pandey AK (2013) Chemistry and biological activities of flavonoids: an overview. TheScientificWorldJournal 2013:162750. https://doi.org/10.1155/2013/162750

Kutchan TM, Gershenzon J, Møller BL, Gang DR (2015) Natural products. In: Buchanan BB, Gruissem W, Jones RL (eds) Biochemistry & molecular biology of plants. American Society of Plants, Rockville, pp 1250–1318

León-Chan RG, López-Meyer M, Osuna-Enciso T, Sañudo-Barajas JA, Heredia JB, León-Félix J (2017) Low temperature and ultraviolet-B radiation affect chlorophyll content and induce the accumulation of UV-B-absorbing and antioxidant compounds in bell pepper (Capsicum annuum) plants. Environ Exp Bot 139:143–151. https://doi.org/10.1016/j.envexpbot.2017.05.006

Leyva-López N, Nair V, Bang WY, Cisneros-Zevallos L, Heredia JB (2016) Protective role of terpenes and polyphenols from three species of Oregano (Lippia graveolens, Lippia palmeri and Hedeoma patens) on the suppression of lipopolysaccharide-induced inflammation in RAW 264.7 macrophage cells. J Ethnopharmacol 187:302–312. https://doi.org/10.1016/j.jep.2016.04.051

Lobiuc A, Vasilache V, Pintilie O, Stoleru T, Burducea M, Oroian M et al (2017) Blue and red LED illumination improves growth and bioactive compounds contents in acyanic and cyanic Ocimum basilicum L. microgreens. Molecules 22(12):14. https://doi.org/10.3390/molecules22122111

López-Alarcón C, Denicola A (2013) Evaluating the antioxidant capacity of natural products: a review on chemical and cellular-based assays. Anal Chim Acta 763:1–10. https://doi.org/10.1016/j.aca.2012.11.051

Ma P, Liu J, Zhang C, Liang Z (2013) Regulation of water-soluble phenolic acid biosynthesis in Salvia miltiorrhiza Bunge. Appl Biochem Biotechnol 170(6):1253–1262. https://doi.org/10.1007/s12010-013-0265-4

Manju K, Jat RK, Anju G (2017) A review on medicinal plants used as a source of anticancer agents. Int J Drug Res Technol 2(2)

Manukyan A (2013) Effects of PAR and UV-B radiation on herbal yield, bioactive compounds and their antioxidant capacity of some medicinal plants under controlled environmental conditions. Photochem Photobiol 89(2):406–414. https://doi.org/10.1111/j.1751-1097.2012.01242.x

Marseglia L, Manti S, D'Angelo G, Nicotera A, Parisi E, Di Rosa G et al (2015) Oxidative stress in obesity: a critical component in human diseases. Int J Mol Sci 16(1). https://doi.org/10.3390/ijms16010378

Martens S, Mithöfer A (2005) Flavones and flavone synthases. Phytochemistry 66(20):2399–2407. https://doi.org/10.1016/j.phytochem.2005.07.013

Nadeem M, Abbasi BH, Younas M, Ahmad W, Zahir A, Hano C (2019) LED-enhanced biosynthesis of biologically active ingredients in callus cultures of Ocimum basilicum. J Photochem Photobiol B Biol 190:172–178. https://doi.org/10.1016/j.jphotobiol.2018.09.011

Nadia Z, Rachid M (2013) Antioxidant and antibacterial activities of Thymus vulgaris L. Med Aromat Plant Res J 1(1):5–11

Nickavar B, Abolhasani L (2013) Bioactivity-guided separation of an α-amylase inhibitor flavonoid from Salvia virgata. Iran J Pharm Res IJPR 12(1):57–61

Nijveldt RJ, Van Nood E, Van Hoorn DE, Boelens PG, Van Norren K, Van Leeuwen PA (2001) Flavonoids: a review of probable mechanisms of action and potential applications. Am J Clin Nutr 74(4):418–425

Opara IE, Chohan M (2014) Culinary herbs and spices: their bioactive properties, the contribution of polyphenols and the challenges in deducing their true health benefits. Int J Mol Sci 15(10). https://doi.org/10.3390/ijms151019183

Pantsi WG, Marnewick JL, Esterhuyse AJ, Rautenbach F, van Rooyen J (2011) Rooibos (Aspalathus linearis) offers cardiac protection against ischaemia/reperfusion in the isolated perfused rat heart. Phytomedicine 18(14):1220–1228. https://doi.org/10.1016/j.phymed.2011.09.069

Peng M, Shahzad R, Gul A, Subthain H, Shen S, Lei L et al (2017) Differentially evolved glucosyltransferases determine natural variation of rice flavone accumulation and UV-tolerance. Nat Commun 8(1):1975. https://doi.org/10.1038/s41467-017-02168-x

Pisoschi AM, Pop A (2015) The role of antioxidants in the chemistry of oxidative stress: a review. Eur J Med Chem 97:55–74. https://doi.org/10.1016/j.ejmech.2015.04.040

Pizzino G, Irrera N, Cucinotta M, Pallio G, Mannino F, Arcoraci V et al (2017) Oxidative stress: harms and benefits for human health. Oxidative Med Cell Longev 2017:13. https://doi.org/10.1155/2017/8416763

Procházková D, Boušová I, Wilhelmová N (2011) Antioxidant and prooxidant properties of flavonoids. Fitoterapia 82(4):513–523. https://doi.org/10.1016/j.fitote.2011.01.018

Radyukina NL, Toaima VIM, Zaripova NR (2012) The involvement of low-molecular antioxidants in cross-adaptation of medicine plants to successive action of UV-B radiation and salinity. Russ J Plant Physiol 59(1):71–78. https://doi.org/10.1134/s1021443712010165

Ralston RA, Schwartz SJ, Hostetler GL (2017) Flavones: food sources, bioavailability, metabolism, and bioactivity. Adv Nutr 8(3):423–435. https://doi.org/10.3945/an.116.012948

Rubió L, Motilva M-J, Romero M-P (2013) Recent advances in biologically active compounds in herbs and spices: a review of the Most effective antioxidant and anti-inflammatory active principles. Crit Rev Food Sci Nutr 53(9):943–953. https://doi.org/10.1080/10408398.2011.574802

Sesso HD, Lee IM, Blumberg JB, Buring JE, Zhang SM, Wang L (2009) Dietary intake of selected flavonols, flavones, and flavonoid-rich foods and risk of cancer in middle-aged and older women. Am J Clin Nutr 89(3):905–912. https://doi.org/10.3945/ajcn.2008.26913

Singh M, Kaur M, Silakari O (2014) Flavones: an important scaffold for medicinal chemistry. Eur J Med Chem 84:206–239. https://doi.org/10.1016/j.ejmech.2014.07.013

Smith C, Swart A (2018) Aspalathus linearis (Rooibos) – a functional food targeting cardiovascular disease. Food Funct 9(10):5041–5058. https://doi.org/10.1039/C8FO01010B

Stagnari F, Di Mattia C, Galieni A, Santarelli V, D'Egidio S, Pagnani G et al (2018) Light quantity and quality supplies sharply affect growth, morphological, physiological and quality traits of basil. Ind Crop Prod 122:277–289. https://doi.org/10.1016/j.indcrop.2018.05.073

Stewart AJ, Bozonnet S, Mullen W, Jenkins GI, Lean MEJ, Crozier A (2000) Occurrence of flavonols in tomatoes and tomato-based products. J Agric Food Chem 48(7):2663–2669. https://doi.org/10.1021/jf000070p

Szymańska R, Ślesak I, Orzechowska A, Kruk J (2017) Physiological and biochemical responses to high light and temperature stress in plants. Environ Exp Bot 139:165–177. https://doi.org/10.1016/j.envexpbot.2017.05.002

Tattini M, Galardi C, Pinelli P, Massai R, Remorini D, Agati G (2004) Differential accumulation of flavonoids and hydroxycinnamates in leaves of Ligustrum vulgare under excess light and drought stress. New Phytol 163(3):547–561. https://doi.org/10.1111/j.1469-8137.2004.01126.x

Taulavuori K, Pyysalo A, Taulavuori E, Julkunen-Tiitto R (2018) Responses of phenolic acid and flavonoid synthesis to blue and blue-violet light depends on plant species. Environ Exp Bot 150:183–187. https://doi.org/10.1016/j.envexpbot.2018.03.016

Tian Y, Sun L-M, Liu X-Q, Li B, Wang Q, Dong J-X (2010) Anti-HBV active flavone glucosides from Euphorbia humifusa Willd. Fitoterapia 81(7):799–802. https://doi.org/10.1016/j.fitote.2010.04.012

Torres-Contreras AM, Senés-Guerrero C, Pacheco A, González-Agüero M, Ramos-Parra PA, Cisneros-Zevallos L et al (2018) Genes differentially expressed in broccoli as an early and late response to wounding stress. Postharvest Biol Technol 145:172–182. https://doi.org/10.1016/j.postharvbio.2018.07.010

Trivellini A, Lucchesini M, Maggini R, Mosadegh H, Villamarin TSS, Vernieri P et al (2016) Lamiaceae phenols as multifaceted compounds: bioactivity, industrial prospects and role of "positive-stress". Ind Crop Prod 83:241–254. https://doi.org/10.1016/j.indcrop.2015.12.039

Tsao R (2010) Chemistry and biochemistry of dietary polyphenols. Nutrients 2(12):1231–1246. https://doi.org/10.3390/nu2121231

Wang T-y, Li Q, K-s B (2018) Bioactive flavonoids in medicinal plants: structure, activity and biological fate. Asian J Pharm Sci 13(1):12–23. https://doi.org/10.1016/j.ajps.2017.08.004

Zaman SB, Hussain MA, Nye R, Mehta V, Mamun KT, Hossain N (2017) A review on antibiotic resistance: alarm bells are ringing. Cureus 9(6):e1403-e. https://doi.org/10.7759/cureus.1403

Zhou Z, Schenke D, Miao Y, Cai D (2017) Investigation of the crosstalk between the flg22 and the UV-B-induced flavonol pathway in Arabidopsis thaliana seedlings. Plant Cell Environ 40(3):453–458. https://doi.org/10.1111/pce.12869

Interactive Biology of Auxins and Phenolics in Plant Environment

5

Ambreen Ahmed, Aqsa Tariq, and Sabiha Habib

Abstract

Plant environment is a complex system where coordinated interactive biology involving various metabolite products and other intermediates determines the overall development and growth of plant. Among the phytohormones, auxins play a fundamental role in various signaling pathways involving other hormones and metabolites affecting cell division and differentiation of plant tissues. Likewise, phenolics are the secondary metabolites secreted by plants that play a key role as defense agents during environmental stress conditions. Biosynthesis of auxins and phenolics follows different metabolic pathways, although shikimate pathway is considered as the root for the production of auxins and phenolics following the synthesis of their corresponding precursors. The interactions between these two compounds may have some physiological and biochemical alterations in plant metabolism, thus affecting plant biology. In addition, the role of soil microbiota is also evident to mediate the communicative behavior of both auxins and phenolics. Phenolic compounds may affect auxin transport and play its role in defense signaling of plants. Some representative examples regarding interactive biology of auxins and phenolic compounds under in vitro conditions are also discussed in this chapter.

Keywords

Biology · Auxins · Phenolics · Environment · Signal pathways

A. Ahmed (✉) · A. Tariq · S. Habib
Department of Botany, University of the Punjab, Lahore, Pakistan

R. Lone et al. (eds.), *Plant Phenolics in Sustainable Agriculture*,
https://doi.org/10.1007/978-981-15-4890-1_5

5.1 Introduction

The living biota over the planet Earth presents a number of solutions to the issues of their survival on account of facing ever-changing and adverse environmental conditions. Animals change their behavior in response to their environment by either fighting or escaping the situation, while plants on the other hand have striking capabilities to optimize their metabolic processes to meet the environmental stresses. This developmental plasticity comprises the production of growth regulators and secretion of secondary metabolites that collectively not only improve the growth but also contribute toward plant defense mechanism in response to external stimuli (Davies 2004; Tanaka et al. 2006). Plants have some mobile signals including primary and secondary metabolites that regulate the physiological and developmental processes in plants. These endogenous hormonal signaling molecules trigger the plant cells for tissue reprogramming, thereby altering the plant physiology in a better way. Crosstalk between hormonal signaling pathways eventually governs the physiological concerns (Davies 2004; Vanneste and Friml 2009).

The flexible behavior of plants in terms of their growth and development and their response to multiple environmental conditions indicates the complex regulatory mechanisms. Although there are many substances known as plant growth regulators, nonetheless, plant development is largely governed by an indolic substance, auxin (Tanaka et al. 2006). The gradients of auxins are involved in several auxin-arbitrated developmental courses in plants and may be attributed to auxin biosynthesis and auxin transport, in response to environmental and developmental indicators. These gradients may have some interactions with plant secondary metabolites including phenolic compounds which together improve plant biology and help in defense system of plants (Cheng et al. 2006, 2007; Stepanova et al. 2008; Tao et al. 2008). Thus, modulation of auxins in plants is expected when they get interacted with phenolic compounds.

The past few decades primarily employed the auxin machinery and phenolic biology in plant systems, but the interactive behavior of phenolic compounds with auxins is neglected for years. Plant developmental plasticity is concerned about the translation of external signals in plant environment in addition to their coordination with plant defense mechanisms. The key aspects discussed in this chapter regarding auxins and phenolics include basic introduction of auxins and phenolics, role of auxins and phenolics in plants, biosynthetic pathways for phenolics and auxins, biosynthesis of phenolics in response to biotic and abiotic stress, crosstalks of phenolics and auxins in plant rhizosphere, role of phenolic compound (flavonoids) in auxin transport, interaction of phenolic hormone (salicylic acid) with auxin signaling during plant defense mechanism, and case studies regarding interactive behavior of auxins with phenolic compounds. Conspicuous and pervasive features of these key points are the unusual self-organizing properties of auxin biology, in relation with plant phenolics, which is comprehensively characterized in this chapter.

5.2 Plant Phenolics

Phenolics are the aromatic compounds containing benzene ring having one or more hydroxyl (OH) group produced from plants for protection under stress conditions. According to Harborne (1980), phenolics are one of the most abundant groups of secondary metabolites and are considered as bioactive substances that are extensively found in plant kingdom. These are closely related with the nutritional quality of fresh as well as processed plant foods. Phenolics have long been considered as allelochemicals for competitive plants, for instance, volatile terpenoids, hydroxycinnamates, hydroquinones, hydroxybenzoates, and 5-hydroxynapthoquinones (Weir et al. 2004; Xuan et al. 2005). Phenolics are generally synthesized when plants recognize some potential pathogens in its proximal areas by pathogen-associated molecular patterns (PAMPs), thereby causing PAMP-triggered immunity (Schuhegger et al. 2006; Newman et al. 2007; Ongena et al. 2007; Tran et al. 2007; Zipfel 2008). Biogenetically, phenolic compounds follow shikimate metabolic pathway that forms phenylpropanoids and acetic acid pathway that forms simple phenol (Sanchez-Moreno 2002). According to Hollman (2001), most of the plants' phenolic compounds are formed through shikimate pathway. However, flavonoids, the most plenteous group of phenolics, are formed by the combination of both pathways (Tomás-Barberán and Espín 2001).

Phenolic compounds contain a large number of heterogeneous structures, from simple to complex ones; hence, they are classified in different ways. On the basis of carbon chains, they are divided into 16 major classes including phenols (C_6), benzoquinones (C_6), phenolic acids (C_6-C_1), acetophenones (C_6-C_2), phenylacetic acid (C_6-C_2), hydroxycinnamic acids (C_6-C_3), phenylpropenes (C_6-C_3), coumarins/isocoumarins (C_6-C_3), chromones (C_6-C_3), naphthoquinones (C_6-C_2), xanthones (C_6-C_1-C_5), stilbenes (C_6-C_2-C_6), anthraquinones (C_6-C_2-C_6), flavonoids (C_6-C_3-C_6), lignans and neolignans ($(C_6$-$C_3)_2$), and lignins ($(C_6$-$C_3)_n$). Conversely, on the basis of distribution of phenolics in nature, they are categorized into three classes. The first class is "shortly distributed" that includes phenols, hydroquinones, resorcinol, aldehydes which are derivatives of benzoic acid, etc. The second class is "widely distributed" that contains flavonoids and derivatives, phenolic acids, and coumarins, whereas the third class is "polymers" having lignin and tannins (Bravo 1998). Lastly, on the basis of their locality in plants, phenolic compounds may be soluble and insoluble. Soluble form consists of low molecular weight tannins, phenol, and flavonoids, while condensed tannins and phenolic acids are present in insoluble form (Sanchez-Moreno 2002).

5.2.1 Role of Phenolics in Plants

Phenolic compounds play their role in plant environment exhibiting interactions with biotic and abiotic factors, but the exact mechanism about the functioning of these phenolic compounds is difficult to estimate (Harborne and Simmonds 1964). The role of phenolics for plant development is evident especially for synthesis of pigment and lignin, resistance to pests, pollination, plant germination, and reproduction. These compounds also maintain the vigor and provide structural integrity to plants and involve in plant physiology and cellular metabolism. In addition, plants get their sensorial characteristics such as color, taste, aroma, and acidity from these phenolic compounds (Tomas-Barberan and Epsin 2001). Essentially, phenolic phytoalexins secreted from wounded areas of plants kill several microorganisms. Therefore, these compounds have been proposed to act as suitable substitutes to the chemical control of pathogens for agricultural plants (Bhattacharya et al. 2010), whereas some pathogens may counteract their effect or can nullify the defenses triggered by phenolic phytoalexins (Boudet 2007). In contrast to beneficial aspects of phenolics in plant system, these compounds also act as natural growth inhibitors and retard different processes like root elongation, stem elongation, opening of buds, seed germination processes, etc. Phenolics can also hinder the growth of other plants when they are released into the environment (Kefeli and Kadyrov 1971). Plant tissue culture is a modern technique for large-scale development of genetically similar plants from different explants (Sarkar and Naik 2000). The calli produced during tissue culture experiments exhibit browning due to the oxidation of exuded phenolic compounds which hinder nutrient uptake and ultimately cause the death of explants (George 1996).

5.3 Auxins: Powerful Phytohormone

The presence of auxin in plants as transportable growth hormone was firstly inferred by Charles and Frances Darwin, as described in their book, *The Power of Movement in Plants* (Darwin and Darwin 1880). However, auxin was first isolated chemically by Went (1926) who first ever mentioned auxins in plant physiology and described general conception of regulators and coordinators in plants. To understand the cast of auxins, i.e., receptors, transporters, synthesizers, and inactivators, a number of reviews have been presented by Chapman and Estelle (2009), Lokerse and Weijers (2009), Petrasek and Friml (2009), and Zhao (2010). There are still some important features missing; however, an obvious shift in understanding the whole plant ecosystem with interactions of auxin machinery brought new insights toward auxin biology in plants.

Auxin, an essential plant hormone, naturally produced by plants, is not only accountable for cell elongation but also helps to repair wounds in plants and is involved in various metabolic pathways. It contains aromatic ring having carboxylic acid group. This phytohormone drastically affects the plant direction in response to sunlight as well as gravity. Auxins comprise of three major classes, i.e., 1-naphthaleneacetic acid (NAA), 3-indole acetic acid (IAA), and indole-3-butyric acid (IBA). However, the most widespread and potent auxin is indole acetic acid (IAA) which has imperative role in growth and developmental processes in plant's life cycle including cardinal role in plant defense mechanism. An amino acid, tryptophan, is responsible for the regulation of IAA production (Bari and Jones 2009; Zhao 2010; Habib et al. 2019).

5.3.1 Role of Auxins in Plants

Auxin, a chief plant growth regulator, is released naturally in plants which is mainly concentrated in meristematic regions and then distributed toward the roots; therefore, plants tend to grow with the gradient of auxin concentrations. The dynamic distribution of auxins within plants controls a variety of developmental processes which not only facilitate the plant growth but also alter the plant morphology in a better way. Ahmed and Hasnain (2010) presented a detailed review on auxins, their role, biosynthesis, and interaction with microbes and plants. They stated that auxins are generally produced in the meristematic regions of plant stems, although it may also produce in shoots, roots, and leaves. The concentration of auxin varies depending on the influx and efflux from the plant tissues and also its biosynthesis from tryptophan (precursor of auxin) and formation of IAA conjugates. Authors also discussed the signaling crosstalks of auxins with other phytohormones and with microbes (Ahmed and Hasnain 2010). Auxins may affect cellular processes, cytoplasmic streaming, and various enzymatic reactions. The proliferative behavior of plant cells assisted by auxins causes stem and root elongation, flower initiation, fruit development, and even tuber or bulb formation. IAA (indole-3-acetic acid) contributes its role to almost every aspect of growth and development of plants (Zhao 2012).

It has been reported that synthesis of auxin and intercellular auxin transport in plants is affected by auxin distribution assimilated by various environmental and plants' endogenous signals. Moreover, auxins are predominantly involved in some sort of cellular regulations and organogenic processes in response to environmental stimuli. During early embryogenic growth of plants, auxin accumulates in apical cells, root pole, and cotyledons, while post-embryogenesis incorporates the auxin accumulation during organogenesis of flowers, floral parts, leaves, and lateral roots (Benková et al. 2003; Friml 2003). Thus, auxin acts as a multipurpose activator of preprogrammed developmental changes in plants.

5.4 Biosynthetic Pathways of Phenolics and Auxins

5.4.1 Biosynthesis of Auxins

Biosynthetic pathways for auxins have been proposed containing one tryptophan-independent pathway and four tryptophan-dependent pathways. These are indole-3-acetamide (IAM) pathway, indole-3-acetaldoxime (IAOx) pathway, tryptamine (TAM) pathway, and indole-3-pyruvic acid (IPA) pathway, respectively (Woodward and Bartel 2005; Vanneste and Friml 2009).

Tryptophan-independent pathway was reported by Wright and his coworkers (1991) during their experimental studies in maize plant and by Normanly et al. (1993) in *Arabidopsis*. In tryptophan-independent pathway, indole and indole-3-glycerol phosphate are the precursors of indole-3-acetic acid (Ouyang et al. 2000; Zhang et al. 2008a, b). Indole-3-glycerol phosphate by the action of indole-3-

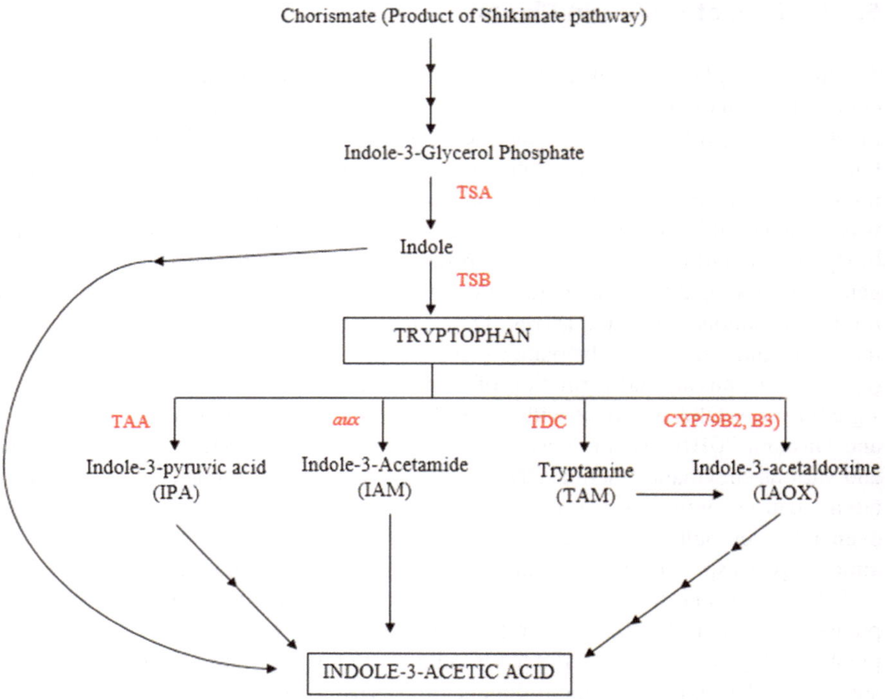

Fig. 5.1 Biosynthetic pathway of auxin

glycerol phosphate lyase (IGL)/indole synthase (INS) converts into indole which through various unknown enzymes and intermediates ultimately forms IAA.

In tryptophan-dependent pathway, indole-3-glycerol phosphate synthesized from chorismate through various steps and enzymes by the action of a-subunit containing tryptophan synthetase (TSA) converts into indole. The a-subunit mediates the removal of side chain from indole-3-glycerol phosphate, and the resultant product, i.e., indole, in turn converts into tryptophan by the activity of enzyme b-subunit containing tryptophan synthetase (TSB). Tryptophan synthesis occurs in chloroplast (Pan et al. 1997; Zhao 2012; Nonhebel 2015). Tryptone converts into indole-3-acetamide (IAM), indole-3-acetaldoximie (IAOX), tryptamine (TAM), and indole-3-pyruvic acid (IPA) by the activity of auxin-synthesizing genes (*aux*), tryptophan-specific P450 monogenase (CYP79B2, B3), tryptophan decarboxylase (TDC), and tryptophan aminotransferase (TAA), respectively, which ultimately by the activity of various enzymes and intermediates results in the formation of IAA (Zhao 2010; Mano and Nemoto 2012) (Fig. 5.1).

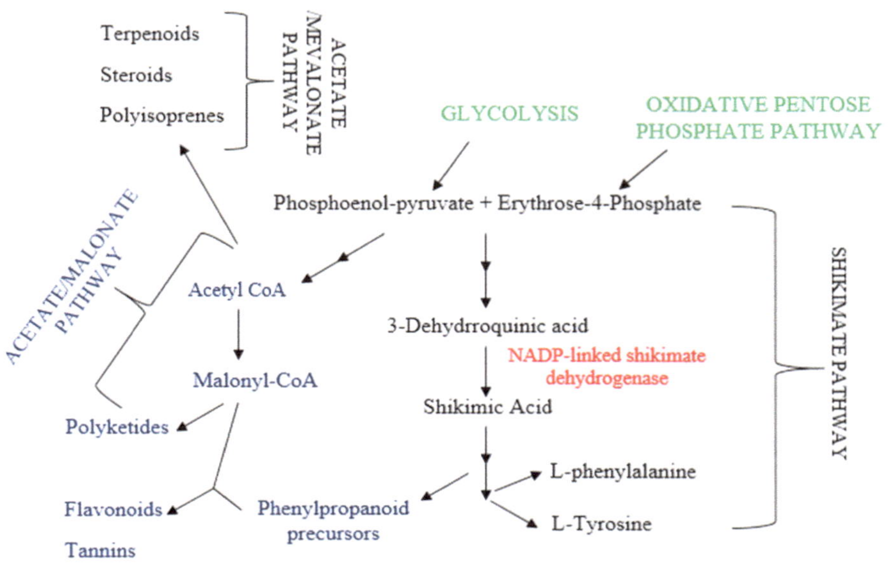

Fig. 5.2 Biosynthetic pathway of phenolics

5.4.2 Biosynthesis of Phenolics

Biosynthetic pathway for phenolics is mainly comprised of three major pathways: (a) shikimate/chorismate or succinylbenzoate pathway producing phenylpropanoid derivatives (C_6–C_3); (b) acetate/malonate or polyketide pathway producing side-chain-elongated phenylpropanoids, including flavonoids (C_6–C_3–C_6) and quinones; and (c) the acetate/mevalonate pathway producing aromatic terpenoids (Bhattacharya et al. 2010).

In shikimate pathway, phosphoenolpyruvate from glycolysis and erythrose-4-phosphate from oxidative pentose phosphate combine to form 3-dehydroquinic acid with ultimately forms shikimic acid that is precursor of L-phenylalanine and L-tyrosine. Shikimic acid forms precursors of phenylpropanoid, i.e., cinnamic acid, *p*-coumaric acid, and *p*-coumaroyl acid. Acetate/malonate pathway involves three main enzymes: phenylalanine ammonia-lyase (PAL), cinnamate 4-hydroxylase (C4H), and 4-coumarate:CoA ligase (4CL). In this pathway, malonyl-CoA produced from acetyl-CoA combined with *p*-coumaroyl acid, formed from shikimic acid, ultimately forms aromatic phenolic compounds, i.e., flavonoids, tannins, etc. Acetate/mevalonate pathway also involves acetyl-CoA that produces dimethylallyl pyrophosphate which acts as precursor to form terpenoids, steroids, and polyisoprenes (Vickery and Vickery 1981; Bhattacharya et al. 2010) (Fig. 5.2).

5.4.2.1 Phenolic Production in Response to Biotic and Abiotic Stress

Phenolics accumulate in subepidermal layers of plant tissues which are facing any biotic or abiotic stress including pathogen attacks (Schmitz-Hoerner and Weissenbock 2003; Clé et al. 2008). Concentrations of phenolics secreted in plant environment may vary according to the seasons and growth stages of plants (Thomas and Ravindra 1999; Ozyigit et al. 2007). Production and accumulation of phenolics are significantly affected by various factors including trauma, drought, extreme environment, wounding, and soilborne or phytopathogen attack. Plants when exposed to light may produce significant amount of phenolics in chloroplasts and vacuoles (Kefeli et al. 2003). Nutrient stress, deficiency in macronutrients (nitrogen, phosphorous, and potassium), deficiency of micronutrients (iron, magnesium, and boron), and photo-inhibition also contribute to the production of phenolic compounds in response to these stresses (Balasundram et al. 2006).

5.4.2.2 Biogenesis of Phenolics and Auxins by Shikimate Pathway

In plants, phenolic compounds are produced due to biotic and abiotic stimuli, e.g., unfavorable environmental conditions like pH, temperature, salinity, or pathogenic attack. There are various pathways for the synthesis of phenolics and auxins (IAA), but primarily both are believed to be synthesized from the precursors produced through the shikimate pathway. It is a significant pathway for the production of various primary and secondary metabolites in plants. Shikimate pathway produces precursors for the biosynthesis of different indole compounds, alkaloids, aromatic amino acids, other aromatic metabolites, and flavonoids. The IAA precursor L-tryptophan is synthesized from chorismate, the end product of the shikimate pathway. Likewise, L-phenylalanine and L-tyrosine produced from the same key compound (chorismate) are the precursor of phenolic compounds. Thus, this pathway is considered the root for biogenesis of auxins and phenolics (Santos-Sánchez et al. 2019) (Fig. 5.3).

5.5 Crosstalks of Phenolics and Auxins in Plant Rhizosphere

Plants acquire all necessary nutrients from soil where they are exposed to biotic and abiotic factors of environment and interact with various microorganisms whether beneficial or harmful. Plants develop various signaling mechanisms to cope with adverse conditions in their environment. To develop better vigor and growth, plants must maintain an equilibrium between growth and defense signaling. The rhizosphere of the plants is the dynamic habitat for a variety of flora and fauna where microorganisms interact with each other in the form of interlocked system (Whipps 2001). These interactions are generally triggered by phenolic compounds secreted from plant roots as root exudates that usually include water, ions, mucilage, enzymes, free oxygen, and various primary and secondary metabolites, importantly, phenolics (Bais et al. 2006; Bertin et al. 2003; Dakora and Phillips 2002). Phenolics occupy their concentration range in soil ecosystem from 2.1 to 4.4% in monocots, while 0.1–0.6% of phenolics were found in dicots (Hartley and Harris 1981). They

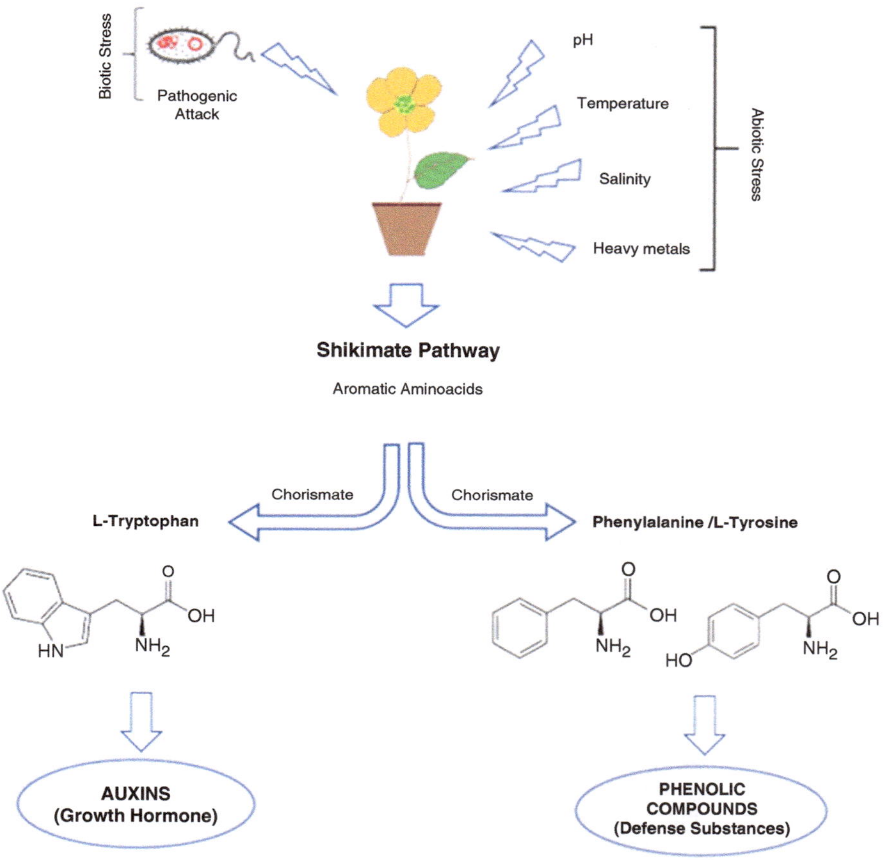

Fig. 5.3 Schematic representation of shikimate pathway for biogenesis of auxins and phenolic compounds. *Arabidopsis* plant when exposed to biotic (pathogenic attack) and abiotic stress (pH, temperature, salinity, and heavy metal stress) initiates various metabolic processes through shikimate pathway, synthesizing different aromatic amino acids that act as precursors, i.e., phenylalanine/L-tyrosine and L-tryptophan which both are formed through chorismate, end product of shikimate pathway, for the synthesis of different compounds including auxins and phenolic compounds, respectively

may have the influence over soil microbiota which are present in the vicinity of plant roots, together with hormonal balance, nutrient competition with neighboring plants, and various enzymatic activities (Hattenschwiler and Vitousek 2000; Kraus et al. 2003).

Some microbes, present in plant rhizosphere, have the potential to produce auxins, an important plant growth regulator which when taken up by plants ultimately enhances plant growth, thus known as auxin-producing rhizobacteria (Ahmed and Hasnain 2010; Aslam and Ahmed 2018; Karamat and Ahmed 2018;

Habib et al. 2019). Moreover, phenolics present in rhizosphere bind with organic matter and are metabolized by rhizobacteria. These bacteria break down phenolic compounds that facilitate the mineralization of nitrogen, thereby enhancing mineral uptake by plants along with auxin uptake synthesized by rhizobacteria (Halvorson et al. 2009). In addition, many phenolic compounds are released in the form of root exudates and act as chemotactic signals for auxin-producing bacterial populations to attract them toward plant rhizosphere that results in the colonization of plant roots by these bacteria contributing toward the growth amelioration of plants (Perret et al. 2000; Taylor and Grotewold 2005). Hence, these crosstalk pathways are not only initiated by soil microbiota but also involve phenolic compounds released in plant rhizosphere.

According to Lattanzio (2013), root exudates contain various substances including phenolics like iso-flavonoids and flavonoids which activate genes of nearby symbiotic microbes for nodulation. The process of nodule formation is regulated by phenolic flavonoid compounds of root exudates of plants that act as transmitters between bacteria and their host plants. These flavonoids tempt bacterial community toward the plant roots, thereby activating nod (nodulation) gene expression, resulting in the formation of nod factors that facilitate the nodule development (Lattanzio 2013). Phenolics also carry redox reactions in soil medium that help to colonize the bacteria with plant roots in rhizospheric region and regulate the hormonal balance of neighboring plants (Bhattacharya et al. 2010) and, depending on bacterial species and type of flavonoid, directly and indirectly influence the expression of bacterial nod genes whether in positive or negative way (Reddy et al. 2007). For instance, daidzein and genistein (isoflavones) stimulate expression of nod genes in *Bradyrhizobium japonicum* but prevent in *Sinorhizobium meliloti*. Likewise, naringenin (flavanone) encourages nod formation in *Rhizobium leguminosarum*, but quercetin (flavonol) suppresses its production (Hartwig et al. 1990; Webster et al. 1998; Abdel-Lateif et al. 2012). Besides, negative interaction between auxins and phenolics can be understood by considering crown gall disease. Briefly, in plants, galls are formed due to uncontrolled cell division causing crown gall disease which is specifically caused by *Agrobacterium tumefaciens*. During infection, these bacteria inject their DNA fragment in plant genome that carries genes for plant hormones. Auxins are reported to promote tumor formation. It has been observed that increase in auxin levels caused the formation of tumor-like structures in plants. Matveeva et al. (2001) stated that auxins' sensitivity in plants is a reason for tumor initiation. Moreover, Hartley (1999) proposed galls in plants are also due to the interactions of phenolic compounds with the IAA or IAA oxidase, which regulates IAA activity.

A number of phenolic compounds have been reported phytotoxic that inhibit germination processes and plant growth. These compounds are coumarin, salicylic acid, parahydroxybenzoic acid, benzoic acid, and syringic acid (Baleroni et al. 2000). Phenolics eventually inhibit the production of phosphatase and prolyl aminopeptidase which are involved in seed germination, thereby causing retardation of seedling growth. The negative impact of these phenolics in plants may be due to the interference of cell divisions and normal enzymatic activities (Madhan et al. 2009).

5.5.1 Role of Flavonoids in Auxin Transport

Flavonoids are the most important group of phenolic compounds that occur naturally and display important role in plant physiology and development. Besides protection against environmental stress, those also act as regulating and signaling molecule. Flavonoids have influence over auxin transport as they promote cell divisions and contribute in development of plants by modifying the activity of indole-3-acetic acid (IAA), which causes tissue differentiation, callus, and tylose formation (Mierziak et al. 2014). Flavonoids act as a link between environment and auxin transport by accumulating as a response to environmental stimuli. They act as natural auxin effluent inhibitors (AEIs) that regulate the activity and localization of auxin transporter proteins (e.g., PIN proteins) (Peer and Murphy 2007). Some flavonoids, e.g., flavonols, inhibit the transportation process of auxins by competing with synthetic AEIs (Bernasconi 1996; Murphy and Taiz 1999).

Studies of Jacobs and Rubery (1988) revealed that under in vitro conditions, flavonoids contend for auxin transporters with 1-naphthylphthalamic acid (NPA) by binding to NPA-interacting proteins. Studies of *Arabidopsis* mutants that contain less flavonoid content displayed increased auxin transport and phenotypic changes, whereas mutants with increased flavonols exhibited decreased auxin transport rates (Brown et al. 2001; Peer et al. 2004). Flavonoids' colonization with auxins helps to control the process of auxin transport (Murphy et al. 2000; Buer et al. 2006). They are considered as effective inhibitors of glycoproteins of PIN and MDR and also for kinases which are tremendously involved in auxin transport. Quercetin (phenolic flavonoid) has been reported as a competent inhibitor for auxin transportation than kaempferol (Lewis et al. 2011). Flavonoid controlled auxin transport has a key role in stress-induced morphogenic response in plants (SIMR). Plant species having high dihydroxy-flavonoids displayed remarkably multi-morphogenic features as compared to those with high monohydroxy-flavonoids (Potters et al. 2007, 2009). It has been also observed that depending on their chemical structures, flavonoids also control the role of IAA oxidase (Jansen et al. 2001).

5.5.2 Interaction of Salicylic Acid with Auxins During Defense Signaling

Plants produce free radicals as a response to unfavorable stimuli that act as a signal to produce various substances including salicylic acid. It is a phenolic hormone affecting plant defense by developing systemic acquired resistance in host plants (Fu and Dong 2013). Wang et al. (2007) studied the treatment of *Arabidopsis* seedlings with salicylic acid analogue benzothiadiazole which caused the suppression of auxin signaling genes, while genes that involve in IAA conjugation (i.e., GH3) were upregulated. Salicylic acid represses the expression of the auxin receptor proteins which maximizes the stabilization of auxin repressor protein and repressed auxin signaling and responses; therefore, reduced plant growth may be ascribed to salicylic acid-interceded repression of auxin signaling (Zhang et al. 2003; Huot et al.

2014). It also promotes the conversion of free auxin to inactive auxin by IAA-conjugating enzyme GH3.5 (Staswick et al. 2005). Studies showed that overexpression of the gene GH3.5 in *Arabidopsis* exhibited high resistance to infection with *Pseudomonas syringae*, but at the same time it also displayed dwarf phenotype due to high expression of PR1 gene (Park et al. 2007; Zhang et al. 2008a, b). Salicylic acid-mediated auxin homeostasis is important for achieving a balance between plant growth and defense (Attaran et al. 2014; Huot et al. 2014). Generally, auxin reduces salicylic acid-mediated plant defenses. For instance, studies of Robert-Seilaniantz et al. (2011) showed that increased auxin signaling due to overexpression of the auxin receptor gene caused substantial reduction in salicylic acid accumulation in *Arabidopsis* after pathogen infection, though overexpression of the gene for auxin biosynthesis gene (YUCCA 1) resulted in high susceptibility without affecting the response of salicylic acid (Mutka et al. 2013).

5.6 Case Studies

Following are some case studies reported by researchers to describe the synergistic interactions between auxins and phenolic compounds.

I. A study was conducted by Jones and Hatfield (1976), who gave insightful demonstration of synergistic association between auxins and phenolic compounds. This report is about the rooting of Apple shoots in the presence of various phenolic compounds. In their study, Apple rootstock shoots were in vitro cultured over auxin media supplemented with phenolic compounds, i.e., phloroglucinol, phloretic acid, caffeic acid, catechol, and pyrogallol, and observed the root and shoot production to propagate the plant. The results showed that phloroglucinol and phloretic acid significantly increased shooting, whereas caffeic acid, catechol, and pyrogallol were not efficient in this respect.
II. Baque and his coworkers (2010) presented a study in which they cultured adventitious roots of *Morinda citrifolia* with different types of auxins and cytokinins. They used naphthalene acetic acid (NAA) and indole butyric acid (IBA) as auxins and kinetin as cytokinins, in different concentrations. After these treatments, growth, production of secondary metabolites (anthraquinone, phenolics, and flavonoid contents), and antioxidant enzymatic activity (catalase, guaiacol peroxidase, and ascorbate peroxidase) were determined. The results suggest that indole butyric acid (IBA), source of auxin supplementation, is expedient for growth and secondary metabolite production including phenolics, for adventitious roots of *Morinda citrifolia* (Baque et al. 2010).
III. Uddin and fellows (2012) conducted a research over herbicidal activity of phenolics produced from hairy root cultures of *Fagopyrum tataricum*. The aim of their study is to optimize the conditions for hairy root cultures using different growth media supplemented with different concentrations of auxins for the production of these compounds and to evaluate the effect of these phenolic compounds to hinder the growth of some weeds in growth chamber and

glasshouse conditions. They revealed the significance of indole-3-butyric acid (IBA) among all auxin treatments, for profound production of phenolic compounds, i.e., rutin, gallic acid, and chlorogenic acid, to get maximal hairy root mass as compared to control treatment. Phenolic compounds (hairy root extracts) were then subjected to check their inhibitory effect on weeds exhibiting that these compounds lowered the germination rate together with inhibition of roots and shoots of tested weeds. Thus, it may be concluded that the production of phenolic compounds can be enhanced with auxin treatments of growth media for respective plant and these phenolics can further be used as herbicide to kill the weeds (Uddin et al. 2012).

5.7 Conclusion

The chapter summarizes and analyzes the role of phenolics, auxins, and their interactive biological behavior in plants. All related data was compiled in order to discuss tripartite association in which microbes play an important role for developing interactions between auxins and phenolic compounds exhibiting crosstalks in plant rhizosphere. Phenolic compounds, importantly, flavonoids and salicylic acid, are involved in auxin transport and defense signaling, thereby causing morphogenic changes in plants. Additionally, some case studies have been presented in this chapter, in which in vitro culturing of plants was done with supplementation of auxins that trigger some response over the production of phenolic compounds. In conclusion, there must be some advance studies comprising the mechanisms behind the interactive biology between auxins and phenolics to have better understanding about these crosstalks.

References

Abdel-Lateif K, Bogusz D, Hocher V (2012) The role of flavonoids in the establishment of plant roots endosymbioses with arbuscular mycorrhiza fungi, rhizobia and *Frankia* bacteria. Plant Signal Behav 7(6):636–641

Ahmed A, Hasnain S (2010) Auxin-producing Bacillus sp.: Auxin quantification and effect on the growth of *Solanum tuberosum*. Pure Appl Chem 82(1):313–319

Aslam T, Ahmed A (2018) Lead-tolerant bacteria can minimize lead toxicity in plants. RADS J Biol Res Appl Sci 9(1):1–7

Attaran E, Major IT, Cruz JA et al (2014) Temporal dynamics of growth and photosynthesis suppression in response to jasmonate signaling. Plant Physiol 165(3):1302–1314

Bais HP, Weir TL, Perry LG et al (2006) The role of root exudates in rhizosphere interactions with plants and other organisms. Annu Rev Plant Biol 57(1):233–266

Balasundram N, Sundram K, Samman S (2006) Phenolic compounds in plants and Agri-industrial by-products: antioxidant activity, occurrence, and potential uses. Food Chem 99(1):191–203

Baleroni CRS, Ferrarese MLL, Souza NE et al (2000) Lipid accumulation during canola seed germination in response to cinnamic acid derivatives. Biol Plant 43(2):313–316

Baque MA, Hahn EJ, Paek KY (2010) Growth, secondary metabolite production and antioxidant enzyme response of *Morinda citrifolia* adventitious root as affected by auxin and cytokinin. Plant Biotechnol Rep 4(2):109–116

Bari R, Jones JD (2009) Role of plant hormones in plant defence responses. Plant Mol Biol 69 (4):473–488

Benková E, Michniewicz M, Sauer M et al (2003) Local, efflux-dependent auxin gradients as a common module for plant organ formation. Cell 115(5):591–602

Bernasconi P (1996) Effect of synthetic and natural protein tyrosine kinase inhibitors on auxin efflux in zucchini (*Cucurbita pepo*) hypocotyls. Physiol Plant 96(2):205–210

Bertin C, Yang XH, Weston LA (2003) The role of root exudates and allelochemicals in the rhizosphere. Plant Soil 256(1):67–83

Bhattacharya A, Sood P, Citovsky V (2010) The roles of plant phenolics in defence and communication during agrobacterium and rhizobium infection. Mol Plant Pathol 11(5):705–719

Boudet A (2007) Evolution and current status of research in phenolic compounds. Phytochemistry 68(22–24):2722–2735

Bravo L (1998) Polyphenols: chemistry, dietary sources, metabolism, and nutritional significance. Nutr Rev 56(11):317–333

Brown DE, Rashotte AM, Murphy AS et al (2001) Flavonoids act as negative regulators of auxin transport in vivo in *Arabidopsis*. Plant Physiol 126(2):524–535

Buer CS, Sukumar P, Muday GK (2006) Ethylene modulates flavonoid accumulation and gravitropic responses in roots of *Arabidopsis*. Plant Physiol 140(4):1384–1396

Chapman EJ, Estelle M (2009) Mechanism of auxin-regulated gene expression in plants. Annu Rev Genet 43(1):265–285

Cheng Y, Dai X, Zhao Y (2006) Auxin biosynthesis by the YUCCA flavin monooxygenases controls the formation of floral organs and vascular tissues in Arabidopsis. Genes Dev 20 (13):1790–1799

Cheng Y, Dai X, Zhao Y (2007) Auxin synthesized by the YUCCA flavin monooxygenases is essential for embryogenesis and leaf formation in *Arabidopsis*. Plant Cell 19(8):2430–2439

Clé C, Hill LM, Niggeweg R et al (2008) Modulation of chlorogenic acid biosynthesis in *Solanum lycopersicum*; consequences for phenolic accumulation and UV-tolerance. Phytochemistry 69 (11):2149–2156

Dakora FD, Phillips DA (2002) Root exudates as mediators of mineral acquisition in low-nutrient environments. In: Food security in nutrient-stressed environments: exploiting plants' genetic capabilities. Springer, Dordrecht, pp 201–213

Darwin C, Darwin F (1880) Sensitiveness of plants to light: it's transmitted effect. The power of movement in plants. pp 574−592

Davies PJ (ed) (2004) Plant hormones: biosynthesis, signal transduction, action! Springer, Dordrecht

Friml J (2003) Auxin transport—shaping the plant. Curr Opin Plant Biol 6(1):7–12

Fu ZQ, Dong X (2013) Systemic acquired resistance: turning local infection into global defense. Annu Rev Plant Biol 64(1):839–863

George EF (1996) Plant propagation by tissue culture. Edington, Springer Science Business Media

Habib S, Fatima H, Ahmed A (2019) Comparative analysis of pre-germination and post-germination inoculation treatments of *Zea mays* L. to mitigate chromium toxicity in Cr-contaminated soils. Pol J Environ Stud 28(2):597–607

Halvorson JJ, Gonzalez JM, Hagerman AE et al (2009) Sorption of tannin and related phenolic compounds and effects on soluble-N in soil. Soil Biol Biochem 41(9):2002–2010

Harborne JB (1980) Plant phenolics. In: Bell EA, Charlwood BV (eds) Encyclopedia of plant physiology, secondary plant products, new series, Springer Verlag, vol 8. Berlin, Heidelberg, New York, pp 329–402

Harborne JB, Simmonds NW (1964) Natural distribution of the phenolic aglycones. In: Harborne JB (ed) Biochemistry of phenolic compounds. Academic Press, London, pp 77–128

Hartley SE (1999) Are gall insects large rhizobia? Oikos 84(2):333–342

Hartley RD, Harris PJ (1981) Phenolic constituents of the cell walls of dicotyledons. Biochem Syst Ecol 9(2–3):189–203

Hartwig UA, Maxwell CA, Joseph CM et al (1990) Chrysoeriol and luteolin released from alfalfa seeds induce nod genes in *Rhizobium meliloti*. Plant Physiol 92(1):116–122
Hättenschwiler S, Vitousek PM (2000) The role of polyphenols in terrestrial ecosystem nutrient cycling. Trends Ecol Evol 15(6):238–242
Hollman PCH (2001) Evidence for health benefits of plant phenols: local or systemic effects? J Sci Food Agric 81(9):842–852
Huot B, Yao J, Montgomery BL et al (2014) Growth–defense tradeoffs in plants: a balancing act to optimize fitness. Mol Plant 7(8):1267–1287
Jacobs M, Rubery PH (1988) Naturally occurring auxin transport regulators. Science 241 (4863):346–349
Jansen MA, van den Noort RE, Tan MY et al (2001) Phenol-oxidizing peroxidases contribute to the protection of plants from ultraviolet radiation stress. Plant Physiol 126(3):1012–1023
Jones OP, Hatfield SGS (1976) Root initiation in apple shoots cultured in vitro with auxins and phenolic compounds. J Horti Sci 51(4):495–499
Karamat M, Ahmed A (2018) Impact of *Arthrobacter mysorens*, *Kushneria avicenniae*, *Halomonas* spp and *Bacillus* sp on *Helianthus annuus* L for growth enhancement. J Anim Plant Sci 28 (6):1629–1634
Kefeli VI, Kadyrov CS (1971) Natural growth inhibitors, their chemical and physiological properties. Annu Rev Plant Physiol 22(1):185–196
Kefeli VI, Kalevitch MV, Borsari B (2003) Phenolic cycle in plants and environment. J Cell Mol Biol 2(1):13–18
Kraus TEC, Dahlgren RA, Zasoski RJ (2003) Tannins in nutrient dynamics of forest ecosystems – a review. Plant Soil 256(1):41–66
Lattanzio V (2013) Phenolic compounds: introduction. In: Ramawat GK (ed) Natural products: phytochemistry, botany and metabolism of alkaloids, phenolics and terpenes, pp 1543–1580
Lewis DR, Negi S, Sukumar P et al (2011) Ethylene inhibits lateral root development, increases IAA transport and expression of PIN3 and PIN7 auxin efflux carriers. Development 138 (16):3485–3495
Lokerse AS, Weijers D (2009) Auxin enters the matrix—assembly of response machineries for specific outputs. Curr Opin Plant Biol 12(5):520–526
Madhan SSR, Girish R, Karthik N et al (2009) Allelopathic effects of phenolics and terpenoids extracted from *Gmelina arborea* on germination of black gram (*Vigna mungo*) and green gram (*Vigna radiata*). Allelopathy J 23(2):323–332
Mano Y, Nemoto K (2012) The pathway of auxin biosynthesis in plants. J Exp Bot 63 (8):2853–2872
Matveeva TV, Lutova LA, Nester Y (2001) Tumor formation in plants. Russ J Genet 37 (9):993–1001
Mierziak J, Kostyn K, Kulma A (2014) Flavonoids as important molecules of plant interactions with the environment. Molecules 19(10):16240–16265
Murphy A, Taiz L (1999) Naphthylphthalamic acid is enzymatically hydrolyzed at the hypocotyl-root transition zone and other tissues of *Arabidopsis thaliana* seedlings. Plant Physiol Biochem 37(6):413–430
Murphy A, Peer WA, Taiz L (2000) Regulation of auxin transport by aminopeptidases and endogenous flavonoids. Planta 211(3):315–324
Mutka AM, Fawley S, Tsao T et al (2013) Auxin promotes susceptibility to *Pseudomonas syringae* via a mechanism independent of suppression of salicylic acid-mediated defenses. Plant J 74 (5):746–754
Newman MA, Dow JM, Molinaro A et al (2007) Priming, induction and modulation of plant defense responses by bacteria lipopolysaccharides. J Endotoxin Res 13(2):69–84
Nonhebel HM (2015) Tryptophan-independent indole-3-acetic acid synthesis: critical evaluation of the evidence. Plant Physiol 169(2):1001–1005

Normanly J, Cohen JD, Fink GR (1993) Arabidopsis thaliana auxotrophs reveal a tryptophan-independent biosynthetic pathway for indole-3- acetic acid. Proc Natl Acad Sci U S A 90 (21):10355–10359

Ongena M, Jourdan E, Adam A et al (2007) Surfactin and fengycin lipopeptides of Bacillus subtilis as elicitors of induced systemic resistance in plants. Environ Microbiol 9(4):1084–1090

Ouyang J, Shao X, Li J (2000) Indole-3-glycerol phosphate, a branchpoint of indole-3-acetic acid biosynthesis from the tryptophan biosynthetic pathway in *Arabidopsis thaliana*. Plant J 24 (3):327–333

Ozyigit II, Kahraman MV, Ercan O (2007) Relation between explant age, total phenols and regeneration response of tissue cultured cotton (Gossypium hirsutum L). Afr J Biotechnol 6 (1):3–8

Pan P, Woehl E, Dunn MF (1997) Protein architecture, dynamics and allostery in tryptophan synthase channeling. Trends Biochem Sci 22(1):22–27

Park JE, Park JY, Kim YS et al (2007) GH3-mediated auxin homeostasis links growth regulation with stress adaptation response in *Arabidopsis*. J Biol Chem 282(13):10036–10046

Peer WA, Murphy AS (2007) Flavonoids and auxin transport: modulators or regulators? Trends Plant Sci 12(12):556–563

Peer WA, Bandyopadhyay A, Blakeslee JJ et al (2004) Variation in expression and protein localization of the PIN family of auxin efflux facilitator proteins in flavonoid mutants with altered auxin transport in *Arabidopsis thaliana*. Plant Cell 16(7):1898–1911

Perret X, Staehelin C, Broughton WJ (2000) Molecular basis of symbiotic promiscuity. Microbiol Mol Biol Rev 64(1):180–201

Petrášek J, Friml J (2009) Auxin transport routes in plant development. Dev 136(16):2675–2688

Potters G, Pasternak TP, Guisez Y et al (2007) Stress-induced morphogenic responses: growing out of trouble? Trends Plant Sci 12(3):98–105

Potters G, Pasternak TP, Guisez Y et al (2009) Different stresses, similar morphogenic responses: integrating a plethora of pathways. Plant Cell Environ 32(2):158–169

Reddy PM, Rendón-Anaya M, de los Dolores Soto del Rio M et al (2007) Flavonoids as signalling molecules and regulators of root nodule development. Dyn Soil Dyn Plant 1(2):83–94

Robert-Seilaniantz A, MacLean D, Jikumaru Y et al (2011) The microRNA miR393 re-directs secondary metabolite biosynthesis away from camalexin and towards glucosinolates. Plant J 67 (2):218–231

Sánchez-Moreno C (2002) Methods used to evaluate the free radical scavenging activity in foods and biological systems. Food Sci Technol Int 8(3):121–137

Santos-Sánchez NF, Salas-Coronado R, Hernández-Carlos B et al (2019) Shikimic acid pathway in biosynthesis of phenolic compounds. In: Plant physiological aspects of phenolic compounds. IntechOpen, London

Sarkar D, Naik PS (2000) Phloroglucinol enhances growth and rate of axillary shoot proliferation in potato shoot tip cultures *in vitro*. Plant Cell Tissue Organ Cult 60(2):139–149

Schmitz-Hoerner R, Weissenbock G (2003) Contribution of phenolic compounds to the UV-B screening capacity of developing barley primary leaves in relation to DNA damage and repair under elevated UV-B levels. Phytochemistry 64(1):243–255

Schuhegger R, Ihring A, Gantner S (2006) Induction of systemic resistance in tomato by N-acyl-L-homoserine lactone-producing rhizosphere bacteria. Plant Cell Environ 29(5):909–918

Staswick PE, Serban B, Rowe M et al (2005) Characterization of an *Arabidopsis* enzyme family that conjugates amino acids to indole-3-acetic acid. Plant Cell 17(2):616–627

Stepanova AN, Robertson-Hoyt J, Yun J et al (2008) TAA1-mediated auxin biosynthesis is essential for hormone crosstalk and plant development. Cell 133(1):177–191

Tanaka H, Dhonukshe P, Brewer PB et al (2006) Spatiotemporal asymmetric auxin distribution: a means to coordinate plant development. Cell Mol Life Sci 63(23):2738–2754

Tao Y, Ferrer JL, Ljung K et al (2008) Rapid synthesis of auxin via a new tryptophan-dependent pathway is required for shade avoidance in plants. Cell 133(1):164–176

Taylor LP, Grotewold E (2005) Flavonoids as developmental regulators. Curr Opin Plant Biol 8 (3):317–323

Thomas P, Ravindra MB (1999) Shoot tip culture in mango: influence of medium, genotype, explant factors, season and decontamination treatments on phenolic exudation, explant survival and axenic culture establishment. J Hortic Sci 72(5):713–722

Tomás-Barberán FA, Espín JC (2001) Phenolic compounds and related enzymes as determinants of quality in fruits and vegetables. J Sci Food Agri 81(9):853–876

Tran H, Ficke A, Aslimwe T et al (2007) Role of the cyclic lipopeptide massetolide a in biological control of Phytophthora infestans and in colonization of tomato plants by Pseudomonas fluorescens. New Phytol 175(4):731–742

Uddin MR, Li X, Won OJ et al (2012) Herbicidal activity of phenolic compounds from hairy root cultures of Fagopyrum tataricum. Weed Res 52(1):25–33

Vanneste S, Friml J (2009) Auxin: a trigger for change in plant development. Cell 136 (6):1005–1016

Vickery ML, Vickery B (1981) The acetate-mevalonate pathway. In: Secondary plant metabolism. Palgrave, London, pp 112–156

Wang D, Pajerowska-Mukhtar K, Culler AH et al (2007) Salicylic acid inhibits pathogen growth in plants through repression of the auxin signaling pathway. Curr Biol 17(20):1784–1790

Webster G, Jain V, Davey MR et al (1998) The flavonoid naringenin stimulates the intercellular colonization of wheat roots by *Azorhizobium caulinodans*. Plant Cell Environ 21(4):373–383

Weir TL, Park SW, Vivanco JM (2004) Biochemical and physiological mechanisms mediated by allelochemicals. Curr Opin Plant Biol 7(4):472–479

Went F (1926) On growth-accelerating substances in the coleoptile of *Avena sativa*. Proc Kon Akad Wetensch Amsterdam 30:10–19

Whipps JM (2001) Microbial interactions and biocontrol in the rhizosphere. J Exp Bot 52 (supp_1)):487–511

Woodward AW, Bartel B (2005) Auxin: regulation, action, and interaction. Ann Bot 95(5):707–735

Wright AD, Sampson MB, Neuffer MG et al (1991) Indole-3-acetic acid biosynthesis in the mutant maize orange pericarp, a tryptophan auxotroph. Science 254(5034):998–1000

Xuan TD, Shinkichi T, Khanh TD (2005) Biological control of weeds and plant pathogens in paddy rice by exploiting plant allelopathy: an overview. Crop Prot 24(3):197–206

Zhang Y, Goritschnig S, Dong X et al (2003) A gain-of-function mutation in a plant disease resistance gene leads to constitutive activation of downstream signal transduction pathways in suppressor of npr1-1, constitutive 1. Plant Cell 15(11):2636–2646

Zhang Z, Wang M, Li Z et al (2008a) *Arabidopsis* GH3.5 regulates salicylic acid-dependent and both NPR1-dependent and independent defense responses. Plant Signal Behav 3(8):537–542

Zhang R, Wang B, Ouyang J et al (2008b) Arabidopsis indole synthase, a homolog of tryptophan synthase alpha, is an enzyme involved in the Trp-independent indole-containing metabolite biosynthesis. J Integr Plant Biol 50(9):1070–1077

Zhao Y (2010) Auxin biosynthesis and its role in plant development. Annu Rev Plant Biol 61 (1):49–64

Zhao Y (2012) Auxin biosynthesis: a simple two-step pathway converts tryptophan to indole-3-acetic acid in plants. Mol Plant 5(2):334–338

Zipfel C (2008) Pattern-recognition receptors in plant innate immunity. Curr Opin Immunol 20 (1):10–16

Bioavailability and Nutritional Analysis of Flavonoids

6

N. L. Meena, Pooja Verma, Rachna Pande, Manoj Kumar, Anshul Watts, and O. P. Gupta

Abstract

Flavonoids are plant-specific dietary components with their promising health-promoting effects in human cell which has been well proven with plethora of experimental and human clinical studies. They own a variety of biological activities and properties as antioxidant, anti-inflammatory, anti-cancerous, anti-allergic and many more which can modulate cell signaling and gene expression-related disease development. Poor bioavailability of flavonoids is a great concern as it can put a check or even can hinder their health effects. Therefore, efforts to improve their bioavailability with the aim of improving the efficacy of flavonoids are subject of current research in this area. This chapter highlights the overall picture of flavonoids, including their beneficial roles, contributing toward improving the human health.

Keywords

Flavonoids · Antioxidant · Anti-inflammatory · Anti-cancerous · Anti-allergic · Human health

N. L. Meena (✉)
ICAR-National Bureau of Plant Genetic Resources, New Delhi, India

P. Verma · R. Pande
ICAR-Central Institute for Cotton Research, Nagpur, Maharashtra, India

M. Kumar
ICAR-Central Institute for Research on Cotton Technology, Mumbai, Maharashtra, India

A. Watts
ICAR-Indian Institute of Plant Biotechnology, New Delhi, India

O. P. Gupta
ICAR-Indian Institute of Wheat and Barley Research, Karnal, Haryana, India

R. Lone et al. (eds.), *Plant Phenolics in Sustainable Agriculture*,
https://doi.org/10.1007/978-981-15-4890-1_6

6.1 Introduction

Flavonoids are one of the plant-specific secondary metabolites with a substantial impact on human health as a dietary component. These are named secondary compounds because they do not involve themselves in the survival of the plant cell. On the contrary, these secondary metabolites are amazingly diverse, with upwards of 10^5 structures reported so far, which are being synthesized at a substantial rate in the plant system (Sabina Passamonti et al. 2009). Flavonoids are known to be at their best while contributing to pigmentation and flavor. These have been first reported owing their biological activity by Rusznyak and Szent-Gyorgyi in 1936. Initially, they were proposed to be required as vitamins, and thus the term "vitamin P" for flavonoids was suggested, although this was later dismissed (Kuo 1997). In chemical terms, they can be primarily characterized by a 15-carbon skeleton, organized as C6-C3-C6, with different substitutions ending up with different subclasses. Approximately 20% of the carbon fixed by photosynthesis is committed toward structuring and building these flavonoids (Ververidis et al. 2007). With more than 8000 compounds, flavonoids are the most abundant polyphenols present in plants (Harborne and Williams 2000; Corcoran et al. 2012) which along with allied polyphenols are mainly derived from different metabolic pathways: pentose phosphate, shikimate, and phenylpropanoid pathways (Randhir et al. 2004). Flavonoids have a wide range of functions including regulation of growth and developmental processes, responding and aiding the plants during biotic and abiotic environmental stimuli, etc. (Hassan et al. 2015).

Since the human body cannot synthesize these flavonoids as their own, the source to access them is only through plants. Flavonoids are widely distributed in foods and beverages of plant origin, such as fruits, vegetables, tea, cocoa, and wine (Ross and Kasum 2002). A plethora of literature exists regarding their content in various foods. The range of flavonoid content in different food ranges from 10 to 10^4 mg kg^{-1} of fresh weight (Manach et al. 2004; Kyle and Duthie 2006). With ups and down in the research trend on flavonoids, they are still ruling because of their useful biological activities improving the human health, such as antioxidant, anti-inflammatory, and anti-cancerous activities. Though, extensive research over 80 years helped us to understand the complex pathways involved in bioavailability of flavonoids in the human body which are now well understood, improvement in their adequate availability approaching the targets is a dire need.

6.2 Classification of Flavonoids

Chemically, flavonoids are 15-carbon frame consisting of 2 benzene rings (A and B) linked via a heterocyclic pyran ring (C) (Kumar and Pandey 2013). Flavonoids can be subdivided into different groups depending on the carbon of the third ring (C ring) on which the second (B ring) ring is attached and the degree of unsaturation and oxidation of the C ring (Panche et al. 2016). They contribute a larger share to plant phenolics with thousands of different described flavonoids falling under six major subclasses, including flavones, flavonols, flavanones, flavanols, anthocyanidins, and isoflavones (Ross and Kasum 2002).

Flavones have a repeat between positions 2 and 3 of the double bonds, in case of ketone double bond position at 4 of the C ring with their wide distribution in leaves, flowers, and fruits. Second subgroup: Flavonols are categorized in the ketone group of flavonoids, which are considered as primary units of proanthocyanins. These are usually ubiquitous in foods (Beecher 2003) and the chief sources are apples, berries, red grapes (Lotito and Frei 2006; Ratnasooriya et al. 2010), onions, red wine, teas, curly kale, leeks, and broccoli (Archivio et al. 2007). Quercetin, the most common flavonol, is abundant in onions (Slimestad et al. 2007). Flavanones (compounds like hesperitin, naringenin, eriodictyol) are another important class of flavonoids which is normally present in citrus fruits such as oranges, lemons, and grapes. These compounds are known to be associated with several health benefits owing to their free radical scavenging properties (Panche et al. 2016). They are also responsible for the bitter taste of the juice and peel of citrus fruits. Flavanols, also referred as flavan-3-ols, are the derivatives of flavonones and termed so because the hydroxyl group is always bound to position 3 of the C ring. These are one of the most commonly consumed flavonoids (Otaki et al. 2009; Bai et al. 2014; Vogiatzoglou et al. 2015), abundantly found in dark chocolates, green tea, and black tea (Rothwell et al. 2013; Kuhnle et al. 2009). Anthocyanidins are one of the major categories of natural pigments in plants exhibiting a blue, purple, or red color (Archivio et al. 2007; Wojdylo et al. 2008). Cherry, easberry, and strawberry are the main source of anthocyanidins (Hertoget al. 1992; Stewart et al. 2000). Isoflavones are very unique subgroup of flavonoids, though a very huge subgroup but with a limited distribution in the plant kingdom. They are predominantly found in soybeans (Reinli and Block 1996) and some other leguminous plants. A few isoflavonoids have also been reported from microbes (Matthies et al. 2008).

6.3 Bioavailability and Metabolism of Flavonoids

Bioavailability of a compound or substance refers to the proportion of that substance which makes itself available to the systemic circulation unchanged, subsequent to a particular course of administration (Ververidis et al. 2007). As far as, dietary flavonoids are concerned, its administration is clearly the oral one. Initially a lot of work was done on flavonoids but not on its bioavailability, metabolism, and subsequent health benefits to human and animals. Then, it was recognized that an increased consumption of vegetables and fruits had positive impact in improving health directly or indirectly which further reduced the risk of numerous chronic diseases (Boeing et al. 2012) and protected the human body against cancer and cardiovascular diseases (Ness and Powles 1997; Research WCRF=AICR 1997; Ness et al. 1999). This laid down a hypothesis that vegetables and fruits may contain some bioactive compounds having a protective effect, responsible for these health benefits. Among these biologically active compounds, flavonoids form a large group of plant-based natural antioxidants which are high in diet of plant foods (Kuhnau 1976). The protective effect might be ascribed to their free radical scavenging and antioxidant activities (Middleton et al. 2000; Lin and Weng 2006), antimicrobial (Proestos et al. 2006; Martini et al. 2004), anticarcinogenic (Deep and Agarwal 2007; Ren et al. 2003), anti-inflammatory (Rao et al. 2005; Narayana et al. 2001), and vasodilatory effects (Calderone et al. 2004). As defined by the US Food and Drug Administration (FDA), bioavailability is "the rate and extent to which the active ingredient or active moiety is absorbed from a drug product and becomes available at the site of action" (FDA 2014). The same principle can be applied to bioavailability of flavonoids present in food for that, rate of its absorption and availability at the site of action is of supreme importance to make it "bioavailable" and for being effective within biological systems.

Absorption and bioavailability of flavonoids have been studied and investigated in human very well (Hollman and Katan 1999). Investigations on animals can be helpful, but since there are differences in their basic endogenous metabolism between human and animal and gut microflora, the same data can't be applied to human studies (Borges et al. 2016). Initially, it was believed that absorption of flavonoids from the food diet is almost negligible, as majority of the food flavonoids are bound to glycosides which on the other hand, can't cross the cell membrane being hydrophilic. Only aglycones were expected to pass freely into the blood stream from the gut wall, because of the absence of any enzyme which is secreted in the gut and able to cleave the glycosidic bonds (Kuhnau 1976). Over the time, with the advancement in research, however, it has been demonstrated that the bioavailability of flavonoids is much higher than previously thought.

Bioavailability of flavonoids depends on the degree of its absorption and the way it has been metabolized inside human body. To be absorbed, they must be released from plant foods by chewing, acted upon by digestive enzymes in the gastrointestinal tract, and then by the microorganisms of the colon. Further, the metabolism of flavonoids involves two phase process including two compartments. The first compartment consists of small intestine, liver, and kidneys, and the second one is colon (Hollman 2004). Majority of the flavonoids are present in the diet as glycosides, except catechins which are too hydrophilic for absorption in the small intestine through passive diffusion. Therefore, they need to undergo deglycosylation prior to be absorbed (Depeint et al. 2002; Cermak et al. 2004; Arts et al. 2004) which is usually performed by hydrolyzing the sugar moiety using intracellular cytoplasmic β-glucosidase (Cermak et al. 2004; Williamson et al. 2000; Day et al. 1998). Studies on human have revealed three different such β-glucosidases so far; a broad specificity cytosolic β-glucosidase, lactase-phlorizin hydrolase (LPH), and glucocerebrosidase (CBG) (Day et al. 1998; Mackey et al. 2002). Different dietary flavonoids differ in their rates of absorption and bioavailability. Isoflavones are known to be the best absorbed dietary flavonoids among all. Flavanols, flavanones, and flavonol glycosides fall under intermediate category, whereas proanthocyanidins, flavanol gallates, and anthocyanins are the poorly absorbed. However, the absorption of dietary flavonoids may differ due to differences in the food matrix in which they are consumed (Viskupicová et al. 2008). Still they are not available in the concentration in which they are required. So, the leading concern is about identifying and resolving the causes of poor bioavailabilities. Efforts have been made in this area which clarify to an extent that the first-pass metabolism involving phase II conjugation, i.e., glucuronidation and/or sulfonation of flavonoids is the principal cause of their poor bioavailabilities (Wu et al. 2011).

6.4 Factors Affecting Flavonoids and Its Bioavailability

Environmental factors, viz., temperature, soil moisture, water status, light, and nitrogen, affect the biosynthesis of flavonoids and its bioaccumulation (Christie et al. 1994; Kubasek et al. 1998; Castellarin et al. 2007; Azuma et al. 2012; Wang et al. 2015; Hernández et al. 2004; Albert et al. 2009; Olsen et al. 2009; Steyn et al. 2009). Flavonoid biosynthesis and its degradation are sensitive to temperature (Olsen et al. 2009). Low temperature and soil moisture favors the total flavonoids content in Ginkgo leaves (Wang et al. 2015). In maize lower temperature changed the anthocyanin content (Christie et al. 1994) and elevated temperature reduced total content of quercetin, proanthocyanidin, and anthocyanin in *Vitis vinifera* (Tarara et al. 2008; Yamamoto et al. 2010). Water stress had no specific trend on biosynthesis of flavonoids (Tattini et al. 2004) as anthocyanin increased under water stress but limited effect was observed on proanthocyanidin and flavonol (Castellarin et al. 2007), but in *Cistus clusii* leaves α-tocopherol and ascorbic acid concentration was increased under drought stress (Hernández et al. 2004).

In humans, bioavailability of dietary flavonoids was mainly found to be affected by Phase -II metabolism (Manach et al. 2005). In vivo in comparison to parent compound flavonoids show reduced bioactivity. Numerous factors affect the bioavailability of dietary flavonoids viz., molecular weight (Scalbert et al. 2002), glycosylation (Hollman et al. 1997, 1999), metabolic conversion (Williamson and Manach 2005; Ishizawa et al. 2011), interaction with colonic microflora, etc. (Spencer et al. 2001)

6.5 Efforts to Improve Bioavailability

Flavonoids have promising effects which can be hinder because of its low bioavailability. Practical applications of flavonoids are limited as flavonoids are low soluble and stable in lipophilic media. Improvement in biological and physicochemical properties of natural compounds after selective modification improved them as promising for industrial purposes. Some of the efforts adopted to improve the bioavailability of flavonoids are discussed in preceding part. Recently in order to eliminate this limitation its enzymatic acylation with fatty acids has been introduced. In plants, biosynthesis of flavonoids the last step is ended with acylation. Acylation is primarily the transfer of either aromatic or aliphatic acyl group from a CoA-donor molecule to hydroxyl residues catalyzed by various acyltransferases (Davies and Schwinn 2006). It improves the bioavailability of flavonoids by altering physicochemical and biological characteristics of maternal compounds (Viskupicova et al. 2009; Ishihara and Nakajima 2003).

Some of the acylating agents present in flavonols, flavones, and anthocyanins including aliphatic and aromatic acids are acetic, malonic, lactic, vinylpropionic, succinic butyric, isobutyric, ferulic, 3-methylbutyric, isoferulic, crotonic, sinapic, n-butanoic, methylsinapic, benzoic, p-hydroxybenzoic, gallic, cinnamic, p-coumaric, caffeic, isovaleric, quinic, tiglic, malic, and tartaric (Williams 2006; Andersen and Jordheim 2006). Most often coumaric, ferulic, malonic, or caffeic acid caused acylation (Bloor and Abrahams 2002; Fujiwara et al. 1998). Enzymatic acylation was influenced by several key factor, viz., type of enzyme, enzyme concentration, the structure and concentration of the substrates (acyl donor, acyl acceptor, and their ratio), nature of reaction and reaction medium, water content of medium, and reaction temperature. Glycosylation, methylation, and acylation are the main factors responsible for huge diversity of flavonoid structures with modifications of its basic skeleton (Schijlen et al. 2004). More than 65% acylation are reported among anthocynins (Andersen and Jordheim 2006).

Remarkable stability to pH changes, heat and exposure to light (Giusti and Wrolstad 2003), and bioavailability (Moussou et al. 2007) to anthocyanin molecules (complex patterns of glycosylation and acylation) has been developed due to discovery of anthocyanin colorants. Malonylation of anthocyanins (aliphatic acylation) stabilize structure, enhance solubility in water, and inhibit enzymatic degradation of glycosides (Nakayama et al. 2003). Flavonoids acylated with aromatic carboxylic acids (specific) improve resistance to light and thermostability (Delazar et al. 2005; Ishihara and Nakajima 2003; Alluis and Dangles 1999; Jungblut et al. 1995). For synthesis of modified flavonoids, its esterification catalyzed by lipase in organic media is a well-studied and well-standardized technique (Chebil et al. 2006, 2007). Esterification of astragalin with p-coumaric acid increases anti-inflammatory activity (Harborne and Williams 2000). Flavonoids are low soluble and stable in lipophilic media during processing and storage (Viskupicova et al. 2009). Acylation modulate the physiological characteristics of modified flavonoids by shifting their interaction with cellular targets, reactivity, stability, and solubility (Ferrer et al. 2008; Sakai et al. 1994).

Several efforts have been made to maintain or increase the bioactivity of flavonoids in vivo, viz., novel delivery systems (Zhang et al. 2011) improving metabolic stability (Walle 2007; Cao et al. 2013) and use of absorption enhancers for improving the intestinal absorption (Shen et al. 2011) and changing the site of absorption (Nielsen et al. 2006).

6.6 Application of Flavonoids

Several researchers have tried to exploit the flavonoids in different applied aspects. Specific flavonoids work as attractants and enhance pollination in flowers and dispersal of seeds impart resistance in plants against drought, ultraviolet (UV) light, cold temperature, and wounding (Bohm 1998; Winkel-Shirley 2002;

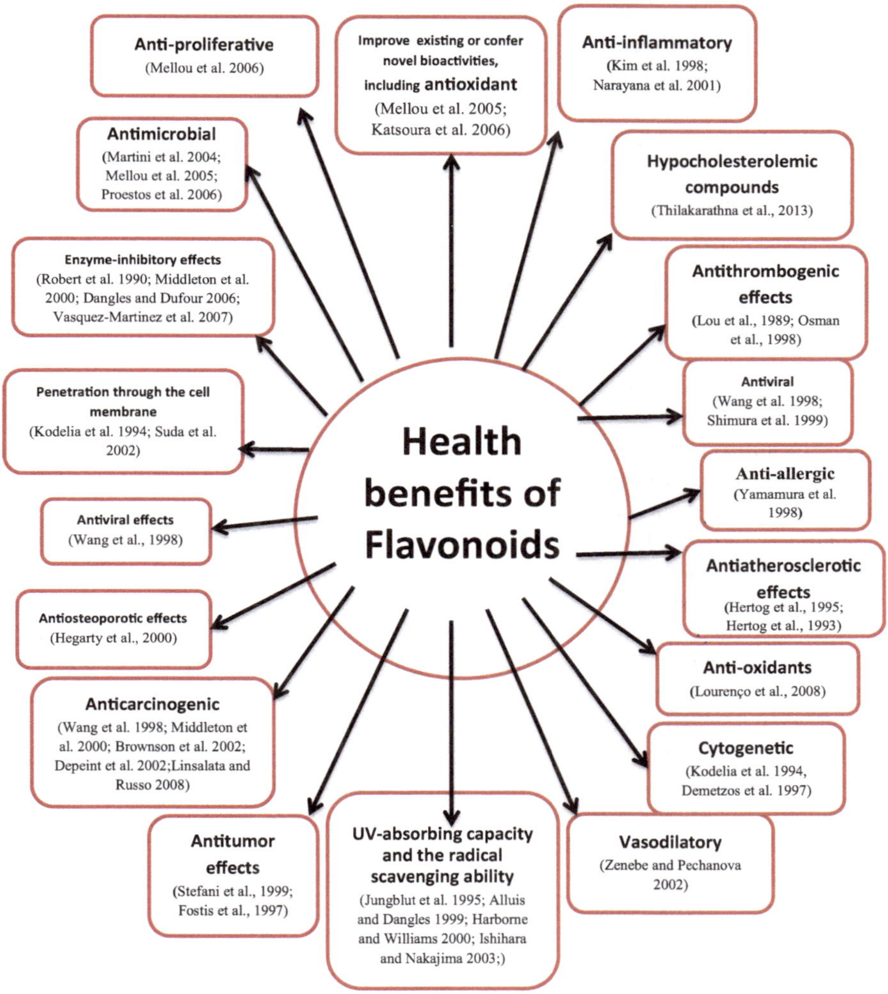

Fig. 6.1 Health benefits of flavonoids

Viskupicova et al. 2009), but the most preferred area is the exploration of physiological properties of flavonoids as biomedicine for improving status of human health (Harborne and Williams 2000; Thilakarathna and Rupasinghe 2013) which is generally attributed by their free radical scavenging and antioxidant activity (Frankel et al. 1993; Rice-Evans et al. 1996; Lin and Weng 2006). Indirectly or directly flavonoids exert promising beneficial effects on health (Fig. 6.1).

Flavonoids are attaining a growing concentration and interest as their use in treatment of various disorders. Its acylated derivatives have promising role in fields of therapy, cosmetics, and dietetic (Nakayama et al. 2003). Some acylated flavonoids act as attractants for pollinator, stimulants for oviposition, phytoalexins (Iwashina 2003), and insect antifeedants (Harborne and Williams 1998) and play crucial role in plant insect interaction. Use of esterified flavonoids in pharmaceutical, dermopharmaceutical, cosmetic, nutritional or agri-foodstuff compositions has been patented also (Fukami et al. 2007; Moussou et al. 2007; Ghoul et al. 2006; Moussou et al. 2004; Bok et al. 2001; Perrier et al. 2001; Otto et al. 2001; Nicolosi et al. 1999; Sakai et al. 1994).

6.6.1 Enhance the Food Solubility and Stability

Acetylated flavonoid improves the solubility and stability of fat-based food as flavonoids are low soluble and less stable in lipophilic media. They also modify the sensory properties which is not desirable. Food contains acetylated flavonoids when consumed may cause changes in their availability and activity and can prevent the disease (Viskupicova et al. 2009). Flavonoid-containing food items generally have some astringent taste. Sensory properties of food can be modified using acylation and glycosylation of flavonoids (Degenhardt et al. 2007). Highly reactive unsaturated fatty acids cause detrimental damages in food. These highly reactive unsaturated fatty acids with free radicals can be stabilized by enzymatic synthesis of flavonoids with unsaturated fatty acids (Mellou et al. 2006). Use of acylated anthocyanins can be used as food colorants which can replace the synthetic additives from food industry (Giusti and Wrolstad 2003; Fox 2000; Asen et al. 1979). Addition of acylated flavonoids also imparts desirable colors with stability in wide range of pH. Best example is stable food colorant isolated from the Heavenly Blue morning glory (*Ipomoea tricolor*) which produced color range from purplish-red to blue in food products at pH values ranging from 2.0 to 8.0 (Asen et al. 1979). Another important example belongs to purple sunflower hulls (anthocyanins acylated with chlorogenic acid) which impart a stable, ruby red natural color in food products, cosmetics, pharmaceuticals, and other materials (Fox 2000). Radishes, red potatoes, red cabbage, black carrots, and purple, sweet potatoes are some examples which can act as a supplier of acylated anthocyanin (Giusti and Wrolstad 2003).

6.6.2 Protective Characteristics of Flavonoids

Disease related with elevated lipid level in blood, e.g., arteriosclerosis, hyperlipidemia, angina pectoris, hepatic, and stroke can be treated and prevented using novel acylated flavanone. Acylated flavonone inhibits acylcolicholesterol acyl transferase activity and the HMG-CoA reductase activity and exert no toxicity or mitogenicity in mice (Bok et al. 2001). Some specific acetylated esters cause cytotoxicity against 4 leukemic cell lines, viz., HUT78, MOLT3, DAUDI, and HL60 which is absent in parent compound (Demetzos et al. 1997). Sugarcane derived flavones shows antiproliferative activity against many human cancer cell lines (Duarte-Almeida et al. 2007). Esterified flavonoids derivatives with polyunsaturated fatty acids are antitumor and antiangiogenic (Mellou et al. 2006). Catechins (fatty acid ester derivatives) have anti-tumorigenesis and antibacterial activity (Fukami et al. 2007). Acylated quercetagetin glycosides with caffeic and p-coumaric acid show high radical scavenging activity (Parejo et al. 2005).

All the cosmetics, especially dermopharmaceutical contains fatty phase which can be easily oxidized at room temperature. This oxidation make the product unusable as it change the properties of original product (Viskupicova et al. 2009). It is clear that antioxidant agent can protect these cosmetic products (N'guyen 1995). Thus, flavonoids can be added as antioxidant to the cosmetic product as they have skin cleansing and protective properties, very effective against skin discoloration and aging (Ghoul et al. 2006). Esterified flavonoids with omega substitute protect the skin from UV radiation. Protection from UV radiation imparts the protection against sun burn, photo-aging, and skin wrinkles (Moussou et al. 2007).

6.6.3 Nutritional Analysis of Flavonoids

Flavonoids are functionally active polyphenolic compound extensively dispersed in plants (Panche et al. 2016; Kumar and Pandey 2013). The flavonoids have raised substantial interestedness recently due to their potential favorable effects on human health. These compounds found to have antiviral, anti-allergic, antiplatelet, anti-inflammatory, antitumor, and also antioxidant functionality (Kumar et al. 2018; Kumar et al. 2019; Ginwala et al. 2019; David et al. 2016; Xiao et al. 2011; Mouradov and Spangenberg 2014; Lin et al. 2018; Tapas et al. 2008; Ballard and Maróstica 2019). The capability of flavonoids to behave as antioxidants relies upon their structure. The location of hydroxyl groups and other characters in the chemical structure of flavonoids are decisive for antioxidant and free radical scavenging properties (Porras et al. 2017; Treml and Šmejkal 2016). Quercetin is an example of the most abundant dietary flavonol that act as a most influential antioxidant because it has all the rightly oriented structural characteristics for free radical scavenging property (Treml and Šmejkal 2016).

Flavonoids also contribute to defense against diseases by devoting along with antioxidant enzymes and vitamins and to the total antioxidant defense system of human body (Van De Wier et al. 2015). The share of these compounds to the antioxidant defense setup may be extraordinary seeing that the overall daily consumption of flavonoids can vary from 50 to 800 mg. This consumption is more in comparison to the average daily consumption of alternative dietary antioxidants like vitamin E (7–10 mg), vitamin C (70 mg), or carotenoids (2–3 mg) (Kumar 2016). Flavonoid intake relies upon the consumption of fruits, vegetables, and certain beverages, such as red wine, beer, and tea (Godos et al. 2017; Nettleton et al. 2006;

Chiva-Blanch and Badimon 2017). The high intake of tea and wine may be most important for total flavonoid intake in certain groups of human beings.

6.6.4 Flavonoids Content in Fruit and Vegetables

It has been found that flavonoids have six major classes of the flavonoids, namely, falvones, flavonols, flavanones, flavan-3-ones, anthocyanins, and isoflavonoids (Tsanova-Savova et al. 2018). Various nutritional programs consider food composition as important information. The nutritional analysis of the fruits and vegetable by analyzing the flavonoid content will help the researchers and food analyst to establish the interrelationship between foods, nutrients, and health. Kühnau (1976), has done the first assessment of flavonoids, and it was found that dietary intake of flavonols and flavone-aglycones was 115 mg/day. Sampson et al. (2000) established an average daily intake of flavonoids of 20–22 mg/day. Flavonoid profile of fruit, vegetable, and other foods are analyzed and characterized using well-established techniques such as high-performance liquid chromatography, fluorescence detection and UV-VIS detection with diode-array detection (Tsanova-Savova and Ribarova 2013; Tsanova-Savova et al. 2005).

The first comprehensive data on the flavonoid nutritional profile in various fruits and vegetables was published in 2003 and later on revised in 2014 and 2018 by USDA. This data reflects the average of flavonoid content based on the previously published work by various researchers. Table 6.1 represents the flavonoid content of fruits and vegetables in mg/100 g product (USDA database 2018). Flavonols are the most omnipresent flavonoids in foodstuff, and the prime members are quercetin and kaempferol. They normally exist at low concentrations of 15–30 mg/kg fresh wt. The richest points of supply are onions, leeks, curly kale, blueberries, broccoli, tea, red wine, and leafy vegetables such as cabbage and lettuce (Hyun et al. 2016). In human foods, flavanones exist in tomatoes and specific aromatic plants such as mint, but they are existing in more amounts in citrus fruit. The main aglycones are hesperetin in oranges, naringenin in grapefruit, and eriodictyol in lemons. Flavanones are normally glycosylated by a disaccharide at seventh position: either a neohesperidose, which imparts a bitter taste or a rutinose, which don't have flavor. Orange juice contains 15–85 mg narirutin/L and 200–600 mg hesperidin/L and a glass of orange juice may contain 40–140 mg flavanone glycosides (Assi et al. 2015). Isoflavones are flavonoids with structural similarities to estrogens. Albeit they are not belongs to steroids, they contain hydroxyl groups at 7 and 4′ positions in a configuration similar to that of the hydroxyl group in the estradiol. Isoflavones are found mainly in leguminous plants. Flavanols present in two forms, i.e., the monomer (catechins) and the polymer (proanthocyanidins). Catechins are found in fruit apricots, red wine and green tea, and chocolate are the richest sources. Catechin and epicatechin are the major flavanols in fruit, whereas gallocatechin, epigallocatechin gallate, and epigallocatechin exist in certain seeds of in grapes, leguminous plants, and more importantly in tea (Pascual-Teresa et al. 2010).

Table 6.1 Flavonoid content of selected fruits and vegetables

Sr. No.	Source	Class	Type of flavonoid	Flavonoid content (units = mg/ 100 g)
1.	Apricots, (*Prunus armeniaca*)	Flavan-3-ols	(−)-Epicatechin	4.74
			(+)-Catechin	3.67
2.	Avocados, raw, all commercial varieties (*Persea americana*)	Anthocyanidins	Cyanidin	0.33
		Flavan-3-ols	(−)-Epicatechin	0.37
			(−)-Epigallocatechin 3-gallate	0.15
3.	Bayberries, raw	Flavonols	Myricetin	3.65
			Quercetin	4.36
4.	Bilberry, raw	Anthocyanidins	Cyanidin	85.26
			Delphinidin	97.59
			Malvidin	39.22
			Peonidin	20.45
			Petunidin	42.69
		Flavonols	Myricetin	1.09
			Quercetin	3.04
5.	Chokeberry, raw	Anthocyanidins	Cyanidin	344.07
			Delphinidin	0.65
6.	Figs, raw (*Ficus carica*)	Anthocyanidins	Cyanidin	0.50
			Pelargonidin	0.01
		Flavan-3-ols	(−)-Epicatechin	0.50
			(+)-Catechin	1.59
		Flavonols	Quercetin	5.47
7.	Gooseberries, raw (*Ribes* spp.)	Anthocyanidins	Cyanidin	8.73
			Delphinidin	0.01
			Peonidin	0.77
8.	Grapes, black (*Vitis vinifera*)	Flavan-3-ols	(−)-Epicatechin	8.68
			(−)-Epicatechin 3-gallate	2.81
			(+)-Catechin	10.14
		Flavonols	Kaempferol	0.09
			Myricetin	0.22
			Quercetin	2.08
9.	Guava, white-fleshed	Flavonols	Quercetin	1.20
10.	Jabuticaba (Brazilian grape), raw (*Myrciaria jaboticaba*)	Flavonols	Quercetin	1.10
11.	Kiwifruit, green, raw (*Actinidia deliciosa*)	Flavan-3-ols	(−)-Epicatechin	0.27
			(−)-Epicatechin 3-gallate	0.01

(continued)

Table 6.1 (continued)

Sr. No.	Source	Class	Type of flavonoid	Flavonoid content (units = mg/ 100 g)
			(−)-Epigallocatechin 3-gallate	0.09
		Flavones	Luteolin	0.74
		Flavonols	Kaempferol	1.03
			Quercetin	0.04
12.	Papayas, raw (*Carica papaya*)	Flavones	Apigenin	0.01
			Luteolin	0.02
		Flavonols	Kaempferol	0.01
			Myricetin	0.02
13.	Pomegranates, raw (*Punica granatum*)	Flavan-3-ols	(−)-Epicatechin	0.08
			(−)-Epigallocatechin	0.16
			(+)-Catechin	0.40
			(+)-Gallocatechin	0.17
14.	Artichokes, (globe or French), raw (*Cynara scolymus*)	Flavanones	Naringenin	12.50
		Flavones	Apigenin	7.48
			Luteolin	2.30
15.	Asparagus, raw (*Asparagus officinalis*)	Flavonols	Isorhamnetin	5.70
			Kaempferol	1.39
			Quercetin	13.98
16.	Beets, raw (*Beta vulgaris*)	Flavones	Luteolin	0.37
17.	Broccoli, raw (*Brassica oleracea* var. *italica*)	Flavones	Luteolin	0.80
		Flavonols	Kaempferol	7.84
			Myricetin	0.06
			Quercetin	3.26
18.	Carrots, raw (*Daucus carota*)	Flavones	Luteolin	0.11
		Flavonols	Kaempferol	0.24
			Myricetin	0.04
			Quercetin	0.21
19.	Drumstick (horseradish tree) leaves, raw (*Moringa oleifera*)	Flavonols	Isorhamnetin	0.44
			Kaempferol	5.95
			Quercetin	16.65
20.	Eggplant, raw (*Solanum melongena*)	Anthocyanidins	Delphinidin	85.69
		Flavonols	Quercetin	0.04
21.	Kale, raw (Brassica oleracea) (Acephala group)	Flavonols	Isorhamnetin	23.60
			Kaempferol	46.80
			Quercetin	22.58
22.	Okra, raw (*Abelmoschus esculentus*)	Flavonols	Quercetin	20.97

(continued)

Table 6.1 (continued)

Sr. No.	Source	Class	Type of flavonoid	Flavonoid content (units = mg/ 100 g)
23.	Onions, raw (Allium cepa)	Flavones	Apigenin	0.01
			Luteolin	0.02
		Flavonols	Isorhamnetin	5.01
			Kaempferol	0.65
			Myricetin	0.03
			Quercetin	20.30
24.	Pumpkin, raw (*Cucurbita* spp.)	Flavones	Luteolin	1.63
25.	Spinach, raw (*Spinacia oleracea*)	Flavones	Luteolin	0.74
		Flavonols	Kaempferol	6.38
			Myricetin	0.35
			Quercetin	3.97

Adapted and modified from USDA Database, 2018. http://www.ars.usda.gov/nutrientdata/flav

6.7 Current Trends on Flavonoids

Time from eighteenth century to 1940s is considered as the age of discovery of flavonoids and its biological activity. By the end of this time scientist were able to elucidate positive effect of flavonoids on human health. From 1950s to 1980s the research in the field of these compounds was limited. Certain pharmaceutical companies made flavonoid-based products, but information about their bioavailability was limited, and very limited attention was paid toward nutritional values of flavonoid-containing foods (Perez-Vizcaino and Fraga 2018). After 1990s, the research in the field of flavonoids has increased, and it can establish by increasing number research publications in the field of flavonoids. The number of papers published per year on flavonoids has risen from 740 in 1991 to more than 9000 in 2015 (Perez-Vizcaino and Fraga 2018). The increased research may be the result of flavonoids has general belief in the population that fruits and vegetable are healthier source then those based on animal products. It has been noticed from the review of literature that maximum research in the field of flavonoid is going in the area of Chemistry, Pharmacology, Food Science, Biochemistry, and Plant Sciences. Oncology, Endocrinology, Neurology, and Cardiology are the other leading areas which directly deals flavonoid research with the human health (Perez-Vizcaino and Fraga 2018).

Over last few years, researchers have become increasingly involved in the potential for distinct dietary flavonoids to illustrate some of the health benefits related to diets rich in fruit and vegetable. These health benefits are being used to encourage the intake of flavonoid-rich foods, dietary supplements, and beverages. These health benefits are mostly related anti-inflammatory effects (e.g., hesperidin, luteolin, and quercetin), antibacterial activity (e.g., apigenin, galangin, flavone and flavonol glycosides, isoflavones, flavanones, and chalcones), antiviral activity (e.g., quercetin, hesperetin, and naringin) and anti-cancer activity of flavonoids (Karak 2019). Hence, researchers are more focused on study of these compounds on human health.

6.8 Conclusion

Flavonoids play a potential role in the prevention of various illness and diseases. Fruits and vegetables are natural source of the flavonoids and considered healthier than animal sources. Due to the beneficial effects of the flavonoids and demand in nutraceuticals and food industry, research has been focused on utilization and biological activity of flavonoids in human system. Further, flavonoids can also be utilized as a therapeutic agent against various diseases without causing ill effects/ side effects.

References

Albert NW, Lewis DH, Zhang H, Irving LJ, Jameson PE et al (2009) Light induced vegetative anthocyanin pigmentation in Petunia. J Exp Bot 60:2191–2202

Alluis B, Dangles O (1999) Acylated flavone glucosides: synthesis, conformational investigation, and complexation properties. Helv Chim Acta 82:2201–2212

Andersen OM, Jordheim M (2006) The anthocyanins. In: Andersen ØM, Markham KR (eds) Flavonoids: chemistry, biochemistry and applications, 1st edn. CRC Press, New York, pp 471–551

Archivio DM, Filesi C, Benedetto DR, Gargiulo R, Giovannini C, Masella R (2007) Polyphenols, dietary sources and bioavailability. Ann Ist Super Sanita 43:348–361

Arts ICW, Sesink ALA, Faassen-Peters M, Hollman PCH (2004) The type of sugar moiety is a major determinant of the small intestinal uptake and subsequent biliary excretion of dietary quercetin glycosides. Br J Nutr 91:841–847

Asen S, Stewart RN, Norris KH (1979) Stable foods and beverages containing the anthocyanin, peonidin-3-(dicaffeylsophoroside)-5-glucosid. Patent US 4172902

Assi RA, Darwis Y, Abdulbaqi IM, Khan AA, Vuanghao L, Laghari MH (2015) *Morinda citrifolia* (Noni): a comprehensive review on its industrial uses, pharmacological activities, and clinical trials. Arabian J Chem 10:691. https://doi.org/10.1016/j.arabjc.2015.06.018

Azuma A, Yakushiji H, Koshita Y, Kobayashi S (2012) Flavonoid biosynthesis related genes in grape skin is differentially regulated by temperature and light conditions. Planta 236:1067–1080

Bai W, Wang C, Ren C (2014) Intakes of total and individual flavonoids by US adults. Int J Food Sci Nutr 165(1):9–20

Ballard CR, Maróstica MR (2019) Health benefits of flavonoids. Bioact Compd:185–201. https://doi.org/10.1016/b978-0-12-814774-0.00010-4

Beecher GR (2003) Overview of dietary flavonoids: nomenclature, occurrence and intake. J Nutr 3:3248S–3254S

Bloor SJ, Abrahams S (2002) The structure of the major anthocyanin in Arabidopsis thaliana. Phytochemistry 59:343–346

Boeing H, Bechthold A, Bub A (2012) Critical review: vegetables and fruit in the prevention of chronic diseases. Eur J Nutr 51:637–663

Bohm B (1998) Introduction of flavonoids. Introduction to flavonoids:1–503

Bok SH, Jeong TS, Lee SK, Kim JR, Moon SS, Choi MS, Hyun BH, Lee CH, Choi YK (2001) Flavanone derivatives and composition for preventing or treating blood lipid level-related diseases comprising same. Patent US 6455577

Borges G, van der Hooft JJJ, Crozier A (2016) A comprehensive evaluation of the [2–14C](−)-epicatechin metabolome in rats. Free Radic Biol Med 99:128–138. https://doi.org/10.1016/j/freeradbiomed.2016.08.00

Brownson DM, Azios NG, Fuqua BK, Dharmawardhane SF, Mabry TJ (2002) Flavonoid effects relevant to cancer. J Nutr 132:3482–3489

Calderone V, Chericoni S, Martinelli C, Testai L, Nardi A, Morelli I et al (2004) Vasorelaxing effects of flavonoids: investigation on the possible involvement of potassium channels. Naunyn-Schmiedeberg's. Arch Pharmacol 370:290–298

Cao H, Jing X, Wu D, Shi Y (2013) Methylation of genistein and kaempferol improves their affinities for proteins. Int J Food Sci Nutr 64:437. https://doi.org/10.3109/09637486.2012.759186

Castellarin SD, Matthews MA, Di Gaspero G, Gambetta GA (2007) Water deficits accelerate ripening and induce changes in gene expression regulating flavonoid biosynthesis in grape berries. Planta 227:101–112

Cermak R, Landgraf S, Wolffram S (2004) Quercetin glucosides inhibit glucose uptake into brush-bordermembrane vesicles of porcine jejunum. Br J Nutr 91:849–855

Chebil L, Humeau C, Falcimaigne A, Engasser J-M, Ghoul M (2006) Enzymatic acylation of flavonoids. Process Biochem 41:2237–2251

Chebil L, Anthoni J, Humeau C, Gerardin C, Engasser J-M, Ghoul M (2007) Enzymatic acylation of flavonoids: effect of the nature of the substrate, origin of lipase, and operating conditions on conversion yield and regioselectivity. J Agric Food Chem 55:9496–9502

Chiva-Blanch G, Badimon L (2017) Effects of polyphenol intake on metabolic syndrome: current evidences from human trials. Oxidative Med Cell Longev 2017:1–18. https://doi.org/10.1155/2017/5812401

Christie PJ, Alfenito MR, Walbot V (1994) Impact of low-temperature stress on general phenylpropanoid and anthocyanin pathways: enhancement of transcript abundance and anthocyanin pigmentation in maize seedlings. Planta 194:541–549

Corcoran MP, McKay DL, Blumberg JB (2012) Flavonoid basics: chemistry, sources, mechanisms of action, and safety. J Nutr Gerontol Geriatr 31:176–189

Dangles O, Dufour C (2006) Flavonoid-protein interactions. In: Andersen ØM, Markham KR (eds) Flavonoids: chemistry, biochemistry and applications, 1st edn. CRC Press, New York, pp 443–469

David AVA, Arulmoli R, Parasuraman S (2016) Overviews of biological importance of quercetin: a bioactive flavonoid. Pharmacogn Rev 10:84–89

Davies KM, Schwinn KE (2006) Molecular biology and biotechnology of flavonoid biosynthesis. In: Andersen ØM, Markham KR (eds) Flavonoids: chemistry, biochemistry and applications, 1st edn. CRC Press, New York, pp 143–218

Day AJ, Dupont MS, Ridley S, Rhodes M, Rhodes MJC, Morgan MRA, Williamson G (1998) Deglycosylation of flavonoid and isoflavonoid glycosides by human small intestine and liver β-glucosidase activity. FEBS Lett 436:71–75

Deep G, Agarwal R (2007) Chemopreventive efficacy of silymarin in skin and prostate cancer. Integr Cancer Ther 6:130–145

Degenhardt A, Ullrich F, Hofmann T, Stark T (2007) Flavonoid sugar addition products, method for manufacture and use thereof. Patent US 20070269570

Delazar A, Celik S, Gokturk RS, Nahar L, Sarker SD (2005) Two acylated flavonoid glycosides from Stachys bombycina, and their free radical scavenging activity. Pharmazie 60:878–880

Demetzos C, Magiatis P, Typas MA, Dimas K, Sotiriadou R, Perez S, Kokkinopoulos D (1997) Biotransformation of the flavonoid tiliroside to 7-methylether tiliroside: bioactivity of this metabolite and of its acetylated derivative. Cell Mol Life Sci 53:587–592

Depeint F, Gee JM, Williamson G, Johnson IT (2002) Evidence for consistent patterns between flavonoid structures and cellular activities. Proc Nutr Soc 61:97–103
Duarte-Almeida JM, Negri G, Salatino A, de Carvalho JE, Lajolo FM (2007) Antiproliferative and antioxidant activities of a tricin acylated glycoside from sugarcane (Saccharum officinarum) juice. Phytochemistry 68:1165–1171
Ferrer JL, Austin MB, Stewart C, Noel JP (2008) Structure and function of enzymes involved in the biosynthesis of phenylpropanoids. Plant Physiol Biochem 46:356–370
Food US (2014) Drug Administration Code of Federal Regulations Title 21. Department of Health and Human Services, ed. 21CFR20157. Washington: US Food and Drug Administration
Fox GJ (2000) Natural red sunflower anthocyanin colorant with naturally stabilized color qualities, and the process of making. Patent US 6132791
Frankel EN, Kanner J, German JB, Parks E, Kinsella JE (1993) Inhibition of oxidation of human low-density lipoprotein by phenolic substances in red wine. Lancet 341:454–457
Fujiwara H, Tanaka Y, Yonekura-Sakakibara K, Fukuchi-Mizutani M, Nakao M, Fukui Y, Yamaguchi M, Ashikari T, Kusumi T (1998) cDNA cloning, gene expression and subcellular localization of anthocyanin 5- aromatic acyltransferase from Gentiana triflora. Plant J 16:421–431
Fukami H, Nakao M, Namikawa K, Maeda M (2007) Esterified catechin, process for producing the same, food and drink or cosmetic containing the same. Patent EP 1849779
Ghoul M, Engasser J-M, Moussou P, Pauly G, Ardhaoui M, Falcimaigne A (2006) Enzymatic production of acyl flavonoid derivatives. Patent US 20060115880
Ginwala R, Bhavsar R, Chigbu DI, Jain P, Khan ZK (2019) Potential role of flavonoids in treating chronic inflammatory diseases with a special focus on the anti-inflammatory activity of apigenin. Antioxidants 8(2):35. https://doi.org/10.3390/antiox8020035
Giusti MM, Wrolstad RE (2003) Acylated anthocyanins from edible sources and their applications in food systems. Biochem Eng J 14:217–225
Godos J, Rapisarda G, Marventano S, Galvano F, Mistretta A, Grosso G (2017) Association between polyphenol intake and adherence to the Mediterranean diet in Sicily, southern Italy. NFS J 8:1–7. https://doi.org/10.1016/j.nfs.2017.06.001
Harborne JB, Williams CA (1998) Anthocyanins and other flavonoids. Nat Prod Rep 15:631–652
Harborne JB, Williams CA (2000) Advances in flavonoid research since 1992. Phytochemistry 55 (6):481–504
Hassan MA et al (2015) Effects of salt and water stress on plant growth and on accumulation of osmolytes and antioxidant compounds in cherry tomato. Not Bot Horti Agrobo 43(1):1–11
Hegarty VM, May HM, Khaw KT (2000) Tea drinking and bone mineral density in older women. Am J Clin Nutr 71:1003–1007
Heo HJ, Lee CY (2005) Epicatechin and catechin in cocoa inhibit amyloid beta protein induced apoptosis. J Agric Food Chem 53:1445–1448
Hernández I, Alegre L, Munné-Bosch S (2004) Drought-induced changes in flavonoids and other low molecular weight antioxidants in Cistusclusii grown under Mediterranean field conditions. Tree Physiol 24:1303–1311
Hertog MGL, Hollman PCH, Katan MB (1992) Content of potentially anticarcinogenic flavonoids of 28 vegetables and 9 fruits commonly consumed in the Netherlands. J Agric Food Chem 40 (12):2379–2383
Hertog MG, Feskens EJ, Hollman PC, Katan MB, Kromhout D (1993) Dietary antioxidant flavonoids and risk of coronary heart disease: the Zutphen Elderly Study. Lancet 342:1007–1011
Hertog MG, Kromhout D, Aravanis C, Blackburn H, Buzina R, Fidanza F et al (1995) Flavonoid intake and long-term risk of coronary heart disease and cancer in the seven countries study. Arch Intern Med 155:381–386
Hollman PCH (2004) Absorption, bioavailability, and metabolism of flavonoids. Pharm Biol 42:74–83

Hollman PC, Katan MB (1999) Dietary flavonoids: intake, health effects, and bioavailability. Food Chem Toxicol 37:937–942

Hollman PCH, van Trijp JMP, Buysman MNCP, van der Gaag MS, Mengelers MJB, de Vries JHM, Katan MB (1997) Relative bioavailability of the antioxidant quercetin from various foods in man. FEBS Lett 418:152–156

Hollman PCH, Buijsman MNCP, van Gameren Y, Cnossen PJ, de Vries JHM, Katan MB (1999) The sugar moiety is a major determinant of the absorption of dietary flavonoid glycosides in man. Free Radic Res 31:569–573

Hyun, TK, Kim HC, Ko YJ, Kim JS (2016) Antioxidant, α-glucosidase inhibitory and anti-inflammatory effects of aerial parts extract from Korean crowberry (Empetrum nigrum var. japonicum). Saudi J Biol Sci 23(2):181–188

Ishihara K, Nakajima N (2003) Structural aspects of acylated plant pigments: stabilization of flavonoid glucosides and interpretation of their functions. J Mol Catal B Enzym 23(2–6):411–417

Ishizawa K, Yoshizumi M, Kawai Y (2011) Pharmacology in health food: metabolism of quercetin in vivo and its protective effect against arteriosclerosis. J Pharmacol Sci 115:466–470

Iwashina T (2003) Flavonoid function and activity to plants and other organisms. Biol Sci Space 17:24–44

Jungblut TP, Schnitzler JP, Hertkorn N, Metzger JW, Heller W, Sandermann H (1995) Acylated flavonoids as plant defense compounds against environmentally relevant ultraviolet-B radiation in Scots pine seedlings. In: Phytochemicals and health proceedings, tenth annual Penn State Symposium in Plant Physiology. American Society of Plant Physiologists, Rockville, MD, pp 266–267

Karak P (2019) Biological activities of flavonoids: an overview. Int J Pharm Sci Res 10 (4):1567–1574

Katsoura MH, Polydera AC, Tsironis L, Tselepis AD, Stamatis H (2006) Use of ionic liquids as media for the biocatalytic preparation of flavonoids derivatives with antioxidant potency. J Biotechnol 123:491–503

Kim HP, Mani I, Iversen L, Ziboh VA (1998) Effects of naturally-occurring flavonoids and bioflavonoids on epidermal cyclooxygenase and lipoxygenase from Guinea-pigs. Prostaglandins Leukot Essent Fatty Acids 58:17–24

Kodelia G, Athanasiou K, Kolisis FN (1994) Enzymatic synthesis of butyryl-rutin ester in organic solvents and its cytogenetic effects in mammalian cells in culture. Appl Biochem Biotechnol 44:205–212

Kubasek WL, Ausubel FM, Shirley BW (1998) A light-independent developmental mechanism potentiates flavonoid gene expression in Arabidopsis seedlings. Plant Mol Biol 37:217–223

Kuhnau J (1976) The flavonoids. A class of semi-essential food components: their role in human nutrition. World Rev Nutr Diet 24:117–191

Kuhnle GG, Dell'Aquila C, Runswick SA, Bingham SA (2009) Variability of phytoestrogen content in foods from different sources. Food Chem 113:1184–1187

Kumar M (2016) Anthocyanins as ingredient for food industry: strategy for extraction, functional characterization and enhanced stability. PhD thesis. ICAR-Indian Agricultural Research Institute, New Delhi, India

Kumar S, Pandey AK (2013) Chemistry and biological activities of flavonoids: an overview. Sci World J 2013:1. https://doi.org/10.1155/2013/162750

Kumar M, Dahuja A, Sachdev A, Kaur C, Varghese E, Saha S, Sairam KVSS (2018) Valorization of Black carrot marc: antioxidant properties and enzyme assisted extraction of flavonoids. Res J Biotechnol 13(11):12–21

Kumar M, Dahuja A, Sachdev A, Kaur C, Varghese E, Saha S, Sairam KVSS (2019) Valorisation of black carrot pomace: microwave assisted extraction of bioactive phytoceuticals and antioxidant activity using Box–Behnken design. J Food Sci Technol 56:995. https://doi.org/10.1007/s13197-018-03566-9

Kuo SM (1997) Dietary flavonoid and cancer prevention: evidence and potential mechanism. Crit Rev Oncog 8:47–69
Kyle JAM, Duthie GG (2006) Flavonoid in Foods. In: Andersen MØ, Markham KR (eds) Flavonoids chemistry, biochemistry and applications. Boca Raton, CRC Taylor & Francis Group, pp 219–262
Lin JK, Weng MS (2006) Flavonoids as nutraceuticals. In: Grotewold E (ed) The science of flavonoids, 1st edn. Springer, New York, pp 213–238
Lin M, Zhang J, Chen X (2018) Bioactive flavonoids in Moringa oleifera and their health-promoting properties. J Funct Foods 47:469–479
Linsalata M, Russo F (2008) Nutritional factors and polyamine metabolism in colorectal cancer. Nutrition 24:382–389
Lotito S, Frei B (2006) Consumption of flavonoid-rich foods and increased plasma antioxidant capacity in humans: cause, consequence, or epiphenomenon? Free Radic Biol Med 41:1727–1746
Lou FQ, Zhang MF, Zhang XG, Liu JM, Yuan WL (1989) A study on teapigment in prevention of atherosclerosis. Chin Med J 102:579–583
Lourenço CF, Gago B, Barbosa RM, de Freitas V, Laranjinha J (2008) LDL isolated from plasma-loaded red wine proanthocyanidins resist lipid oxidation and tocopherol depletion. J Agric Food Chem 56:3798–3804
Mackey AD, Henderson N, Gregory JF (2002) Enzymatic hydrolysis of pyridoxine-5′-β-D-glucoside is catalyzed by intestinal lactase-phlorizin hydrolase. J Biol Chem 277:26858–26864
Manach C, Scalbert A, Morand C, Remesy C, Jimenez L (2004) Polyphenols: food sources and bioavailability. Am J Clin Nutr 79(5):727–747
Manach C, Williamson G, Morand C, Scalbert A, Rémésy C (2005) Bioavailability and bioefficacy of polyphenols in humans. I. Review of 97 bioavailability studies. Am J Clin Nutr 81:230S–242S
Martini ND, Katerere DRP, Eloff JN (2004) Biological activity of five antibacterial flavonoids from Combretum erythrophyllum (Combretaceae). J Ethnopharmacol 93:207–212
Matthies A, Clavel T, Gütschow M et al (2008) Conversion of daidzein and genistein by an anaerobic bacterium newly isolated from the mouse intestine. Appl Envrion Microbiol 74:4847–4852
Mellou F, Lazari D, Skaltsa H, Tselepis AD, Kolisis FN, Stamatis H (2005) Biocatalytic preparation of acylated derivatives of flavonoid glycosides enhances their antioxidant and antimicrobial activity. J Biotechnol 116:295–304
Mellou F, Loutrari H, Stamatis H, Roussos C, Kolisis FN (2006) Enzymatic esterification of flavonoids with unsaturated fatty acids: effect of novel esters on vascular endothelial growth factor release from K562 cells. Process Biochem 41:2029–2034
Middleton EJR, Kandaswami C, Theoharides TC (2000) The effects of plant flavonoids on mammalian cells: implications for inflammation, heart disease, and cancer. Pharmacol Rev 52:673–751
Mouradov A, Spangenberg G (2014) Flavonoids: a metabolic network mediating plants adaptation to their real estate. Front Plant Sci 5,620
Moussou P, Falcimaigne A, Pauly G, Ghoul M, Engasser JM, Ardhaoui M (2004) Preparation of flavonoid derivatives. Patent EP1426445
Moussou P, Falcimaigne A, Ghoul M, Danoux L, Pauly G (2007) Esters of flavonoids with w-substituted C6-C22 fatty acids. Patent US 20070184098
N'guyen QL (1995) Cosmetic or dermopharmaceutical composition containing, in combination, a lauroylmethionate of a basic amino acid and at least one polyphenol. Patent US 5431912
Nakayama T, Suzuki H, Nishino T (2003) Anthocyanin acyltransferases: specificities, mechanism, phylogenetics, and applications. J Mol Catal B Enzym 23:117–132
Narayana KR, Reddy MS, Chaluvadi MR, Krishna DR (2001) Bioflavonoids classification, pharmacological, biochemical effects and therapeutic potential. Indian J Pharmacol 33(1):2–16

Ness AR, Powles JW (1997) Fruit and vegetables, and cardiovascular disease: a review. Int J Epidemiol 26:1–13

Ness AR, Egger M, Powles JW (1999) Fruit and vegetables, and ischaemic heart disease: systematic review or misleading meta-analysis? [letter]. Eur J Clin Nutr 53:900–902

Nettleton JA, Harnack LJ, Scrafford CG, Mink PJ, Barraj LM, Jacobs DR (2006) Dietary flavonoids and flavonoid-rich foods are not associated with risk of type 2 diabetes in postmenopausal women. J Nutr 136(12):3039–3045. https://doi.org/10.1093/jn/136.12.3039

Nicolosi G, Piattelli M, Lambusta D, Patti A (1999) Biocatalytic process for the preparation of 3-O-acyl-flavonoids. Patent WO/1999/066062

Nielsen IL, Chee WS, Poulsen L, Offord-Cavin E, Rasmussen SE, Frederiksen H, Enslen M et al (2006) Bioavailability is improved by enzymatic modification of the citrus flavonoid hesperidin in humans: a randomized, double-blind, crossover trial. J Nutr 136:404–408

Olsen KM, Slimestad R, Lea US, Brede C, Løvdal T et al (2009) Temperature and nitrogen effects on regulators and products of the flavonoid pathway: experimental and kinetic model studies. Plant Cell Environ 32:286–299

Osman HE, Maalej N, Shanmuganayagam D, Folts JD (1998) Grape juice but not orange or grapefruit juice inhibits platelet activity in dogs and monkeys. J Nutr 128:2307–2312

Otaki N, Kimira M, Katsumata S, Uehara M, Watanabe S, Suzuki K (2009) Distribution and major sources of flavonoid intakes in the middle-aged Japanese women. J Clin Biochem Nutr 44 (3):231–238

Otto R, Geers B, Weiss A, Petersohn D, Schlotmann K, Schroeder KR (2001) Novel flavone glycoside derivatives for use in cosmetics, pharmaceuticals and nutrition. Patent EP 1274712

Panche AN, Diwan AD, Chandra SR (2016) Flavonoids: an overview. J Nutr Sci 5:1–15

Parejo I, Bastida J, Viladomat F, Codina C (2005) Acylated quercetagetin glycosides with antioxidant activity from Tagetes maxima. Phytochemistry 66:2356–2362

Pascual-Teresa SD, Moreno DA, GarcíaViguera C (2010) Flavanols and anthocyanins in cardiovascular health: a review of current evidence. Int J Mol Sci 11:1679–1703

Passamonti S, Terdoslavich M, Franca R, Vanzo A, Tramer F, Braidot E, Petrussa E, Vianello A (2009) Bioavailability of flavonoids: a review of their membrane transport and the function of Bilitranslocase in animal and plant organisms. Curr Drug Metab 10:369–394

Perez-Vizcaino F, Fraga CG (2018) Research trends in flavonoids and health. Arch Biochem Biophys 646:107–112. https://doi.org/10.1016/j.abb.2018.03.022

Perrier E, Mariotte AM, Boumendjel A, Bresson-Rival D (2001) Flavonoide esters and their use notably in cosmetics. Patent US 6235294

Porras D, Nistal E, Martínez-Flórez S, Pisonero-Vaquero S, Olcoz JL, Jover R, González-Gallego J, García-Mediavilla MV, Sánchez-Campos S (2017) Protective effect of quercetin on high-fat diet-induced non-alcoholic fatty liver disease in mice is mediated by modulating intestinal microbiota imbalance and related gut-liver axis activation. Free Radic Biol Med 102:188–202

Proestos C, Boziaris IS, Nychas GJE, Komaitis M (2006) Analysis of flavonoids and phenolic acids in Greek aromatic plants: investigation of their antioxidant capacity and antimicrobial activity. Food Chem 95:664–671

Randhir R, Lin YT, Shetty K (2004) Phenolics, their antioxidant and antimicrobial activity in dark germinated fenugreek sprouts in response to peptide and phytochemical elicitors. Asia Pac J Clin Nutr 13:295–307

Rao YK, Fang SH, Tzeng YM (2005) Antiinflammatory activities of flavonoids isolated from Caesalpinia pulcherrima. J Ethnopharmacol 100(3):249–253

Ratnasooriya C, Rupasinghe HPV, Jamieson A (2010) Juice quality and polyphenol concentration of fresh fruits and pomace of selected Nov Scotia-grown grape cultivars. Can J Plant Sci 90:193–205

Reinli K, Block G (1996) Phytoestrogen content of foods: a compendium of literature values. Nutr Cancer 26(2):123–148

Ren W, Qian Z, Wang H, Zhu L, Zhang L (2003) Flavonoids: promising anticancer agents. Med Res Rev 23:519–534

Research WCRF-AICR (1997) Food, nutrition and the prevention of cancer: a global perspective. American Institute for Cancer Research, Washington, DC
Rice-Evans CA, Miller NJ, Paganga G (1996) Structure-antioxidant activity relationships of flavonoids and phenolic acids. Free Radic Biol Med 20:933–956
Robert L, Godeau G, Gavignet-Jeannin C, Groult N, Six C, Robert AM (1990) The effect of procyanidolic oligomers on vascular permeability. A study using quantitative morphology. Pathol Biol (Paris) 38:608–616
Ross JA, Kasum CM (2002) Dietary flavonoids: bioavailability, metabolic effects, and safety. Annu Rev Nutr 22:19–34
Rothwell JA, Perez-Jimenez J, Neveu V, Medina-Remon A, M'hiri N, García-Lobato P, Manach C, Knox C, Eisner R, Wishart DS, Scalbert A (2013) Phenol-Explorer 3.0: a major update of the Phenol-Explorer database to incorporate data on the effects of food processing on polyphenol content. Database 2013(0):bat070-bat070
Rusznyak SP, Szent-Gyorgyi A (1936) Vitamin P: flavonols as vitamins. Nature 138:27
Sakai M, Suzuki M, Nanjo F, Hara Y (1994) 3-O-acylated catechins and methods of producing same. Patent EP 0618203
Sampson L, Rimm E, Hollman PCH, de Vries JH, Katan MB (2000) Flavonol and flavone intakes in US health professionals. J Am Diet Assoc 102(10):1414–1420
Scalbert A, Morand C, Manach C, Remesy C (2002) Absorption and metabolism of polyphenols in the gut and impact on health. Biomed Pharmacother 56:276–282
Schijlen EGWM, Ric de Vos CH, van Tunen AJ, Bovy AG (2004) Modification of flavonoid biosynthesis in crop plants. Phytochemistry 65:2631–2648
Shen Q, Li X, Li W, Zhao X (2011) Enhanced intestinal absorption of daidzein by borneol/menthol eutectic mixture and microemulsion. AAPS Pharm Sci Tech 12:1044–1049
Shimura MZY, Asada Y, Yoshikawa T, Hatake K, Takaku F, Ishizaka Y (1999) Inhibition of Vpr-induced cell cycle abnormality by quercetin: a novel strategy for searching compounds targeting Vpr. Biochem Biophys Res Commun 261:308–316
Slimestad R, Fossen T, Vagen IM (2007) Onions: a source of unique dietary flavonoids. J Agric Food Chem 55(25):10067–10080
Spencer JP, Schroeter H, Rechner AR, Rice-Evans C (2001) Bioavailability of flavan-3-ols and procyanidins: gastrointestinal tract influences and their relevance to bioactive forms in vivo. Antioxid Redox Signal 3:1023–1039
Stefani ED, Boffetta P, Deneo-Pellegrini H, Mendilaharsu M, Carzoglio JC, Ronco A, Olivera L (1999) Dietary antioxidants and lung cancer risk: a case-control study in Uruguay. Nutr Cancer 34:100–110
Stewart A, Bozonnet S, Mullen W, Jenkins GI, Lean ME, Crozier A (2000) Occurrence of flavonols in tomatoes and tomato-based products. J Agric Food Chem 48(7):2663–2669
Steyn WJ, Wand SJ, Jacobs G, Rosecrance RC, Roberts SC (2009) Evidence for a photoprotective function of low-temperature-induced anthocyanin accumulation in apple and pear peel. Physiol Plant 136:461–472
Suda I, Oki T, Masuda M, Nishiba Y, Furuta S, Matsugano K, Sugita K, Terahara N (2002) Direct absorption of acylated anthocyanin in purplefleshed sweet potato into rats. J Agric Food Chem 50:1672–1676
Tapas AR, Sakarkar DM, Kakde RB (2008) Flavonoids as nutraceuticals: a review. Trop J Pharm Res 7(3):1089–1099
Tarara JM, Lee J, Spayd SE, Scagel CF (2008) Berry temperature and solar radiation alter acylation, proportion, and concentration of anthocyanin in Merlot grapes. Am J Enol Viticult 59:235–247
Tattini M, Galardi C, Pinelli P, Massai R, Remorini D, Agati G (2004) Differential accumulation of flavonoids and hydroxycinnamates in leaves of Ligustrumvulgare under excess light and drought stress. New Phytol 163:547–561
Thilakarathna SH, Rupasinghe HP (2013) Flavonoid bioavailability and attempts for bioavailability enhancement. Nutrients 5(9):3367–3387

Treml J, Šmejkal K (2016) Flavonoids as potent scavengers of hydroxyl radicals. Compr Rev Food Sci Food Saf 15(4):720–738. https://doi.org/10.1111/1541-4337.12204

Tsanova-Savova S, Ribarova F (2013) Flavonols and flavones in some Bulgarian plant foods. Pol J Food Nutr Sci 63(3):173–177

Tsanova-Savova S, Ribarova F, Gerova M (2005) (+)-cathechin and (−)-epicathechin in Bulgarian fruits. J Food Compos Anal 18(7):691–698

Tsanova-Savova S, Denev P, Ribarova F (2018) Flavonoids in foods and their role in healthy nutrition. Diet Microb Health:165–198. https://doi.org/10.1016/b978-0-12-811440-7.00007-7

U.S. Department of Agriculture, Agricultural Research Service. USDA National Nutrient Database for the Isoflavone Content of Selected Foods, Release 2.0, 2008. Available online: http://www.ars.usda.gov/SP2UserFiles/Place/12354500/Data/isoflav/Isoflav_R2.pdf. Accessed on 2 June 2013

USDA Database for the Flavonoid Content of Selected Foods, Release 3.3 (March 2018) Available from: Nutrient Data Laboratory Home Page. http://www.ars.usda.gov/nutrientdata/flav

Van De Wier B, Koek GH, Bast A, Haenen GRMM (2015) The potential of flavonoids in the treatment of non-alcoholic fatty liver disease. Crit Rev Food Sci Nutr 57(4):834–855. https://doi.org/10.1080/10408398.2014.952399

Vasquez-Martinez Y, Ohri RV, Kenyon V, Holman TR, Sepu'lveda-Boza S (2007) Structure–activity relationship studies of flavonoids as potent inhibitors of human platelet 12-hLO, reticulocyte 15-hLO-1, and prostate epithelial 15-hLO-2. Bioorg Med Chem 15:7408–7425

Ververidis F, Trantas E, Douglas C, Vollmer G, Kretzschmar G, Panopoulos N (2007) Biotechnology of flavonoids and other phenylpropanoid- derived natural products. Part I: Chemical diversity, impacts on plant biology and human health. Biotechnol J 2(10):1214–1234

Viskupicová J, Ondrejovič M, Sturdík E (2008) Bioavailability and metabolism of flavonoids. J Food Nutr Res 47(4):151–162

Viskupicova J, Ondrejovic M, Sturdık E (2009) The potential and practical applications of acylated flavonoids. Pharmazie 64(6):355–360

Vogiatzoglou A, Mulligan AA, Lentjes M, Luben RN, Spencer JPE, Schroeter H, Khaw KT, Kuhnle GGC (2015) Flavonoid intake in European adults (18 to 64 years). PLoS One 10 (5):128–132

Walle T (2007) Methylation of dietary flavones greatly improves their hepatic metabolic stability and intestinal absorption. Mol Pharm 4:826–832

Wang HK, Xia Y, Yang ZY, Natschke SL, Lee KH (1998) Recent advances in the discovery and development of flavonoids and their analogues as antitumor and anti-HIV agents. Adv Exp Med Biol 439:191–225

Wang G, Cao F, Wang G, Yousry A, Kassaby E (2015) Role of temperature and soil moisture conditions on flavonoid production and biosynthesis-related genes in Ginkgo (Ginkgo biloba L.) leaves. Nat Prod Chem Res 3:162. https://doi.org/10.4172/2329-6836.1000162

Williams CA (2006) Flavone and flavonol O-glycosides. In: Anderse ØM, Markham KR (eds) Flavonoids: chemistry, biochemistry and applications, 1st edn. CRC Press, New York, pp 397–441

Williamson G, Manach C (2005) Bioavailability and bioefficacy of polyphenols in humans. II. Review of 93 intervention studies. Am J Clin Nutr 81:243S–255S

Williamson G, Day AJ, Plumb GW, Couteau D (2000) Human metabolic pathways of dietary flavonoids and cinnamates. Biochem Soc Trans 28:16–22

Winkel-Shirley B (2002) Biosynthesis of flavonoids and effects of stress. Curr Opin Plant Biol 5:218–223

Wojdylo A, Oszmianski J, Laskowski P (2008) Polyphenolic compounds and antioxidant activity of new and old apple varieties. J Agric Food Chem 56:6520–6530

Wu B, Kulkarni K, Basu S, Zhang S, Hu M (2011) First-pass metabolism via UDP-glucuronosyltransferase: a barrier to oral bioavailability of phenolics. J Pharm Sci 100 (9):3655–3681

Xiao ZP, Peng ZY, Peng MJ, Yan WB, Ouyang YZ, Zhu HL (2011) Flavonoids health benefits and their molecular mechanism. Mini-Rev Med Chem 11:169–177. https://doi.org/10.2174/138955711794519546
Yamamoto GN, Mori K, Numata M, Koyama K, Kitayama M (2010) Effects of temperature and water regimes on flavonoid contents and composition in the skin of red-wine grapes. J Int SciVigne Vin (special issue Macrowine):75–80
Yamamura S, Ozawa K, Ohtani K, Kasai R, Yamasaki K (1998) Antihistaminic flavones and aliphatic glycosides from Mentha spicata. Phytochemistry 48:131–136
Zhang Z, Huang Y, Gao F, Gao Z, Bu H, Gu W, Li Y (2011) A self-assembled nanodelivery system enhances the oral bioavailability of daidzein: *in vitro* characteristics and *in vivo* performance. Nanomedicine (London) 6:1365–1379

Newly Identified Phenolic Compounds from Different Plant Families

7

Gulab Khan Rohela, Mohammad Yaseen Mir, Pawan Shukla, and Aftab Ahmad Shabnam

Abstract

In plants among the different phytochemicals, phenolic and polyphenolic compound-based research has gained rapid momentum during the present decade. Even though the phenolic compounds are not required in primary processes such as growth and development of a plant, these compounds which are considered as secondary metabolites have gained a lot of attention of researchers across the world for their role in diversified functions of a plant system. Majorly phenolic compound-based research is being carried out to understand the role of various classes of phenols right from their role in adapting a plant species to different environments which encompasses biotic and abiotic factors to understanding their role in reproduction, in interaction with other biomolecules, and in specified function of every individual phenolic compound. More advanced research is also being carried to know the action of phenolic compounds on the enzymes involved in the epigenetic regulation of gene expression. From the 1970s, the plant phenolic-based research has gained importance, and researchers have started identifying the new phenolic compounds from several plant species. Now, through this book chapter, we put together the newly identified phenolic

G. K. Rohela (✉) · P. Shukla
Biotechnology Section, Moriculture Division, Central Sericultural Research & Training Institute, Central Silk Board, Ministry of Textiles, Government of India, Pampore, Jammu and Kashmir, India

M. Y. Mir
Centre of Research for Development, University of Kashmir, Srinagar, Jammu and Kashmir, India

A. A. Shabnam
Mulberry Breeding & Genetics Section, Moriculture Division, Central Sericultural Research & Training Institute, Central Silk Board, Ministry of Textiles, Government of India, Pampore, Jammu and Kashmir, India

R. Lone et al. (eds.), *Plant Phenolics in Sustainable Agriculture*,
https://doi.org/10.1007/978-981-15-4890-1_7

compounds of the present decade (i.e., 2010–April 2019) which were identified from different plant families.

Keywords

Phenolic compounds · Chemoattractant · Pollinator · Families

7.1 Introduction

Phenolic compounds are the main phytochemicals which have gained a lot of importance in the modern era of plant-based chemical research for understanding the diversified roles of phenolic compounds. In plants, even though these compounds may not be utilized as the primary metabolites, their relative content, site of production, and type of phenolic do play a major role in several activities of a plant system such as the following: as a chemoattractant for pollinators (Nicolson et al. 2015; Zhang et al. 2016); as a plant protectant against abiotic (physical damage, salinity, UV damage, drought, etc.) (Crozier et al. 2007) and biotic stresses (pathogens and herbivore animals) (Liu et al. 2007a, b; Redman et al. 2003); as a molecule of communication within a plant system (Bhattacharya et al. 2010); and as a molecule of interaction between plant-plant (Elizabeth and John 1998; Vittorio and Clay 2013), plant-microbe (Siqueira et al. 1991; Baker et al. 1997; Hammer 2005; Mandal et al. 2010), and plant-animal interactions (Liu et al. 2004, 2007a, b). Thus, by the virtue of above functions, these phenolic compounds represent a striking example for enabling the plants to adapt to changing environments under the influence of biotic and abiotic factors.

Phenolic compounds also enable the plants to produce their products (flowers, fruits) with diversified color, taste, putativeness, palatability, and odor (Naczk and Shahidi 2012; Dykes and Rooney 2007) which either attract other organisms or provide economic importance by enhancing the commercial values of plant products. In agro-industrial and food sectors, the plant-based phenolic has a wide influence over the final product by exerting either positive or negative impacts during the processing of food products and thus ultimately effecting the quality and economic value of food and medicinal products (Dong et al. 2017). Due to this, in the modern era, the plant phenolic-based research has gained lot of importance.

Due to advancements in molecular biology and genomics with regard to tracing out the genes responsible for a specific step of a biochemical pathway, now it is possible to make a gene to undergo either upregulation or downregulation by CRISPR (clustered regularly interspaced short palindromic repeats)- and CAS (CRISPR-associated genes)-mediated modification in gene expression pattern (Gilbert et al. 2013; Cheng et al. 2013; Shalem et al. 2014), and therefore one can understand the response of plants in defense and other functions in relation to altered expression of phenolic compounds. In recent days, researchers have focused on the metabolic pathways of phenolic and polyphenolic compounds to understand their mechanism of synthesis (Cheynier et al. 2013) and their relative content in different parts of a plant (Fialova et al. 2012). Studies regarding the detailed structures and properties of newly identified phenolic compounds also gained rapid interest in recent decades to elucidate specificity of individual phenolic compounds with regard to their function.

Across the globe, from the past five decades, there are several reports about the new phenolic compounds identified from several plant species belonging to different families. Now, through this review paper, we put together the newly identified phenolic compounds of the present decade (i.e., 2010–April 2019) from various plant species of different families.

7.2 Asteraceae

In the present decade (2010–April 2019), from family Asteraceae, there is a report (Zhizhen et al. 2010) on identification of two new phenolic compounds, viz., p-hydroxyphenylferulate and 5,30-dihydroxy-4,40-dimethoxy-2,70-cycloligna-7,70-diene-9,90-lactone, from *Liatris elegans* which is a perennial plant, distributed widely in the southeastern parts of the United States. Through high-resolution electrospray ionization mass spectrometry (HRESIMS), the chemical formula and mass of p-hydroxyphenylferulate was elucidated as $C_{16}H_{14}O_5$ and 287.0968 (Zhizhen et al. 2010). For 5,30-dihydroxy-4,40-dimethoxy-2,70-cycloligna-7,70-diene-9,90-lactone, the chemical formula and mass was elucidated as $C_{20}H_{16}O_6$ and 353.1066 (Zhizhen et al. 2010). The structure of both the compounds was deduced through the H1-NMR studies (Fig. 7.1a, b) (Zhizhen et al. 2010).

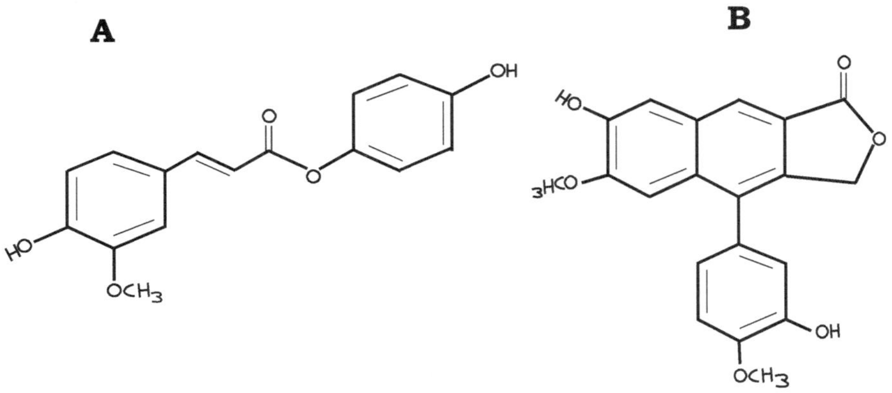

Fig. 7.1 Structure of new phenolic compounds of *Liatris elegans* deduced through H1-NMR studies (Zhizhen et al. 2010)
(**a**) Structure of p-hydroxyphenylferulate
(**b**) Structure of 5,30-dihydroxy-4,40-dimethoxy-2,70-cycloligna-7,70-diene-9,90-lactone

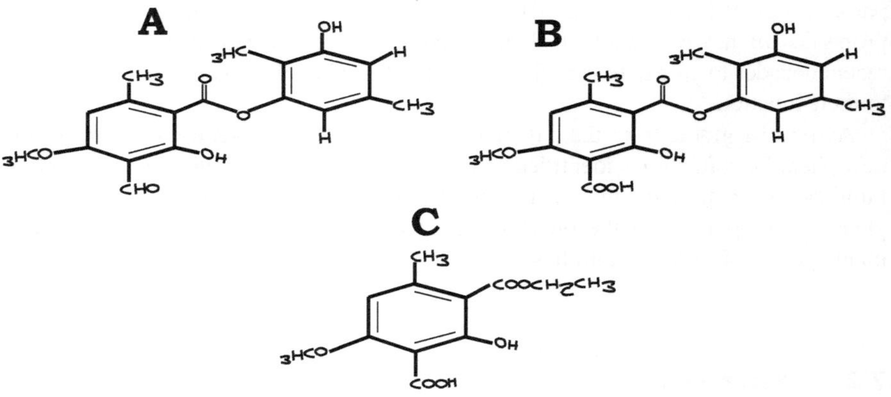

Fig. 7.2 Structure of new phenolic compounds of *Thamnolia vermicularis* deduced through 2D-NMR studies (Jia et al. 2011)
(**a**) Structure of thamnoliadepside-A, (**b**) structure of thamnoliadepside-B, and (**c**) structure of thamnolic acid-A

7.3 Icmadophilaceae

In 2011, Jia et al. identified three new phenolic compounds from lichen *Thamnolia vermicularis* belonging to the family Icmadophilaceae. Newly identified phenolic compounds are thamnoliadepside-A, thamnoliadepside-B, and thamnolic acid-A. Earlier also several other phenolic compounds have been isolated from this lichen species (Sun and Shen 1985; Jiang and Sun 2002). A phenolic compound known as baeomycesic acid was also reported from this plant for anticancer activity against different human tumor cell lines (Haraldsdottir et al. 2004). Jia et al. (2011) have carried the high-resolution electrospray ionization mass spectrometry (HRESI-MS) and identified the chemical formula and mass of thamnoliadepside-A ($C_{20}H_{16}O_6$ and 353.1066), thamnoliadepside-B ($C_{18}H_{18}O_6$ and 329.1028), and thamnolic acid-A ($C_{12}H_{14}O_6$ and 253.0716). Structures of all the three newly identified phenolic compounds were identified by two-dimensional nuclear magnetic resonance spectroscopy (2D-NMR) and high-resolution mass spectrometry (HR-MS) (Fig. 7.2a, b, c) (Jia et al. 2011). Based on NMR studies, Jia et al. (2011) also reported that thamnoliadepside-A has the ability to bond with G-quadruplex DNA, thereby exhibiting antiprostate cancer activity.

7.4 Oxalidaceae

A new phenolic compound known as 2,4-dihydroxy-6-(hydroxymethylene)benzaldehyde was reported from *Averrhoa bilimbi* Linn of Oxalidaceae family (Gunawan and Anamy 2013). Leaf extracts of this plant species have been in use since ancient

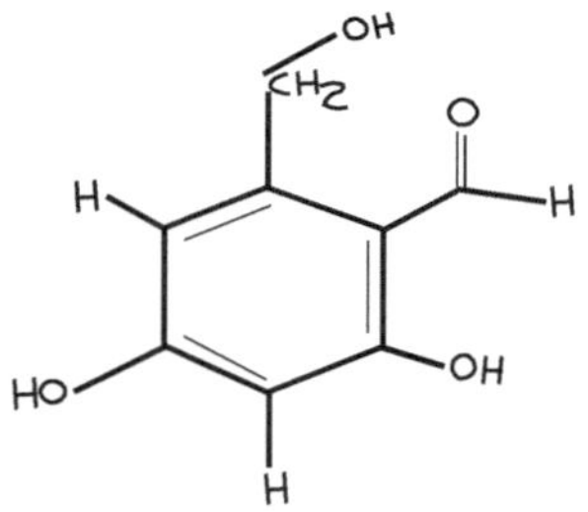

Fig. 7.3 Structure of new phenolic compound (2,4-dihydroxy-6-(hydroxymethylene) benzaldehyde) of *Averrhoa bilimbi* deduced through 2D-NMR studies (Gunawan and Anamy 2013)

days as a traditional medicine for treatment of hypertension, pain, inflammation, and pathogenic infections, as antidiabetic, and as an antioxidant. By electrospray mass spectrometry (ES-MS), electron ionization (EI-MS), and infrared spectroscopy (IR-S), the chemical formula and mass of 2,4-dihydroxy-6-(hydroxymethylene)benzaldehyde was elucidated (Gunawan and Anamy 2013). The structure of this compound was deduced through 2D-NMR studies (Fig. 7.3) (Gunawan and Anamy 2013).

7.5 Orchidaceae

Gastrodia elata is an important medicinal plant in Orchidaceae family; the rhizome of this plant is being used as herbal medicine for treatment of rheumatism, epilepsy, and headaches (Sin et al. 1997). The extracts of the rhizome were also reported to be used as an anti-asthmatic (Jang et al. 2010), anti-inflammatory (Hwang et al. 2009), antidepressant (Chen et al. 2008b), anti-osteoporotic (Seo et al. 2010), and a neuroprotective agent (Kim et al. 2001; Yong et al. 2009; Yu et al. 2010; An et al. 2010). In 2011, Ah-Reum et al. have isolated and identified 12 phenolic compounds from the rhizomes of *Gastrodia elata.* Among the 12, two were identified as new phenolic compounds, viz., 4-hydroxybenzyl vanillyl ether and 4-{{4-[4-(methoxymethyl) phenoxy]benzyl}oxy}benzyl methyl ether (Ah-Reum et al. 2011). By investigating through high-resolution electrospray ionization mass spectrometry (HR-ESI-MS), the chemical formula and mass of new phenolic compounds were identified as $C_{15}H_{16}O_4$ and 260.1041 for the first compound and $C_{23}H_{24}O_4$ and 364.1695 for the second compound (Ah-Reum et al. 2011). The structure of both compounds was deduced through the NMR (1H, 1H-COSY, NOESY, HSQC, and HMBC) studies (Fig. 7.4a, b) (Ah-Reum et al. 2011).

In 2015, there was another report from family Orchidaceae about isolation and identification of five new phenolic compounds from *Dendrobium aphyllum* (Dan et al. 2015). Genus *Dendrobium* with nearly about 1100 species is widely distributed among diversified climatic regions of Asia, Australia, and Europe. In China, *Dendrobium* has wide applications in traditional medicine against chronic diseases and as well as health food source (Zhang et al. 2003). Phenolic compounds, alkaloids, terpenoids, and carbohydrates are the main phytochemicals of *Dendrobium* which provide them medicinal values (Chen and Guo 2001; Ng et al. 2012). Hence, several researchers have carried the extensive research to identify

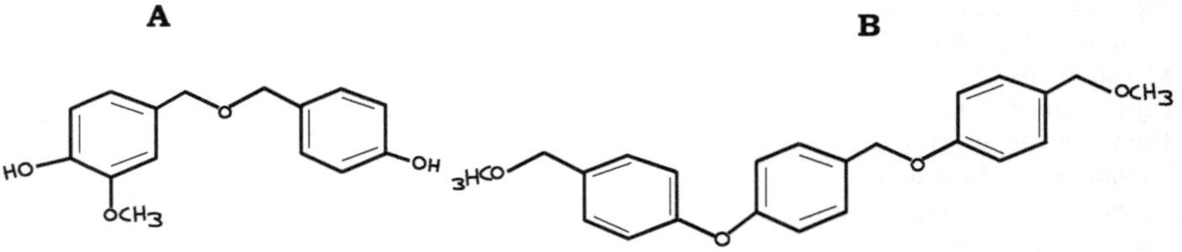

Fig. 7.4 Structure of new phenolic compounds of *Gastrodia elata* deduced through 2D-NMR (Ah-Reum et al. 2011)
(**a**) Structure of hydroxybenzyl vanillyl ether
(**b**) Structure of 4-{{4-[4-(methoxymethyl) phenoxy]benzyl}oxy}benzyl methyl ether

these phytochemicals for their specific identification along with their detailed structures and activity (Majumder and Sen 1987; Zhao et al. 1994; Zhang et al. 2008, Shao et al. 2008; Chen et al. 2008a, b).

Recently, by carrying out thorough investigations on chemical constituents from stem portion of *Dendrobium aphyllum*, Dan et al. (2015) identified five new phenolic compounds, viz., aphyllone-A, aphyllone-B, aphyllal-C, aphyllal-D, and aphyllal-E. Techniques such as X-ray diffraction, quantum calculations, optical rotations, circular dichroism (CD) spectra, infrared spectra, NMR spectra, and API QSTAR time-of-flight spectrometer were applied for identification of chemical formulae, mass, and detailed structures (Dan et al. 2015) (Fig. 7.5a–e).

7.6 Fabaceae

In family Fabaceae, *Glycyrrhiza yunnanensis* species is known for possessing pharmaceutically important phytochemicals such as alkaloids (Hu and Shen 1995), flavonoids (Hu et al. 1994; Gao and Zhang 1994), and saponins (Zeng et al. 1990; Ohtani et al. 1992; Kazuhiro et al. 1994). Due to these pharmaceutically important compounds, *G. yunnanensis* have also been used as an alternative source of *G. uralensis* in preparation of folk medicine (Gao and Zhange 1993). In 2013, Qing et al. have identified three new phenolic compounds from the roots of *Glycyrrhiza yunnanensis*. Roots of *Glycyrrhiza yunnanensis* were air-dried, powdered, and extracted with different solvent systems (Qing et al. 2013).

These compounds were separated on silica gel by thin-layer chromatography technique, and individual compounds were eluted from silica gel by using different solvents (Qing et al. 2013). The separated individual compounds were subjected to their identification in terms of their chemical formulae, mass, and structures by using various techniques of spectroscopy (UV, IR, NMR [1D and 2D], and HR-ESI-MS) (Qing et al. 2013). After thorough investigations of Qing et al. (2013), out of 19 phenolic compounds, three were identified as new compounds. Three phenolic compounds are 2-(2′-methoxy-4′-hydroxy)-aryl-3-methy-6-hydroxy-benzofuran ($C_{16}H_{14}O_4$ and 271.0951), (2S)-6,7-(2,2-dimethyl dihydropyrano)-8-prenyl-4′-hydroxyflavanone ($C_{25}H_{28}O_4$ and 393.2060), and

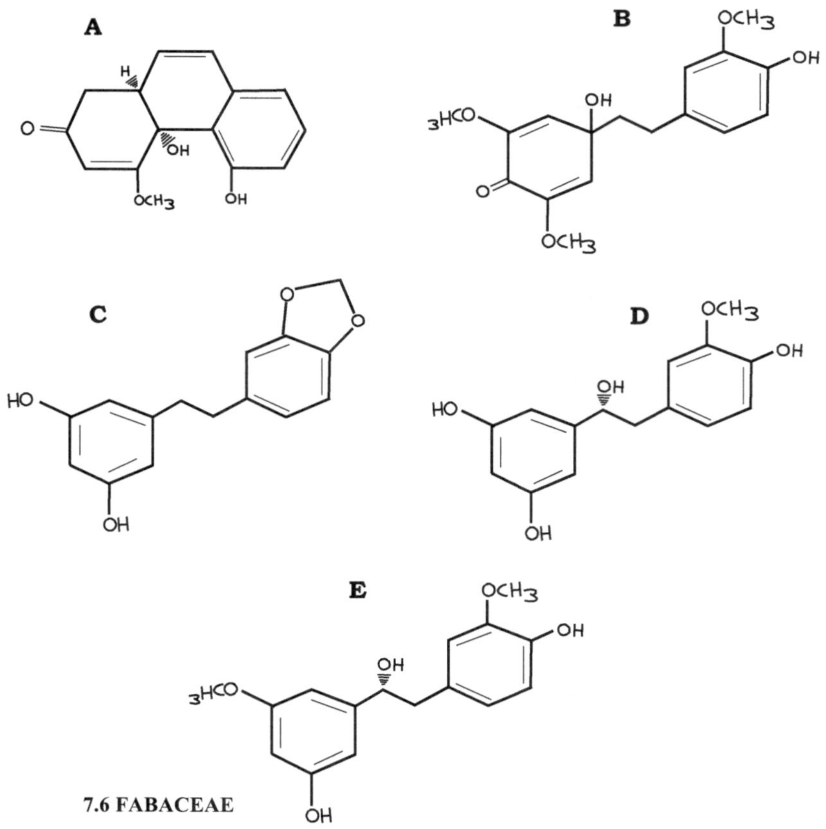

Fig. 7.5 Structure of new phenolic compounds identified from *Dendrobium aphyllum* (Dan et al. 2015)
(**a**) Structure of aphyllone-A, (**b**) structure of aphyllone-B
(**c**) Structure of aphyllal-C, (**d**) structure of aphyllal-D
(**e**) Structure of aphyllal-E

6-prenyl-7,3′,4′-trihydroxyflavone ($C_{20}H_{18}O_5$ and 339.1227) with their chemical formula and mass in parenthesis (Qing et al. 2013) (Fig. 7.6a–c).

7.7 Theaceae

In 2013, Ayaka et al. identified two new phenolic compounds, i.e., teasperol and teasperin, from fermented tea products of plant species *Camellia sinensis* L. Ayaka et al. (2013) identified these two new phenolic compounds after carrying out fermentation of dry leaves of *Camellia sinensis* L. with fungal organism *Aspergillus* sp. Out of two phenolic compounds, teasperol was identified from Chinese post-fermented tea product, and teasperin was identified from Japanese post-fermented

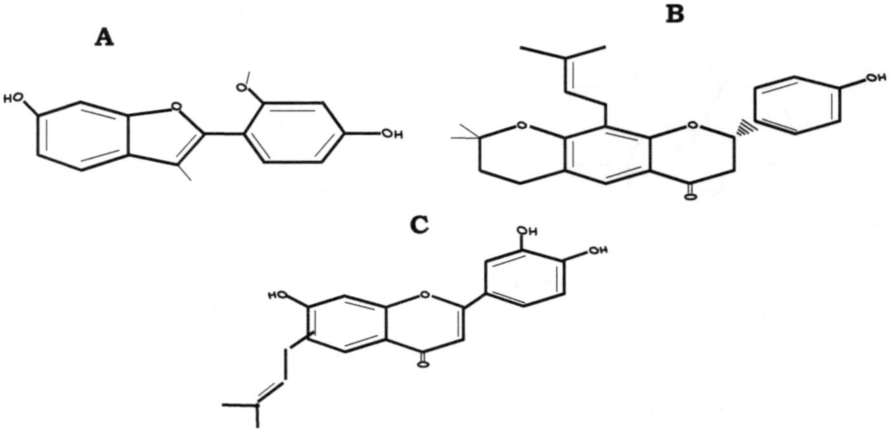

Fig. 7.6 Structure of new phenolic compounds identified from *Glycyrrhiza yunnanensis* (Qing et al. 2013)
(**a**) Structure of 2-(2′-methoxy-4′-hydroxy)-aryl-3-methy-6-hydroxy-benzofuran
(**b**) Structure of (2S)-6,7-(2,2-dimethyl dihydropyrano)-8-prenyl-4′-hydroxyflavanone
(**c**) Structure of 6-prenyl-7,3′,4′-trihydroxyflavone

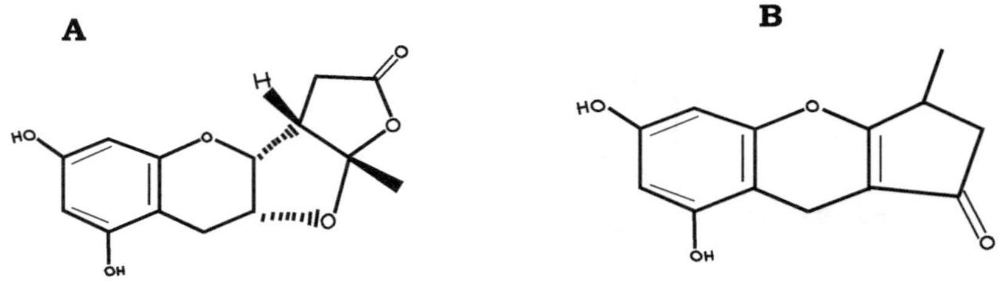

Fig. 7.7 Structure of new phenolic compounds identified from *Camellia sinensis* L. (Ayaka et al. 2013)
(**a**) Structure of teasperol, (**b**) structure of teasperin

tea product. Initially, they have extracted these compounds along with known phenolic compounds by carrying out the extraction procedure with 60% ethanol or hot water, followed by their separation through column chromatography.

After separation of compounds by chromatographic technique, individual compounds were subjected to identification in terms of their chemical formulae, mass, and structures by using various techniques of spectroscopy such as HPLC-QTOF-MS, ^{1}H-NMR, ^{13}C-NMR, and HMBC (Ayaka et al. 2013). Chemical formulae and mass of teasperol were identified as $C_{14}H_{14}O_6$ and 277.0719, whereas for teasperin it was identified as $C_{13}H_{12}O_4$ and 231.0665 (Ayaka et al. 2013) (Fig. 7.7a, b).

7.8 Moraceae

7.8.1 *Broussonetia papyrifera*

In 2014, Yang et al. have identified five new phenolic compounds from the leaves of paper mulberry (*Broussonetia papyrifera*) along with 15 known phenolic compounds. The newly identified five phenolic compounds are broussoside-A, broussoside-B, broussoside-C, broussoside-D, and broussoside-E. Dried leaves of *B. papyrifera* were used for the extraction of phenolic compounds by using methanol, and later the extracts were purified by suspending in distilled water and partitioned by using solvents such as chloroform and n-butanol (Yang et al. 2014). The soluble fraction of chloroform and n-butanol was then subjected to purification through silica gel column-based chromatography and was separated into individual compounds by preparative high-performance column chromatography (PHPLC) (Yang et al. 2014).

After separation of compounds by various chromatographic techniques, individual compounds were subjected to their identification in terms of chemical formulae, mass, and structures by using various techniques of spectroscopy such as ER-ESI-MS, ^{1}H-NMR, ^{13}C-NMR, and HSQC (Yang et al. 2014). Chemical formulae and mass of broussoside-A were identified as $C_{32}H_{44}O_{13}$ and 659.2674, whereas for broussoside-B it was identified as $C_{32}H_{44}O_{13}$ and 659.2674, for broussoside-C it was identified as $C_{32}H_{46}O_{13}$ and 673.2831, for broussoside-D it was identified as $C_{26}H_{34}O_{09}$ and 513.2095, and for broussoside-E it was identified as $C_{32}H_{42}O_{13}$ and 657.2518 (Yang et al. 2014) (Figs. 7.8a, 7.8b, 7.8c, 7.8d, and 7.8e).

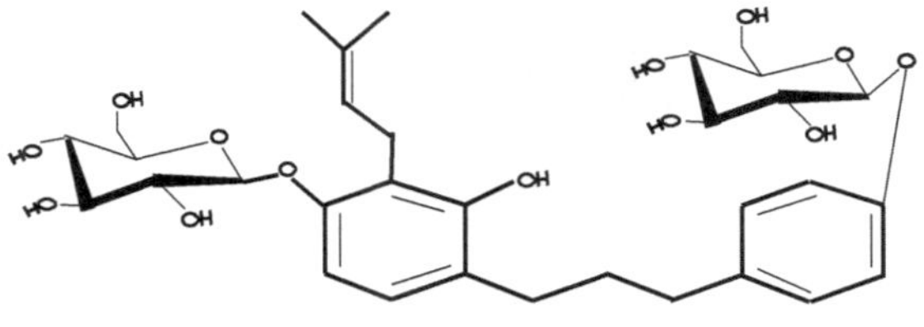

Fig. 7.8a Structure of new phenolic compound *broussoside-A* identified from *Broussonetia papyrifera* (Yang et al. 2014)

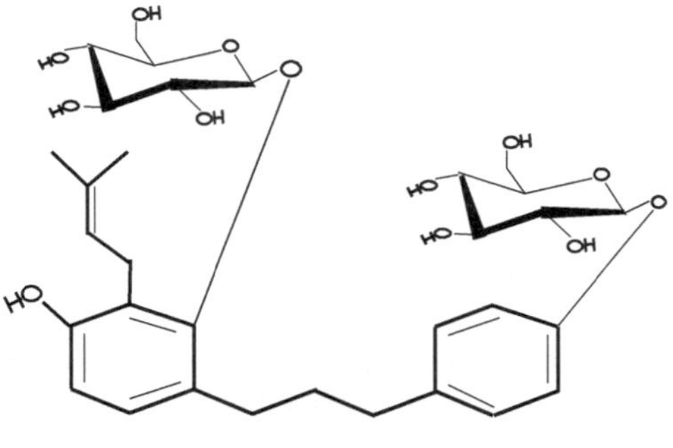

Fig. 7.8b Structure of new phenolic compound *broussoside-B* identified from *Broussonetia papyrifera* (Yang et al. 2014)

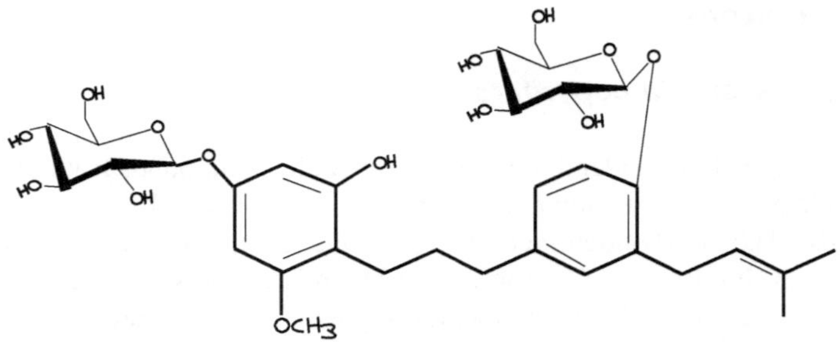

Fig. 7.8c Structure of new phenolic compound *broussoside-C* identified from *Broussonetia papyrifera* (Yang et al. 2014)

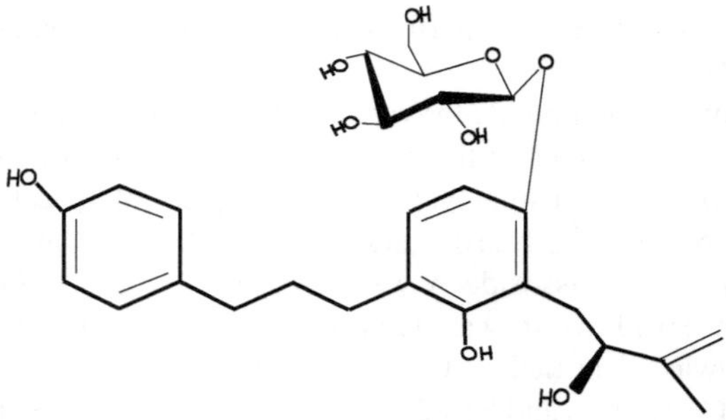

Fig. 7.8d Structure of new phenolic compound *broussoside-D* identified from *Broussonetia papyrifera* (Yang et al. 2014)

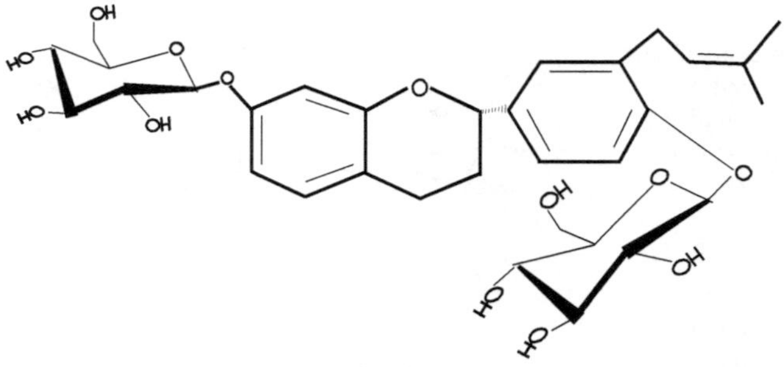

Fig. 7.8e Structure of new phenolic compound *broussoside-E* identified from *Broussonetia papyrifera* (Yang et al. 2014)

7.9 Lamiaceae

From family Lamiaceae, there are two reports in the resent decade about the identification of new phenolic compounds, i.e., from the plant species *Schizonepeta tenuifolia* and *Lavandula angustifolia*.

7.9.1 *Schizonepeta tenuifolia*

Schizonepeta tenuifolia is a widely spread herbaceous plant species from family Lamiaceae (Yoo et al. 2011). This plant species is widely used from ancient days in treatment of common cold, cough, fever, and headache (Oshima et al. 1989). Medicinal properties of this plant species are majorly due to the presence of phytochemicals like oils, phenolic compounds, etc. Several investigations were carried out to identify the phytoconstituents of this plant species (Lee et al. 2008; Zhang et al. 2006; Yang and Lee 2013). Recently, Xu et al. (2016) have isolated and identified two new phenolic compounds, i.e., schitenoside-A and schitenoside-B, along with other six known phenolic compounds from dried parts of whole plants of *S. tenuifolia.* Ethanol solvent-based extractions were carried from dried parts of *S. tenuifolia*, and it was reduced under pressure to form a crude extract (Xu et al. 2016).

The obtained crude extract was initially suspended in water and later partitioned sequentially with three solvents of petroleum ether, ethyl acetate, and n-butanol to yield three portions (Xu et al. 2016). These three portions were subjected to silica gel-based chromatography with different solvent systems and eluting solutions for the separation and purification of the individual compounds (Xu et al. 2016). The separated individual compounds were subjected to their identification through spectroscopic analysis (ER-ESI-MS, ^{1}H-NMR, ^{13}C-NMR, and HMBC) for their chemical formulae, mass, and structural details (Xu et al. 2016).

Through the generated spectroscopic data, the chemical formulae and mass of schitenoside-A were elucidated as $C_{21}H_{24}O_{10}$ and 459.1267, whereas for schitenoside-B it was elucidated as $C_{24}H_{26}O_{12}$ and 529.1322 (Xu et al. 2016) (Figs. 7.9a and 7.9b). Antibacterial activity of newly identified schitenoside-A and schitenoside-B was also carried out by inhibition zone assay by using four strains of bacteria which includes both Gram-positive and Gram-negative stains, viz., *Staphylococcus aureus*, *Pseudomonas aeruginosa*, *Salmonell*a *typhimurium*, and *Escherichia coli* (Xu et al. 2016). Through the assay, it was identified that both the newly identified compounds showed moderate antibacterial activity (Xu et al. 2016).

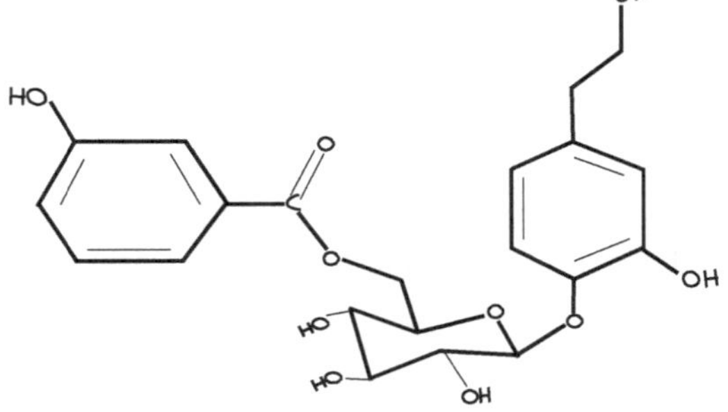

Fig. 7.9a Structure of new phenolic compound *schitenoside-A* identified from *Schizonepeta tenuifolia*

Fig. 7.9b Structure of new phenolic compound *schitenoside-B* identified from *Schizonepeta tenuifolia*

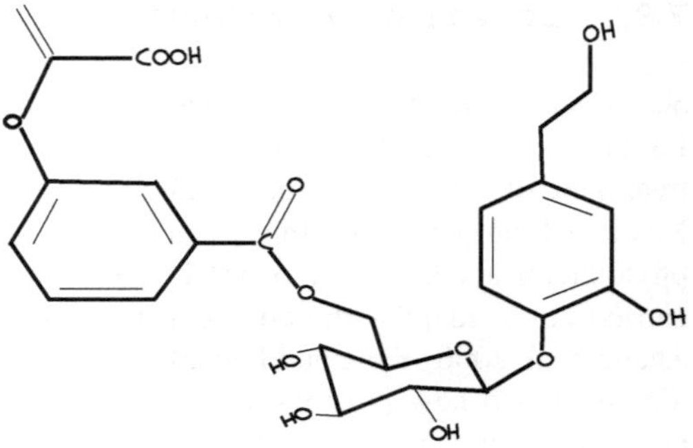

7.9.2 *Lavandula angustifolia*

Lavandula angustifolia of Lamiaceae family is an important medicinal plant which has been in intense cultivation across the world for its pharmaceutical values and for essential oil components (Yang and Gao 2010). This plant species (common name: lavender) has been used as traditional medicine for its antibacterial nature, spasmolysis, sedative, and neuroprotective function (Chinese Pharmacopoeia Commission, 1999). The aromatic and biological activities of lavender oil are majorly due to the presence of several compounds such as β-ocimene, camphor, cineole, linalool, linalyl acetate, lavender alcohol, p-lavender acetate, terpene-4-alcohol, etc. (Shi 2012; Danh et al. 2012). Many types of phenolic compounds were isolated and identified earlier in *L. angustifolia* (Nitzsche et al. 2004; Shan et al. 2005; Wu et al. 2007a, b; Castro et al. 2014). Few nonvolatile compounds like apigenin, caffeic acid, chlorogenic acid, protocatechuic acid, and rosmarinic acid were identified earlier from the lavandula waste which was obtained after the extraction of essential oil (Torras et al. 2007).

In 2018, Nigary et al. have isolated and identified seven new phenolic compounds (i.e., lavandunat (1), lavandufurandiol (2), lavandufluoren (3), lavandupyrone-A (4), lavandupyrone-B (5), lavandudiphenyl-A (6), and lavandudiphenyl-B (7)) from the ethyl acetate extracts from the waste of *L. angustifolia* as mentioned above. During isolation, ethanolic extracts of waste of lavandula were fractionized by using different solvents of dichloromethane, ethyl acetate, n-butanol, and petroleum ether, and compounds were separated and purified by silica gel and HPLC chromatographic (Nigary et al. 2018).

After separation of compounds by chromatographic techniques, individual new compounds (seven in number) were identified with regard to their chemical formulae, mass, and structures (Figs. 7.10a, 7.10b, 7.10c, 7.10d, 7.10e, 7.10f, and 7.10g) by various spectroscopy techniques such as ER-ESI-MS, ^{1}H-NMR, ^{13}C-NMR, ^{1}H-^{1}H COSY, and HMBC (Nigary et al. 2018). Chemical formulae and mass of lavandunat (1) were identified as $C_{12}H_{13}O_6$ and 253.0790; for lavandufurandiol (2), it was identified as $C_{25}H_{20}O_9$ and 463.1107, for lavandufluoren (3), it was identified as $C_{26}H_{22}O_9$ and 477.1264; for lavandupyrone-A (4), it was identified as $C_{26}H_{19}O_{10}$ and 491.1056; for

A

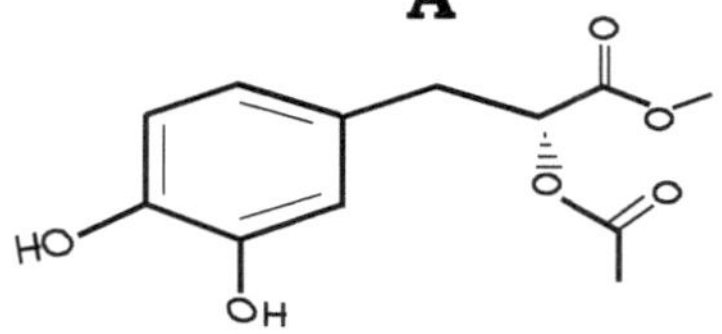

Fig. 7.10a Structure of new phenolic compound *lavandunat* identified from *Lavandula angustifolia* (Nigary et al. 2018)

B

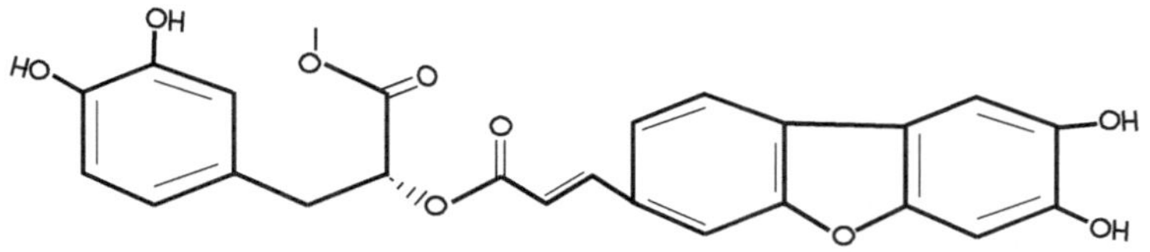

Fig. 7.10b Structure of new phenolic compound *lavandufurandiol* identified from *Lavandula angustifolia* (Nigary et al. 2018)

C

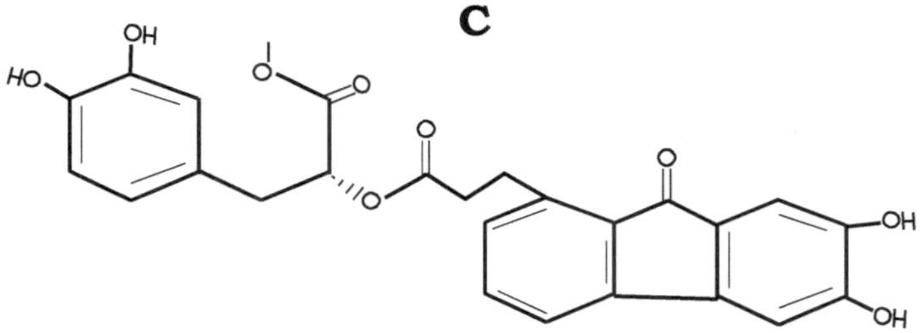

Fig. 7.10c Structure of new phenolic compound *lavandufluoren* identified from *Lavandula angustifolia* (Nigary et al. 2018)

D

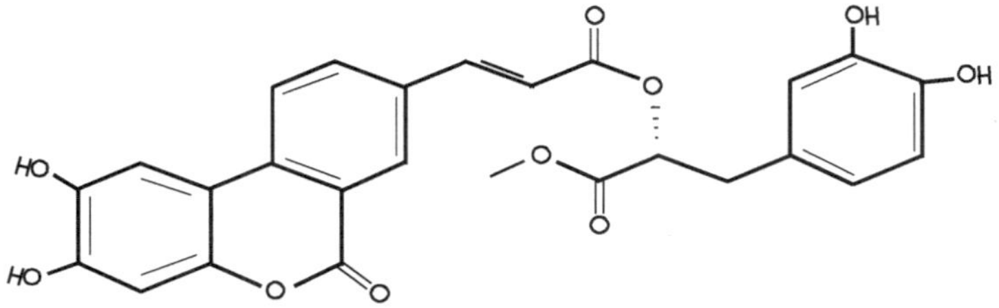

Fig. 7.10d Structure of new phenolic compound *lavandupyrone-A* identified from *Lavandula angustifolia* (Nigary et al. 2018)

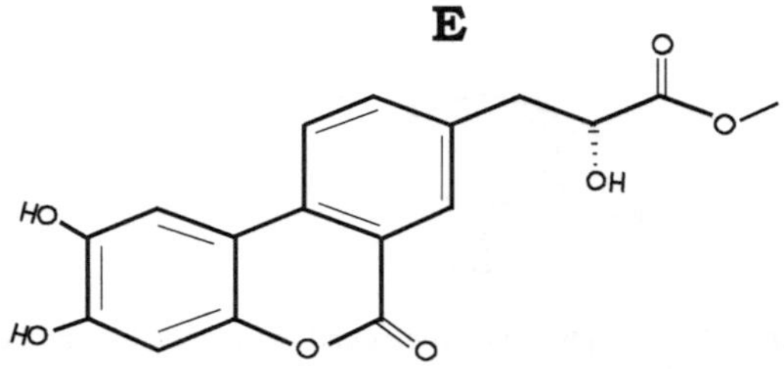

Fig. 7.10e Structure of new phenolic compound *lavandupyrone-B* identified from *Lavandula angustifolia* (Nigary et al. 2018)

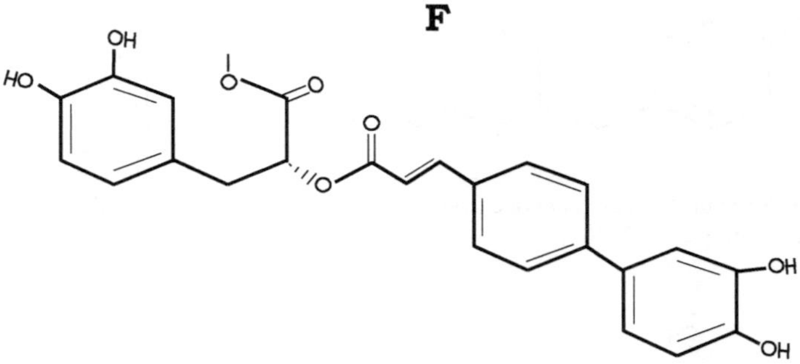

Fig. 7.10f Structure of new phenolic compound *lavandudiphenyl-A* identified from *Lavandula angustifolia* (Nigary et al. 2018)

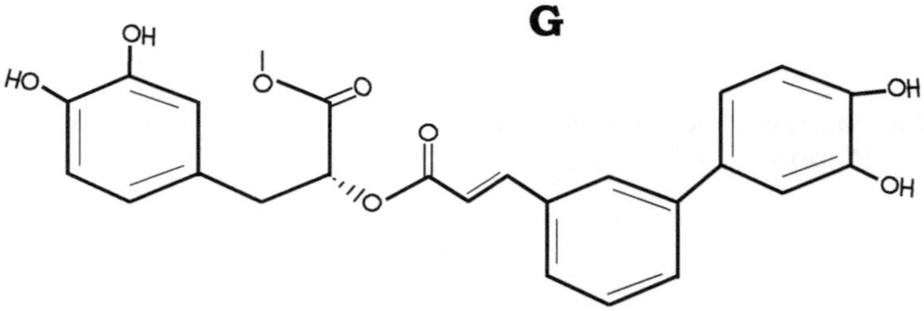

Fig. 7.10g Structure of new phenolic compound *lavandudiphenyl-B* identified from *Lavandula angustifolia* (Nigary et al. 2018)

lavandupyrone-B (5), it was identified as $C_{17}H_{13}O_7$ and 329.0740; and for lavandudiphenyl-A (6) and lavandudiphenyl-B (7), it was identified as $C_{25}H_{22}O_8$ and 439.1315 (Nigary et al. 2018).

7.10 Pandanaceae

Pandanus tectorius is an important medicinal plant of Pandanaceae family which is widely distributed in China. Previously from the fruits of *P. tectorius*, several phytochemicals of phenolic and flavonoid nature such as tangeretin, ethyl caffeate, and dihydroconiferyl alcohol have been isolated (Zhang et al. 2012). In 2016, Xiaopo et al. have isolated and identified two new phenolic compounds, i.e., pandanusphenol-A and pandanusphenol-B, from the fruits of *P. tectorius*. For isolation, the fruits were dried and macerated with 95% ethanol for 48-h duration, and then the filtrate is concentrated under vacuum to form crude extract (Xiaopo et al. 2013).

The crude extract was subjected to silica gel-based column chromatography to separate the individual compounds, and the separated individual compounds were subjected to their identification in terms of chemical formulae, mass, and their structures (Figs. 7.11a and 7.11b) through spectroscopic analysis of 2D NMR experiments (1H-1H COSY, HMBC, and NOESY), HR-ESI-MS, and CD (Xiaopo et al. 2013). Through the generated spectroscopic and CD data, the chemical formulae and mass of pandanusphenol-A were elucidated as $C_{21}H_{24}O_{06}$ and 395.1471, whereas for pandanusphenol-B it was elucidated as $C_{12}H_{16}O_{05}$ and 263.0895 (Xiaopo et al. 2013).

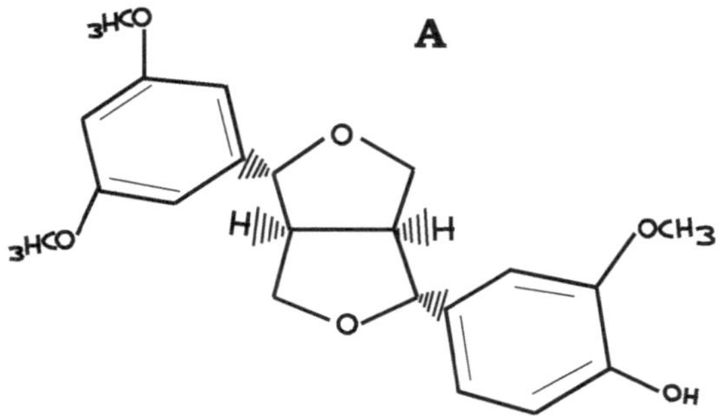

Fig. 7.11a Structure of new phenolic compound *pandanusphenol-A* identified from *Pandanus tectorius* (Xiaopo et al. 2013)

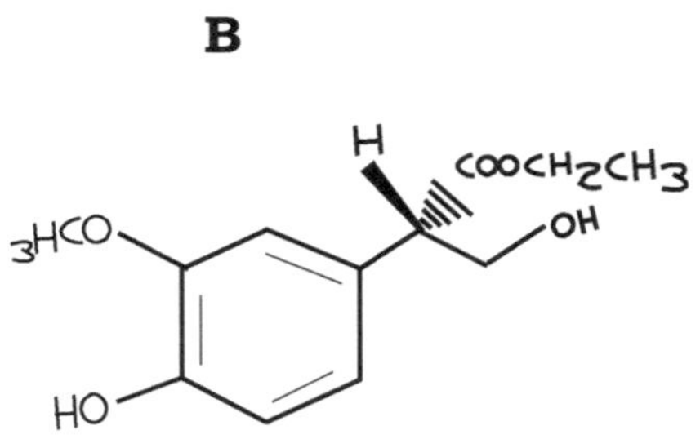

Fig. 7.11b Structure of new phenolic compound *pandanusphenol-B* identified from *Pandanus tectorius* (Xiaopo et al. 2013)

7.11 Asparagaceae

In family Asparagaceae, the plant species *Asparagus officinalis* is famously known as a vegetable crop across the world (Caporali et al. 1994). Majorly, stem portion of this plant is consumed as a food material and also used to prepare extracts for the treatment of various ailments (Jang et al. 2004). Medicinal properties of stem portion of this plant are majorly due to the presence of pharmaceutically important phytochemicals such as flavonoids, saponins (Huang et al. 2008), sugars (Fukushi et al. 2000), sulfur compounds (Terada et al. 1995), nitrogen compounds (Yanagawa et al. 1973), and acetylenic compounds (Jiang et al. 2014). Investigations in the stem portion of *Asparagus officinalis* have led to the isolation and identification of two new phenolic compounds (asparoffin-C and asparoffin-D) along with four known phenolic compounds (asparinin-A, asparenyol, gobicusin B, and 1-methoxy-2-hydroxy-4-[5-(4-hydroxyphenoxy)-3-penten-1-ynyl] phenol) (Xue et al. 2016).

After separation of compounds by various chromatographic techniques, individual compounds were subjected to their identification in terms of chemical formulae, mass, and structures (Fig. 7.12a, b) by using various techniques of spectroscopy such as IR, ER-ESI-MS, ^{1}H-NMR, ^{13}C-NMR, DEPT spectra, 1H-1H COSY, and HMBC (Xue et al. 2016). Chemical formulae and mass of asparoffin-C were identified as $C_{20}H_{22}O_4$ and 365.1150, whereas for asparoffin-D it was identified as $C_{18}H_{16}O_3$ and 279.1027 (Xue et al. 2016).

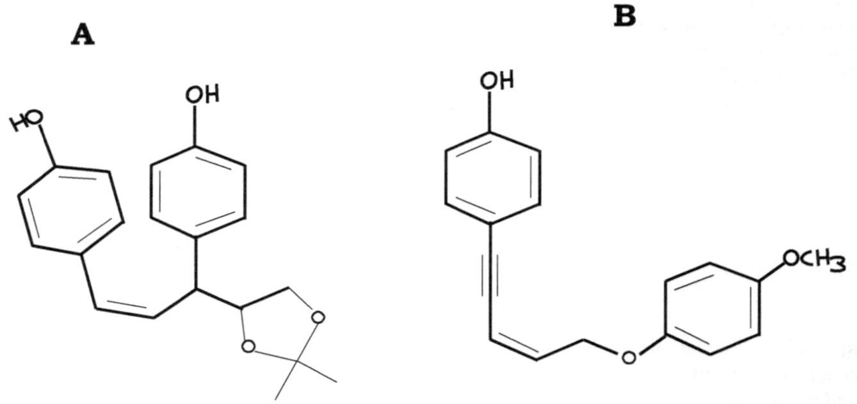

Fig. 7.12 Structure of new phenolic compounds identified from *Asparagus officinalis* (Xue et al. 2016)
(**a**) Structure of asparoffin-C, (**b**) structure of asparoffin-D

7.12 Lauraceae

Machilus yunnanensis is a tree species in family Lauraceae with main distribution in Sichuan and Yunnan provinces of China (The Flora of China, 1982). This extract of this plant species is traditionally used as a medicine for the treatment of skin-related disorders, sore, bone fractures, mumps, burns, and rheumatism (Chinese Medicine Resources Annales, 1994). Several types of phytochemicals such as terpenoids (Cheng et al. 2009), lignans (Bu et al. 2013), glycosides (Gan et al. 2012), lactones (Liu et al. 2012), and alkaloids (Liu et al. 2007a, b) were investigated in the past from this plant species. Recently, Hui et al. (2016) have carried further phytochemical investigations of this plant and identified two new phenolic compounds (i.e., 8-*O*-acetyl-phenylethanoid-4-*O*-*β*-D-glucopyranoside and (*E*)-2,3-bis(4-hydroxyphenyl) acrylaldehyde) along with 16 other known compounds from seed extracts with extraction procedure employing 70% aqueous acetone.

Initially, seeds *M. yunnanensis* were air-dried and extracted with 70% aqueous acetone, and seed extracts were concentrated under vacuum and partitioned with water and ethyl acetate (Hui et al. 2016). Compounds further purified by using silica gel and HPLC chromatographic techniques and individual compounds were subjected to their identification in terms of chemical formulae, mass, and structures (Fig. 7.13a, b) by using various techniques of spectroscopy such as ER-EI-MS, ^{1}H-NMR, ^{13}C-NMR, HM, QC, and HMBC (Hui et al. 2016). Chemical formulae and mass of compound-1 were identified as $C_{16}H_{22}O_8$ and 365.1207, whereas for compound-2 it was identified as $C_{15}H_{12}O_3$ and 263.0679 (Hui et al. 2016).

7.13 Fagaceae

There is a recent report about the identification of one new phenolic compound from *Castanea mollissima* of Fagaceae family (Wu et al. 2019). This plant species is commonly known as Chinese chestnut and is native to China and possibly to Korea (Jia et al. 2010). It has been in use as a traditional medicine against epistaxis, hemafecia, pertussis, and regurgitation (Jia et al. 2010). Extracts of this plant are also reported for its antimicrobial activity (You et al. 2014), for triggering necrosis and arresting the cell cycle of cultured animal cells (Hep G_2 cells) (Zhang et al.

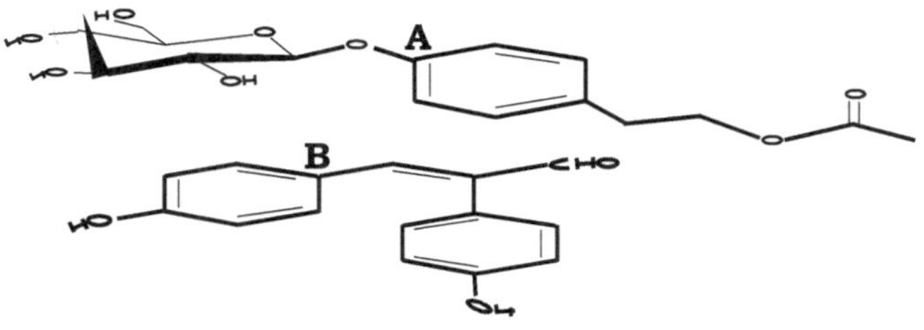

Fig. 7.13 Structure of new phenolic compounds identified from *Machilus yunnanensis* (Hui et al. 2016)

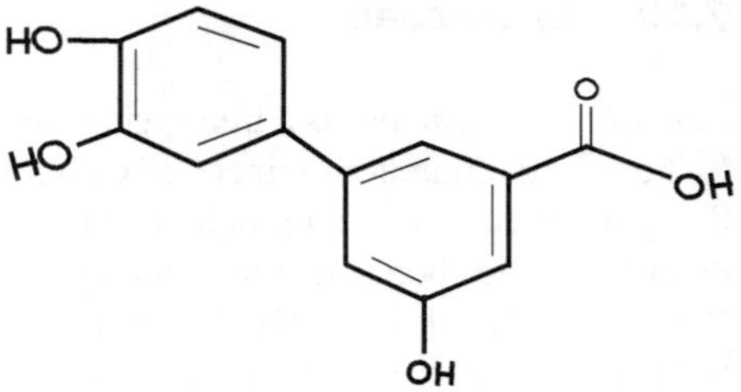

Fig. 7.14 Structure of new phenolic compound *castanol-b* identified from *Castanea mollissima* (Wu et al. 2019)

2014). After processing of chestnuts by agricultural industries, the shell of *C. mollissima* is often considered as a waste product, but now it's been considered for its medicinal values as the shells possess pharmaceutically important compounds. From the shells of *C. mollissima*, recently, Wu et al. (2019) have identified one new phenolic compound (castanol-B) along with other six known compounds.

In their study, Wu et al. (2019) have isolated the phenolic compounds from methanolic extracts of shells of *C. mollissima*. Phenolic compounds were separated into individual compounds and purified further by preparative HPLC chromatographic technique. The separated individual compounds were subjected to identification in terms of their chemical formulae, mass, and structure by carrying various spectroscopic investigations such as ER-ESI-MS, ^{1}H-NMR, ^{13}C-NMR, DEPT Spectra, and HMBC (Wu et al. 2019). Chemical formulae and mass of newly identified compound, i.e., castanol-b, were identified as $C_{13}H_{10}O_5$ and 245.0462 (Wu et al. 2019) (Fig. 7.14).

7.14 Conclusion

Plant phenolic-based research should be given much more importance for understanding their synthesis, structure, physiological functions, and their role in interaction with other organisms and environment. Newly identified phenolic compounds of different plant families should also be screened for their medicinal properties, so that on research and commercial basis every individual identified phenolic compound could be isolated at large scales and can be utilized as a phytomedicinal compound in treatment of specific disorder or a disease (Table 7.1).

Table 7.1 Newly identified phenolic compounds from several plant species during the present decade (2010–April 2019)

Sl. No	New phenolic compound	Chemical formula	Mol.wt (gr)	Plant species	Family	References
01	p-Hydroxyphenylferulate	$C_{16}H_{14}O_5$	287.0968	*Liatris elegans*	Asteraceae	Zhizhen et al. (2010)
02	5,30-Dihydroxy-4,40-dimethoxy-2,70-cyclolignа-7,70-diene-9,90-lactone	$C_{20}H_{16}O_6$	353.1066			
03	Thamnoliadepside-A	$C_{18}H_{18}O_6$	329.1028	*Thamnolia vermicularis*	Icmadophilaceae	Jia et al. (2011)
04	Thamnoliadepside-B	$C_{18}H_{18}O_7$	345.0975			
05	Thamnolic acid-A	$C_{12}H_{14}O_6$	253.0716			
06	3-(6,10,14-Trimethyl pentadecan-2-yl) furan-2	–	–	*Averrhoa bilimbi* Linn	Oxalidaceae	Gunawan and Anamy (2013)
07	2,3-bis(2,6,10-Trimethylundeca-1,5,9-trienyl) oxirane	–	–			
08	2,4-Dihydroxy-6-((4-methylpentyloxy)methyl) benzaldehyde	–	–			
09	4-Hydroxybenzyl vanillyl ether	$C_{15}H_{16}O_4$	260.1041	*Gastrodia elata*	Orchidaceae	Ah-Reum et al. (2011)
10	4-{4-[4-(Methoxymethyl) phenoxy]benzyl}oxy} benzyl methyl ether	$C_{23}H_{24}O_4$	364.1695			
11	2-(2′-Methoxy-4′-hydroxy)-aryl-3-methy-6-hydroxy-benzofuran	$C_{16}H_{14}O_4$	271.0951	*Glycyrrhiza yunnanensis*	Fabaceae	Qing et al. (2013)
12	(2S)-6,7-(2,2-Dimethyl dihydropyrano)-8-prenyl-4′-hydroxyflavanone	$C_{25}H_{28}O_4$	393.2060			
13	6-Prenyl-7,3′,4′-trihydroxyflavone	$C_{20}H_{18}O_5$	339.1227			
14	Teasperol	$C_{14}H_{14}O_6$	277.0719	*Camellia sinensis*	Theaceae	Ayaka et al. (2013)
15	Teasperin	$C_{13}H_{12}O_4$	231.0665			
16	Broussoside-A	$C_{32}H_{44}O_{13}$	659.2674	*Broussonetia papyrifera*	Moraceae	Yang et al. (2014)
17	Broussoside-B	$C_{32}H_{44}O_{13}$	659.2674			
18	Broussoside-C	$C_{33}H_{46}O_{13}$	673.2831			
19	Broussoside-D	$C_{26}H_{34}O_9$	513.2095			
20	Broussoside-E	$C_{32}H_{42}O_{13}$	657.2518			

(continued)

Table 7.1 (continued)

Sl. No	New phenolic compound	Chemical formula	Mol.wt (gr)	Plant species	Family	References
21	Aphyllone-A	$C_{15}H_{14}O_4$	259.0965	*Dendrobium aphyllum*	Orchidaceae	Dan et al. (2015)
22	Aphyllone-B	$C_{17}H_{20}O_6$	320.1260			
23	Aphyllal-C	$C_{15}H_{14}O_4$	259.0965			
24	Aphyllal-D	$C_{15}H_{16}O_5$	275.0924			
25	Aphyllal-E	$C_{16}H_{18}O_5$	313.1052			
26	Schitenoside-A	$C_{21}H_{24}O_{10}$	459.1267	*Schizonepeta tenuifolia*	Lamiaceae	
27	Schitenoside-B	$C_{24}H_{26}O_{12}$	529.1322			
28	Pandanusphenol A	$C_{21}H_{24}O_6$	395.1471	*Pandanus tectorius*	Pandanaceae	Xiaopo et al. (2013)
29	Pandanusphenol B	$C_{12}H_{16}O_5$	263.0895			
30	Asparoffin-C	$C_{20}H_{22}O_4$	365.1150	*Asparagus officinalis*	Asparagaceae	Xue et al. (2016)
31	Asparoffin-D	$C_{18}H_{16}O_3$	279.1027			
32	8-*O*-acetyl-phenylethanoid-4-*O*-*β*-D-glucopyranoside	$C_{16}H_{22}O_8$	365.1207	*Machilus yunnanensis*	Lauraceae	Hui et al. (2016)
33	2,3-bis(4-Hydroxyphenyl) acrylaldehyde	$C_{15}H_{12}O_3$	263.0679			
34	Lavandunat	$C_{12}H_{13}O_6$	253.0790	*Lavandula angustifolia*	Lamiaceae	Nigary et al. (2018)
35	Lavandufurandiol	$C_{25}H_{20}O_9$	463.1107			
36	Lavandufluoren	$C_{26}H_{22}O_9$	477.1264			
37	Lavandupyrone-A	$C_{26}H_{19}O_{10}$	491.1056			
38	Lavandupyrone-B	$C_{17}H_{13}O_7$	329.0740			
39	Lavandudiphenyl-A	$C_{25}H_{22}O_8$	449.1315			
40	Lavandudiphenyl-B	$C_{25}H_{22}O_8$	449.1315			
41	Castanol-B	$C_{13}H_{10}O_5$	245.0462	*Castanea mollissima*	Fagaceae	Wu et al. (2019)

References

Ah-Reum H, Hyun JS, Ha RJ, Je-Hyun L, Dongho L, Eun KS (2011) Two new phenolic compounds from the rhizomes of *Gastrodia elata* Blume. Helvetica 1310 Chimica Acta 94

An H, Kim IS, Koppula S, Kim BW, Park PJ, Lim BO, Choi WS, Lee KH, Choi DK (2010) Protective effects of Gastrodia elata Blume on MPP+-induced cytotoxicity in human dopaminergic SH-SY5Y cells. J Ethnopharmacol 130:290

Ayaka K, Arisa S, Hideyuki N, Yoko N, Ichiro Y, Yosuke M, Takashi T, Kanji I (2013) New phenolic compounds from *Camellia sinensis* L. fermented leaves. J Nat Med 67:652–656. https://doi.org/10.1007/s11418-012-0704-5

Baker B, Zambryski P, Staskawicz B, Dinesh Kumar SP (1997) Signaling in plant-microbe interactions. Science 276:726–733

Bhattacharya A, Sood P, Citovsky V (2010) The roles of plant phenolics in defence and communication during *Agrobacterium* and *Rhizobium* iñfection. Mol Plant Pathol 11(5):705–719. https://doi.org/10.1111/j.1364-3703.2010.00625.x

Bu PB, Li YR, Jiang M, Wang F, Wang XL, Lin S, Zhu CG, Shi JG (2013) Glycosides from the bark of Machilus robusta. J Asian Nat Prod Res 15:482

Caporali E, Carboni A, Galli MG, Rossi G, Spada A, Longo GPM (1994) Development of male and female flower in Asparagus officinalis. Search for point of transition from hermaphroditic to unisexual developmental pathway. Sex Plant Reprod 7:239

Castro VL, Leon RV, Alanon ME, Perez Coello MS, Gonzalez-Porto AV (2014) Floral origin markers for authenticating Lavandin honey (*Lavandula angustifolia* x *latifolia*). Discrimination from Lavender honey (Lavandula latifolia). Food Control 37:362–370

Chen XM, Guo SX (2001) Advances in the research of constituents and pharmacology of Dendrobium. Nat Prod Res Dev 13:70–75

Chen PJ, Hsieh CL, Su KP, Hou YC, Chiang HM, Lin IH, Sheen LY (2008a) The antidepressant effect of Gastrodia elata Bl. on the forced-swimming test in rats. Am J Chin Med 36:95

Chen YG, Li JT, Wang LQ, Liu Y (2008b) Aromatic compounds from *Dendrobium aphyllum*. Biochem Syst Ecol 36:458–460

Cheng W, Zhu CG, Xu WD, Fan XN, Yang YC, Li Y, Chen XG, Wang WJ, Shi JG (2009) Chemical constituents of the bark of Machilus wangchiana and their biological activities. J Nat Prod 72:2145

Cheng AW, Wang H, Yang H, Shi L, Katz Y, Theunissen TW, Rangarajan S, Shivalila CS, Dadon DB, Jaenisch R (2013) Multiplexed activation of endogenous genes by CRISPR-on, an RNA-guided transcriptional activator system. Cell Res 23:1163–1171. https://doi.org/10.1038/cr.2013.122

Cheynier V, Comte G, Davies K, Lattanzio V, Martens S (2013) Plant phenolics: recent advances on their biosynthesis, genetics, and ecophysiology. Plant Physiol Biochem Societe francaise de physiologie vegetale 72:1–20. https://doi.org/10.1016/j.plaphy.2013.05.009

Chinese Herbal Medicine Company, Chinese Medicine Resources Annales (1994) J Asian Nat Prod Res:295–312. Science Press, Beijing

Chinese Pharmacopoeia Commission (1999) Drug standard of Ministry of Public Health of the Peoples Republic of China (Uygur Medicine Branch). Xinjiang Science and Technology Health Publishing House

Crozier A, Jaganath I , Clifford M (2007) Phenols, polyphenols and tannins: an overview. https://doi.org/10.1002/9780470988558.ch1

Dan Y, Liang YL, Zhong QC, Feng QX, Wei WF, Cheng TZ, Fa WD, Jun Z, Zhong TD, Jiang MH (2015) Five new phenolic compounds from *Dendrobium aphyllum*. Fitoterapia 100:11–18. https://doi.org/10.1016/j.fitote.2014.11.004

Danh LT, Triet NDA, Han LTN, Zhao J, Mammucari R, Foster N (2012) Antioxidant activity, yield and chemical composition of lavender essential oil extracted by supercritical CO2. J Supercrit Fluids 70:27–34

Dong PX, Ya L, Xiao M, Tong Z, Yue Z, Jie Z, Jiao JZ, Hua B (2017) Natural antioxidants in foods and medicinal plants: extraction, assessment and resources. Int J Mol Sci 18(1):96. https://doi.org/10.3390/ijms18010096

Dykes L, Rooney LW (2007) Phenolic compounds in cereal grains and their health benefits. Cereal Foods World 52(3):105–111

Editorial Board of the Flora of China (1982) The flora of China, vol 31. Science Press, Beijing, p 14

Elizabeth ME, John IY (1998) Plant-plant communications: rhizosphere signaling between parasitic angiosperms and their hosts. Plant Physiol 116:1–7

Fialova S, Tekelova D, Grancai D (2012) The content of phenolic compounds in underground and aerial parts of different *Mentha* species. Acta Facultatis Pharmaceuticae Universitatis Comenianae 59:30–38. https://doi.org/10.2478/v10219-012-0013-9

Fukushi E, Onodera S, Yamamori A, Shiomi N, Kawabata J (2000) NMR analysis of tri- and tetrasaccharides from asparagus. Magn Reson Chem 38:1005

Gan ML, Liu MT, Gan LS, Lin S, Liu B, Zhang YL, Zi JC, Song WX, Shi JG (2012) Dammarane glycosides from the root of Machilus yaoshansis. J Nat Prod 75:1373

Gao DY, Zhang RY (1993) Studies of flavonoids from *Glycyrrhiza yunnanensis*. J Beijing Med Univ 25:302

Gao DY, Zhang RY (1994) Studies on the chemical constituents of malay licorice. Chin Tradit Herb Drug 25:507–513

Gilbert LA, Larson MH, Morsut L, Liu Z, Brar GA, Torres SE, Stern Ginossar N, Brandman O, Whitehead EH, Doudna JA, Lim WA, Weissman JS, Qi LS (2013) CRISPR-mediated modular RNA-guided regulation of transcription in eukaryotes. Cell 154:442–451. https://doi.org/10.1016/j.cell.2013.06.044

Gunawan AC, Anamy P (2013) Structure elucidation of two new phytol derivatives, a new phenolic compound and other metabolites of *Averrhoa bilimbi*. In: Presented at the research congress, De La Salle University Manila March 7–9, 2013

Hammerschmidt R (2005) Phenols and plant–pathogen interactions: the saga continues. Physiol Molec Plant Pathol 66:77–78. https://doi.org/10.1016/j.pmpp.2005.08.001

Haraldsdottir S, Guolaugsdottir E, Ingolfsdottir K, Ogmundsdottir HM (2004) Anti-proliferative effects of lichen-derived lipoxygenase inhibitors on twelve human cancer cell lines of different tissue origin in vitro. Planta Med 70:1098–1100

Hu JF, Shen FJ (1995) Structure of a new alkaloid from the roots of *Glycyrrhiza yunnanensis*. Chem J Chin Univ 16:1245–1247

Hu JF, Ye ZL, Shen FJ (1994) Constituents from the roots of *Glycyrrhiza yunnanensis* (I). Nat Prod Res Dev 6:6–8

Huang XF, Lin YY, Kong LY (2008) Steroids from the roots of Asparagus officinalis and their cytotoxic activity. J Integr Plant Biol 50:717

Hui L, Dan L, Yan Y, Ling Z, De BP, Jie L, Xue JC, Xu Z, Shuang L, Jun BG, Yong MW, Xu J, Wei LX (2016) Two new phenolic compounds from the seeds of *Machilus yunnanensis*. J Asian Nat Prod Res 2016:1–7. https://doi.org/10.1080/10286020.2016.1187139

Hwang SM, Lee YJ, Kang DG, Lee HS (2009) Anti-inflammatory effect of Gastrodia elata rhizome in human umbilical vein endothelial cells. Am J Chin Med 37:395

Jang DS, Cuendet M, Fong HHS, Pezzuto ZM, Kinghorn AD (2004) Constituents of Asparagus officinalis evaluated for inhibitory activity against cyclooxygenase-2. J Agric Food Chem 52:2218–2222

Jang YW, Lee JY, Kim CJ (2010) Anti-asthmatic activity of phenolic compounds from the roots of Gastrodia elata Bl. Int Immunopharmacol 10:147

Jang MS, Park HY, Nam KH (2014) Whitening effects of 4-hydroxyphenethyl alcohol isolated from water boiled with *Hizikia fusiformis*. Food Sci Biotechnol 23:555–560

Jia L, Xi F, Wang N, Jing LL, Kong DY (2010) Chemical constituents of Castanea mollissima Blume Shell. Chin J Pharm 41:98–102

Jia G, Zhan-lin L, Ao-li W, Xiao-qiu L, JianW XG, Yong-kui J, Hui-ming H (2011) Three new phenolic compounds from the lichen *Thamnolia vermicularis* and their antiproliferative effects in prostate cancer cells. Planta Med 77:2042–2046

Jiang B, Sun HD (2002) Constituents from Thamnolia vermicularis. Acta Bot Yunn 24:525–530

Jiang D, Tao FY, Li YQ, Zhao W (2014) Sci Technol Food Ind 35:357

Kazuhiro O, Katsuki O, Oleanane RK (1994) Glycosides from *Glycyrrhiza yunnanensis* roots. Phytochemistry 36:139–145

Kim HJ, Moon KD, Oh SY, Kim SP, Lee SR (2001) Ether fraction of methanol extracts of Gastrodia elata, a traditional medicinal herb, protects against kainic acid-induced neuronal damage in the mouse hippocampus. Neurosci Lett 314:65

Lee IK, Kim MA, Lee SY, Hong JK, Lee JK, Lee KR (2008) Phytochemical constituents of *Schizonepeta tenuifolia* Briquet. Nat Prod Sci 14:100–106

Liu FL, Fu WJ, Yang DR, Peng YQ, Zhang XW, He JZ (2004) Reinforcement of bee-plant interaction by phenolics in food. J Apicult Res 43:155–157

Liu F, Chen J, Chai J, Zhang X, Bai X, He D (2007a) Adaptive functions of defensive plant phenolics and a non-linear bee response to nectar components. Funct Ecol 21:96–100. https://doi.org/10.1111/j.1365-2435.2006.01200.x

Liu MT, Lin S, Wang YH, He WY, Li S, Wang SJ, Yang YC, Shi JG (2007b) Yaoshanenolides A and B: new spirolactones from the bark of Machilus yaoshansis. Org Lett 9:129

Liu MT, Lin S, Gan ML, Liu B, Zi JC, Song WX, Zhang YL, Fan XN, Liu Y, Tan W, Wang SJ, Yang YC, Shi JG (2012) Butanolide derivatives from the bark of Machilus yaoshansis. J Asian Nat Prod Res 14:713

Majumder PL, Sen RC (1987) Structure of moscatin a new phenanthrene derivative from the orchid Dendrobium moscatum. India J Chem 26B:18–20

Mandal SM, Chakraborty D, Dey S (2010) Phenolic acids act as signaling molecules in plant-microbe symbioses. Plant Signal Behav 5(4):359–368

Naczk M, Shahidi F (2012) Extraction and analysis of phenolics in food. J Chromatogr A 1054:1–2

Ng TB, Liu J, Wong JH, Ye XJ, StephenCWS TY (2012) Review of research on Dendrobium, a prized folk medicine. Appl Microbiol Biotechnol 93:1795–1803

Nicolson SW, Lerch-henning S, Welsford M, Johnson SD (2015) Nectar palatability can selectively filter bird and insect visitors to coral tree flowers. Evol Ecol 29:405–417. https://doi.org/10.1007/s10682-014-9718-7

Nigary Y, Khayrulla B, Gen L, Haji AA (2018) Seven new phenolic compounds from Lavandula angustifolia. Phytochem Lett 23:149–154. https://doi.org/10.1016/j.phytol.2017.12.005

Nitzsche A, Tokalov SV, Gutzeit HO, Ludwig MJ (2004) Chemical and biological characterization of cinnamic acid derivatives from cell cultures of lavender (Lavandula officinalis) induced by stress and jasmonic acid. J Agric Food Chem 52(10):2915–2923

Ohtani K, Ogawa K, Kasai R (1992) Oleanane glycosides from *Glycyrrhiza yunnanensis* roots. Phytochemistry 31:1747–1752

Oshima Y, Takata S, Hikino H (1989) Schizonodiol, schizonol, and schizonepetosides D and E, monoterpenoids of schizonepeta tenuifolia Spikes1. Planta Med 55(2):179–180

Qing W, Shuai J, Si-wang Y, Hui-xia W, Xiong-hao L, Tao-tao M, Xue Q, Cheng X, Min Y, Guo D-a (2013) Three new phenolic compounds from the roots of *Glycyrrhiza yunnanensis*. Fitoterapia 85:35–40. https://doi.org/10.1016/j.fitote.2012.12.027

Redman RS, Freeman S, Clifton DR, Morrel J, Brown G, Rodriguez RJ (2003) Biochemical analysis of plant protection afforded by a nonpathogenic endophytic mutant of colletotrichum magna. Plant Physiol 119:795–804

Seo BI, Joo SJ, Park JH, Roh SS, Kim SM, Choi HS, Ku SK (2010) The antiosteoporotic effect of aqueous extracts of Gastrodiae Rhizoma in vitro and in vivo. J Health Sci 56:422–433

Shalem O, Sanjana NE, Hartenian E, Shi X, Scott DA, Mikkelsen TS, Heckl D, Ebert BL, Root DE, Doench JG, Zhang F (2014) Genome-scale CRISPR-Cas9 knockout screening in human cells. Science 343:84–87

Shan B, Cai YZ, Sun M, Corke H (2005) Antioxidant capacity of 26 spice extracts and characterization of their phenolic constituents. J Agric Food Chem 53:7749–7759

Shao L, Huang WH, Zhang CF, Wang L, Zhang M, Wang ZT (2008) Study on chemical constituents from stem of *Dendrobium aphyllum*. China J Chin Mater Med 33:1693–1695

Shi YG (2012) Lavender volatile composition analysis and its quality control, vol. MEM. Xinjiang University

Sin MK, Kim CM, Ahn DK, Lee KS (1997) Dictionary of Chinese Materia Medica (in Korean). Jungdam Publishing, Seoul

Siqueira JO, Nair MG, Hammerschmidt R, Safir GR (1991) Significance of phenolic compounds in plant-soil-microbial systems. Crit Rev Plant Sci 10:63–121

Sun HD, Shen XY (1985) The structure of vermicularin. Acta Bot Yunn 7:109–113

Terada K, Honda C, Suwa K, Takeyama S, Oku H, Kamisako W (1995) Acetylenic compounds isolated from cultured cells of Asparagus officinalis. Chem Pharm Bull 43:564

Torras CL, Jauregui O, Bastida J, Codina C, Viladomat F (2007) Antioxidant activity and phenolic composition of lavandin (Lavandula x intermedia Emeric ex Loiseleur) waste. J Agric Food Chem 55:8436–8443

Vittorio V, Clay F (2013) Chemical Signaling between plants and plant-pathogenic Bacteria. Annu Rev Phytopathol 51:17–37

Wu X, Liu J, Yu Z, Ye Y, Zhou Y (2007a) Chemical constituents of *Lavandula angustifolia*. Acta Chim Sin 65:1649–1653

Wu X, Liu J, Yu Z, Ye Y, Zhou Y (2007b) Studies on flavones in of *Lavandula angustifolia*. China J Chin Mater Med:821–823

Wu F, Yao X, Xu J, Wu Y, Yang Y, Jin Y, Xie H, Liu Y, Yang Y, Zheng X (2019) One new phenolic compound from Castanea mollissima shells and its suppression of hepatoma cell proliferation and inflammation by inhibiting NF-κβ pathway. Int J Mol Sci 20:466. https://doi.org/10.3390/ijms20030466

Xiaopo Z, Haifeng W, Chongming W, Peng G, Xudong X, Yang M (2013) Pandanusphenol a and B: two new phenolic compounds from the fruits of *Pandanus tectorius* Soland. Rec Nat Prod 7 (4):359–362

Xu HH, Jia HS, Chun CZ, Jia C, Xiu HS, Xiao QX, Wen TZ (2016) Two new phenolic compounds from *Schizonepeta tenuifolia* (Benth.) Briq. Rec Nat Prod 10(5):604–608

Xue ML, Jin LC, Le W, Wen XW, Hong LA, Zi CM (2016) Two new phenolic compounds and antitumor activities of asparinin a from *Asparagus officinalis*. J Asian Nat Prod Res. https://doi.org/10.1080/10286020.2016.1206529

Yanagawa H, Kato T, Kitahara Y (1973) Asparagusic acid-S-oxides, new plant growth regulators in etiolated young asparagus shoots. Tetrahedron Lett 14:1073

Yang J, Gao FL (2010) Study on antioxidant activities of total flavonoids from lavandula angustifolia of Xinjiang. China Food Addit 2:162–165

Yang JY, Lee HS (2013) Changes in acaricidal potency by introducing functional radicals and an acaricidal constituent isolated from *Schizonepeta tenuifolia*. J Agric Food Chem 61:11511–11516

Yang C, Li F, Du B, Chen B, Wang F (2014) Isolation and characterization of new phenolic compounds with Estrogen biosynthesis-inhibiting and Antioxidation activities from *Broussonetia papyrifera* leaves. PLoS One 9(4):e94198. https://doi.org/10.1371/journal.pone.0094198

Yong W, Xing TR, Wang S, Chen L, Hu P, Li CC, Wang HI, Wang M, Chen JT, Ruan DT (2009) Protective effects of gastrodin on lead-induced synaptic plasticity deficits in rat hippocampus. Planta Med 75:1112–1117

Yoo JS, Kim DK, Sangjin K, Shin TY (2011) Anti-allergic effects of *Schizonepeta tenuifolia* on mast cell-mediated allergy model. Nat Prod Sci 17:239–244

You TT, Zhou SK, Wen JL, Ma C, Xu F (2014) Chemical composition, properties, and antimicrobial activity of the water soluble pigments from *Castanea mollissima* shells. J Agric Food 62:1936–1944

Yu SS, Zhao J, Zheng WP, Zhao Y (2010) Brain Res 1308:167

Zeng L, Zhang RY, Wang D (1990) Two triterpenoids from roots of *Glycyrrhiza yunnanensis*. Phytochemistry 29:3605–3659

Zhang GN, Bi ZM, Wang ZT, Xu LS, Xu GJ (2003) Advances in studies on chemical constituents from plants of *Dendrobium* Sw. Chin Tradit Herb Drug 6:102–105

Zhang YH, Zhou L, Shi RB, Guo YJ, Dong Y (2006) Studies on chemical constituents in spikes of *Schizonepeta tenuifolia*. Chin J Chin Mater med 31:1247–1249

Zhang CF, Shao L, Huang WH, Wang L, Wang ZT, Xu LS (2008) Phenolic components from herbs of *Dendrobium aphyllum*. China J Chin Mater Med 33:2922–2925

Zhang XP, Guo P, Sun GL, Chen SL, Yang MH, Fu NG, Wu HF, Xu XD (2012) Phenolic compounds and flavonoids from the fruits of *Pandanus tectorius* Soland. J Med Plant Res 6:2622–2626

Zhang H, Ke J, ShaoT LJ, Duan Y, He Y, Zhang C, Chen G, Sun G, Sun X (2014) Cytotoxic effects of procyanidins from *Castanea mollissima* Bl. shell on human hepatoma G2 cells in vitro. Food Chem Toxicol 64:166–176

Zhang FP, Yang QY, Zhang SB (2016) Dual effect of phenolic nectar on three floral visitors of *Elsholtzia rugulosa* (Lamiaceae) in SW China. PLoS One 11(4):e0154381. https://doi.org/10.1371/journal.pone.0154381

Zhao YL, Wang SL, Li XY (1994) Studies on the polysaccharides of *Dendrobium aphyllum*. Acta Bot Yunnan, 16:392–396

Zhizhen Z, Stacy O, Ping W, Wei Y, Wanli Z, Shiyou L (2010) New phenolic compounds from *Liatris elegans*. Nat Prod Res Form Nat Prod Lett 24(11):1079–1085. https://doi.org/10.1080/14786410903027789

Phenolic Allelochemicals from Crops and Weed Management

8

Chella Perumal Palanisamy, Vinoth Prasanna Gunasekaran, Sophia Dominic, and Tran Dang Xuan

Abstract

Plants play a major role in communicating and channelizing the growth of other plants through chemicals, which is referred to as allelopathy. The chemicals involved in the process of allelopathy are known as allelochemicals. The majority of secondary metabolites are involved in allelopathic interactions and are considered to be allelochemicals. Some well-known allelochemicals are phenolic compounds, alkaloids, terpenes, benzoxazinoids, tannins, etc. Allelopathic efficiency of crops may be utilized for weed management without adding or spraying any synthetic weed control agents as these synthetic weed control agents may pose a threat to the field crops and henceforth the ecosystem. In the natural ecosystem, phenolic compounds have been proved to be the most important and efficient plant allelochemical. Here, this chapter focuses on the classification, extraction, isolation, and purification of phenolic compounds and its role in weed management.

C. P. Palanisamy
Department of Plant Sciences, University of the Free State, QwaQwa Campus, Phuthaditjhaba, South Africa

V. P. Gunasekaran
Department of Biomedical Science, Bharathidasan University, Tiruchirappalli, Tamil Nadu, India

S. Dominic
Department of Biochemistry, Jyoti Nivas Degree College, Bangalore, India

T. D. Xuan (✉)
Graduate School for Advanced Science and Engineering, Transdisciplinary Science and Engineering Program, Hiroshima University, Hiroshima, Japan
e-mail: tdxuan@hiroshima-u.ac.jp

R. Lone et al. (eds.), *Plant Phenolics in Sustainable Agriculture*,
https://doi.org/10.1007/978-981-15-4890-1_8

Keywords

Allelopathy · Phenolic allelochemicals · Secondary metabolites · Weed management

8.1 Introduction

Due to the fast growing of world population in recent decades, alternative technologies are warranted for large-scale and industrial farming production (Wezel et al. 2014). Environmental and sustainability problems, overuse of chemical pesticides and fertilizers, other synthetic chemicals in the farming, and its negative health effects are the basic key factors for creating environmentally friendly integrated production approach (Zhao et al. 2006). Synthetic weedicides or herbicides are acutely toxic and can be lethal to the people who are in contact with it for a long term. Thus, the impressions for organically developed products are constantly rising (Smith-Spangler et al. 2012).

The plants are able to synthesize massive amount of chemicals or by-products in response to stress and other unpleasant conditions like harsh climate, grazing animals, disease conditions, drought conditions, nutrition deficiency, etc. (Baranski et al. 2014). These by-products or chemical substances are called as "secondary metabolites." Initially, these secondary metabolites were considered to be insignificant to the plant and the animal kingdom. But later on, a number of extensive studies and investigations by ecologists, phytochemists, and pharmacologists have unraveled its potential biological importance (Veberic 2016). These compounds formed from the plants and microorganisms also known as alleleochemicals have a major role in the natural habitat of plants (Pagare et al. 2015). These allelochemicals play an important role in crop production and weed management. The concentration of allelochemicals varies from one plant to the other plant (Piasecka et al. 2015). The synthesis of allelochemicals occurs in several or all parts of the plant tissues. Plants communicate to their surrounding ecosystem through these allelochemicals. This can be achieved through the release of allelochemicals from the plants' root or from the other parts of the plant by volatilization (Akula and Ravishankar 2011).

Allelopathy has the capability to slow down or to stimulate other plant growth in the surroundings by chemical resources (Kalt 2005). Consumption of allelopathy may increase the crop production through prevention of negative impacts, exploitation of stimulatory effects, and management and development of allelopathic crops to inhibit weeds, in addition to the use of allelochemicals as herbicides and growth regulators (Kohli et al. 1998). Allelochemicals might have destructive properties on crop production, and some weeds could reveal allelopathic action in the field of agriculture (Khanh et al. 2005). There are numerous factors that may affect the synthesis and release of allelochemicals from plants. The plants releasing allelochemicals into the environment and the ones receiving these allelochemicals are called as "donor" plants and "receiver or target species," respectively (Xuan and

Tsuzuki 2002). Usually, the allelopathic activities are highly expressed in the aerial parts of the plants. The expression of defense genes in the target plants can be regulated by the allelochemicals from donor plants (Khanh et al. 2007).

Phenolics are very essential and widespread in plant allelochemicals in the ecosystem. It could be formed by different biosynthetic pathways, which are succinylbenzoate, polyketide, and mevalonate pathways (Batish et al. 2001). These pathways produce the phenylpropanoids, quinines, aromatic terpenoids, etc. The main role of phenolic compounds in plant physiology and relations with biotic and abiotic environments are intricate to overvalue (Bhattacharya et al. 2010). Phenolic compounds are giving structural reliability and scaffolding support to plants, and they are implicated in biosynthesis of lignin and a variety of pigments such as anthocyanidins, anthocyanins, etc. (Ferrer et al. 2008).

In plant tissues, many simple and complex phenolics are accumulated to act as phytoanticipins, phytoalexins, and nematicides for soilborne pathogens and phytophagous insects. Phenolic phytoalexins (secreted by wounded and perturbed plants) significantly deter and/or kill a variety of microorganisms; some pathogens could reverse these defenses or challenge them to their own advantage (Akhtar and Malik 2000). As long as phenolic compounds are recognized as phytoestrogens (in animals) and allelochemicals (in crop production and weed management), consequently, phenolics are proposed to serve as constructive substitutes for the chemical control of pathogens in agricultural products (Langcake et al. 1981). Here, this chapter focuses on the classes of phenolic compounds, extraction, isolation, and purification of phenolics and phenolics in crops and effects on weed management (Quideau 2009).

8.2 Phenolic Compound Classification

Generally, the phenolic compounds are a vast class of secondary metabolites (Fig. 8.1) which possess a crucial role in defense and survival of the plants, and their structures differ extensively. Therefore, the phenolic compounds are classified into groups according to their basic skeleton pattern (Luthria et al. 2006). Moreover, the phenolic compound classifications are numerous, and a newly identified compound does not match with the structural criteria to be incorporated in the accessible classes. Hence, Harbone (1984) modified the classification system. They are divided into three major classes, and each class is further separated into subcategories. The group of phenolic compounds to fully hierarchize their structural differences (Naczk and Shahidi 2004).

8.2.1 Simple Phenols (Phenolic Acid)

Simple phenols are compounds that have at least one hydroxyl group attached with aromatic ring in basic skeleton (Fig. 8.1) and the portion of entire family serving as generic name, the phenol is also having specific name for its simplest member,

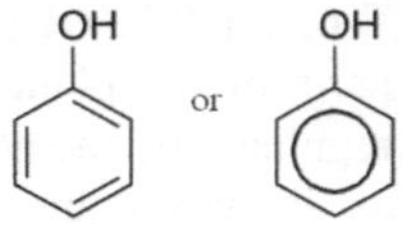

Fig. 8.1 Basic skeleton of phenolic compounds

monohydroxy benzene (Luthria 2006). Some of the rare simple phenolic compounds such as phloroglucinol, catechol, and hydroquinone were identified from aromatic and medicinal plants. The O-glycoside of hydroquinone and arbutin are recognized from different *Vaccinium sp.* leaves (such as blueberry, cranberry, cowberry, and pear trees (*Pyrus communis* L., Rosaceae)) (Zeng et al. 2008). The richest source of arbutin (compound ranging from 15% to 23% per dry weight of the plant) has been identified from *Bergenia crassifolia* (L.) when compared with other plants. Catechol has been identified from leaves of the *Gaultheria species* (Singh et al. 2001).

8.2.1.1 Hydroxycinnamic Acid and Derivatives

Hydroxycinnamic acid derivatives are a significant class of phenolic compounds initiated from the mevalonate-shikimate biosynthesis pathway in plants (Callaway and Aschehoug 2000). These groups of compounds overflowing in food that might account for about one-third of the phenolic compounds in our diet are supposed to function in some growth and developmental progression, such as floral induction, flower formation, sexual differentiation, tuberization, cell division, and cytomorphogenesis. Cinnamic acid, p-coumaric acid, ferulic acid, caffeic acid, chlorogenic acid, and rosmarinic acid are also belonging to this class (Ridenour and Callaway 2001).

8.2.1.2 Hydroxybenzoic Acid and Derivatives

Hydroxybenzoic acids (C_6–C_1 type of structure) are the component of a complex structure, and they are directly derivative compounds from benzoic acid. The occurrence ratio of hydroxybenzoic acids in foods of plant origin is usually very less, but in some berries and vegetables such as onions and horseradish, the substance of hydroxybenzoic acids could be very high. Some hydroxybenzoic acids are found in olive products (Capasso 1997).

8.2.2 Polyphenols

Polyphenols may refer as a group of component that takes place in plants that might be characterized by the presence of about one phenol unit or building block per molecule. Usually, they are involved in defense activity against ultraviolet radiation or aggression by pathogens. Moreover, they can be divided into two broad classes such as tannins and flavonoids (Capasso et al. 1992).

8.2.2.1 Tannins

Tannins are a major class of mordant polyphenolic compounds that binds to precipitate proteins and other different natural compounds. Tannins could be defined as high-molecular-weight (ranging from 500 to 3000), water-soluble compounds which form an insoluble complex by the property of combining with proteins, cellulose, gelatin, and pectin. Tannins are further classified into two main groups; they are hydrolyzable and condensed tannins (Hartmann 1996).

8.2.2.1.1 Hydrolyzable Tannins

Hydrolyzable tannins contain a polyhydric alcohol at their core; the hydroxyl groups are moderately or completely esterified among the gallic or hexahydroxydiphenic acids. They may embrace the long chains of gallic acid from the central glucose core. During the hydrolysis process with acid or enzymes, the hydrolyzable tannins go down into their constituent phenolic acids and carbohydrates (Franz 2002).

8.2.2.1.2 Condensed Tannins

The molecules possess 2–50 flavonoid units which are joined by carbon-carbon bonds, and they are not cleaved by hydrolysis called hydrolyzable tannins. It is also known as proanthocyanidins. Normally, they do not contain sugar residues (Santana et al. 2009). Various types of condensed tannins exist that are widely distributed in plants such as the procyanidins, propelargonidins, prodelphinidins, profisetinidins, proteracacinidins, proguibourtinidins, or prorobinetidins (Chou and Leu 1992).

8.2.2.2 Flavonoids

Flavonoids are a major class of secondary metabolites, widely distributed in plants, and they contain 15 carbon atoms (C6-C3-C6) in the basic skeleton which fulfills many functions. Currently, more than 5000 flavonoid compounds have been discovered from the plants (Chou and Lin 1976). In the plant, flavonoids play a significant role in plant pigments (such as flower coloration, producing yellow, red, and blue pigmentation in petals intended to attract pollinator animals), and they are concerned in floral pigmentation, symbiotic nitrogen fixation, and UV filtration. Some flavonoids could also serve as cell cycle inhibitors, chemical messengers, and physiological regulators; moreover, some flavonoids possess inhibitory activities for microorganisms causing plant diseases (Chou and Lee 1991).

Flavonoids are further classified into multiple groups such as chalcones, dihydrochalcones, aurones, flavones, flavonols, flavadiols, anthocyanidins, dihydroxyflavonols, flavanones, isoflavonoids, and biflavonoids. Flavonoid components might arise as free aglycones or as conjugated forms with other substituents such as aliphatic acids, methoxyl, prenyl, methylenedioxy, glycosyl, and isoprenyl (Tang 1986).

8.2.3 Other Groups

Other phenolic compounds of lignans, lignins, coumarins, stilbene derivatives such as resveratrol, and other compounds are not further classified into the separate subgroups (Fig. 8.2). This group of compounds may also arise as free aglycones or may be conjugated with one or more substituents (Waller 1987).

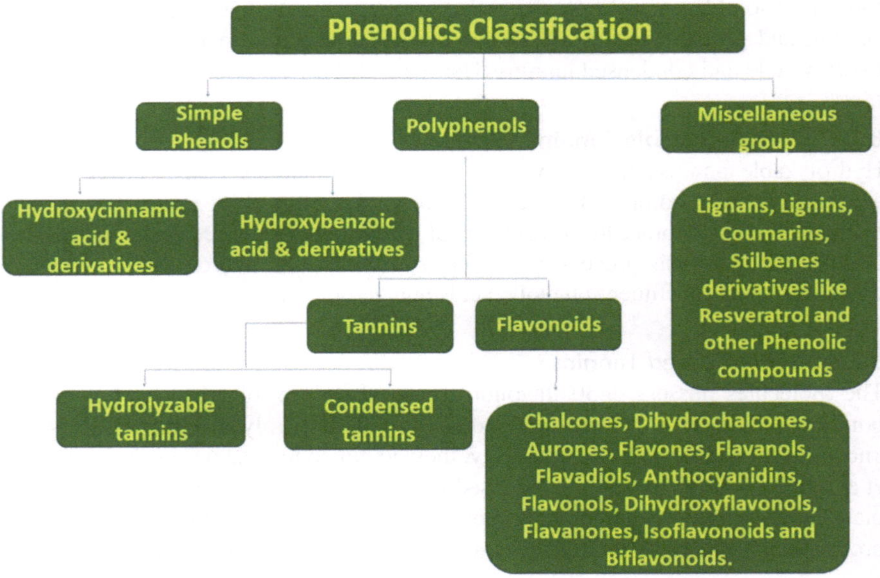

Fig. 8.2 Major classification of plant phenolic compounds

8.3 Process of Extraction and Isolation of Phenolic Compounds

The process of extraction and isolation of phenolic compounds (Fig. 8.3) is greatly complicated, because no distinct standardized procedure could be suggested for phenolics (Junqueira-Gonçalves et al. 2015). Therefore, procedures might be optimized based on the nature of sample, target analyses, and objectives of the study (Batish et al. 2008a, b).

8.3.1 Preparation of Sample

It is the initial step of extraction and isolation process, and the main objectives of sample preparation are:

(i) To develop the sample stability
(ii) To improve the efficiency of the extraction process
(iii) To eradicate or decrease possible interferences
(iv) To enhance the analyses or to transform them into derivatives which could be easily detected or quantified (Mahall and Callaway 1992)

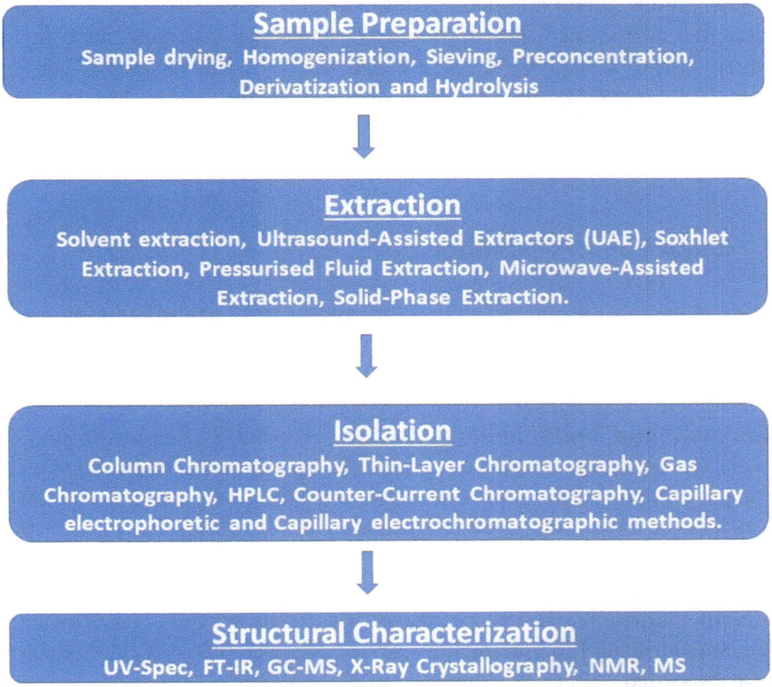

Fig. 8.3 The steps involved in isolation of active compounds

Preparation of the sample contains several steps: they are (i) sample drying, (ii) homogenization, (iii) sieving, (iv) preconcentration, (v) derivatization, and (vi) hydrolysis. Even though the significance of sample preparation process has long been documented, it has received relatively less attention than separation and detection, although it is essential to ensure the quality and consistency of the acquired results (Ghafar et al. 2000). The significance of sample preparations is:

A. Phenolic content and composition significantly diverge, depending on the part of the plant (i.e., roots, bark, leaves, flowers, and fruits).
B. Samples must be sufficiently preserved up to the analysis, so that no changes in their chemical composition are produced.
C. Samples could be stored in fresh, frozen, and dried form.
D. Freezing and drying are the preferred options for long storage (Sasikumar et al. 2002).

Hydrolysis process could shorten the composition of the sample, helps to extract the compounds, and develops the potential of detection and quantification. Furthermore, they could be used in structural elucidation and characterization of phenolic compounds and are also required to discharge insoluble phenolics bounce to sample matrix compounds which are directly not extracted through organic solvents. Nevertheless, this is a choice that desires to be charily considered since hydrolysis techniques are not constantly proficient and might demolish several compounds. However, if the aglycones are the target analytes, hydrolysis could be commenced (Willis 1985).

8.3.2 Extraction of Phenolics

Extraction is the most important step in the process of phenolic isolation from plant materials and other organisms. The extraction of phenolics is influenced through their chemical nature, the extraction method employed, sample particle size, and the presence of interfering substances (Hostettmann et al. 1997).

Solid-liquid and liquid-liquid extraction procedure is frequently used to analyze the polyphenolics and simple phenolics in the plants. Because of efficiency, a wide range of applicability, and user-friendly still, they are most widely used techniques in the extraction process (Djurdjevic et al. 2007).

Solvent extraction: Commonly, polar solvents such as ethyl acetate, acetone, ethanol, and methanol are combined with water to extract the phenolic compounds. The solvent extraction has been achieved by using special apparatus including Soxhlet, supercritical fluid extraction device, accelerated solvent extractor, and microwave-assisted extractors (Putnam and Tang 1986).

Ultrasound-assisted extractors can be accomplished by ultrasonic probes or ultrasonic baths in either separate or constant mode, and most of the applications are static type. The use of ultrasonic bath is cheaper and simpler to handle the ultrasonic probes, and that could provide better outcome (Inderjit 1996).

Soxhlet extraction is a frequently used method for phenolic compound isolation from the solid samples. Generally, aqueous methanol or acetonitrile is used as solvents (Hadacek 2002).

Pressurized fluid extraction employs predictable solvents at controlled temperatures and pressure to extract the bioactive compounds (Zeng et al. 2001).

A microwave-assisted extraction procedure was initially utilized for the continuous determination of isoflavonoids in *Radix astragali*. This procedure archives highest extraction efficiency compared to Soxhlet, ultrasonic extraction, and reflux methods (Leslie and Romani 1998).

Solid-phase extraction is the standard procedure used for purification of plant crude extracts or other biological samples. It is a rapid and reproducible method and fairly purifies the extracts. Solid-phase extraction is essentially emulsion-free and is used for small volume of samples. A very simple solid-phase extraction method is essential for all acidic and basic analyte isolation from the plant crude extract, and high recoveries are achieved with these procedures (Ni 2000).

8.3.3 Isolation of Phenolics

Acquired extracts from the plants and other sources could be further fractioned for the isolation of compounds or preparation of simpler mixtures. Typically, the process of fractionation has been executed by column chromatography, although countercurrent chromatography has expanded place in recent times (Clé et al. 2008). The range of chromatographic resins with distinct separation methods is commercially accessible, and most of them are used in more or less extent for the phenolic compound isolation from plant extracts (Haribal and Enwick 1998). Most of the standard column chromatography separations of phenolic compounds are based on adsorption, which is mostly directed by the formation of hydrogen bonding between phenolic proton donors and acceptor groups in the resin and also by hydrophobic interactions with the aromatic rings of the phenolic compound. Consequently, the affinity in the direction of the stationary phase could be expected to develop with numerous hydroxyl phenolic groups and aromatic rings (Einhellig et al. 2004).

8.3.3.1 Thin-Layer Chromatography (TLC)

TLC is the standard method to analyze the phenolics from the plant extracts, and they play a significant part in determining the phenolic acids in natural products. Mainly it is used as a rapid screening method for extracting plant derivatives for pharmacologically active substances before the detailed analysis by instrumental techniques (Chou and Muller 1972).

8.3.3.2 Gas Chromatography (GC)

GC is one of the important chromatographic techniques used for the analysis of phenolic compounds in plants. Preparation of samples for GC comprises the removal of lipids from the extract, liberation of phenolics from ester, and glycosidic bonds by alkali, acid, or enzymatic hydrolysis. Commonly, gas phase analysis involves a chemical derivatization step, in addition to sample extraction, isolation, and cleanup (Watson and Renney 1974).

8.3.3.3 High-Performance Liquid Chromatography (HPLC)

In the past two decades, HPLC has been dominated as analytical technique for separation and characterization of phenolics from the plants. As a result of comparatively high-molecular-mass and hydrophilic flavonoid glycosides, intrinsic features of hydrophobic flavonoid aglycones and the crushing mass of chromatographic techniques in the literature drop in the realm of HPLC and associated technologies (Tyser and Key 1988). HPLC techniques might propose an exclusive possibility to separate continuously all analyzed components together with their possible derivatives or degradation products. The main advantages of HPLC are (i) the broad range of commercially available columns and (ii) the possibility of joining two or more columns in a switching mode (Callaway and Ridenour 2004).

8.3.3.4 Countercurrent Chromatography (CCC)

CCC has been emerged as a powerful isolation tool for the compounds at preparative and semi-preparative scale, competing as the beneficial option with column chromatography (Bais et al. 2002). Unlike other chromatographic methods, CCC does not use the solid support as stationary phase, but it is the all-liquid chromatography method in which the separation of compounds is due to their partitioning between two immiscible solvents; one of them is stationary phase, while another one is the mobile phase. A supplementary solvent, miscible in both phases, might be used to aid the partitioning of the analytes between the two immiscible phases (Liebler et al. 1996).

8.3.3.5 Capillary Electrophoretic and Capillary Electrochromatographic Methods

The electromigration modes are generally capillary zone electrophoresis, capillary electrophoresis, and micellar electrokinetic chromatography which uses phosphate or borate buffers, capillaries of 50–100 lmid, injection volumes of 10–50 nL, and voltages of 10–30 kV. Detection is commonly carried out with UV, and also it can be attained by electrochemical and MS detectors. Mostly, the capillary electrophoretic techniques are utilized for the analysis of phenolics that fall in the field of natural product research such as analysis of plants (Fig. 8.3), vegetables, herbs, and fruit-derived products (Durant et al. 2002).

8.4 Phenolics in Crops and Effects on Weed Management

8.4.1 Allelopathy in Crops and Weed Management

The present agriculture is productivity-oriented and relies primarily on synthetic inputs to attempt the weeds and other pest problems. The use of rigorous herbicide to manage weeds is posing severe ecological and environmental threats to the planet and its inhabitants in the last few decades (Clé et al. 2008). Herbicide residues in agricultural products, soil and groundwater, evolution of resistant weed biotypes, shifts in weed populations, and associated health hazards have forced to discover and establish alternative weed management strategies. Worldwide, allelopathy is one of the important ecological facts that makes clear interference among species through biochemical pathways. It is the tool that can be utilized to manage weeds in agroecosystems. Utilization of allelopathic properties of native plant/crop species provides ample opportunities for this purpose (Ferrer et al. 2008). Allelopathy could control the plant biodiversity via its impact on adaptation, survival, and community organization of plants. Allelopathic crop residues could be exploited for weed suppression and helpful in reducing reliance on herbicides (Van Pelt et al. 1998).

8.4.2 Role of Phenolic Allelochemicals in Crops and Weed Management

The phytotoxic nature of the plants might be due to its higher total phenolic content. It has been reported that the phenolic compounds are the largest class of allelochemicals inhibiting various crops of weed species. Several studies indicate that phenolic allelochemicals are identified in both natural and managed ecosystems which cause numerous ecological and economic problems including reduced crop yield due to soil sickness, replanting problems in orchards, and regeneration failure of natural forests (Langcake et al. 1981). The structure of phenolic compounds and its modes of action vary and provide potential lead compounds for the development of potential herbicides or pesticides. In soil, phenolics can exist in three forms such as free, reversibly bound, and bound form. Commonly, the first two forms are considered as important in the process of allelopathy. Thus, the structural diversity and intraspecific variability represents the important characteristics of phenolic compounds (Seal et al. 2004).

Phenolic compounds have the ability to influence the nutrient uptake, cell membrane permeability, enzyme activity, photosynthesis, respiration, etc., which in turn might increase the process of lipid peroxidation. This may lead to slow growth of the plants or death of plant tissue. Phenolic compounds are generally made up of tannins, hydroxy and substituted cinnamic acids, simple aromatic phenols, hydroxy and substituted benzoic acids, and aldehydes and coumarins. Among the phenolic compounds, flavonoids are the major chemical studied in chemical, biological, agricultural, and medical studies. (Politycka 1997).

Recent reports have shown that phenolic compounds are involved in plant allelopathy, and they are universally distributed in plants and very common in plant decomposition products, and they are important precursors of humic substances in soils (Bais et al. 2003). Ortho-substituted phenolics, such as o-coumaric acids and salicylic, and dihydro-substituted phenolics, such as caffeic acids and protocatechuic, are absorbed by clay minerals by forming chelate complexes with metals. Free phenolic compounds may accumulate in rhizosphere soils, mostly in soils flooded with vegetable wastewaters, thus influencing the accumulation of nutrient cycling and availability of soil nutrients, which both eventually distress plant growth (Perry et al. 2005a, b).

Recent studies have observed the biotechnology, chemistry, and ecotoxicology of naturally occurring polyphenols in waste of vegetable. The best example of the protective effect of phenolic compound is that of protocatechuic acid and catechol from onion, which is known to treat infection of *Colletotrichum circinans*. Water-soluble phenolic compounds diffused out from the dead cell layers of the scales and decreased the germination of spore and hyphal penetration of the pathogen (Cruz et al. 1998).

8.4.3 Phenolic Allelochemical Mode of Action in Crops

Changes in Membrane Permeability and Inhibition of Plant Nutrient Uptake Phenolic allelochemicals could improve the cell membrane permeability. Finally, there is slowdown in the growth or death of plant tissue (Li et al. 1993). Further, phenolic allelochemicals also might inhibit the plants from absorbing nutrients from surroundings and affect the normal growth of plants. Recently, cucumber (*Cucumis sativus*) was treated with benzoic acid and cinnamic acid derivatives for 7 days and the result proved that declined in phenol glycosylation and phenyl-ß-glucosyltransferase activity decrease were linked with raises of membrane permeability (Patterson 1981).

Inhibition of Cell Division, Elongation, and Submicroscopic Structure Phenolic allelochemicals possibly decrease the plant root elongation and cell division, change cell ultrastructure, and further interfere the progression and normal growth of whole plant (Yu et al. 2003).

Effects on Plant Photosynthesis and Respiration The impact of phenolic allelochemicals on the respiration of plants has mainly been shown to involve weakened oxygen absorption capacity, while the photosynthesis has generally been to decrease the chlorophyll content and photosynthetic rate (Rice 1974).

Effects on Synthesis of Plant Endogenous Hormones Phenolic allelochemicals can inhibit or inactivate the hormone's physiological functions in the crops, which may further inhibit the normal physiological process of the plants. Recent study reported that polyphenols, hydroxyl benzoic acid, and other compounds can affect the decomposition process of gibberellins and indoleacetic acid (Politycka 1998).

Effects on Protein Synthesis Some phenolic compounds such as cinnamic acid and ferulic acid can decrease the protein synthesis, amino acid transport, and the subsequent growth of the plants. All phenolic compounds can decrease the integrity of DNA and RNA, but in the vast majority of cases, phenolic compounds showed as a mixture and not as a single substance. Hence, the contribution made to allelopathy by phenolic compounds is probably not possible due to single substance (Batish et al. 2008a, b).

8.5 Discussion and Future Prospects

The allelopathic characteristics of plants can be utilized successfully as a tool for pathogen and weed reduction. In last few decades, preliminary surveys of some hundred allelopathic plants in the Southeast Asia and Japan ecosystems have been done, and more than 30 species including crops which demonstrated the maximum allelopathic potential have been selected and examined for their impacts on emergence of pathogens and weed management (Bhattacharya et al. 2010).

Allelochemicals could be helpful to reduce serious problems in the present agricultural production such as environmental pollution, human health concerns, soil sickness, unsafe product depletion of crop diversity, and reduction of crop productivity. Numerous crops such as alfalfa, sorghum, buckwheat, wheat, maize, sunflower, rice, etc., are affected either by phytotoxin exudates or their own toxicity during their decomposed residues in the soil that show strong reduction on weed emergences (Thelen et al. 2005). Interpretations confirmed that application of these plant materials at 1–2 ton ha^{-1} can decrease weed biomass by about 70% and increase the rice yield by about 20%. Some species show strong inhibition on many plant pathogens, and they might become potential tools in dropping plant pathogens and weeds (Weir et al. 2006). Conversely, application of 1–2 ton ha^{-1} of plant material to the field makes heavy fieldwork. Several growth inhibitors identified from these allelopathic plants are accountable for their allelopathic properties and could be an efficient source for the future development of bio-herbicides and pesticides (Weir et al. 2003).

Discovery of phenolic allelochemicals from plants and other sources could offer a huge and nearly unexploited reservoir of chemical compounds which possess a variety of potential uses including pest control and weed management in the field of agriculture production. They are having very less risk when compared with synthetic compounds which are toxicologically and environmentally redundant (Perry et al. 2005a, b). Therefore, isolation and identification of allelopathic phenolic compounds for pest control and weed management tools might produce toxicologically benign pest control and weed management tools that might have more improved nutraceutical value in crop production. Some of the potential phenolic allelochemicals were identified, and their mode of action from the plants such as 1, 8-cineole, p-coumaric acid, and cyperin isolated from *Eucalyptus* spp., *O. sativa*, and plant pathogens, respectively, was analyzed (Won et al. 2013).

8.6 Conclusion

In the modern era, significant and speedy analysis technologies made it possible to isolate and employ the sophisticated structural characterization of phenolic allelochemicals. Furthermore, an extensive diversity of allelopathy mechanisms was well characterized. Due to the broad research in the field of allelopathy, the great number of phenolic allelochemicals has been illustrated and it plays a significant role in complex interaction with the living organisms in the environment that is slowly unwinding. Research in phenolic allelochemical mixture showed that the individual compounds might be preserved when being anticipated for phytotoxic affects. However, further analyses are needed to confirm the current findings of phenolic activities under the field condition. To recognize the mechanism of phenolic compounds in allelopathy more clearly, further studies are mandatory on the role and production of phenolic compounds in the ecosystem. Significant developments in this field ultimately lead to the progress of biorational pest control and weed management by using phenolic compounds.

References

Akhtar M, Malik A (2000) Roles of organic soil amendments and soil organisms in the biological control of plant-parasitic nematodes: a review. Bioresour Technol 74:35–47

Akula R, Ravishankar GA (2011) Influence of abiotic stress signals on secondary metabolites in plants. Plant Signal Behav 6:1720–1731

Bais HP, Walker TS, Stermitz FR, Hufbauer RA, Vivanco JM (2002) Enantiomeric-dependent phytotoxic and antimicrobial activity of (±)-catechin. A rhizosecreted racemic mixture from spotted knapweed. Plant Physiol 128:1173–1179

Bais HP, Vepachedu R, Gilroy S, Callaway RM, Vivanco JM (2003) Allelopathy and exotic plant invasion: from molecules and genes to species interactions. Science 301:1377–1380

Baranski M, Srednicka-Tober D, Volakakis N, Seal C, Sanderson R, Stewart GB, Benbrook C, Biavati B, Markellou E, Giotis C (2014) Higher antioxidant and lower cadmium concentrations and lower incidence of pesticide residues in organically grown crops: a systematic literature review and meta-analyses. Br J Nutr 112:794–811

Batish DR, Singh HP, Kohli RK, Kaur S (2001) Crop allelopathy and its role in ecological agriculture. J Crop Prod 4:121–162

Batish DR, Kaur S, Singh HP, Kohli RK (2008a) Role of root-mediated interactions in phytotoxic interference of *Ageratum conyzoides* with rice (Oryza sativa). Flora 204:388–395

Batish DR, Singh HP, Kaur S, Kohli RK, Yadav SS (2008b) Caffeic acid affects early growth, and morphogenetic response of hypocotyl cuttings of mung bean (*Phaseolus aureus*). J Plant Physiol 165:297–305

Bhattacharya A, Sood P, Citovsky V (2010) The roles of plant phenolics in defence and communication during agrobacterium and rhizobium infection. Mol Plant Pathol 11(5):705–719

Callaway RM, Aschehoug ET (2000) Invasive plants versus their new and old neighbors: a mechanism for exotic plant invasion. Science 290:521–523

Callaway RM, Ridenour WM (2004) Novel weapons: invasive success and the evolution of increased competitive ability. Front Ecol Environ 2:436

Capasso R (1997) The chemistry, biotechnology and ecotoxicology of the polyphenols naturally occurring in vegetable wastes. Curr Top Phytochem 1:145–156

Capasso R, Cristinzio G, Evidente A, Scognamiglio F (1992) Isolation, spectroscopy and selective phytotoxic effects of polyphenols from vegetable waste waters. Phytochemistry 31:4125–4128

Chou CH, Lee YY (1991) Allelopathic dominance of Miscanthus transmorrisonensis in an alpine grassland community in Taiwan. J Chem Ecol 17:2267–2281

Chou CH, Leu LL (1992) Allelopathic substances and interactions of Delonix regia (BOJ) RAF. J Chem Ecol 18:2285–2303

Chou CH, Lin HJ (1976) Autointoxication mechanism of *Oryza sativa* I. Phytotoxic effects of decomposing rice residues in soil. J Chem Ecol 2:353–367

Chou CH, Muller CH (1972) Allelopathic mechanisms of *Arctostaphylos glandulosa* var. *zacaensis*. Am Midl Nat 88:34–347

Clé C, Hill LM, Niggeweg R, Martin CR, Guisez Y, Prinsen E, Jansen MAK (2008) Modulation of chlorogenic acid biosynthesis in *Solanum lycopersicum*; consequences for phenolic accumulation and UV-tolerance. Phytochemistry 69:2149–2156

Cruz OR, Anaya AL, Hernandez-Bautista BE (1998) Effects of allelochemical stress produced by sicyos deppei on seedling root ultrastructure of *Phaseolus vulgaris* and *Cucurbita ficifolia*. J Chem Ecol 24:2039–2057

Djurdjevic L, Mitrovic M, Pavlovic P (2007) Methodology of allelopathy research: 2. Forest ecosystems. Allelopath J 20:1–24

Durant AA, Fente CA, Franco CM, Vazquez BI, Cepeda A (2002) Gas-chromatography-tandem mass spectrometry determination of 17 α-ethinylestradiol residue in the hair of cattle, application to treated animals. J Agric Food Chem 50:436–440

Einhellig FA, Galindo JCG, Molinillo JMG, Cutler HG (2004) Mode of allelochemical action of phenolic compounds. In: Allelopathy: chemistry and mode of action of allelochemicals. CRC Press, Boca Raton, pp 217–238

Ferrer JL, Austin MB, Stewart C, Noelb JP (2008) Structure and function of enzymes involved in the biosynthesis of phenylpropanoids. Plant Physiol Biochem 46(3):356–370

Franz H (2002) Secondary metabolites as plant traits: current assessment and future perspectives. Plant Sci 21:273–322

Ghafar A, Saleem B, Qureshi MJ (2000) Allelopathic effects of sunflower on germination and seedling growth of wheat. Pak J Biol Sci 3:1301–1302

Hadacek F (2002) Secondary metabolites as plant traits: current assessment and future perspectives. Crit Rev Plant Sci 21:273–322

Harbone JB (1984) Phytochemical methods: a guide to modern techniques of plant analysis. Springer, Dordrecht. 37–99 pp

Haribal M, Enwick JAA (1998) Isovitexin 6‴-O-β-D-glucopyranoside: a feeding deterrent to *Pieris napi oleracea* from *Alliaria petiolata*. Phytochemistry 47:1237–1240

Hartmann T (1996) Diversity and variability of plant secondary metabolism: a mechanistic view. Entomol Gen Appl 80:177–188

Hostettmann K, Marston A, Hostettmann M (1997) Preparative chromatography techniques: application in natural product isolation. Springer, Berlin

Inderjit (1996) Plant phenolics in allelopathy. Bot Rev 62:186–202

Junqueira-Gonçalves MP, Yanez L, Morales C, Navarro M, Contreras RA, Zuniga GE (2015) Isolation and characterization of phenolic compounds and anthocyanins from murta (*Ugni molinae* Turcz.) fruits. Assessment of antioxidant and antibacterial activity. Molecules 20:5698–5713

Kalt W (2005) Effects of production and processing factors on major fruit and vegetable antioxidants. J Food Sci 70:R11–R19

Khanh TD, Hong NH, Xuan TD, Chung IM (2005) Paddy weed control by medical and leguminous plants from Southeast Asia. Crop Prot 24:421–431

Khanh TD, Elzaawely AA, Chung IM, Ahn JK, Tawata S, Xuan TD (2007) Role of allelochemicals for weed management in rice. Allelopath J 19:76–85

Kohli RK, Batish D, Singh HP (1998) Allelopathy and its implication in agroecosystems. J Crop Prod 1:169–202

Langcake P, Irvine JA, Jeger MJ (1981) Alternative chemical agents for controlling plant disease. Philos Trans R Soc Lond B: Biol Sci 295:83–101

Leslie CA, Romani RJ (1998) Inhibition of ethylene biosynthesis by salicylic acid. Plant Physiol 88:833–837

Li HH, Inoue M, Nishimura H, Mizutani J, Tsuzuki E (1993) Interaction of trans-cinnamic acid, its related phenolic allelochemicals, and abscisic-acid in seedling growth and seed-germination of lettuce. J Chem Ecol 19:1775–1787

Liebler DC, Burr JA, Philips L, Ham AJL (1996) Gas chromatography-mass spectrometry analysis of vitamin E and its oxidation products. Anal Biochem 263:27–34

Luthria DL (2006) Significance of sample preparation in developing analytical methodologies for accurate estimation of bioactive compounds in functional foods. J Sci Food Agric 86:2266–2272

Luthria DL, Mukhopadhyay S, Krizek DT (2006) Content of total phenolics and phenolic acids in tomato (*Lycopersicon esculentum* Mill.) fruits as influenced by cultivar and solar UV radiation. J Food Compos Anal 19:771–777

Mahall BE, Callaway RM (1992) Root communication mechanisms and intercommunity distributions of 2 mojave desert shrubs. Ecology 73:2145–2151

Naczk M, Shahidi F (2004) Extraction and analysis of phenolics in food. J Chromatogr A 1054:95–111

Ni HW (2000) Present status and prospect of crop allelopathy in China. In: Kim KU, Shin DH (eds) Rice allelopathy. Kyunpook National University, Kyungpook, pp 41–48

Pagare S, Bhatia M, Tripathi N, Pagare S, Bansal YK (2015) Secondary metabolites of plants and their role: overview. Curr Trends Biotechnol Pharm 9(3):293–304
Patterson DT (1981) Effects of allelopathic chemicals on growth and physiological response of soybean (Glycine max). Weed Sci 29:53–58
Perry LG, Johnson C, Alford ÉR, Vivanco JM, Paschke MW (2005a) Screening of grassland plants for restoration after spotted knapweed invasion. Restor Ecol 13:725–735
Perry LG, Thelen GC, Ridenour WM, Weir TL, Callaway RM, Paschke MW, Vivanco JM (2005b) Dual role for an allelochemical (±)-catechin from *Centaurea maculosa* root exudates regulates conspecific seedling establishment. J Ecol 93:1126–1135
Piasecka A, Jedrzejczak-Rey N, Bednarek P (2015) Secondary metabolites in plant innate immunity: conserved function of divergent chemicals. New Phytol 206:948–964
Politycka B (1997) Free and glucosylated phenolics, phenol-beta-glucosyltransferase activity and membrane permeability in cucumber roots affected by derivatives of cinnamic and benzoic acid. Acta Physiol Plant 19:311–317
Politycka B (1998) Phenolics and the activities of phenylalanine ammonia-1yase, phenol-beta-glucosyltransferase and beta-glucosidase in cucumber roots as affected by phenolic allelochemicals. Acta Physiol Plant 20:405–410
Putnam AR, Tang CS (1986) The science of allelopathy. Wiley, New York, pp 1–19
Quideau S (2009) Why bother with polyphenols? In: Groupe polyphenols: the international society dedicated to the promotion of research on polyphenols. Bordeaux
Rice EL (1974) Allelopathy, vol 353. Academic, New York
Ridenour WM, Callaway RM (2001) The relative importance of allelopathy in interference: the effects of an invasive weed on a native bunchgrass. Oecologia 126:444–450
Santana CM, Ferrera ZS, Padrón MET, Rodríguez JJS (2009) Methodologies for the extraction of phenolic compounds from environmental samples: new approaches. Molecules 14:298–320
Sasikumar K, Vijayalakshmi C, Parthiban KT (2002) Allelopathic effects of Eucalyptus on blackgram (*Phaseolus mungo* L.). Allelopath J 9:205–214
Seal AN, Pratley JE, Haig T, An M (2004) Identification and quantitation of compounds in a series of allelopathic and non-allelopathic rice root exudates. J Chem Ecol 30:1647–1662
Singh HP, Kohli RK, Batish DR (2001) Allelopathy in agroecosystems: An overview. J Crop Prod 4:1–41
Smith-Spangler C, Brandeau ML, Hunter GE, Bavinger C, Pearson M, Eschbach PJ, Sundaram V, Liu H, Schirmer P, Stave C (2012) Are organic foods safer or healthier than conventional alternatives? Ann Intern Med 157:348–366
Tang CS (1986) Continuous trapping techniques for the study of allelochemicals from higher plants. In: Putnam AR, Tang CS (eds) The science of allelopathy. Wiley, New York, pp 113–131
Thelen GC, Vivanco JM, Newingham B, Good W, Bais HP, Landers P, Caesar A, Callaway RM (2005) Insect herbivory stimulates allelopathic exudation by an invasive plant and the suppression of natives. Ecol Lett 8:209–217
Tyser RW, Key CH (1988) Spotted knapweed in natural area fescue grasslands: An ecological assessment. Northwest Sci 62:981–987
Van Pelt CD, Haggarty H, Brenna JT (1998) Quantitative subfemtomole analysis of α-tocopherol and deuterated isotopomers in plasma using tabletop GC/MS/MS. Anal Chem 70:4369–4375
Veberic R (2016) The impact of production technology on plant phenolics. Horticulturae 2(3):8
Waller GR (1987) Allelochemicals: role in agriculture and forestry, ACS symposium series 330. American Chemical Society, Washington, DC, p 606
Watson AK, Renney AJ (1974) The biology of Canadian weeds. 6. *Centaurea diffusa* and *Centaurea maculosa*. Can J Plant Sci 54:687–701
Weir TL, Bais HP, Vivanco JM (2003) Intraspecific and interspecific interactions mediated by a phytotoxin, (−)-catechin, secreted by the roots of *Centaurea maculosa* (spotted knapweed). J Chem Ecol 29:2379–2394

Weir TL, Bais HP, Stull VJ, Callaway RM, Thelen GC, Ridenour WM, Bhamidi S, Stermitz FR, Vivanco JM (2006) Oxalate contributes to the resistance of *Gaillardia grandiflora* and *Lupinus sericeus* to a phytotoxin produced by *Centaurea maculosa*. Planta 223:785–795

Wezel A, Casagrande M, Celette F, Vian JF, Ferrer A, Peigne J (2014) Agroecological practices for sustainable agriculture. A review. Agron Sustain Dev 34:1–20

Willis RJ (1985) The historical bases of the concept of allelopathy. J Hist Biol 18:71–102

Won OJ, Uddin MR, Park KW, Pyon JY, Park SU (2013) Phenolic compounds in sorghum leaf extracts and their effects on weed control. Allelopath J 31((1)L):147–156

Xuan TD, Tsuzuki E (2002) Varietal difference in allelopathic potential of alfalfa. J Agron Crop Sci 188:2–7

Yu JQ, Ye SF, Zhang MF, Hu WH (2003) Effects of root exudates and aqueous root extracts of cucumber (*Cucumis sativus*) and allelochemicals, on photosynthesis and antioxidant enzymes in cucumber. Biochem Syst Ecol 31:129–139

Zeng RS, Luo SM, Shi YH (2001) Physiological and biochemical mechanism of allelopathy of secalonic acid on higher plants. Agron J 93:72–79

Zeng RS, Mallik AU, Luo SM (2008) Allelopathy in sustainable agriculture and forestry. Springer, New York

Zhao X, Carey EE, Wang WQ, Rajashekar CB (2006) Does organic production enhance phytochemical content of fruit and vegetables? Current knowledge and prospects for research. HortTechnology 16:449–456

Phenolic Compounds Against Fungal and Viral Plant Diseases

9

Younes Rashad, Dalia Aseel, and Saad Hammad

Abstract

Plants are subjected to attack by various pathogens including fungi, bacteria, viruses, nematodes, protozoa, and insects causing serious and economically important diseases. To defend themselves against invading pathogens, plants have developed varied defense mechanisms including preformed structures and metabolites and induced cascade of immune responses. Plant defense lines start from the pattern recognition receptors, proceed locally at the first, and then become systemic via the triggered signaling molecules which activate different defense mechanisms along the entire plant. One of the main induced hypersensitive responses is stimulating the phenolic compounds production in the plant. In addition, phenolic compounds have different important roles in seed germination, plant growth and resistance against various stress factors such as sun exposure, injuries, and heavy metal stress. Most of them have antioxidant and antimicrobial properties; some repel herbivores, while others are involved in the lignin and suberin biosynthesis for cell wall strengthening. In this chapter, we will discuss the biosynthesis, gene regulation, and defense mechanisms of the different categories of phenolic compounds in response to plant fungal and viral diseases.

Keywords

Phenolic compounds · Fungi · Viruses · Diseases · Pathogen

Y. Rashad (✉) · D. Aseel · S. Hammad
Plant Protection and Biomolecular Diagnosis Department, Arid Lands Cultivation Research Institute, City of Scientific Research and Technological Applications, Alexandria, Egypt

R. Lone et al. (eds.), *Plant Phenolics in Sustainable Agriculture*,
https://doi.org/10.1007/978-981-15-4890-1_9

9.1 Introduction

Phenolic substances are diverse groups of phytochemicals which structurally have an aromatic ring with at least one hydroxyl group substituent and produced by plants as secondary metabolites. These widely occurring compounds play many important physiological, morphological, and protective roles throughout all plant life stages. They have considerable involvements in plant growth, proliferation, pigmentation, attraction of pollinators, UV-screening, allelopathy, defense signaling, and protection against attacking pathogens and herbivores (Mandal et al. 2010; Agati et al. 2013; Lucas-Barbosa et al. 2016; Šežiene et al. 2017; Bernardi et al. 2018). In order to defend themselves against invading pathogens, plants have released a diverse set of defense-related phenolic compounds as induced systemic resistance responses. Many of them are derived from phenylpropanoid biosynthesis pathway which derived from shikimate pathway (Deng and Lu 2017). The defensive roles of plant phenolic compounds against fungal and viral plant diseases have been extensively studied by many researchers in the last decades. Wang et al. (2018) investigated the antifungal activity of the phenolic monoterpenes (carvacrol and thymol) against some phytopathogenic fungi in vitro. Although both monoterpenes showed a high antifungal activity, their ester derivatives were more toxic against the tested fungi. In another study, the fungitoxic effect of the seed extract of *Ammi visnaga* against Rhizoctonia root rot, caused by *Rhizoctonia solani*, in maize was attributed to their content of coumarins (Rashad et al. 2018). Various fungitoxic mechanisms were reported for different phenolic compounds including distortion of the cell wall integrity, altering of the cell membrane permeability, enzymes suppression, oxidative bursts elicitation (free radical formation), DNA damage, inhibition of protein synthesis, and/or repression of virulence/toxin genes (Ansari et al. 2014; Upadhyay et al. 2015; Negritto et al. 2017; Fernandes et al. 2019).

On the other hand, Dunkić et al. (2010) reported the antiviral activity of carvacrol and thymol from the essential oil of *Satureja montana* L. ssp. *Variegata* against *Tobacco mosaic virus* and *Cucumber mosaic virus*. A considerable reduction in the number of local lesions was noticed (29.2% and 24.1%, respectively). Hu et al. (2013) reported the antiviral activity of different phenolic compounds isolated from *Arundina graminifolia* against *Tobacco mosaic virus*. Li et al. (2015) isolated and characterized a new antiviral phenolic compound gramniphenol I (1) which exhibited 16.8% inhibition against *Tobacco mosaic virus*. The main antiviral mechanisms exerted by the phenolic compounds are suppression of the infection process and/or repression of viral replication. In this regard, the reported antiviral modes of action include viral DNA/protein damage and/or inhibition of the viral enzymes (Kumar and Pandey 2013).

Based on the complexity of the carbonaceous skeleton, substitution, modification degree, and linkage with other molecules, phenolic compounds can be classified into simple phenols (C_6), phenolic acids (C_6-C_1 or C_6-C_3), flavonoids, lignins, tannins, and coumarins. However, diversity of phenolic compounds in the plant is more complicated than mentioned here. In this chapter, we will explore these various categories of phenolic compounds in response to plant fungal and viral diseases discussing their biosynthesis, gene regulation, and defense mechanisms.

9.2 Simple Phenols

Simple phenols are compounds which contain at least one hydroxyl group attached to an aromatic ring such as phenol and cresol, or multihydroxyl groups such as catechol and pyrogallol. Many of them have significant functions in plants as defense lines against attacking pathogens (Bhattacharya et al. 2010), or they may be released into the rhizospheric soils as roots exudates performing allelopathic effects on the growth of surrounding plants and microbiota (Makoi and Ndakidemi 2012). Moreover, some phenolic compounds released in the rhizosphere act as molecules for cross talk and communication between the plant and their surrounding plants, mircobiota, and fauna in the rhizosphere (Bhattacharya et al. 2010).

Although simple phenols are synthesized normally in the healthy plant tissues, their accumulation is a typical characteristic in case of biotic or abiotic stresses (Arici et al. 2014). Simple phenols and their derivatives are produced in the plant tissues via two main pathways. Carbohydrates developed via the Calvin cycle are degraded and transformed into pyruvate by glycolysis or into glyceraldehyde-3-phosphate via oxidative pentose phosphate pathway leading to acetate-malonate and shikimate pathways. Both of these pathways are crucial for simple phenols biosynthesis, especially via phenylpropanoid precursors using phenylalanine ammonia-lyase (PAL) and chalcone synthase (CHS) (Cheynier et al. 2013). Moreover, modifications of different phenolic compounds via methylation, acylation, and glycosylation are essential in many biological processes in plant. These modifications lead to altering various biochemical properties and biological activities of these compounds (Kim et al. 2017).

Early and fast accumulation of phenolic compounds is an important characteristic in resistant plants encountering invading pathogens. At infection, induced resistance responses are activated including accumulation of toxic substances at the infection site including simple phenols, phenolic phytoalexins, hydroxycinnamic acids, and their derivatives. These antifungal molecules interfere with cell metabolism inhabiting primary enzymes, damage its cellular structure or integrity, or disrupt DNA leading to rapid cell death (Daayf et al. 2012). The elevated levels of these simple phenols provide appropriate conditions for the oxidative enzymes such as polyphenol oxidases and peroxidase that catalyze the oxidation reactions of these toxic phenols into quinones, which are more toxic to the pathogens causing oxidative damages in their tissues (Shabana et al. 2008). The relative toxicity of these simple phenols is correlated with the site and number of the hydroxyl group(s) on the phenol ring as well as the oxidation state. The Highly oxidation of simple phenols leds to more toxicity to the pathogens (Scalbert 1991).

9.3 Phenolic Acids

Phenolic acids are compounds which contain at least one carboxylic acid group attached to a phenol ring. According to the attached carbon chain, they are divided into C_6-C_1 (hydroxybenzoic acids) or C_6-C_3 (hydroxycinnamic acids). In addition, the phenolic acids are varied based on the number and site of the hydroxyl group (s) on their aromatic ring (Goleniowski et al. 2013). Most of phenolic acids in the plant are attached via ester, ether, or acetal bonds to other structural components, or natural molecules such as cellulose, flavonoids, glucose, maleic, or tartaric acids (de Oliveira et al. 2015; Alam et al. 2016). Phenolic acids include two categories: hydroxybenzoic acids, which derived from benzoic acid such as *p*-hydroxybenzoic acid, salicylic acid, gallic acid, vanillic acid, syringic acid, and protocatechuic acid, or hydroxycinnamic acids which derived from cinnamic acid such as *o*-, *m*-, or *p*-coumaric acid, ferulic acid, sinapic acid, and caffeic acid. Ferulic acid and *p*-coumaric acid form the precursors of lignins (Goleniowski et al. 2013). They have important roles in physical and chemical defense against attacking pathogens, structural functions, enzymes activities, nutrient uptake, protein synthesis, photosynthesis, and allelopathy (Leváková and Lacko-Bartošová 2017).

Biosynthesis of hydroxycinnamic acids derived from L- phenylalanine via the phenylpropanoid pathway. Cinnamic acid formation is catalyzed by PAL from L-phenylalanine and transformed to *p*-coumaric acid by cinnamate 4-hydroxylase (C4H). Addition of another (OH) group to *p*-coumaric acid is catalyzed by *p*-coumarate 3-hydroxylase (C3H) to give caffeic acid which is then methylated to give ferulic acid by caffeic acid *O*-methyltransferase (COMT). Sinapic acid is produced as a result of methylation of the hydroxyferulic acid (El-Seedi et al. 2012). On the other hand, some of hydroxybenzoic acids may be produced from 3-dehydroshikimic acid (gallic acid) or degradation of hydroxycinnamic acids and flavonoids (Chen et al. 2009; Goleniowski et al. 2013).

Upon pathogen attack, plants tend to defend themselves by triggering different defense responses via accumulation of signaling molecules. Phenolic acids such as salicylic acid act as signaling molecules which activate the plant-induced defense mechanisms at the site of infection (hypersensitivity reaction) and along the plant tissues to prevent the infection spread (Mandal et al. 2010). Various phenolic acids are known to have potent fungitoxic and antiviral activities which accumulated in the plant tissue and act as phytoalexins and phytoanticipins (Hammerschmidt 2014; Alves et al. 2014; Li et al. 2017). In addition, the pathogenic stress leds to induce the biosynthesis of cinnamic acid and benzoic acid derivatives which incorporate in the cell wall strengthening (de Ascensao and Dubery 2003).

9.4 Flavonoids

Flavonoids are low-molecular-weight phenolic compounds which have chemical structure which depends on 15-carbon skeleton, containing two aromatic rings linked by a heterocyclic pyran ring (C6-C3-C6). They are a large and diverse category of polyphenolic compounds that found in higher plants as aglycones, glycosides, and methylated derivatives (Tohge et al. 2017). Now, more than 6000 compounds of flavonoids have been isolated and identified (Zakaryan et al. 2017). Due to their antioxidative, anti-inflammatory, anti-mutagenic, and anti-carcinogenic properties, they have many pharmaceutical, medicinal, and cosmetic uses. In plants, they play many important roles and biological functions. In addition to their responsibility for the color and odor of flowers and fruits, flavonoids have a protection role against different biotic (fungal, bacterial, and viral pathogenic attack) and abiotic stresses (UV, drought, heat acclimatization, and freezing). Moreover, they act as signal molecules, allelopathic compounds, phytoalexins, detoxifying agents, and antimicrobial defensive compounds (Panche et al. 2016). Based on the position of the C ring on which the B ring is linked, the degree of unsaturation, and the level of oxidation of their heterocyclic ring, flavonoids are classified into different subgroups including flavones, flavonols, flavanones, flavanonols, flavanols, anthocyanins, and chalcones (Falcone Ferreyra et al. 2012).

9.4.1 Types of Flavonoids

9.4.1.1 Flavones

Flavones are flavonoid compounds which have a double bond between positions 2 and 3 and a ketone in position 4 of the C ring. They are widely present in fruits, leaves, and flowers as glucosides. The major sources of flavones include red peppers, parsley, celery, chamomile, ginkgo biloba, and mint.

9.4.1.2 Flavonols

Flavonols are flavonoids with a ketone group which have a hydroxyl group in position 3 of the C ring. They are the building blocks of proanthocyanins. They are the most common subgroup of flavonoids in fruits and vegetables such as kale, onions, berries, tomatoes, lettuce, grapes, apples, and tea.

9.4.1.3 Flavanones

Flavanones are flavonoids which have the C ring saturated (the double bond between positions 2 and 3 is saturated). They are found in oranges, lemons and grapes, and responsible for the bitter taste of the juice and peel of citrus fruits. Examples of flavanones include hesperitin, naringenin, and eriodictyol.

9.4.1.4 Isoflavonoids

The rich sources of isoflavonoids, such as genistein and daidzein, include soya beans and other leguminous plants. They act as precursors for the development of phytoalexins showing immense potentials against various diseases.

9.4.1.5 Neoflavonoids

Neoflavonoids are polyphenolic compounds which have a 4-phenylchromen backbone with no hydroxyl group substitution at position 2. The most common neoflavone "calophyllolide" is found in *Calophyllum inophyllum* seeds and *Mesua thwaitesii.*

9.4.1.6 Flavanols

Flavanols, which called catechins, are 3-hydroxy derivatives of flavanones. They are found in bananas, apples, blueberries, peaches, and pears.

9.4.1.7 Anthocyanins

Anthocyanins are responsible for the colors in plants (fruits and flowers). Their color depends on the methylation or acylation at the hydroxyl groups on the rings and also pH (Iwashina 2013). The most commonly studied anthocyanins are delphinidin, malvidin, cyanidin, and peonidinpelargonidin. They are found in cranberries, bilberries, raspberries, blueberries, black currants, strawberries, and blackberries.

9.4.1.8 Chalcones

Chalcones are open-chain flavonoids which do not have C ring in their structure. The most common examples are phloridzin, arbutin, phloretin, and chalconaringenin. They are found in tomatoes, pears, strawberries, bearberries, and certain wheat products.

9.4.2 Flavonoid Biosynthesis

Flavonoids are synthesized via the phenylpropanoid pathway where phenylalanine is converted to 4-coumaroyl- CoA, which is then used in the flavonoid biosynthesis pathway. Initially, chalcone, from which all flavonoids are derived, is produced by condensation of 4-coumaroyl- CoA with malonyl-CoA using CHS. Chalcone is then converted to flavanone by chalcone flavanone isomerase (CHI). Several side branches are derived from this central intermediate producing diverse set of flavonoids. A set of enzymes, such as isomerases, reductases, and hydroxylases, play a role in altering the main flavonoid skeleton, resulting in the biosynthesis of various flavonoid subclasses (Martens et al. 2010). These enzymes are located on the cytosolic side of the endoplasmic reticulum (Jorgensen et al. 2005). Substitution by glycosylation, malonylation, methylation, hydroxylation, acylation, prenylation, or polymerization results in differentiation of these compounds and modifies their function, solubility, and degradation (Zhang et al. 2009). Generally, flavonoids have been found in the nucleus, the vacuole, cell wall, cell membrane, and the cytoplasm (Naoumkina and Dixon 2008). In the root, they are found at the root tip and cap cells from where they can be exuded into the soil (Weston and Mathesius 2013).

9.4.3 Regulation of Flavonoid Biosynthesis

Various sets of transcription factor families are involved in the regulation of flavonoid biosynthesis. Many of these transcription factors have been identified in different plant species such as MYB proteins (Table 9.1), which represent one of the most important transcription factor families in higher plants (Li 2014). In this regard, Zhu et al. (2019) reported the R2R3-MYB transcription factor (*VvMYBC2L2*) as a negative regulator of anthocyanin biosynthesis in grapevine. However, numerous R2R3-MYB transcription factors are isolated from strawberry, tobacco, potato, apple, pear, and mangosteen (Hichri et al. 2011). In *Arabidopsis thaliana*, flavonoid biosynthetic pathway encoding genes are transcriptionally regulated by different transcription factors such as *transparent testa* (*TT4*, *TT5*, *TT6*, and *TT7)* which regulate *CHS, CHI,* flavanone 3-hydroxylase (*F3H*), and flavanone 3′-hydroxylase (*F3′H*) encoding genes, respectively (Appelhagen et al. 2011; Li 2014). Three R2R3-MYB proteins (MYB11, MYB12, and MYB111) positively regulate the flavonoid biosynthetic genes *CHS*, *CHI*, *F3H*, and flavonol synthase (*FLS1*) (Falcone Ferreyra et al. 2012; Ravaglia et al. 2013), while, three classes of regulatory proteins (R2R3-MYBs, bHLHs, and Transparent Testa Glabra 1) regulate dihydroflavonol 4-reductase (*DFR*), leukoanthocyanidin dioxygenase (*LDOX*), anthocyanidin reductase (*ANR*), and *TT12* (Gou et al. 2011). Biosynthesis of anthocyanin is regulated by a MBW complex includes TTG1, one R2R3-MYB protein from production of anthocyanin pigments (PAP1), PAP2, MYB113, one bHLH protein from TT8, GLABROUS3 (GL3), or enhancer of GLABRA3 (EGL3) (Gou et al. 2011). In addition, recent researches have reported other regulators that participate in flavonoid production such as R3-MYB protein MYBL2, squamosa promoter-binding protein-like 9 (SPL9), the WIP-type zinc finger protein TT1, and the class II CIN-TCP protein TCP3 (Albert et al. 2014; Zheng et al. 2019).

9.4.4 Defensive Mechanisms of Flavonoids

The antifungal activity of flavonoids, such as isoflavones, flavanes, and flavanones, has been extensively reported against a plethora of phytopathogenic fungi in the last years (Junior et al. 2014; Chepkirui et al. 2014). The fungitoxic mechanisms of flavonoids comprise cell walls disruption, cytoplasmic membrane damage, enzyme inhibition, cell death induction, chelation of metal ions, inhibition of efflux pump, and/or binding with extracellular and soluble proteins (Mierziak et al. 2014). Serpa et al. (2012) reported the antifungal activity of the flavone compound baicalein inhibiting the efflux pump and inducing apoptosis in *Candida albicans*. In another study, it was found that the isoflavone sedonan A, isolated from *Dalea formosa*, has a fungitoxic potentiality against *C. albicans* and *C. glabrata* inhibiting the efflux-mediated pumps and affecting the intracellular transcription targets (Belofsky et al. 2013). The chalcone carvacrol has also been found to disrupt the cellular cytoplasmic membrane and induce cell apoptosis in several *Candida* spp. (Zuzarte et al. 2012).

Table 9.1 Examples of MYB transcription factors involved in the genes regulation of flavonoid biosynthesis in different plants

Plant name	Transcription factor	Target genes/functions	Reference
Zea mays	ZmP1	Regulation of 3-deoxyanthocyanin, phlobaphene and flavonol biosynthesis	Grotewold et al. (1998), Bruce et al. (2000), Ferreyra et al. (2010)
Sorghum bicolor	SbY1	CHS, CHI, DFR regulation of 3-deoxyflavonoid biosynthesis	Du et al. (2009)
Antirrhinum majus	AmMYB305 AmMYB340	PAL, CHI, F3H Regulation of flavonol biosynthesis	
Solanum tuberosum	StD	F3H, DFR, F3-5-H Regulation of anthocyanin biosynthesis in tuber skin	Jung et al. (2009)
Nicotiana tabacum	NtAN2	Regulation of anthocyanin biosynthesis in flowers	Pattanaik et al. (2010)
Brassica oleracea var. *botrytis*	Pr-D	F3H, DFR, ANS Regulation of anthocyanin biosynthesis	Chiu et al. (2010)
Vitis vinifera	VvMYBF1	Regulation of flavonol biosynthesis in developing grape berries	Czemmel et al. (2009)
	VvMYBPA1 VvMYB5a VvMYB5b	LAR1, ANR Regulation of proanthocyanidin biosynthesis in developing grape berries	Bogs et al. (2007)
	VvMYB5b VvMYBA1 VvMYBA2	Regulation of anthocyanidin biosynthesis in ripening grape berries	Deluc et al. (2008)
Pinus pinaster	PpNAC1	Positively regulate SCW formation	Pascual et al. (2018)
Populus trichocarpa	PtrMYB3/20	Activate the biosynthetic pathways of cellulose, xylan, and lignin and are directly target of PtrWND2.	McCarthy et al. (2010)
Quercus suber	QsMYB1	Related to secondary growth and cork biosynthesis.	Almeida et al. (2013)
Populus tomentosa Carr.	PtoMYB156	Repressor, repress phenylpropanoid biosynthesis and negatively regulate SCW formation.	Yang et al. (2017)
Malus	MdMYB9 MdMYB11	Proanthocyanidin biosynthesis in leaves	Gesell et al. (2014)
Myrica rubra	MrMYB1	Anthocyanin biosynthesis in fruit	Niu et al. (2010)
Prunus persica	PpMYB10	Anthocyanin biosynthesis in fruit skin	Ravaglia et al. (2013)

The antiviral activity of flavonoids has also been reported against viral plant diseases (Krcatović et al. 2008; Zhao et al. 2013; Wan et al. 2015). The antiphytoviral modes of action of flavonoids include inhibition of the viral protein synthesis, disruption of the viral RNA translation, suppression of viral DNA synthesis, interference with the viral structural protein, and inhibition of transcription of the viral genome and enzymes (Zakaryan et al. 2017). Binding of the ring of the flavonoids with viral nucleic acid bases via formation of hydrogen bonds can lead to inhibition of the viral DNA and RNA synthesis and their polymerases. In this regard, Zhao et al. (2013) reported a high anti-*Tobacco mosaic virus* activity of two flavonoids, fistula flavonoids B and C, isolated from the bark and stems of *Cassia fistula*, achieving inhibition rates of 28.5% and 31.3%, respectively compared to 24.7% for ningnanmycin.

9.5 Lignin

Lignin is one of the most common natural polymers on the earth ranking second after cellulose. This polymer (**C6**-**C3**)n is formed from three basic phenolic molecules (monolignols) in the cell wall (Alejandro et al. 2012). Deposition of lignin in the cell wall has an important defensive function in the plant forming a physical barrier against the invading pathogens. In addition, it plays another important role enhancing plant growth and development, cell wall rigidity, and hydrophobic properties (Schuetz et al. 2014).

9.5.1 Lignin Biosynthesis

Biosynthesis of lignin can be divided into three stages: biosynthesis of monolignols, translocation from cytoplasm to the cell wall, and polymerization of monolignols. Biosynthesis of the monolignols (*p*-coumaryl alcohol, coniferyl alcohol, and sinapyl alcohol) occurs in the cell cytoplasm via the phenylpropanoid pathway from the *p*-coumaric acid using a subset of enzymes, and then these monolignols are actively translocated into the cell walls via an ATP-binding cassette transporter. In the cell wall, lignin is formed by polymerization of the three monolignols by peroxidase and laccase (Liu et al. 2018). However, other compounds such as hydroxycinnamaldehydes, tricin flavones, hydroxystilbenes, and xenobiotics have also been known as lignin subunits (Del Río et al. 2017).

9.5.2 Regulation of Lignin Biosynthesis

It was found that lignin metabolism is regulated by transcription factors such as MYB and NAC families, which represent their main transcriptional regulatory mechanism (Nakano et al. 2015). In *A. thaliana*, overexpression of the transcription factors MYB46, MYB58, MYB63, MYB83, and MYB85 activates the transcriptional expression of most phenylpropanoid and lignin biosynthetic pathways genes (Zhong and Ye 2012; Kim et al. 2014). Moreover, NAC transcription factors, SND1, and NST1 are also able to regulate lignin biosynthesis-related genes in *Arabidopsis* (Zhong et al. 2007). On contrary, the transcription factors MYB4, MYB32, and MYB7 suppress the expression of the lignin biosynthesis-related genes (Wang and Dixon 2012). The transcription factor WRKY12 was found also to suppress lignin biosynthesis in *Arabidopsis* (Wang et al. 2010). In *Medicago sativa*, downregulation of a WRKY transcription factor enhanced the lignin biosynthesis (Gallego-Giraldo et al. 2016).

9.5.3 Defensive Mechanisms of Lignin

Lignin plays a well-recognized defensive role against attacking phytopathogenic fungi. It has antifungal activity, acts as a physical barrier to suppress fungal invasion, and represses the diffusion of their toxins (Sattler and Funnell-Harris 2013; Xie et al. 2018). Lignin deposition during the attack of many pathogenic fungi has been widely reported. In this regard, Marques et al. (2018) studied the cell wall-associated defense responses in smut-resistant and susceptible sugarcane cultivars against infection with *Sporisorium scitamineum*, the causal agent of smut. Lignin and phenolic compounds accumulation was detected at the infection site in the resistant cultivars during the early stages of infection to prevent the initial penetration of the fungus. Later, cell wall thickening as a physical barrier was also observed in the resistant cultivars but not in the susceptible ones. Moreover, lignin exhibited the most potent antifungal activity when tested in vitro against the fungal pathogen *Diplodia pinea*, the causal agent of tip blight and canker of Austrian pine, compared to the other tested phenolic compounds (Sherwood and Bonello 2013). However, downregulation of some lignin biosynthesis-related genes was found to improve the plant immunity. In *Arabidopsis* and *Medicago*, suppression of *HCT* led to accumulation of salicylic acid and upregulation of some pathogenesis-related genes, resulting in the plant immunity enhancement (Gallego-Giraldo et al. 2011a, b).

Regard to plant viral diseases, Kofalvi and Nassuth (1995) reported no change in the lignin content of wheat leaves when infected with mosaic virus. However, the antiviral activity of lignin and its derivatives has been reported against many human and animal viruses (Spiridon 2018; Srisapoome et al. 2018). In this regard, a strong antiviral potentiality of low-molecular-weight lignin was observed against hepatitis C virus in the cultured cells (Matsuhisa et al. 2015). In another study, Lee et al. (2011) also reported the antiviral activity of lignin-carbohydrate-protein complexes extracted from *Pimpinella anisum* against HSV 1 and 2.

9.6 Tannins

Tannins, the most complex group of polyphenols (C6-C3-C6)n, are oligomeric compounds with free phenolic groups, of a high molecular weight ranging from 500 to 3000 Da, which can bind and precipitate proteins, starch, cellulose, and minerals. There are two main categories of tannins; the first is the hydrolysable tannins such as gallotannins and ellagitannins which contain a carbohydrate as a central core, and the second is the condensed tannins (proanthocyanidins) which consist of flavan-3-ol units linked by carbon-carbon bonds (Crozier et al. 2006). They are responsible for the astringent taste of some leaves, fruits and wines.

9.6.1 Tannins Biosynthesis

Tannins are biosynthesized from various branches of the flavonoid pathway, sharing the same anthocyanin pathway. As previously described in the flavonoid biosynthesis, chalcone are converted to flavanone in the cytoplasm of plant cells using CHI, which then hydroxylated by F3′H and F3′5′H (flavonoid 3′,5′-hydroxylase) into eriodictyol or pentahydroxyflavanone, respectively, from which leucoanthocyanidins are produced. Leucoanthocyanidins are then oxidized by ANS to give colored anthocyanidins. In another branch, it can be converted by leucoanthocyanidin reductase into (2R,3S)-flavan-3-ols which is the potential precursor of tannins (He et al. 2008).

9.6.2 Regulation of Tannins Biosynthesis

Multiple regulatory genes, belonging to six different families, are contributed in the control of the proanthocyanidin biosynthesis (Myc transcriptional factors, Myb transcriptional factors, WD40-like protein, WRKY transcription factors, MADS homeodomain genes, and TFIIIA-like proteins) (Marles et al. 2003). Typically, *Myc* and *Myb* genes have the widely control of the proanthocyanidin biosynthesis (Gonzalez et al. 2016; Escaray et al. 2017). WD40-like protein regulates the accumulation of proanthocyanidin, while, WRKY transcription factor acts downstream of WD40-like protein. *MADS* genes regulate BAN and may act upstream of other regulatory genes. WIP proteins specifically control the assembly of proanthocyanidin polymer, as well as the biosynthesis of monomeric flavan-3-ol molecules (He et al. 2008). In addition, other different regulatory genes can directly or indirectly affect the proanthocyanidin biosynthetic-related genes such as Anthocyaninless2 and AtDOF4;2 (Skirycz et al. 2007).

9.6.3 Defensive Mechanisms of Tannins

The antifungal activity of hydrolysable tannins and proanthocyanidins has been widely reported against phytopathogenic fungi. In this regard, Anttila et al. (2013) studied the fungitoxicity of the cone and bark extracts of conifer tannins against 8 brown-rot fungi, 3 white-rot fungi, and 4 soft-rot fungi in liquid cultures. A high growth inhibition for the tested brown-rot fungi, but not the white-rot or soft-rot fungi, was achieved even at low concentrations. In temperate trees, tannins and related phenolic compounds prevent the fungal decay of the heartwood and inhibit extracellular hydrolases from invading pathogens. An antifungal activity of tannins extract from *Acacia mearnsii* against *Aspergillus niger* and *Candida* sp. was also reported (Dos Santos et al. 2017). Their antifungal mechanisms of action comprise inhibition of the extracellular enzymes (cellulase, pectinase, laccase, xylanase, etc.), nutrient privation of substrates (metal complexation, protein insolubilization), and inhibition of oxidative phosphorylation (Ogawa and Yazaki 2018).

9.7 Coumarins

Coumarins are an important widely distributed group of polyphenolic compounds. They are a class of benzopyrones (1,2-benzopyrones or 2H-1-benzopyran-2-ones) which will be produced by plants as a defense response against herbivores and attacking pathogens. However, they have a wide range of applications as perfumes, cosmetics, aroma enhancers, medicinal drugs, and industrial additives (Kontogiorgis et al. 2012; Rohini and Srikumar 2014). They can be subdivided into simple coumarins, furanocoumarins, pyranocoumarins, and phenylcoumarins (Jain and Joshi 2012).

9.7.1 Coumarins Biosynthesis

Simple coumarins are biosynthesized via the phenylpropanoid biosynthesis pathway from *p*-coumaric acid. The biosynthesis process comprises C-2 hydroxylation, chain isomerization, and lactonization producing the umbelliferone. Pyrano and furocoumarins are also produced from *p*-coumaric acid. These coumarins are subdivided into two classes (lineal and angular) according to the condensation position of isopentenyl pyrophosphate forming the heterocycle. Cyclization of a simple coumarin can also results in production of these coumarins (Dewick 2002).

9.7.2 Defensive Mechanisms of Coumarins

Coumarins have a strong fungitoxicity against a wide range of phytopathogens (Montagner et al. 2008; Guerra et al. 2015). Their modes of action include altering the mitochondrial matrix making it thick, induction of cell death, and cell wall perforation leading to release of the cell cytoplasm (Widodo et al. 2012). In this regard, Al-Barwani and Eltayeb reported a strong antifungal activity of psoralen and fouranocoumarin against *Cercospora carotae*, *Sclerotinia sclerotiorum,* and *Alternaria brassicicola*. In addition, Al-Amiery et al. (2012) reported a strong antifungal activity of some newly synthesized coumarins against *A. niger* and *C. albicans* in vitro. Inoculation of celery and parsnip roots with *S. sclerotiorum* led to upregulation of furanocoumarins expression level (Rahman 2000).

The antiviral activities of coumarins against many plant viruses have also been reported. Liu et al. (2016) studied the antiviral potential of seven coumarins including two new coumarins 6-hydroxy-7-methyl-3-(4′-methoxyphenyl)-coumarin and 6-hydroxy-5-methoxy-7-methyl-3-(4′-methoxyphenyl)-coumarin, isolated from the leaves of *Nicotiana tabacum*, against *Tobacco mosaic virus*. Among the tested coumarins, both new compounds exhibited anti-TMV activity ranging from 13.7 to 28.6%. The antiviral mechanism of coumarins includes inhibition of reverse transcriptase, integrase, and protease enzymes (Shokoohinia et al. 2014).

References

Agati G, Brunetti C, Di Ferdinando M, Ferrini F, Pollastri S, Tattini M (2013) Functional roles of flavonoids in photoprotection: new evidence, lessons from the past. Plant Physiol Biochem 72:35–45

Alam MA, Subhan N, Hossain H, Hossain M, Reza HM, Rahman MM, Ullah MO (2016) Hydroxycinnamic acid derivatives: a potential class of natural compounds for the management of lipid metabolism and obesity. Nutr Metab (Lond) 13:27

Al-Amiery AA, Kadhum AA, Mohamad AB (2012) Antifungal activities of new coumarins. Molecules 17(5):5713–5723

Al-Barwani FM, Eltayeb EA (2004) Antifungal compounds from induced *Conium macultum* L. plants. Biochem Syst Ecol 32:1097–1108

Albert NW, Davies KM, Lewis DH, Zhang H, Montefiori M, Brendolise C, Boase MR, Ngo H, Jameson PE, Schwinn KE (2014) A conserved network of transcriptional activators and repressors regulates anthocyanin pigmentation in eudicots. Plant Cell 26(3):962–980

Alejandro S, Lee Y, Tohge T, Sudre D, Osorio S, Park J, Bovet L, Geldner N, Fernie AR, Martinoia E (2012) AtABCG29 is a monolignol transporter involved in lignin biosynthesis. Curr Biol 22:1207–1212

Almeida T, Menéndez E, Capote T, Ribeiro T, Santos C, Gonçalves S (2013) Molecular characterization of Quercus suber MYB1, a transcription factor up-regulated in cork tissues. J Plant Physiol 170(2):172–178

Alves CT, Ferreira IC, Barros L, Silva S, Azeredo J, Henriques M (2014) Antifungal activity of phenolic compounds identified in flowers from North Eastern Portugal against Candida species. Future Microbiol 9(2):139–146

Ansari MA, Fatima Z, Hameed S (2014) Sesamol: a natural phenolic compound with promising anticandidal potential. J Pathog 2014:895193

Anttila A-K, Pirttilä AM, Häggman H, Harju A, Venäläinen M, Haapala A, Holmbom B, Julkunen-Tiitto R (2013) Condensed conifer tannins as antifungal agents in liquid culture. Holzforschung 67(7):825–832

Appelhagen I, Lu GH, Huep G, Schmelzer E, Weisshaar B, Sagasser M (2011) TRANSPARENT TESTA1 interacts with R2R3-MYB factors and affects early and late steps of flavonoid biosynthesis in the endothelium of *Arabidopsis thaliana* seeds. Plant J 67:406–419

Arici SE, Kafkas E, Kaymak S, Koc NK (2014) Phenolic compounds of apple cultivars resistant or susceptible to *Venturia inaequalis*. Pharm Biol 52(7):904–908

Belofsky G, Kolaczkowski M, Adams E, Schreiber J, Eisenberg V, Coleman CM, Zou Y, Ferreira D (2013) Fungal ABC transporter-associated activity of isoflavonoids from the root extract of *Dalea formosa*. J Nat Prod 76(5):915–925

Bernardi J, Stagnati L, Lucini L, Rocchetti G, Lanubile A, Cortellini C, De Poli G, Busconi M, Marocco A (2018) Phenolic profile and susceptibility to *Fusarium* infection of pigmented maize cultivars. Front Plant Sci 9:1189

Bhattacharya A, Sood P, Citovsky V (2010) The roles of plant phenolics in defence and communication during *Agrobacterium* and *Rhizobium* infection. Mol Plant Pathol 11(5):705–719

Bogs J, Jaffe FW, Takos AM, Walker AR, Robinson SP (2007) The grapevine transcription factor VvMYBPA1 regulates proanthocyanidin synthesis during fruit development. Plant Physiol 143:1347–1361

Bruce W, Folkerts O, Garnaat C, Crasta O, Roth B, Bowen B (2000) Expression profiling of the maize flavonoid pathway genes controlled by estradiol-inducible transcription factors CRC and P. Plant Cell 12:65–80

Chen Z, Zheng Z, Huang J, Lai Z, Fan B (2009) Biosynthesis of salicylic acid in plants. Plant Signal Behav 4(6):493–496

Chepkirui C, Matasyoh JC, Wagara IN, Nakavuma J (2014) Antifungal activity of flavonoids isolated from Monanthotaxis littoralis against mycotoxigenic fungi from maize. Am J Chem Appl 1(4):54–60

Cheynier V, Comte G, Davies KM, Lattanzio V, Martens S (2013) Plant phenolics: recent advances on their biosynthesis, genetics, and ecophysiology. Plant Physiol Biochem 72:1–20

Chiu LW, Zhou X, Burke S, Wu X, Prior RL, Li L (2010) The purple cauliflower arises from activation of a MYB transcription factor. Plant Physiol 154:1470–1480

Crozier A, Jaganath IB, Clifford MN (2006) Phenols, polyphenols and tannins: an overview. In: Crozier A, Clifford MN, Ashihara H (eds) Plant secondary metabolites: occurrence, structure and role in the human diet. Blackwell Publishing, Oxford, pp 1–24

Czemmel S, Stracke R, Weisshaar B, Cordon N, Harris NN, Walker AR, Robinson SP, Bogs J (2009) The grapevine R2R3-MYB transcription factor VvMYBF1 regulates flavonol synthesis in developing grape berries. Plant Physiol 151:1513–1530

Daayf F, El Hadrami A, El Bebany A, Henriquez M, Yao Z, Derksen H, El Hadrami I, Adam LR (2012) Phenolic compounds in plant defense and pathogen counter-defense mechanisms. In: Cheynier V, Sarni-Manchado P, Quideau S (eds) Recent advances in polyphenols research, vol 3. Wiley-Blackwell, Oxford

de Ascensao AR, Dubery IA (2003) Soluble and wall bound phenolic polymers in Musa acuminata roots exposed to elicitors from Fusarium oxysporum f.sp. cubense. Phytochemistry 63:679–686

de Oliveira DM, Finger-Teixeira A, Mota TR, Salvador VH, Moreira-Vilar FC, Molinari HB, Mitchell RA, Marchiosi R, Ferrarese-Filho O, dos Santos WD (2015) Ferulic acid: a key component in grass lignocellulose recalcitrance to hydrolysis. Plant Biotechnol J 13 (9):1224–1232

Del Río JC, Rencoret J, Gutiérrez A, Kim H, Ralph J (2017) Hydroxystilbenes are monomers in palm fruit endocarp lignins. Plant Physiol 174:2072–2082

Deluc L, Bogs J, Walker AR, Ferrier T, Decendit A, Merillon JM, Robinson SP, Barrieu F (2008) The transcription factor VvMYB5b contributes to the regulation of anthocyanin and proanthocyanidin biosynthesis in developing grape berries. Plant Physiol 147:2041–2053

Deng Y, Lu S (2017) Biosynthesis and regulation of phenylpropanoids in plants. Crit Rev Plant Sci 36(4):257–290

Dewick PM (2002) Medicinal natural products: a biosynthetic approach, 2nd edn. Wiley, Hoboken

Dos Santos C, Vargas Á, Fronza N, Dos Santos JHZ (2017) Structural, textural and morphological characteristics of tannins from *Acacia mearnsii* encapsulated using sol-gel methods: applications as antimicrobial agents. Colloids Surf B Biointerfaces 151:26–33

Dunkić V, Bezić N, Vuko E, Cukrov D (2010) Antiphytoviral activity of *Satureja montana* L. ssp. *variegata* (Host) P. W. Ball essential oil and phenol compounds on CMV and TMV. Molecules 15(10):6713–6721

Du H, Zhang L, Liu L, Tang X-F, Yang W-J, Wu Y-M, Huang Y-B, Tang Y-X (2009) Biochemical and molecular characterization of plant MYB transcription factor family. Biochemistry (Mosc) 74:1–11

El-Seedi HR, El-Said AM, Khalifa SA, Göransson U, Bohlin L, Borg-Karlson AK, Verpoorte R (2012) Biosynthesis, natural sources, dietary intake, pharmacokinetic properties, and biological activities of hydroxycinnamic acids. J Agric Food Chem 60(44):10877–10895

Escaray FJ, Passeri V, Perea-García A, Antonelli CJ, Damiani F, Ruiz OA, Paolocci F (2017) The *R2R3-MYB TT2b* and the *bHLH TT8* genes are the major regulators of proanthocyanidin biosynthesis in the leaves of *Lotus* species. Planta 246(2):243–261

Falcone Ferreyra ML, Rius SP, Casati P (2012) Flavonoids: biosynthesis, biological functions, and biotechnological applications. Front Plant Sci 3:222

Fernandes KRP, Bittercourt PS, Souza ADL, Souza AQL, Silva FMA, Lima ES, Acho LDR, Nunomura RCS, Teixeira AF, Koolen HHF (2019) Phenolic compounds from *Virola venosa* (Myristicaceae) and evaluation of their antioxidant and enzyme inhibition potential. Acta Amaz 49(1):48–53

Ferreyra MLF, Rius SP, Emiliani J, Pourcel L, Feller A, Morohashi K, Casati P, Grotewold E (2010) Cloning and characterization of a UV-B-inducible maize flavonol synthase. Plant J 62:77–91

Gallego-Giraldo L, Escamilla-Trevino L, Jackson LA, Dixon RA (2011a) Salicylic acid mediates the reduced growth of lignin down-regulated plants. Proc Natl Acad Sci U S A 108:20814–20819

Gallego-Giraldo L, Jikumaru Y, Kamiya Y, Tang Y, Dixon RA (2011b) Selective lignin downregulation leads to constitutive defense response expression in alfalfa (*Medicago sativa* L.). New Phytol 190:627–639

Gallego-Giraldo L, Shadle G, Shen H, Barrosrios J, Fresquet CS, Wang H, Dixon RA (2016) Combining enhanced biomass density with reduced lignin level for improved forage quality. Plant Biotechnol J 14:895–904

Gesell A, Yoshida K, Tran LT, Constabel CP (2014) Characterization of an apple TT2-type R2R3 MYB transcription factor functionally similar to the poplar proanthocyanidin regulator PtMYB134. Planta 240(3):497–511

Goleniowski M, Bonfill M, Cusido R, Palazón J (2013) Phenolic acids. In: Ramawat K, Mérillon JM (eds) Natural products. Springer, Berlin/Heidelberg

Gonzalez A, Brown M, Hatlestad G, Akhavan N, Smith T, Hembd A, Moore J, Montes D, Mosley T, Resendez J, Nguyen H, Wilson L, Campbell A, Sudarshan D, Lloyd A (2016) TTG2 controls the developmental regulation of seed coat tannins in *Arabidopsis* by regulating vacuolar transport steps in the proanthocyanidin pathway. Dev Biol 419(1):54–63

Gou JY, Felippes FF, Liu CJ, Weigel D, Wang JW (2011) Negative regulation of anthocyanin biosynthesis in Arabidopsis by a miR156-targeted SPL transcription factor. Plant Cell 23:1512–1522

Grotewold E, Chamberlin M, Snook M, Siame B, Butler L, Swenson J, Maddock S, Clair GS, Bowen B (1998) Engineering secondary metabolism in maize cells by ectopic expression of transcription factors. Plant Cell 10:721–740

Guerra FQ, de Araújo RS, de Sousa JP, Pereira Fde O, Mendonça-Junior FJ, Barbosa-Filho JM, de Oliveira LE (2015) Evaluation of antifungal activity and mode of action of new coumarin derivative, 7-Hydroxy-6-nitro-2H-1-benzopyran-2-one, against *Aspergillus* spp. Evid Based Complement Alternat Med 2015:925096

Hammerschmidt R (2014) Chlorogenic acid: a versatile defense compound. Physiol Mol Plant Pathol 88:3–6

He F, Pan QH, Shi Y, Duan CQ (2008) Biosynthesis and genetic regulation of proanthocyanidins in plants. Molecules 13(10):2674–2703

Hichri I, Barrieu F, Bogs J, Kappel C, Delrot S, Lauvergeat V (2011) Recent advances in the transcriptional regulation of the flavonoid biosynthetic pathway. J Exp Bot 62:2465–2483

Hu Q-F, Zhou B, Huang J-M, Gao X-M, Shu L-D, Yang G-Y, Che C-T (2013) Antiviral phenolic compounds from *Arundina gramnifolia*. J Nat Prod 76(2):292–296

Iwashina T (2013) Flavonoid properties of five families newly incorporated into the order Caryophyllales (review). Bull Natl Mus Nat Sci 39:25–51

Jain PK, Joshi H (2012) Coumarin: chemical and pharmacological profile. J Appl Pharm Sci 2 (6):236–240

Jorgensen K, Rasmussen AV, Morant M, Nielsen AH, Bjarnholt N, Zagrobelny M, BAK S, Moller BL (2005) Metabolon formation and metabolic channeling in the biosynthesis of plant natural products. Curr Opin Plant Biol 8:280–291

Jung CS, Griffiths HM, De Jong DM, Cheng S, Bodis M, Kim TS, De Jong WS (2009) The potato developer (D) locus encodes an R2R3 MYB transcription factor that regulates expression of multiple anthocyanin structural genes in tuber skin. Theor Appl Genet 120:45–57

Junior HMS, Campos VAC, Alves DS, Cavalheiro AJ, Souza LP, Botelho DMS, Chalfoun SM, Oliveira DF (2014) Antifungal activity of flavonoids from *Heteropterys byrsonimifolia* and a commercial source against *Aspergillus ochraceus*: in silico interactions of these compounds with a protein kinase. Crop Protect 62:107–114

Kim WC, Kim JY, Ko JH, Kang H, Han KH (2014) Identification of direct targets of transcription factor MYB46 provides insights into the transcriptional regulation of secondary wall biosynthesis. Plant Mol Biol 85:589–599

Kim KH, Dutta T, Walter ED, Isern NG, Cort JR, Simmons BA, Singh S (2017) Chemoselective methylation of phenolic hydroxyl group prevents quinone methide formation and repolymerization during lignin depolymerization. ACS Sustain Chem Eng 5(5):3913–3919

Kofalvi SA, Nassuth A (1995) Influence of wheat streak mosaic virus infection on phenylpropanoid metabolism and the accumulation of phenolics and lignin in wheat. Physiol Mol Plant Pathol 47:365–377

Kontogiorgis C, Detsi A, Hadjipavlou-Litina D (2012) Coumarin-based drugs: a patent review (2008 – present). Expert Opin Ther Pat 22(4):437–454

Krcatović E, Rusak G, Bezić N, Krajacić M (2008) Inhibition of tobacco mosaic virus infection by quercetin and vitexin. Acta Virol 52(2):119–124

Kumar S, Pandey AK (2013) Chemistry and biological activities of flavonoids: an overview. Sci World J 2013:162750

Lee JB, Yamagishi C, Hayashi K, Hayashi T (2011) Antiviral and immunostimulating effects of lignin-carbohydrate-protein complexes from *Pimpinella anisum*. Biosci Biotechnol Biochem 75 (3):459–465

Leváková L, Lacko-Bartošová M (2017) Phenolic acids and antioxidant activity of wheat species: a review. Agriculture (Poľnohospodárstvo) 63(3):92–101

Li S (2014) Transcriptional control of flavonoid biosynthesis: fine-tuning of the MYB-bHLH-WD40 (MBW) complex. Plant Signal Behav 9(1):e27522

Li L, Xu W-X, Liu C-B, Zhang C-M, Zhao W, Shang S-Z, Deng L, Guo Y-D (2015) A new antiviral phenolic compounds from *Arundina graminifolia*. Asian J Chem 27:3525–3526

Li ZJ, Liu M, Dawuti G, Dou Q, Ma Y, Liu HG, Aibai S (2017) Antifungal activity of gallic acid in vitro and in vivo. Phytother Res 31(7):1039–1045

Liu CB, Shen QP, Wang Y, Zhang FM (2016) Coumarins from the leaves of *Nicotiana tabacum* and their anti-tobacco mosaic virus activities. Chem Nat Compd 52:992–995

Liu Q, Luo L, Zheng L (2018) Lignins: biosynthesis and biological functions in plants. Int J Mol Sci 19(2):pii: E335

Lucas-Barbosa D, Sun P, Hakman A, van Beek TA, van Loon JJA, Dicke M (2016) Visual and odour cues: plant responses to pollination and herbivory affect the behaviour of flower visitors. Funct Ecol 30(3):431–441

Makoi JH, Ndakidemi PA (2012) Allelopathy as protectant, defence and growth stimulants in legume cereal mixed culture systems. N Z J Crop Hort Sci 40(3):161–186

Mandal SM, Chakraborty D, Dey S (2010) Phenolic acids act as signaling molecules in plant-microbe symbioses. Plant Signal Behav 5(4):359–368

Marles MAS, Ray H, Gruber MY (2003) New perspectives on proanthocyanidin biochemistry and molecular regulation. Phytochemistry 64:367–383

Marques JPR, Hoy JW, Appezzato-da-Glória B, Viveros AFG, Vieira MLC, Baisakh N (2018) Sugarcane cell wall-associated defense responses to infection by *Sporisorium scitamineum*. Front Plant Sci 9:698

Martens S, Preuss A, Matern U (2010) Multifunctional flavonoid dioxygenases: flavonol and anthocyanin biosynthesis in *Arabidopsis thaliana* L. Phytochemistry 71:1040–1049

Matsuhisa K, Yamane S, Okamoto T, Watari A, Kondoh M, Matsuura Y, Yagi K (2015) Anti-HCV effect of *Lentinula edodes* mycelia solid culture extracts and low-molecular-weight lignin. Biochem Biophys Res Commun 462:52–57

McCarthy RL, Zhong R, Fowler S, Lyskowski D, Piyasena H, Carleton K, Spicer C, Ye Z-H (2010) The poplar MYB transcription factors, PtrMYB3 and PtrMYB20, are involved in the regulation of secondary wall biosynthesis. Plant Cell Physiol 51(6):1084–1090

Mierziak J, Kostyn K, Kulma A (2014) Flavonoids as important molecules of plant interactions with the environment. Molecules 19(10):16240–16265

Montagner C, de Souza SM, Groposoa C, Delle Monache F, Smânia EF, Smânia A Jr (2008) Antifungal activity of coumarins. Z Naturforsch C 63:21–28

Nakano Y, Yamaguchi M, Endo H, Rejab NA, Ohtani M (2015) NAC-MYB-based transcriptional regulation of secondary cell wall biosynthesis in land plants. Front Plant Sci 6:288

Naoumkina M, Dixon RA (2008) Subcellular localization of flavonid natural products: a signalling function? Plant Signal Behav 3:573–575

Negritto MC, Valdez C, Sharma J, Rosenberg C, Selassie CR (2017) Growth inhibition and DNA damage induced by X-phenols in yeast: a quantitative structure-activity relationship study. ACS Omega 2(12):8568–8579

Niu SS, Xu CJ, Zhang WS, Zhang B, Li X, Lin-Wang K, Ferguson IB, Allan AC, Chen KS (2010) Coordinated regulation of anthocyanin biosynthesis in Chinese bayberry (Myrica rubra) fruit by a R2R3 MYB transcription factor. Planta 231(4):887–899

Ogawa S, Yazaki Y (2018) Tannins from *Acacia mearnsii* De wild. Bark: tannin determination and biological activities. Molecules 23(4):E837

Panche AN, Diwan AD, Chandra SR (2016) Flavonoids: an overview. J Nutr Sci 5:e47

Pascual MB, Llebrés MT, Craven-Bartle B, Cañas RA, Cánovas FM, Ávila C (2018) PpNAC1, a main regulator of phenylalanine biosynthesis and utilization in maritime pine. Plant Biotechnol J 16(5):1094–1104

Pattanaik S, Kong Q, Zaitlin D, Werkman JR, Xie CH, Patra B, Yuan L (2010) R2R3 MYB regulator from tobacco. Isolation and functional characterization of a floral tissue-specific. Planta 231:1061–1076

Rahman AU (2000) Studies in natural product chemistry, vol 24. Elsevier, Amesterdam

Rashad YM, Aseel DG, Hafez EE (2018) Antifungal potential and defense gene induction in maize against Rhizoctonia root rot by seed extract of *Ammi visnaga* (L.) Lam. Phytopathol Mediterr 57 (1):73–88

Ravaglia D, Espley RV, Henry-Kirk RA, Andreotti C, Ziosi V, Hellens RP, Costa G, Allan AC (2013) Transcriptional regulation of flavonoid biosynthesis in nectarine (*Prunus persica*) by a set of R2R3 MYB transcription factors. BMC Plant Biol 13:68

Rohini K, Srikumar PS (2014) Therapeutic role of coumarins and coumarin-related compounds. J Thermodyn Catal 5:2

Sattler SE, Funnell-Harris DL (2013) Modifying lignin to improve bioenergy feedstocks: strengthening the barrier against pathogens? Front Plant Sci 4:70

Scalbert A (1991) Antimicrobial properties of tannins. Phytochemistry 30:3875–3883

Schuetz M, Benske A, Smith RA, Watanabe Y, Tobimatsu Y, Ralph J, Demura T, Ellis B, Samuels AL (2014) Laccases direct lignification in the discrete secondary cell wall domains of protoxylem. Plant Physiol 166:798–807

Serpa R, Franca EJ, Furlaneto-Maia L, Andrade CG, Diniz A, Furlaneto MC (2012) In vitro antifungal activity of the flavonoid baicalein against *Candida* species. J Med Microbiol 61:1704–1708

Šežiene V, Baležentienė L, Maruška A (2017) Identification and allelochemical activity of phenolic compounds in extracts from the dominant plant species established in clear-cuts of Scots pine stands. iForest 10:309–314

Shabana YM, Abdel-Fattah GM, Ismail AE, Rashad YM (2008) Control of brown spot pathogen of rice (*Bipolaris oryzae*) using some phenolic antioxidants. Braz J Microbiol 39(3):438–444

Sherwood P, Bonello P (2013) Austrian pine phenolics are likely contributors to systemic induced resistance against *Diplodia pinea*. Tree Physiol 33(8):845–854

Shokoohinia Y, Sajjadi SE, Gholamzadeh S, Fattahi A, Behbahani M (2014) Antiviral and cytotoxic evaluation of coumarins from *Prangos ferulacea*. Pharm Biol 52(12):1543–1549

Skirycz A, Jozefczuk S, Stobiecki M, Muth D, Zanor MI, Witt I, Mueller-Roeber B (2007) Transcription factor AtDOF4;2 affects phenylpropanoid metabolism in *Arabidopsis thaliana*. New Phytol 175:425–438

Spiridon I (2018) Biological and pharmaceutical applications of lignin and its derivatives: a mini-review. Cellulose Chem Technol 52(7–8):543–550

Srisapoome P, Hamano K, Tsutsui I, Iiyama K (2018) Immunostimulation and yellow head virus (YHV) disease resistance induced by a lignin-based pulping by-product in black tiger shrimp (*Penaeus monodon* Linn.). Fish Shellfish Immunol 72:494–501

Tohge T, de Souza LP, Fernie AR (2017) Current understanding of the pathways of flavonoid biosynthesis in model and crop plants. J Exp Bot 68:4013–4028

Upadhyay A, Mooyottu S, Yin H, Nair M, Bhattaram V, Venkitanarayanan K (2015) Inhibiting microbial toxins using plant-derived compounds and plant extracts. Medicines 2:186–211

Wan Z, Hu D, Li P, Xie D, Gan X (2015) Synthesis, antiviral bioactivity of novel 4-thioquinazoline derivatives containing chalcone moiety. Molecules 20(7):11861–11874

Wang HZ, Dixon RA (2012) On-off switches for secondary cell wall biosynthesis. Mol Plant 5:297–303

Wang H, Avci U, Nakashima J, Hahn MG, Chen F, Dixon RA (2010) Mutation of WRKY transcription factors initiates pith secondary wall formation and increases stem biomass in dicotyledonous plants. Proc Natl Acad Sci U S A 107:22338–22343

Wang K, Jiang S, Pu T, Fan L, Su F, Ye M (2018) Antifungal activity of phenolic monoterpenes and structure-related compounds against plant pathogenic fungi. Nat Prod Res 15:1–8

Weston LA, Mathesius U (2013) Flavonoids: their structure, biosynthesis and role in the rhizosphere, including allelopathy. J Chem Ecol 39(2):283–297

Widodo GP, Sukandar EY, Adynyana IK (2012) Mechanism of action of coumarin against *C. albicans* by SEM/TEM analysis. ITB J Sci 44A:145–151

Xie M, Zhang J, Tschaplinski TJ, Tuskan GA, Chen J-G, Muchero W (2018) Regulation of lignin biosynthesis and its role in growth-Defense Tradeoffs. Front Plant Sci 9:1427

Yang L, Zhao X, Ran L, Li C, Fan D, Luo K (2017) PtoMYB156 is involved in negative regulation of phenylpropanoid metabolism and secondary cell wall biosynthesis during wood formation in poplar. Sci Rep 7:41209

Zakaryan H, Arabyan E, Oo A, Zandi K (2017) Flavonoids: promising natural compounds against viral infections. Arch Virol 162(9):2539–2551

Zhang J, Subramanian S, Stacey G, Yu O (2009) Flavones and flavonols play distinct critical roles during nodulation of *Medicago truncatula* by *Sinorhizobium meliloti*. Plant J 57:171–183

Zhao W, Zeng XY, Zhang T, Wang L, Yang GY, Chen YK, Hu QF, Miao MM (2013) Flavonoids from the bark and stems of *Cassia fistula* and their anti-tobacco mosaic virus activities. Phytochem Lett 6:179–182

Zheng T, Tan W, Yang H, Zhang L, Li T, Liu B, Zhang D, Lin H (2019) Regulation of anthocyanin accumulation via MYB75/HAT1/TPL-mediated transcriptional repression. PLoS Genet 15(3): e1007993

Zhong R, Ye ZH (2012) MYB46 and MYB83 bind to the SMRE sites and directly activate a suite of transcription factors and secondary wall biosynthetic genes. Plant Cell Physiol 53(2):368–380
Zhong R, Richardson EA, Ye ZH (2007) Two NAC domain transcription factors, SND1 and NST1, function redundantly in regulation of secondary wall synthesis in fibers of Arabidopsis. Planta 225:1603–1611
Zhu Z, Li G, Liu L, Zhang Q, Han Z, Chen X, Li B (2019) A R2R3-MYB transcription factor, VvMYBC2L2, functions as a transcriptional repressor of anthocyanin biosynthesis in grapevine (*Vitis vinifera* L.). Molecules 24(1):92
Zuzarte M, Vale-Silva L, Goncalves MJ, Cavaleiro C, Vaz S, Canhoto J, Pinto E, Salgueiro L (2012) Antifungal activity of phenolic-rich *Lavandula multifida* L. essential oil. Eur J Clin Microbiol Infect Dis 31(7):1359–1366

Phenolic Compounds from Medicinal Herbs: Their Role in Animal Health and Diseases – A New Approach for Sustainable Welfare and Development

10

Younis Ahmad Hajam, Seema Rai, Rajesh Kumar, Mudasir Bashir, and Javid Ahmad Malik

Abstract

Polyphenolic compounds have received a special attention from the scientific community, because of their pleiotropic properties. These polyphenolic compounds are the secondary metabolites synthesized in various components of the plants. They possess significant capability to influence different processes in biological systems. So, these are taken as supplementary foodstuffs, energizers, dietary sources, and medicines. These natural compounds attribute therapeutic and medicinal efficiencies. Phenolic compounds are of great interest in food manufacturing industries as they suppress the oxidative deprivation of lipids and therefore improve the nutritional value of foods. Ethnoveterinary medicine is a word usually applied for folk talent, beliefs, knowledge, practices, and methods associated to animal healthiness and cure of different health problems in the rural areas. Animal welfare is speedily becoming foremost aspect of production of animals worldwide, in the industrialized world and also in developing countries. Livestock plays an essential role in sustaining livelihood, nutritional and environmental security, and augmentation of agriculture. The enormous stride made in the livestock sector in the past decades is one of the main causes for positive growth rates recorded in agricultural sector. There is enormous need to attain vertical growth in terms of improving productivity. Animal husbandry and agriculture are associated with each other; rural communities utilize ethanobotanical culture for the curing of different animal diseases and in the biological control. Diseases in animals remain most important

Y. A. Hajam (✉) · R. Kumar
Department of Biosciences, Division Zoology, Career Point University, Hamirpur, Himachal Pradesh, India

S. Rai · J. A. Malik
Department of Zoology, Guru Ghasidas Vishwavidyalaya, Bilaspur, Chhattisgarh, India

M. Bashir
Department of School Education, Jammu and Kashmir, India

R. Lone et al. (eds.), *Plant Phenolics in Sustainable Agriculture*,
https://doi.org/10.1007/978-981-15-4890-1_10

causes of poor performance in livestock field, gap between the supply of animals, and the requirement of livestock products. Phenolic compounds carry various pleiotropic functions including antibacterial, antioxidant, anti-inflammatory, antiviral, hepatoprotective, anti-allergic, antithrombotic, oncostatic, and immune-modulating activities. Present review paid some attention on the importance of bioactive compounds from plants and their role in animal welfare, to make progress and recover the agricultural productivity.

Keywords

Phenolic compounds, Animal health, Diseases, Herbs, Ethnoveterinary

10.1 Introduction

Nature is at all times a golden indication to illustrate the well-known phenomena of coexistence. Natural products derived from flora are the source for treating animal diseases (Firenzuoli and Gori 2007). In the current era, products obtained from medicinal plants are in demand, and their recognition is expanding progressively. Certainly, plants contribute a very important role by providing essential services in ecosystems. Without plants, humans as well as different other organisms cannot live in standard way. Medicinal herbs are continuously acted as an in general sign of ecosystem health (Singh 2002). Historical employments of plants used for the fitness and restoration, for both human and animal, are centuries old. It has been documented that plants have the ability to fight against the various diseases.

Ethnoveterinary medicine is a term typically applied for traditional skills, values, information, practices, and methods associated to animal well-being and treatment of different health issues in the rural areas (McCorkle and Green 1998). Ethnoveterinary practices have attained enormous significance for the last decade owing to the invention of some effective ethnoveterinary products (Singh 2002). The use of traditional therapies poses a low cost, easier, and sustainable option to synthetic remedy and pharmaceutical product (Dilshad et al. 2010). Thus, it is essential to investigate and estimate the biological efficacy from the phytochemical, pharmacological, toxicological, and clinical perspectives for wider application. Ethnoveterinary medicine has come forward as a demanding field in the recent past (Mathias-Mundy and McCorkle 1989). Medicinal plants have been used for millennia in virtually all cultures and serve both as a source of income and a source of affordable healthcare (Lambert et al. 1997). Plants contain the various bioactive components which are used in the devising of ethnoveterinary medicament. That 70–95% of the human beings living in emergent countries relies on medicinal plants for their primary healthcare needs (Robinon and Zhang 2011). Plant-derived medicine used in veterinary is one of the major methods for curing different diseases and diseases of the livestock. It has reported that 55 medicinal plants have 44 species, 40 genera, and 30 families (Kone and Atindehou 2008).

The fundamental oils and plant extracts revealed different natural actions such as antimicrobial, membrane interference by terpenoids and phenolics, metal chelation by phenols and flavonoids, and target on genetic material by coumarin as well as alkaloids that are considered to slow down the growth of microorganisms (Cowan 1999). Fundamental oils are slightly more efficient next to Gram-positive in contrast to Gram-negative food pathogens (Burt 2004) as they encompass an outer membrane boundary of the cellular wall which restricts the invasion of hydrophobic multipart through its lipopolysaccharide structures (Nazzaro et al. 2013; Vaara 1992). Various essential oils stimulate the development of helpful microbes and limit quantity of pathogenic bacteria in poultry (Wenk 2000). Apparent efficiency of small doses of fundamental oils and sodium butyrate has already been tested adjacent to *Salmonella* in broilers (Cerisuelo et al. 2014). The blend of thymol and cinnamaldehyde is confirmed to have careful antibacterial possessions that slow down the growth of yeast and fungi (Bento et al. 2013). Majority complement declared a substituent to antibiotics has influence on the microflora, either straight or indirectly (Taylor 2001).

10.2 Biological Importance of Secondary Metabolites

Phytochemicals are the secondary metabolites produced in special part of the plants. They contain the significant capability to influence different body processes and functions. Therefore, these bioactive components are taken as foodstuffs, stimulants, nutritional plants, and drugs. These natural components obtained from the plants contain therapeutic and medicinal properties. Among the bioactive components, polyphenolic compounds are the mainly significant group having medicinal value. The most important polyphenolic compounds include phenols, phenolic acids, flavonoids, and phenylpropanoids. Polyphenolic compounds act as antioxidants and free radical scavengers and therefore function to decrease oxidative stress and their damaging effects. Chemically term "phenolic" or "polyphenolic" defined a substance possessing aromatic ring having one or more than hydroxy substituent, together with functional derivatives (methyl ethers, esters, glycosides, etc.) (Lattanzio et al. 2006). Majority of phenolics possess more than two hydroxyl groups and are bioactive substances distributed extensively in food plants that are eaten regularly by large numbers of people.

Therefore, phenols assist in avoidance and control of different dreadful diseases and premature aging. Polyphenolic compounds are also responsible for anti-inflammatory, antibiotic, and antiseptic properties. The distinctive molecular structure of these bioactive components, with definite position of OH groups, possesses strong bioactivities. Polyphenolic compounds are generally present in the plants. Now-a-days plant polyphenols are given more attention, owing to their strong antioxidant properties and their well-known potential to prevent different oxidative stress-related diseases including cancer.

10.3 Properties of Plant Phenolics

10.3.1 Antioxidant Mechanisms of Phenolics

Antioxidant mechanisms of phenolics compounds acquired due to their capability to act as powerful antioxidants in different ways. Therefore, phenolic compounds demonstrate antioxidant properties as their potential to hunt free radicals produced from lipids, proteins, and oligonucleic acids. The free radical hunting capacity of phenolic compounds has been widely studied and related with several biological effects. Polyphenolic compounds contain OH groups, which are excellent hydrogen donors. These hydrogen-donating antioxidative compounds can counter ROS and NOS (Valentao et al. 2003; Valentao et al. 2002) in the execution reaction. These compounds split the cycle of production of fresh radicals. Subsequent contact of these phenolic complexes with primary reactive species results in the production of a radical form of the antioxidant. These are having much more chemical steadiness than the earlier radical. These hydroxyl groups of the phenolic compounds interact with the p-electrons of the benzene ring. The antioxidant capability of polyphenolic compounds are also credited owing to their potential to interlink metal ions involved in the fabrication of free radicals (Yang et al. 2001). Nevertheless, polyphenolic function as pro-oxidants having metal-chelating property that maintains or enhances catalytic action or by decreasing metals, therefore raising their capacity to produce free radicals (Croft 1998). Phenolic compounds very frequently have the possible potential to powerfully interact with proteins. This provides polyphenolic the capability function antioxidants also by virtue of their capacity to slow down some enzymes involved in radical generation, such as various cytochrome P450 isoforms, lipoxygenases, cyclooxygenases, and xanthine oxidases (Parr and Bolwell 2002; Cos et al. 1988).

10.3.2 Health Benefits of Polyphenolic Compounds

Nowadays dietary polyphenols have gained immense importance among the nutritionists, scientists, and customers because they play significant role in sustaining the animal health. Bioactive components extracted from plants and considered the most important ones are phenolic compounds, having ability to suppress the absorption of amylase in the curing of carbohydrate absorption. Fruits and vegetables comprise polyphenolic compounds, including phenolic acids and flavonoids, which could support health benefits by decreasing the risk of metabolic diseases. Nevertheless, many groups of phenolic compounds possess different biological features. Reactive oxygen species (ROS) and reactive nitrogen species (RNS) are highly reactive oxidized molecules, produced continuously through normal cellular metabolism, for example, the function of respiratory chain in mitochondria and inflammation, which can cause damage to other biological molecules, like proteins and DNA. The antioxidant enzymes includes SOD (superoxide dismutase), GPx (glutathione peroxidase), and CAT (catalase), which play a fundamental role in getting relieved from oxidants and protecting the cells from injury.

10.3.2.1 Antibacterial Effect

There are many phenolics that reveal antibacterial activities which belong to the nonflowering medicinal plants to the flowering ones. Polyphenolic compounds are produced by different plant groups together with numerous medicinal plants. Dzoyem et al. reported that flavonoids derived from this plant demonstrated the highest antibacterial activity against both strains. Plant species used in traditional medicine or dietary consumption, e.g., coli, *Klebsiella pneumoniae* (Schroeter) Trevisan, *Pseudomonas aeruginosa* (Schroeter) Migula, and *Enterococcus faecalis*. Polyphenolics are used as antibacterial agent against *P. acnes* are the main cause of skin acne problems.

10.3.2.2 Anticancer Activity

Cancer is frequently related with enhanced risk of death and the toxic side effects caused by the allopathic medicine; cancer patients are always in search of alternative and complementary methods of treatment, and their best alternative is phytomedicine (Kim and Park 2002). In present era, chemotherapy is considered an innovative approach for the treatment of cancer. Different cancer patients use herbal therapies. Medicinal plants are well identified for neutralizing free radical and antioxidant activities (Agarwal et al. 2001). Free radical generation that causes attack results in the oxidative stress to various biomolecules such as lipids, proteins, and DNA. Plants contain different bioactive components having strong antioxidant properties which might inhibit and control the cancer and other diseases by shielding the cells from the damaging effects of the free radicals.

10.4 Role of Livestock in Economy

Previous studies reported inappropriate animal husbandry practices; approximately 30–35% loss in animal breeding sectors has occurred, particularly in emerging countries (FAO 2002). However, the rural people are greatly reliant on livestock farming for their livelihood activities (Pica-Ciamarra et al. 2011). People in rural and mountainous regions in developing countries believe that livestock is fundamental source for financial system. Social security and food are also considered symbol of prestige for a particular family. Formulation of medicines using plant sources have been continuously growing since ancient time for the treatment of animal diseases.

Approximately 25% of the therapeutic preparations are derivative of plants and still many plants are unexplored. The traditional drugs for animals based on both plant products have received less attention. In well-developed countries, veterinary practices and animal well-being in rural population relied on ethnomedical remedies, mainly when access to allopathic veterinary products is hard or too costly for the local farmer (Nyamanga et al. 2008). Therefore, to maintain the health conditions of animals, traditional plants and their products are used from centuries and have been passed down orally from generation to generation (Martínez and Luján 2011). Traditional veterinary knowledge includes compilation of beliefs and practices about animal welfare that involved in the consumption of natural resources (plant and animals) and other materials. This knowledge is usually transferred orally from decade to decade and, as other traditional beliefs, is presently vulnerable by technological development, sociocultural changes, and environmental alterations (McCorkle and Martin 1998). Within the industrialized and urban society, there is an increasing attention in alternative or complementary medicine which, collectively with other natural therapies, is based on the utilization of plant-based therapeutics. Therefore, utilization of homeopathic and herbal therapy medicines in veterinary is given more attention, due to growing demands on the value of meat and milk products that are required for producing organic food goods (Pieroni et al. 2004).

Livestock subsector contributes 56% in the agriculture as a source of income and about 11% toward the GDP (gross domestic product), and 30 million rural people are associated with livestock subsector directly or indirectly (Sarwar et al. 2002). Thus, livestock plays a significant role in poverty-decreasing tactics. People living in the far-flung areas use medicinal plants for livestock's health. Mainly, the predictable lifestyle of nomadic and pastoralists makes it hard for them to reach veterinary extension services owing to high expenses and least accessibility of allopathic medicines (Raziq et al. 2010). Livestock plays a central role in sustaining livelihood, nutritional and environmental security, and growth of Indian agriculture. Enormous stride made in the livestock sector in the past decades is one of the main causes for positive growth rates recorded in agricultural sector. There is immense need to attain vertical growth in terms of improving productivity. Deficiency of food resources in India has been recognized by different organizations. But various locally accessible feed resources used for feeding milch animals have not been given any consideration. These comprise industrial by-products, horticulture and vegetable wastes, local grasses, tree leaves, weeds, and other non-conventional feed resources. The accessible feed resources are not fed in right amount as per the necessity of animals, leading to imbalance of nutrients in the ration.

10.5 Livestock Diseases and Role of Herbal Medicines

Diseases in animals remained primary causes of poor domestic animal performance, foremost to a still creating gap among the supply and the requirement for livestock products (Agrawal 1995). Local healers are aware and having adequate knowledge as well as experienced in traditional systems of treatment is important. But these information are unpublished and hence dwindling fast (Bekele 2007). It has been also seen and observed that the knowledge of medicinal plants is on the verge of irretrievable loss and dilapidated to deterioration. The reasons might be oral passage of herbal heritage from one age group to generation rather than in writings, despite their very important position for the health of human and domestic animal population (Mesfin et al. 2009). Environmental deprivation, agricultural development, farming of insignificant lands, and urbanization are also posing a considerable warning to the future well-being of human. Animal populations are dependent on these sources to overcome different ailments for generations (Lulekal et al. 2008; Giday et al. 2009. Hence, it is appropriate time to document and publish, promote, and preserve the country's medicinal plant diversity. Such publications might be extremely important to sustain cultural uniqueness of the country (Cetinkaya 2009; Lynam et al. 2007). This may also create a great enthusiasm and new thought for scientific discovery to develop new drugs (Cos et al. 2006). The livestock subsector contributes only 30% of the agricultural sector GDP compared to 55% from crops. Farming absolutely contributes 18% of the countrywide GDP (URT 1998). The limited contribution of the sector to the economy is explained by many factors including animal diseases, poor nutrition, lack of water, and poor animal genetic base. Animal diseases are serious problems for livestock development. Ethnoveterinary medicines give conventional medicines, which are locally accessible and usually low cost than typical treatment. Domestic animal holders can arrange and utilize homemade remedies with least amount of expense. Some domestic animals owners in rural areas. There are comparatively few veterinarians and shortages of other services; conventional therapeutic plants are the only choice to treat different diseases (McCorkle and Martin 1998) (Table 10.1).

10.6 Therapeutic Role of Medicinal Plants

Animal disease disaster may occur when there are sudden outbreaks of epidemic diseases or other animal health-associated events which have the possibility to cause serious socioeconomic results for a country. These emergencies are occurring recurrently by eruption of transboundary animal diseases (TADs), which are of significant economic, trade, and/or food security importance for different countries. Such diseases can increase without difficulty and attain epidemic proportions; management/organization, including exclusion, requires collaboration between numerous countries. These diseases affects the country in different ways such as; conciliation food security by serious loss of animal protein and/or loss of draught animal. They cause main losses in production of domestic animal products including meat, milk, and dairy products, wool and fibers, and skins. They cause loss of valuable domestic animals of high genetic potential. They might limit chance for improving the manufacturing potential of local livestock industries by building it hard to import exotic high-producing breeds that are enormously vulnerable to TADs. They adjoin considerably to the cost of domestic animal production since valuable disease control measures need to be applied and critically interrupt or slow down trade of livestock, germplasm, and livestock items inside a country or internationally. This might cause main losses in nationwide export income in important livestock-farming countries. Continuous reduction in livestock, thus trap livestock

Table 10.1 Plant species used to treat livestock, their habit, and disease treated, plant part used, preparation and administration

Family and botanical names, voucher number	Common names	Habit	Disease	Part used	Preparation, administration
Anacardiaceae, *R. lancea* L. f. (EV0030LT)	Karre (Eng.), Mushakaladza (V)	T	LSD in cattle	L.	Leaves are boiled, 1 liter to adults and ½ liter to calves
Apocynaceae, *Carissa bispinosa* (L.) Desf. *Ex* Brenan subsp. *bispinosa* (EV0040LT)	Num-num (Eng.), Tshirungulu (V)	S	Calving difficulties in cattle	R and B.	Take bulb, grind them, and give 1 l to cow
Asparagaceae, *Asparagus falcatus* L. (EV0025LT)	Sickle thorn (Eng.), Lufhaladzamakole (V)	H	Constipation in cattle	W and P.	Cut the whole plant including roots, immerse for 24 h, 1 l once a day for 3 days
Asteraceae, *Tagetes minuta* L. (EV0022LT)	Khaki weed (Eng.), Mushushathuri (V)	H	Tick control in cattle	L.	Take the leaves, mix with peri-peri (*Capsicum frutescens*), grind and apply the mixture on the ticks
Asteraceae, *V. colorata* (Wild.) Drake subsp. *colorata* (EV0002LT)	Phathane (V)	H	Diarrhea Worms	R	Take the roots, boil, give 1 l to cow, and ½ liter to young ones
					Worms take the roots and soak until the color change to dark brown (like coke), give the animal in 1 L
Asteraceae, *Vernonia corymbosa* Less. (EV0034LT)	Phathaphathane (V)	H	Worms in cattle	R	Take the roots, grind them, and mix it with water. Liters and half a liter to young calves
Boraginaceae, *Ehretia rigida* (Thunb.) Druce (EV0012LT)	Puzzle bush (Eng.), Mutepe/Murovherovhe (V)	T	Eating problems (tshiuthwane) in cattle	R	Boil the roots; give the cow in 1 l and ½ a liter to young animals
Capparaceae, *Maerua angolensis* DC. (EV0033LT)	Bead-bean tree (Eng.), Mutambanamme (V)	T	Acting problem drought tonic	L	Take leaves, grind them, mix with water, and give to cows in 1 l and calves in ½ liter

Celastraceae, *Elaeodendron transvaalensis* (Burt Davy) RH Archer (EV0031LT)	Bushveld Saffron (Eng.), Mulumanama (V)	T	Worms in cattle	F	Take the fruits, grind them, mix with water, and give 1 L to cows and ½ liter to calves
Clusiaceae, *Garcinia livingstonei* T. Anders. (EV0029LT)	Lowveld mangosteen (Eng.), Mupimbi (V)	T	Eye problems in cattle	L	Take the fresh leaves and squeeze the juice into the eye of the animal
Combretaceae, *T. sericea* Burch. *ex* DC. (EV0005LT)	Silver cluster-leaf (Eng.), Mususu (V)	T	Diarrhea (utshuluwa) in cattle	R	Take the roots, boil, and give the animal in 1 l, ½ a liter to young ones (mix with milk).
			Ticks and wounds		Grind the roots, mix with water, apply on the ticks and wounds
			Gut conditions – diarrhea		
Combretaceae, *C. molle* R. Br. ex G. Don (EV0007LT)	Velvet bushwillow (Eng.), Mugwiti (V)	T	Worm infestation Breeding problems, for example, difficult calving	L	An infusion is administered: 1 L to cows and ½ liter to calves
Ebenaceae, *Diospyros lycioides* Desf. subsp. *lyciodes* (EV0017LT)	Karoo bluebush (Eng.), Muthala (V)	S	Ticks in cattle	I	Grind leaves, mix with water, and apply on the affected area
Euphorbiaceae, *Pseudolachnostylis maprouneifoloia* Pax (EV0021LT)	Kudu berry (Eng.), Mutondowe (V)	S	Drought tonic	B	Grind the bark, mix with water, sieve the liquid, and give 1 l to cows and ½ liter to calves
Euphorbiaceae, *Synadenium cupulare* (Boiss.) L.C. Wheeler (EV0001LT)	Dead-man's tree (Eng.), Muswoswo (V)	S	Eye problems (infections) Stems Black quarter LSD	Stems	Apply latex on the side of the eyelid after applying pig or cattle fat
					Strike with latex branch on the affected area
					Cut the branch, apply the oozing latex on the limb
Fabaceae, *B. speciosus* (Bolus) Harms (EV0048LT)	Nkohlwane	T	Retained placenta in cattle	R	Pounded roots are immersed for 12 h, 2 liters for 3 days
Fabaceae, *Elephantorrhiza burkei* Benth. (EV0024LT)	Sumach bean (Eng.), Gumululo (V)	S	Diarrhea in cattle	R and B	Take the bulb, gr water, and give to animal ind it, mix with
Fabaceae, *P. angolensis* DC. (EV0009LT)	Transvaal teak (Eng.), Mutondo (V)	T	Mali" and not eating in cattle	B	Soak the bark in water; give to the cows in 1 l and ½ a liter to calves

(continued)

Table 10.1 (continued)

Family and botanical names, voucher number	Common names	Habit	Disease	Part used	Preparation, administration
Fabaceae, *Senna petersiana* (Bolle) Lock (EV0019LT)	Monkey pod (Eng.), Munembenembe (V)	T	Diarrhea in cattle	B and L	Take the bark, grind it, mix with salt, and give to cattle or leaves are soaked for 12 h; 2 liters given to animal
			Eating problem		Take the bark, grind it, boil it, and give it in 1 l for cows and ½ liters for calves
Iridaceae, *Gladiolus dalenii* Van Geel (EV0035LT)	Wild gladiolus (Eng.), Phende-phende (V)	H	Eye problems in goats, sheep	R and B	Grind fresh bulb and put it in a sack, squeeze the juice on the infected eye
Lauraceae, *Cassytha filiformis* L. (EV0010LT)	Love vine (Eng.), Luangalala (V)	H	Eating problems (tshiuthwane) in cattle, goats, and sheep. Calving difficulties	Stem	Take the stem and mix with leaves of *D. eriocarpum*, crush the mixture, boil, give in 1 l bottle
					Take the stem and mix with leaves of *D. eriocarpum*, crush the mixture, boil, give in 1 L
Liliaceae, *A. marlothii* Berger	Mountain aloe (Eng.), Tshikhopha (V)	H	Liver problems in chickens/ Newcastle disease	L	Take the broad leaves, grind them, squeeze the juice in water, and let the chickens drink
Asphodelaceae, *Aloe marlothii* Berger (EV0020LT)					
Meliaceae, *Turrae obtusifolia* Hochst (EV0042LT)	Mbhovane	S	Wounds in goats, sheep, and cattle	L	Crush leaves, apply directly on the wounds
Ochnaceae, *Ochna holstii* Engl. (EV0026LT)	Red ironwood (Eng.), Tshipfure (V)	S	Not eating (tshiuthwane) in cattle	Shoots	Mix shoots with water let the animal drink with 1 l.
Olacaceae, *Ximenia americana* L. var. *microphylla* Welw. *ex* Oliv. (EV0032LT)	Blue sourplum (Eng.), Muthanzwa (V)	T	Wounds in goats, sheep and cattle	L	Boil leaves and branches for 2 h, 1 l once a day for 3 days.

Pedaliaceae, *D. eriocarpium* (Decne) Abels (EV0003LT)	Devil's thorn (Eng.), Museto (V)	H	Calving difficulties in cattle	Stem and L	Grind aerial parts, mix with water, and give 1 l to cows
			Worms in cattle		Grind aerial part, mix it with water, and give it to animals (1 l bottle)
Rosaceae, *Prunus persica* (L.) Batsch (EV0011LT)	Peach tree (E), Muberegisi (V)		Wounds in cattle		Take the leaves, grind them, squeeze the juice, and apply to the wound/eye
			Eye problems in cattle		
Rubiaceae, *Cephalanthus natalensis* Oliv. (EV0014LT)	Strawberry bush (Eng.), Murondo (V)	S	Eye problem in cattle	L	Take leaves, grind them, and give the animal in 1 l bottle
Rubiaceae, *Hyperacanthus amoena* (Sims) Bridson (EV0016LT)	Thorny gardenia (Eng.), Murombe (V)	T	Eye problems	R	Take the fresh roots, crush them, and squeeze the juice in the eye
Rubiaceae, *Rothmannia capensis* Thunb (EV0023LT)	Wild gardenia (Eng.), Murathamapfene/ Murathambila (V)	T	Eating problem in cattle	R	Decoction is administered in 2 l bottle
Solanaceae, *Solanum incanum* L. (EV0028LT)	Bitter apple (Eng.), Mututulwa muhulwane (V)	S	Eye problems in goats, sheep, and cattle	F	Take the fruits (thuthulwa khulwane), grind them, and apply to the eyes
Ulmaceae, *Trema orientalis* (L.) Blume (EV0013LT)	Pigeonwood (Eng.), Mukurukuru (V)	T	Eye problems in goats, sheep, and cattle	L	Grind leaves, mix with water, and give to an animal
			Gall sickness		Take leaves, grind them, mix them with water, and give them to animals

Luseba and Tshisikhawe (2013), Journal of Medicinal Plants Research producer in uneconomic agriculture. They also cause public health issues where diseases can be transmitted to humans (i.e., zoonoses).

Domestic animal production is a fundamental part of the way of life for the people of the world. Every year millions of rupees are costed to control the animal diseases. Unhealthy animals need more food and take much more time for growth than healthy animals. In general animals get birth free of diseases or parasites. However, they generally acquire these problems either during their contact with infected animals or because of inappropriate sanitation, feeding, care, and management. Animals are affected by a variety of diseases. Awareness regarding the animal diseases is compulsory from community healthiness point. Keeping animals healthy by employing proper sound values of sanitation, feeding, and management by sensible use of suitable and reliable vaccines is the practical and economical sound to avoid losses from the disease. Good accommodation to sustain the health of the herd but judicious feeding not only builds up body fight against the disease but also helps in speedy recovery in case there is a disease attack.

10.7 Role of Phytochemicals in Animal Health and Welfare

Plants were once the principal source of medicines in the world. Now-a-days plants are used as therapeutic medicine because of the presence of polyphenolic compounds. Plants act usually to revive and strengthen the body's restorative forces. Remarkably, huge quantity of drugs comes from an insubstantial number of plants that are already investigated pharmacologically. Medicinal herbs and their derivatives act as main source of bulk of modern pharmacological drugs. During recent years, natural polyphenolic compounds have been given great attention; these polyphenolic compounds are widespread in nature among the group of secondary metabolites. There are 100,000–200,000 approximately secondary bioactive metabolites having ubiquitous distribution (Metcalf 1987). Phenolic compounds possess pleiotropic functions such as antibacterial, antiviral, anti-inflammatory, hepatoprotective, anti-allergic, antioxidant, antithrombotic, oncostatic, and immune-modulating functions (Cushnie and Lamb 2005). Their modulatory function in different biological processes has been well known such as oxidation, enzymatic detoxification, apoptosis, host immune system, and some other functions. These bioactive compounds, present in all fruits, vegetables, and medicinal plants that are taken daily, might be exhibiting potential for homeostatic metabolism in favorable way. Polyphenolic compounds are now given keen attention in the present era as they are having multifaceted health benefits on animal health and welfare. These compounds are known by important functional ingredients and health-promoting biomolecules in current literature due to their potential in supporting health and suppressing the chronic ailments (Nijveldt et al. 2001). Latest literature survey signified that polyphones and flavonoids might be interacting synergistically with xenobiotic compounds such as drugs vitamins, as well as other phenolic compounds (Mikstacka et al. 2010). These interactions may lead to the antagonistic effects as well. These functional activities are associated interactions (synergism, additive, or antagonism) of polyphenols with other compounds and co-antioxidants is a aspect grab the attention scientific community. Recent reports suggested that synergistic or antagonistic antioxidative effects depend upon concentration; small concentration exhibits synergistic antioxidative effects; however the high concentration exhibits additive effects (Mikstacka et al. 2010). Earlier reports revealed the synergistic roles of flavonoids for antioxidation (Mikstacka et al. 2010), platelet aggregation (Pignatelli et al. 2000), antimicrobial action, and life span of meat and improved value (Kamboh and Zhu 2013). Flavonoids slow down oxidation through different mechanisms, and their defensive roles in living systems are attributed because of their capability to transfer electrons to free radicals, act as interchelate agents in metal ions, stimulate antioxidative enzymatic machinery, decrease alpha-tocopherol radicals, and slow down oxidases (Middleton et al. 2000; Nijveldt et al. 2001). Latest work has proposed that the cellular defensive importance of polyphenolic compounds might be mediated by their interactions with precise proteins that are essential to some intracellular signaling cascades. In particular, flavonoids might be operating selectively with various components of protein kinase signaling cascades, such as phosphoinositide 3-kinase, protein kinase-C, Akt/protein kinase-B, etc. (Hou and Kumamoto 2010).

10.7.1 Role of Secondary Metabolites in Immune Regulation

All vertebrates possess the well-distinct immune systems that protect the host from infectious agents present in the environment (e.g., viruses, bacteria, fungi) and from other harmful abuses. Dietary phytoconstituents are having antioxidant potential like flavonoids are well known to positively modulate the immune functions in all taxa of vertebrates. The polyphenols in diet not only stimulate the immune coordination but also amend the detoxifying capability of enzymes and free radical scavenging potential and also control the gene expression. Previous studies reported that epigallocatechin gallate and cyanidin glycoside-rich juices could improve the IL-2 secretion, lymphocyte proliferation, and pathogen-eliminating action of natural killer (NK) cells. Leaves could produce a significant increase in proliferation responsiveness of peripheral blood mononuclear cells by means of improved synthesis and secretion of immunoreactive IL-2 and IL-4 cytokines. Also, amplifies eliminating activity of natural killer (NK) cells and salivary IgA secretion has been also found. The working mechanism behind this phenomenon is not understood yet; however, generally it is conjectured that supplementation of antioxidants might be having potential to decrease the detrimental roles of ROS on the immune response. That might result in the improved functions of the immune system and act as immune cocktails. Different parts from three categories of plants, i.e., herbs, shrubs, and trees as well as their parts (root, stem, leaves, fruits, flowers, and bud). These are very rich in polyphenolic compounds and essential oils having therapeutic importance. Conventional practitioners prepare crude extracts from various plant parts containing different medicinal values and ues whole plants or their parts [solid (powders), semisolid (paste, gel or colloid), and aqueous (fluid)]. Medicinal importance of some shrubs and herbs (tulsi, aloe vera, fenugreek, and haldi) is well explained. Confirmations from scientific reports revealed that plant-derived bioactive components particularly the polyphenols might be helpful to decrease the oxidative stress at the cellular level. Therefore, it can recover the genomic constancy and cellular integrity. These polyphenols also branded to improve immunity and gut function that eventually trim down chance of communicable diseases and increase animal performance. Moreover, these might be producing the synergisms with one another and with other organic and synthetic growth promoters. Therefore, phenolic compounds, both in purified and plant extracts, could be having potential source to increase the production of domestic animal farming.

10.8 Role of Plants and Their Bioactive Components in Nutrition

Good nutrition is the basic obligation for all farm animals; it is principal contributory factor to maintain the animal welfare and healthiness. Insufficient nutrition influences not only productivity but also health, behavior, and well-being of an animal. Simultaneously, protection and value of the foodstuff chain are ultimately affected by the welfare of farming of domestic animals, because they are closely associated among animal welfare, animal health, and food-borne diseases including *Escherichia coli*, *Salmonella* and *Campylobacter*, etc. Stress factors, poor well-being, and imbalanced nutrition can augment vulnerability to diseases in animals, thus raising the requirement for veterinary management, creating risks to food consumers, declining productivity, and endangering environmental sustainability of the domestic animal rearing systems and its interconnected animal food chains. Polyphenolic compounds influence on animal health, nutrition, performance, and environmental sustainability. The diversity of polyphenolic compounds prevent a widespread summary of their effects; impact is influenced by the type, concentration, and astringency of the dietary phenolic compound, animal physiological status, sex, exposure to pathogens, predictable productive performance, and environmental circumstances. Nowadays, there is emerging need of nutraceuticals globally. They hold superiority over other therapeutic molecules because they are pure and natural in origin, safe to use, and easy to access. However different confront at the forefront to nutraceutical industrial establishment. These challenges include the requirement to accurate and precise tests to check the efficacy and safety of such chemical compounds, identify their correct mode of action, estimate their bioavailability and study possible mode of communications in the body. This strongly specifies these polyphenolic compounds are valuable gifts to focus nutraceutical importance and their roles in the research. Therefore, roles in the nutraceuticals should be given more and more emphasis so that their outstanding importance in the pharmacological actions can be benefited. Cattle husbandry is generally practice in traditional systems (sedentary) and faces serious challenge about the growth of the manufacture which rely on the well-being of the animals. Animal diseases result in the significant decrease in all animal productions (meat, milk, skin, etc.) and their main source of income. Besides, different dairy products are processed and are component for household utilization. Animal husbandry and agriculture are associated with each other. Rural people utilize ethanobotanical culture for the curing of animal diseases and in the biological control.

10.8.1 Dietary Efficacy of Phenolic Compounds

Secondary metabolites ubiquitously present in plants play an essential role in human diet as well (Gazor et al. 2017). Most of polyphenolic compounds might be decreasing the risk of vulnerable health issues, having antioxidant properties. Biochemical examination revealed that these polyphenolic compounds have practical uses (Hamedi et al. 2017). Polyphenolic compounds are synthesized during various biochemical pathways that are not essential for primary growth, but are used for defensive purposes. These polyphenolic compounds are having biological and pharmaceutical importance. Antioxidant molecules present in plants include flavonoids and phenolic acids. Numerous studies suggested that these polyphenolic compounds possess various biological actions, such as strong antioxidants and free radical scavengers, oncostatic, anti-inflammatory and immune modulating (Shahidi and Ambigaipalan 2015; Selamoglu et al. 2016). In general, the capability of phenolic compounds that are efficient antioxidants depends on three factors. These act as metal-interchelating having potential that is powerfully dependent on the alignment of hydroxyls and carbonyl group approximately the molecule. Hydrogen ions (electron donating) are able to decrease free radicals, and the capability of the polyphenolic compounds to delocalize the unpaired electron causes the production of a stable ion. Mechanisms regarding the different roles of polyphenolic compounds such as antioxidant effect, preventive mechanism, and chain-breaking mechanism are well known. These studies explained the antiradical activity such as chain-breaking action (Shahidi and Ambigaipalan 2015; Ozgen et al. 2016; Selamoglu et al. 2016). These polyphenolic compounds inhibit the oxidation of biomolecules in the living systems, so that these molecules are now given more preference as dietary items.

10.9 Importance of Animal Products

Animal products are in great demand through the world which produces enormous pressure on the natural wealth and on the farm animals. In exhaustive production systems, a secondary pressure is pushed on animals to give maximum productivity. However, widespread and smallholder systems in growing nations animal farming and welfare are compromised by improper nutrition. Still in demanding production systems where animals get plentiful nutritious diets, animal welfare could be affected due to extreme or unsuitable feeding. A number of management- linked factors – including accommodation and bedding, preventive systems, space and crowding, transport conditions, dramatic and slaughter system, castration of males, and tail docking – affect welfare. Vast literature is available till date which suggests that these factors affect animal well-being, productivity, and product quality. However, least consideration and attention have been paid to understand the association between animal nutrition and animal welfare.

Farmers are in great trouble to take up such practices that support animal welfare without having good information on the effect of such practices on animal productivity and on their income. In the present era, the world is suffering great extent of food crisis, diseases, which compel animal scientists and farmers to search alternative feedstuffs for animal production. Therefore, importance of plants and compounds extracted from them are considered as potential natural alternatives to increase the livestock productivity. The livestock industry must continuously look for and employ alternative animal nutritional approach. Which can fulfill consumer wish for food products that are produced in a hygienic, green, and ethical conduct? Hence, the significance of plant products and their antioxidative activity not usually present in the diets of ruminants warrants search. The supplementation of the diet with plants, distinguished for its antioxidant activity, has the capability to improve the health and production performances of animals. Grazing might provide ruminants health benefits from some vitamins and minerals. While clean forages are classically considered ability for supplying sufficient levels of antioxidants for dairy cattle, the accessibility of these compounds for lactating grazing cows is decreased when pasture availability is not enough to meet their energy necessities. Assessment of oxidative constraint in ruminants is contributing considerably to understand the basic processes concerned with metabolic disorders. It is thought that oxidative constraint plays a considerable function in the balancing of the metabolic action of different organs and efficiency in farm animals (Celi 2011). Milk production is linked with oxidative strain and, thus, supplementation of cows' diets with antioxidant which might affect development of their oxidative status and productive performances. Animal welfare is speedily becoming one of the foremost aspects of animal production worldwide, at the intensity of the industrialized world and in developing countries.

In fact, there are sufficient husbandry and managing practices, transport and slaughtering conditions, have a major affect on product quality and farmer income. It was practical observed that animals were not subjected to nutritional stress showed a quieter approach when compared with the animals that were losing weight.

10.10 Conclusion

Significance of traditional medicines (TM) and their pleiotropic roles in sustaining the animal health worldwide cannot be underemphasized. In the recent era, plant polyphenolic complexes have been broadly investigated. These compounds present in different foods have been coupled with health-promoting and economic importance. The association between the veterinary utilization and biological activity is investigated in various species or practices, and great variety of native resources still remains unfamiliar from a pharmacological approach of view. Nonetheless, information of novel therapeutic implications of plants shall support to study. The application of plant therapeutic based formulations are promoting the development of of alternative medicines. The use of natural products as therapeutic remedies decreasing the dependency on allopathic drugs, all of which entails rising in the value and price of animal foodstuffs in accordance with new market demands.

In this sense, it is important to unfold and record therapeutic plant uses in veterinary medicine, within an ethnoscientific perspective as approached in this review. Therefore it is necessary to understands the association human beings the nature and their culture. It is very importance because it allows the characterization of social plan by their meticulous environmental awareness, and provides helpful tools for the development of conservation policies". A systematic recording is important concern to use biodiversity for different purposes is required.

References

Agarwal SK, Chatterjee S, Misra SK (2001) Immune potentiating activity of a poly herbal formulation 'Immu-21'. Phytomedica 2:1–22

Agrawal A (1995) Indigenous and scientific knowledge: some critical comments. Indig Knowl Dev Monit 3:3–35

Bekele E (2007) Actual situation of medicinal plants in Ethiopia. Prepared for Japan Association for International Collaboration of Agriculture and Forestry (JAICAF)

Bento TS, Torres LM, Fialho MB, Bononi VL (2013) Growth inhibition and Antioxidative response of wood decay Fungi exposed to plant extracts of Casearia species. Lett Appl Microbiol 58(1):79–86

Burt S (2004) Essential oils: their antibacterial properties and potential applications in foods-a review. Int J Food Microbiol 94(3):223–253

Celi P (2011) Biomarkers of oxidative stress in ruminant medicine. Immunopharm Immunot 33:233–240

Cerisuelo A, Marin C, Sanchez-Vizcaino F, Gómez EA, de la Fuente JM, Durán R, Fernández C (2014) The impact of a specific blend of essential oil components and sodium butyrate in feed on growth performance and Salmonella counts in experimentally challenged broilers. Poultry Sci 93:599–606

Cetinkaya G (2009) Challenges for the maintenance of traditional knowledge in the Satoyama and Satoumi ecosystems, Noto peninsula. Japan Hum Ecol Rev 16:27–40

Cos P, Ying L, Calomme M, Hu JP, Cimanga K, Poel BV, Pieters L, Vlietinck AJ, Berghe DV (1988) Structure-activity relationship and classification of flavonoids as inhibitors of xanthine oxidase and superoxide scavengers. J Nat Prod 61:71–76

Cos P, Vlietinck AJ, Vanden Berghe D, Maes L (2006) Anti-infective potential of natural products: how to develop a stronger in vitro 'proof – of concept'. J Ethnopharmacol 106(3):290–302

Cowan MM (1999) Plant products as antimicrobial agents. Clin Microbiol Rev 12(4):564–582

Croft KD (1998) The chemistry and biological effects of flavonoids and phenolic acids. Ann N Y Acad Sci 854(1):435–442

Cushnie TPT, Lamb AJ (2005) Antimicrobial activity of flavonoids. Int J Antimicrob Agents 26 (5):343–356

Dilshad SR, Rehman NU, Ahmad N, Iqbal A (2010) Documentation of ethnoveterinary practices for mastitis in dairy animals in Pakistan. Pak Vet J 30(3):167–171

FAO (2002) Genetics and animal health spotlight. FAO, Rome
Gazor R, Asgari M, Pasdaran A, Mohammad Ghasemi F, Nasiri E, Roushan ZA (2017) Evaluation of hepatoprotective effect of Acantholimon gilliati aerial part methanolic extract. Iran J Pharm Res 16:135–141
Giday M, Asfaw Z, Woldu Z (2009) Medicinal plants of the Meinit ethnic group of Ethiopia: an ethanobotanical study. J Ethnopharmacol 124(3):513–521
Gori FF (2007) Herbal medicine today: clinical and research issues. Evid Based Complement Alternat Med 4:37–40
Hamedi A, Pasdaran A, Zebarjad Z, Moein M (2017) A survey on chemical constituents and ındications of aromatic waters soft drinks (Hydrosols) used in Persian nutrition culture and folk medicine for neurological disorders and mental health. J Evid Based Complement Alternate Med 22(4):744–752
Hou DX, Kumamoto T (2010) Flavonoids as protein kinase inhibitors for cancer chemoprevention: direct binding and molecular modeling. Antioxid Redox Signal 13(5):691–719
Kamboh AA, Zhu WY (2013) Individual and combined effects of genistein and hesperidin supplementation on meat quality in meat-type broiler chickens. J Sci Food Agric 93 (13):3362–3367
Kim J, Park EJ (2002) Cytotoxic anticancer candidates from natural resources. Curr Med Chem Anticancer Agent 20(4):485–537
Kone WM, Atindehou KK (2008) Ethnobotanical inventory of medicinal plants used in traditional veterinary medicine in Northern Côte d'Ivoire (West Africa). S Afr J Bot 74(1):76–84
Lambert J, Srivastava JP, Vietmeyer N (1997) Medicinal plants: rescuing a global heritage. The World Bank
Lattanzio V, Lattanzio VM, Cardinali A (2006) Role of phenolics in the resistance mechanisms of plants against fungal pathogens and insects. Phytochem Adv Res 661(2):23–67
Lulekal E, Kelbessa E, Bekele T, Yineger H (2008) An ethanobotanical study of medicinal plants in Mana Angetu District, southeastern Ethiopia. J Ethnobiol Ethnomed 4(1):1–10
Luseba D, Tshisikhawe MP (2013) Medicinal plants used in the treatment of livestock diseases in Vhembe region, Limpopo province, South Africa. J Med Plants Res 7(10):593–601
Lynam T, De Jong W, Sheil D, Kusumanto T, Evans K (2007) A review of tools for incorporating community knowledge, preferences, and values into decision making in natural resources management. Ecol Soc 12(2):5
Martínez GJ, Luján MC (2011) Medicinal plants used for traditional veterinary in the Sierras de Córdoba (Argentina): an ethnobotanical comparison with human medicinal uses. J Ethnobiol Ethnomed 7(1):23
Mathias-Mundy E, McCorkle CM (1989) Ethnoveterinary medicine: an annotated bibliography. Bibliographies in technology and social change, technology and social change program. Iowa State University, Ames, p 199
McCorkle CM, Green EC (1998) Intersectoral healthcare delivery. Agric Hum Values 15 (2):105–114
McCorkle C, Martin M (1998) Parallels and potentials in animal and human ethnomedical technique. Agric Hum Values 15(2):139–144
Mesfin F, Demissew S, Teklehaymanot T (2009) An ethnobotanical study of medicinal plants in Wonago Woreda, SNNPR, Ethiopia. J Ethnobiol Ethnomed 5(1):1–18
Metcalf RL (1987) Plant volatiles as insect attractants. CRC Crit Rev Plant Sci 5(3):251–301
Middleton E, Kandaswami C, Theoharides T (2000) The effect of plant flavonoids on mammalian cells: implications for inflammation, heart disease and cancer. Pharmacol Rev 52(4):673–751
Mikstacka R, Rimando A, Ignatowicz E (2010) Antioxidant effect of trans-resveratrol, pterostilbene, quercetin and their combinations in human erythrocytes in vitro. Plant Foods Hum Nutr 65(1):57–63
Nazzaro F, Fratianni F, Martino LD, Coppola R, Feo VD (2013) Effect of essential oils on pathogenic bacteria. Pharmaceuticals 6(12):1451–1474

Nijveldt RJ, van Nood E, van Hoorn DE, Boelens PG, van Norren K, van Leeuwen PA (2001) Flavonoids: a review of probable mechanisms of action and potential applications. Am J Clin Nutr 74(4):418–425

Nyamanga PA, Suda C, Aagaard-Hansen J (2008) The socio-cultural context and practical implications of ethnoveterinary medical pluralism in western Kenya. Agric Hum Values 25 (4):513–527

Ozgen S, Kilinc OK, Selamoglu Z (2016) Antioxidant activity of quercetin: a mechanistic review. Turkish J Agric Food Sci Technol 4(12):1134–1138

Parr AJ, Bolwell JP (2002) Phenols in the plant and in man. The potential for possible nutritional enhancement of the diet by modifying the phenols content or profile. J Sci Food Agric 80 (7):985–1012

Pica-Ciamarra U, Baker D, Morgan N, Zezza A, Meas (2011) Contribution livest. househ. livelihoods: a livest. module multi-topic househ. surv

Pieroni A, Howard P, Volpato G, Santoro RF (2004) Natural remedies and nutraceutical used in ethnoveterinary practices in Inland Southern Italy. Vet Res Commun 28(1):55–80

Pignatelli P, Pulcinelli FM, Celestini A, Lenti L, Ghiselli A, Gazzaniga PP, Violi F (2000) The flavonoids quercetin and catechin synergistically inhibit platelet function by antagonizing the intracellular production of hydrogen peroxide. Am J Clin Nutr 72(5):1150–1155

Raziq A, Verdier K, Younas M (2010) Ethnoveterinary treatments by dromedary camel herders in the Suleiman Mountainous Region in Pakistan; an observation and questionnaire study. J Ethnobiol Ethnomed 6(1):16

Robinon MR, Zhang X (2011) The world medicine situation (traditional medicines: global situation, issues and challenges). World Health Organization, Geneva

Sarwar M, Khan MA, Nisa M, Iqbal Z (2002) Dairy industry in Pakistan: a scenario. Int J Agric Biol 3(1):420–428

Selamoglu Z, Ustuntas HE, Ozgen S (2016) Traditional and complementary alternative medicine practices of some aromatic plants in the human health. Res J Biol 4:52–54

Shahidi F, Ambigaipalan P (2015) Phenolics and polyphenolics in foods, beverages and spices: antioxidant activity and health effects – a review. J Funct Foods 18:820–897

Singh JS (2002) The biodiversity crisis: a multifaceted review. Curr Sci 82(6):638–647

Tabuti JR, Dhillion SS, Lye KA (2003) Ethnoveterinary medicines for cattle (Bos indicus) in Bulamogi county, Uganda: plant species and mode of use. J Ethnopharmacol 88(2–3):279–286

Taylor DJ (2001) Effects of antimicrobials and their alternatives. Br Poult Sci 42(l):67–68

URT (1998) Expanded agricultural survey 1996/97 Tanzania Mainland Vol II main report, Statistics Unit and National Bureau of Statistics

Vaara M (1992) Agents that increase the permeability of the outer membrane. Microb Rev 56 (3):395–411

Valentao P, Fernandes E, Carvalho F, Andrade PB, Seabra RM, Bastos ML (2002) Antioxidative properties of cardoon (Cynara cardunculus L.) infusion against superoxide radical, hydroxyl radical, and hypochlorous acid. J Agric Food Chem 50(17):4989–4993

Valentao P, Fernandes E, Carvalho F, Andrade PB, Seabra RM, Bastos ML (2003) Hydroxyl radical and hypochlorous acid scavenging activity of small centaury (Centaurium erythraea) infusion. A comparative study with green tea (Camellia sinensis). Phytomedicine 10 (6–7):517–522

Wenk C (2000) Recent advances in animal feed additives such as metabolic modifiers, antimicrobial agents, probiotics, enzymes and highly available minerals-review. Asian Aus J Anim Sci 13 (1):86–95

Yang CS, Landau JM, Huang MT, Newmark HL (2001) Inhibition of carcinogenesis by dietary polyphenolic compounds. Annu Rev Nutr 21(1):381–406

Phenolics: A Game Changer in the Life Cycle of Plants

11

Nasir Aziz Wagay, Rafiq Lone, Shah Rafiq, and Sheikh Ummar Bashir

Abstract

Phenolics are ubiquitously present across the kingdom Plantae ranging from simple phenolic acids to very large and complex polymers. Phenolics are vital for plant growth and development as they are important in almost every process of plants including senescence, flower development, extracellular linking, pathogen and herbivore resistance, UV protection and photodamage, etc. Phenolics also act as chemical messengers or internal physiological regulators which regulate processes like transcription, trafficking of vesicles, signal transduction, and permeability of membrane. They inhibit or induce oxidative bursts and affect the respiration and photosynthesis rates. Plant phenolics also play an important role by acting as both underground signaling (mycorrhizal symbiosis, actinorhizal symbiosis, plant growth-promoting rhizobacteria (PGPR)) and as aboveground signaling molecules. Phenolics also act as potential antioxidant compounds during stressful conditions as they can donate electrons for the compound detoxification. Also, phenolics are accumulated in response to ROS under stress conditions. Phenolic compounds are also produced by plants as response to environmental stresses including wounding, salt, drought, heavy metal,

N. A. Wagay (✉)
Botany Research Laboratory, Department of Botany, Vidyabharati Mahavidyalya College, Amravati, Maharashtra, India

R. Lone
Department of Botany, Sant Baba Bhag Singh University, Jalandhar, Punjab, India

S. Rafiq
Plant Tissue Culture Laboratory, Department of Botany, University of Kashmir, Srinagar, Jammu and Kashmir, India

S. U. Bashir
Department of Botany, SGB Amravati University, Amravati, Maharashtra, India

R. Lone et al. (eds.), *Plant Phenolics in Sustainable Agriculture*,
https://doi.org/10.1007/978-981-15-4890-1_11

temperature, etc. Phenolics provide resistance to the host plant by providing resistance against pathogen attack, herbivores, etc., and they also act as indicators of mineral deficiencies during various stages of plant growth and development. The action of resisting plants is apparently to increase either phenolic synthesizing enzyme levels or their activities, leading to increased phenolic production. Phenolics have also played a role in the plant invasion to land as the ability to synthesize phenolics has been selected throughout the course of evolution. Such compounds addressed the special needs to cope up in a constantly changing environment. It can be said that plant life without phenolics is impossible.

Keywords

Phenolics · Plant · Life cycle · Growth and development · Environment

11.1 Introduction

"Plant polyphenols" or "plant phenolics" are secondary metabolites which are synthesized by the plant kingdom by phenylpropanoid/shikimate pathway from the aromatic amino acids (phenylalanine, tryptophan, and tyrosine) which provides simple phenylpropanoids, or the malonate/"polyketide" acetate pathway producing monomeric, polymeric, and polyphenols or simple phenols, sometimes both. Phenolics are widely found across the kingdom Plantae with nearly 10,000 structures identified possessing an aromatic ring bearing one or more hydroxyl substituent (Boudet 2007; Harborne 1980). They range from simple phenolic acids to complex polymers (Hopkins 1997). Phenolic compounds tend to be soluble in water, as they most frequently occur as glycosides (combined with sugars). Glycosylated phenols are common in metabolically active sites (Hopkinson 1969). Several thousand natural phenolic compounds have been identified till date; among these the flavonoid group is the largest, but also considerable amount of phenylpropanoids, phenolic quinones, and simple monocyclic phenols have been identified. Several important phenolic groups present in plants are lignins, melanins, tannins, etc. and are made from phenolic units, and occasionally these polyphenolic compounds are encountered in proteins, alkaloids, and the terpenoids (Harborne 1998).

Phenols are mainly classified on the basis of carbon atom number they contain and the pathway of their biosynthesis by which each group is derived (Harborne and Simmonds 1964).

Simple phenols (C_6/C_7) containing simple aromatic ring include phenolic alcohols, aldehydes, ketones, and their glycosides. An example of simple phenol is arbutin (hydroquinone b-D-glucoside, I) which is found in the various *Vaccinium* spp. leaves, such as cowberry, cranberry, blueberry, and pear trees (*Pyrus communis* L., Rosaceae) (Clifford 2000). *Phenolic acids* (C_7, C_8, C_9) include both benzoic acids and cinnamic acids (phenylpropanes). *Acetophenones* (C_6–C_2 ketones) are aromatic ketones and occur as glycosides. It has been found as the main component of spruce needles (*Picea abies* (L.) Karst.) and also occurs in *Larix deciduas* Mill.,

Populus balsamifera L., and *Salix* spp. (Cheynier et al. 2013). *Phenyl propanes* (C_6–C_3) are compounds of volatile nature which are associated with essential oils. They occur as both open and closed chain forms with or without combining with sugars. *Xanthones* (C_6–C_1–C_6) resemble benzophenones but differ in the presence of a heterocyclic oxygen ring in between the two benzene rings. Only few higher plant families (Podostemaceae, Guttiferae, Gentianaceae, Loganiaceae, and Polygalaceae) have this class of plant phenolics group; therefore this group is highly valuable for chemotaxonomic classification. Mangiferin (VII) is a unique example of natural xanthones as this compound has much wider natural occurrence than that of any other xanthone. This 2-C-glucoside of 1,3,6,7-tetrahydroxyxanthone was first found in the leaves of *Mangifera indica* L. (Bennett and Lee 1988; Peres and Nagem 1977; Peres et al. 2000). *Coumarins* are formed by the lactonization of o-hydroxycinnamic acids. Almost all the naturally occurring compounds are O-substituted at C_7. Furano- and pyranocoumarins have a furan or pyran ring fused with the benzene ring of coumarins. Coumarins can be found in free form or in combined form with sugars as heterosides and glycosides in many families of eudicots (eudicotyledons), including the Fabaceae, Apiaceae, Rosaceae, Asteraceae, Moraceae, Solanaceae, and Rubiaceae (Petersen et al. 1999). *Stilbenes* (C_6–C_2–C_6) are compounds occurring mainly as heartwood constituents and some of its derivatives are condensed tannins. They are found in many eudicot angiosperms, gymnosperms, ferns, and in some liverworts, ranging from the unsubstituted trans-stilbene from *Petiveria* and *Alnus* to the hexasubstituted combretastatin A-1 (VIII) from *Combretum caffrum* (Eckl. & Zeyh.) Kuntze, and also include various acylated and glycosylated derivatives (Gorham 1989, 1995). *Lignans* are dimers formed by the condensation of two cinnamic acids/cinnamyl alcohols through the β-carbons of their aliphatic side chains. The aromatic rings are often oxygenated, and additional rings are also formed in the molecule. Gymnosperm lignin is comprised of coniferyl alcohol subunits, while angiosperm lignin is mixture of sinapyl and coniferyl alcohol subunits (Hopkins 1997). *Flavonoids* (C_6–C_3–C_6) include the compounds formed by the skeleton of flavone (2-phenyl chromone) as well as its related or derived compounds like anthocyanins, isoflavones, neoflavones, etc. The different classes of these flavonoids are recognized chiefly by the oxidation pattern of the C_3 fragment, the additional oxygen heterocyclic rings, and the glycosylation. They often occur as glycosides in plants, but a number of them are present in free state too. *Quinones* are the aromatic diketones which form the largest class of naturally coloring matters. Quinones may be benzo-, naphtho-, or anthraquinones, depending on the mono-, bi-, or tricyclic ring system they possess. *Tannins* are polyphenols having an astringent taste and in plants are divided in two groups: hydrolysable and condensed. Hydrolysable tannins are soluble in water and contain simple phenolic acids (gallic, digallic, and ellagic acids) esterified with one or more sugar molecules and are hydrolyzed by dilute acids to component compounds. Condensed tannins (flavolans) are polyphenols insoluble in water, which, on treatment with acids, yield complex products of unknown composition called "tannin reds" or "phlobaphenes" (Daneil 2006). More recently a third class of tannins, the phlorotannins, has been isolated from several genera of algae. The phlorotannins are composed almost entirely of phloroglucinol subunits linked in some cases by C–C and in others by C–O–C(aryl ether) bonds (Quideau 2009; Ferreira and Slade 2002; Haslam 1998; Khanbabaee and Van Ree 2001; Singh et al. 2010).

Phenolic derivatives are synthesized as they act as visual cues for the attraction of pollinators as well as for other animals which help in the seed dispersal, and they also act as molecular cues for the protection of plants during stress conditions (Mol et al. 1998) such as low temperatures (photo-oxidation and photo-inhibition) allowing more efficient nutrient resorption especially nitrogen ("photoprotection theory"). During winter, the combination of dry, cold, and bright sunlight conditions can result in excess energy capture relative to processing, photosynthetic inhibition, increase in the production of oxidative species (ROS), and free radicals, greater photooxidative damage, etc. (Lee and Gould 2002; Gould 2004; Archetti 2009; Archetti et al. 2009; Nikiforou and Manetas 2010; Hughes 2011). Phenolics also play important roles in the development of plant particularly in scaffolding support to plants and structural integrity, pigment and lignin biosynthesis, nutrient deficiency, and infection by pathogens. Perturbed or wounded plants secrete phenolic phytoalexins as defense molecules to repel or kill many microorganisms, and some pathogens can nullify or counteract these defenses or even subvert them to their own advantage (Bhattacharya et al. 2010). Plant phenolics also take part in the modulation of physiologically essential processes such as signal transduction, transcriptional regulation, vesicle trafficking, and membrane permeability. They can decrease or increase the respiration and photosynthesis rates. We can say that, at the whole organism level, the processes such as development, adaptation, symbiosis, diseases, and male sterility can be better understood on the basis of phenolics (Bidel et al. 2010). The diverse roles of phenolic compounds during interactions of plant with their abiotic and biotic environments and also in plant physiology are difficult to overestimate, and some of them have been discussed in this chapter.

11.2 Cell Wall and Extracellular Linking

Phenolic compounds, precisely hydroxycinnamates, act as molecular bridges by oxidative coupling or inter-polymeric cross-linking reactions between polymer-bound phenolic groups in plant cell walls where they are present in significant levels. Such coupling may result in strengthening cell walls that are less susceptible to turgor-driven cell expansion (Fry 1986; Macadam et al. 1992).The phenolic (tyrosine) residues of cell wall glycoproteins form isodityrosine, pulcherosine, and di-isodityrosine by oxidative coupled reactions, thereby cross-linking the glycoproteins (Brady and Fry 1997). Phenolic coupling also contribute to cell–cell adhesion in the middle lamella (Parr et al. 1996). For example, in grasses, ferulic acid and 4-coumaric acid link polysaccharide polymers to lignin, such as glucuronoarabinoxylan, through ether bonds and/or labile esters (Jung et al. 1993; Wallace and Fry 1994), and in bamboo the α-D-Xyl (D-xylose) residues of xyloglucans are linked with lignin by these two phenols (Ishii 1997). The composition and amount differences of lignin-bound arabinoxylans in the plant cell walls determine their rigidity and also affect the biomass degradability. The major determinants for biomass degradation in the *Miscanthus* were found directly related to lignin content, cellulose/lignin, and cellulose/xylan ratios (Lygin et al. 2011).

In cultured maize cells, feruloyl coupling product, a specific trimer of ferulate, has been identified (Rouau et al. 2003).The oxidative coupling of polysaccharide-bound feruloyl groups yields several distinct dimers (e.g., 5–5¢-diferulate, 8-O-4-diferulate, etc.), which could represent inter-polysaccharide cross-links (Ralph et al. 1994; Waldron et al. 1997). The total composition and content of ester- and ether-bound phenolics (hydroxycinnamic acid) also differ in cell walls of plants with predominant type of diferulate acids (Lygin et al. 2011). The controlled formation of diferuloyl bridges in vivo could theoretically govern wall extensibility and thus growth of plant (Fry 1979; Wakabayashi et al. 1997a, b).

Dimerized hydroxycinnamic acids (diferulic acid) cross-link pectins to noncellulosic polysaccharides or with other pectins by forming ester linkages in between them (Fry 1986; Parr et al. 1996; Waldron et al. 1997). It has been confirmed in the plant family Chenopodiaceae, in which hydroxycinnamic acids present in cell walls are attached to pectic polysaccharides (Wallace and Fry 1994).

Although it is difficult to evaluate the role of cross-linking (Fry 1983), there is evidence that cross-linking of matrix polysaccharides through diferulic acid bridges in oats play a protective role, in which it acts as a barrier to pathogen ingress by increasing its incompatible interaction with the crown rust pathogen. Also, phenolic coupling in response to pathogenic attack may provide resistance against fungal penetration (Brady and Fry 1997).The enzymatic or microbial degradability is correlated negatively with both p-hydroxycinnamic acids and lignins present in cell wall (Jung and Deetz 1993; Besle et al. 1994). The cell wall hydroxycinnamic acids could also determine recalcitrance of lignocellulosic feedstocks to chemical or enzymatic saccharification during processing to biofuel (Grabber et al. 1995, 1998a, b). Further studies are needed to clarify the role and effects of phenolics on composition of cell wall, ether, and ester cross-linking.

11.3 Phenolics During Flower Maturation and Development

Development of flower involves various interrelated processes, viz., growth, abscission, and senescence (Sood and Nagar 2003). During flower senescence and bud opening, a well-defined sequence of various events take place such as cell division, cellular differentiation, shifts in membrane permeability, cell elongation, and a wide range of gene expression (Kumar et al. 2008a) in association with the changes in concentration of secondary metabolites and endogenous plant growth regulators (Kumar et al. 2008a; Mayak and Halevy 1972; Sood and Nagar 2003).

Little information is available on phenolic content of developing flowers. A sharp increase in the phenolic content has been detected during flower development in *Rosa bourboniana* Desport. and *Rosa damascena* Mill. (Sood and Nagar 2003). Some phenolics such as gallic acid and protocatechuic acid and catechin, chlorogenic, caffeic, and p-coumaric acid which have been identified in rose petals

(Cai et al. 2005; Kumar et al. 2008b; Velioglu and Mazza 1991) showed constant decrease in these phenolics during progression of various developmental stages from bud to senescent stages in rose flowers (Schmitzer et al. 2009). Flower development in miniature rose 'KORcrisett' also revealed a significant negative trend in analyzed phenolic acids from bud to senescent stage (Schmitzer et al. 2009). Decline in the concentration of phenylalanine (PAL) has been also reported in *Malus domestica* Borkh. Flowers during developmental stages (Dong et al. 1998) however, in *Rosmarinus officinalis* L., the concentration of phenolics remained constant (Del Bano et al. 2003).

The later stages of flower development may limit the function of the phenolics/peroxidase/ascorbic acid system in antioxidant defense and make the flower more vulnerable to oxidative stress due to the decline in phenolics concentration (Takahama and Oniki 1997). The possible role of phenolics in flower development regulation is still an unexplored field of plant developmental biology.

11.4 Senescence

The process of senescence occurs in photosynthetic prokaryotes, algae, and non-angiosperm land plants. Senescence is a highly dynamic, integral, and genetically regulated program that involves an ordered set of events at physiological and biochemical levels, including activation of hydrolytic enzymes, production of reactive oxygen species (ROS), loss of plasma membrane integrity, degradation of proteins, lipids, nucleic acids, and carbohydrates (Ahmad and Tahir 2015, 2016; Rahmani et al. 2015).

Senescence is an essential part of growth and development in plants and is a cell death program phenomenon. However, it is still unknown whether ROS increment during senescence does play a role in the regulation of the phenylpropanoid metabolism in vegetables and fruits by acting as signal mediator as reported under stress conditions (Reyes et al. 2007).

During the process of senescence, phytochemicals (secondary metabolites) could be accumulated or diminished, depending on the tissue of plant (Dangl et al. 2000). In this period there is an increment in the reactive oxygen species (ROS) formation and also an increasing homeostatic imbalance. In order to continue the cell repair mechanisms, it is possible that secondary metabolism may be activated depending on whether these compounds are needed in the process or not. Little information is found in relation to this point.

Polyphenol oxidase activity increases during senescence of detached leaves or leaf disks (Farkas et al. 1964; Kisban et al. 1964; Sacher 1967) as well as with the physiological age of the attached leaves (Maraite 1973). The oxidation of compounds during senescence is due to either peroxidase or polyphenol oxidase activity but rarely both (Bonner 1957). Peroxidase activity acts as a catalyzer for oxidation of compounds, and absence of polyphenol oxidase activity has also been reported in *Oryza sativa* leaves (Sridhar 1972; Toyoda and Suzuki 1960).

Polyphenol oxidase enzyme employs pyrogallol as the substrate at pH optima between pH 6.5 and 7 (Varga 1970).

Polyphenols contribute to the taste, color, and nutritional properties of most climacteric tree fruits (Cheynier 2005). So far, little is known whether/how endogenous polyphenols may play a role in senescence and postharvest ripening of fruits, despite of remarkable changes in phenolic content and components during ripening of various fruits (Aaby et al. 2012; Fernández et al. 1993). At present, it is difficult to determine endogenous functions of polyphenols in fruit by overexpressing or silencing synthesis of the phenolic compounds in tree fruits. A typical example of climacteric fruit is *Pyrus malus* (apple) which is also rich in polyphenols (Ceymann et al. 2012). Among polyphenols, chlorogenic acid is a principal component in most climacteric fruits, especially in tree fruits (Tsao et al. 2003; Cui et al. 2005; Scattino et al. 2014; Cao et al. 2015); therefore, it was used as a representative of endogenous polyphenols, and it was experimentally demonstrated that chlorogenic acid could suppress apple pulp disk senescence (Xi et al. 2016).

Xi et al. (2016) noticed that treatment with *Pyrus malus* phenolic extract can effectively reduce loss of red color and browning of pericarp in *Litchi* (Zhang et al. 2015), despite this physiological role of endogenous phenolic compounds in fruits being limited. Chlorogenic acid (CHA) is the principle phenolic compound in *Pyrus malus* pulp, inhibiting increase in rate of respiration in apple pulp disks during senescence. In addition, NADP-malic enzyme (NADP-ME) was also inhibited by CHA (Xi et al. 2016). The mechanism for the characterization of phytochemical synthesis or degradation during the over-ripening or advanced senescence stage of climacteric fruit is still unclearly explained.

Tamagnone et al. (1998a, b) experimentally confirmed that *Antirrhinum* transcription factor AmMYB308 inhibited phenolic acid (and monolignol) biosynthesis in *Nicotiana tabacum* (tobacco) plants by reducing level of hydroxycinnamic acid and monolignol metabolism gene transcripts. In Myb308 *Nicotiana tabacum* plants, phenolic derivatives of monolignols and hydroxycinnamic acids play a significant role in the development of leaf palisade cells in tobacco, probably through the production of dehydrodiconiferyl alcohol glucosides (DCGs) phenolic signaling molecules as they can modify cell expansion and cell division. Myb308 plants had paler leaves than the control plants, and cells in mature leaves undertook precocious cell death which evaluated the importance of secondary metabolism in the normal growth and development of plants. Overexpression of AmMYB308 results in a general repression of phenolic acid and monolignol production, while precise phenolic acid or its derivative has not been identified (Tamagnone et al. 1998a, b).

A very similar phenotype has been reported for *Nicotiana tabacum* plants showing reduced activity of phenylalanine ammonia-lyase (PAL) and consequently reduced monolignol or hydroxycinnamic acid metabolism (Elkind et al. 1990). This suggested that premature leaf cell death and paler leaves arise as a direct consequence of reduced phenolic acid metabolism.

Phenolics play an important role during leaf senescence by normally controlling the processes of cell death inhibition. Phenolic acid metabolism controls senescence by providing important compounds which enhance the ability of buffer cells against the catalytic damage invoked by ROS. Therefore, they act as natural antioxidants and have a significant biological action (Tamagnone et al. 1998a, b). However, determination of process for cell death is difficult as it could be triggered by intrinsic signaling in case of programmed cell death (PCD) pathway or from the direct effect of elevated ROS (Korsmeyer 1995; Heath 1998).

The oxidation of phenolics to o-quinones triggers browning of senescent tissues which result in membrane damage (Thompson et al. 1987). Lipoxygenase released during membrane damage oxidizes membrane lipids (Thompson et al. 1987), and polyphenol oxidases which oxidize cellular phenolics are released by plastids (Thompson et al. 1987). Phenolic oxidation by peroxidases in cell walls is also stimulated during membrane damage process (Noodén et al. 1997). The leaf tissue in which phenol biosynthesis was inhibited (Myb308 transcript plants) did not become brown as they eventually became completely dead and white (Tamagnone et al. 1998a, b). This study suggests that oxidation of phenolics act as a protection mechanism during senescence by reducing the impact of increased levels of ROS and also allows withdrawal of cellular contents in a programmed manner from aging leaves via controlled senescence. It has been explained and supported by the enzymatic evidence; as H_2O_2 is scavenged by chlorogenic acid after cell damage (Takahama and Oniki 1997) and in tobacco leaves, this function has been related to abundant phenolic acid derivatives such as chlorogenic acid.

The cells of aging leaves in the absence of antioxidant phenolics (either bound to the cell wall or soluble in the vacuole) are less protected against the oxidative stresses generated by senescence process and die rapidly. Similarly, the lipid peroxidation levels in mature leaves of Myb308 cell death is observed directly due to the programmed pathway activation which lead to death of senescent cells of control leaves in a programmed manner (Noodén et al. 1997; Pennell and Lamb 1997). Therefore it can be said that phenolics may play both general as well as specific roles in cell death modulation.

In *Zea mays*, the Lls1gene (cell death suppresser gene) encode a protein with structural similarities to dioxygenases that metabolize phenolics in bacteria (Gray et al. 1997). In this case, an interpretation is that Lls1 gene product (which suppresses cell death) removes the cell death causing phenolic mediator. Possibly, Lls1 mutant plants fail in the synthesis of phenolic derivative, important in buffering reactive oxygen species and tempering cell death.

11.5 Photodamage Stress

When light energy is absorbed above the capacity of photosynthetic and photoprotecting processes, the excess energy causes the formation of highly reactive chemical species – particularly triplet chlorophyll, singlet oxygen, and hydroxyl radicals (Niyogi 1999). These reactive intermediates and by-products can cause oxidative damage (Asada 1994; Foyer and Harbinson 1994). Specifically, they can oxidize lipids, proteins, pigments (Knox and Dodge 1985), and Calvin cycle enzymes (Asada 1994; Kaiser 1979). If conditions are severe, they can cause fatal damage to leaf tissues and whole plants (Wise and Naylor 1987), termed photodamage.

In order to minimize potential photodamage, plants produce a large range of antioxidant molecules and scavenging enzymes that can quench reactive chemical species. These important molecules include carotenoids (Frank and Cogdell 1993, 1996; Demmig-Adams et al, 1996; Havaux 1998), superoxide dismutases (Asada 1994; Foyer et al. 1994), tocopherols (Foyer et al. 1994), ascorbate (Conklin et al. 1996, 1997), and glutathione (Asada 1994; Foyer et al. 1994).

Phenolics also quench reactive chemical species as they have antioxidant capacity and may protect plants from photodamage. Flavonoids, a class of phenolic compounds with lower molecular weight, may act as plant physiological antioxidants (Rice-Evans et al. 1996). Similarly, phenolic acids (Rice-Evans et al. 1996) and anthocyanins (Yamasaki et al. 1996; Wang et al. 1997) have also been demonstrated to act as antioxidants. Condensed and hydrolyzable tannins are compounds with higher molecular weight and are more effective than simple phenolics in quenching peroxyl radicals (Hagerman et al. 1998). These tannins are characterized by many phenolic hydroxyl groups and/or high degree of polymerization, both of which contribute to antioxidant quenching efficacy (Hodnick et al. 1988; Ariga and Hamano 1990).

Many phenolics vary in plants under different light conditions; i.e., leaves in sunny parts (high light) of a tree need more antioxidants for protection from photodamage than those in the shade and hence have higher phenolic levels. Plants may increase phenolic production directly in response to oxidative pressure produced from excess light energy and as a physiological response to quench reactive chemical species. Patterns of phenolic levels between species also reflect different selective pressure from potential risk of photodamage.

The concentration of leaf phenolics including tannins has frequently shown to increase in plants grown under conditions of high light or nutrient limitation (Newbery and de Foresta 1985; Bryant et al. 1987; Mole et al. 1988; Iason and Hester 1993; Iason et al. 1996). For example, sun-adapted leaves of *Mahonia repens* have more than three times the amount of chlorogenic acid than shade-adapted leaves (Grace et al. 1998). Similarly, sun-adapted foliage of *Eucalyptus nitens* seedlings has more than twice the level of hydrolyzable tannins (di-, tri-, tetra-, and pentagalloylglucose) than shade-adapted foliage (Close et al. 2001). Leaves from the sunny side of the Victorian plum tree (*Prunus domestica* var. Victoria) had significantly higher phenolic levels than leaves from the shady side (Hillis and Swain 1959). Induction of condensed and hydrolyzable tannins within plant species varies under various incident light regimes and also depends on their position in the plant (Mole et al. 1988). Phenolic levels in plants are also dependent on the nutrient concentration present in the vicinity of its roots. Hydrolyzable tannin levels were also about twofold higher in foliage of nitrogen-deficient, compared with nitrogen-sufficient, *E. nitens* seedlings (Close et al. 2001). These responses to light and nutrient availability are similar to those of other antioxidant compounds, such as ascorbate and glutathione.

Leaves from slow-growing trees adapted to nutrient-poor boreal forest environments are subjected to large amounts of oxidative pressure during periods of low temperature coupled with high light. Furthermore, because they are long-lived, they are subject to a high potential accumulation of free radical damage relative to shorter-lived leaves. As a result of selective pressure from both the daily and cumulative potential costs of photodamage, these slow-growing plants have evolved higher constitutive phenolic levels for their antioxidant capacity. The life span of animals can also be extended by consuming these phenolic antioxidants, presumably through lowering the oxidative damage to tissue that can occur from biochemical processes (Perez-Campo et al. 1998). Production of high phenolic levels in plants may in a sense "enable" the viability of long-lived leaves similarly through their antioxidant capacity.

Low temperature (which decreases the turnover rates of photosynthetic enzymes) or drought/water logging (which induces closure of leaf stomata and in turn photosynthetic carbon dioxide limitation) both limit photosynthesis, increase oxidative stress, and therefore could induce an increase in phenolic levels. Consistent with this prediction, Grace et al. (1998) found that the phenolic chlorogenic acid was about twofold greater in *M. repens* foliage in winter compared to summer. In *Eucalyptus nitens* foliage, phenolic levels are higher in winter than in summer, and this is associated with higher levels of cold-induced photoinhibition (measured by chlorophyll fluorescence and xanthophyll cycle pigment dynamics), indicative of higher oxidative pressure (Close et al. 2001). These patterns explain consistent predictions that oxidative pressure and risk of photodamage cause an increase in phenolic levels.

11.6 Role of Plant Phenolics in Rooting

Rooting in plants takes place when the total phenol content increases three times than the original level of phenol concentration in the cutting of various plants. Some of the phenolic compounds which have been identified for root formation stimulators in cuttings are catechol (Hackett 1970), phloretic acid (Jones and Hatfield 1976), chlorogenic acid (Hammerschlag 1982), and phloroglucinol (James and Thurbon 1981; Zimmerman 1984). The amount of total phenol levels increased in the cuttings of *Protea cynaroides* during root formation (Wu 2006). Rooting in plants has both stimulatory as well as inhibitory results depending on the dose of phenolics. The root stimulation causing range of phenol concentrations is usually narrow at lower levels, while higher phenol concentrations affect root inhibition. The effects of phenolic substances have been extensively studied in plant species which are difficult to root, and the results concluded that difficult-to-root stem cuttings have endogenous rooting inhibitors in higher amounts, which delay or inhibit root formation, compared to easy-to-root stems which have high content of rooting promoters (Richards 1964; Fadl and Hartmann 1967; Taylor and Odom 1970; Biran and Halevy 1973; Reuveni and Adato 1974). The phenolics are also linked to poor rooting of *Protea* cuttings as these cuttings are inherently difficult to root. The effect of phenolics on growth regulation has been assessed by dose response bioassay of aqueous stem extracts from *P. cynaroides* on lettuce seeds which suggested that phenolic compounds are directly contributing to rooting inhibition and/or stimulation during in vitro and in vivo propagation (Wu et al. 2007a, b).

Furthermore, the root stimulation by phenolic compound coumarins such as caffeic, ferulic, and gallic acid at higher concentrations has been reported in many plant cuttings individually (San Antonio 1952; Dhawan and Nanda 1982) or in association with 3,4-dihydroxybenzoic acid in *C. lanceolata* (Zhiqun et al. 2002), *R. japonicus* (Elzaawely et al. 2005), and *Phytolacca* spp. cuttings (Kim et al. 2005). Moreover, salicylic acid individually has stimulatory effect for root initiation of *Phaseolus aureus* cuttings (Kling and Meyer 1983). It can therefore be assumed that this group "coumarin" of phenolic compounds has an important role to play in regulating formation of roots, particularly in terms of stimulation. Phenolic concentrations play pivotal role in the rooting of numerous plant species which likely relates these compounds as endogenous promoters or inhibitors of adventitious rooting.

Correlation between phenomenon of allelopathy and phenolic compound concentrations has been reported in various plant species. 3,4-Dihydroxybenzoic acid is a widespread and common allelopathic agent which can influence growth during different plant developmental stages (Rice 1984). Phenolic bioassay reports have widely suggested the root inhibition by 3,4-dihydroxybenzoic acid and some other phenolic compounds present in plant species such as *Arctostaphylos glandulosa* (Chou and Muller 1972), *Chrysanthemum morifolium* (Kil and Lee 1987), *Pennisetum clandestinum* and *Cunninghamia lanceolata* (Chou et al. 1987), *Rumex japonicus* (Elzaawely et al. 2005) and *Vulpia myuros* (An et al. 2000), *Tilia americana* (Morsink and Smith 1975), etc. However, Bär et al. (1997) have reported that 3,4-dihydroxybenzoic acid has no observable effect on the root growth of *Kalanchoe* cuttings at lower concentrations while root growth was inhibited at higher levels.

Few studies have used in vitro conditions to determine the effects of plant-extracted phenolics on root growth. In vitro studies on *P. cynaroides* (Wu et al. 2007a, b) explants suggested that the lower concentrations of 3,4-dihydroxybenzoic acid has no observable effect, while stimulation of root growth was apparent when the concentration reached 100 mg L^{-1} and 3,4-dihydroxybenzoic acid at higher concentration (500 mg L^{-1}) was toxic to the *P. cynaroides* explants, which was clearly demonstrated by root inhibition and browning of explants. Nevertheless, similar type of results in *Nicotiana tabacum* explants have been reported also (Mucciarelli et al. 2000). Pellissier (1994) has reported that root formation is stimulated by higher concentrations of 3,4-dihydroxybenzoic acid in *Picea abies*, it can be concluded that 3,4-dihydroxybenzoic acid does play a primary role in root formation, depending largely on its endogenous concentration which increases after the cutting plantation and is maintained during the period of rooting. Until now, change in concentration of 3,4-dihydroxybenzoic acid during rooting of cuttings has not been reported.

Auxin induces root formation and is a known fact from more than half century. The severe reductions are caused due to interactions between bioactive molecules (isoflavonoids like genistein, formononetin, and quercetin) and plant hormones during commitment phase of root formation in in vitro root formation model of legume *M. truncatula*. This effect is related to an auxin transport inhibition and/or redox cell status regulation (Imin et al. 2006). Tannic acid and rutin are also inhibitors of root formation (Still et al. 1976).

11.7 Plant Phenolics as Signaling Molecules

The interaction of plant with its environment is via secondary metabolites which are often diffusible. The alteration in the physicochemical properties of soil by root exudations is called rhizosphere (Hiltner 1904). The root exudates are used as nutrient sources by the terrestrial organisms as these mostly include enzymes, ions, and molecules rich in organic carbon (primary and secondary metabolites). This environment with the soil exudates make ups a conducive zone for the survival of microorganisms (Bertin et al. 2003). Also, the root exudates (secondary metabolites) are mostly used by the selected organisms (commensals or pathogens) as a source of chemotaxis. This chemical environment is favorable to the establishment of various interactions in particular with symbiotic microorganisms.

11.8 Phenols in Signaling of *Rhizobium*/Legume Symbiosis

In the case of *Rhizobium*/legume mutualistic symbiosis, root exudates are synthesized by the plant and diffused in soil (rhizosphere) in response to stress, whether biotic or abiotic (Nicholson and Hammerschmidt 1992). These exudates include phenolic derivatives in particular isoflavonoids and flavones (flavonoids) and act as signaling molecules and are recognized (oxidation, position of the substituent) by the nucleotide-binding and oligomerization domain D receptors of the bacteria which in turn control nod (A, B, and C) gene expression and also Nod factor (lipochitin oligosaccharide) production (Zhang et al. 2009).

Rhizobium/legume symbiosis induces the synthesis of various phenolic derivatives for signaling and has been confirmed in various plants such as soybean (Estabrook and Sengupta-Gopalan 1991), *M. sativa* (Savoure et al. 1994), *Arachis hypogea* L. (Devi and Reddy 2002), peanut (Chakraborty and Mandal 2008; Mandal et al. 2009), alfalfa (Dakora et al. 1993), *Phaseolus vulgaris* L. (Dardanelli et al. 2008), and *Lotus subbiflorus* Lag. (Prinsen et al. 1991). The interaction of bacteria modifies the plant metabolite synthesis by inducing phenylalanine ammonia-lyase gene expression, a major enzyme of the phenylpropanoid pathway (PHP) and a major route for the synthesis of phenolic derivatives (Estabrook and Sengupta-Gopalan 1991). The signals induced due to Nod factors, many secondary metabolites, and cosubstrates (sinapoyl CoA feruloyl CoA) perceive signals for the synthesis of enzymes for phenylpropanoid pathway (Zhang et al. 2009). Increased synthesis of phenolic compounds such as cis- and trans-caffeic acid, p-coumaric acid, trans- and cis-ferulic acid, chlorogenic acid, vanillic acid, etc. is due to the induction of Nod factors which results in the overexpression of phenylpropanoid pathway enzymes (Van Rossum et al. 1995) and especially flavonoids such as kaempferol, aglycone and medicarpin glucoside, formononetin, etc. (Dakora et al. 1993; Zhang et al. 2009; Devi and Reddy 2002).

11.9 Phenolics in the Actinorhizal Symbiotic Association

Some soil filamentous bacteria, microaerophilic or aerobic actinomycetes of genus *Frankia* (Family *Frankiaceae*), are associated with eudicot angiosperm families (Coriariaceae, Betulaceae, Casuarinaceae, Datiscaceae, Rhamnaceae, Myricaceae, Rosaceae, and Elaeagnaceae) for the nitrogen fixation and are called as the actinorhizal plants (Normand et al. 1988). The different strains have host specificity abilities that allow to distinguish compatible or incompatible strains to the symbiosis with a plant species (Huguet et al. 2005). The specificity of interaction seems here to be related to the biosynthesis of specific phenylpropanoids and flavonoids (Hammad et al. 2003). The increase or decrease of hydroxycinnamic acids (HCA) during actinorhizal symbiotic associations is modulated by the inoculated bacterial strain (Hückelhoven 2007). An increase of the HCA level limits and/or prevents the symbiosis, while decrease of HCA biosynthesis facilitates the colonization of host plant tissues by *Frankia* which is, being related to the modification for the production of coumarins or lignin (lignin biosynthesis), a common mechanism of allowance to regulate the bacterial penetration in plant tissues (Hückelhoven 2007). Actinorhizal symbiosis has been studied in *Frankia* and *Alnus* (Hammad et al. 2003), *Frankia* and *Myricaceae* sp. (Popovici et al. 2011), etc.

11.10 Phenols as Signaling Molecules in AMF

In mycorrhizal symbiotic association first step is the recognition of signal molecules between the two partners (plant roots and fungi). The beginning of interaction is by perception of signal compounds (more precisely flavonoids) secreted by the plant and specifically recognizable by the fungus (arbuscular mycorrhizal fungi). These chemicals are present in high concentrations in roots and exudates of plants and they can stimulate AMF growth. Due to the reactivity of these phenolic derivatives, there is an initiation of interaction through recognition mechanisms of these signaling molecules by the different partners allowing thus a molecular dialog.

Increase in the phenolic content or its derivatives due to mycorrhization have been confirmed in plants such as date palm (Jaiti et al. 2008), *Begonia* (Selvaraj et al. 2008), *Rudbeckia* (Araim et al. 2009), *Echinacea purpurea* (Araim et al. 2009), etc. The concentration of tannins is also increased in presence of AMF, and these tannins contribute to different resistance-related mechanisms in plants against microorganisms (Cheynier et al. 2013).

11.11 Drought Stress

The "phenolic molecules" act as signal trigger leading to protective mechanisms against drought stress in leaf cells (Akula and Ravishankar 2011). The plant responds to ultraviolet radiations and drought stress by the accumulating flavonoids and phenolic acids which act as sunshields and antioxidants (Nichols et al. 2015). Under non-watered conditions, the rice plants often tolerate drought stress by delaying the process of reduced leaf drying and delayed leaf rolling (Hu et al. 2006). Large amount of phenolic acids and flavonoids is synthesized in response to cell-damaging oxidants during water stress in *Triticum aestivum* leaves, (Ma et al. 2014), rice (Quan et al. 2016), *Brassica napus* cultivar (Rezayian et al. 2018), and tobacco plants (Torras-Claveria et al. 2012), high increase of quercetin and kaempferol in drought-resistant tomato cultivars (Sánchez-Rodríguez et al. 2011), and increase in some phenylpropanoids such asp-coumaric acid and caffeic acid content in Maize (Alvarez et al. 2008). The mechanism of drought tolerance is controlled by application of phenolic compounds exogenously such as SA which strengthens drought tolerance in *Oryza sativa* (Farooq et al. 2009), *Hordeum vulgare* (Fayez and Bazaid, 2014), etc., but phenolic compound does vary among different species (Akula and Ravishankar 2011). The drought stress induces metabolic alterations and production of free radicals in flowers and leaves of *Tridax procumbens* resulting in the shift of metabolic processes by triggering the biosynthetic processes which increase the production of highly reduced phenolics for better adaptation to drought stress and was confirmed by significant increase in the sucrose content, total phenolics, and decrease of relative leaf water content, as these phenolics prevented oxidative damage of the cells and also increased drought tolerance (Gnanasekaran and Kalavathy 2017). The accumulation and increase of flavonoids and phenolics during drought stress is proportional to drought resistance levels. The drought tolerance ability via phenolic acids greatly varies among plant genotypes (Sabar and Arif 2014). These phenolic compounds may be used as suitable markers of stress at high level of drought.

11.12 Heavy Metal Stress

Heavy metals are defined as that group of elements having specific weights higher than 5 g/cm^3. A number of them (Cu, Co, Ni, Fe, Mo, Zn, Mn) are essential micronutrients, and they are necessary for normal development and growth by functioning in electron transfers, reduction-oxidation reactions, and other important metabolic processes in plants. Metals which are considered nonessential (Hg, Pb, Cr, Cd, etc.) are potentially highly toxic for plants (Sebastiani et al. 2004; Devi and Prasad 1998; Rai et al. 2004). Large land areas are heavy metal (the main group of inorganic contaminants) contaminated resulting from urban activities, agricultural practices, and industry (Khan et al. 2000; Clemens 2001). The processes leading to decrease in biomass, growth inhibition, toxicity, and death of plants have also been correlated to the excessive concentrations of trace elements such as Cd, Hg, Co, Mn,

Cr, Pb, Ni, and Zn (Zenk 1996). Physiological processes such as plant-water relationship, mineral nutrition, N-metabolism, photosynthesis, respiration, and cell elongation are also inhibited by heavy metals (Zornoza et al. 2002).

Heavy metal contamination results mainly from anthropogenic activities such as mining and metal smelting menacing humans and whole ecosystems. Exposure of humans and wildlife to heavy metals is through several pathways that may embed contaminated food and drinking water, polluted soil health, and inhaling particulates (Ikenaka et al. 2010; Qu et al. 2012). Gruca-Królikowska and Waclawski (2006) stated that the principal reason for reduction in productivity of plants growing on heavy metal-polluted regions is a significant reduction in the photosynthesis efficiency, dependent to interference of photosynthetic pigment biosynthesis.

Stress conditions and different environmental factors enhance the phenolic compound content and phenylpropanoid metabolism (Lavola et al. 2000; Grace and Logan 2000; Díaz et al. 2001; Sakihama and Yamasaki 2002). High concentration of phenolics is exuded by the roots of many plants when exposed to heavy metals (Winkel-Shirley 2001). It has been reported that the biosynthesis of phenolic acids in *Triticum aestivum* is induced in response to nickel toxicity, in *Zea mays* in response to aluminum (Winkel-Shirley 2001), and in *Phaseolus vulgaris* when exposed to Cd^{2+} (Díaz et al. 2001). Copper sulfate sprays increased the amount of phenolics in *Phyllanthus tenellus* leaves (Bors et al. 1990). Lead (Pb) caused a significant increase in the total amounts of phenols and flavonoids in wheat (Kaimoyo et al. 2008). Lead in the soil has the ability to complex other plant elements such as phosphorus and both access for uptake by plant roots (Pazoki 2015).

The increase in the amount of phenolic compounds and their de novo synthesis under heavy metal stress has been correlated with the increased activity of phenolic acid metabolism enzymes. In contrast, some evidence indicates that the increase in flavonoid concentration is mainly the result of conjugate hydrolysis and not due to de novo biosynthesis of phenolics (Parry et al. 1996). Stress induces an increase in the content of intermediates in lignin biosynthesis (soluble phenolics) and also some anatomical changes which prevent cells against harmful action of heavy metals (Díaz et al. 2001).

Phenolics, especially phenylpropanoids and flavonoids, can be oxidized by peroxidase and act in the hydrogen peroxide scavenging, phenolic/POX/ASC system against heavy metal contamination (Michalak 2006). Heavy metal toxicity creates reduction of reactive oxygen species using Fenton reaction (typical for transfer metals such as iron or copper) and autoxidation, mostly non-redox reactive heavy metals such as cadmium and mercury block, essential functional groups in biomolecules like proteins (by the inactivation of the SH groups in enzyme active centers) and polynucleotides (Baranowska-Morek 2003; MithÖfer et al. 2004), and substitution of essential metal ions by other incorrect ones in biomolecules (Rai et al. 2004; Schützendübel and Polle 2002). Phenolic compounds can act as metal chelators within heavy metal stress (Michalak 2006).

These evidences suggest that plants respond to the heavy metal contamination by phenolic metabolism. The decrease in dry weight might be due to decline in photosynthesis and pigment contents stated by Okhi (1978), Joshi et al. (1999), and Sinhal (2005). However, the metabolic inducers of such effects are still unknown (Kennedy et al. 2002).

11.13 Influence of Salinity Stress on Phenolic Acids

Among the different environmental constraints, salt stress is considered as an important abiotic factor, limiting productivity and plant growth especially in semi-arid and arid regions (Zhu 2001). The specific ion effect (salt stress) is thought to have detrimental effects on plant physiology, nutritional imbalance, and redirection of energy from growth to extracting pure water from the saline water and to producing defensive chemicals or a combination of these different factors (Munns and Tester 2008). The oxidative stress and disturbance in balance between the elimination of reactive oxygen species (ROS) and rates of production is mainly associated with the salt stress (Turkan and Demiral 2009; Tounekti et al. 2011).

Plants have evolved different defense systems to avoid the oxidative damage caused by salt including overproduction of antioxidant metabolites which stop the propagation of oxidative chain reactions. The abiotic/biotic stresses especially salt stress conditions trigger response for the synthesis of polyphenols (Souza and Devaraj 2010). Polyphenolic compounds such as flavonoids, phenolic acids, anthocyanins, and proanthocyanidins are well-known antioxidant compounds having powerful radical scavenging ability, and the prevention of stress-induced oxidative damage by producing ROS scavenging agents is due to the increase in phenolic content of plant tissues under salinity stress (Hichem et al. 2009; Bourgou et al. 2010). The ROS scavenging ability of plants mainly depends on the defense system by antioxidants including non-enzymatic components (Cuin and Shabala 2008) consisting of various secondary metabolites, such as hydrophilic phenolics and flavonols, organic acids, lipophilic carotenoids, and water-soluble ascorbate, and the enhancement of the phenolics metabolism leading to an increase in phenolic compounds by inducing disturbances in the secondary metabolic pathways is considered one of the responses to abiotic stresses (Close and McArthur 2002; Ksouri et al. 2007). The ROS produced in the course of salt stress are detoxified by several antioxidants from the photosynthetic electron transport chain within the chloroplasts (Tounekti et al. 2011). Under saline conditions, plants produce excess amount of ROS as a result of oxidative stress, and if they are poorly protected, these reactive molecules may damage macromolecules such as DNA, proteins, and membrane lipids resulting in cell death.

Plants may vary widely in their phenolic contents and compositions, with both genetic and environment affecting the type and level of these compounds (Awika and Rooney 2004; De Abreu and Mazzafera 2005). Additionally, these compounds are accumulated to response in the increases of ROS under salt stress (Dixon and Paiva 1995; Roberts and Paul 2006; Julkunen-Tiito et al. 2015) by exhibiting antioxidant activity in tissues to inactivate lipid-free radicals or prevent decomposition of hydroperoxides into free radicals (Krishnaiah et al. 2011; Agati et al. 2012; Brunetti et al. 2013). Higher buildup of phenolics and flavonoids in the plant under salt stress may assist the plant to lighten the salinity-induced oxidative stress (Wahid and Ghazanfar 2006). Increase in phenolic content in different tissues under increasing salinity has been reported in *Cakile maritima* leaves (Ksouri et al. 2007); *Artichoke* leaves (Hanen et al. 2008); red matured paper fruits (Navarro et al. 2006); mangrove plants (Parida et al. 2004); *Oryza sativa* (Minh et al. 2016); virgin olive oil (Ahmed et al. 2009); *Cordyline fruticosa* leaves (Plaza et al. 2009); *Cynara carduncolus* leaves (Hanen et al. 2008), hyacinth bean leaves (Souza and Devaraj 2010), *Echinacea angustifolia* roots (Montanari et al. 2008); salt-tolerant *Medicago ciliaris* line leaves and roots (Salah et al. 2011); *Zea mays* and *Saccharum officinarum* (Wahid and Ghazanfar 2006); and *H. pruinatum* plantlets (Caliskan et al. 2017). Based on these studies, we can say that in plant tissues enhanced phenolic production is good indicator for salt resistance under salt stress conditions. The decrease in phenolic content of the aforesaid plant species in response to salt treatments could probably be due to their sensitivity to saline.

Ferulic acid and p-coumaric acid might play a certain role in salinity tolerance mechanism as accumulation of p-coumaric acid assists to decrease oxidative pressure by expressing high radical scavenging activity due to its hydroxyl nature (Jamalian et al. 2013). Ferulic acid strengthens plant cell wall and helps in the overall cell elongation under osmotic stress (Wakabayashi et al. 1997a, b). Besides, ferulic acid copes with dehydration stress by decreasing lipid peroxidation due to activation of antioxidant enzymes and increasing proline and soluble sugar content (Li et al. 2013).

However, decrease in total carotenoids and flavonoids content in *Calendula officinalis* (Khalid and da Silva 2010) and total phenolic content in radish sprouts (Yuan et al. 2010) and *Nigella sativa* (Bourgou et al. 2010) has been also reported. Salt-resistant varieties and species accumulate differential fractions of phenolic compounds, i.e., phenolic acids and flavonoids as an adaptation to contrasting salinities (Mahmoudi et al. 2010).

11.14 Temperature Stress

Most plants suffer from both biochemical and physiological damage by exposure to lower or higher temperatures than for optimal growth (Lyons 1973; Grace et al. 1998). The results of these injuries are reflected in most metabolic processes (Anderson et al. 1994; Prasad et al. 1994a, b), such as reduced growth capacity of the crops thereby lowering commercial yield (Wang 1982). Plant growth is stunted by heat and cold stress as these induce the production of soluble phenolics such as phenylpropanoids and flavonoids which in turn increase phenylalanine ammonia-lyase (PAL) activity, and against stress it is considered as one of the main lines of cell acclimation in plants (Nozolillo et al. 1990; Christie et al. 1994, b; Dixon and Paiva 1995; Bharti and Khurana 1997; Kacperska 1993; Levine et al. 1994; Leyva et al. 1995). The development of acclimated mechanisms has been reported by Rivero et al. (2001) in tomato and watermelon plants as they accumulate phenolic compounds which may act as manipulating factors to trigger acclimated mechanisms in plants under temperature stress. Phenols are oxidized by polyphenol oxidase (PPO) and peroxidase (POD) which catalyze the oxidation as well as hydroxylation of monophenols (Vaughn and Duke 1981; Shöderhäll 1995; Thypyapong et al. 1995; Lafuente and Martınez-Tellez 1997). The response to different types of stress, both abiotic and biotic enzymes, results in enzyme increase (Pandolfini et al. 1992; Kwak et al. 1996, b; Smith-Becker et al. 1998; Ruiz et al. 1998, 1999). Polyphenol oxidase (PPO) and peroxidase (POD) enzymes have been specifically related to the appearance of physiological injuries caused in plants by thermal stress (Lyons 1973; Grace et al. 1998; Leyva et al. 1995; Lafuente and and Martınez-Tellez 1997; Cohen et al. 1988; Ke and Salveit 1988).

Phenolic compounds (SA, ASA, SSA) significantly enhanced plant growth attributes, i.e., plant height, number of branches, number of leaves, root and shoot length, and total dry biomass of plant at different growth stages. The ability of SA which ameliorates the adverse effects of cold stress has significant implications in increasing the plant growth and development. Among the phenolic compounds, SA is most effective in low-temperature stress conditions followed by ASA and then SSA as foliar application of these phenolics increased the level of cell division by stimulating the mitotic system of the apical meristem in tomato seedling roots which caused an increase in plant growth (Orabi et al. 2015). Similar reports about SA application which enhanced growth under low-temperature stress conditions have been reported by AOAC (1990), Boroum and Jazi et al. (2011), and Khandaker et al. (2011) in plants *Brassica napus* and *Amaranthus tricolor*; Orabi et al. (2015) in tomato; Saltveit and Morris (1990) in watermelon; Imami et al. (2011) in chickpea; and Gharib (2006) in basil. Hussein et al. (2007) and Hayat et al. (2010) also reported increased productivity due to an improvement in growth attributes with the application of sulfosalicylic acid and SA under stress conditions in maize plants. Król et al. (2015) demonstrated that phenolic acids and total content of phenolics, antioxidant, and reducing power activity decrease in the leaves of *Vitis vinifera* under long-term cold stress as some elements of the secondary metabolism in plants slow down during long-term cold stress.

Rodriǵuez et al. (2015) has reported that extreme temperatures affect the photosynthetic activity and yield of *B. oleracea* cultivars by causing oxidative stress. The modification in the phenolic compound composition during low temperature-induced stress has been reported in different anatomical parts of *Pisum sativum* (Rudikovskaya et al. 2008), rapeseed (Solecka et al. 1999; Solecka and Kacperska 2003; Stefanowska et al. 2002), soybean (Janas et al. 2000, b), maize (Christie et al. 1994, b), elderberry (Thomas et al. 2008), winter wheat (Olenichenko et al. 2006), *Ammopiptanthus mongolicus* (Liu et al. 2007), and *Cotinus coggygria* (Oren-Shamirand Levi-Nissin 1997).

Khan et al. (2015) reported significant decrease in content of lycopene and β-carotene under low-temperature stress in tomato, while phenolic compounds such as SA significantly enhanced the content of these non-enzyme antioxidants (Hafeznia et al. 2014, Huang et al. 2008). Salicylic acid can activate lycopene biosynthesis pathway by upregulating the encoding of enzymes relating to lycopene level during fruit development. Plants also synthesize lipophilic antioxidant known as α-tocopherol or vitamin E which acts as scavenger of free radicals in combination with other antioxidants (Massacci et al. 2008). Munne-Bosch and Dixon (2003) reported that α-tocopherol helps in membrane stabilization and alleviates tolerance of plants during oxidative stress caused by the temperature.

11.15 Phenolics in Wounding and Hyperoxia

It is well-known that wound healing is induced by the production of suberin and lignin during wounding stress (Rittinger et al. 1987). Lignin is composed of monolignol residues that are synthesized from hydroxycinnamic acid precursors such as FA (Boerjan et al. 2003). FA is synthesized at the similar rate as it is utilized for suberin and lignin synthesis.

Phenolic oxidation usually occurs because of wounding (tissue damage) which immediately activates both enzymes (NADPH oxidase, SOD), NADPH oxidase generates superoxide radical ($O_2^{\bullet-}$), and SOD transforms superoxide radicals into transformed H_2O_2. These ROS ($O_2^{\bullet-}$ and H_2O_2) act as a signal that increase the mitochondrial respiration rate in the wounded tissue (Mittler 2002;Murphy and DeCoursey 2006) inducing a subsequent oxidative burst in shredded carrots (Jacobo-Velázquez et al. 2011) and potato tubers (Razem and Bernards 2003). The lower H_2O_2 detoxification ability also induces higher PC (Jacobo-Velázquez et al. 2011). ROS activate plant defense genes and play a major role in plants subjected to environmental stresses (Dixon and Paiva 1995; Chen et al. 2008).

ROS induce PAL activation which accumulates phenolic compounds during stress induced by wounding and hyperoxia in carrots (Jacobo-Velázquez et al. 2011). The accumulation of total PC intensifies with increasing wounding intensity and hyperoxia stress. Hyperoxia increases PAL activity and respiration rate and accumulates higher concentrations of H_2O_2 at the wound site which induces activation of the phenylpropanoid metabolism, thereby accumulating soluble phenolics associated with oxidative cross-linking of suberin poly(phenolics) (Razem and Bernards 2003). This increase of PAL activity and total phenolic content increases the ROS concentration, thereby increasing the respiration rate of the tissue (Mittler 2002; Murphy and DeCoursey 2006) which partially inhibits the ascorbate peroxidase (APX) activated by wound induction and catalase (CAT) by chemical signaling (extracellular ATP). As ROS higher levels are toxic for plant cells, CAT and APX activities increase to finely tune ROS levels (Mittler 2002).

During the wounding stress process, cell disruption occurs inducing the liberation of cytosolic ATP into the extracellular matrix. This chemical signal synthesized at wounding site diffuses from the injury site into adjacent cells where it is recognized by its receptor at the plasma membrane. Once ATP binds to its receptor, the cytosolic Ca^{2+} concentration is increased, triggering the NADPH oxidase activation and thus $O_2^{\bullet-}$ production (Song et al. 2006; Kang and Saltveit 2003). This signal may be partially removed from the site of injury by the water dips which in turn decreases PAL activity and signaling compounds that trigger the NADPH oxidase-mediated $O_2^{\bullet-}$ and wound-induced activation of PAL (Song et al. 2006). The exact nature of the wound signal that activates PAL remains unknown (Jacobo-Velazquez et al. 2011).

Similar reports on higher accumulation of phenolic compounds in hyperoxia-treated blueberries (Zheng et al. 2003), strawberries (Ayala-Zavala et al. 2007), and Chinese bayberries (Yang et al. 2009) suggest that both stresses, wounding and hyperoxia, may induce the activation of the phenylpropanoid metabolism through a common signal. An increase in phenolic content has also been observed in poplar and oak trees in response to wounding (Hammerschmidt and Schultz 1996).

The stress-induced production of phenolics resulted in an enhancement of the AOX (antioxidant capacity) value in fruits and vegetables (Heo et al. 2007). Each phenolic compound has a particular AOX based on its chemical structure; the AOX value of fruits and vegetables not only depends on their total PC, it is also affected by their phenolic profiles (type of phenolics present and their relative amounts or proportions) (Rice-Evans et al. 1996; Jacobo-Velázquez and Cisneros-Zevallos 2009). The increased PC amounts generated by wounding and hyperoxia stresses also result in enhancement of their AOX values.

The application of wounding and hyperoxia stresses accumulated phenols such as3-O-caffeoylquinic acid (3-CQA), FA (ferulic acid), 3,5-dicaffeoylquinic acid (3,5-diCQA), and 4,5-dicaffeoylquinic acid (4,5-diCQA) in carrot tissue (Clifford et al. 2003). The concentrations of CQAs during stress in carrot tissues have been found higher as compared to coffee beans and artichoke (Ky et al. 1997; Lattanzio et al. 2009).

11.16 Role of Phenolics in Plant Architecture and Auxin Transport

The plant phenolics act as chemical messengers or internal physiological regulators, and there effects are associated with the growth hormone auxin (IAA). The functioning of IAA oxidase is destroyed by the enzyme peroxidase in which monohydroxy B-ring flavonoids act as its cofactor while IAA-degrading activity is inhibited by dihydroxy B-ring forms (Stafford 1991; Mathesius 2001). The inhibition of the IAA transport through plasma membrane is also implicated by the binding of these two monohydroxy and dihydroxy flavonoids to receptor of N-1-naphthylphthalamic acid (NPA) protein. NPA is a synthetic compound, which probably binds a regulatory protein that is associated with the transmembrane efflux of IAA anions mediated by carrier. Some flavonoids, such as apigenin, kaempferol, and quercetin, also block the polar auxin transport in the plasma membrane of plant cell by competing with the NPA receptor of IAA. However, these flavonoids do not compete directly with IAA receptor but act through their own receptor of NPA. These effects upon auxin transport could influence plant architecture. The studies on flavonoid defective mutants have shown a wide range of alterations to shoot and root development which confirms the link between flavonoids and the plant architecture (Murphy et al. 2002; Buer and Muday 2004; Imin et al. 2006; Peer and Murphy 2007; Buer and Djordjevic 2009).

11.17 Role of Phenolics in Pollen Development and Sterility

The role of Phenolic derivatives (flavonoids) in functional pollen development has been confirmed in *Zea mays* and *Petunia* mutants. The functionality of pollens in gymnosperms and angiosperms (monocots and dicots) has been correlated with the flavonol requirements of the pollens which suggest that it might have arisen in early ancestors of land plants (Mo et al. 1992; Van der Meer et al. 1992; Ylstra et al. 1996; Taylor and Grotewold 2005).

11.18 Role of Phenolics in Nyctinasty

The phenolic derivatives (p-coumaroyl moiety) as an endogenous bioactive substance play its role in nyctinastic leaf movement (Ueda and Nakamura 2010).

References

Aaby K, Mazur S, Nes A, Skrede G (2012) Phenolic compounds in strawberry (Fragaria x ananassa Duch.) fruits: composition in 27 cultivars and changes during ripening. Food Chem 132:86–97

Agati G, Azzarello E, Pollastri S, Tattini M (2012) Flavonoids as antioxidants in plants: location and functional significance. Plant Sci 196:67–76

Ahmad SS, Tahir I (2015) Storage protocol for improving the postharvest performance in cut scapes of Iris versicolor. Acta Hortic 1060:71–79

Ahmad SS, Tahir I (2016) Putrescine and jasmonates outplay conventional growth regulators in improving postharvest performance of Iris germanica L. cut scapes. Proc Natl Acad Sci India Sec B Biol Sci. (Online). https://doi.org/10.1007/s40011-016-0767-2

Ahmed CB, Rouina BB, Sensoy S (2009) Saline water ırrigation effects on fruit development, quality, and phenolic composition of virgin olive oils, cv. Chemlali. J Agric Food Chem 57:2803–2811

Akula R, Ravishankar GA (2011) Influence of abiotic stress signals on secondary metabolites in plants. Plant Signal Behav 6:1720–1731

Alvarez S, Marsh EL, Schroeder SG, Schachtman DP (2008) Metabolomic and proteomic changes in the xylem sap of maize under drought. Plant Cell Environ 31:325–340

An M, Haig T, Pratley JE (2000) Phytotoxicity of Vulpia residues: II. Separation, identification, and quantification of allelochemicals from Vulpia myuros. J Chem Ecol 26(6):1465–1476

Anderson MD, Prasad TK, Martin BA, Steward CR (1994) Differential gene expression in chilling acclimated maize seedlings and evidence for the involvement of abscisic acid in chilling tolerance. Plant Physiol 105:331–339

AOAC (1990) Methods of analysis, 10th edn. Washington DC

Araim G, Saleem A, Arnason JT, Charest C (2009) Root colonization by an arbuscular mycorrhizal (AM) fungus increases growth and secondary metabolism of purple coneflower, *Echinacea purpurea* (L.) Moench. J Agric Food Chem 57(6):2255–2258

Archetti M (2009) Decoupling vigour and quality in the autumn colours game: weak individuals can signal, cheating can pay. J Theor Biol 256:479–484

Archetti M, Döring TF, Hagen SB, Hughes NM, Leather SR, Lee DW, Lev-Yadun S, Manetas Y, Ougham H, Schaberg P, Thomas H (2009) Unravelling the evolution of autumn colours: an interdisciplinary approach. Trends Ecol Evol 24:166–173

Ariga T, Hamano M (1990) Radical scavenging action and its mode in proanthocyanidins B-1 and B-3 from azuki beans to peroxyl radicals. Agric Biol Chem 54:2499–2504

Asada K (1994) Production and action of active oxygen species in photosynthetic tissues. In: Foyer CH, Mullineaux PM (eds) Causes of photooxidative stress and amelioration of defense systems in plants. CRC Press, Boca Raton, pp 77–104

Awika JM, Rooney LM (2004) Sorghum phytochemicals and their potential impact on human health. Phytochemistry 65:1199–1221

Ayala-Zavala JF, Wang SY, Wang CY, Gonzalez-Aguilar GA (2007) High oxygen treatment increases antioxidant capacity and postharvest life of strawberry fruit. Food Technol Biotechnol 45:166–173

Bär W, Pfeifer P, Dettner K (1997) Intra- and inter-specific Allelochemical effects in three Kalanchoe-species (Crassulaceae). J Biosci 52(7–8):441–449

Baranowska-Morek A (2003) Roślinnemechanizmytolerancjinatoksycznedziałaniemetalicieżkich. Kosmos 52:283

Bennett GJ, Lee HH (1988) Xanthones from Guttiferae. Phytochemistry 28:967–998

Bertin C, Yang X, Weston A (2003) The role of root exudates and allelochemicals in the rhizosphere. Plant Soil 256:67–83

Besle JM, Cornu A, Jouany JP (1994) Roles of structural phenyl propanoids in forage cell wall digestion. J Sci Food Agric 64:171–190

Bharti AK, Khurana JP (1997) Mutant of *Arabidopsis* as tools to understand the regulation of phenylpropanoids pathway and UVB protection mechanism. J Photochem Photobiol 65:765–776

Bhattacharya A, Sood P, Citovsky V (2010) The roles of plant phenolics in defence and communication during agrobacterium and rhizobium infection. Mol Plant Pathol 11(5):705–719

Bidel LPR, Coumans M, Baissac Y, Doumas P, Jay-Allemand C (2010) Biological activity of phenolics in plant cells. In: Santos-Buelga C, Escribano-Bailon MT, Lattanzio V (eds) Recent advances in polyphenol research, vol 2. Wiley-Blackwell, Oxford, pp 163–205

Biran I, Halevy AH (1973) Endogenous levels of growth regulators and their relationship to the rooting of dahlia cuttings. Physiol Plant 28:244–247

Boerjan W, Ralph J, Baucher M (2003) Lignin biosynthesis. Annu Rev Plant Biol 54:519–546

Bonner WD Jr (1957) Soluble oxidases and their inhibitors. Annu Rev Plant Physiol 8:427–452

Boroumand JS, Lariyazdi H, Ranjbar M (2011) Effect of salicylic acid on some plant growth parameters under lead stress in *Brassica napus* var. okapi. Int J Plant Physiol 1(3):117–185

Bors W, Heller W, Michel C, Saran M (1990) Flavonoids as antioxidants: determination of radical-scavenging efficiencies. Methods Enzymol 186:343–355

Boudet AM (2007) Evolution and current status of research in phenolic compounds. Phytochemistry 68:2722–2735

Bourgou S, Bettaieb I, Saidani M (2010) Fatty acids, essential oil, and phenolics modifications of black cumin fruit under NaCl stress conditions. J Agric Food Chem 58:12399–12406

Brady JD, Fry SC (1997) Formation of di-isodityrosine and loss of isodityrosine in the cell walls of tomato cell-suspension cultures treated with fungal elicitors or H_2O_2. Plant Physiol 115:87–92

Brunetti C, Di Ferdinando M, Fini A, Pollastri S, Tattini M (2013) Flavonoids as antioxidants and development regulators: relative significance in plants and humans. Int J Mol Sci 14:3540–3555

Bryant JP, Chapin FS, Reichardt PB, Clausen TP (1987) Response of winter chemical defense in Alaska paper birch and green alder to manipulation of plant carbon/nutrient balance. Oecologia 72(4):510–514

Buer CS, Djordjevic MA (2009) Architectural phenotypes in the transparent testa mutants of *Arabidopsis thaliana*. J Exp Bot 60:751–763

Buer CS, Muday GK (2004) The transparent testa4 mutation prevents flavonoid synthesis and alters auxin transport and the response of *Arabidopsis* roots to gravity and light. Plant Cell 16:1191–1205

Cai YZ, Xing J, Sun M, Zhan ZQ, Corke H (2005) Phenolic antioxidants (hydrolysable tannins, flavonols, and anthocyanins) identified by LC-ESI-MS and MALDI-QIT-TOF MS from *Rosa chinensis* flowers. J Agric Food Chem 53:9940–9948

Caliskan O, Radusiene J, Temizel KE, Staunis Z, Cirak C, Kurt D, Odabas MS (2017) The effects of salt and drought stress on phenolic accumulation in greenhouse-grown Hypericum pruinatum. Ital J Agron 12(3):29

Cao J, Jiang Q, Lin J, Li X, Sun C, Chen K (2015) Physicochemical characterisation of four cherry species (Prunus spp.) grown in China. Food Chem 173:855–863

Ceymann M, Arrigoni E, Schärer H, Nising AB, Hurrell RF (2012) Identification of apples rich in health promoting flavan-3-ols and phenolic acids by measuring the polyphenol profile. J Food Compos Anal 26:128–135

Chakraborty D, Mandal SM (2008) Fractional changes in phenolic acids composition in root nodules of *Arachis hypogaea* L. Plant Growth Regul 55(3):159

Chen JY, He LH, Jiang YM, Wang Y, Joyce DC, Ji ZL, Lu WJ (2008) Role of phenylalanine ammonia-lyase in heat pretreatment-induced chilling tolerance in banana fruit. Physiol Plant 132:318–328

Cheynier V (2005) Polyphenols in foods are more complex than often thought. Am J Clin Nutr 81 (1):223S–229S

Cheynier V, Comte G, Davies KM, Lattanzio V, Martens S (2013) Plant phenolics: recent advances on their biosynthesis, genetics, and ecophysiology. Plant Physiol Biochem 72:1–20

Chou CH, Muller CH (1972) Allelopathic mechanisms of Arctostaphylos glandulosa var. zacaensis. Am Midl Nat 88(2):324–347

Chou CH, Hwang SY, Peng CI (1987) The selective allelopathic interaction of a pasture-forest intercropping in Taiwan. Plant Soil 98:31–41

Christie PJ, Alfenito MR, Walbot V (1994) Impact of low temperature stress on general phenylpropanoid and anthocyanin pathways: enhancement of transcript abundance and anthocyanin pigmentation in maize seedlings. Planta 194:541–549

Clemens S (2001) Molecular mechanisms of plant metal tolerance and homeostasis. Planta 212 (4):475–486

Clifford MN (2000) Miscellaneous phenols in foods and beverages–nature, occurrence and dietary burden. J Sci Food Agric 80(7):1126–1137

Clifford MN, Johnston KL, Knight S, Kuhnert N (2003) Hierarchical scheme for LC-MSn identification of chlorogenic acids. J Agric Food Chem 51:2900–2911

Close DC, McArthur C (2002) Rethinking the role of many plant phenolics protection from photodamage not herbivores? Oikos 99:166–172

Close D, Davies NW, Beadle CL (2001) Temporal variation of tannins (galloylglucoses), flavonols and anthocyanins in leaves of *Eucalyptus nitens* seedlings: implications for light attenuation and antioxidant activities. Aust J Plant Physiol 28:1–10

Cohen E, Laurie S, Yavin S (1988) Prevention of red blotch in degreened lemon fruit. HortScience 23:864–866

Conklin PL, Williams EH, Last RL (1996) Environmental stress sensitivity of an ascorbic acid deficient *Arabidopsis* mutant. Proc Natl Acad Sci USA 93:970–9974

Conklin PL, Pallanca JE, Last RL, Smirnoff N (1997) L-ascorbic acid metabolism in the ascorbate-deficient Arabidopsis mutant vtc1. Plant Physiol 115(3):1277–1285

Cui T, Nakamura K, Ma L, Li JZ, Kayahara H (2005) Analyses of arbutin and chlorogenic acid, the major phenolic constituents in oriental pear. J Agric Food Chem 53:3882–3887

Cuin TA, Shabala S (2008) Compatible solutes mitigate damaging effects of salt stress by reducing the impact of stress-induced reactive oxygen species. Plant Signal Behav 3:207–208

Dakora FD, Joseph CM, Phillips DA (1993) Alfalfa (*Medicago sativa* L.) root exudates contain isoflavonoids in the presence of *Rhizobium meliloti*. Plant Physiol 101(3):819–824

Daneil M (2006) Medicinal plants: chemistry and properties. Oxford & IBH Publishing, Oxford, p 140

Dangl JL, Dietrich RA, Thomas H (2000) Senescence and programmed cell death. In: Biochemistry and molecular biology of plants, 1st edn. American Society of Plant Physiologist, Rockville

Dardanelli MS, de Cordoba FJ, Espuny MR, Carvajal MA, Díaz ME, Serrano AM, Okon Y, Megías M (2008) Effect of *Azospirillum brasilense* coinoculated with *Rhizobium* on *Phaseolus vulgaris* flavonoids and nod factor production under salt stress. Soil Biol Biochem 40(11):2713–2721

De Abreu IN, Mazzafera P (2005) Effect of water and temperature stress on the content of active constituents of *Hypericum brasilienne* Choisy. Plant Physiol Biochem 43:241–248

de Fernández SB, Hernández T, Estrella I (1993) Phenolic composition of white grapes (Var. Airen). Changes during ripening. Food Chem 47:47–52

Del Bano MJ, Lorente J, Castillo J, Benavente-Garcıa O, Del Rıo JA, Ortuno A, Quirin KW, Gerard D (2003) Phenolic diterpenes, flavones, and rosmarinic acid distribution during the development of leaves, flowers, stems, and roots of *Rosmarinus officinalis.* Antioxidant activity. J Agric Food Chem 51:4247–4253

Demmig-Adams B, Gilmore AM, Adams WW III (1996) In vivo functions of carotenoids in higher plants. FASEB J 10:403–412

Devi SR, Prasad MN (1998) Copper toxicity in *Ceratophyllum demersum* L. (Coontail), a free floating macrophyte: response of antioxidant enzymes and antioxidants. Plant Sci 138 (2):157–165

Devi MC, Reddy MN (2002) Phenolic acid metabolism of groundnut (*Arachis hypogaea* L.) plants inoculated with VAM fungus and *Rhizobium.* Plant Growth Regul 37(2):151–156

Dhawan RS, Nanda KK (1982) Stimulation of root formation on *Impatiens balsamina* L. cuttings by coumarin and the associated biochemical changes. Biol Plant 24(3):177–182

Díaz J, Bernal A, Pomar F, Merino F (2001) Induction of shikimate dehydrogenase and peroxidase in pepper (*Capsicum annuum* L.) seedlings in response to copper stress and its relation to lignification. Plant Sci 161(1):179–188

Dixon RA, Paiva N (1995) Stress-induced phenylpropanoid metabolism. Plant Cell 7:1085–1097

Dong YH, Beuning L, Davies K, Mitra D, Morris B, Kootstra A (1998) Expression of pigmentation genes and photoregulation of anthocyanin biosynthesis in developing Royal Gala apple flowers. Aust J Plant Physiol 25:245–252

Elkind Y, Edwards R, Mavandad M, Hedrick SA, Ribak O, Dixon RA, Lamb CJ (1990) Abnormal plant development and down regulation of phenylpropanoid biosynthesis in transgenic tobacco containing a heterologous phenylalanine ammonia lyase gene. Proc Natl Acad Sci USA 87:9057–9061

Elzaawely AA, Xuan TD, Tawata S (2005) Allelopathic activity and identification of allelochemicals from Rumex japonicus Houtt. Allelopathy J 16(2):209–216

Estabrook EM, Sengupta-Gopalan C (1991) Differential expression of phenylalanine ammonia-lyase and chalcone synthase during soybean nodule development. Plant Cell 3(3):299–308

Fadl MS, Hartmann HT (1967) Relationship between seasonal changes in endogenous promoter and inhibitors in pear buds and cutting bases and the rooting of pear hardwood cuttings. Proc Am Soc Hortic Sci 91:96–112

Farkas GL, Dezsi L, Horvath M, Kisban K, Udvardy J (1964) Common pattern of enzymatic changes in detached leaves and tissues attacked by parasites. Phytopathol Z 49:343–354

Farooq M, Basra SMA, Wahid A, Ahmad N, Saleem BA (2009) Improving the drought tolerance in rice (*Oryza sativa* L.) by exogenous application of salicylic acid. J Agron Crop Sci 195:237–246

Fayez KA, Bazaid SA (2014) Improving drought and salinity tolerance in barley by application of salicylic acid and potassium nitrate. J Saudi Soc Agric Sci 13:45–55

Ferreira D, Slade D (2002) Oligomeric proanthocyanidins: naturally occurring O-heterocycles. Nat Prod Rep 19:517–541

Foyer CH, Harbinson J (1994) Oxygen metabolism and the regulation of photosynthetic electron transport. In: Foyer CH, Mullineaux PM (eds) Causes of photooxidative stress and amelioration of defense systems in plants. CRC Press, Boca Raton, pp 1–42

Foyer CH, Lelandais M, Kunert KJ (1994) Photooxidative stress in plants. Physiol Plant 63:257–264

Frank HA, Cogdell RJ (1993) The photochemistry and function of carotenoids in photosynthesis. In: Young A, Britton G (eds) Carotenoids in photosynthesis. Chapman and Hall, London, pp 252–326
Frank HA, Cogdell RJ (1996) Carotenoids in photosynthesis. Photochem Photobiol 63:257–264
Fry SC (1979) Phenolic components of the primary cell wall and their possible role in the hormonal regulation of growth. Planta 146:343–351
Fry SC (1983) Feruloylated pectins from the primary cell wall: their structure and possible functions. Planta 157:111–123
Fry SC (1986) Cross-linking of matrix polymers in the growing cell walls of angiosperms. Annu Rev Plant Physiol 37:165–186
Gharib FA (2006) Effect of salicylic acid on the growth, metabolic activities and oil content of basil and marjoram. Int J Agric Biol 4:485–492
Gnanasekaran N, Kalavathy S (2017) Drought stress signal promote the synthesis of more reduced phenolic compounds (chloroform insoluble fraction) in *Tridax procumbens*. Free Radicals Antioxid 7(1):128–136
Gorham J (1989) Stilbenes and phenanthrenes. In: Harborne JB (ed) Methods in plant biochemistry, Plant Phenolics, vol 1. Academic Press, London, pp 159–196
Gorham J (1995) The biochemistry of stilbenoids (with Contributions by M. Toriand Y. Asakawa). Chapman & Hall, London
Gould KS (2004) Nature's Swiss army knife: the diverse protective roles of anthocyanins in leaves. J Biomed Biotechnol 5:314–320
Grabber JH, Hatfield RD, Ralph J, Zon J, Amrhein N (1995) Ferulate cross-linking in cell walls isolated from maize cell suspensions. Phytochemistry 40:1077–1082
Grabber JH, Hatfield RD, Ralph J (1998a) Diferulate cross-links impede the enzymatic degradation of non-lignified maize walls. J Sci Food Agric 77:193–200
Grabber JH, Hatfield RD, Ralph J (1998b) Ferulate cross-links limited the enzymatic degradation of synthetically lignified primary walls of maize. J Agric Food Chem 46:2609–2614
Grace SC, Logan BA (2000) Energy dissipation and radical scavenging by the plant phenylpropanoid pathway. Philos Trans R Soc Lond B Biol Sci 355(1402):1499–1510
Grace SC, Logan BA, Adams WW III (1998) Seasonal differences in foliar content of chlorogenic acid, a phenylpropanoid antioxidant, in *Mahonia repens*. Plant Cell Environ 21:513–521
Gray J, Close PS, Briggs SP, Johal GS (1997) A novel suppressor of cell death in plants encoded by the Lls1 gene of maize. Cell 89:25–31
Gruca-Krolikowska S, Waclawek W (2006) Metale w środowisku.Cz. II. Wpływmetaliciężkichnarośliny. Chemia-Dydaktyka-Ekologia-Metrologia 11(1–2):41–54
Hackett WP (1970) The influence of auxin, catechol and methanolic tissue extracts on root initiation in aseptically cultured shoot apices of the juvenile and adult forms of Hedera helix. J Am Soc Hortic Sci 95:398–402
Hafeznia M, Mashayekhi K, Ghaderifar F, Mousavizadeh SJ (2014) Tomato morphological and biochemical characteristics in response to foliar applying of salicylic acid. Int J Biosci 5 (9):237–243
Hagerman AE, Riedl KM, Jones GA, Sovik KN, Ritchard NT, Hartzfeld PW, Riechel TL (1998) High molecular weight plant polyphenolics (tannins) as biological antioxidants. J Agric Food Chem 46(5):1887–1892
Hammad Y, Nalin R, Marechal J, Fiasson K, Pepin R, Berry AM, Normand P, Domenach AM (2003) A possible role for phenyl acetic acid (PAA) on *Alnus glutinosa* nodulation by *Frankia*. In: *Frankia* symbiosis. Springer, Dordrecht, pp 193–205
Hammerschlag F (1982) Factors influencing *in vitro* multiplication and rooting of the plum rootstock 'Myrobalan' (Prunus cerasifera Ehrh.). J Am Soc Hortic Sci 107:44–47
Hammerschmidt R, Schultz JC (1996) Multiple defences and signals in plant defence against pathogens and herbivores. In: Romeo JT et al (eds) Phytochemical diversity and redundancy in ecological interactions. Plenum Press, New York, pp 121–154

Hanen F, Ksouri R, Megdiche W, Trabelsi N, Boulaaba M, Abdelly C (2008) Effect of salinity on growth, leaf phenolic content and antioxidant scavenging activity in *Cynara cardunculus* L. In: Abdelli C, Öztürk M, Ashraf M, Grignon YC (eds) Biosaline agriculture and high salinity tolerance. Basel, Birkhauser Verlag, pp 335–343

Harborne AJ (1998) Phytochemical methods A guide to modern techniques of plant analysis, 3rd edn. Springer, Dordrecht. ISBN 978-0-412-57260-9

Harborne JB (1980) Plant phenolics. In: Bell EA, Charlwood BV (eds) Encyclopedia of plant physiology, Secondary plant products, vol 8. Springer, Berlin, pp 329–402

Harborne JB, Simmonds NW (1964) The natural distribution of the phenolic aglycones. In: Biochemistry of phenolic compounds, pp 77–127

Haslam E (1998) Practical polyphenolics: from structure to molecular recognition and physiological function. Cambridge University Press, Cambridge

Havaux M (1998) Carotenoids as membrane stabilisers in chloroplasts. Trends Plant Sci 3:147–151

Hayat Q, Hayat S, Irfan M, Ahmad A (2010) Effect of exogenous salicylic acid under changing environment: A review. Environ Exp Bot 68:14–25

Heath M (1998) Apoptosis, programmed cell death and the hypersensitive response. Eur J Plant Pathol 104:117–124

Heo HJ, Kim YJ, Chung D, Kim DO (2007) Antioxidant capacities of individual and combined phenolics in a model system. Food Chem 104:87–92

Hichem H, Mounir D, Naceur EA (2009) Differential responses of two maize (*Zea mays* L.) varieties to salt stress: changes on polyphenols composition of foliage and oxidative damages. Ind Crop Prod 30:144–151

Hillis WE, Swain T (1959) The phenolic constituents of *Prunus domestica*. II. The analysis of tissues of the Victorian plum tree. J Sci Food Agric 10:135–144

Hiltner L (1904) Uber neure Erfahrungen und probleme auf dem gebeit der bodenbackteriologie und unter besonderer berucksichtigung der grundungung und brache. Arb Deut Landwirsch Ges 98:59–78

Hodnick WF, Milosavljevic EB, Nelson JH et al (1988) Electrochemistry of flavonoids. Biochem Pharmacol 37:2607–2611

Hopkins WG (1997) Introduction to plant physiology, 2nd edn. Wiley, Hoboken, pp 267–284

Hopkinson SM (1969) Quaternary Reviews 23, p 98. In: Daneil M (2006) Medicinal plants: chemistry and properties. Oxford & IBH Publishing, p 140

Hu H, Dai M, Yao J, Xiao B, Li X, Zhang Q, Xiong L (2006) Overexpressing a NAM, ATAF, and CUC (NAC) transcription factor enhances drought resistance and salt tolerance in rice. Proc Natl Acad Sci USA 103:12987–12992

Huang J, Wang MM, Bao YM, Sun SJ, Pan LJ, Zhang HS (2008) SRWD: a novel WD40 protein subfamily regulated by salt stress in rice (*Oryza sativa* L.). Gene 424(1–2):71–79

Hückelhoven R (2007) Cell wall–associated mechanisms of disease resistance and susceptibility. Annu Rev Phytopathol 45:101–127

Hughes NM (2011) Winter leaf reddening in 'evergreen' species. New Phytol 190:573–581

Huguet V, Gouy M, Normand P, Zimpfer JF, Fernandez MP (2005) Molecular phylogeny of Myricaceae: a reexamination of host–symbiont specificity. Mol Phylogenet Evol 34(3):557–568

Hussein MM, Balbaa LK, Gaballah MS (2007) Salicylic acid and salinity effects on growth of maize plants. Res J Agric Biol Sci 3(4):321–328

Iason GR, Hester AJ (1993) The response of heather (*Calluna vulgaris*) to shade and nutrients – predictions of the carbon-nutrient balance hypothesis. J Ecol 81:75–80

Iason GR, Duncan AJ, Hartley SE, Staines BW (1996) Feeding behaviour of red deer (*Cervuse laphus*) on Sitka spruce (*Picea sitchensis*): the role of carbon-nutrient balance. For Ecol Manag 88(1–2):121–129

Ikenaka Y, Nakayama SM, Muzandu K, Choongo K, Teraoka H, Mizuno N, Ishizuka M (2010) Heavy metal contamination of soil and sediment in Zambia. Afr J Environ Sci Technol 4 (11):729–739

Imami S, Jamshidi S, Shahrokhi S (2011) Salicylic acid foliar and soil application effect on chickpea resistance to chilling stress. In: International conference on biology, environment and chemistry. IPCBEE. IACSIT Press, Singapore, p 24

Imin N, Nizamidin M, Wu T, Rolfe BG (2006) Factors involved in root formation in *Medicago truncatula*. J Exp Bot 58:439–451

Ishii T (1997) Structure and functions of feruloylated polysaccharides. Plant Sci 127:111–127

Jacobo-Velázquez DA, Cisneros-Zevallos L (2009) Correlations of antioxidant activity against phenolic content revisited: a new approach in data analysis for food and medicinal plants. J Food Sci 74(9):R107–R113

Jacobo-Velázquez DA, Martínez-Hernández GB, del C Rodríguez S, Cao CM, Cisneros-Zevallos L (2011) Plants as biofactories: physiological role of reactive oxygen species on the accumulation of phenolic antioxidants in carrot tissue under wounding and hyperoxia stress. J Agric Food Chem 59(12):6583–6593

Jaiti F, Kassami M, Meddich A, Hadrami EI (2008) Effect of arbuscular mycorrhization on the accumulation of hydroxycinnamic acid derivatives in date palm seedlings challenged with *Fusarium oxysporum* f. sp. albedinis. J Phytopathol 156(11–12):641–646

Jamalian S, Gholami M, Esna-Ashari M (2013) Abscisic acid-mediated leaf phenolic compounds, plant growth and yield is strawberry under different salt stress regimes. Theor Exp Plant Physiol 25:291–299

James DJ, Thurbon IJ (1981) Shoot and root initiation *in vitro* in the apple rootstock M9 and the promotive effects of phloroglucinol. J Hortic Sci 56:15–20

Janas KM, Cvikrova M, Pagiewicz A, Eder J (2000) Alterations in phenylpropanoid content in soybean roots during low temperature acclimations. Plant Physiol Biochem 38:587–593

Jones OP, Hatfield SGS (1976) Root initiation in apple shoots cultured *in vitro* with auxins and phenolic compounds. J Hortic Sci 51:495–499

Joshi VN, Rathore SS, Arora SK (1999) Effect of chromium on growth and development of cowpea (*Vigna unguicalata* L.). India J Environ Prot 19:745–749

Julkunen-Tiito R, Nenadis N, Neugart S, Robson M, Agati G, Vepsäläinen J, Zipoli G, Nybakken L, Winkler B, Jansen M (2015) Assessing the response of plant flavonoids to UV radiation: an overview of appropriate techniques. Phytochem Rev 14:273–279

Jung HG, Deetz DA (1993) Cell wall lignification and degradability. In: Forage cell wall structure and digestibility. Jan(foragecellwalls), pp 315–346

Jung HG, Buxton DR, Hatfield RD, Ralph J (1993) Forage cell wall structure and digestibility. American Society of Agronomy, Madison

Kacperska A (1993) Water potential alteration: a prerequisite or a triggering stimulus for the development of freezing tolerance in overwintering herbaceous plants. In: Advances in plant cold hardiness, vol 73, p 91

Kaimoyo E, Farag MA, Sumner LW, Wasmann C, Cuello JL, VanEtten H (2008) Sublethal levels of electric current elicit the biosynthesis of plant secondary metabolites. Biotechnol Prog 24 (2):377–384

Kaiser WM (1979) Reversible inhibition of the Calvin cycle and activation of oxidative pentose phosphate cycle in isolated intact chloroplasts by hydrogen peroxide. Planta 145:377–382

Kang H, Saltveit ME (2003) Wound-induced increases in phenolic content of fresh-cut lettuce is reduced by a short immersion in aqueous hypertonic solutions. Postharvest Biol Technol 29:271–277

Ke D, Salveit E Jr (1988) Plant hormone interaction and phenolic metabolism in the regulation of russet spotting in iceberg lettuce. Plant Physiol 88:1136–1140

Kennedy JA, Matthews MA, Waterhouse AL (2002) Effect of maturity and vine water status on grape skin and wine flavonoids. Am J Enol Vitic 53:268–274

Khalid KA, da Silva JAT (2010) Yield, essential oil and pigment content of Calendula officinalis L. flower heads cultivated under salt stress conditions. Sci Hortic 126:297–305

Khan AG, Kuek C, Chaudhry TM, Khoo CS, Hayes WJ (2000) Role of plants, mycorrhizae and phytochelators in heavy metal contaminated land remediation. Chemosphere 41(1–2):197–207

Khan TA, Fariduddin Q, Yusuf M (2015) *Lycopersicon esculentum* under low temperature stress: an approach toward enhanced antioxidants and yield. Environ Sci Pollut Res 22 (18):14178–14188

Khanbabaee K, Van Ree T (2001) Tannins: classification and definition. Nat Prod Rep 18:641–649

Khandaker L, Masumakond ASMG, Oba S (2011) Foliar application of salicylic acid improved the growth, yield and leaves bioactive compounds in red Amaranth (*Amaranthus tricolor* L.). veg. Crops Res Bull 74:77–86

Kil BS, Lee SY (1987) Allelopathic effects of *Chyrsanthemum morifolium* on germination and growth of several herbaceous plants. J Chem Ecol 13(2):299–308

Kim YO, Johnson JD, Lee EJ (2005) Phytotoxicity effects and chemical analysis of leaf extracts from three Phytolaccaceae species in South Korea. J Chem Ecol 31(5):1175–1186

Kisban G, Horvath M, Dezsi L, Udvardy J, Farkas GL (1964) Role of root system in the regulation of enzyme levels in leaf tissues. Acta Bot Acad Sci Hung 10:275–287

Kling GJ, Meyer MM Jr (1983) Effects of phenolic compounds and indoleacetic acid on adventitious root initiation in cuttings of *Phaseolus aureus*, *Acer saccharinum*, and *Acer griseum*. Hortic Sci 18(3):352–354

Knox JP, Dodge AD (1985) Singlet oxygen and plants. Phytochemistry 24:889–896

Korsmeyer SJ (1995) Regulators of cell death. Trends Genet 11:101–105

Krishnaiah D, Sarbatly R, Nithyanandam R (2011) A review on the antioxidant potential of medicinal plant species. Food Bioprod Process 89:217–233

Król A, Amarowicz R, Weidner S (2015) The effects of cold stress on the phenolic compounds and antioxidant capacity of grapevine (*Vitis vinifera* L.) leaves. J Plant Physiol 189:97–104

Ksouri R, Megdiche W, Debez A (2007) Salinity effects on polyphenol content and antioxidant activities in leaves of the halophyte *Cakile maritima*. Plant Physiol Biochem 45:244–249

Kumar N, Srivastava GC, Dixit K (2008a) Flower bud opening and senescence in roses (*Rosa hybrida* L.). Plant Growth Regul 55:81–99

Kumar N, Bhandari P, Singh B, Gupta AP, Kaul VK (2008b) Reversed phase-HPLC for rapid determination of polyphenols in flowers of rose species. J Sep Sci 31:262–267

Kwak SS, Kim SK, Park IJ, Lui JR (1996) Enhancement of peroxidase activity by stress related chemicals in sweet potato. Phytochemistry 43:565–568

Ky CL, Noirot M, Hamon S (1997) Comparison of five purification methods for chlorogenic acids in green coffee beans (Coffea sp.). J Agric Food Chem 42:786–790

Lafuente MT, Martínez-Téllez MA (1997) Effect of the high temperature conditioning on ethylene, phenylalanine ammonia-lyase, peroxidase and polyphenol oxidase activities in flavedo of chilled Fortune mandarin fruit. J Plant Physiol 150:674–678

Lattanzio V, Kroon PA, Linsalata V, Cardinali A (2009) Globe artichoke: A functional food and source of nutraceutical ingredients. J Funct Foods 1:132–144

Lavola A, Julkunen-Tiitto R, de la Rosa TM, Lehto T, Aphalo PJ (2000) Allocation of carbon to growth and secondary metabolites in birch seedlings under UV-B radiation and CO2 exposure. Physiol Plant 109(3):260–267

Lee DW, Gould KS (2002) Why leaves turn red. Am Sci 90:524–531

Levine A, Tenhaken R, Dixon R, Lamb C (1994) H_2O_2 from the oxidative burst orchestrates the plant hypersensitive disease resistance response. Cell 79:583–593

Leyva A, Jarrillo JA, Salinas J, Martınez-Zapater M (1995) Low temperature induces the accumulation of phenylalanine ammonia-lyase and chalcone synthase mRNA of *Arabidopsis thaliana* in light-dependent manner. Plant Physiol 108:39–46

Li DM, Nie YX, Zhang J, Yin JS, Li Q, Wang XJ, Bai JG (2013) Ferulic acid pretreatment enhances dehydration-stress tolerance of cucumber seedlings. Biol Plant 57:711–717

Liu MQ, Chen YY, Lu CF, Zhang H, Yin WL (2007) Cold acclimation induced accumulation of phenolic compounds and freezing tolerance in *Ammopiptanthus mongolicus*. For Stud China 9:203–207

Lygin AV, Upton J, Dohleman FG, Juvik J, Zabotina OA, Widholm JM, Lozovaya VV (2011) Composition of cell wall phenolics and polysaccharides of the potential bioenergy crop–Miscanthus. GCB Bioenergy 3(4):333–345
Lyons JM (1973) Chilling injury in plants. Annu Rev Plant Physiol 24:445–466
Ma D, Sun D, Wang C, Li Y, Guo T (2014) Expression of flavonoid biosynthesis genes and accumulation of flavonoid in wheat leaves in response to drought stress. Plant Physiol Biochem 80:60–66
Macadam JW, Nelson CJ, Sharp RE (1992) Peroxidase activity in the elongation zone of tall fescue. II. Spatial distribution of apoplastic peroxidase activity in genotypes differing in length of elongation zone. Plant Physiol 99:879–885
Mahmoudi H, Huang J, Gruber MY (2010) The impact of genotype and salinity on physiological function, secondary metabolite accumulation, and antioxidative responses in lettuce. J Agric Food Chem 58:5122–5130
Mandal S, Mandal M, Das A, Pati B, Ghosh A (2009) Stimulation of indoleacetic acid production in a rhizobium isolate of *Vigna mungo* by root nodule phenolic acids. Arch Microbiol 191(4):389
Maraite H (1973) Changes in polyphenol oxidases and peroxidases in muskmelon (Cucumis melo L.) infected by Fusarium oxysporum f. sp. melonis. Physiol. Plant Pathol 3:29–49
Massacci A, Nabiev SM, Pietrosanti L, Nematov SK, Chernikova TN, Thor K, Leipner J (2008) Response of the photosynthetic apparatus of cotton (*Gossypium hirsutum*) to the onset of drought stress under field conditions studied by gas-exchange analysis and chlorophyll fluorescence imaging. Plant Physiol Biochem 46(2):189–195
Mathesius U (2001) Flavonoids induced in cells undergoing nodule organogenesis in white clover are regulators of auxin breakdown by peroxidase. J Exp Bot 52:419–426
Mayak S, Halevy AH (1972) Interrelationship of ethylene and abscisic acid in the control of rose petal senescence. Plant Physiol 50:341–346
Michalak A (2006) Phenolic compounds and their antioxidant activity in plants growing under heavy metal stress. Rev Pol J of Environ Stud 15(4):523–530
Minh LT, Khang DT, Ha PT, Tuyen PT, Minh TN, Quan NV, Xuan TD (2016) Effects of salinity stress on growth and phenolics of rice (*Oryza sativa* L.). Int Lett Nat Sci 57. https://doi.org/10.18052/www.scipress.com/ILNS.57.1
MithÖfer A, Schulz EB, Bolan DW (2004) Biotic and heavy metal stress response in plants: evidence for common signals. FEBS Lett 566:1
Mittler R (2002) Oxidative stress, antioxidants and stress tolerance. Trends Plant Sci 7:405–410
Mo Y, Nagel C, Taylor LP (1992) Biochemical complementation of chalcone synthase mutants defines a role for flavonols in functional pollen. Proc NatlAcad Sci USA 89:7213–7217
Mol J, Grotewold E, Koes R (1998) How genes paint flowers and seeds. Trends Plant Sci 3:212–217
Mole S, Ross JAM, Waterman PG (1988) Light-induced variation in phenolic levels in foliage of rain-forest plants. I Chemical changes. J Chem Ecol 14:1–21
Montanari M, Deglnnocenti E, Maggini R (2008) Effect of nitrate fertilization and saline stress on the contents of active constituents of *Echinacea angustifolia* DC. Food Chem 107:1461–1466
Morsink WAG, Smith VG (1975) The effect of some monohydroxybenzoic and dihydrobenzoic acids as auxin synergists on rooting softwood cuttings of basswood (*Tilia americana*) under mist. Can J For Res 5:500–502
Mucciarelli M, Scannerini S, Gallino M, Maffei M (2000) Effects of 3,4-dihydroxybenzoic acid on tobacco (*Nicotiana tabacum* L.) cultured *in vitro*. Growth regulation in callus and organ cultures. Plant Biosyst 134(2):185–192
Munné-Bosch S, Alegre L (2003) Drought-induced changes in the redox state of α-tocopherol, ascorbate, and the diterpene carnosic acid in chloroplasts of *Labiatae* species differing in carnosic acid contents. Plant Physiol 131(4):1816–1825
Munns R, Tester M (2008) Mechanisms of salinity tolerance. Annu Rev Plant Biol 59:651–681
Murphy R, DeCoursey TE (2006) Charge compensation during the phagocyte respiratory burst. Biochim Biophys Acta 1757:996–1011

Murphy AS, Hoogner KR, Peer WA, Taiz L (2002) Identification, purification, and molecular cloning of N-1-naphthylphthalmic acid-binding plasma membrane-associated aminopeptidases from *Arabidopsis*. Plant Physiol 128:935–950

Navarro JM, Flores P, Garrido C, Martinez V (2006) Changes in the contents of antioxidant compounds in pepper fruits at different ripening stages, as affected by salinity. Food Chem 96:66–73

Newbery DC, de Foresta H (1985) Herbivory and defense in pioneer, gap and understory trees of tropical rain forest in French Guiana. Biotropica 17:238–244

Nichols SN, Hofmann RW, Williams WM (2015) Physiological drought resistance and accumulation of leaf phenolics in white clover interspecific hybrids. Environ Exp Bot 119:40–47

Nicholson RL, Hammerschmidt R (1992) Phenolic compounds and their role in disease resistance. Annu Rev Phytopathol 30:369–389

Nikiforou C, Manetas Y (2010) Strength of winter leaf redness as an indicator of stress vulnerable individuals in *Pistacia lentiscus*. Flora 205:424–427

Niyogi KK (1999) Photoprotection revisited: genetic and molecular approaches. Annu Rev Plant Physiol Plant Mol Biol 50:333–359

Noodén LD, Guiament JJ, John I (1997) Senescence mechanisms. Physiol Plant 101:746–753

Normand P, Simonet P, Bardin R (1988) Conservation of nif sequences in *Frankia*. Mol Gen Genet MGG 213(2–3):238–246

Nozolillo C, Isabelle P, Das G (1990) Seasonal changes in phenolics constituents of jack pine seedlings (*Pinus banksiana*) in relation to the purpling phenomenon. Can J Bot 68:2010–2017

Okhi K (1978) Pb concentration in soybean as related to growth, photosynthesis and carbonic anhydrase activity. Physiol Plant 18:79–82

Olenichenko NA, Ossipow VI, Zagoskina NV (2006) Effect of cold gardening on the phenolic complex of winter wheat leaves. Russ J Plant Physiol 53:554–559

Orabi SA, Dawood MG, Saleem SR (2015) Comparative study between the physiological role of hydrogen peroxide and salicylic acid in alleviating low temperature stress on tomato plants grown under sand- ponic culture. Sci Agic 9(1):49–59

Oren-Shamir M, Levi-Nissin A (1997) Temperature effects on the leaf pigmentation of *Cotinus coggygria* "Royal Purple". J Hortic Sci 72:425–432

Pandolfini T, Gabrielli R, Comparining C (1992) Nickel toxicity and peroxidase activity in seedlings of *Triticum aestivum* L. Plant Cell Environ 15:719–725

Parida AK, Das AB, Sanada Y, Mohanty P (2004) Effects of salinity on biochemical components of the mangrove, *Aegiceras corniculatum*. Aquat Bot 80:77–87

Parr AJ, Waldron KW, Ng A, Parker ML (1996) The wall-bound phenolics of Chinese water chestnut (*Eleocharis dulcis*). J Sci Food Agric 71:501–507

Pazoki A (2015) Evaluation of flavonoids and phenols content of wheat under different Lead, PGPR and mycorrhiza levels. Biol Forum Res Trend 7(1):309

Peer WA, Murphy AS (2007) Flavonoids and auxin transport: modulators or regulators. Trends Plant Sci 12:556–563

Pellissier F (1994) Effect of phenolic compounds in humus on the natural regeneration of spruce. Phytochemistry 36(4):865–867

Pennell RI, Lamb C (1997) Programmed cell death in plants. Plant Cell 9:1157–1168

Peres V, Nagem TJ (1977) Tryoxigenated naturally occurring xanthones. Phytochemistry 44:191–214

Peres V, Nagem TJ, de Oliveira FF (2000) Tetraoxygenated naturally occurring xanthones. Phytochemistry 55:683–710

Perez-Campo R, Lopez-Torres M, Cadenas S, Rojas C, Barja G (1998) The rate of free radical production as a determinant of the rate of aging: evidence from the comparative approach. J Comp Physiol B 168(3):149–158

Petersen M, Strack D, Matern U (1999) Biosynthesis of phenylpropanoids and related compounds. In: Wink M (ed) Biochemistry of plant secondary metabolism, vol 2. Sheffield Academic Press/ CRC Press, Sheffield/Boca Raton, pp 151–222

Plaza BM, Jimenez S, Segura ML (2009) Physiological stress caused by salinity in *Cordyline fruticosa* and its indicators. Commun Soil Sci Plant 40:473–484

Popovici J, Walker V, Bertrand C, Bellvert F, Fernandez MP, Comte G (2011) Strain specificity in the *Myricaceae–Frankia* symbiosis is correlated to plant root phenolics. Funct Plant Biol 38 (9):682–689

Prasad TK, Anderson MD, Marin BA, Steward CR (1994a) Evidence for chilling-induce oxidative stress in maize seedlings and a regulatory role of hydrogen peroxide. Plant Cell 6:65–74

Prasad TK, Anderson MD, Steward CR (1994b) Acclimation, hydrogen peroxide and abcisic acid protect mitochondria against irreversible chilling injury in maize seedlings. Plant Physiol 105:619–627

Prinsen E, Chauvaux N, Schmidt J, John M, Wieneke U, De Greef J, Schell J, Van Onckelen H (1991) Stimulation of indole-3-acetic acid production in *Rhizobium* by flavonoids. FEBS Lett 282(1):53–55

Qu CS, Ma ZW, Yang J, Liu Y, Bi J, Huang L (2012) Human exposure pathways of heavy metals in a lead-zinc mining area, Jiangsu Province, China. PLoS One 7(11):e46793

Quan N, Anh LA, Khang D, Tuyen P, Toan N, Minh T, Minh L, Bach D, Ha P, Elzaawely A, Khanh T (2016) Involvement of secondary metabolites in response to drought stress of rice (*Oryza sativa* L.). Agriculture 6(2):23

Quideau S (2009) Chemistry and biology of ellagitannins, an underestimated class of bioactive plant polyphenols. World Scientific, London

Rahmani I, Ahmadi N, Ghanati F, Sadeghi M (2015) Effects of salicylic acid applied pre- or post-transport on post-harvest characteristics and antioxidant enzyme activity of gladiolus cut flower spikes. NZ J Crop Hortic 43(4):294–305

Rai V, Vajpayee P, Singh SN, Mehrotra S (2004) Effect of chromium accumulation on photosynthetic pigments, oxidative stress defense system, nitrate reduction, proline level and eugenol content of *Ocimum tenuiflorum* L. Plant Sci 167(5):1159–1169

Ralph J, Quideau S, Grabber JH, Hatfield RD (1994) Identification and synthesis of new ferulic acid dehydrodimers present in grass cell walls. J Chem Soc Perkin Trans 11994:3485–3498

Razem FA, Bernards MA (2003) Reactive oxygen species production in association with suberization: evidence for an NADPH-dependent oxidase. J Exp Bot 54:935–941

Reuveni O, Adato I (1974) Endogenous carbohydrates, root promoters and root inhibitors in easy- and difficult-to-root date palm. (*Phoenix dactylifera* L.) offshoots. J Am Soc Hortic Sci 99:361–363

Reyes LR, Villarreal JE, Cisneros-Zevallos L (2007) The increase in antioxidant capacity after wounding depends on the type of fruit or vegetable tissue. Food Chem 101:1254–1262

Rezayian M, Niknam V, Ebrahimzadeh H (2018) Differential responses of phenolic compounds of *Brassica napus* under drought stress. Plant Physiol 8(3):2417–2425

Rice EL (1984) Allelopathy. Academic Press, Orlando, pp 272–274

Rice-Evans CA, Miller NJ, Paganga G (1996) Structure-antioxidant activity relationships of flavonoids and phenolic acids. Free Radic Biol Med 20:933–956

Richards M (1964) Root formation on cuttings of Camellia reticulate var. 'Capt. Rawes'. Nature 204:601–602

Rittinger PA, Biggs AR, Peirson DR (1987) Histochemistry of lignin and suberin deposition in boundary layers formed after wounding in various plant species and organs. Can J Bot 65:1886–1892

Rivero RM, Ruiz JM, Garcıa PC, Lopez-Lefebre LR, Sánchez E, Romero L (2001) Resistance to cold and heat stress: accumulation of phenolic compounds in tomato and watermelon plants. Plant Sci 160(2):315–321

Roberts MR, Paul ND (2006) Seduced by the dark side: integrating molecular and ecological perspectives on the influence of light on plant defense against pests and pathogens. New Phytol 170:677–699

Rodríguez VM, Soengas P, Alonso-Villaverde V, Sotelo T, Cartea ME, Velasco P (2015) Effect of temperature stress on the early vegetative development of L. BMC Plant Biol 15(1):145

Rouau X, Cheynier V, Surget A, Gloux D, Barron C, Meudec E, Louis-Montero J, Criton M (2003) A dehydrotrimer of ferulic acid from maize bran. Phytochemistry 63:899–903

Rudikovskaya EG, Fedorova GA, Dudareva LV, Makarova LE, Rudokovskij AV (2008) Effect of growth temperature on the composition of phenols in pea roots. Russ J Plant Physiol 55:712–715

Ruiz JM, Bretones G, Baghour M, Ragala L, Belakbir A, Romero L (1998) Relationship between boron and phenolic metabolism in tobacco leaves. Phytochemistry 48:269–272

Ruiz JM, Garcıa PC, Rivero RM, Romero L (1999) Response of phenolic metabolism to the application to the carbendazim plus boron in tobacco leaves. Physiol Plant 106:151–157

Sabar M, Arif M (2014) Phenotypic response of rice (*Oryza sativa*) genotypes to variable moisture stress regimes. Int J Agric Biol 16:32–40

Sacher JA (1967) Studies of permeability, RNA and protein turnover during aging of fruit and leaf tissues. In: Woolhouse HW (eds) Aspects of the biology of aging. Symposia of the society for experimental biology. XXI. Cambridge University Press, London, pp 269–303

Sakihama Y, Yamasaki H (2002) Lipid peroxidation induced by phenolics in conjunction with aluminum ions. Biol Plant 45(2):249–254

Salah IB, Mahmoudi H, Gruber M (2011) Phenolic content and antioxidant activity in two contrasting *Medicago ciliaris* lines cultivated under salt stress. Biologia 66:813–820

Saltveit ME, Morris LL (1990) Overview on chilling injury of horticultural crops. In: Chilling injury of horticultural crops, vol 20, pp 3–15

San Antonio JP (1952) The role of coumarin in the growth of roots of *Melilotus alba*. Bot Gaz 114:79–95

Sánchez-Rodríguez E, Moreno DA, Ferreres F, del Mar Rubio-Wilhelmi M, Ruiz JM (2011) Differential responses of five cherry tomato varieties to water stress: changes on phenolic metabolites and related enzymes. Phytochemistry 72:723–729

Savoure A, Magyar Z, Pierre M, Brown S, Schultze M, Dudits D, Kondorosi A, Kondorosi E (1994) Activation of the cell cycle machinery and the isoflavonoid biosynthesis pathway by active *Rhizobium meliloti* Nod signal molecules in *Medicago microcallus* suspensions. EMBO J 13 (5):1093–1102

Scattino C, Castagna A, Neugart S, Chan HM, Schreiner M, Crisosto CH, Tonutti P, Ranieri A (2014) Post-harvest UV-B irradiation induces changes of phenol contents and corresponding biosynthetic gene expression in peaches and nectarines. Food Chem 163:51–60

Schmitzer V, Veberic R, Osterc G, Stampar F (2009) Changes in the phenolic concentration during flower development of rose 'KORcrisett'. J Am Soc Hortic Sci 134(5):491–496

Schützendübel A, Polle A (2002) Plant responses to abiotic stresses: heavy metal-induced oxidative stress and protection by mycorrhization. J Exp Bot 53:1351–1365

Sebastiani L, Scebba F, Tognetti R (2004) Heavy metal accumulation and growth responses in poplar clones Eridano (*Populus deltoides* × *maximowiczii*) and I-214 (P. × euramericana) exposed to industrial waste. Environ Exp Bot 52(1):79–88

Selvaraj T, Rajeshkumar S, Nisha MC, Wondimu L, Tesso M (2008) Effect of Glomus mosseae and plant growth promoting rhizomicroorganisms (PGPR's) on growth, nutrients and content of secondary metabolites in *Begonia malabarica* Lam. Maejo Int J Sci Technol 2(3):516–525

Shöderhäll I (1995) Properties of carrot polyphenol oxidase. Phytochemistry 39:33–38

Singh IP, Sidana J, Bharate SB, Foley WJ (2010) Phloroglucinol compounds of natural origin: synthetic aspects. Nat Prod Rep 27:393–416

Sinhal VK (2005) Phytotoxic, Cytogenetic and Biochemical effects of Pb^{2+} & Pb^{2+} in Vigna mungo (L).Hepper. PhD thesis. M.J.P. Rohilkhand University Bareilly, India

Smith-Becker J, Marois E, Huguet EJ, Midlands SL, Sims JJ, Keen NT (1998) Accumulation of salicylic acid and 4-hydroxybenzoic acid in phloem fluids of cucumber during systemic acquired resistance in preceded by a transient increase in phenylalanine ammonia-lyase activity in petioles and stem. Plant Physiol 116:231–238

Solecka D, Kacperska A (2003) Phenylpropanoid deficiency affects the course of plant acclimation to cold. Physiol Plant 119:253–262

Solecka D, Boudet AM, Kacperska A (1999) Phenylpropanoid and anthocyanin changes in low-temperature treated winter oilseed rape leaves. Plant Physiol Biochem 37:491–496

Song CJ, Steinebrunner I, Wang X, Stout SC, Roux SJ (2006) Extracellular ATP induces the accumulation of superoxide via NADPH oxidases in *Arabidopsis*. Plant Physiol 140:1222–1232

Sood S, Nagar PK (2003) Changes in abscisic acid and phenols during flower development in two diverse species of rose. Acta Physiol Plant 25:411–416

Souza MRD, Devaraj VR (2010) Biochemical responses of hyacinth bean (*Lablab purpureus*) to salinity stress. Acta Physiol Plant 32:341–353

Sridhar R (1972) Influence of nitrogen fertilization and *Pyricularia oryzae* development on some oxidases. Their substrates and respiration of rice plant. Acta Phytopathol Acad Sci Hung 7:57–70

Stafford HA (1991) Flavonoid evolution: an enzymic approach. Plant Physiol 96:680–685

Stefanowska M, Kuras M, Kacperska A (2002) Low temperature-induced modifications in cell ultrastructure and localization of phenolics in winter oilseed rape (*Brassica napus* L. var. oleifera L.) leaves. Ann Bot 90:637–645

Still SM, Dirr MA, Gartner JB (1976) Phytotoxic effects of several bark extracts on mung bean and cucumber growth. J Am Soc Hortic Sci 101:34–37

Takahama U, Oniki T (1997) A peroxidase/phenolics/ascorbate system can scavenge hydrogen peroxide in plant cells. Physiol Plant 101:845–852

Tamagnone L, Merida A, Parr A, Mackay S, Culianez-Macia FA, Roberts K, Martin C (1998a) The AmMYB308 and AmMYB330 transcription factors from Antirrhinum regulate phenylpropanoid and lignin biosynthesis in transgenic tobacco. Plant Cell 10:135–154

Tamagnone L, Merida A, Stacey N, Plaskitt K, Parr A, Chang CF, Lynn D, Dow JM, Roberts K, Martin C (1998b) Inhibition of phenolic acid metabolism results in precocious cell death and altered cell morphology in leaves of transgenic tobacco plants. Plant Cell 10(11):1801–1816

Taylor LP, Grotewold E (2005) Flavonoids as developmental regulators. Curr Opin Plant Biol 8:317–323

Taylor GG, Odom RE (1970) Some biochemical compounds associated with rooting of *Carya illinoensis* stem cuttings. J Am Soc Hortic Sci 95:146–151

Thomas AL, Chen Y-C, Rottinghaus GE, Malone AM, Byers PL, Applequist WL, Finn CE (2008) Occurrence of rutin and chlorogenic acid in elderberry leaf, flower, and stem in response to genotype, environment, and season. Acta Hortic 765:197–206

Thompson JE, Legge RL, Barber RF (1987) The role of free radicals in senescence and wounding. New Phytol 105:317–344

Thypyapong P, Hunt MD, Steffens JC (1995) Systemic wound induction of potato (*Solanum tuberosum*) polyphenol oxidase. Phytochemistry 40:673–676

Torras-Claveria L, Jáuregui O, Codina C, Tiburcio AF, Bastida J, Viladomat F (2012) Analysis of phenolic compounds by high-performance liquid chromatography coupled to electrospray ionization tandem mass spectrometry in senescent and water-stressed tobacco. Plant Sci 182:71–78

Tounekti T, Vadel AM, Onate M, Khemira H, Munne-Bosch S (2011) Salt-induced oxidative stress in rosemary plants: damage or protection? Environ Exp Bot 71:298–305

Toyoda S, Suzuki N (1960) Histochemical studies on rice blast lesions caused by Pyricularia oryzae Cav. IV Changes in the activities of oxidases in infected tissues. Ann Phytopathol Soc Jpn 25:172–177

Tsao R, Yang R, Young JC, Zhu H (2003) Polyphenolic profiles in eight apple cultivars using high-performance liquid chromatography (HPLC). J Agric Food Chem 51:6347–6353. 14518966

Turkan I, Demiral T (2009) Recent developments in understanding salinity tolerance. Environ Exp Bot 67:2–6

Ueda M, Nakamura Y (2010) Plant phenolic compounds controlling leaf-movement. In: Santos-Buelga C, Escribano-Bailon MT, Lattanzio V (eds) Recent advances in polyphenol research, vol 2. Wiley-Blackwell, Oxford, pp 226e–237e

Van der Meer IM, Stam ME, Van Tunen AJ, Mol JN, Stuitje AR (1992) Antisense inhibition of flavonoid biosynthesis in petunia anthers results in male sterility. Plant Cell 4:253–262

Van Rossum DI, Schuurmans FP, Gillis M, Muyotcha A, Van Verseveld HW, Stouthamer AH, Boogerd FC (1995) Genetic and phenetic analyses of *Bradyrhizobium* strains nodulating peanut (*Arachis hypogaea* L.) roots. Appl Environ Microbiol 61(4):1599–1609

Varga M (1970) Correlations between the phenol content, polyphenolase, and peroxidase activity and the growth of rice seedlings with particular respect to flooding conditions. II. Riso 19:353–360

Vaughn KC, Duke SO (1981) Tentoxin-induced loss of plastidic polyphenol oxidase. Physiol Plant 53:421–428

Velioglu YS, Mazza G (1991) Characterisation of flavonoids in petals of *Rosa damascena* by HPLC and spectral analysis. J Agric Food Chem 39:463–467

Wahid A, Ghazanfar A (2006) Possible involvement of some secondary metabolites in salt tolerance of sugarcane. J Plant Physiol 163:723–730

Wakabayashi K, Hoson T, Kamisaka S (1997a) Abscisic acid suppresses the increases in cell wall-bound ferulic and diferulic acid levels in dark-grown wheat (*Triticum aestivum* L.) coleoptiles. Plant Cell Physiol 38:811–817

Wakabayashi K, Hoson T, Kamisaka S (1997b) Osmotic stress suppresses cell wall stiffening and the increase in cell wall bound ferulic and diferulic acids in wheat coleoptiles. Plant Physiol 113:967–973

Waldron KW, Ng A, Parker ML, Parr AJ (1997) Ferulic acid dehydrodimers in the cell walls of Beta vulgaris and their possible role in texture. J Sci Food Agric 74:221–228

Wallace G, Fry SC (1994) Phenolic compounds of the plant cell wall. Int Rev Cytol 151:229–267

Wang CY (1982) Physiological and biochemical response of plant to chilling stress. HortScience 17:173–186

Wang H, Cao G, Prior RL (1997) Oxygen radical absorbing capacity of anthocyanins. J Agic Food Chem 45:304–309

Winkel-Shirley B (2001) Flavonoid biosynthesis: a colorful model for genetics, biochemistry, cell biology and biotechnology. Plant Physiol 126:485–493

Wise RR, Naylor AW (1987) Chilling enhanced photooxidation. Plant Physiol 83: 278–282

Wu HC (2006) Improving *in vitro* propagation of *Protea cynaroides* L. (King Protea) and the roles of starch and phenolic compounds in the rooting of cuttings. PhD thesis. University of Pretoria, Pretoria

Wu HC, Du Toit ES, Reinhardt CF (2007a) Micrografting of *Protea cynaroides*. Plant Cell Tissue Organ Cult 89(1):23–28

Wu HC, Du Toit ES, Reinhardt CF, Rimando AM, Van Der Kooy F, Meyer JJM (2007b) The phenolic, 3, 4-dihydroxybenzoic acid, is an endogenous regulator of rooting in *Protea cynaroides*. Plant Growth Regul 52(3):207–215

Xi Y, Cheng D, Zeng X, Cao J, Jiang W (2016) Evidences for chlorogenic acid – a major endogenous polyphenol involved in regulation of ripening and senescence of apple fruit. PLoS One 11(1):e0146940

Yamasaki H, Uefuji H, Sakihama Y (1996) Bleaching of the red anthocyanin induced by superoxide radical. Arch Biochem Biophys 332:183–186

Yang Z, Zheng Y, Cao S (2009) Effect of high oxygen atmosphere storage on quality, antioxidant enzymes, and DPPH-radical scavenging activity of Chinese bayberry fruit. J Agric Food Chem 57:176–181

Ylstra B, Muskens M, van Tunen AT (1996) Flavonols are not essential for fertilization in *Arabidopsis thaliana*. Plant Mol Biol 32:1155–1158

Yuan G, Wang X, Guo R (2010) Effect of salt stress on phenolic compounds, glucosinolates, myrosinase and antioxidant activity in radish sprouts. Food Chem 121:1014–1019

Zenk MH (1996) Heavy metal detoxification in higher plants-a review. Gene 179(1):21–30

Zhang J, Subramanian S, Stacey G, Yu O (2009) Flavones and flavonols play distinct critical roles during nodulation of *Medicago truncatula* by *Sinorhizobium meliloti*. Plant J 57(1):171–183

Zhang Z, Huber DJ, Qu H, Yun Z, Wang H, Huang Z, Huang H, Jiang Y (2015) Enzymatic browning and antioxidant activities in harvested litchi fruit as influenced by apple polyphenols. Food Chem 171:191–199
Zheng Y, Wang CY, Wang SY, Zheng W (2003) Effect of high oxygen atmospheres on blueberry phenolics, anthocyanins, and antioxidant capacity. J Agric Food Chem 51:7162–7169
Zhiqun H, Haig T, Silong W, Sijie H (2002) Autotoxicity of Chinese fir on seed germination and seedling growth. Allelopathy J 9(2):187–193
Zhu JK (2001) Plant salt tolerance. Trends Plant Sci 6:66–71
Zimmerman RH (1984) Rooting apple cultivars *in vitro*: interactions among light, temperature, phloroglucinol and auxin. Plant Cell Tissue Organ Cult 3:301–311
Zornoza P, Vázquez S, Esteban E, Fernández-Pascual M, Carpena R (2002) Cadmium-stress in nodulated white lupin: strategies to avoid toxicity. Plant Physiol Biochem 40(12):1003–1009

Phenolics as Plant Protective Companion Against Abiotic Stress

12

Manoj Kumar, Yamini Tak, Jayashree Potkule, Prince Choyal, Maharishi Tomar, Nand Lal Meena, and Charanjit Kaur

Abstract

Abiotic stress has become a major risk to food security and is predominantly the leading cause of extensive crop and agriculture produce loss worldwide. It has been estimated that about 50% of major agriculture produce is lost due to various abiotic stress factors. Plants perceive risk alarm by virtue of their receptors and activate protective mechanism to sustain against abiotic stresses. These protective mechanisms include accumulation of protective metabolites such as phenolics, terpenes and alkaloids, out of which phenolics play a vital role in the survival of the plant under various abiotic stresses. Enhanced synthesis of phenolics assures the survival, persistence, endurance and competitiveness of the plant against abiotic stress.

Keywords

Abiotic stress · Phenolics · Elicitors · Oxidative stress

M. Kumar (✉) · J. Potkule
ICAR- Central Institute for Research on Cotton Technology, Mumbai, Maharashtra, India

Y. Tak
Agriculture University Kota, Kota, Rajasthan, India

P. Choyal
Directorate of Medicinal and Aromatic Plants Research, Anand, Gujarat, India

M. Tomar
Indian Grassland and Fodder Research Institute, Jhansi, Uttar Pradesh, India

N. L. Meena
National Bureau of Plant Genetic Resources, New Delhi, India

C. Kaur
ICAR- Indian Agricultural Research Institute, New Delhi, India

R. Lone et al. (eds.), *Plant Phenolics in Sustainable Agriculture*,
https://doi.org/10.1007/978-981-15-4890-1_12

12.1 Introduction

Biologically, stress can be defined as a temporary or permanent aberration in the natural development, function and physiology of plants which could be detrimental and has the ability to impose irreversible harm to the plant system. This stress can be classified under two broad categories, viz. biotic and abiotic. Biotic stress occurs as a result of damage done by living organisms like virus, bacteria, fungi and insects, whereas abiotic stress is due to nonliving factors such as temperature, drought, salinity, light and heavy metal toxicity. These factors adversely influence plant development and crop productivity. Abiotic stress has become a major risk to food security and is predominantly the leading cause of extensive crop and agriculture produce loss worldwide (Boyer 1982; Bray et al. 2000). It has been estimated that about 50% of major agriculture produce is lost due to various abiotic stress factors (Boyer 1982; Bray et al. 2000). Every plant species has a peculiar way to acclimate under specific abiotic stress. Various biochemical, molecular and physiological processes align themselves in order to enable the plant to deal with a specific stress condition which in turn determines the growth and yield of the plant (Devi and Prasad 1998). In this chapter, we have comprehensively reviewed how phenolic compounds act as protective companion to various types of abiotic stresses, viz. heavy metal stress, temperature stress, salinity stress, light stress, drought stress and mineral stress.

12.2 Phenolic Compounds and Abiotic Stress

Phenolic compounds are vast group of secondary metabolites and play a crucial role in plant defence mechanism as these compounds possess strong antioxidant properties. Antioxidant properties of these compounds help the plant to scavenge the reactive oxygen species (ROS) generated during the abiotic stress. A large number of researchers have reported that there is an accumulation of phenolic compounds whenever plant faces stress conditions. For example, it was stated by Janas and co-workers (2010) in dark-grown lentil roots that under Cu^{2+} stress, ROS could serve as a common signal for accumulation of total phenolic compounds. In another example, transgenic potato plant with an elevated concentration of flavonoids showed enhanced antioxidant capacity (Lukaszewicz et al. 2004). Further, drought stress, light stress, temperature stress, etc. also showed the accumulation of phenolics and helped the plant to acclimatize in stress condition.

12.3 Abiotic Stress Elicitor

Abiotic elicitation is the signalling process in which phytochemicals or metabolites are synthesized in response to physical or chemical stimulant (Owolabi et al. 2018). Abiotic stimulants or elicitors are the substances of nonliving source which includes physical factors like pH, heavy metals, salt stress, water deficiency, high or low light intensity, high or low temperature and mineral deficiencies (Veersham 2004). These abiotic stress elicitors along with phenolic as protective compounds are detailed below.

12.4 Heavy Metals

Heavy metals are one of the major abiotic stress factors that cause various biochemical and molecular changes within the plants. By definition, heavy metals are a group of elements having a specific weight higher than 5 g/cm^3. These are classified into two major groups, viz. essential and non-essential heavy metals. Essential heavy metals are composed of cobalt, iron, manganese, molybdenum, nickel, zinc and copper, as these are necessary for growth and cofactors of various metabolic enzymes and participate in several electron transport chain reactions within the plant system. The non-essential heavy metal group is composed of lead, cadmium, chromium, mercury, antimony, selenium, etc. that are toxic to the plant system (Rai et al. 2004; Khan et al. 2000; Clemens 2001).

Heavy metals are identified as major inorganic, persistent and non-biodegradable group of elements having severe mutagenic and cytotoxic effects on plants (Rascio and Navari-Izzo 2011; Hall 2002; Farid et al. 2013). Due to the unsupervised and rapid expansion of industries, urban activities, metal-contaminated waste disposal and intensive agricultural practices, a very large section of usable land has come under heavy metal contamination (Zenk 1996; Zornoza et al. 2002). Plants uptake heavy metals through their root systems which are later transported and immobilized inside the cell walls by the apoplastic system (Ciriakova 2009). Based on physiological and chemical properties, the harmful actions of these heavy metals are classified under three broad categories: (a) reactive oxygen species (ROS) generation by autoxidation and Fenton reaction, (b) inactivation of essential metabolic enzymes by blocking their active sites through covalent modifications of the amino acids present in them and (c) metal ion substitution from biomolecules (Schutzendubel et al. 2002). After entering the plant system, they cause inhibition and retardation of physiological processes like photosynthesis, respiration, cell division, cell elongation, nitrogen metabolism, mineral nutrition and plant-water relationship (Ciriakova 2009; Sytar et al. 2013). Toxicity concentration of heavy metal and site of action or stage of action are summarized in Table 12.1. Heavy metals induce the generation of reactive oxygen species (ROS) like hydroxyl free radicals (OH^-), superoxide free radicals (O_2^-) also non-free radicals like singlet oxygen (O_2*) and hydrogen peroxide (H_2O_2) and cytotoxic compounds like methylglyoxal which causes oxidative stress by disturbing the antioxidant homeostasis in the plant cell (Zengin and Munzuroglu 2005; Hossain et al. 2012; Sytar et al. 2013). This disturbance in homeostasis leads to several repercussions like protein and lipid oxidation, ion leakage from cells through damaged membranes and DNA and RNA damage, finally leading to programmed cell death (Farid et al. 2013; Rellan-Alvarez et al. 2006; Hatata and Abdel-Aal 2008; Zengin and Munzuroglu 2005; Sharma et al. 2012; Flora 2009; Valko et al. 2005).

Table 12.1 Major heavy metals along with their toxic effects on plant system

Sr. no.	Heavy metal	Toxicity	Reference
1	Chromium	Toxic to majority of higher plants at 100 μM/kg plant tissue dry weight. It has an indirect influence on soil pH making it both acidic and alkaline. During the early stages of plant development, it inhibits cell elongation and division leading to reduction in dry matter production and yield	Hawley et al. (2004), Nematshahi et al. (2012), Shanker et al. (2005)
2	Aluminium	Toxic at a concentration of about 2–3 μg/g soil at soil pH below 5.5. It hinders root growth and respiration nutrient uptake leading to acute mineral deficiency in the plant. It also causes reduction in stomatal aperture and hampers photosynthetic activity	Ma et al. (2001), Steiner et al. (2012), Vardar and Unal (2007), Bennet et al. (1985)
3	Manganese	It causes decreases in CO2 assimilation and chlorophyll depletion leading to crinkling, chlorosis of leaves and black specs on stem	Schubert (1992), Rezai and Farboodnia (2008), Izaguirre-Mayoral and Sinclair (2005)
4	Nickel	Toxicity ranges from 25 to 246 μg/g dry weight of plant tissue. It has the ability to compete with various cations like Zn^{+2} and Fe^{+2} preventing their absorption, thus causing their deficiency which ultimately results in germination inhibition and seedling growth retardation	Khan and Khan (2010), Aydinalp and Marinova (2009), Sethy and Ghosh (2013)
5	Copper	Excess copper links to the sulfhydryl groups of cell membrane proteins and causes membrane lipid peroxidation resulting in membrane damage and free radical production	Yruela (2009), Doncheva and Stoyanova (2007), Chen et al. (2000)
6	Zinc	Toxicity concentration from 150 to 300 μg/g soil. Excess zinc produces ROS which alters membrane permeability. It hampers photosynthetic efficiency and functionality by reducing the content of various essential photosynthetic pigments and inhibiting photosystems	Hosseini and Poorakbar (2013), Vassilev et al. (2011), Mirshekali et al. (2012), Tsonev and Lidon (2012), Truta et al. (2013)

12.5 Phenolic as a Protective Companion

In order to protect the plant against ROS produced as a result of heavy metal stress, plants employ a non-enzymatic defence mechanism which is composed of phenols and flavonoids. Under copper stress in red cabbage, phenolic compounds served as an effective line of defence (Posmyk et al. 2009). Some phenolic compounds that serve as precursors of flavonoids and lignin have been found to be important antioxidants and can scavenge harmful ROS. Polyphenols also serve as antioxidants by donating their electron to guaiacol-type peroxidases; these enzymes detoxify H_2O_2 produced under heavy metal stress (Sakihama et al. 2002). Phenolic compound accumulation was observed by Shemet and Fedenko (2005) in maize roots under cadmium stress. This induction of phenolic metabolism under heavy metal stress can be due to increase in phenylalanine ammonia lyase activity (Schutzendubel et al. 2001). Similarly, when the roots of *Panax ginseng* were cultured and exposed to copper sulphate, a significant upregulation of enzymes responsible for flavonoids and other phenol biosynthesis was observed (Ali et al. 2006). List of various phenolic compounds and their potential to combat heavy metal toxicity is summarized in the Table 12.2.

There is a satisfactory information that advocate crucial functions of non-enzymatic and enzymatic antioxidants under heavy metal stress. Molecules like flavonoids and phenols constitute the major group of non-enzymatic antioxidants. There is a need to explore various signalling pathways operating under heavy metal stresses, crosstalk between these pathways and involvement of various phytohormones. Numerous molecular switches that are regulated to protect the plant under heavy metal stress need to be explored. Several questions like thiol regulation during stress, gene expression studies of various pathways involved in production of various phenolic compounds and phenolic contents in various crop species that grow well under heavy metal stress are yet to be answered.

12.6 Temperature Stress

Every plant species prefer a particular ambient temperature for their growth and development, and a slight deviation from this temperature optima adversely affects the plant growth potential. This temperature deviation results into the physiological, biochemical and molecular changes within the plants and makes the plants sturdy to maintain the cellular homeostasis and allow growing under such adverse environment. Both high and low temperature create stressful condition for plants to maintain the normal growth and development and referred as temperature stress.

A set of temperature range, which is optimum for one species, can be stressful for others. Therefore, a temperature range below which plant's normal growth and development process become restricted can be referred as low temperature stress. Generally, plant feels cold or chilling stress at temperature ranging from 0 to 15 °C. This cold stress creates either chilling or freezing or both to the plants. Chilling injury happens at temperature above the freezing point of water and can be

Table 12.2 Phenolics as plant protective companion against heavy metals

Plant/crop	Heavy metal elicitor	Phenolic compound	Mode of action/ signalling	References
Glycine max	$HgCl_2$	Glyceollins, trihydroxypterocarpan (THP) precursor of glyceollin	Presence of $HgCl_2$ increased the rate of synthesis and decrease rate of hydrolysis of glyceollins	Moesta and Grisebach (1980), Yoshikawa (1978)
Lupinus albus	$CuCl_2$	Genistein, 2-OH-genistein	$CuCl_2$ leads to an increase in the amounts of genistein and 2′-hydroxygenistein monoprenyls in root tissues, and a major increase in all isoflavonoids was observed in the exudates	Gagnon and Ibrahim (1997)
Medicago sativa	$CuCl_2$	Medicarpin, vestitol, sativan, (medicarpin 3-*O*-glucoside-6'-*O*-malonate (MGM), formononetin 7-*O*-glucoside-6''-*O*-malonate (FGM)	Elevation in the concentration of phenolic compounds was observed both in roots and leaves under $CuCl_2$ presence	Dewick and Martin (1979), Parry et al. (1994)
Pisum sativum	$CuCl_2$	Pisatin	$CuCl_2$ elicits the accumulation of pisatin and other phytoalexins	Nasu et al. (1992)
Trifolium repens	$HgCl_2$	Medicarpin	$HgCl_2$ leads to oxidation of free sulfhydryl groups of some key membrane proteins leading to their structural alteration. This causes Ca^{+2} influx which further results in a significant elicitation of phytoalexins like medicarpin	Devlin and Gustine (1992)
Trifolium pratense	$CuCl_2$	(−)-Maakiain, formononetin	Copper ions trigger defence reactions and cause cell membrane disintegration leading to the release of isoflavonoid phytoalexins	Tebayashi et al. (2001)
Brassica sp.	$CuCl_2$	Indole phytoalexins	$CuCl_2$ induces the indole phytoalexins	Rouxel et al. (1991)
Daucus carota	$CuCl_2$, $HgCl_2$	6-Methoxymellein	Treatment of carrot cells with $HgCl_2$ triggers a hypersensitive reaction leading to their death.	Marinelli et al. (1991),

(continued)

Table 12.2 (continued)

Plant/crop	Heavy metal elicitor	Phenolic compound	Mode of action/ signalling	References
			This causes the release of the endogenous elicitors present in them, which in turn induces the production of phytoalexins like 6-methoxymellein in their neighbouring alive tissues or cells	Guo et al. (1993)
Sorbus aucuparia	$CuCl_2$	Aucuparin	Treatment of Rosaceae leaves by copper ions leads to production of several phytoalexins like aucuparin	Kokubun and Harborne (1994)
Helianthus tuberosus	$CuCl_2$	7-OH-Coumarines	$CuCl_2$ induces the release of endogenous elicitors following cellular damage	Cabello-Hurtado et al. (1998)
Oryza sativa	$CuCl_2$	Sakuranetin, volatiles	Copper increases the levels of various plant secondary metabolites like sakuranetin through induction of jasmonic acid	Rakwal et al. (1996), Obara et al. (2002)
Brassica oleracea L. var. *rubrum*	$CuSO_4$	Anthocyanin, sinapoyl esters	There was an enhanced accumulation of anthocyanins and sinapoyl esters in red cabbage seedlings under Cu^{+2} stress	Posmyk et al. (2009)
Lens culinaris	$CuSO_4$	Flavonoids, hydroxycinnamic esters (ferulic acid and p-coumaric acid)	A higher accumulation of these phenolics was found in vacuoles and cell walls and may be involved in the scavenging of ROS produced due to copper stress	Janas et al. (2010)
Matricaria chamomilla	$NiCl_2$	Soluble phenolics and flavonoid	Enhancement of phenolic metabolite accumulation was due to high activities of the enzymes like phenylalanine ammonia lyase and shikimate dehydrogenase which are the pivotal enzymes in phenylpropanoid metabolism	Kovacik et al. (2009)

(continued)

Table 12.2 (continued)

Plant/crop	Heavy metal elicitor	Phenolic compound	Mode of action/ signalling	References
Nymphaea	Cd $(NO_3)_2$, CrO_3, Pb $(NO_3)_2$, $HgCl_2$	Polyphenols mainly hydrolysable tannins, tannic acid derivatives and gallic acid	Under heavy metal stress, polyphenols accumulate in the epidermal glands of *Nymphaea* leaf lamina. Polyphenols lead to sequestration of heavy metals in the epidermal glands	Lavid et al. (2001)
Panax ginseng	$CuSO_4$	Flavonoids	Cu^{+2} induces the synthesis of polyphenols by increasing the activities of four major enzymes in polyphenol biosynthetic pathway	Ali et al. (2006)
Vaccinium myrtillus L.	Zn, Pb, Cd	Flavones, flavonols and anthocyanins	The protective mechanisms of these phenolics are based on their antioxidant property which allows them to scavenge free radicals produced due to metal ions and their metal chelating ability	Bialonska et al. (2007)
Erica andevalensis	$CdSO_4$	Cinnamic acid derivate	Cadmium toxicity leads to increase in the levels of total phenolics, flavonoid compounds and total antioxidant capacity	Marquez-Garcia et al. (2012)
		Epicatechin		
		Rutin		
Brassica juncea	$HgCl_2$	Total phenols	There was an enhancement in total phenols due to heavy metal stress. They also serve as metal chelators with ROS scavengers	Singh and Malik (2017)
Zea mays	$CuSO_4$, Pb $(NO_3)_2$, $CdCl_2$	Cinnamic acid, chlorogenic acid, ferulic acid, caffeic acid, vanillic acid and rutin	Concentration of all the phenolic compounds increased under Cd, Cu and Pb toxicity, thus helping the plant to overcome stress	Kisa et al. (2016)

reversible, but prolonged exposure can lead to cell death. Freezing injury occurs at a temperature below freezing point of water. Intracellular ice crystals forms, which grow and finally resulted into cell disruption. Ice crystal forms in the apoplastic region of the cell. These ice crystals create a water potential difference between apoplastic region and cytoplasmic region. As a result, water moves from cytoplasm to apoplastic region and adds to existing crystals and thus creates big ice crystal which ultimately ruptures the plant cell (Olien and Smith 1997) and ultimately leads to cell death.

Temperature above the optimum range creates stressful condition for plant to survive and grow. It was estimated that there may be approximately 17% yield losses for every degree increase in temperature above the average growing season temperature (Lobell and Asner 2003). Heat stress results into decreased seed germination, reduction in photosynthesis, changes in plant phenology, oxidative stress, poor quality of seeds and ultimately reduced crop yield. High temperature decreases enzyme activity and less functional proteins. At higher temperature, plant's photosynthetic machinery gets inhibited resulting into the production of reactive oxygen species, which further damage the plant structures.

Both high and low temperature generates cellular reactive oxygen species (ROS), and these ROS damage photosynthetic machinery, cellular compartments, etc. and damage the cells (Asada 2006; Hasanuzzaman et al. 2013). Moreover, temperature stress abolishes water potential gradients and creates dehydration stress. To combat with dehydration stress generated by temperature, plant accumulates primary metabolites as osmoprotectants such as proline, glycine betaine and soluble sugars which maintain cell water balance and provide the buffering capacity to cell redox potential (Sakamoto and Murata 2000). Plants also activate the antioxidative machinery against ROS. This machinery includes various antioxidative enzymes and antioxidants, which scavenge the ROS and provide defence against them (Balla et al. 2009).

During the stress condition, to balance the energy, plants shift their primary metabolism towards secondary metabolism and synthesize some high-molecular-weight secondary plant products, known as secondary metabolites (Selmar and Kleinwächter 2013). These secondary metabolites include terpenes, phenolics and alkaloids. These secondary metabolites get accumulated during the temperature stresses and provide the stress tolerance capacity to plants.

12.7 Phenolic as Protective Companion

Plant phenolics are the secondary compounds that contain a phenol group and mainly synthesized by shikimic acid pathway. Flavonoids, lignins, tannins, anthocyanins, etc. are few broad classes of phenolics. Temperature stress acts as elicitor for the increased synthesis of phenolics and imparts tolerance to cold stress and also increases the nutraceutical value in fruit crops (Rana and Bhushan 2016) (Table 12.3). During cold stress, these phenolics get accumulated in plants and subsequently incorporated in cell wall either as lignin or as suberin and protect

Table 12.3 Phenolics as plant protective companion against temperature stress

Plant/crop	Temperature elicitor	Phenolic compound	Mode of action/ signalling (flow diagram can be given)	Reference
Tomato (*Lycopersicon esculentum*) and watermelon (*Citrullus lanatus*)	Cold and heat stress	Total soluble phenols	Total soluble phenols get accumulated under temperature stresses through the induction of their biosynthesis and their reduced oxidation	Rivero et al. (2001)
Grapevine roots (*Vitis vinifera* L.)	Low-temperature stress followed by recovery period	*p*-Coumaric acid, caffeic acid and ferulic acid	The content of the phenolics gets decreased during the chilling period but significantly increased after the recovery from the stress	Weidner et al. (2009)
Maize (*Zea mays* L.)	Low-temperature stress	Phenylpropanoid and anthocyanin	Leaf sheath of maize seedling showed increased production of anthocyanin under short-term low-temperature treatment. This increase was correlated with transcript abundance of PAL and chalcone synthase	Christie et al. (1994)
Apple fruits (*Malus domestica*)	Cold stress	Catechin, epicatechin, procyanidins, chlorogenic acid, *p*-coumaroylquinic acid, quercitrin, etc.	Apple peel and pulp behaved differently in terms of phenolics accumulation under cold storage condition. Phenolic content increased in peel under cold storage condition	Perez-Ilzarbe et al. (1997)
Tomato (*Solanum lycopersicon*)	Heat stress	Kaempferol, 3-O-glucoside, naringenin, naringenin chalcone, quercetin-3-hexoside	Phenolic compounds get accumulated under heat stress and help the plant to protect from oxidative damage	Martinez et al. (2016)

(continued)

Table 12.3 (continued)

Plant/crop	Temperature elicitor	Phenolic compound	Mode of action/ signalling (flow diagram can be given)	Reference
Carrot (*Daucus carota* L.)	Heat stress	Coumaric acid, caffeic acid and anthocyanins	The phenolic metabolites protected microfilaments cytoskeletons from reactive oxygen species generated during heat episodes, and plants showed less heat damage under carrot cell culture	Commisso et al. (2016)
Grapevine (*Vitis vinifera*)	Low-temperature stress	Gallic acid, caffeic acid and ferulic acid	Low-temperature stress reduced the total phenolics, but some individual phenolics like gallic acid, caffeic acid and ferulic acid were increased under the stress, and caffeic acid was the dominant phenolic compound elicited during the low-temperature stress	Amarowicz et al. (2010)
Sugarcane (*Saccharum* spp. hybrid)	Low-temperature stress	Total phenols	Low-temperature stress increases the total phenols in buds of sugarcane stubbles. High total phenols along with IAA increase the toxicity level in buds which results into the increased dormancy of bud	Jain et al. (2007)
Kale (*Brassica oleracea* var. *acephala*)	Short-term low temperature	Total phenols, caffeic acid, ferulic acid, and kaempferol	Short-term low-temperature treatment followed by recovery in a controlled environment resulted into	Lee and Oh (2015)

(continued)

Table 12.3 (continued)

Plant/crop	Temperature elicitor	Phenolic compound	Mode of action/ signalling (flow diagram can be given)	Reference
			increase in total as well as individual phenolic compounds in kale	
Red orange (*Citrus sinensis* L.)	Low temperature	Anthocyanins	Long-term low-temperature storage (4 °C) induced the accumulation of anthocyanins in juice vesicles of red oranges. This increase was correlated with the increased expression of PAL, CHS, DFR, etc. phenolic biosynthetic genes	Lo Piero et al. (2005)
Papaya (*Carica papaya* L.)	Low temperature	Ferulic acid and caffeic acid	Papaya, stored at 1 °C, had more content of ferulic acid and caffeic acid compared to storage at 25 °C, while the total carotenoids were decreased under low-temperature storage	Rivera-Pastrana et al. (2010)
Soybean (*Glycine max* L.)	Low temperature	Total phenolic acids, genistein, daidzein and genistin	Low-temperature treatment increases the total phenolic content in the roots of soybean. This increase was attributed to availability of preformed conjugates of phenylpropanoids	Janas et al. (2002)
Sweet potato (*Ipomoea batatas* L.)	Low temperature	Chlorogenic acid, 3,5-dicaffeoylquinic acid	Sweet potato roots accumulated more phenolic compounds when stored at 5 °C for 4 weeks and	Padda and Picha (2008)

(continued)

Table 12.3 (continued)

Plant/crop	Temperature elicitor	Phenolic compound	Mode of action/ signalling (flow diagram can be given)	Reference
			subsequently exposed to ambient temperature. The highest total phenolic content 7.55 g kg^{-1} was observed in the root periderm	
Lettuce (*Lactuca sativa* L.)	Heat shock and chilling stress	Chicoric acid and chlorogenic acid, quercetin-3-O-glucoside and luteolin-7-O-glucoside	Mild heat and chilling stress increased the phenolic content of lettuce. Chilling stress induces the synthesis of PAL, L-GalDH and g-TMT but not the heat stress, while GalDH was consistently increased under both the stresses	Oh et al. (2009)
Petunia × *hybrida*	Cold stress	Gentisic acid and total phenols	Cold pretreatment-induced accumulation of gentisic acid, total phenolics and antioxidants resulted into the increased tolerance to low-temperature stress in petunia	Pennycooke et al. (2005)
Winter oilseed rape (*Brassica napus*)	Low temperature	Soluble phenolic acids (caffeic, *p*-coumaric, ferulic and sinapic acids) and anthocyanins	Brief freezing followed by low-temperature treatment resulted into the accumulation of total soluble phenolics and anthocyanins in rape leaves, and the accumulation was associated with the esterification of phenolics under cold stress	Solecka et al. (1999)

(continued)

Table 12.3 (continued)

Plant/crop	Temperature elicitor	Phenolic compound	Mode of action/ signalling (flow diagram can be given)	Reference
Fairy primrose (*Primula malacoides*)	Freezing stress	Farinose flavonoids	Accumulation of farinose flavonoids during freezing stress in primula leaves protects the plants through inhibition of ice crystallization	Isshiki et al. (2014)

from cellular injury and increase stress tolerance potential (Chalker-Scott and Fuchigami 2018). This accumulation of phenolics in the plant tissue is attributed to the increased activity of phenylalanine ammonia lyase (Oh et al. 2009; Lo Piero et al. 2005; Christie et al. 1994). Cold-induced phenolics accumulations are able to decrease the freezing point of cell or tissues and thus maintain the water potential gradient in the cell and protect from cell disruption (Kasuga et al. 2008). Isshiki et al. (2014) shown that accumulation of farinose flavonoids on aerial part of primula during the freezing stress resulted into higher stress tolerance ability of plant through inhibiting the ice crystallization. It was suggested that deposition of suberin and lignin in cell wall acts as barrier for water and thus maintains the water potential of the tissue (Griffith et al. 1985; Paroschy et al. 1980). Further, Bartolo and Wallner (1986) emphasized the role of cell wall-associated phenolic in cold stress tolerance. They demonstrated that change in the composition and content of phenolic in cell wall could increase the adhesion of cell membrane to the cell wall and, thus, reduces the membrane disruption during the freezing stress. When the roots of grapevine were given the cold treatment, the root extract showed increased amount of tannins and other soluble phenols after the recovery period (Weidner et al. 2009). Rivero et al. (2001) reported that both heat and cold stress act as elicitor for the increased accumulation of total soluble phenolics in watermelon and tomato. It increased the biosynthesis of phenolics and reduced their oxidation. Perez-Ilzarbe et al. (1997) showed that cold storage of apple could result into the higher chlorogenic acid content in the leaf, while the pulp phenolic content decreased under the storage. In carrot cell culture, the biosynthesis of different phenolics like coumaric, caffeic acid and anthocyanins gets stimulated during the heat episodes. These phenolics protected the microfilaments cytoskeletons from reactive oxygen species (Commisso et al. 2016). Amarowicz et al. (2010) reported that the total phenolic content was decreased under low-temperature stress in grapevine but the gallic acid, caffeic acid and ferulic acid were increased significantly, which shows their prime role in the cold tolerance capacity as compared to other phenolics. A few examples reporting the role of phenolics in temperature stress tolerance are given in Table 12.3.

Temperature stress, both heat and cold stress, functions as inducers for the synthesis and accumulation of phenolics in plants. These phenolics function as stress protectant secondary metabolites and provide additional capacity to plant to tolerate the extreme temperature.

12.8 Salinity Stress

Growth and productivity of plants are mostly affected by salt stress conditions than plants propagated under favourable environment. Research on salt stress proved that accumulation of phenols, polyphenols, flavonoids, anthocyanin, phenolic acids, lignin and other secondary metabolites occurs in salt stress condition in diversity of plants. Deliberately applying abiotic stress like salinity on plants increases the quality and ability of the plant to survive in the stressed condition. This will finally result in enhancing the yield of the crops in stressed condition (Khan et al. 2011). Abiotic and environmental factors such as salinity stress cause limiting plant growth and productivity (Parida and Das 2005). Salt-induced osmotic stress is responsible for the formation of reactive oxygen species (ROS) which causes oxidative stress in plants (Saiema et al. 2013). Salt stress leads to production of ROS in plant cell and results in harmful effects on cell integrity. Salinity stress causes various physiological changes, such as nutrient imbalance, interruption of membranes, differences in the antioxidant enzymes, formation of ROS and decrease in stomatal aperture (Rahnama et al. 2010). ROS formation causes oxidative damages in various cellular components such as ion level, antioxidative enzymes and antioxidants, lipids, proteins and nucleic acids.

Salt stress negatively affects physiological and metabolic processes like photosynthetic pathway, photosynthetic pigments such as chlorophyll and total carotenoid contents and soluble protein. Other processes, viz. respiration, glycolysis pathway and nitrogen fixation, disrupt the electron transport system in chloroplasts, and mitochondria are affected during salinity stress. All these effects lead to decreased crop productivity and economic losses to farmers. Further, increase in the salt concentration in the rhizosphere of the plants leads to reduced relative water content, leaf water potential, transpiration rate, ion uptake, water uptake, water use efficiency and water retention (Fig. 12.1) (Parida and Das 2005; Gupta and Huang 2014; Sharma et al. 2005; Zhang et al. 2010).

Salinity stress causes drastic reduction in the productivity of various crops. The effect can be measured in terms of reduced photosynthetic efficiency due to accumulation of salt in the chloroplast of the plant cell. Deteriorative effects of salt can also be demarked by degradation of chlorophyll in crop such as potato, pea, tomato and *Phaseolus vulgaris* (Sudhir and Murthy 2004; Khaled et al. 2016). Figure 12.2 shows salt stress is responsible for reducing the crop productivity by damaging the photosynthetic machinery.

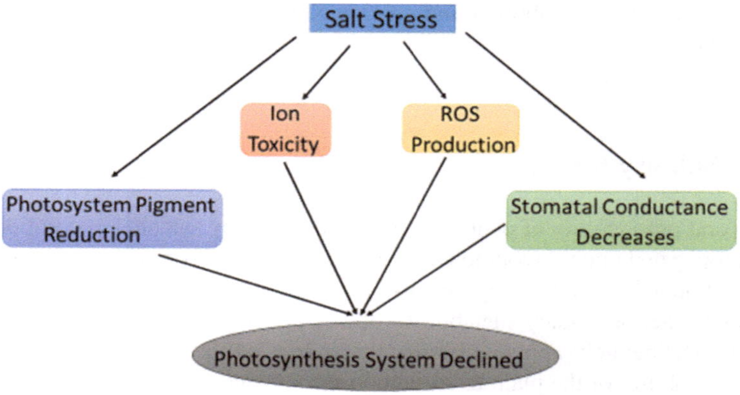

Fig. 12.1 Harmful effects of salt stress on plant at cellular level. (Modified from Farooq et al. 2015)

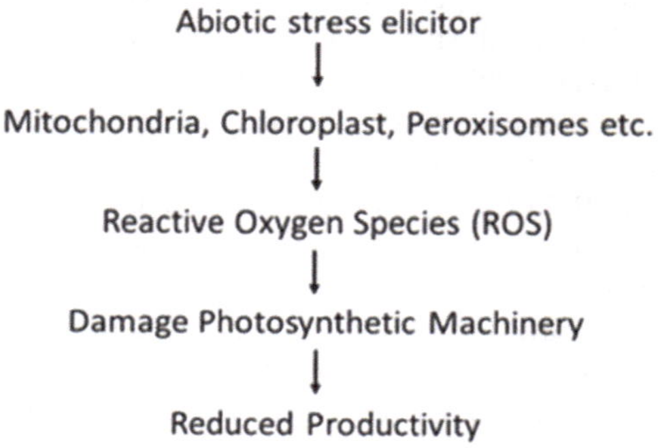

Fig. 12.2 Salt stress affecting the productivity of crops/plants

12.8.1 Phenolic as Protective Companion Against Salinity Stress

Plant phenols, phenolic acids and polyphenols such as flavonoids, lignin and tannins are secondary metabolites arising from malonic acid pathway and shikimic acid pathway/phenylpropenoid pathway. Polyphenols and monomeric and polymeric phenols fulfil a very broad range of metabolic and physiological roles in plants. In case of abiotic stress, condition like salt stress disturbs either the enzymes/genes involved or whole secondary metabolite synthesis pathway, and accumulation of secondary metabolites such as phenols, polyphenol, flavonoids, phenolic acids, tannin and lignin occurs in high amounts which result in decrease in salt stress and osmotic stress and increase in growth and development and survival of plants (Lattanzio 2013).

Salt stress causes oxidative stress in plants. Hence, plants have evolved different defence systems against oxidative damage caused by salinity including antioxidant metabolites which stop oxidative chain reaction. In this case, phenolic compounds

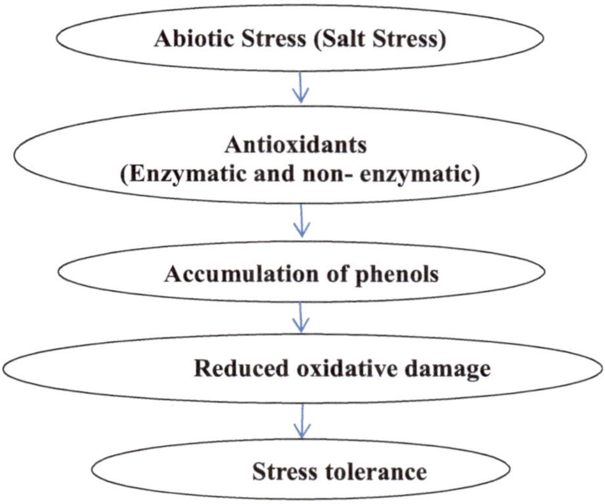

Fig. 12.3 Phenolic salinity stress tolerance system

like polyphenols, anthocyanins, flavonoids, proanthocyanidins, phenolic diterpene and phenolic acid play an important role in quenching singlet oxygen, absorbing and neutralizing free radicals, decomposing peroxides and reducing the detrimental effect of salinity shown in Fig. 12.3. Phenolic compounds protect plants from biotic attacks by pathogens and herbivores (Hichem et al. 2009; Valifard et al. 2014; Omer et al. 2017). Plants such as buckwheat and barley under salt stress have confirmed positive changes occur in phenolic components content. Plant species such as Jerba and Tabarka with accumulation of polyphenols significantly increase by 56% and 30% in response to 100 Mm and 400 mM NaCl treatment repectively. Phenolic compound in *F. esculentum* increased up to 57%, 121% and 153% as compared to control. Four major compounds rutin, vitexin, isoorientin and orientin cause accumulation of phenolic compounds in plants facing salinity stress (Yang et al. 2018).

Salicylic acid is an endogenous growth regulator/signalling molecule and belongs to a group of phenolic compound that plays very important role in plant defence mechanism during stress conditions shown in Fig. 12.4. Many crops including *Vicia faba, Brassica juncea, Medicago sativa and V. radiata* are examples of salicylic acid's role in strengthening salinity tolerance mechanism (Jini and Joseph 2017; Khan et al. 2015). Various examples of the phenolic compounds acting as salinity stress tolerant are presented in Table 12.4.

Salinity is a major abiotic factor that negatively affects growth and development of plants. Salt stress affect on cellular activities of plants leads to decreased crop productivity and economic losses to farmer. Recent research on salt stress proved that phenolic compounds such as phenols, polyphenols, flavonoids, anthocyanin and phenolic acids occur in salt stress condition in diversity of plants. Phenolic compounds protect plants from salt stress through different course, including enzymatic and non-enzymatic antioxidant activity, detoxification of reactive oxygen species, physiological and metabolic process regulation and maintaining cell integrity. Phenols like salicylic acid play a key role in defence system of plant/signalling

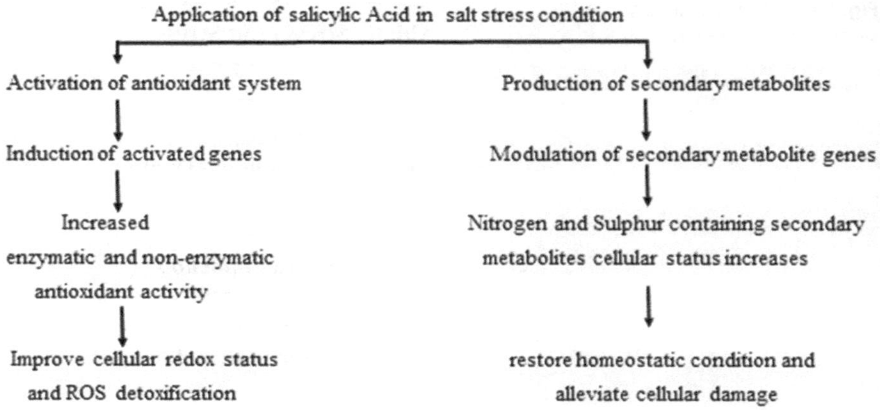

Fig. 12.4 Salicylic acid role in stress plants. (Adopted and modified from Khan et al. 2015)

mechanism. Phenols also act as an osmoprotectant that helps maintain water-soil balance in plant.

12.8.2 Light Stress

Higher plants require visible solar radiation to provide energy for photosynthetic mechanism. During light harvesting, the photosynthetic system is inevitably exposed to relatively high doses of harmful solar radiation/UV radiation. Plants adapt a number of self-sustaining mechanisms such as accumulation of phenolic compounds which enhances antioxidant activity of plants and results in growth and development of plants/crops in reaction to increased intensity of light radiation, gamma rays, all types of UV radiation, blue light and red and far-red light. High light intensity causes enzymatic/non-enzymatic changes in secondary metabolite biosynthesis pathway. Release and accumulation of various secondary metabolites including phenolic compounds, phenolic acids, anthocyanin, triterpenoids and flavonoids have high utilization and economic value due to the well-known antioxidant property. Light plays an irreplaceable role in promoting plant growth and development as well as plays critical role in biosynthesis of both primary and secondary metabolites in plants (Xu et al. 2014; Koyama et al. 2012).

Solar UV radiation exposure may lead to photoinduced DNA and chlorophyll degradation. Growth of plants is inhibited by disrupting different reactions of photosynthetic system. These radiations affect electron transport chain, reaction centre of photosystem II, light-harvesting complex and acceptor and donor site of photosystem II. Gamma rays resulted in dissolution of the pectin through increased activity of pectinmethylesterase and polygalacturonase. Very low light intensity at the critical stage for bud abscission can increase the chances of flower bud abortion (Vantuyl et al. 1985; Gill et al. 2015).

Table 12.4 Phenolic compounds as plant protective companion against salt stress

Sr. no.	Plant/crop	Phenolic compound	Mode of action/signalling	References
1.	*Colubrina asiatica*	Polyphenols, flavonoid	Enzymatic and non-enzymatic antioxidant defence system	Sonar et al. (2011)
2.	*Hypericum pruinatum*	Chlorogenic acid, isoquercetin, hyperoside, quercitrin, rutin and quercetin	Non-enzymatic antioxidant defence system accumulates phenolic compounds in plant	Omer et al. (2017)
3.	Bean (*Phaseolus vulgaris*)	Flavonoid, phenols	Non-enzymatic antioxidant defence system increases flavonoid and total phenol content	Arkadiusz et al. (2008)
4.	*Vicia faba* L.	α-Tocopherol	α-Tocopherol acts as an antioxidant	Semid et al. (2014)
5.	Strawberry (*Fragaria ananassa* Duch.)	Anthocyanins, gallic acid, catechin, caffeic acid, *p*-coumaric acid, ellagic acid, *p*-hydroxycinnamic acid	Enzymatic activity such as oxidase, polyphenol oxidase and Phenylalanine ammonia lyase and Non-enzymatic activity increase the phenolic compound content in strawberry plant	Galli et al. (2016)
6.	Indica rice varieties (*Oryza sativa* L. spp. indica cv.)	Flavonoids	Accumulation of compatible solute (proline) which acts as osmoprotectant, flavonoid acts as an antioxidant which results in growth and development of plant	Chutipaijit et al. (2009)
7.	Rosemary (*Rosmarinus officinalis* L.)	Monoterpene, phenolic diterpene, carnosic acid, carnosol, rosmarinic acid	Accumulation of phenolic compounds by adding specific fertilizers in salt stress condition	Tounekti et al. (2011)
8.	Safflower (*Carthamus tinctorius* L.)	Salicylic acid, carotenoid, flavonoids, anthocyanin	Accumulation of compatible solutes and salicylic acid plays a key role in defence system of plant/signalling mechanism	Fatemeh et al. (2018)
9.	Garlic (*Allium sativum* L.)	Total phenolic content	Effect of selenium on enzymatic antioxidants which increases total phenolic compounds	Rozita et al. (2018)

(continued)

Table 12.4 (continued)

Sr. no.	Plant/crop	Phenolic compound	Mode of action/signalling	References
10.	*Nigella sativa*	Vanillic acid, *p*-hydroxybenzoic acid, catechin, syringic acid, epicatechin, amentoflavone	Antioxidant activity leads to salinity tolerance and plant's growth	Bourgou et al. (2010)
11.	Barley grains (*Hordeum vulgare*)	Vanillic acid, *p*-hydroxybenzoic acid, protocatechuic acid, gallic acid, caffeic acid, *p*-coumaric acid, syringic acid, ferulic acid, sinapic acid	GABA accumulates phenolic compounds by regulating gene expression of enzymes involve in phenylpropanoid pathway as well as enhances antioxidant activity	Yan et al. (2019)
12.	*Salvia mirzayanii*	Total phenolic content	Accumulation of phenolic compound and antioxidant activity	Valifard et al. (2014)
13.	Rice (*Oryza sativa* L.)	Salicylic acid	Salicylic acid salt tolerance activity	Jini and Joseph (2017)
14.	Thai rice lines	Anthocyanin, cyanidin-3-glucoside	Antioxidant activity increases total phenolic content in plant and lipid peroxidation activity, and anthocyanin cyanidin-3-glucoside helps to overcome salt stress	Daiponmak et al. (2010)
15.	Artichoke (*Cynara scolymus*)	Phenols, flavonoids, caffeic and chlorogenic acids	Accumulation of phenolic compounds and antioxidant activity	Rezazadeh et al. (2012)
16.	Onion (*Allium cepa* L.)	Phenols and flavonoids	α-Tocopherol and α-tocopherol + KH_2PO_4 treatment	Mohamed and Amina (2008)
17.	Rapeseed (*Brassica napus*)	Total phenolic, non-flavonoids, tannins, phenolic acids	Accumulation of phenolic compounds and antioxidant activity	Falcinelli et al. (2017)
18.	Scented indica rice varieties	Polyphenols	Accumulation of polyphenols increases growth of plant	Danai-Tambhale et al. (2011)
19.	Rice (*Oryza sativa* L.)	Vanillin, cinnamic acid, protocatechuic acid, ferulic acid, and *p*-coumaric acid	Accumulation of phenolic compounds helps to overcome plant stress	Minh et al. (2016)
20.	*Catharanthus roseus*	Salicylic acid	Reactive oxygen species defence mechanism, enzymatic antioxidant activity and salicylic acid defence mechanism	Misra et al. (2014)

(continued)

Table 12.4 (continued)

Sr. no.	Plant/crop	Phenolic compound	Mode of action/signalling	References
21	Fennel (*Foeniculum vulgare* Mill.)	Polyphenols	Antioxidant activity leads to salinity stress tolerance	Rebey et al. (2017)
22	Canola (*Brassica napus* var. *oleifera* Del.)	Glutathione and polyadenylic acid	Oxidative defence system	Kattab (2007)
23	*Salvia macrosiphon*	Vanillin and protocatechuic acid	Accumulation of phenols and antioxidant activity	Valifard et al. (2017)
24	Honeysuckle (*Lonicera japonica* Thunb.)	Total phenolic compound	Accumulation of total phenols helps to overcome plant stress	Yan et al. (2017)

12.9 Phenolics as Plant Protective Companion Against Light Stress

Light is a crucial factor which stimulates the plant to produce secondary metabolites. Light can stimulate the production of secondary metabolites like gingerol and zingiberene in the callus culture of *Zingiber officinale* (Anasori and Asghari 2008). Ultraviolet (UV) radiation stimulates the production of phenolic compounds (Table 12.5). UV-B exposure to field-grown plants leads to an increase in yield of essential oils and phenolic compounds and a decrease in amount of toxic beta-asarone (Kumari et al. 2009). Plants/crops produce phenolics due to the exposure to all types of light stresses, and phenolics protect them from ill effects of light intensity/stress. Flavonoids and hydroxycinnamic acids found in fruits and berries determine the aroma, colour, astringency and antioxidant properties (He and Giusti 2010). The effect of light on anthocyanin synthesis modify endogenous signals, environmental factors, developmental stages and previous light exposure. Pollination and seed dispersal are visual functions and protect tissues against photoinhibition; these are general functions of anthocyanin in plants. Flavonoids and anthocyanin protect cells and tissues from exposure of UV-B irradiation in many plants (Steyn et al. 2002). Few examples of the phenolics that can act as plant protective companion against light stress are given in Table 12.5.

12.10 Drought Stress

It can be well comprehended from the above sections that abiotic stresses such as temperature, heavy metals, light and salinity cause an accumulation of the phenolic compounds and aid the plants for their survival in such adverse situations. Similarly,

Table 12.5 Phenolics as plant protective companion against light stress

Sr. no.	Plant/crop	Type of light stress (high intensity/ irradiance or UV light)	Phenolic compound	Mode of action/ signalling	References
1.	Young ginger varieties (*Zingiber officinale*)	Different light intensities	Total phenolic and flavonoids	Accumulation of phenols and flavonoids enhances medicinal components and antioxidant activities	Ghasemzadeh et al. (2010)
2.	Cucumber (Cucumis sativus L.)	UV-B irradiation	Total phenolic compound	Enzymatic and non-enzymatic activity enhance light tolerance	Kondo and Kawashima (2000)
3.	Ginkgo (*Ginkgo biloba* L.)	100% light intensity	Flavonoids	Accumulation of flavonoids enhances growth and development of plant	Xu et al. (2014)
4.	Immature radish microgreen (*Raphanus raphanistrum*)	Different light spectrum (white and blue UV-A and darkness)	Anthocyanin	Accumulation of anthocyanin content enhances antioxidative activity of immature radish microgreen which leads to growth and development of plant	Zhang et al. (2019)
5.	Red wine grapes (*Vitis vinifera* cv.)	UV and visible light	Flavonoid, proanthocyanidin and flavonols	Accumulation of phenolic compounds and photoprotection activity	Koyama et al. (2012)
6.	Bilberry (*Vaccinium myrtillus* L.)	Red light, long and short light intensity	Hydroxycinnamic acids, flavanols and quinic acid	Accumulation of phenolic compound increases the level of antioxidant activity	Uleberg et al. (2012)

(continued)

Table 12.5 (continued)

Sr. no.	Plant/crop	Type of light stress (high intensity/ irradiance or UV light)	Phenolic compound	Mode of action/ signalling	References
7.	Buckwheat (*Fagopyrum esculentum and F. tataricum*)	UV-B	Flavonoids and Anthocyanin	Accumulation of flavonoids and anthocyanin leads to stress tolerance	Regvar et al. (2012)
8.	Basil (*Ocimum basilicum*), arugula (*Eruca sativa*) and bloody dock (*Rumex sanguineus*)	Blue and blue-violet light	Flavonoids and phenolic acids	Accumulation of phenolic compound leads to light stress tolerance	Taulavuori et al. (2018)

under drought stress, plants accumulate phenolic compound for their survival. It was observed by Larson (1988) in willow leaves that drought conditions often cause oxidative stress in the plant system and also lead to an increase in the amounts of phenolic acids and flavonoids. An increased accumulation of anthocyanins was reported under drought stress and at cold temperatures. Plant tissues having anthocyanins are usually resistant to drought (Chalker-Scott 1999). For example, a purple cultivar of chilli can more effectively tolerate water stress than a green cultivar (Bahler et al. 1991). Flavonoids have protective role during drought stress. Del Moral (1972) and Kirakosyan et al. (2004) reported that there was ten- and sixfold enhancement in chlorogenic acid in case of Helianthus annuus and Crataegus spp., respectively, under drought stress. Nogues et al. (1998), De Abreu and Mazzafera (2005); Hernandez et al. (2006), Liu et al. (2011) and Gray et al. (2003) also reported the accumulation of phenolic compounds under drought stress.

12.11 Mineral Deficiency

Deficiency of specific mineral nutrient in plant system results in poor growth and development of the plant/crop. In the current chapter, it is studied in depth that abiotic environmental stresses can induce phenolic compound biosynthesis and accumulation in plants/crops. Mineral deficiency is also said to be an abiotic stress to the plants. Higher level of phenolic compounds that is accumulated as induced by nitrogen deficiency was found in many plants, such as *Potamogeton amplifolius*, *Nuphar advena* and *Arabidopsis* (Cronin and Lodge 2003; Lillo et al. 2008). Kovacik and Klejdus (2014) reported phenolic compounds are accumulated in the plant cell in nitrogen-deficient conditions. Co-deprivation of photosynthetic electron transport mineral pairs significantly enhanced phenolic compounds in *Arthrospira (Spirulina) platensis* (Hifney et al. 2019). Hence, it can be said that mineral deficiencies lead to the accumulation of the phenolics in the plants and lead to improved survival of the plants in mineral deficiencies.

12.12 Conclusion

From the above discussion in various section, we can conclude that abiotic stresses, viz. heavy metals, temperature, light, salinity, drought and mineral deficiencies, lead to accumulation of phenolic compounds and can support plant to fight against abiotic stress. All abiotic stresses lead to synthesis and accumulation of phenolics and provide additional capacity to plant to tolerate the extreme environmental conditions. There is a convincing evidence reported in the chapter that suggest the important roles of non-enzymatic and enzymatic antioxidants under various abiotic stresses in plant. Schematic diagram of interrelationship between abiotic stress elicitor and phenolics as stress companions to plants can be seen in Fig. 12.5.

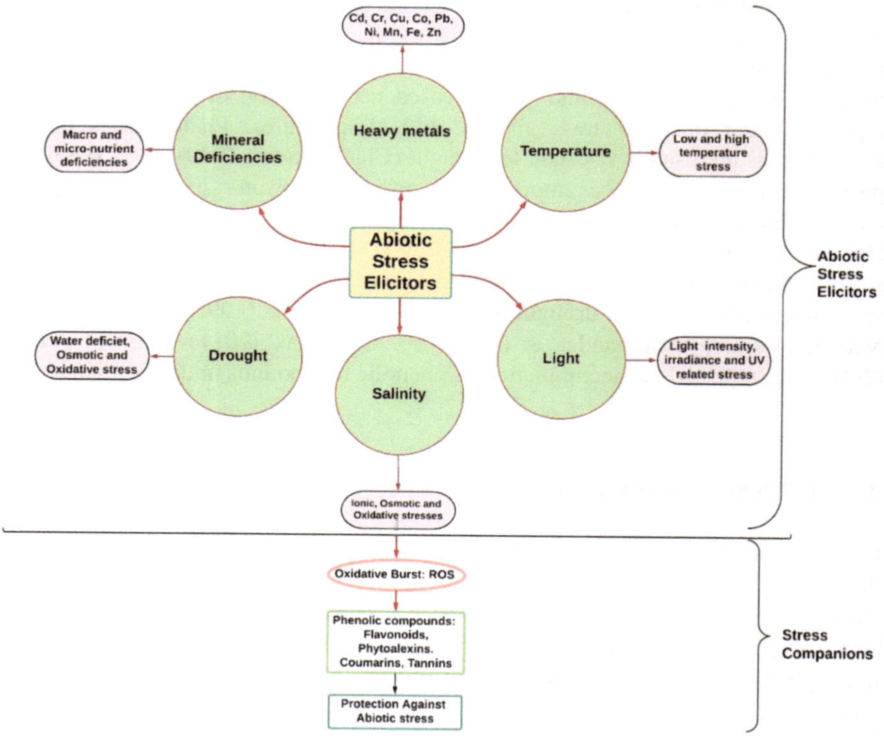

Fig. 12.5 Diagram showing abiotic stress elicitors and downstream stress companions (phenolic compounds) of plants

References

Ali MB, Singh N, Shohael AM, Hahn EJ, Paek KY (2006) Phenolics metabolism and lignin synthesis in root suspension cultures of *Panax ginseng* in response to copper stress. Plant Sci 171(1):147–154

Amarowicz R, Weidner S, Wójtowicz I, Karmac M, Kosinska A, Rybarczyk A (2010) Influence of low-temperature stress on changes in the composition of grapevine leaf phenolic compounds and their antioxidant properties. Funct Plant Sci Biotechnol 4:90–96

Anasori P, Asghari G (2008) Effects of light and differentiation on gingerol and zingiberene production in callus culture of *Zingiber officinale* Rosc. Res Pharm Sci 3:59–63

Asada K (2006) Production and scavenging of reactive oxygen species in chloroplasts and their functions. Plant Physiol 141(2):391–396

Aydinalp C, Marinova S (2009) The effects of heavy metals on seed germination and plant growth on alfalfa plant (*Medicago sativa*). Bulgarian J Agr Sci 15(4):347–350

Bahler BD, Steffen KL, Orzolek MD (1991) Morphological and biochemical comparison of a purple-leafed and a green-leafed pepper cultivar. Hort Sci 26:736

Balla K, Bencze S, Janda T, Veisz O (2009) Analysis of heat stress tolerance in winter wheat. Acta Agronomica Hungarica 57(4):437–444

Bartolo ME, Wallner SJ (1986) Cold hardiness and cellulase resistance induced by wounding. Plant Physiol Suppl 80:122

Bennet RJ, Breen C, Fey MV (1985) Aluminium induced changes in the morphology of the quiescent centre, proximal meristem and growth region of the root of *Zea mays*. S Afr J Bot 51(5):355–362

Bialonska D, Zobel AM, Kuras M, Tykarska T, Sawicka-Kapusta K (2007) Phenolic compounds and cell structure in bilberry leaves affected by emissions from a Zn–Pb smelter. Water Air Soil Pollut 181(1–4):123

Bourgou S, Kchouk ME, Bellila A, Marzouk B (2010) Effect of salinity on phenolic composition and biological activity of tunician *Nigella sativa* shoots and roots. C R Biol 331(1):48–55

Boyer JS (1982) Plant productivity and environment. Science 218(4571):443–448

Bray EA, Bailey-Serres J, Weretilnyk E (2000) Biochemistry and molecular biology of plants. American Society of Plant Physiol, Rockville, pp 1158–1203

Cabello-Hurtado F, Durst F, Jorrin JV, Werck-Reichhart D (1998) Coumarins in *Helianthus tuberosus*: characterization, induced accumulation and biosynthesis. Phytochemistry 49 (4):1029–1036

Chalker-Scott L (1999) Environmental significance of anthocyanins in plant stress responses. Photochem Photobiol 70:1–9

Chalker-Scott L, Fuchigami LH (2018) The role of phenolic compounds in plant stress responses. In: Low temperature stress physiology in crops. CRC Press, Boca Raton, pp 67–80

Chen LM, Lin CC, Kao CH (2000) Copper toxicity in rice seedlings: changes in antioxidative enzyme activities, H_2O_2 level, and cell wall peroxidase activity in roots. Bot Bull Acad Sin 41:99–103

Christie PJ, Alfenito MR, Walbot V (1994) Impact of low-temperature stress on general phenylpropanoid and anthocyanin pathways: enhancement of transcript abundance and anthocyanin pigmentation in maize seedlings. Planta 194(4):541–549

Chutipaijit S, Cha-Um S, Sompornpailin K (2009) Differential accumulations of proline and flavonoids in indica rice varieties against salinity. Pak J Bot 41(5):2497–2506

Ciriakova A (2009) Heavy metals in the vascular plants of Tatra Mountains. Oecologia Montana 18 (1–2):23–26

Clemens S (2001) Molecular mechanisms of plant metal tolerance and homeostasis. Planta 212 (4):475–486

Commisso M, Toffali K, Strazzer P, Stocchero M, Ceoldo S, Baldan B, Levi M, Guzzo F (2016) Impact of phenylpropanoid compounds on heat stress tolerance in carrot cell cultures. Front Plant Sci 7:1439

Cronin G, Lodge DM (2003) Effects of light and nutrient availability on the growth, allocation, carbon/nitrogen balance, phenolic chemistry, and resistance to herbivory of two freshwater macrophytes. Oecologia 137:32–41

Daiponmak W, Theerakulpisut P, Thanonkao P, Vanavichit A, Prathepha P (2010) Changes of anthocyanin cyanidin-3-glucoside content and antioxidant activity in Thai rice varieties under salinity stress. Sci Asia 36:286–291

Danai-Tambhale S, Vinay KV, Shriram V (2011) Differential response of two scented indica rice (*Oryza sativa*) cultivars under salt stress. J Stress Physiol Biochem 7(4):387–397

De Abreu IN, Mazzafera P (2005) Effect of water and temperature stress on the content of active constituents of Hypericum brasiliense Choisy. Plant Physiol Biochem 43:241–248

Del Moral R (1972) On the variability of chlorogenic acid concentration. Oecologia 9:289–300

Devi SR, Prasad MNV (1998) Copper toxicity in *Ceratophyllum demersum* L. (Coontail), a free floating macrophyte: response of antioxidant enzymes and antioxidants. Plant Sci 138 (2):157–165

Devlin WS, Gustine DL (1992) Involvement of the oxidative burst in phytoalexin accumulation and the hypersensitive reaction. Plant Physiol 100(3):1189–1195

Dewick PM, Martin M (1979) Biosynthesis of pterocarpan and isoflavan phytoalexins in Medicago sativa: the biochemical interconversion of pterocarpans and 2′ hydroxyisoflavans. Phytochemistry 18(4):591–596

Doncheva S, Stoyanova Z (2007) Plant response to copper and zinc hydroxidesulphate and hydroxidecarbonate used as an alternative copper and zinc sources in mineral nutrition. Rom Agric Res (7–8):15–23

Falcinelli B, Lutts S, Sileoni V, Marconi O, Perretti G, Quinet M, Benincasa P (2017) Germination under moderate salinity increases phenolic content and antioxidant activity in rapeseed (*Brassica napus var oleifera* del.) sprouts. Molecules 22:1–13

Farid M, Shakoor MB, Ehsan S, Ali S, Zubair M, Hanif MS (2013) Morphological, physiological and biochemical responses of different plant species to Cd stress. Int J Chem Biochem Sci 3:53–60

Farooq M, Hussain M, Wakeel A, Siddique HM (2015) Salt stress in maize: effects, resistance mechanisms and management: a review. Agron Sustain Dev 35:461–481

Fatemeh S, Hasan EM, Vahid N (2018) Growth enhancement and salt tolerance of safflower by salicylic acid. Curr Plant Biol 13:16–22

Flora SJ (2009) Structural, chemical and biological aspects of antioxidants for strategies against metal and metalloid exposure. Oxidative Med Cell Longev 2(4):191–206

Gagnon H, Ibrahim RK (1997) Effects of various elicitors on the accumulation and secretion of isoflavonoids in white lupin. Phytochemistry 44(8):1463–1467

Galli V, Borowski JM, Messias RS, Perin EC, Bamberg AL, Rombaldi CV (2016) Mild salt stress improves strawberry fruit quality. Food Sci Technol 73:693–699

Ghasemzadeh A, Jaafar HZE, Rahmat A (2010) Antioxidant activities, total phenolics and flavonoids content in two varieties of Malaysia young ginger (*Zingiber officinale* Roscoe). Molecules 15:4324–4333

Gill SS, Anjum NA, Gill R, Jha M, Tuteja N (2015) DNA damage and repair in plants under ultraviolet and ionizing radiations. Sci World J 2015:1–11

Gray DE, Pallardy SG, Garrett HE, Rottinghaus G (2003) Acute drought stress and plant age effects on alkamide and phenolic acid content in purple coneflower roots. Planta Med 69(1):50–55

Griffith M, Huner NPA, Espelie KE, Kolattukudy PE (1985) Lipid polymers accumulate in the epidermis and mestome sheath cell walls during low temperature development of winter rye leaves. Protoplasma 125(1–2):53–64

Guo ZJ, Nakagawara S, Sumitani K, Ohta Y (1993) Effect of intracellular glutathione level on the production of 6-methoxymellein in cultured carrot (*Daucus carota*) cells. Plant Physiol 102 (1):45–51

Gupta B, Huang B (2014) Mechanism of salinity tolerance in plants: physiological, biochemical and molecular characterization. Int Gen Genomics 1:1–18

Hall JL (2002) Cellular mechanisms for heavy metal detoxification and tolerance. J Exp Bot 53 (366):1–11
Hasanuzzaman M, Nahar K, Alam M, Roychowdhury R, Fujita M (2013) Physiological, biochemical, and molecular mechanisms of heat stress tolerance in plants. Int J Mol Sci 14(5):9643–9684
Hatata MM, Abdel-Aal EA (2008) Oxidative stress and antioxidant defense mechanisms in response to cadmium treatments. Am Eurasian J Agric Environ Sci 4(6):655–669
Hawley EL, Deeb RA, Kavanaugh MC, Jacobs JA (2004) Treatment technologies for chromium (VI). In: Chromium (VI) handbook. CRC Press, Boca Raton, pp 275–309
He J, Giusti M (2010) Anthocyanins: natural colorants with health- promoting properties. Annu Rev Food Sci Technol 1:163–187
Hernandez I, Alegre L, Munne-Bosch S (2006) Enhanced oxidation of flavan-3-ols and proanthocyanidin accumulation in water-stressed tea plants. Phytochemistry 67:1120–1126
Hichem H, Mounir D, Naucer EA (2009) Differential responses of two maize *(zea mays l)* varieties to salt stress: changes on polyphenol composition of foliage and oxidative damages. Ind Crop Prod 30:144–151
Hifney AF, El-Shazoly RM, Abdel-Baset R (2019) Co-deprivation of photosynthetic electron transport mineral pairs significantly enhanced antioxidant contents in a local isolate of *Arthrospira (Spirulina) platensis*. S Afr J Bot 121:1–6
Hossain MA, Piyatida P, Silva D, JAT & Fujita M (2012) Molecular mechanism of heavy metal toxicity and tolerance in plants: central role of glutathione in detoxification of reactive oxygen species and methylglyoxal and in heavy metal chelation. J Bot 872875:1–35
Hosseini Z, Poorakbar L (2013) Zinc toxicity on antioxidative response in (*Zea mays L.)* at two different pH. J Stress Physiol Biochem 9(1):66–73
Isshiki R, Galis I, Tanakamaru S (2014) Farinose flavonoids are associated with high freezing tolerance in fairy primrose (*Primula malacoides*) plants. J Integr Plant Biol 56(2):181–188
Izaguirre-Mayoral ML, Sinclair TR (2005) Soybean genotypic difference in growth, nutrient accumulation and ultrastructure in response to manganese and iron supply in solution culture. Ann Bot 96(1):149–158
Jain R, Shrivastava AK, Solomon S, Yadav RL (2007) Low temperature stress-induced biochemical changes affect stubble bud sprouting in sugarcane (*Saccharum spp.* hybrid). Plant Growth Regul 53(1):17–23
Janas KM, Cvikrova M, Pałagiewicz A, Szafranska K, Posmyk MM (2002) Constitutive elevated accumulation of phenylpropanoids in soybean roots at low temperature. Plant Sci 163 (2):369–373
Janas KM, Zielińska-Tomaszewska J, Rybaczek D, Maszewski J, Posmyk MM, Amarowicz R, Kosińska A (2010) The impact of copper ions on growth, lipid peroxidation, and phenolic compound accumulation and localization in lentil (*Lens culinaris* Medic.) seedlings. J Plant Physiol 167(4):270–276
Jini D, Joseph B (2017) Physiological mechanism of salicylic acid for alleviation of salt stress in rice. Rice Sci 24(2):97–108
Kasuga J, Hashidoko Y, Nishioka A, Yoshiba M, Arakawa K, Fujikawa S (2008) Deep supercooling xylem parenchyma cells of katsura tree (*Cercidiphyllum japonicum*) contain flavonol glycosides exhibiting high anti-ice nucleation activity. Plant Cell Environ 31 (9):1335–1348
Kattab H (2007) Role of glutathione & polyadenylic acid on the oxidative defense systems of two different cultivars of canola seedlings grown under saline condition. Aust J Basic Appl Sci 1 (3):323–334
Khaled T, Fadhila T, Moulay B, Leila AA, Amel E, Jose MM (2016) Effect of salt stress on growth, chlorophyll content, lipid peroxidation and antioxidant defence systems in *phaseolus vulgaris L.* S Afr J Bot 105:306–312
Khan MR, Khan MM (2010) Effect of varying concentration of nickel and cobalt on the plant growth and yield of chickpea. Aust J Basic Appl Sci 4(6):1036–1046

Khan AG, Kuek C, Chaudhry TM, Khoo CS, Hayes WJ (2000) Role of plants, mycorrhizae and phytochelators in heavy metal contaminated land remediation. Chemosphere 41(1–2):197–207
Khan TA, Mazid, Mohammad F (2011) Status of secondary plant products under abiotic stress: an overview. J Stress Physiol Biochem 7(2):75–98
Khan MIR, Fatma M, Per TS, Anjum NA, Khan NA (2015) Salicylic acid induce abiotic stress tolerance and underlying mechanisms in plants. Front Plant Sci 6:462
Kirakosyan A, Kaufman P, Warber S, Zick S, Aaronson K, Bolling S, Chul Chang S (2004) Applied environmental stresses to enhance the levels of polyphenolics in leaves of hawthorn plants. Physiol Plant 121(2):182–186
Kisa D, Elmastas M, Ozturk L, Kayir O (2016) Responses of the phenolic compounds of *Zea mays* under heavy metal stress. Appl Biol Chem 59(6):813–820
Kokubun T, Harborne JB (1994) A survey of phytoalexin induction in leaves of the *Rosaceae* by copper ions. Zeitschrift fur Naturforschung C 49(9–10):628–634
Kondo N, Kawashima M (2000) Enhancement of the tolerance to oxidative stress in cucumber (*Cucumis sativus* L.) seedlings by UV-B irradiation: possible involvement of phenolic compounds and Antioxidative enzymes. J Plant Res 113:311–317
Kovacik J, Klejdus B (2014) Induction of phenolic metabolites and physiological changes in chamomile plants in relation to nitrogen nutrition. Food Chem 142:334–341
Kovacik J, Klejdus B, Backor M (2009) Phenolic metabolism of *Matricaria chamomilla* plants exposed to nickel. J Plant Physiol 166(13):1460–1464
Koyama K, Ikeda H, Poudel PR, Goto-Yamamoto N (2012) Light quality affects flavonoid biosynthesis in young berries of cabernet sauvignon grape. Phytochemistry 78:54–64
Kumari R, Agrawal SB, Singh S, Dubey NK (2009) Supplemental ultraviolet-B induced changes in essential oil composition and total phenolics of *Acorus calamus* L (sweet flag). Ecotox Environ Safe 72(7):2013–2019
Larson RA (1988) The antioxidants of higher plants. Phytochemistry 27:969–978
Lattanzio V (2013) Phenolic compound: introduction. Nat Prod:1543–1580
Lavid N, Schwartz A, Yarden O, Tel-Or E (2001) The involvement of polyphenols and peroxidase activities in heavy-metal accumulation by epidermal glands of the waterlily (*Nymphaeaceae*). Planta 212(3):323–331
Lee JH, Oh MM (2015) Short-term low temperature increases phenolic antioxidant levels in kale. Hortic Environ Biotechnol 56(5):588–596
Lillo C, Lea US, Ruoff P (2008) Nutrient depletion as a key factor for manipulating gene expression and product formation in different branches of the flavonoid pathway. Plant Cell Environ 31:587–601
Liu H, Wang X, Wang D, Zou Z, Lianga Z (2011) Effect of drought stress on growth and accumulation of active constituents in Salvia miltiorrhiza Bunge. Ind Crop Prod 33:84–88
Lo Piero AR, Puglisi I, Rapisarda P, Petrone G (2005) Anthocyanins accumulation and related gene expression in red orange fruit induced by low temperature storage. J Agric Food Chem 53 (23):9083–9088
Lobell DB, Asner GP (2003) Climate and management contributions to recent trends in U.S. agricultural yields. Science 299:1032
Lukaszewicz M, Matysiak-Kata I, Skala J, Fecka I, Cisowski W, Szopa J (2004) Antioxidant capacity manipulation in transgenic potato tuber by changes in phenolic compounds content. J Agric Food Chem 52(6):1526–1533
Ma JF, Ryan PR, Delhaize E (2001) Aluminium tolerance in plants and the complexing role of organic acids. Trends Plant Sci 6(6):273–278
Marinelli F, Di Gregorio S, Ronchi VN (1991) Phytoalexin production and cell death in elicited carrot cell suspension cultures. Plant Sci 77(2):261–266
Marquez-Garcia B, Fernandez-Recamales M, Cordoba F (2012) Effects of cadmium on phenolic composition and antioxidant activities of *Erica andevalensis*. Environ Exp Bot 75(1):159–166

Martinez V, Mestre TC, Rubio F, Girones-Vilaplana A, Moreno DA, Mittler R, Rivero RM (2016) Accumulation of flavonols over hydroxycinnamic acids favors oxidative damage protection under abiotic stress. Front Plant Sci 7:838

Minh LT, Minh TN, Khang DT, Thu Ha PT, Tuyen PT, Quan NV, Xuan TD (2016) Effects of salinity stress on growth and phenolics of rice (*Oryza sativa L*). Int Lett Nat Sci 57:1–10

Mirshekali H, Hadi HASHEM, Amirnia R, Khodaverdiloo H (2012) Effect of zinc toxicity on plant productivity, chlorophyll and Zn contents of sorghum (*Sorghum bicolor*) and common lambsquarter (*Chenopodium album*). Int J Agric 2(3):247–254

Misra N, Misra R, Mariam A, Yusuf K, Yusuf L (2014) Salicylic acid alters antioxidant and phenolics metabolism in *Catharanthus roseus* grown under salinity stress. Afr J Tradit Complement 11(5):118–125

Moesta P, Grisebach H (1980) Effects of biotic and abiotic elicitors on phytoalexin metabolism in soybean. Nature 286(5774):710

Mohamed AA, Amina AA (2008) Alterations of some secondary metabolites and enzymes activity by using exogenous antioxidant compound in onion plants grown under sea water stress. Am Euras J Sci Res 3(2):139–146

Nasu K, Shiraishi T, Yoshioka H, Hori N, Ichinose Y, Yamada T, Oku H (1992) An endogenous suppressor of the defense response in *Pisum sativum*. Plant Cell Physiol 33(5):617–626

Nematshahi N, Lahouti M, Ganjeali A (2012) Accumulation of chromium and its effect on growth of (*Allium cepa* cv. Hybrid). Eur J Exp Bio 2(4):969–974

Nogues S, Allen DJ, Morison JIL, Baker NR (1998) Ultraviolet-B radiation effects on water relations, leaf development and photosynthesis in droughted pea plants. Plant Physiol 117:173–181

Obara N, Hasegawa M, Kodama O (2002) Induced volatiles in elicitor-treated and rice blast fungus-inoculated rice leaves. Biosci Biotechnol Biochem 66(12):2549–2559

Oh MM, Carey EE, Rajashekar CB (2009) Environmental stresses induce health-promoting phytochemicals in lettuce. Plant Physiol Biochnol 47(7):578–583

Olien CR, Smith MN (1997) Ice adhesions in relation to freeze stress. Plant Physiol 60:499–503

Omer C, Jolita R, Kadir ET, Zydrunas S, Cuneyt C, Dursun K, Mehmat SO (2017) The effect of salt and drought stress on phenolic accumulation in greenhouse-grown *hypericum pruinatum*. Ital J Agron 12(918):271–275

Owolabi IO, Yupanqui CT, Siripongvutikorn S (2018) Enhancing secondary metabolites (emphasis on phenolics and antioxidants) in plants through elicitation and metabolomics. Pak J Nutr 17:411–420

Padda MS, Picha DH (2008) Effect of low temperature storage on phenolic composition and antioxidant activity of sweet potatoes. Postharvest Biol Technol 47(2):176–180

Parida AK, Das AB (2005) Salt tolerance and salinity effects on plants: a review. Ecotoxicol Environ Safe 60:324–349

Paroschy JH, Meiering AG, Peterson RL, Hostetter G, Neff A (1980) Mechanical winter injury in grapevine trunks. Am J Enol Vitic 31(3):227–232

Parry AD, Tiller SA, Edwards R (1994) The effects of heavy metals and root immersion on isoflavonoid metabolism in alfalfa (*Medicago sativa L)*. Plant Physiol 106(1):195–202

Pennycooke JC, Cox S, Stushnoff C (2005) Relationship of cold acclimation, total phenolic content and antioxidant capacity with chilling tolerance in petunia (*Petunia*× *hybrida*). Environ Exp Bot 53(2):225–232

Perez-Ilzarbe J, Hernandez T, Estrella I, Vendrell M (1997) Cold storage of apples (cv. Granny Smith) and changes in phenolic compounds. Zeitschrift fur Lebensmitteluntersuchung und-Forschung A 204(1):52–55

Posmyk MM, Kontek R, Janas KM (2009) Antioxidant enzymes activity and phenolic compounds content in red cabbage seedlings exposed to copper stress. Ecotoxicol Environ Safe 72 (2):596–602

Rahnama A, James RA, Poustini K, Munns R (2010) Stomatal conductance as a screen for osmotic stress tolerance in durum wheat growing in saline soil. Funct Plant Biol 37:255–263

Rai V, Vajpayee P, Singh SN, Mehrotra S (2004) Effect of chromium accumulation on photosynthetic pigments, oxidative stress defense system, nitrate reduction, proline level and eugenol content of *Ocimum tenuiflorum* L. Plant Sci 167(5):1159–1169

Rakwal R, Tamogami S, Kodama O (1996) Role of jasmonic acid as a signaling molecule in copper chloride-elicited rice phytoalexin production. Biosci Biotechnol Biochem 60(6):1046–1048

Rana S, Bhushan S (2016) Apple phenolics as nutraceuticals: assessment, analysis and application. J Food Sci Technol 53(4):1727–1738

Rascio N, Navari-Izzo F (2011) Heavy metal hyperaccumulating plants: how and why do they do it? And what makes them so interesting? Plant Sci 180(2):169–181

Rebey B, Bourgou S, Marzouk B, Ml F, Ksouris R (2017) Salinity impact on seed yield, polyphenols composition and antioxidant activity of fennel (*Foeniculum vulgare* mill) extracts. CSIEA 3:2610–2619

Regvar M, Bukovnik U, Likar M, Kreft I (2012) UV-B radiation affects flavonoids and fungal colonisation in *Fagopyrum esculentum* and *F. tataricum*. Open Life Sci 7:275–283

Rellan-Alvarez R, Ortega-Villasante C, Alvarez-Fernandez A, Del Campo FF, Hernandez LE (2006) Stress responses of *Zea mays* to cadmium and mercury. Plant Soil 279(1–2):41–50

Rezai K, Farboodnia T (2008) Manganese toxicity effects on chlorophyll content and antioxidant enzymes in pea plant (*Pisum sativum* L. cv Qazvin). Agric J 3(6):454–458

Rezazadeh A, Ghasemnezhad A, Barani M, Telmadarrehei T (2012) Effect of salinity on phenolic composition and antioxidant activity of artichoke (*Cynara scolymus L*) leaves. Res J Med Plant 6(3):245–252

Rivera-Pastrana DM, Yahia EM, González-Aguilar GA (2010) Phenolic and carotenoid profiles of papaya fruit (*Carica papaya* L.) and their contents under low temperature storage. J Sci Food Agric 90(14):2358–2365

Rivero RM, Ruiz JM, Garcıa PC, Lopez-Lefebre LR, Sanchez E, Romero L (2001) Resistance to cold and heat stress: accumulation of phenolic compounds in tomato and watermelon plants. Plant Sci 160(2):315–321

Rouxel T, Kollmann A, Boulidard L, Mithen R (1991) Abiotic elicitation of indole phytoalexins and resistance to *Leptosphaeria maculans* within Brassicaceae. Planta 184(2):271–278

Rozita KA, Sahebali B, Fariborz ZN, Shahin O (2018) Effect of selenium application on phenylalanine ammonia-lyase (pal) activity, phenol leakage and total phenolic content in garlic (*allium sativum L.*) under nacl stress. Inform Process Agric 5:339–344

Saiema R, Asiya H, Azooz MM, Muneeb R, Siddhiqi OT, Parvaiz A (2013) Salt stress: causes, type and responses of plant. In: Ecophysiology and responses of plant under salt stress. Springer, New York, pp 1–25

Sakamoto A, Murata N (2000) Genetic engineering of glycine betaine synthesis in plants: current status and implications for enhancement of stress tolerance. J Exp Bot 51(342):81–88

Sakihama Y, Cohen MF, Grace SC, Yamasaki H (2002) Plant phenolic antioxidant and prooxidant activities: phenolics-induced oxidative damage mediated by metals in plants. Toxicology 177 (1):67–80

Schubert TS (1992) Manganese toxicity of plants in Florida, vol 353. Florida Department of Agriculture and Consumer Services Division of Plant Industry, Gainesville

Schutzendubel A, Schwanz P, Teichmann T, Gross K, Langenfeld-Heyser R, Godbold DL, Polle A (2001) Cadmium-induced changes in antioxidative systems, hydrogen peroxide content, and differentiation in Scots pine roots. Plant Physiol 127(3):887–898

Schutzendubel A, Nikolova P, Rudolf C, Polle A (2002) Cadmium and H_2O_2-induced oxidative stress in *Populus canescens* roots. Plant Physiol Biochem 40(6–8):577–584

Selmar D, Kleinwächter M (2013) Influencing the product quality by deliberately applying drought stress during the cultivation of medicinal plants. Ind Crop Prod 42:558–566

Semid WM, Taha RS, Abdelhamid MT, Rady MT (2014) Foliar applied α-tocopherol enhances salt tolerance in *Vicia faba L.* S Afr J Bot 95:24–31

Sethy SK, Ghosh S (2013) Effect of heavy metals on germination of seeds. J Nat Sci Biol Med 4 (2):272–275

Shanker AK, Cervantes C, Loza-Tavera H, Avudainayagam S (2005) Chromium toxicity in plants. Environ Int 31(5):739–753

Sharma N, Gupta NK, Gupta S, Hasegawa H (2005) Effect of salinity on photosynthetic rate, transpiration rate, and oxidative stress tolerance in contrasting wheat genotype. Photosynthetica 43(4):609–613

Sharma P, Jha AB, Dubey RS, Pessarakli M (2012) Reactive oxygen species, oxidative damage, and antioxidative defense mechanism in plants under stressful conditions. J Bot 217037:1–26

Shemet SA, Fedenko VS (2005) Accumulation of phenolic compounds in maize seedlings under toxic cadmium influence. Fiziologiia i biokhimiia kul'turnykh rastenii 37(6):505

Singh Y, Malik CP (2017) Phenols and their antioxidant activity in Brassica juncea seedlings growing under HgCl2 stress. J Microbiol Biotechnol Res 1(4):124–130

Solecka D, Boudet AM, Kacperska A (1999) Phenylpropanoid and anthocyanin changes in low-temperature treated winter oilseed rape leaves. Plant Physiol Biochem 37(6):491–496

Sonar BA, Desai NM, Gaikwad DK, Chavan PD (2011) Assessment of salinity-induced antioxidative defense system in *colubrina asiatica* brong. J Stress Physiol Biochem 7 (3):193–200

Steiner F, Zoz T, Junior ASP, Castagnara DD, Dranski JAL (2012) Effects of aluminum on plant growth and nutrient uptake in young physic nut plants. Semina: Ciencias Agrarias 33 (5):1779–1788

Steyn WJ, Wand SJE, Holcroft DM, Jacobs G (2002) Anthocyanins in vegetative tissues: a proposed unified function in photoprotection. New Phytol 155:349–361

Sudhir P, Murthy SDS (2004) Effects of salt stress on basic processes of photosynthesis. Photosynthetica 42(4):481–486

Sytar O, Kumar A, Latowski D, Kuczynska P, Strzałka K, Prasad MNV (2013) Heavy metal-induced oxidative damage, defense reactions, and detoxification mechanisms in plants. Acta Physiol Plant 35(4):985–999

Taulavuori K, Pyysalo A, Taulavuori E, Julkunen-Tiitto E (2018) Responses of phenolic acid and flavonoid synthesis to blue and blue-violet light depends on plant species. Environ Exp Bot 3414:1–18

Tebayashi SI, Ishihara A, Iwamura H (2001) Elicitor-induced changes in isoflavonoid metabolism in red clover roots. J Exp Bot 52(357):681–689

Tounekti T, Vadel AM, Ennajeh M, Khemira H, Bosch SM (2011) Ionic interactions and salinity affect monoterpene and phenolic diterpene composition in rosemary *(rosmarinus officinalis)*. Plant Nutr Soil Sci 00:1–11

Truta E, Gherghel D, Bara ICI, Vochita GV (2013) Zinc-induced genotoxic effects in root meristems of barley seedlings. Notulae Botanicae Horti Agrobotanici Cluj-Napoca 41 (1):150–156

Tsonev T, Lidon FJC (2012) Zinc in plants--an overview. Emirates J Food Agric 24(4):322–334

Uleberg E, Rohloff J, Jaakola L, Trost K, Junttila O, Haggman H, Martinussen I (2012) Effects of temperature and photoperiod on yield and chemical composition of northern and southern clones of bilberry (*Vaccinium myrtillus* L). J Agric Food Chem 60:10406–10414

Valifard V, Mohsenzadeh S, Kholdebarin B, Rowshan V (2014) Effect of salt stress on volatile compound, total phenolic compound and antioxidant activities of *Salvia mirzayanii*. South African J Bot 93:92–97

Valifard M, Mohsenzadeh S, Kholdebarin B (2017) Salinity effects on phenolic content and antioxidant activity of salvia macrosiphon. Iran J Sci Technol 41:295–300

Valko MMHCM, Morris H, Cronin MTD (2005) Metals, toxicity and oxidative stress. Curr Med Chem 12(10):1161–1208

Vantuyl JM, Vangroenestijin JE, Toxopeus SJ (1985) Low light intensity and flower bud abortion in Asiatic hybrid lilies. I Genetic variation among cultivars and progenies of a DTALLEL cross. Euphytica 34:83–92

Vardar F, Unal M (2007) Aluminum toxicity and resistance in higher plants. Adv Mol Bio 1:1–12

Vassilev A, Nikolova A, Koleva L, Lidon F (2011) Effects of excess Zn on growth and photosynthetic performance of young bean plants. J Phytology 3(6):58–62
Veersham C (2004) In elicitation: medicinal plant biotechnology. CBS Publisher, pp 270–293
Weidner S, Karolak M, Karamac M, Kosinska A, Amarowicz R (2009) Phenolic compounds and properties of antioxidants in grapevine roots [*Vitis vinifera* L.] under drought stress followed by recovery. Acta Soc Bot Pol 78(2):97–103
Xu YBD, Wang G, Cao F, Zhu C, Wang G, Yousry A (2014) Light intensity affects the growth and flavonol biosynthesis of Ginkgo (*Ginkgo biloba* L). Nanjing Forestry Univ 30:51–54
Yan K, Zhao S, Bian L, Chen X (2017) Saline stress enhanced accumulation of leaf phenolics in honeysuckle (*Lonicera japonica* Thunb.) without induction of oxidative stress. Plant Physiol Biochem 112:326–334
Yan M, Pei W, Mian W, Maomao S, Zhenxin U, Runqiang Y (2019) GABA mediates phenolic compound accumulation and the antioxidant system enhancement in germinating Hulless Barley under Nacl stress. Food Chem 270:593–601
Yang L, Wen K, Ruan X, Zhao Y, Wei F, Wang Q (2018) Response of plant secondary metabolites to environmental factors. Molecules 23:276
Yoshikawa M (1978) Diverse modes of action of biotic and abiotic phytoalexin elicitors. Nature 275(5680):546
Yruela I (2009) Copper in plants: acquisition, transport and interactions. Funct Plant Biol 36 (5):409–430
Zengin FK, Munzuroglu O (2005) Effects of some heavy metals on content of chlorophyll, proline and some antioxidant chemicals in bean (*Phaseolus vulgaris L.*) seedlings. Acta Biol Cracov Ser Bot 47(2):157–164
Zenk MH (1996) Heavy metal detoxification in higher plants-a review. Gene 179(1):21–30
Zhang S, Song J, Wang H, Feng G (2010) Effect of salinity on seed germination, ion content and photosynthesis of cotyledons in halophytes or xerophyte growing in Central Asia. J Plant Eco 3 (4):259–267
Zhang X, Wei J, Tian J, Li N, Jia L, Shen W, Cui J (2019) Enhanced anthocyanin accumulation of immature radish microgreens by hydrogen-rich water under short wavelength light. Sci Hortic 247:75–85
Zornoza P, Vazquez S, Esteban E, Fernández-Pascual M, Carpena R (2002) Cadmium-stress in nodulated white lupin: strategies to avoid toxicity. Plant Physiol Biochem 40(12):1003–1009

Phenolics: A Key Defence Secondary Metabolite to Counter Biotic Stress

13

Yamini Tak and Manoj Kumar

Abstract

Polyphenols or plant phenolics are the secondary metabolites which have diverse functions to mitigate various abiotic (heat, drought, cold, salt or heavy metal) and biotic (bacterial, fungal, viral, insect or weed) stresses. Phenolic compounds derived via the common phenylpropanoid pathway perform as signalling molecule and can act as agents in plant shielding. Phenolic compounds play important role in the regulation of seed germination and cooperate in regulating the growth of plants, also taking part in defence responses during viral or bacterial infection, excessive sun or light exposure, wounds and heavy metal stresses. One of the crucial role of phenolic compounds is their antimicrobial activity in plants is to act as safeguarding compounds opposed to disease agents such as fungi, bacteria, nematodes, weeds and other pathogens. In the present chapter, we will try to understand how plants fight against the different types of biotic stresses by employing phenolic compounds.

Keywords

Biotic · Stress · Phenolics · Phenylpropanoid pathway · Defence

13.1 Introduction

God has stated equal right to each and every living soul in this universe. Every living being has been provided with their own defence system for survival against various stresses they face in their life period. "Survival of the fittest" is a phrase that arose from Darwinian evolutionary theory as a way of narrating the mechanism of natural

Y. Tak (✉) · M. Kumar
Agriculture University Kota, Kota, Rajasthan, India

ICAR- Central Institute for Research on Cotton Technology, Mumbai, India

R. Lone et al. (eds.), *Plant Phenolics in Sustainable Agriculture*,
https://doi.org/10.1007/978-981-15-4890-1_13

selection. The crops/plants/animals that are able to save themselves from the different types of stresses (biotic or abiotic) and can be able to produce their offspring can lead to successful natural selection. Plants have their own mechanism to fight against the abiotic and biotic stresses. They have antioxidants and various phenolic compounds so that they can face the above-said stresses.

13.2 Stresses in Plants

Plant growth and development is influenced by a wide range of environmental stresses that are key constraints for sustainable agricultural productivity in soil health and fertility (Kumar and Verma 2018). Abiotic and biotic stress contributes 50% and 30%, respectively, to losses in agricultural productivity worldwide. Various biotic stresses like viruses, bacteria, fungi, insects, nematodes, etc. and abiotic stress factors such as salinity, frost, chilling, flood, drought, heavy metals, temperature and nutrient deficiency have a huge negative impact on agriculture (Dresselhaus and Hückelhoven 2018). Considering all the biotic stressors, fungi cause the most severe threat to plant species as 85% of diseases in plants are caused by fungus or fungal-like organisms (Behmann et al. 2014). Furthermore, viruses are treated as serious agent for destroying crop plants. These biotic organisms cause numerous symptoms in plants such as necrosis, leaf spots, blights, blasts, wilting, mottling, tumour formation and many more (Saddique et al. 2018). Another biotic agent, namely, weeds, also affects the growth and yield of economically important plants by direct destruction of plant or by aggravating competition for area and nutrient availability (Melvin et al. 2017).

13.3 Biotic Stress and Phenolics

Plants machinery give rise to the various metabolites as a response biotic stress in particular stage of development. These metabolites can be phenolic compounds, and their biosynthesis is originated from a specific primary metabolite (Guo et al. 2018). These phenolics are secondary metabolites that are composed of one of the most common and widespread groups of substances in plants (Saltveit 2009). These compounds are synthesized by the phenylpropanoid pathway and concentrated in subepidermal layer of plant tissue when plants are attacked by the pathogens. Additionally, phenolics are also found covalently associated with plant cell wall and other types of phenolics present in waxes or on the peripheral surfaces of plant organs (Vishwanath et al. 2014). Besides the above-said factors, natural factors such as soil type, moisture content, temperature and ultraviolet ray's incidence can affect the accumulation and biosynthesis of phenolics (Mukherjee et al. 2018).

13.4 Biotic Stress

Usually, biotic agents like insects, nematodes and vertebrate pests, viz. rodents, cause 25%, 20% and 6–8% losses in crops, respectively (Bentur 2011). Bacteria are microorganisms and can hinder the ability of the plant to transport nutrients and water to other parts of plants, and as a result, plant starts to die (Jones et al. 2004). Bacterial infection can lead to mosaic-like symptoms which are very similar to the viral diseases and can also lead gigantic plant deformities such as galls or tumours (Ludwig-Müller 2015). The other symptoms are spots on leaves or fruit, blights or deadening of tissue on leaves, stems or tree trunks and rots of any part of the plant.

The infection caused by viruses displays many physiological and metabolic abnormalities in the plants (Culver and Padmanabhan 2007). The infection disseminates via lesions or injured plant parts because they provide direct gateway to viral genetic material to take entry inside the plant cells. The symptoms of viral affected plant include mottling, mosaic, leaf spots, curling of leaf, puckering of leaf and other symptoms like stunting and reduction in the yield. These symptoms may involve direct or indirect effects that disrupt the host physiology (Thakur et al. 2018).

Another group of biotic stress agent is fungi which have as many as 1.5 million of species. Fungi are biotic agents which can be classified into two categories, i.e. biotrophs and necrotrophs (Lattanzio et al. 2006). As fungi are non-photosynthetic and cannot be able to synthesize their own food, they adapted to evolve approaches to get their food from either living or lifeless organisms. Biotrophic fungal pathogens acquire nutrients from living host tissues, often via specialized cells called haustoria that form inside host cells. Necrotrophic pathogens kill host tissue through production of toxins and get nutrient from dead cells (Delaye et al. 2013).

Nematodes such as root-knot nematode do most damage below ground which causes root-knot by feeding on root of plants and reduces the plants' ability to absorb nutrient and water (Bais et al. 2006). Nematode infestation accounts for about 14% global loss amounting to ~100 billion dollars annually (Chitwood 2003). Nematodes act as parasites in the plants and cause mechanical damage to plant tissues with their stylets (Holbein et al. 2016). Supposedly, nematodes may also secrete digesting enzymes which cause dissolution of cell walls enabling intracellular and intercellular movement of endoparasites, and they act as "digestive enzymes" digesting solid components of cells which can be imbibed by the parasite. Hence, nematodes digest the internal content of plant cells and kill them (Kyndt et al. 2017). Typical root symptoms by nematode attack are root-knots or galls, root lesions, excessive root branching, injured root tips and stunted root systems. Plant shows wilting even with ample soil moisture, yellowing of leaves and stunted growth (Marahatta 2018).

Insects injure growing crops by two major ways, i.e. indirect and direct: Direct damage caused to the crop/plant by the insects by directly feeding on them and indirect damage by transmitting a bacterial, viral or fungal infection into crop plants (Fried et al. 2017). After the insect attacks, general indications are flower malformation, abnormal shoot branching, chlorotic or yellow patterns of mosaic or even stunted growth of plant. Among crops, the total global potential drop due to pests varied from about 50% in wheat to more than 80% in cotton production (Ueda et al. 2018).

Agricultural crops have highest potential loss due to weeds (34%), with animal pests and pathogens being less important (Christos 2016). Loss in yield due to weeds is almost each time caused by a group of different weed species, and these can differ substantially in competitive ability (Jabran et al. 2017). Due to distinct growth habits and adaptations, weeds can grow well in many ecological niches where other plants are unable to grow. Weeds compete not only with themselves but also with agricultural or horticultural crop plants (Wallace et al. 2017). Weeds affect the plant growth and yield by direct destruction of crop or by aggravating competition for area and nutrient availability.

13.5 Phenolics

Phenolics are the compounds playing critical functions in plant protection and indispensable to fight against different types of stresses (Wani et al. 2016). Their structure has aromatic ring attached to one or more hydroxyl group. Phenolics are very important in the development of plants as they contribute to the lignin biosynthesis and pigment biosynthesis and safeguards the plants from external infections (Asgari Lajayer et al. 2017). It also provides structural and scaffolding integrity to plants. Phenolic phytoalexins, secreted by injured or otherwise damaged plants, repel many microorganisms. but other pathogens can prevent these protections or even use them for their own advantage (Dar et al. 2017). Plant phenolics may be categorized in two classes:

1. Preformed phenolics are formed at the time of the usual growth and development of plant tissues.
2. Induced phenolics are formed during physical or mechanical injury, during microbial or pathogen attack or when stressed by external component such as heavy metal, drought, salt, UV-irradiation, temperature, etc.

13.6 Biosynthesis of Phenolics

Phenolics are cosmopolitan compounds in the plant kingdom but are not very common among the bacteria, algae and fungi (Ma et al. 2015). For phenolic biosynthesis, the first step is participation of glucose in the hexose monophosphate shunt and converting it into glucose-6-phosphate and later on irreversible transformation into ribulose-5-phosphate. Glucose-6-phosphate dehydrogenase performed the first committed step in the conversion to ribulose-5-phosphate (Lin et al. 2016). On the other side, hexose monophosphate shunt produces erythrose-4-phosphate along with phosphoenolpyruvate from glycolysis, which is then utilized through the phenylpropanoid pathway to generate phenolic compounds after being channelled to the shikimic acid pathway to produce phenylalanine. The shikimic acid pathway is involved in the synthesis of a majority of the phenolic compounds in plants, bacteria and fungi (Santos-Sánchez et al. 2019). Another pathway is malonic acid pathway which is less significant in the formation of phenolic acids in higher plants as compared to fungi and bacteria (Cheynier et al. 2013).

Flavonoids are the biggest single group of phenolic C15 compounds containing two phenolic rings connected by a three-carbon unit, and these are considered as derivatives of phenolic acids which are biosynthetically derived from malonyl CoA and p-coumaroyl CoA, derived from acetate and shikimate, respectively (Patel et al. 2017).

Phenylalanine ammonia lyase is the key enzyme which mediates the formation of cinnamic acid from phenylalanine and is a crucial branch point of primary and secondary metabolism (Saltveit 2017). It is considered as the most important regulatory step in the formation of phenolic acids. After biosynthesis, phenolic acids are integrated to the cell wall of plants to compensate for biotic stress with increased variability through synthesis of cinnamic and benzoic acid derivatives and help the plant to fight against the stress (Ledesma-Escobar et al. 2019).

13.7 Functions of Phenolic Compounds in General

Phenolics have antibiotic, antimicrobial, anti-nutritional and have unpalatable properties. These properties help the plant to save itself from biotic and abiotic stresses. Plant phenolics have key roles as pigments, antioxidants, metal chelators and plant growth regulators signalling agents and function as UV light screens (Bhattacharya et al. 2010). Phenolics usually accumulate in the central vacuoles of guard and epidermal cells or subepidermal cells in plants (Radice and Arena 2015). Plant phenolics have the following major functions:

13.7.1 UV Sunscreens Protection

When agricultural crops are exposed to solar ultraviolet-β radiation (280–320 nm), it negatively affects genetic material of cell, lipids and proteins (Silva et al. 2018). UVR have many roles in evolution-generating mutations, leading to new traits and leading to the diversity among the species. Phenolic compounds prevent mutagenesis (hydroxycinnamic and hydroxybenzoic acids) and programmed cell death by dimerization of thymine units in the DNA. Flavonoids regulate auxin transport and are found to be promising in regulation of growth and development of plant under different intensities of light (Czemmel et al. 2017). Red clover (*Trifolium pratense*) contains isoflavones (a subclass of flavonoids), and some metabolically related compounds are found to be very helpful in providing protection from UV-induced inflammation and immune reppression in mice (Danciu et al. 2018).

13.7.2 Attracting Pollinators and Seeds Dispersers

Anthocyanins are important class of flavonoids which gives colour to various flowers and performs as pollinator attractants (Kong et al. 2003). Different pigments (cyanidin, pelargonidin, petunidin, delphinidin, malvidin and peonidin) present in different parts of a flower to attract numerous type of pollinators (Miller et al. 2011). Generally younger flowers are bee pollinated, and the older are visited by birds for pollination. As flower ages, plant switches to birds visitation by increasing the sugar concentration in the nectar and flower colour by anthocyanin pigments. Colours of the flowers are contingent on possible complexes with Fe^{+3} and Al^{+3} as well as pH. Anthocyanins are stable at acidic pH and give beautiful brilliant red shades, and under basic pH it gives blue colour (Hubbermann 2016). These colours play very salient role in attraction of pollinators. Phenylpropanoids act as key chemical modulators of plant communication with insects and microbes, either as attractants or repellants, to attract the pollinator by flower colour (Robert and Junker Amy 2015).

13.7.3 Phenolics as Signal Compounds

Based on the nature and type of root-originated chemical substances, both positive and negative cross-talk routes are initiated. These pathways may happen between rhizome and insects or roots and pathogens. A variety of organisms can be repelled or attracted to the specific chemical signal, which then elicits responses in recipient microbe or organism (Bhattacharya et al. 2010). Phenolics function as a warning molecule to trigger interplay among legumes and rhizobia. In legumes, phenolic acids are secreted rapidly from growing roots (nodule development) throughout seed germination and seedling growth (Ferdous et al. 2017). Rhizobia species have the special power to make use of phenolic acids as a carbon source. Root exudates from legumes which contain phenolics act as chemoattractant and signalling compounds for numerous soil microorganisms (Huang et al. 2014). These organisms identify them and progress towards plant roots in the carbon-rich rhizosphere. Flavonoids and isoflavonoids from soybean, vanillin from peanut and betaines and aldonic acids in legume seed initiate nodule formation and function as signals to the microbial symbiont (Janczarek et al. 2015). In relation to legumes, phenolic compounds control *nod* gene expression by the *rhizobium* and modify the legume-rhizonium symbiosis by secreting these phenolics that act as signalling molecule during expression of various symbiotic plasmid encoded *nod* (nodulation) genes (Gourion et al. 2015). Glycolipids (chito-lipo-oligosaccharides) trigger root hair deformations and cortical cell divisions within the root leading to nodule formation. The *Rhizobium* quorum sensing enhances nodulation efficiency, symbiosome development, exopolysaccharide production, nitrogen fixation and adaptation to various types of stress (Pérez-Montaño et al. 2011). Quorum sensing guides rhizobia to synchronize themselves to phenolic signals on a population-wide scale and to do duty as multicellular organisms for successful symbiosis (Garg et al. 2013).

13.7.4 Role in Allelopathy

Allelopathy is defined as the negative effect of one plant species on neighbouring plants via the release of allelochemicals into the surroundings (Del Fabbro and Prati 2015). Phenolic allelochemicals result in increased cell membrane permeability. Due to increased permeability, cell contents spill, resulting in raised lipid peroxidation. Phenolic allelochemicals can impede the absorption of nutrients from the soil and can lead to dwarfing in the plants or may lead to death of the whole plant (Scavo et al. 2018). Treatment of cucumber (*Cucumis sativus*) for 7 days with benzoic and cinnamic acid derivatives results in decreased phenolic glycosylation and phenyl-ß-glucosyltransferase activity, and this decrease was linked with increased membrane permeability (Li et al. 2010). Gallic acid, p-coumaric acid, p-hydroxybenzoic acid and ferulic acid isolated from *Ageratum conyzoides* L (Scognamiglio et al. 2013) are the major allelochemicals in the rhizosphere. Syringic or hydroxybenzoic acids at 2 mM concentration and gallic acid at 1 mM concentrations significantly hindered the germination in various agricultural crops (Li et al. 2010).

13.8 Role of Phenolic Compounds in Biotic Stress

Phenolics act as safeguarding agents, inhibitors, pesticides and natural animal toxicants hostile to various biotic agents, i.e. herbivores, nematodes, insects and fungal, viral and bacterial pathogen (Ghosh et al. 2017). Phenolics also perform as phytoestrogens in animals and as allelochemicals for competitive plants and weeds. Mainly, the vaporous terpenoids, the toxic water-soluble hydroquinones, escopoletins, hydroxybenzoates, caffeic acids, hydroxycinnamates and the 5-hydroxynapthoquinones are broadly effectual allelochemicals (Bhattacharya et al. 2010). Phenolics are the best natural shielding agents and can offer an alternative to the chemical control of pathogens on agricultural crops. Plant phenolic compounds produced during host-pathogen interactions work by several mechanisms in plant protection (Tripathi 2004).

As soon as a plant host is infected by the pathogen, it shows accumulation of phenolics as an initial response, and it may lead to general increase in host metabolism (Mayer et al. 2001). It has been perceived that in maize, there is a significant increase in the two phenolic caffeic acid esters after infected with *Glomerella graminicola* or *C. heterostrophus* (Pusztahely et al. 2017). Although these polyphenols or phenolic compounds are not toxic to fungi, their rapid accumulation and sudden decline in the concentration proposed that these phenols may serve as pool for synthesis of other protective compounds. The intercommunication between the infected plant and infecting agent can be built by the capability of the infecting agent to metabolize the phytoalexins formed by the infected plant (Hardoim et al. 2015). These compounds reduce the dissemination of biotic agent infection. The bacteriostatic and bactericidal properties of phytoalexin are identical to the antibiotics (Pina-Pérez and Ferrús Pérez 2018). Phytoalexins limit sporulation development and growth of hyphae of plant pathogenic fungi. Defence by phytoalexins against microbes or pathogens depends on the rate of synthesis and their accumulation in plant tissues (Duke 2018).

Quorum sensing leads to exponential growth and regulates many other activities in bacteria; polyphenols can disrupt this process and restrict the growth of these pathogens, and therefore phytoalexins have potent role in controlling harmful activities of the biotic stressors (Raman et al. 2015). Phenolic derivatives such as quinones are very valuable for the plants, as they produce complex derivatives by combining with proteins and inhibit the maceration of proteomic substance by the herbivores (Easwar Rao et al. 2017). Quantity of these compounds in amino acids leads to decreased digestion of proteins in the insects that hampers pest growth. Phenols also reduce the incidence of reactive oxygen species such as peroxides, superoxide and singlet oxygen involved in a series of biotic and abiotic stresses. These phenolic compounds activate the beginning of defensive enzymes and hence protect plants from these stresses (Saddique et al. 2018).

Phenolics accumulate at the infected site of plant and inhibit the overall growth and development of the microorganism which occurs as the result of cell death due to hypersensitive reaction (Lincoln et al. 2018). To overcome the biotic stresses, biosynthesis of lignin proceeds from L-phenylalanine via 4-coumaric acid and the CoA-esters of 4-coumaric, ferulic and sinapic acids to the corresponding alcohols which supposedly polymerize under the action of a peroxidase and help in fighting against biotic stress agents (Varbanova et al. 2011; Peperidou et al. 2017; Bi et al. 2017). Scientific studies by various researchers suggest that low molecular weight phenols, like the benzoic acids and the phenyl propanoids, are formed in the initial response to infection. It is strongly suggested by the evidence that esterification of phenols to cell wall materials is especially for the coping of the plants against various stresses (Mandal et al. 2010).

Many microbes, such as fungi, bacteria, viruses and herbivore insects have threat of harm and danger to plants (War et al. 2012). Plants have established a variety of resistance mechanisms for protection and survival, several of which are generated as a consequence of pathogen attack.

13.8.1 Bacteria

Bacteria can infect the plant and spread in a number of ways including herbivores insects, swashing water or other infected plants or plant parts. They enter plants through tiny openings either through damage or cuts but also through natural openings in the plant itself (Hirano and Upper 2000). Bacteria usually use Type III secretion system which is a complex bacterial structure that provides virulence to gram-negative bacteria to inject chemicals or effector protein directly into plant cells. This system resembles a syringe and plunger system that bacteria use to inject effector proteins to cause diseases and trigger defence responses in plants (Green and Mecsas 2015). Bacterial pathogens utilize quorum sensing to ensure that virulence genes are only expressed after their population has reached a critical size.

Among all bacterial pathogens that spread disease to plants, *Pseudomonas syringae* pathogens are the most familiar and well-investigated ones. *Pseudomonas syringae* is a gram-negative bacterium that shows various symptoms in plants such as blights, cankers, leaf spots and wilting of plant (Sun et al. 2017).

Control Through Phenolic Compounds

- Due to innate immunity of plants, plant pattern recognition receptors recognize potential pathogens by conserved pathogen-associated molecular patterns (PAMPs), and they are responsible for PAMP-triggered immunity. As a result, the progress of the infection is restricted long before the pathogen gains complete hold of the host (Postel and Kemmerling 2009).
- Polyphenols such as catechins can affect different bacterial species including *Pseudomonas aeruginosa*, *Serratia marcescens*, *Salmonella choleraesuis*, *Escherichia coli*, *Bordetella bronchiseptica*, *Klebsiella pneumoniae*, *Bacillus subtilis* and *Staphylococcus aureus* by changing the permeability and properties of plasma membranes and the production of reactive oxygen species like hydrogen peroxide (Wang et al. 2018).
- In *E. coli* by exogenous application of trans-cinnamaldehyde and tannic acid heterologous expression of quorum sensing proteins (LasI and RhlI) have been observed. Eugenol and carvacrol specifically interfere with the quorum sensing regulator in pectobacteria (Joshi et al. 2016).

13.8.2 Fungus

Fungus is a eukaryotic creature which digests food externally and absorbs nutrients directly to its cell walls. Most fungi reproduce by spores or hyphae; a vegetative growth has a body (thallus) composed of microscopic tubular cells (Dube 2015). The spores of some fungi pass through the air and attack leaves produce dead spots or even death of the whole leaf. Some fungi reside in the soil and enter in plant through roots. Fungi cause wilting in plant by blocking the water-conducting cells or kill them. In many cases the plant is seriously damaged or may even die (Carris et al. 2012).

Magnaporthe oryzae, a filamentous ascomycete fungus, is the causative agent of rice blast disease, the most devastating disease of rice worldwide. All foliar tissues of the host plant are subject to infection; however, infection of the panicle can lead to complete loss of grain (Hajano et al. 2011).

Control Through Phenolic Compounds

- Resveratrol-O-methyltransferase and resveratrol synthase 3 increase the hostility to soybean when it was exposed to the pathogen *Rhizoctonia solani* (Zernova et al. 2014). For halting growth of the fungus, relative rate of formation as well as

the quality of cell wall appositions is thought to be first line of the "complementary defence mechanisms".
- Hydroxycinnamic derivatives, oleuropein derivatives, flavonol monoglucosides derivatives and tyrosol derivatives are exceptionally responsible for olive tree hostility to the leaf spot disease caused by *Fusicladium oleagineum* (Talhaoui et al. 2015).
- The fungus which infects onion crop is a soil saprophyte which begins its parasitic existence on the dead outer scales of the onion bulb. Later it perforates into the lower succulent scales and deep into the bulb. Some varieties of onion are resistant to the fungus, the resistance being correlated with catechol, protocatechuic acid and red or yellow pigmentation of the dead outer bulb scales (Levin 1971). These compounds are soluble in water and diffuse from the dead cell layers of the scales to the infection drop, where they inhibit germination and penetration.

13.8.3 Virus

Viruses are infectious particles made up of nucleic acid core and protein. Once it infects a susceptible cell or wounded cell, the infectious virus particle sheds its protein coat and the nucleic acid then directs the formation of several copies of itself and related proteins principally development of new virus particles (Maccheroni et al. 2005). Movement of plant virus occurs through cell-to-cell or by cytoplasmic "bridges". Plant viruses are generally transmitted from plant to plant by a vector, but mechanical and seed transmission are also observed (Congdon et al. 2017). Vector transmission is often by an insect (aphids, white fly), but some nematodes, fungi and protozoa have been observed to be viral vectors. Plant viruses cause disease in different plant parts by causing a reallocation of photosynthates and a disruption of normal cellular processes as they replicate (Burch-Smith and Zambryski 2016).

The first plant virus discovered was tobacco mosaic virus (TMV), which attacks members of the nightshade, or Solanaceae, family.

Control Through Phenolic Compounds

- Increases in the concentration of kaempferol, quercetin, caffeic acid and chlorogenic acid have been linked with virus activity in various host plants (Parr and Bolwell 2000).
- In *Matthiola incana* the level of kaempferol increased, while the level of anthocyanin declined (Baskar et al. 2018).

13.8.3.1 Nematode

Plant parasitic nematodes are more destructive plant pathogens causing worldwide losses exceeding $ 125 billion annually (Chitwood 2003). To feed on living plant tissue plant parasitic nematodes rupture host cells by using an oral. For partial digestion of the cell contents, plant nematodes inject enzymes into a host cell

(Escobar et al. 2015). Root-knot nematodes parasitize plant root systems and thus directly inhibit the consumption of water and nutrients needed for normal plant growth, development and reproduction of the plant (Davies and Spiegel 2011).

About 2000 plants worldwide are susceptible to infection by root-knot nematodes, and they cause approximately 5% of global crop loss. Potato cyst nematodes (*Globodera pallida* and *Globodera rostochiensis*) are widely distributed in Europe and North and South America and cause $300 million worth of damage in Europe every year (Minnis et al. 2002).

Control Through Phenolic Compounds

- Phenolics play role by different mechanisms for resistance of nematode: (i) Browning and slow formation of wide necrosis in plants (ii) Speedy browning and emergence of non- expandable necrosis (iii) IAA-oxidase inhibition favours auxin accumulation and as a result of galls or giant cell formation (iv) IAA-oxidase stimulation which favours auxin decomposition and of necrosis in plants (Giebel 1982).
- Nematodes are able to do activation of phenols by decompose bound phenol to free secreting fl-glycosidases into the host tissue (Ohri and Pannu 2010).
- Glucoside amygdaline present in peach roots is hydrolysed by 3-glucosidase of *Pratylenchus penetrans* to hydrocyanic acid and benzaldehyde (Giebel 1982).

13.8.4 Insects and Parasites

Phytophagous insects produce stresses to plant by salivary secretions which contain various enzymes, playing an elemental role in food digestion of sucking-piercing insects (Lattanzio et al. 2006). Plant parasites directly couple to the vascularized system of host plants thereby reducing the uptake of water, nutrients and carbohydrates as a result leading to immensely reduced biomass and losses in seed yields of the contaminated host plants (Hegenauer et al. 2017).

Control Through Phenolic Compounds

- Juglone is a phenolic produced by *Carya ovata*, which is not palatable for bark beetle *Scolytus multistriatus* (Byers 1995).
- Glands on the leaves, sepals and petals of *Lypericum* spp. secrete a crimson-coloured phenolic quinone referred to as hypericin. Ingestion of this compound causes intensive photosensitivity and skin irritation, leading sometimes to blindness, skin lesions, refusal to eat, and sometimes death, and it is toxic against insects as well as mammals (Levin 1971).
- Formation of gossypol in leaves and flower buds presumably defends against grazing by mammalian herbivores and has been observed to defend against infestation by bollworm and tobacco budworm (Levin 1971).

13.8.5 Weeds

The annual global economic loss caused by weeds has been estimated at more than 100 billion US dollars. Weeds can be a harbour to pest, vector of virus or reservoir of a pathogen; weeds can remarkably influence disease prevalence (Swanton et al. 2015).

Phalaris minor is the major weed of wheat crop in rice. Witch weeds (*Striga* species), for example, generally attack cereal crops, including maize, millet, sorghum and rice. Strigolactone, a type of phytohormone exuded from the host root, triggers germination of the parasite seed (Qasem 2006).

Control Through Phenolic Compounds

- High concentrations of catechin reduce *C. maculosa* germination and development, thereby performing as an autoinhibitor (Li et al. 2010). It has been observed that application of a bur cucumber seed extract and their phenolic allelochemical (2-linoleoyl glycerol) triggers the accumulation of abscisic acid, salicylic acid and jasmonic acid and inhibits the gibberellin pathway, so seed germination will be halted (Lee et al. 2015).
- p-hydroxybenzoic, p-coumaric, gallic, syringic, gentisic and vanillic acids are allelopathic compounds in the bark, fresh leaves or litter of *Eucalyptus polycarpa*, *E. microtheca* and *E. camaldulensis* which have injurious effects on the crops in the ecosystem resulting in the delaying of germination, mortality of seedling and reduction in growth and yield of the other plants (John 2012).

  Various biotic stresses and phenolics as shield against biotic stresses are introduced in Table 13.1.

13.9 Conclusion and Future Prospective

Due to global warming, and potential climate abnormalities associated with it, crops typically encounter increased various abiotic and biotic stress combinations, which critically affect their germination, growth and yield.

Combination of various biotic stresses can cause serious threats to plant growth and development. Main causative agents of biotic stress are bacteria, viruses, fungi, weeds, nematodes, insects and parasites in the plants.

Biotic stresses are caused by various living organisms, such as bacteria, fungi, insects and weeds (Lattanzio et al. 2006). They are very disastrous for plant growth and development. From the different biotic organisms, fungi pose a major serious threat as compared with others. An enormous species of fungi, i.e. about 8100 fungal species, have been found to perform as plant pathogens (Tarkowski and Vereecke 2014). Similarly, viruses are also precarious pathogen which demolish agricultural crops much more than fungi world over (Bai et al. 2002) These pathogens may cause various types of injurious symptoms in plants, such as rotting, yellowing or wilting

Table 13.1 Major biotic stresses in plant, their effects and mitigation by phenolic compounds

Example	Mode of action	Phenolic compound	Function	References
Pseudomonas savastanoi – olive knot disease	In fresh wounds of olive trees, pathogen initially colonizes the tissues around the infection point and, through pectolytic and hemicellulolytic enzymes, disrupts the integrity of the host cells, producing cavities that are filled with the bacterium	Secoiridoids oleuropein and verbascoside (accumulate inside vacuoles) catechol, coumarins	Catechins change the properties and permeability of plasma membranes and increase formation of hydrogen peroxide	Young (2004), Maddox et al. (2010), Buonaurio et al. (2015)
Xylella fastidiosa – Pierce's disease in grape	Attacks and gives rise to toxic chemicals that destroy the surrounding plant cells		Polyphenols disrupt the process of quorum sensing	
	Block the plants' ability to uptake water and nutrients to the whole plant			
	Eventually the plant begins to wilt or droop			
Phaeomoniella chlamydospora, Phaeoacremonium aleophilum – Petri disease in grape vine	Fungal mycelium sticks to the xylem and tyloses formation in the xylem vessels by enzymatic activities softening of the xylem walls	p-Coumaric acid, catechin, caffeic acid, tannin	p-Coumaric acid inhibits radial growth of fungus and prevents the rate of sporulation	Río et al. (2004), Olivier et al. (2018), Schnee et al. (2008)
Botrytis cinerea – downy mildew in grape vine	Rapid synthesis of callose in the stomata of grapevine leaves	Pterostilbene and resveratrol, piceide, ε-viniferin (leaf epidermal cell)	Stilbene induces development of the infectious structure in the penetrated epidermal cell and produces resistance	

(continued)

Table 13.1 (continued)

Example	Mode of action	Phenolic compound	Function	References
Tomato leaf curl disease in tomato – transmitted by whitefly	Stunted growth, shoots and leaves are smaller than normal with an erect position, chlorotic leaves	Rutin and chlorogenic acid	THT gene catalyses the synthesis of p-coumaroyltyramine from the amine tyramine and the thioester p-coumaroyl-CoA and synthesis of phenylalanaine ammonia lyase	Singh et al. (2014), Sade et al. (2014), Mandhania et al. (2018)
Cotton leaf curl disease transmitted by white fly	Upward or downward curling of leaf margins, swelling and darkening of veins, develop into cup-shaped and leaflike outgrowths called 'enations'	Gossypol		
Meloidogyne javanica (root-knot nematode) infection in mungbean	Formation of typical root galls on the roots	Chlorogenic acid (leaves and roots), transcinnamic acid	Browning and slow formation of wide necrosis in plants vulnerable to the migratory nematodes	Ahmed et al. (2009), Ohri and Pannu (2010)
	Water and nutrient uptake are substantially reduced and stunted and chlorotic plants and reduce the photosynthetic rates		Browning and slow formation of wide necrosis in plants, speedy browning and emergence of non-expandable necrosis, IAA-oxidase inhibition favours auxin accumulation and as a result of galls or giant cell formation, IAA-oxidase stimulation which favours auxin decomposition and of necrosis in plants	

Trichobaris mucorea attacks on *Nicotiana attenuata* leaves	Plant cell wall destruction by creating an osmotically fragile cell	Chlorogenic acid (caffeic acid, quinic acid), catechin	Chlorogenic acid can be oxidized by plant polyphenol oxidase and generates reactive quinone species that limits microbial growth and tissue damage by cross-linking cell wall components	Lee et al. (2017), Niggeweg et al. (2004)
Canavalia ensiformis attacks on *Beta vulgaris* subsp. cicla	Compete with crop and less availability of water and nutrient to the plant	Cinnamic acid, naringin and rutin	Phenolic compounds of herbal extracts decrease the amylase activity in weeds, which slows starch hydrolysis process and seed germination	Soltys et al. (2013), Li et al. (2010), Mendes and Rezende (2014)
			Reduces cell division and retards weed growth	

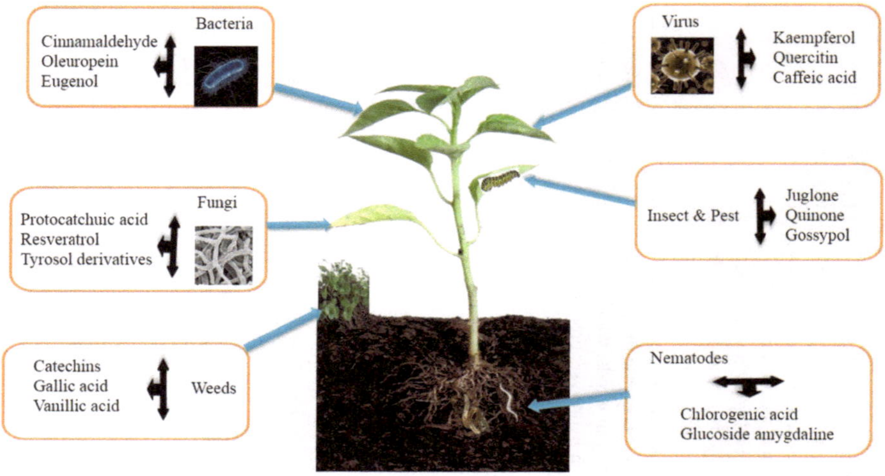

Fig. 13.1 Responses of phenolic compounds to various biotic stresses

of plant, spots on leaves, formation of galls and damaging of seeds (Anderson et al. 2004). They can harm all plant parts, including roots, stem, bark, leaves and flowers (Walling 2008) and transmit the disease caused by bacteria and viruses from infected part to uninfected plants parts (Mann et al. 2012). Weeds are the self-growing wild plants that restrict or inhibit the growth of other plants either directly by destruction or indirectly by increased competition with the host plant for area, light and nutrient availability. Weeds are fast-growing plants and their germination and growth is also high, so sometimes their rapid growth and adaptability to the environment helps them dominate to the host plants (Dass et al. 2016). The interrelationship of biotic stresses and phenolic compounds has been presented in Fig. 13.1.

References

Ahmed N, Abbasi MW, Shaukat SS, Zaki MJ (2009) Physiological changes in leaves of mungbean plants infected with Meloidogyne javanica. Phytopathol Mediterr 48:262–268

Anderson PK, Cunningham AA, Patel NG, Morales FJ, Epstein PR, Daszak P (2004) Emerging infectious diseases of plants: pathogen pollution, climate change and agrotechnology drivers. Trends Ecol Evol 19(10):535–544

Asgari Lajayer B, Ghorbanpour M, Nikabadi S (2017) Heavy metals in contaminated environment: Destiny of secondary metabolite biosynthesis, oxidative status and phytoextraction in medicinal plants. Ecotoxicol Environ Saf 145:377–390

Bai FW, Yan J, Qu ZC, Zhang HW, Xu J, Ye MM, Shen DL (2002) Phylogenetic analysis reveals that a dwarfing disease on different cereal crops in China is due to rice black streaked dwarf virus (RBSDV). Virus Gene 25(2):201–206

Bais HP, Weir TL, Perry LG, Gilroy S, Vivanco JM (2006) The role of root exudates in rhizosphere interactions with plants and other organisms. Annu Rev Plant Biol 57(1):233–266

Baskar V, Venkatesh R, Ramalingam S (2018) Flavonoids (antioxidants systems) in higher plants and their response to stresses. In: Antioxidants and Antioxidant Enzymes in Higher Plants. Springer, Cham, pp 253–268
Behmann J, Mahlein AK, Rumpf T, Römer C, Plümer L (2014) A review of advanced machine learning methods for the detection of biotic stress in precision crop protection. Precis Agric 16 (3):239–260
Bentur JS (2011) Insect pests of Rice in India and their management. In: Pests and pathogens: management strategies. CRC Press, Boca Raton, pp 1–42
Bhattacharya A, Sood P, Citovsky V (2010) The roles of plant phenolics in defence and communication during agrobacterium and rhizobium infection. Mol Plant Path 11(5):705–719
Bi B, Tang J, Han S, Guo J, Miao Y (2017) Sinapic acid or its derivatives interfere with abscisic acid homeostasis during Arabidopsis thaliana seed germination. BMC Plant Biol 17(1):99
Buonaurio R, Moretti C, da Silva DP, Cortese C, Ramos C, Venturi V (2015) The olive knot disease as a model to study the role of interspecies bacterial communities in plant disease. Front Plant Sci 6:434
Burch-Smith TM, Zambryski PC (2016) Regulation of plasmodesmal transport and modification of Plasmodesmata during development and following infection by viruses and viral proteins. Plant Virus Int:87–122
Byers JA (1995) Host-tree chemistry affecting colonization in bark beetles. Chem Ecol Insect 2:154–213
Carris LM, Little CR, Stiles CM (2012) Introduction to fungi. The plant health instructor
Cheynier V, Comte G, Daviesc MK, Lattanziod V, Martense S (2013) Plant phenolics: recent advances on their biosynthesis, genetics, and ecophysiology. Plant Physiol Biochem 72:1–20
Chitwood DJ (2003) Research on plant-parasitic nematode biology conducted by the United States Department of Agriculture-Agricultural Research Service. Pest Manag Sci 59:748–753
Christos A (2016) Damalas safe food production with minimum and judicious use of pesticides. In: Food safety. Springer, pp 43–55
Congdon BS, Coutts BA, Renton M, Flematti GR, Jones RAC (2017) Establishing alighting preferences and species transmission differences for pea seed-borne mosaic virus aphid vectors. Virus Res 241:145–155
Culver JN, Padmanabhan MS (2007) Virus-induced disease: altering host physiology one interaction at a time. Annu Rev Phytopathol 45(1):221–243
Czemmel S, Höll J, Loyola R, Arce-Johnson P, Alcalde JA, Matus JT, Bogs J (2017) Transcriptome-wide identification of novel UV-B- and light modulated flavonol pathway genes controlled by VviMYBF1. Front Plant Sci 8. https://doi.org/10.3389/fpls.2017.01084
Danciu C, Avram S, Pavel IZ, Ghiulai R, Dehelean CA, Ersilia A, Soica C (2018) Main isoflavones found in dietary sources as natural anti-inflammatory agents. Curr Drug Target 19(7):841–853
Dar SA, Rather BA, Wani AR, Ganie MA (2017) Resistance against insect pests by plant Phenolics and their derivative compounds. Chem Sci Rev Lett 6(23):1941–1949
Dass A, Shekhawat K, Choudhary AK, Sepat S, Rathore SS, Mahajan G, Chauhan BS (2016) Weed management in rice using crop competition a review. Crop Prot 95:45–52
Davies K, Spiegel Y (2011) Biological control of plant-parasitic nematodes. https://doi.org/10.1007/978-1-4020-9648-8
Del Fabbro C, Prati D (2015) The relative importance of immediate allelopathy and allelopathic legacy in invasive plant species. Basic Appl Ecol 16(1):28–35
Delaye L, García-Guzmán G, Heil M (2013) Endophytes versus biotrophic and necrotrophic pathogens—are fungal lifestyles evolutionarily stable traits? Fungal Divers 60(1):125–135
Dresselhaus T, Hückelhoven R (2018) Biotic and abiotic stress responses in crop plants. Agron 8:267
Dube HC (2015) An introduction to fungi, 4th edn. Scientific Publishers, pp 978–981
Duke SO (2018) Interaction of chemical pesticides and their formulation ingredients with microbes associated with plants and plant pests. J Agric Food Chem 66(29):7553–7561

Easwar Rao D, Divya K, Prathyusha IVSN, Rama Krishna C, Chaitanya KV (2017) Insect-resistant plants. In: Current developments in biotechnology and bioengineering. Elsevier, Amsterdam, pp 47–74

Escobar C, Barcala M, Cabrera J, Fenoll C (2015) Overview of root-knot nematodes and Giant cells. In: Plant nematode interactions – a view on compatible interrelationships. Academic, Amsterdam, pp 1–32

Ferdous AS, Islam MR, Khan H (2017) Transkingdom signaling systems between plant and its associated beneficial microbes in relation to plant growth and development. In: Mycorrhiza – nutrient uptake, biocontrol, ecorestoration. Springer, Cham, pp 451–472

Fried G, Chauvel B, Reynaud P, Sache I (2017) Decreases in crop production by non-native weeds, pests, and pathogens. In: Impact of biological invasions on ecosystem services. Springer, Cham, pp 83–101

Garg N, Manchanda G, Kumar A (2013) Bacterial quorum sensing: circuits and applications. Antonie Van Leeuwenhoek 105(2):289–305

Ghosh P, Bhattacharya A, Char B (2017) Manipulating disease and pest resistance pathways in plants for enhanced crop improvement. Biosci Biotech Res Commun 10(4):631–644

Giebel J (1982) Mechanism of resistance to plant nematodes. Annu Rev Phytopathol 20 (1):257–279

Gourion B, Berrabah F, Ratet P, Stacey G (2015) Rhizobium–legume symbioses: the crucial role of plant immunity. Trends Plant Sci 20(3):186–194

Green ER, Mecsas J (2015) Bacterial secretion systems: an overview. In: Virulence mechanisms of bacterial pathogens, 5th edn. ASM Press, Washington, DC, pp 215–239

Guo Q, Major IT, Howe GA (2018) Resolution of growth–defense conflict: mechanistic insights from jasmonate signaling. Curr Opin Plant Biol 44:72–81

Hajano JUD, Pathan MA, Rajput AQ, Lodhi MA (2011) Rice blast-mycoflora, symptomatology and pathogenicity. Int J Agro Vet Med Sci 5(1):53–63

Hardoim PR, Van Overbeek LS, Berg G, Pirttilä AM, Compant S, Campisano A, Sessitsch A (2015) The hidden world within plants: ecological and evolutionary considerations for defining functioning of microbial endophytes. Microbiol Mol Biol Rev 79(3):293–320

Hegenauer V, Körner M, Albert M (2017) Plants under stress by parasitic plants. Curr Opin Plant Biol 38:34–41

Hirano SS, Upper CD (2000) Bacteria in the leaf ecosystem with emphasis on pseudomonas syringae-a pathogen, ice nucleus, and epiphyte. Microbiol Mol Biol Rev 64(3):624–653

Holbein J, Grundler FMW, Siddique S (2016) Plant basal resistance to nematodes: an update. J Exp Bot 67(7):2049–2061

Huang XF, Chaparro JM, Reardon KF, Zhang R, Shen Q, Vivanco JM (2014) Rhizosphere interactions: root exudates, microbes, and microbial communities. Botany 92(4):267–275

Hubbermann EM (2016) Coloring of low-moisture and gelatinized food products. In: Handbook on natural pigments in food and beverages, pp 179–196

Jabran K, Mahmood K, Melander B, Bajwa AA, Kudsk P (2017) Weed dynamics and management in wheat. AdvAgron:97–166

Janczarek M, Rachwał K, Marzec A, Grządziel J, Palusińska-Szysz M (2015) Signal molecules and cell-surface components involved in early stages of the legume–rhizobium interactions. Appl Soil Ecol 85:94–113

John J (2012) Role of phenolics in allelopathic interactions. Allelopathy J 29(2):215–230

Jones DL, Hodge A, Kuzyakov Y (2004) Plant and mycorrhizal regulation of rhizodeposition. New Phytol 163(3):459–480

Joshi JR, Khazanov N, Senderowitz H, Burdman S, Lipsky A, Yedidia I (2016) Plant phenolic volatiles inhibit quorum sensing in pectobacteria and reduce their virulence by potential binding to ExpI and ExpR proteins. Sci Rep 6(1)

Kong JM, Chia LS, Goh NK, Chia TF, Brouillard R (2003) Analysis and biological activities of anthocyanins. Phytochemistry 64(5):923–933

Kumar A, Verma JP (2018) Does plant—microbe interaction confer stress tolerance in plants: a review? Microbiol Res 207:41–52

Kyndt T, Zemene HY, Haeck A, Singh R, De Vleesschauwer D, Denil S, Gheysen G (2017) Below-ground attack by the root knot nematode meloidogyne graminicola predisposes Rice to blast disease. Mol Plant Microb Int 30(3):255–266

Lattanzio V, Lattanzio VMT, Cardinali A (2006) Role of phenolics in the resistance mechanisms of plants against fungal pathogens and insects. Phytochem Adv Res 81:23–67

Ledesma-Escobar CA, Priego-Capote F, Luque de Castro MD (2019) Relevance and analysis of Citrus flavonoids. In: Polyphenols in plants. Elsevier, London, pp 133–150

Lee SM, Radhakrishnan R, Kang SM, Kim JH, Lee IY, Moon BK, Lee IJ (2015) Phytotoxic mechanisms of bur cucumber seed extracts on lettuce with special reference to analysis of chloroplast proteins, phytohormones, and nutritional elements. Ecotoxicol Environ Saf 122:230–237

Lee G, Joo Y, Kim SG, Baldwin IT (2017) What happens in the pith stays in the pith: tissue -localized defense responses facilitate chemical niche differentiation between two spatially separated herbivores. Plant J 92:414–425

Levin DA (1971) Plant phenolics: an ecological perspective. Am Nat 105(942):157–181

Li ZH, Wang Q, Ruan X, Pan C, Jiang D (2010) Phenolics and plant allelopathy. Molecules 15 (12):8933–8952

Lin D, Xiao M, Zhao J, Li Z, Xing B, Li X, Chen S (2016) An overview of plant phenolic compounds and their importance in human nutrition and management of Type 2 diabetes. Molecules 21(10):1374

Lincoln JE, Sanchez JP, Zumstein K, Gilchrist DG (2018) Plant and animal PR1 family members inhibit programmed cell death and suppress bacterial pathogens in plant tissues. Mol Plant Path 19(9):2111–2123

Ludwig-Müller J (2015) Bacteria and fungi controlling plant growth by manipulating auxin: balance between development and defense. J Plant Physiol 172:4–12

Ma XH, Ma Y, Tang JF, He YL, Liu YC, Ma XJ, Huang LQ (2015) The biosynthetic pathways of Tanshinones and phenolic acids in Salvia miltiorrhiza. Molecules 20(9):16235–16254

Maccheroni W, Alegria MC, Greggio CC, Piazza JP, Kamla RF, Zacharias PRA, da Silva ACR (2005) Identification and genomic characterization of a new virus (Tymoviridae family) associated with citrus sudden death disease. J Virol 79(5):3028–3037

Maddox CE, Laur LM, Tian L (2010) Antibacterial activity of phenolic compounds against the Phytopathogen *Xylella fastidiosa*. Curr Microbiol 60(1):53–58

Mandal SM, Chakraborty D, Dey S (2010) Phenolic acids act as signaling molecules in plant-microbe symbioses. Plant Signal Behav 5(4):359–368

Mandhania S, Sangwan RS, Siwach SS, Pundir SR, Sangwan O, Janu A (2018) Role of biochemical constituents and minerals against cotton leaf curl disease in cotton. J Environ Biol 39:221–227

Mann RS, Ali JG, Hermann SL, Tiwari S, Pelz-Stelinski KS, Alborn HT, Stelinski LL (2012) Induced release of a plant-defense volatile 'deceptively' attracts insect vectors to plants infected with a bacterial pathogen. PLoS Pathog 8(3):111–124

Marahatta SP (2018) Nematode community associated with recently planted breadfruit, Artocarpus Altilis, rhizosphere on Kaua'I, Hawai'I. Nematropica 48(2)

Mayer AM, Staples RC, Gil-ad NL (2001) Mechanisms of survival of necrotrophic fungal plant pathogens in hosts expressing the hypersensitive response. Phytochemistry 58(1):33–41

Melvin P, Bankapalli K, D'Silva P, Shivaprasad PV (2017) Methylglyoxal detoxification by a DJ-1 family protein provides dual abiotic and biotic stress tolerance in transgenic plants. Plant Mol Biol 94(4–5):381–397

Mendes IDS, Rezende MOO (2014) Assessment of the allelopathic effect of leaf and seed extracts of Canavalia ensiformis as post emergent bioherbicides: a green alternative for sustainable agriculture. J Environ Sci Health Part B 49(5):374–380

Miller R, Owens SJ, Rørslett B (2011) Plants and colour: flowers and pollination. Opt Laser Technol 43(2):282–294

Minnis ST, Haydock PPJ, Ibrahim SK, Grove IG, Evans K, Russell MD (2002) Potato cyst nematodes in England and Wales – occurrence and distribution. Ann App Biol 140(2):187–195

Mukherjee S, Kutty NN, Bera P, Mitra A (2018) Impact of light and sucrose supplementation on cellular differentiation, metabolic shift and modulation of gene expression in hairy roots of Daucus carota. In: Plant cell, tissue and organ culture. Springer, Berlin, pp 1–15

Niggeweg R, Michael AJ, Martin C (2004) Engineering plants with increased levels of the antioxidant chlorogenic acid. Nat Biotechnol 22:746–754

Ohri P, Pannu SK (2010) Effect of phenolic compounds on nematodes- a review. J Appl Nat Sci 2 (2):344–350

Olivier V, Spring JL, Gindro K (2018) Stilbenes: biomarkers of grapevine resistance to fungal diseases. OENO One 52(3):235–241

Parr AJ, Bolwell GP (2000) Phenols in the plant and in man. The potential for possible nutritional enhancement of the diet by modifying the phenols content or profile. J Sci Food Agric 80 (7):985–1012

Patel A, Panchal I, Parmar I, Mishtry B (2017) Synthesis of new flavonoid and chalcone derivatives as antimicrobial agent by green chemistry approach. Int J Pharm Sci Res 8(6):2725–2730

Peperidou A, Pontiki E, Hadjipavlou-Litina D, Voulgari E, Avgoustakis K (2017) Multifunctional cinnamic acid derivatives. Molecules 22(8):1247

Pérez-Montaño F, Guasch-Vidal B, González-Barroso S, López-Baena FJ, Cubo T, Ollero FJ, Espuny MR (2011) Nodulation-gene-inducing flavonoids increase overall production of autoinducers and expression of N-acyl homoserine lactone synthesis genes in rhizobia. Res Microbiol 162(7):715–723

Pina-Pérez MC, Ferrús Pérez MA (2018) Antimicrobial potential of legume extracts against foodborne pathogens: a review. Trends Food Sci Technol 72:114–124

Postel S, Kemmerling B (2009) Plant systems for recognition of pathogen-associated molecular patterns. Semin Cell Dev Biol 20(9):1025–1031

Pusztahely T, István H, Pócsi IJ (2017) Plant-fungal interactions: special secondary metabolites of the biotrophic, necrotrophic, and other specific interactions. In: Fungal metabolites. Academic, London, pp 1–58

Qasem JR (2006) Parasitic weeds and allelopathy: from the hypothesis to the proofs. Allelopathy:565–637

Radice S, Arena ME (2015) Environmental effect on the leaf morphology and anatomy of Berberis microphylla G. Forst Int J Plant Biol 6:5677

Raman N, Marchillo K, Lee MR, Rodríguez López A, Andes DR, Palecek SP, Lynn DM (2015) Intraluminal release of an antifungal β-peptide enhances the antifungal and anti-biofilm activities of multilayer-coated catheters in a rat model of venous catheter infection. ACS Biomater Sci Eng 2(1):112–121

Río JAD, Gómez P, Báidez A, Fuster MD, Ortuño A, Frías V (2004) Phenolic compounds have a role in the defence mechanism protecting grapevine against the fungi involved in Petri disease. Phytopathol Mediterr 43:87–94

Robert R, Junker Amy L (2015) Working towards a holistic view on flower traits—how floral scents mediate plant–animal interactions in concert with other floral characters. J Ind Ins Sci 95 (1):43–67

Saddique M, Kamran M, Shahbaz M (2018) Differential responses of plants to biotic stress and the role of metabolites. In: Plant metabolites and regulation under environmental stress. Elsevier, San Diego, pp 69–87

Sade D, Shriki O, Cuadros-Inostroza A, Tohge T, Semel Y, Haviv Y, Brotman Y (2014) Comparative metabolomics and transcriptomics of plant response to tomato yellow leaf curl virus infection in resistant and susceptible tomato cultivars. Metabolomics 11(1):81–97

Saltveit ME (2009) Synthesis and metabolism of phenolic compounds. In: Fruit and vegetable phytochemicals: chemistry, nutritional value, and stability. Wiley-Blackwell, Ames, pp 89–100

Saltveit ME (2017) Synthesis and metabolism of phenolic compounds. In: Fruit and vegetable phytochemicals. Wiley-Blackwell, Ames, pp 89–100

Santos-Sánchez NF, Salas-Coronado R, Hernández-Carlos B, Villanueva-Cañongo C (2019) Shikimic acid pathway in biosynthesis of phenolic compounds. In: Plant physiological aspects of phenolic compounds. https://doi.org/10.5772/intechopen.83815

Scavo A, Restuccia A, Mauromicale G (2018) Allelopathy: principles and basic aspects for agroecosystem control. In: Sustainable agriculture reviews. Springer, Cham, pp 47–101
Schnee S, Viret O, Gindro K (2008) Role of stilbenes in the resistance of grapevine to powdery mildew. Physiol Mol Plant Path 72(4–6):128–133
Scognamiglio M, D'Abrosca B, Esposito A, Pacifico S, Monaco P, Fiorentino A (2013) Plant growth inhibitors: allelopathic role or phytotoxic effects? Focus on Mediterranean biomes. Phytochem 12(4):803–830
Silva S, Santos C, Serodio J, Silva AMS, Dias MC (2018) Physiological performance of drought-stressed olive plants when exposed to a combined heat–UV-B shock and after stress relief. Func Plant Biol. https://doi.org/10.1071/FP18026
Singh RK, Rai N, Singh M, Singh SN, Srivastava K (2014) Selection of tomato genotypes resistant to tomato leaf curl virus disease using biochemical and physiological markers. J Agric Sci 153 (04):646–655
Soltys D, Krasuska U, Bogatek R, Gniazdow A (2013) Allelochemicals asbio-herbicides -present and perspectives. In: Price AJ, Kelton JA (eds) Herbicides -current research and case studies in use. InTech, Croatia, pp 517–542
Sun S, Zhi Y, Zhu Z, Jin J, Duan C, Wu X, Xiaoming W (2017) An emerging disease caused by pseudomonas syringae pv. phaseolicola threatens mung bean production in China. Plant Dis 101 (1):95–102
Swanton CJ, Nkoa R, Blackshaw RE (2015) Experimental methods for crop–weed competition studies. Weed Sci 63:2–11
Talhaoui N, Taamalli A, Gómez-Caravaca AM, Fernández-Gutiérrez A, Segura-Carretero A (2015) Phenolic compounds in olive leaves: analytical determination, biotic and abiotic influence, and health benefits. Food Res Int 77:92–108
Tarkowski P, Vereecke D (2014) Threats and opportunities of plant pathogenic bacteria. Biotechnol Adv 32(1):215–229
Thakur H, Jindal SK, Sharma A, Dhaliwal MS (2018) Chilli leaf curl virus disease: a serious threat for chilli cultivation. J Plant Dis Protect 125(3):239–249
Tripathi P (2004) Exploitation of natural products as an alternative strategy to control postharvest fungal rotting of fruit and vegetables. Postharv Biol Technol 32(3):235–245
Ueda H, Kugimiya S, Tabata J, Kitamoto H, Mitsuhara I (2018) Accumulation of salicylic acid in tomato plant under biological stress affects oviposition preference of Bemisia tabaci. J Plant Int 14(1):73–78
Varbanova M, Porter K, Lu F, Ralph J, Hammerschmidt R, Jones AD, Day B (2011) Molecular and biochemical basis for stress-induced accumulation of free and bound p-coumaraldehyde in cucumber. Plant Physiol 157(3):1056–1066
Vishwanath SJ, Delude C, Domergue F, Rowland O (2014) Suberin: biosynthesis, regulation, and polymer assembly of a protective extracellular barrier. Plant Cell Rep 34(4):573–586
Wallace KJ, Laughlin DC, Clarkson BD (2017) Exotic weeds and fluctuating microclimate can constrain native plant regeneration in urban forest restoration. Ecol Appl 27(4):1268–1279
Walling LL (2008) Avoiding effective defenses: strategies employed by phloem-feeding insects. Plant Physiol 146(3):859–866
Wang L, Sun R, Zhang Q, Luo Q, Zeng S, Li X, Liu Z (2018) An update on polyphenol disposition via coupled metabolic pathways. In: Expert opinion on drug metabolism & toxicology. Ashley Publications, London, pp 1–15
Wani SH, Kumar V, Shriram V, Sah SK (2016) Phytohormones and their metabolic engineering for abiotic stress tolerance in crop plants. Crop J 4(3):162–176
War AR, Paulraj MG, Ahmad T, Buhroo AA, Hussain B, Ignacimuthu S, Sharma HC (2012) Mechanisms of plant defense against insect herbivores. Plant Signal Behav 7(10):1306–1320
Young JM (2004) Olive knot and its pathogen. Aust Plant Path 33:33–49
Zernova O, Lygin A, Pawlowski M, Hill C, Hartman G, Widholm J, Lozovaya V (2014) Regulation of plant immunity through modulation of phytoalexin synthesis. Molecules 19(6):7480–7496

Phenolics from Agro-industrial By-Products 14

Gabriela Vázquez-Olivo, Luis A. Cabanillas-Bojórquez,
C. Alicia Elizalde-Romero, and J. Basilio Heredia

Abstract

Phenolics are organic compounds which contain a phenol ring and one or more hydroxyl groups. These molecules are produced as secondary metabolites and are ubiquitous in all plant tissues. Besides, the consumption of phenolic compounds has been associated with the promotion of human health. Therefore, the extraction of these molecules has been of great interest for many researches and could be used for food and pharmaceutical purposes. Agro-industrial by-products are rich in many phytochemicals, such as phenolic compounds. These by-products are generated from fruit and vegetable processing and have been mainly regarded as wastes. Thus, there is a need to revalorize these by-products. Recently, the research have been focusing of the identification and quantification of phenolic content from agro-industrial by-products, the maximization of the yield of recovery of phenolic compounds assessed by different methods of extraction, and the enhancement of the concentration of phenolic compounds. Thus, this chapter describes the extraction of phenolics from agro-industrial by-products by different methodologies, as well as the enhancement of the phenolic content and their biological properties.

Keywords

Agro industry · By-products · Phenolics · Extraction · Methods

G. Vázquez-Olivo · L. A. Cabanillas-Bojórquez · C. A. Elizalde-Romero · J. B. Heredia (✉)
Department of Food Science and Technology, Centro de Investigación en Alimentación y Desarrollo, A.C, Culiacán, Sinaloa, Mexico
e-mail: jbheredia@ciad.mx

R. Lone et al. (eds.), *Plant Phenolics in Sustainable Agriculture*,
https://doi.org/10.1007/978-981-15-4890-1_14

14.1 Introduction

Phenolic compounds are ubiquitous in plants, present in almost all plant tissues, and they have been studied to present a variety of chemical structures (Santana-Gálvez and Jacobo-Velázquez 2018). Their consumption has been associated to a wide range of bioactivities mainly due to its radical scavenging capacity which is relevant for the treatment and/or prevention of many diseases (de Camargo et al. 2016).

Agro-industrial by-products have been evaluated for their bioactivity since there is a need for a further utilization of this plant material. They represent an abundant and cheap source of phenolic compounds with many potential applications, being in some cases richer in phenolic compounds than the main product. Furthermore, adding value to agro-industrial by-products can enhance profits and reduce the costs and the negative environmental effect of its disposition. Hence, the evaluation of the biological effect and utilization of agro-industrial by-products have become of scientific interest (Oskoueian et al. 2011). This chapter aims to review the phenolic compounds present in some agro-industrial by-products and discuss the methods of recovery of this molecules, besides, the biological effect for their potential application in many industries.

14.2 Phenolic Compounds from Agro-industrial By-Products

Phenolic compounds come from plant's secondary carbon metabolism from the shikimic and the malonic acid pathways. Over 8000 of different phenolic compounds have been identified, with different structures and metabolic functions (Santana-Gálvez and Jacobo-Velázquez 2018). They all have in common a phenyl ring and a hydroxyl group (Fig. 14.1).

Phenolic compounds are known by their antioxidant activity which is mainly related to its ability to chelate metals (Decker 1997). Their synthesis is genetically regulated in each plant, so the type of phenolics and the amounts of them may vary. There are some environmental factors that can have an influence in their production, such as biotic and abiotic factors (Gimeno Creus 2004).

14.3 Classification of Phenolic Compounds

There are many forms to classify phenolic compounds; for instance, phenolics can be classified according to its structure (Table 14.1).

Fig. 14.1 Structure of phenol

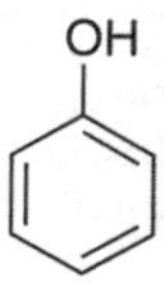

Table 14.1 Classification of phenolic compounds

Class	Basic skeleton
Simple phenols	C_6
Benzoquinones	C_6
Phenolic acids	$C_6–C_1$
Acetophenones	$C_6–C_2$
Phenylacetic acids	$C_6–C_2$
Hydroxycinnamic acids	$C_6–C_3$
Phenylpropenes	$C_6–C_3$
Coumarins, isocoumarins	$C_6–C_3$
Chromones	$C_6–C_3$
Naphtoquinones	$C_6–C_4$
Xanthones	$C_6–C_1–C_6$
Stilbenes	$C_6–C_2–C_6$
Athraquinones	$C_6–C_2–C_6$
Flavonoids	$C_6–C_3–C_6$
Lignans, neolignans	$(C_6–C_3)_2$
Lignins	$(C_6–C_3)_n$

Source: (Balasundram et al. 2006)

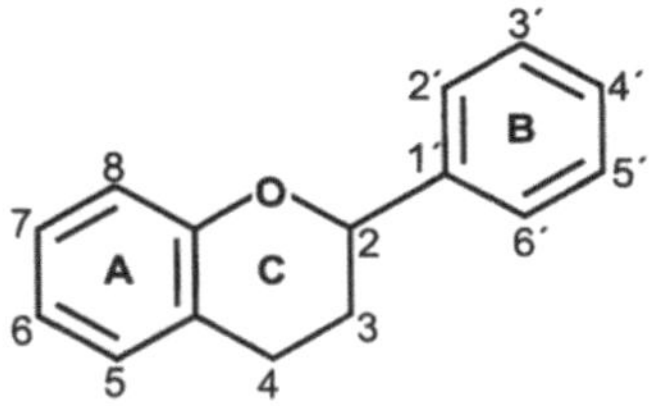

Fig. 14.2 Basic chemical structure of flavonoids

Flavonoids are known for being the most important group of phenolic compounds with more than 5000 structures described in literature. The basic structure of flavonoids consists in two rings (A and B) inked by a 3-carbon unit (ring C) (Fig. 14.2). This group of polyphenols is subdivided into 13 classes: chalcones, dihydrochalcones, flavones, aurones, flavonols, dihydroflavonol, flavanones, flavanol, flavandiol, anthocyanidin, isoflavonoids, biflavonids, and condensed tannins (Santana-Gálvez and Jacobo-Velázquez 2018).

In nature, flavonoids are usually present in its glycosylated form mainly with glucose or rhamnose. They can also be linked with galactose, arabinose, xylose, and other sugars. The classification of flavonoids is summarized in Table 14.2.

The concentration of phenolic compounds vary greatly, even among cultivars if the same species. Their synthesis depends mostly on light, which explains the higher concentrations of some phenolics in leaves. Other factors, such as genetics and environmental conditions can change phenolic quantities and composition. Ripeness, germination, procession, storage, and other factors have to be considered too (Bravo 1998).

Table 14.2 Classification of flavonoids

Group	Structure	Examples
Flavone	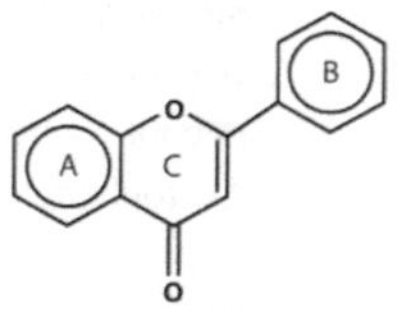	Apigenin Luteolin
Flavonol	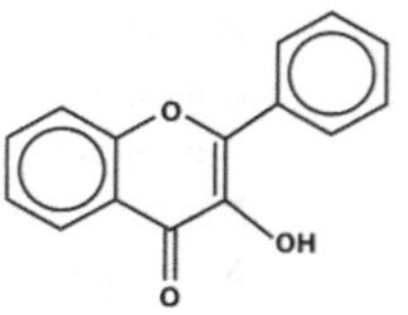	Kaempferol Quercetin
Flavanone	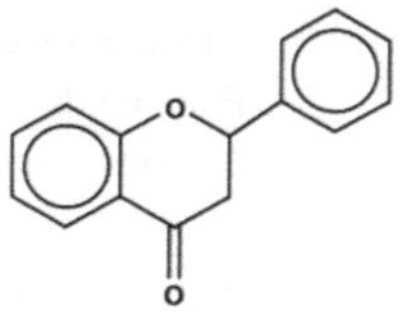	Hesperitin Naringenin
Flavanol	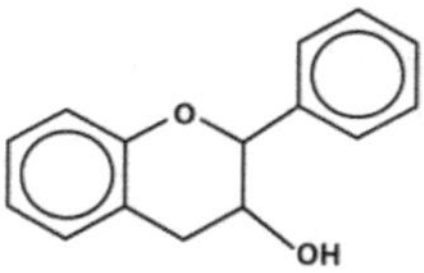	Epicatechin Catechin
Anthocyanidin	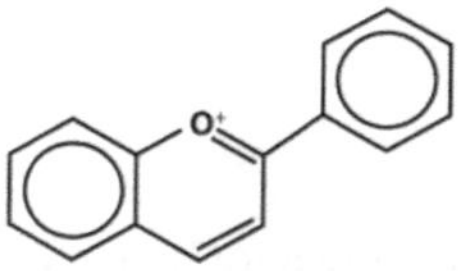	Cyanidin Peonidin

14.4 Biological Properties of Phenolic Rich Extracts from Agro-industrial By-Products

Phenolic compounds have attracted scientific attention ought to their biological activity which could be beneficial for human health as demonstrated in in vitro and in vivo assays. As stated earlier in this chapter, phenolic compounds possess a plethora of structural characteristics which can in turn affect their biological effect.

Agro-industrial by-products are derived from plant sources and can therefore be useful for phenolic compound extraction, as well as for the application in many industries such cosmetic, food, and pharmaceutic; hence, the study of their bioactive properties is important. Agro-industrial by-products have been reported to have potential in the treatment and/or prevention of several diseases or conditions, such as thrombosis, hepatic steatosis, diabetes, obesity, and among others (de Camargo et al. 2016; Meshkini and Tahmasbi 2017; Vergani et al. 2016). Some cellular assays have been performed, for example, in olive by-products. In this sense, olive industry generates by-products such as olive pomace and olive leaf. Olive pomace was tested for its potential effect on hepatic steatosis. For this aim, FaO cells, from rat hepatoma, were overloaded of lipids. Olive pomace (0.05 mg caffeic acid equivalents/mL) significantly reduced the content of triglycerides in FaO cells. Furthermore, 0.1 mg caffeic acid equivalents/mL of olive pomace extract reduced the 3,4-methylenedioxyamphetamine levels, which is used to evaluate lipid peroxidation through the thiobarbituric acid reactive substances assay (Vergani et al. 2016).

In another study, polyphenols from walnut hull showed to have antiplatelet aggregation properties through modulation of the intracellular redox status which in turn modulates the caspase activation. This is important for the treatment of thrombotic disorders (Meshkini and Tahmasbi 2017). Rapeseed meal, cottonseed meal, and soybean meal are by-products produced from oil industry. These by-products are scarcely studied; nonetheless they are rich in polyphenols which can be used for food and pharmaceutical industries.

Other in vitro studies have shown radical scavenging activity, xanthine and tyrosinase inhibitory activity, as well as anti-inflammatory potential. Rapeseed meal, cottonseed meal, and soybean meal were reported to inhibit DPPH radical scavenging activity and nitric oxide in a dose-dependent manner. Also, they showed xanthine oxidase inhibitory activity and tyrosinase activity, in a concentration-dependent manner. However, rapeseed meal showed the best results. Furthermore, the inhibition on nitric oxide production in RAW 264.7 cells induced by lipopolysaccharide and interferon gamma was evaluated, and rapeseed meal extract (62.5 μg/mL) inhibits the nitric oxide production comparable to N-nitro-L-arginine methyl ester (Oskoueian et al. 2011). *Panax notoginseng* stem leaf is rich in saponins and flavonoid glycosides; furthermore, it has potential to be used in cosmetic industry as it can inhibit tyrosinase activity resulting in a reduction in the melatonin synthesis (Dai et al. 2018).

Apart from the by-products obtaining from the processing of foods, there are by-products such as the pods and stems obtained from the pruning of the olive trees that can be exploited for phenolic compound extraction (Salido et al. 2015). Olive wood shavings were extracted with dichloromethane and ethyl acetate for the DPPH assay. Moreover, simple phenols, flavonoids, and terpenoids were also identified.

Furthermore, it has been reported that phenolic compounds correlate positively with antioxidant activity; for instance, broccoli by-products such as florets, stalks, and leaves, which are rich in bioactive compounds such as glucosinolates, phenolic acids, and flavonoids, showed a positive correlation with antioxidant activity ($r^2 = 0.906$) (Domínguez-Perles et al. 2010).

Table 14.3 Biological effect of agro-industrial by-products

By-product	Biological effect	Reference
Olive pomace	Anti-steatotic effect	Vergani et al. (2016)
Walnut hull	Antiplatelet aggregation property	Meshkini and Tahmasbi (2017)
Broccoli by-products	In vitro radical scavenging capacity	Domínguez-Perles et al. (2010)
Rapeseed meal, cottonseed, and soybean meal	Antioxidant and anti-inflammatory activity	Oskoueian et al. (2011)
Olive wood shavings	Radical scavenging capacity	Salido et al. (2015)
Panax notoginseng stem leaf	Free radical scavenging activity and superoxide anion removal ability	Dai et al. (2018)
Winemaking by-products	Free radical scavenging activity, reducing power, inhibition of the activity of α-amylase and lipase enzymes	de Camargo et al. (2016)
Olive leaf	Radical scavenging capacity	Şahin et al. (2018)
Red and purple rice bran	Oxygen radical absorbance capacity	Chen et al. (2016)
Black rice bran	Oxygen radical absorbance capacity	Zhang et al. (2010)

In addition, agro-industrial by-products have shown to possess higher antioxidant capacity than the main product. For instance, pomegranate peel has strongest radical scavenging capacity than juice (Fawole and Opara 2016). Table 14.3 summarizes the biological effects studied in some agro-industrial by-products.

Although there is great evidence of the antioxidant properties of some agro-industrial by-products, there is still need for evaluating their stability and activity after a simulated in vitro digestion and in vivo metabolism utilizing the whole by-product (Fawole and Opara 2016). Another important fact is that bioactivity can be improved or reduced by the method of extraction of phenolic compounds; for instance, winemaking by-products showed to increase their bioactivity after an enzymatic treatment. Therefore, some of the traditional and emerging methods for the extraction of phenolics from agro-industrial by-products are reviewed later in this chapter (de Camargo et al. 2016).

14.5 Methods of Recovery of Phenolics from Agro-industrial By-Products

Conventional methods of extraction such as reflux extraction and Soxhlet are still being used for the recovery of phenolics; however, due to the concern for enhancing the yield of phenolic recovery and the maintenance of their bioactivity, emerging technologies have been assessed; among this, the most reported are pressure-driven membrane process, microwave-assisted extraction, supercritical fluid extraction, ultrasound-assisted extraction, and subcritical water extraction (Giannuzzo et al. 2003).

14.5.1 Conventional Extraction Techniques

Conventional extraction techniques have been reported to be efficient for the recovery of flavonoids and phenolic acids from agro-industrial by-products (Al-Jabari 2002; da Silva et al. 2016).

14.5.2 Maceration and Reflux Extraction

Grapefruit peel is rich in naringin, a glycosylated flavonoid. Therefore, different methods were used to recover this bioactive compound. By means of maceration and reflux, the yields were 13.5 ± 0.5 g/Kg grapefruit peel and 11.1 ± 0.6 g/Kg grapefruit peel, respectively (Giannuzzo et al. 2003). Differences observed in the yield of extraction might be due to the extraction time.

Olive tree by-products such as olive tree pruning and olive mill leaf were evaluated for the efficient solvent to extract phenolic acids and flavonoids. Gullón et al. (2018) studied a bioactive compound extraction with different solvents from olive tree pruning and olive mill leaf, among these the highest total phenolic and flavonoids content was with 50% ethanol an extraction conditions of extraction time of 90 min, extraction temperature of 55 °C, and a liquid to solid ratio of 6 mL/g.

Similarly, in order to increase the phenolic compound extraction of non-pomace residue from grape processing, Haas et al. (2018) reported that conventional extraction (methanol-water mixture extraction with mechanical orbital agitation) with 5 min of time extraction, solvent to water ratio of 80:20%, and solid/solvent ratio of 1:20 (w/v) were the most efficient conditions to polyphenol extraction. They agree with other authors who report that the water addition in organic solvents facilitates the bioactive compound extraction such as phenolics; this could be due to the polarity being affected by the addition of water to the mixture, therefore increasing the phenolic extraction.

14.5.3 Soxhlet Extraction

Giannuzzo et al. (2003) reported the flavonoid glycosylated naringin extraction from grapefruit peel; they obtained a yield naringin extraction of 15.2 ± 0.5 g/kg grapefruit peel to the extraction conditions of 8 h and 200 mL of ethanol. The Soxhlet extraction was the highest naringin yield in comparison with other extraction methods; nevertheless the extraction time was long, and a significant amount of solvents was used. On the other hand, Chanioti and Tzia (2017) extracted phenolics compounds from olive pomace with a Soxhlet method using 250 mL of hexane for 8 h. They reported a yield extraction of 12.38 ± 0.07% and total phenolic content of 0.207 ± 0.009 mg gallic acid/g oil. In comparison with ultrasound-assisted extraction, yield and total phenolics content by Soxhlet method was lower, therefore the importance of using alternative methods for bioactive compound extraction from industry by-products such as olive pomace.

Similarly, Louli et al. (2004) studied the phenolic compound extraction of red grape by-products with different solvents (aqueous KOH (3%), methanol, and ethyl acetate). They reported that the best solvent for total phenolics compound extraction was ethyl acetate; nevertheless, the extracts must be purified, and therefore they used the extraction of supercritical fluids as purification method.

On the other hand, Da Porto et al. (2013) have reported the polyphenol extraction (89.47 ± 2.29 mg GAE/g flour) from grape seeds using maceration and Soxhlet extraction; they found that despite obtaining a significant amount using both successive extractions, the highest polyphenol content extraction was using ultrasound and Soxhlet (105.81 ± 3.21 mg GAE/g flour); also, the polyphenol extraction was influenced to the de-oiling method from grape seeds, which can interfere into the polyphenol extraction.

"Horchata" by-products are rich sources of phenolic compounds; therefore different extraction techniques are reported; in this sense, Roselló-Soto et al. (2019) studied the supercritical CO_2 and conventional extractions on phenolic compound recovery. They reported an important amount of phenolic compounds using conventional extraction; besides, *p*-coumaric and ferulic acids were identified with conventional extraction.

Although these methods of extractions have been widely used, many disadvantages have been reported, such as the use of large amounts of solvents, which disposition generates a negative impact in the environment. Also, these techniques require long extraction times and need to further processes to remove solvents.

14.6 Emerging Technologies

Emerging technologies for bioactive compound extraction have been studied in the last years; due to climate change, environmentally friendly extraction methods have been studied, which have focused on nonhazardous extraction processes, with a minimum consumption of energy, reduction of extraction times, and the use of organic solvents, as well as high yields and purity of bioactive compounds (Chanioti and Tzia 2017; da Silva et al. 2016; Galanakis 2012).

14.6.1 Pressure-Driven Membrane Processing

Membrane separation processes have been studied for the recovery of bioactive compounds (such as phenolic compounds) from food by-products because it is considered as a clean and economical technology and they can eliminate solids in suspension as well as concentrating macromolecules (Nunes et al. 2019; Syed et al. 2017). There is different pressure-driven membrane process such as nanofiltration, ultrafiltration, and microfiltration (Cassano et al. 2011; Castro-Muñoz et al. 2016). Likewise, some bioactive compounds can be recovered using membranes from agroindustry by-products such as phenolic compounds. In this sense, different agroindustry by-products have been studied for phenolic compound extraction, among these are the following: orange press liquor, nixtamalization wastewater, olive mill wastewater, grape seeds, fermented grape pomace, red grapes, pomace, artichoke wastewater, mate tree residues, and soy processing waste, which have been reported the phenolic compound extraction using different membranes (Castro-Muñoz et al. 2016). Recently, Nunes et al. (2019) reported an increase of total phenolics (1030–1284 mg gallic acid equivalent/L) and flavonoids (398.9–467.3 mg epicatechin equivalent/L) of olive pomace concentrate utilizing membrane separation process in comparison with initial aqueous extract.

14.6.2 Microwave-Assisted Extraction

Microwave-assisted extraction is a technology that has been increasing in recent years due to the multiple reports that have proven their efficiency in obtaining bioactive compounds, microwave energy is able to accelerate the mass transfer of matrix bioactive compounds to the solvent through the solvent heating, and on the other hand, its main characteristic is its easy handling and requires moderate amounts of solvent (Galanakis 2012). According to Balasundram et al. (2006), citrus industry by-products contain important amounts of polyphenols; therefore they have become a study model. In this sense, mandarins are an important citrus crop that industrially produce large amounts of by-product in the form of seed, leaves, and peel waste; also, it has been reported that it contains bioactive compounds; therefore different extraction methods have been studied. Ateş et al. (2019) demonstrated that microwave-assisted extraction has a significant effect on the yield of total phenolics and flavonoids of mandarin leaves, because as irradiation power and time extraction increased, the yield increased; however at higher conditions, these decreased. The total phenolics and flavonoid (17.22 mg gallic acid equivalent/g dried leaf and 1.71 mg catechin equivalent/g dried leaf, respectively) content of mandarin leaves extracted in optimum conditions of microwave-assisted was found at extraction time of 45 s and microwave irradiation power of 43 W.

On the other hand, it has been reported that juice processing generates high by-products with important bioactive compounds which could be used to give added value; in this sense, important quantities of sea buckthorn pomace are produced during the berry processing for juice extraction. Due to the sea buckthorn berries which are a rich source of bioactive compounds, sea buckthorn pomace is also a study model for the recovery of bioactive compounds (Rösch et al. 2004). Therefore, Périno-Issartier et al. (2011) studied the effect of a novel technology (microwave hydro-diffusion and gravity) on total phenolics content of sea buckthorn pomace; they found that similarly amounts of flavonoids were obtained by microwave hydro-diffusion and gravity in comparison with conventional solid-liquid extraction; in fact, quercetin 3-O-glucoside and isorhamnetin were higher in microwave extracts.

14.6.3 Supercritical Fluid Extraction

Supercritical fluid extraction is a technology that has been studied for the recovery of phenolic compounds in industry by-products; carbon dioxide (CO_2) is the most used gas for supercritical fluids extractions, because its critical pressure and temperature point are relatively easy to reach; also, it is considered as harmless to humans. Supercritical fluid extraction is based on carrying a gas above its critical temperature and pressure point, which modifies its characteristics; in this sense it increases the fluid diffusion and penetrates through the material pores increasing the mass transfer and therefore the compound extraction (Ameer et al. 2017; Galanakis 2012; Herrero et al. 2010). Similar to microwave-assisted extraction, the effect of supercritical fluids on the increase of phenolic compounds and flavonoids of mandarin by-products (specifically in the leaf) has been studied. Ateş et al. (2019) reported that total phenolic content increases with pressure; it could be due to the modification in the fluid density as well as to solvation power; on the other hand when increasing the extraction temperature, the phenolic content decreased until certain point, because there is a temperature and pressure relation that modifies the properties of supercritical fluid. It was found a total phenolic and flavonoid content of 79.9 mg gallic acid equivalent/g dry leaf and 7.56 mg catechin equivalent/g dry leaf, respectively, at the optimum temperature, flow, and pressure extraction of 50 ° C, 0.39 mL/min, and 200 bar. It is important to knowledge that these values were found up to four times higher than polyphenols obtained by microwave-assisted extraction; therefore, supercritical fluid extraction is widely studied.

On the other hand, the flavonoid extraction by supercritical fluids in grapefruit peel has been reported; in this sense, Giannuzzo et al. (2003) reported the highest naringin yield (14.4 g/kg) at supercritical CO_2 extraction of 58.6 °C, 95 bar and ethanol 15% as co-solvent. Besides, the naringin supercritical CO_2 yield was higher in fresh peel than dried; also naringin yield by supercritical CO_2 was higher than conventional extraction methods; therefore supercritical CO_2 extraction can be used for the bioactive compound extraction from fruit by-products.

As for phenolic compound extraction from red grape by products, Louli et al. (2004) studied the effect of supercritical extraction on ethyl acetate extract of red grape by products; in this sense, they reported that supercritical extract possessed high total phenolic compounds (18.1 equivalent gallic acid % w/w) at pressure (150 bar) and temperature (45 °C) moderates, which in turn showed a high antioxidant activity; therefore, the use of supercritical extraction could be used for increase phenolic compound extraction of grape by-product extracts.

Recently, Roselló-Soto et al. (2019) studied the supercritical CO_2 extraction effect on phenolic compound of oil from "horchata" by-products (by-products obtaining after the horchata product from tiger nuts). They reported that phenolic profile is linked with the different supercritical conditions; likewise, an increase of pressure increases the phenolic compound amounts, being at 40 MPa the highest phenolic compound recovery. Besides, the predominant phenolic compounds found in supercritical CO_2 extract were caffeic acid, scopoletin, and lignin isohydroxymatairesinol.

14.6.4 Ultrasound-Assisted Extraction

Ultrasound-assisted extraction is based on the damage to the cell wall caused by ultrasound wave acceleration through the effect of cavitation and therefore accelerates mass transfer, and metabolites such as polyphenols are available to be separated. In this sense, it has been reported that ultrasound-assisted extraction is a promising technique for bioactive compound extraction such as phenolics and flavonoids, due to some advantage than conventional extraction techniques such as low-energy consumption, low-temperature extractions, low-period extractions, and solvent consumption and high repeatability (Görgüç et al. 2019; Khan et al. 2010).

On the other hand, industry by-products have important amounts of bioactive compounds as polyphenols; thus different techniques have been studied for recovery, among them is ultrasound-assisted extraction. Khan et al. (2010) optimized the polyphenol extraction by ultrasound-assisted extraction of orange peels. They reported a naringin (70.3 mg/100 g of fresh weight) and hesperidin (205.2 mg/100 g of fresh weight) increase by ultrasound extraction at optimum conditions (150 W ultrasound power, 80% ethanol: water ratio and 40 °C extraction temperature). In addition, phenolic compound extraction was up to three times faster using ultrasound.

Likewise, Chanioti and Tzia (2017) optimized the olive pomace oil by ultrasound-assisted extraction. They reported that ultrasound affected the total phenolic compounds of olive pomace oil; in this sense, temperature, solid/liquid ratio, and particle seize were studied for the enhancement of the total phenolic compounds; the highest total phenolic compounds were 0.261 mg gallic acid/g oil at optimized conditions of extraction temperature of 50 °C, solid/liquid ratio of 1.8 g/mL, and 0.5 mm of particle seize. However, Haas et al. (2018) reported a less efficiency extraction of phenolic compounds and flavonoids of non-pomace residue by ultrasound-assisted with hydroalcoholic solvent; likewise, solvent extraction was the most efficient.

On the other hand, Fernández et al. (2018) reported the effectiveness of ultrasound-assisted extraction in the phenolic compound recovery from agro-industrial by-products (olive cake, dried scapes and umbel onion, peels, seeds, and cull fruits of tomato and pear) after a combination process of lactic acid, glucose, and 15% water pretreatment. They reported different optimal conditions of ultrasound-assisted extraction for each material studied; also, the process variables for extraction were extraction time and material/solvent ratio; likewise high phenolic compound extraction efficiency was found compared with traditional solvents.

Goula et al. (2016) optimized the phenolic compound extraction from grape pomace; the variables studied were extraction temperature, pulse duration/pulse interval ratio, amplitude level, and solvent/pomace ratio; the highest yield was 10.14 mg gallic acid equivalents/g dry pomace and was found under the optimum conditions of 34.5 °C extraction temperature, ratio of pulse duration/pulse interval of 8/6, amplitude level of 39%, and solid/solvent ratio of 18.1 mL/g.

Similarly, it has been reported that an important by-product grape processing is seeds; these are a rich source of phenolic compounds; in this sense, Da Porto et al. (2013) reported an improvement in the polyphenol extraction (105.81 ± 3.21 mg GAE/g flour) from the grape seed using ultrasound-assisted and Soxhlet extraction; they found that using both coupled extractions increased the total content of polyphenols and cinnamic acids; likewise, they attributed the increase ultrasound conditions (30 min, 20 kHz, 150 W, and 30 °C, respectively) used in the extraction by ultrasound, as well as to the Soxhlet extraction conditions (12 h extraction time at room temperature and nitrogen atmosphere), and prevented the polyphenolic compound degradation.

14.6.5 Subcritical Water Extraction

Subcritical water or hot-compressed water is a novel technique for the extraction of bioactive compounds, which is related to changes in water properties. It has been reported that subcritical water increased the mass transfer rates due to lower viscosity and surface tension (Yu et al. 2008). Likewise, there is a relationship between the temperature increase and hydrogen bound dissociation, which produces a greater solubility of hydrophobic compounds; on the other hand, the polarity of the water is also increased when the temperature increases; therefore, it has been used to the extraction of bioactive compounds such as phenolic compounds (Möller et al. 2011; Pedras et al. 2017).

In this sense, Pedras et al. (2017) studied the effect of subcritical water extraction on phenolic compounds from white grape pomace; they reported that the highest

Table 14.14 Advantages and disadvantages of phenolic compound extraction techniques

Extraction technique	Advantages	Disadvantages	Reference
Maceration and reflux extraction	Easy to replicate	High solvent use Long extraction times	Giannuzzo et al. (2003)
Soxhlet extraction	High yields	High solvent use Long extraction times	Giannuzzo et al. (2003)
Microwave-assisted extraction	Short extraction times Small solvent volumes Performance heating	Thermolabile degradation compounds Operational control variables	Galanakis (2012), Lucchesi et al. (2004) and Périno-Issartier et al. (2011)
Pressure-driven membrane extraction	Clean and economical technology High separation efficiency Simple operation	Large time separations Is not selective for phenolic compounds	Castro-Muñoz et al. (2016), Nunes et al. (2019) and Syed et al. (2017)
Subcritical water extraction	Simple operation Don't require the use of polar solvents	High-temperature requirements	Möller et al. (2011), Pedras et al. (2017) and Yu et al. (2008)
Supercritical fluid extraction	High selectivity High bioactive compound concentration Low extraction time Low solvents consumption	High equipment, security, and maintenance costs Low yield of extraction of polar compounds	Ameer et al. (2017), da Silva et al. (2016), Galanakis (2012) and Herrero et al. (2010)
Ultrasound-assisted extraction	High reproducibility Low extraction time Low solvent consumption	Need a filtration steep	Galanakis (2012)

phenolic compound yield extraction (113.4 ± 5.1 mg gallic acid equivalent/g extract) was at 210 °C and 10 mL/min of water flow rate. Besides, they identified six phenolic compounds (vanillin, caffeic acid, epicatechin, chlorogenic acid, gallic acid, and catechin). Table 14.14 shows the advantages and disadvantages of phenolic compound extraction techniques.

14.7 Conclusions

Agro-industrial by-products are a potential source of bioactive compounds as demonstrated in many researches; however, there is still a need of the evaluation of the energy consumption of technologies to the recovery of phenolic compounds as well as the metabolism in in vivo models. Furthermore, the use of the whole biomass generated by the plant material processing has not been fully evaluated, for example, the use of the pruning and wastes left in the field after the harvest of the fruits.

References

Al-Jabari M (2002) Kinetic models of supercritical fluid extraction. 25:477–489. https://doi.org/10.1002/1615-9314(20020601)25:8<477::AID-JSSC477>3.0.CO;2-C

Ameer K, Shahbaz HM, Kwon J-H (2017) Green extraction methods for polyphenols from plant matrices and their byproducts: a review. Compr Rev Food Sci Food Saf 16:295–315. https://doi.org/10.1111/1541-4337.12253

Ateş F, Şahin S, İlbay Z, Kırbaşlar ŞİJ (2019) A green valorisation approach using microwaves and supercritical CO2 for high-added value ingredients from mandarin (Citrus deliciosa Tenore) leaf waste. Waste Biomass Valoriz 10:533–546. https://doi.org/10.1007/s12649-017-0074-z

Balasundram N, Sundram K, Samman S (2006) Phenolic compounds in plants and Agri-industrial by-products: antioxidant activity, occurrence, and potential uses. Food Chem 99:191–203. https://doi.org/10.1016/j.foodchem.2005.07.042

Bravo L (1998) Polyphenols: chemistry, dietary sources, metabolism, and nutritional significance. Nutr Rev 56:317–333. https://doi.org/10.1111/j.1753-4887.1998.tb01670.x

Cassano A, Conidi C, Drioli E (2011) Comparison of the performance of UF membranes in olive mill wastewaters treatment. Water Res 45:3197–3204. https://doi.org/10.1016/j.watres.2011.03.041

Castro-Muñoz R, Yáñez-Fernández J, Fíla V (2016) Phenolic compounds recovered from agro-food by-products using membrane technologies: an overview. Food Chem 213:753–762. https://doi.org/10.1016/j.foodchem.2016.07.030

Chanioti S, Tzia C (2017) Optimization of ultrasound-assisted extraction of oil from olive pomace using response surface technology: oil recovery, unsaponifiable matter, total phenol content and antioxidant activity. LWT Food Sci Technol 79:178–189. https://doi.org/10.1016/j.lwt.2017.01.029

Chen M-H, McClung AM, Bergman CJ (2016) Concentrations of oligomers and polymers of proanthocyanidins in red and purple rice bran and their relationships to total phenolics, flavonoids, antioxidant capacity and whole grain color. Food Chem 208:279–287

Da Porto C, Porretto E, Decorti D (2013) Comparison of ultrasound-assisted extraction with conventional extraction methods of oil and polyphenols from grape (Vitis vinifera L.) seeds. Ultrason Sonochem 20:1076–1080. https://doi.org/10.1016/j.ultsonch.2012.12.002

da Silva RPFF, Rocha-Santos TAP, Duarte AC (2016) Supercritical fluid extraction of bioactive compounds. TrAC Trends Anal Chem 76:40–51. https://doi.org/10.1016/j.trac.2015.11.013

Dai C-Y, Liu P-F, Liao P-R, Qu Y, Wang C-X, Yang Y, Cui X-M (2018) Optimization of flavonoids extraction process in panax notoginseng stem leaf and a study of antioxidant activity and its effects on mouse melanoma B16 cells. Molecules 23:2219

de Camargo AC, Regitano-d'Arce MAB, Biasoto ACT, Shahidi F (2016) Enzyme-assisted extraction of phenolics from winemaking by-products: antioxidant potential and inhibition of alpha-glucosidase and lipase activities. Food Chem 212:395–402

Decker A (1997) Phenolics: prooxidants or antioxidants? 55:396–398. https://doi.org/10.1111/j.1753-4887.1997.tb01580.x

Domínguez-Perles R, Martínez-Ballesta MC, Carvajal M, García-Viguera C, Moreno DA (2010) Broccoli-derived by-products—a promising source of bioactive ingredients. J Food Sci 75: C383–C392. https://doi.org/10.1111/j.1750-3841.2010.01606.x

Fawole OA, Opara UL (2016) Stability of total phenolic concentration and antioxidant capacity of extracts from pomegranate co-products subjected to in vitro digestion. BMC Complement Altern Med 16:358

Fernández MÁ, Espino M, Gomez FJV, Silva MF (2018) Novel approaches mediated by tailor-made green solvents for the extraction of phenolic compounds from agro-food industrial by-products. Food Chem 239:671–678. https://doi.org/10.1016/j.foodchem.2017.06.150

Galanakis CM (2012) Recovery of high added-value components from food wastes: conventional, emerging technologies and commercialized applications. Trends Food Sci Technol 26:68–87. https://doi.org/10.1016/j.tifs.2012.03.003

Giannuzzo AN, Boggetti HJ, Nazareno MA, Mishima HT (2003) Supercritical fluid extraction of naringin from the peel of Citrus paradisi. Phytochem Anal 14:221–223. https://doi.org/10.1002/pca.706

Gimeno Creus E (2004) Compuestos fenólicos. Un análisis de sus beneficios para la salud. Offarm 23:80–84

Görgüç A, Bircan C, Yılmaz FM (2019) Sesame bran as an unexploited by-product: effect of enzyme and ultrasound-assisted extraction on the recovery of protein and antioxidant compounds. Food Chem 283:637–645. https://doi.org/10.1016/j.foodchem.2019.01.077

Goula AM, Thymiatis K, Kaderides K (2016) Valorization of grape pomace: drying behavior and ultrasound extraction of phenolics. Food Bioprod Process 100:132–144. https://doi.org/10.1016/j.fbp.2016.06.016

Gullón B et al (2018) Valorisation of olive agro-industrial by-products as a source of bioactive compounds. Sci Total Environ 645:533–542. https://doi.org/10.1016/j.scitotenv.2018.07.155

Haas ICDS, Toaldo IM, Burin VM, Bordignon-Luiz MT (2018) Extraction optimization for polyphenolic profiling and bioactive enrichment of extractives of non-pomace residue from grape processing. Ind Crop Prod 112:593–601. https://doi.org/10.1016/j.indcrop.2017.12.058

Herrero M, Mendiola JA, Cifuentes A, Ibáñez E (2010) Supercritical fluid extraction: recent advances and applications. J Chromatogr A 1217:2495–2511. https://doi.org/10.1016/j.chroma.2009.12.019

Khan MK, Abert-Vian M, Fabiano-Tixier A-S, Dangles O, Chemat F (2010) Ultrasound-assisted extraction of polyphenols (flavanone glycosides) from orange (Citrus sinensis L.) peel. Food Chem 119:851–858. https://doi.org/10.1016/j.foodchem.2009.08.046

Louli V, Ragoussis N, Magoulas K (2004) Recovery of phenolic antioxidants from wine industry by-products. Bioresour Technol 92:201–208. https://doi.org/10.1016/j.biortech.2003.06.002

Lucchesi ME, Chemat F, Smadja J (2004) Solvent-free microwave extraction of essential oil from aromatic herbs: comparison with conventional hydro-distillation. J Chromatogr A 1043:323–327. https://doi.org/10.1016/j.chroma.2004.05.083

Meshkini A, Tahmasbi M (2017) Antiplatelet aggregation activity of walnut hull extract via suppression of reactive oxygen species generation and caspase activation. J Acupunct Meridian Stud 10:193–203. https://doi.org/10.1016/j.jams.2017.02.007

Möller M, Nilges P, Harnisch F, Schröder U (2011) Subcritical water as reaction environment: fundamentals of hydrothermal biomass transformation. ChemSusChem 4:566–579. https://doi.org/10.1002/cssc.201000341

Nunes AM, Pawlowski S, Costa ASG, Alves RC, Oliveira MBPP, Velizarov S (2019) Valorization of olive pomace by a green integrated approach applying sustainable extraction and membrane-assisted concentration. Sci Total Environ 652:40–47. https://doi.org/10.1016/j.scitotenv.2018.10.204

Oskoueian E, Abdullah N, Hendra R, Karimi E (2011) Bioactive compounds, antioxidant, xanthine oxidase inhibitory, tyrosinase inhibitory and anti-inflammatory activities of selected agro-industrial by-products. Int J Mol Sci 12. https://doi.org/10.3390/ijms12128610

Pedras B, Salema-Oom M, Sá-Nogueira I, Simões P, Paiva A, Barreiros S (2017) Valorization of white wine grape pomace through application of subcritical water: analysis of extraction, hydrolysis, and biological activity of the extracts obtained. J Supercrit Fluids 128:138–144. https://doi.org/10.1016/j.supflu.2017.05.020

Périno-Issartier S, Zill-e-Huma A-VM, Chemat F (2011) Solvent free microwave-assisted extraction of antioxidants from sea buckthorn (Hippophae rhamnoides) food by-products. Food Bioprocess Technol 4:1020–1028. https://doi.org/10.1007/s11947-010-0438-x

Rösch D, Krumbein A, Kroh LW (2004) Antioxidant gallocatechins, dimeric and trimeric proanthocyanidins from sea buckthorn (Hippophaë rhamnoides) pomace. Eur Food Res Technol 219:605–613. https://doi.org/10.1007/s00217-004-1002-6

Roselló-Soto E, Barba FJ, Lorenzo JM, Munekata PES, Gómez B, Moltó JC (2019) Phenolic profile of oils obtained from "horchata" by-products assisted by supercritical-CO2 and its relationship with antioxidant and lipid oxidation parameters: triple TOF-LC-MS-MS characterization. Food Chem 274:865–871. https://doi.org/10.1016/j.foodchem.2018.09.055

Şahin S, Elhussein E, Bilgin M, Lorenzo JM, Barba FJ, Roohinejad S (2018) Effect of drying method on oleuropein, total phenolic content, flavonoid content, and antioxidant activity of olive (Olea europaea) leaf. J Food Process Preserv 42:e13604

Salido S, Pérez-Bonilla M, Adams RP, Altarejos J (2015) Phenolic components and antioxidant activity of wood extracts from 10 main Spanish olive cultivars. J Agric Food Chem 63:6493–6500

Santana-Gálvez J, Jacobo-Velázquez DA (2018) Classification of phenolic compounds. In: Nollet LM, Gutierrez-Uribe JA (eds) Phenolic compounds in food: characterization and analysis, vol 1, 1st edn. CRC Press, Boca Raton, pp 3–20. https://doi.org/10.1201/9781315120157

Syed UT, Brazinha C, Crespo JG, Ricardo-da-Silva JM (2017) Valorisation of grape pomace: fractionation of bioactive flavan-3-ols by membrane processing. Sep Purif Technol 172:404–414. https://doi.org/10.1016/j.seppur.2016.07.039

Vergani L et al (2016) Antioxidant and hepatoprotective potentials of phenolic compounds from olive pomace. Chem Eng Trans 49:475–480

Yu Y, Lou X, Wu H (2008) Some recent advances in hydrolysis of biomass in hot-compressed water and its comparisons with other hydrolysis methods. Energy Fuel 22:46–60. https://doi.org/10.1021/ef700292p

Zhang MW, Zhang RF, Zhang FX, Liu RH (2010) Phenolic profiles and antioxidant activity of black rice bran of different commercially available varieties. J Agric Food Chem 58:7580–7587

Plant Phenolics and Postharvesting Technologies

15

Leticia X. López-Martínez, Ofelia Marquez-Molina, Erick P. Gutiérrez-Grijalva, and J. Basilio Heredia

Abstract

Phenolic compounds have a crucial role in the defense mechanisms of plants against factors that generate stress (biotic or abiotic). In fruits and vegetables, the exposure to postharvest conditions such as storage, controlled atmospheres, phytohormones (ethylene and methyl jasmonate), radiation (ultraviolet and gamma), thermal shocks, edible coatings, and minimal processing induce the synthesis of these phenolic compounds. The synthesis of phenolic compounds is through the phenylpropanoid pathway. The purpose of this chapter focuses on analyzing how the postharvest treatments of fruits and vegetables exert influence on the total phenolic content, anthocyanins, and other flavonoids compounds.

Keywords

Plant · Phenolics · Anthocyanins · Flavonoids · Postharvesting technology

L. X. López-Martínez
Cátedras CONACYT – Centro de Investigación en Alimentación y Desarrollo, AC, Hermosillo, Sonora, México

O. Marquez-Molina
Centro Universitario UAEM Amecameca, Universidad Autónoma del Estado de México, Toluca, Estado de México, Mexico

E. P. Gutiérrez-Grijalva
Cátedras CONACYT – Centro de Investigación en Alimentación y Desarrollo, AC, Culiacán, Sinaloa, México

J. B. Heredia (✉)
Department of Food Science and Technology, Centro de Investigación en Alimentación y Desarrollo, Culiacán, Sinaloa, Mexico
e-mail: jbheredia@ciad.mx

R. Lone et al. (eds.), *Plant Phenolics in Sustainable Agriculture*,
https://doi.org/10.1007/978-981-15-4890-1_15

15.1 Introduction

Several postharvest treatments have been assessed to minimize the losses in quality attributes of fruits and vegetables, which have shown efficacy in maintaining quality. These plant foods contain phytochemical compounds that are non-nutrients but have demonstrated beneficial effects on human health. Among them, glucosinolates, betalains, carotenoids, and phenolics are the main groups. Most postharvest treatments imply the alteration of the standard conditions of the fruit or vegetable to extend the postharvest life. Modified atmospheres (CO_2 and O_2 changes), gases, and irradiation cause changes on some biochemical processes in the metabolism of fruits and vegetables (Charles et al. 2009). Thermal treatments affect ethylene synthesis, respiration, softening, and cell-wall metabolism (Zhang et al. 2009).

Other postharvest treatments could induce some mechanisms that result in changes in the metabolic activity of vegetables and fruits, such as the stimulation of the antioxidant mechanism. Postharvest treatments have received attention over the years due to their impact on the content of phytochemical compounds that have health benefits during storage. In this way, the objective of this chapter is to get a better knowledge of the impact and changes that occur in the phenolic compounds of fruits and vegetables due to the application of postharvest technologies.

15.2 Technologies to Increase the Postharvest Life of Fruits and Vegetables

Fruits and vegetables are exposed to quality changes after harvesting, where large losses have been reported due to lack of packaging, storage facilities, and poor means of transportation. The postharvest handling of horticultural products is an important area of research, and there are many studies conducted in this area in different crops. Therefore, it is necessary to establish technologies that would help to extend the shelf life of fruits and vegetables in addition to studying its effects on total phenolic compounds, flavonoids, and anthocyanins.

15.2.1 Minimal Processing

"Minimal processing" refers to the physical alteration of fruits or vegetables from its original state but remaining in a fresh "unprocessed" state. Within this context, "fresh-cut" is fruits or vegetables that the consumer can acquire in a state that allows direct and immediate consumption without prior preparation or transformation. To prepare the minimally processed products, the whole fruit or vegetables are divide into smaller parts, and the shape has to be given to the product depending on its final use. However, these processes cause tissue damage, make its storage life shorter, and induce changes in the phenolic metabolism, mainly affecting the phenolic antioxidant composition (Barry-Ryan and O'Beirne 2000).

In general, the wounding caused during minimal processing induces phenolic compound biosynthesis, which is accumulated during the storage of processed products. Du et al. (2012) evaluated the variation in total phenolic content (TPC) of commercial fresh-cut carrot products (baby carrots, carrot sticks, shredded carrots, crinkle-cut coins, and chips of carrots). The content of TPC found in fresh-cut carrot products at day 0 varied from 12.9 to 24.2 mg chlorogenic acid equivalents/100 g, while wounding alone contributed to a large increase in TPC for sliced carrots at day 3 (crinkle-cut coins and oblong chips).

On the contrary, smaller increases in total phenolic were determined in baby carrots, carrot sticks, and shredded carrot samples. These behaviors indicate that cutting style influences the accumulation of total phenolic in fresh carrot products. Similarly, Heredia and Cisneros-Zevallos (2009) and Surjadinata and Cisneros-Zevallos (2012) also reported that the accumulation of total phenolic compounds in carrots depends on the intensity of the wound. Kenny and O'Beirne (2010) concluded that machine-peeled carrots showed a higher accumulation of phenolic compounds compared to hand-peeled carrots. This may be due to the synthesis of PAL, an enzyme that generates an increase in phenolic compounds.

Furthermore, hand peeling onions caused a decrease of total phenolic compounds (Makris and Rossiter 2001; Gennaro et al. 2002). This could be due to the accumulation of phenolic compounds on the skin of the onions in place of the stress caused by peeling. Li et al. (2017) researched the effect of different cutting styles on the TPC on storage at 15 °C for 4 days of pitaya fruit. The pitaya was prepared in three various forms: in a slice, half a slice, and the fourth slice with 1 cm thick each. The TPC and PAL activity increased significantly with cutting wounding intensity at 2 days of storage in comparison with the whole fruit. The content of TPC was the fourth slice > half a slice > in a slice > whole fruit. These results reveal that depending on the cutting styles, the biosynthesis of phenolic compounds and the antioxidant activity of the pitayas could increase.

15.2.2 Storage

Storage at low temperatures is one of the postharvest treatments utilized to maintain quality, reduce metabolism, and prolong the shelf life of fruits and vegetables. It is considered as a non-chilling sensitive, slows the development of microorganisms that cause spoilage during the storage times, and decreases respiration rates (Lamikanra and Richard 2002). The ripening, softening, and senescence processes can be reflected in changes in the profiles of TPC on the fruits and vegetables on storage.

According to Mishra et al. (2012), the shelf life of fresh-cut products usually extends up to 16 days at 4 °C, 12 days at 10 °C, and 5 days at 26 °C storage temperatures, although the type of matrix determines this. Different studies about the changes in bioactive compounds during the time of cold storage have shown no general behavior (Table 15.1).

Table 15.1 Effect of storage on the phenolic, flavonoid, and anthocyanin content of fruits and vegetables

Storage	Fruit or vegetables	Effects	References
Fresh	Apples	Total phenolics	Pérez-Ilzarbe et al. (1997)
4 °C, 30 days		↑ 23%	
4 °C, 60 days		↑ 117%	
Fresh	Lowbush blueberry	Anthocyanins (354 mg/ 100 g)	Kader et al. (1996)
L °C, 15 days		↑18%	
Fresh	Golden Delicious apple	Total flavonoids (168 mg/ 100 g)	Lattanzio et al. (2001)
2 °C, 60 days		↑ 19%	
2 °C, 100 days		– 11%	
Fresh	Blueberries	Total phenolics (335 mg/ 100 g)	Connor et al. (2002)
5 °C, 21 days		Total phenolics 7%	
5 °C, 35 days		13↑	
Fresh	Blueberries	Anthocyanins (132 mg/ 100 g)	Connor et al. (2002)
5 °C, 21 days		– 0.78%	
5 °C, 35 days		↑ 8.5%	
Air 0 days	Plums	Total phenolics (140 mg/ kg)	Singh and Singh (2013)
0–1 °C, 35 days		– 10%	
0–1 °C, 56 days		– 15%	
0 °C, 7 days	Strawberry	Total phenolics (650 mg/ 100 g)	Ayala-Zavala et al. (2004)
5 °C, 7 days		↑16.6%	
10 °C, 7 days		↑ 25.7%	
Fresh	Lettuce	Total phenolics (5.79 mg/ 100 g)	Tavarini et al. (2007)
4 °C, 1 day		↑11.5%	
4 °C, 3 days		↑12.3%	
Blade cut	Eggplant	Total phenolics (1.23 mg/g)	Mishra et al. (2012)
4 ± 2 °C, 1 day		↑11.2%	
4 ± 2 °C, 5 days		↑10.3%	
4 ± 2 °C, 10 days		– 13.5%	
4 ± 2 °C, 16 days		– 10.9%	

(↑) Increase; (–) decrease

15.2.3 Controlled (CA) and Modified Atmospheres (MA)

The CA and MA alter the atmosphere composition different from regular air, varying the CO_2 and O_2 concentration. Among the effects caused by CA and MA in fruits or vegetables are inhibition of ethylene action or production, reduction of respiration rate, and retardation of ripening (Yahia 2009). The positive effects depend on the type of the food matrix (fruit or vegetable), cultivar, concentration of gases in the atmosphere, temperature, and storage duration (Pesis 2005; Singh and Pal 2008).

The most common atmosphere consists of elevated CO_2 and reduced O_2 levels. However, other gases are sometimes included (ethylene, carbon monoxide, propylene, and acetylene); the concentration and combination of gases used are based on the variety, origin, and season of fruit or vegetable. Some fruits and vegetables are stored under atmospheres reduced in O_2 and with high levels of CO_2. These storage conditions could show effects on phenolic accumulation and phenolic composition. Several researchers indicated that CA has the benefit of controlling the postharvest deterioration of fruit. However, a CO_2-enriched atmosphere with low O_2 concentration can affect anthocyanin content detrimentally, with negative effects in fruit color and nutritional composition (Holcroft and Kader 1999).

Another study compared the effects of CA and MA packaging (MAP) on the content of phenolic compounds in mature "Blackamber" Japanese plums stored at 1 °C during 5 and 8 days in air, CA-1 (1% O_2 + 3% CO_2), CA-2 (2.5% O_2 + 3% CO_2), and MAP (10% O_2 and 3.8% CO_2). Overall, they observed that CA-2 was more effective than CA-1 and MAP in the retention of TPC during cold storage (Singh and Singh 2013).

Limited information is available on the results of elevated O_2 concentrations on phytochemicals. Day (1996) reported that high O_2 levels did not decrease the antioxidant content in fresh-cut lettuce compared with low O_2 levels. Montero-Calderón et al. (2010) found that a concentration of O_2 (38%) with 1% CO_2 in comparison to 12% O_2 with 1% CO_2 provided a decrease in TPC on fresh-cut pineapple. In contrast, Duan et al. (2011) investigated the effect of 100% O_2 (pure oxygen) and humidified air (control) on metabolism, TPC, and anthocyanins of litchi fruit. The exposition to pure oxygen retards the reduction of TPC and anthocyanins in litchi fruit pericarp. There was no appreciable difference for the contents of TPC and anthocyanins between control and pure-oxygen-exposed fruit at 2 days. After 4 and 6 days of storage, the content of TPC and anthocyanins in oxygen (100%) revealed that litchi fruits were higher than those from control. Stress such as altered O_2 and CO_2 levels in CA and modified atmosphere packaging (MAP), or C_2H_4 gassing for ripening and greening, could affect phytochemical accumulation too. For example, PAL generated an increase in TPC in certain fresh-cut products (Kenny and O'Beirne 2010).

Some studies have found a detrimental correlation between phenolic content and antioxidant capacity during CA storage compared to standard atmospheric storage. It is presumed that a prolonged time of CA storage may cause the oxidation of the phenolic compounds (Schotsmans et al. 2007). Zheng et al. (2005) reported that high-oxygen treatments might improve the antioxidant capacity in fruit but only to a limited storage period.

15.2.4 Light

Light is an important factor that affects the phytochemical content in plant tissue (Massa et al. 2008). Several studies have demonstrated that postharvest light-emitting diode (LED) treatment could enhance antioxidant capacity in vegetables such as lettuce, barley leaf, spinach, and komatsuna (Ohashi-Kaneko et al. 2007; Samuolienė et al. 2012).

The effect of light treatment (fluorescent and LED green light) on phytochemicals in broccoli florets was investigated (Jin et al. 2015). The results showed that LED green light treatments increased TPC during all storage time. On the second day, the TPC was found 40.4% and 29.6% higher in LED green light treated than those broccoli florets treated with fluorescent light. This agreed with the report by Li and Kubota (2009); they investigated the different LED light quality effects on baby red leaf lettuce; they reported a significant increase of phenolic compounds under fluorescent lighting supplemented with red LEDs and did not notice supplemental blue or green LED's influence on phenolic accumulation. Moreover, Lee et al. (2010) showed that the treatment of barley leaves with green or blue LEDs during recultivation and the second harvest caused an increase in the level of TPC.

Ko et al. (2015) studied the effect of different types of light (fluorescent, blue, red, and ultraviolet) on onion after harvest and storage, which influenced the content and concentration of quercetin and quercetin glucosides. All the light treatments increased quercetin content in peeled onion bulbs. Among them, especially, the fluorescent light stimulates the highest synthesis of quercetin in onion. In the case of whole onion bulbs, skin and pulp showed different responses to light treatment. The pulp had the maximum content of quercetin glucosides under blue light, whereas the minimum was produced under fluorescent light. Onion skin showed the opposite behavior when it is compared to the pulp. The light treatment proved to be more useful to increase the level of quercetin content. These results could be due to the induction of PAL activity, which is the key enzyme of the phenylpropanoid pathway, in fruits and vegetables treated with fluorescent and LED light.

15.2.5 UV Radiation

The application of UV-B or UV-C radiation during postharvest storage can affect the accumulation of TPC in fruits and vegetables. Ultraviolet radiation is an abiotic physical inducer of resistance mechanisms, leading to a rapid increase in phenolic acids, flavonoids, and anthocyanins in response to stress (Oufedjikh et al. 2000; Higashio et al. 2005). The biosynthesis of TPC is affected by UV irradiation owing to the increased activity of PAL. Besides, the increased activity of other enzymes such as chalcone synthase, chalcone isomerase, and dihydroflavonol-4-reductase that are involved in flavonoid synthesis (Tomás-Barberán and Espín 2001).

Low doses of irradiation, ultraviolet light (UV), and UV-C irradiation (240–280 nm) stimulated the hormesis (beneficial reactions in biological organs) (Shama 2007). Hormetic doses of UV-C can delay the senescence process and fruit ripening (Gonzalez-Aguilar et al. 2007). Also, the induction of natural defense and elicitors against fungi and bacteria (Alothman et al. 2009) prolongs the postharvest life. This defense is showed through the induction and accumulation of phytoalexins, carotenoids, vitamin C, and phenolic compounds (El-Ghaouth et al. 1998). UV-C treatment (3.7 kJ m^{-2}) is known to delay the ripening and senescence in fruits (Liu et al. 1993; Maharaj et al. 1999). However, the exact mode of action of UV treatment needs to be studied in more detail.

Li et al. (2009) studied the changes in trans-resveratrol and other phenolic compounds in the seed and skin of red grape (*Vitis amurensis* "Beiquan") during the temperature of cold storage ($-1° \pm 0.5$ °C) after postharvest irradiation (UV-B and UV-C) at 25 °C. Total levels of anthocyanins in grape skins after either UV treatment were significantly higher than skin and seeds without treatments, during the first week (7 days) of storage, whereas only those in UV-B-treated skins were significantly higher than controls during 21–28 days of storage. Total flavan-3-ol and flavonol contents remained constant in UV-B- and UV-C-treated grape skins during cold storage and did not differ from controls. The total flavan-3-ol and flavonol contents in control and UV-B-treated grape seeds declined from day 2 to day 28 of cold storage, while those of UV-C-treated grape seeds remained constant. Changes in catechin and epicatechin were responsible for the differences observed in total flavan-3-ols and flavonols during cold storage.

Sari et al. (2016) evaluated the TPC in the peel and pulp of untreated and UV-C-treated pineapple. The exposure of pineapple to UV-C did not affect TPC content in pulp, whereas UV-C irradiation significantly increased TPC in the peel of pineapple during storage for up to 28 days. The longer UV-C exposure time showed higher TPC. Additionally, the TPC content in all treatments increased along with the storage time. These results indicated that a formation or accumulation of TPC content in the peel is a response of pineapple to oxidative stress caused by UV-C radiation (Shama and Alderson 2005). However, the TPC in pulp was not significantly different among the light treatments, suggesting UV-C radiation cannot penetrate the pineapple peel to generate stress to the pulp.

Similar to TPC, UV-C irradiation significantly increased the total flavonoid content (TFC) of pineapple peel. The increase of TFC in UV-C-treated tissue could be a defense mechanism against UV-C irradiation (Liu et al. 2012). Gonzalez-Aguilar et al. (2007) studied the effect on TPC and TFC of fresh-cut "Tommy Atkins" mango stored for 15 days at 5 °C and irradiated for 0, 1, 3, 5, and 10 min. UV-C irradiation for 10 min induced a fast defense response resulting in the TFC and TFC accumulation. Similarly, mature green tomatoes (*Solanum lycopersicum*) were UV-C radiated, increasing the TFC (Liu et al. 2012).

Furthermore, Liu et al. (2012) exposed the tomato fruits to different doses of UV-C irradiation (2, 4, 8, and 16 kJ m^{-2}), and after irradiation, the tomato fruits were stored in the absence of light at 14 °C and 95% relative humidity for 35 d. UV-C irradiation at 4 or 8 kJ m^{-2} increased the content of gallic acid, chlorogenic acid, syringic acid, p-coumaric acid, and quercetin. They concluded that the optimum doses of UV-C irradiation to increase the TPC and enhance antioxidant activity were 4 or 8 kJ m^{-2}. Gallic acid was the major phenolic compound determined in tomato. In contrast, previous studies reported that UV-C light treatment had no significant effect on the anthocyanin content of grapes and pomegranate (Cantos et al. 2002; López-Rubira et al. 2005). In a different study, Marais et al. (2001) demonstrated that UV-B + incandescent light had no effect on the accumulation of anthocyanins in apples and pears.

Plant anthocyanin absorbs UV radiation in a short time, while flavonols and phenolic acids have an impact on protection of the tissue of fruits and vegetables from the possible damage of UV-B irradiation; the accumulation or induction of flavonoids is considered to be a specific acclimation response. These data support the hypothesis of the existence of different UV-signaling pathways in tissues of fruits and vegetables (Kucera et al. 2003).

15.2.6 Gamma Radiation

Mostafavi et al. (2012) evaluated the effect of different irradiation doses (0, 300, 600, 900, and 1200 Gy) on TPC and antioxidant activity of Red Delicious apple stored at 1 °C during 3-month intervals. The results showed that dose range of 900 Gy significantly decreased phenolic content and antioxidant activity. According to Benoit et al. (2000), the total phenol content was directly related to the dose of irradiation applied (1.17, 1.28, and 1.38 μg/g for mushrooms treated at 0.5, 1.5, and 2.5 kGy, respectively). Ionizing treatments increased PAL activity and total phenolic concentration between days 1 and 4. From days 4 to 12 (end of the storage period), both PAL activity and total phenols in the irradiated mushrooms significantly decrease to lower values. In a different study, the effect of gamma irradiation on the phenolic compounds in rice grains of three genotypes (white, black, and red) was investigated. Ferulic acid, ρ-coumaric acid, and sinapic acid were identified as the major phenolic acids, while cyanidin-3-glucoside and peonidin-3-glucoside were identified in pigmented grain samples.

In general, gamma irradiation in all samples at most of the doses could decrease the phenolic acid content and total anthocyanin content in the black rice. On the contrary, 6 and 8 kGy increased the total content of anthocyanins and phenolic acids in black rice (Zhu et al. 2010). Gamma irradiation at 60 kGy was capable of changing membrane permeability of some vegetables and fruits, favoring water loss. The changes in the localization of anthocyanins are presumed to be related to the water content. An increase in oxidative stress produced by free radicals generated by dehydration during postharvest can decrease the intracellular pH. Therefore, it would favor processes such as the synthesis of anthocyanins and decomposition of lignin (Schreiner and Huyskens-Keil 2006). Irradiation can increase PAL activity of plant tissue, generating an accumulation of phenolic compounds. On the other hand, some results showed that gamma irradiation could significantly decrease total phenolic contents in plant materials (Zhu et al. 2010). Previous studies indicate that adequate doses of irradiation should be selected to minimize the loss of antioxidant phenolic compounds in grain rice during storage.

15.2.7 Edible Coatings

The MAC combined with storage at low temperature is the most common way to prolong the shelf life of fresh-cut products and increase the quality of fresh-cut or minimally processed fruits and vegetables. Besides, it acts as moisture and gas barriers and can serve as carriers of flavorings, vitamins, and minerals. This is in addition to other food-grade additives such as antimicrobial, antibrowning, or antisoftening agents for controlling microbial growth and preserving color, moisture, and texture (Olivas et al. 2008). Edible coatings have been designed to extend shelf life and maintain the quality of minimally processed fruits and vegetables by preventing changes in aroma, taste, texture, and appearance (Barry-Ryan and O'Beirne 2000). These coatings are generally made from lipids, proteins, and polysaccharides (cellulose, pectins, alginates, chitosan, gums, carrageenan, etc.) (Tay and Perera 2004).

The effect of different edible coatings alginate (1.5%), alginate (1.5%) + chitosan (1%), and quitosan (2%) on total phenolic content and total anthocyanin in highbush blueberry (*Vaccinium corymbosum* L. cv. Berkeley and cv. O'Neal) was analyzed (Chiabrando and Giacalone 2015). Highbush blueberries were stored at 0 °C and collected at 15-day periods for 45 days; the application of chitosan coating retards the decrease in anthocyanin content and phenolic content. In both studied cultivars, anthocyanin content was higher in 0 days compared to harvest with significant differences among coatings. In the case of cv. Berkeley berries, anthocyanin content of the alginate-coated and control berries was 42.23 and 34.63 mg cyanidin 3-gluc/100 g FW (the highest and lowest, respectively). At the end of 45 days (storage time), control berries showed a minor concentration of anthocyanins compared with all the coated blueberries. Total phenolic content revealed varied behavior between berries during the postharvest storage period. In alginate-coated samples, TPC showed a slight increase during the next 30 days and then declined at the end of storage. In alginate plus chitosan samples and chitosan-coated berries, TPC decreased during the first 15 days of storage and then increased to its maximum on day 30, after which it declined moderately during the last storage days. In control blueberries, the total phenolic content remains constant during storage.

Between treatments, chitosan-coated samples showed the highest total phenolic content values. Considerable changes were found in flavonoid content in berries treated with several concentrations of chitosan (0.5, 1.0, and 1.5 g/100 mL) during 5 min at 20 °C and subsequently stored at two different temperatures (5 °C and 10 °C); most of the flavonoids increased slowly but steadily at 5 or 10 °C of storage temperatures (Wang and Gao 2013). Flavonoids of the treated berries showed no decrease, but not the control at the end of storage time. The TPC and total anthocyanin content increased substantially and reached a maximum value at 5 °C for 6 days and 3 days at 10 °C after these storage times then decreased in untreated strawberries. In those berries treated with chitosan coating, total anthocyanins and total phenolics also increased slowly and did not decrease at the end of the storage time as the control samples. As a result, chitosan-treated fruit maintained higher anthocyanin and phenolic compounds than the control samples at the end of storage.

Strawberries stored at temperatures higher than 0 °C showed an increase in anthocyanins, and total phenolic content and the magnitude of these increases were related to storage temperature (Kalt et al. 1999). In a similar study, Pen and Jiang (2003) treated fresh-cut Chinese water chestnut with an aqueous solution (0.5, 1, or 2 g chitosan/100 mL), placed it into trays overwrapped with plastic films, and stored it at 4 °C. Changes in PAL and TPC were determined. Application of chitosan coating retards discoloration produced by reduction of PAL activity, including lower TPC compared to the control. Increasing the concentration of chitosan coating enhanced the beneficial effects. The results demonstrated that the application of chitosan coating increased the shelf life and maintained quality of fresh-cut chestnut. PAL activity increased in control slices during the first 6 days, before decreasing in the latter period of storage. Treatment with chitosan coating inhibited PAL activity, and the inhibition enhanced as the treatment concentrations increased. However, higher PAL activity of the fresh-cut chestnut treated with 2 g chitosan was observed at the end of the storage. Phenolic content increased, reached a maximum value after 7 days of storage, and, after this time, then deceased. Treatment with chitosan coating inhibited the variation in the content of total phenols in a concentration-dependent manner. As trans-cinnamate (Ke and Saltveit Jr 1989), a lower level of phenolics of the chitosan-treated slices may be due to the inhibition of PAL activity.

Fresh-cut carrots, apples, and potatoes were coated by a blended whey protein–pectin film–transglutaminase, and the phenolic content of the coated and uncoated samples was analyzed during storage (4 °C for 10 days). Uncoated and coated carrot, apple, and potato phenolic content was determined at 0 and 10 days, where a significant decrease in uncoated apple (about 45%), potato, and carrot (about 20%) samples was found. This decrease was prevented at 50% levels when the samples were coated with whey protein/pectin/transglutaminase-based films.

They have obtained promising results to maintain or increase the content of TPC in fruits and vegetables with edible coating. However, sensory implications may be the main problem for commercialization and consumption. Therefore, more research is needed to develop edible coatings with adequate sensory performance.

15.2.8 Heat Treatments

Heat tolerance of different fruits and vegetables depends on the species, the genotype, the stage of maturity, the type, and the intensity of the heat treatments applied in addition to the conditioning treatment of the postharvest fruit or vegetable (Jacobi et al. 2001). Several investigations relate heat tolerance to the increase in heat shock-resistant proteins, antioxidant enzymes, and phytochemicals such as phenolic compounds. Tomato plants were grown at 15, 25, and 35 °C during 30 days, where the heat stress took place at 35 °C, causing the highest PAL activity, lowest polyphenol oxidase (PPO) and peroxidase activity, and also an accumulation of phenolic compounds (Rivero et al. 2001).

Reyes and Cisneros-Zevallos (2003) reported a different behavior; they stored potatoes at 2, 10, or 20 °C and concluded that treatments did not significantly affect the content of phenolic compounds when compared to potatoes without treatments. Talcott et al. (2005) evaluated mangos using hot water immersion at 50 °C for 60 min and storage at 5 °C and 20 °C against mangos without treatments. After 8 days of storage, the treated fruit at 5 °C was transferred at 20 ° C to complete ripening. Phytochemical and antioxidant capacity were analyzed every 4 days during 20 days. In the study, they concluded that the hot water treatment was not a determinant factor that would affect the content of phenolic compounds or the antioxidant activity of the fruits. However, the storage temperature showed an important role in these changes. These results indicate that temperature plays a fundamental role in the accumulation of phenolic compounds in fruits, activating their accumulation. This could be considered an acclimatization mechanism, which helps to prolong the postharvest life of certain types of fruits and promotes an increase in bioactive compounds.

Hot water treatment such as 35 °C for 12 h and 42 °C for 24 h in different fruits and vegetables is reported to inhibit polyphenol oxidase activities leading to delayed anthocyanin synthesis protective of pigment color changes by maintaining the anthocyanins in their red-pigmented form with high antioxidative activity in post-harvest (Civello et al. 2001; Fallik 2004). However, the potential effects of thermal treatments on fruits and vegetables still need to be studied in greater depth.

15.2.9 Low-Temperature Storage

The temperature has a marked influence on the phytochemicals present during the ripening stage of the product. Research focusing on chilling injury prevention techniques during postharvest conservation of some fruits such as mangos and papaya is abundant. However, little is known about the changes in the antioxidant compounds, especially phenolics and the result of critical storage temperature conditions on antioxidant levels.

For example, Rivera-Pastrana et al. (2010) induced chilling stress in mature green papaya by storing at low temperature (1 °C) and assessed the effect on phenolic compounds in comparison with storage temperature (25 °C). Ranges for ferulic acid content (1.33–1.62 g kg^{-1} dry weight), caffeic acid (0.46–0.68 g/kg), and rutin (0.10–0.16 g/kg) were established in papaya fruit, which tends to decrease during ripening at 25 °C. Low temperature (1 °C) maintained or increased ferulic and caffeic acid levels concerning higher temperatures (25 °C).

Berries (strawberries, red currants, and raspberries) and cherries (*Prunus avium* and *Prunus cerasus*) were stored at 4 °C and 25 °C; the total phenolic compounds, flavonoids, and anthocyanins were evaluated (Piljac-Žegarac and Šamec 2011). Red currants and strawberries showed the highest content of total phenolics at 4 °C (322.40 and 335.47 mg GAE/100 g, respectively) and maintained the highest content during storage at both temperatures. Berries such as strawberries, raspberries, and blueberries stored at (0–6) >15 °C have a lower anthocyanin and phenolic contents as compared to berries stored at higher temperatures (15 °C) (Kalt et al. 1999; Cordenunsi et al. 2005). Furthermore, Wang and Stretch (2001) found the highest anthocyanin and phenolic contents in cranberries stored at 15 °C. The accumulation of anthocyanins is perhaps due to the formation of carbon skeletons from organic acids, through interconversion with carbohydrates, for the synthesis of phenolics, including anthocyanins (Kalt et al. 1999). Fruits and vegetables that are not sensitive to chilling the flavonoid content manifest an increase at low temperatures.

Concellón et al. (2012) evaluated the effect of two temperatures of storage in dark-purple American eggplant (0 or 10 °C) for 0, 3, 5, 10, or 14 days, on Folin–Ciocalteu reacting substances and the content of chlorogenic acid. The Folin–Ciocalteu reagent increased during the first 3 days of storage at 0 °C, but later, significant degradation was found. The major phenolic detected was chlorogenic acid and an ester between caffeic (CA) and quinic acids (QA), which accumulated in fruit maintained at 10 °C, increasing by 60% after 14 days of storage.

In blushed pear "Rosemarie," low temperatures increased the red color as well as the activity of both PAL and flavonoid 3-O-glucosyltransferase. In contrast, the phenolic compounds and enzyme activity in the red pear "Bon Rouge" showed a very low response to low temperatures (Concellón et al. 2012). These observations indicated an additional genetic effect (Steyn et al. 2004). According to Lattanzio (2003), the postharvest metabolic changes of phenolic compounds are often coupled with the activity of phenylalanine ammonia-lyase enzyme that is induced by low temperatures in plant tissue.

15.2.10 Ethylene (C2H4)

Fruits and vegetables minimally processed promote large amounts of ethylene, which results in a shorter life of the products. Ethylene accelerates ripening, softening, and senescence, which provokes changes in the cell membrane of the products (Siddiqui et al. 2011) and hence could promote the accumulation of the phytochemical compounds at postharvest (Kader 2002).

Endogenous sensitivity to C_2H_4 varies during plant development, as well as its rate of synthesis and loss by diffusion from the plant. The responses to exogenous C2H4 applied to fruits and vegetables are numerous and varied and could be beneficial or detrimental in the production of phenolic compounds (Table 15.2). In non-climacteric fruits, ethylene induced senescence associated with a high decomposition rate of secondary metabolites (Kader 2002), or as in the case of bell pepper, no significant differences were noted for phenolics due to postharvest ethylene exposition (Fox et al. 2005).

Ethylene treatment of grape berries was associated with an increase in gene expression of chalcone synthase, flavanone 3-hydroxylase, leucoanthocyanidin

Table 15.2 Effects of ethylene on phytochemicals from fruit and vegetables

Fruit or vegetables	Treatment	Effects	References
		Anthocyanins	Reyes and Cisneros-Zevallos (2003)
Purple-flesh potatoes	Fresh	152 mg/kg	
	Air	183 mg/kg (↑)	
	Ethylene	90 mg/kg (–)	
		Total phenolic	Reyes and Cisneros-Zevallos (2003)
Purple-flesh potatoes	Fresh	1000 mg/kg	
	Air	1600 mg/kg (↑)	
	Ethylene	1500 mg/kg (–)	
		Anthocyanins	Costa et al. (2018)
Blueberries	Control, 1 day	1450 mg/kg	
	Ethylene, 1 day	2100 mg/kg (↑)	
	Ethylene, 28 days	2700 mg/kg (↑)	
	Ethylene, 56 days	1000 mg/kg (–)	
		Anthocyanins	Anastasiadi et al. (2016)
Gooseberries	Control, 1 day	600 mg/kg	
	Ethylene, 1 day	800 mg/kg (↑)	
	Control, 7 days	900 mg/kg (↑)	
	Ethylene, 7 days	750 mg/kg (–)	
	Control, 11 days	1090 mg/kg (↑)	
	Ethylene, 11 days	1050 mg/kg (–)	
		Total phenolics	Ma et al. (2017)
Pears	Control, 5 days	70 mg/g	
	Ethylene, 5 days	00 mg/g (↑)	
	Control, 15 days	05 mg/g (↑)	
	Ethylene, 15 days	25 mg/g (↑)	
	Control, 25 days	95 mg/g (–)	
	Ethylene, 25 days	2.25 mg/g (↑)	
		Total phenolics	Massolo et al. (2019)
Cut celery	Control, 5 days	7 mg/kg	
	Ethylene, 5 days	3 mg/kg (–)	
	Control, 15 days	4 mg/kg (–)	
	Ethylene, 15 days	4 mg/kg (↑)	

(continued)

Table 15.2 (continued)

Fruit or vegetables	Treatment	Effects	References
	Control, 25 days	9 mg/kg (↑)	
	Ethylene, 25 days	86.1 mg/kg (−)	
		Total phenolics	Persic et al. (2019)
Persimmon	Control, 1 day	900 mg/kg	
	Ethylene, 1 day	700 mg/kg (−)	
	Control, 5 days	690 mg/kg (−)	
	Ethylene, 5 days	900 mg/kg (↑)	

(↑) Increase; (−) decrease

dioxygenase, and UDP glucose–flavonoid 3-O-glucosyltransferase leading to an increased concentration of phenolic compounds. However, as shown in apples, other phenolics such as flavonoids and chlorogenic acid were not affected by ethylene treatment (Awad and de Jager 2002), suggesting a specific ethylene susceptibility of anthocyanin biosynthesis. The treatment with ethylene of fruits and vegetables shows an important effect on its antioxidant properties and on the content of some of its phenolic acids. The effect of prolonged-term storage is apparent. However, it can only be evaluated in the presence of ethylene treatment.

15.2.11 Methyl Jasmonate (MJ)

The results of the postharvest use of (−)- and (+)-MJ enantiomers, grapes treated with MJ racemic mixture, and grapes without treatment (control) on TPC and total anthocyanin concentration were determined on and 7 days after application (Flores et al. 2015). Exogenous treatment of MJ induced enhancement in TPC when compared to control grapes. This increase was higher when the enantiomers were used individually (up to 42%TCP), rather than the mixture was applied. Similarly, a significant increase was reported in the total anthocyanin content when the grapes were exposed to the enantiomers. This increase varied from 14% to 42% after 5 days of storage and from 22% to 64% when grapes were preserved for 7 days. Postharvest treatment of grapes with MJ enantiomers is suggested to increase the antioxidant activity and anthocyanin concentration in grapes.

Flores and del Castillo (2014) evaluated the effect of postharvest treatments with MJ on the content of myricetin, quercetin, and ellagic acid in raspberries. Different raspberry varieties and different MJ concentrations were analyzed in the study. Postharvest MJ treatment did not cause an increase in the concentration of myricetin, quercetin, and ellagic acid during storage, which was constant between 60 and 100 mg/g. Enzyme studies reflected an increase in PAL activity after preharvest MJ treatment. Yang et al. (2011) studied the application of MJ for regulation of pericarp browning of harvested lychees concerning to the content of anthocyanins and (−)-epicatechin. Lychees were dipped for 3 min in 0.6 μg/L MJ, 1 μg/L MJ, 5 μg/L MJ + 0.05% Sportak, 25 μg/L MJ, or 0.05% Sportak (control).

The treatment with MJ significantly maintained the contents of anthocyanins and (−)-epicatechin and extended the shelf life of lychees. Furthermore, MJ treatment increased PAL activity, which mainly explains the anthocyanin or (−)-epicatechin content. After these treatments, 1 μg/L MJ was effective in inhibiting pericarp browning and in maintaining the contents of anthocyanins and (−)-epicatechin of lychees that did not suffer any variation. MJ induces anthocyanin biosynthesis by induction of chalcone synthase and dihydroflavonol-4-reductase gene expression (Tamari et al. 1995; Saniewski et al. 1998). Therefore, after postharvest MJ application, increased anthocyanin contents and phenolic acids were found in apple, guava, and papaya (Rudell et al. 2002; Kondo et al. 2005). Application of MJ also enhanced the activity of PAL and thus the concentration of other flavonoids (Tomás-Barberán and Espín 2001).

15.3 Conclusion

The efficacy and efficiency of postharvest treatments have been assessed mainly by the quality maintenance of harvested fruit and vegetables. However, with the increasingly growing consumer interest in foods that are beneficial to health, attention has been focused on quality maintenance and sensorial attributes. Also, there is a powerful emphasis on the enhancement of health-promoting phytochemicals. Most postharvest treatments involve tissue stress resulting in an accumulation of TPC via the stimulation of the PAL, the key enzyme that induces the flavonoid metabolism. Therefore, to obtain products of fruit and vegetables enriched with this type of phytochemicals, it can be realized through treatments and postharvest technologies. However, it is necessary to study these treatments deeply to maintain and increase the content of phytochemicals as phenolic compounds.

References

Alothman M, Bhat R, Karim AA (2009) UV radiation-induced changes of antioxidant capacity of fresh-cut tropical fruits. Innov Food Sci Emerg Technol 10:512–516

Anastasiadi M, Mwangi PM, Ordaz-Ortiz JJ, Redfern SP, Berry M, Simmonds MS, Terry LA (2016) Tissue biochemical diversity of 20 gooseberry cultivars and the effect of ethylene supplementation on postharvest life. Posth Biol Technol 117:141–151

Awad MA, de Jager A (2002) Formation of flavonoids, especially anthocyanin and chlorogenic acid in 'Jonagold' apple skin: influences of growth regulators and fruit maturity. Sci Hort 93:257–266

Ayala-Zavala JF, Wang SY, Wang CY, González-Aguilar GA (2004) Effect of storage temperatures on antioxidant capacity and aroma compounds in strawberry fruit. LWT-Food Sci Technol 37:687–695

Barry-Ryan C, O'Beirne D (2000) Effects of peeling methods on the quality of ready-to-use carrot slices. Int J Food Sci Technol 35:243–254

Benoit MA, D'Aprano G, Lacroix M (2000) Effect of γ-irradiation on phenylalanine ammonia-lyase activity, total phenolic content, and respiration of mushrooms (*Agaricus bisporus*). J Agric Food Chem 48:6312–6316

Cantos E, Espin JC, Tomás-Barberán FA (2002) Varietal differences among the polyphenol profiles of seven table grape cultivars studied by LC–DAD–MS–MS. J Agric Food Chem 50:5691–5696

Charles MT, Tano K, Asselin A, Arul J (2009) Physiological basis of UV-C induced resistance to *Botrytis cinerea* in tomato fruit. V. Constitutive defence enzymes and inducible pathogenesis-related proteins. Postharvest Biol Technol 51:414–424

Chiabrando V, Giacalone G (2015) Anthocyanins, phenolics and antioxidant capacity after fresh storage of blueberry treated with edible coatings. Int J Food Sci Nutr 66:248–253

Civello PM, Vicente AR, Martinez GA, Chaves R (2001) Refrigerated storage of heat treated strawberry fruit. Acta Hort 553:551–553

Concellón A, Zaro MJ, Chaves AR, Vicente AR (2012) Changes in quality and phenolic antioxidants in dark purple American eggplant (*Solanum melongena* L. cv. Lucía) as affected by storage at 0 °C and 10 °C. Postharvest Biol Technol 66:35–41

Connor AM, Luby JJ, Hancock JF, Berkheimer S, Hanson EJ (2002) Changes in fruit antioxidant activity among blueberry cultivars during cold-temperature storage. J Agric Food Chem 50:893–898. https://www.ncbi.nlm.nih.gov/nlmcatalog?term=%22J+Agric+Food+Chem%22%5Bta%5D

Cordenunsi BR, Genovese MI, do Nascimento JRO, Hassimotto NMA, dos Santos RJ, Lajolo FM (2005) Effects of temperature on the chemical composition and antioxidant activity of three strawberry cultivars. Food Chem 91:113–121

Costa DV, Almeida DP, Pintado M (2018) Effect of postharvest application of ethylene on the profile of phenolic acids and anthocyanins in three blueberry cultivars (*Vaccinium corymbosum*). J Sci Food Agric 98:5052–5061

Day BPF (1996) High oxygen modified atmosphere packaging for fresh prepared produce. Postharvest News Inf 7:31N–34N

Du WX, Avena-Bustillos RJ, Breksa AP III, McHugh TH (2012) Effect of UV-B light and different cutting styles on antioxidant enhancement of commercial fresh-cut carrot products. Food Chem 134:1862–1869

Duan X, Liu T, Zhang D, Su X, Lin H, Jiang Y (2011) Effect of pure oxygen atmosphere on antioxidant enzyme and antioxidant activity of harvested litchi fruit during storage. Food Res Int 44:1905–1911

El-Ghaouth A, Wilson CL, Wisniewski M (1998) Ultrastructural and cytochemical aspects of the biological control of *Botrytis cinerea* by *Candida saitoana* in apple fruit. Phytopathology 88:282–291

Fallik E (2004) Pre-storage hot water treatments (immersion, rinsing and brushing)-a review. Posth Biol Technol 32:125–134

Flores G, del Castillo MLR (2014) Influence of preharvest and postharvest methyl jasmonate treatments on flavonoid content and metabolomic enzymes in red raspberry. Postharvest Biol Technol 97:77–82

Flores G, Blanch GP, del Castillo MLR (2015) Postharvest treatment with (−) and (+)-methyl jasmonate stimulates anthocyanin accumulation in grapes. LWT-Food Sci Technol 62:807–812

Fox AJ, Del Pozo-Insfran D, Lee JH, Sargent SA, Talcott ST (2005) Ripening-induced chemical and antioxidant changes in bell peppers as affected by harvest maturity and postharvest ethylene exposure. Hort Sci 40:732–736

Gennaro L, Leonardi C, Esposito F, Salucci M, Maiani G, Quaglia G, Fogliano V (2002) Flavonoid and carbohydrate contents in Tropea red onions: effects of homelike peeling and storage. J Agric Food Chem 50:1904–1910

Gonzalez-Aguilar GA, Villegas-Ochoa MA, Martinez-Tellez MA, Gardea AA, Ayala-Zavala JF (2007) Improving antioxidant capacity of fresh-cut mangoes treated with UV-C. J Food Sci 72: S197–S202

Heredia JB, Cisneros-Zevallos L (2009) The effect of exogenous ethylene and methyl jasmonate on pal activity, phenolic profiles and antioxidant capacity of carrots (*Daucus carota*) under different wounding intensities. Postharvest Biol Technol 51:242–249

Higashio H, Hirokane H, Sato F, Tokuda S, Uragami A (2005) Enhancement of functional compounds in Allium vegetables with UV radiation. In: I International symposium on human health effects of fruits and vegetables, vol 744, pp 357–362

Holcroft DM, Kader AA (1999) Controlled atmosphere-induced changes in pH and organic acid metabolism may affect color of stored strawberry fruit. Postharvest Biol Technol 17:19–32

Jacobi KK, MacRae EA, Hetherington SE (2001) Postharvest heat disinfestation treatments of mango fruit. Sci Hort 89(3):171–193

Jin P, Yao D, Xu F, Wang H, Zheng Y (2015) Effect of light on quality and bioactive compounds in postharvest broccoli florets. Food Chem 172:705–709

Kader AA (2002) Postharvest technology of horticultural crops, vol 3311. University of California Agriculture and Natural Resources

Kader F, Rovel B, Girardin M, Metche M (1996) Fractionation and identification of the phenolic compounds of highbush blueberries (*Vaccinium corymbosum*, L.). Food Chem 55:35–40

Kalt W, Forney CF, Martin A, Prior RI (1999) Antioxidant capacity, vitamin C, phenolics, and anthocyanins after fresh storage of small fruits. Postharvest Biol Technol 47:4638–4644

Ke D, Saltveit ME Jr (1989) Wound-induced ethylene production, phenolic metabolism and susceptibility to russet spotting in iceberg lettuce. Physiol Plant 76:412–418

Kenny O, O'Beirne D (2010) Antioxidant phytochemicals in fresh-cut carrot disks as affected by peeling method. Postharvest Biol Technol 58:247–253

Ko EY, Nile SH, Sharma K, Li GH, Park SW (2015) Effect of different exposed lights on quercetin and quercetin glucoside content in onion (*Allium cepa* L.). Saudi J Biol Sci 22:398–403

Kondo S, Kittikorn M, Kanlayanarat S (2005) Preharvest antioxidant activities of tropical fruit and the effect of low temperature storage on antioxidants and jasmonates. Postharvest Biol Technol 36:309–318

Kucera B, Leubner-Metzger G, Wellmann E (2003) Distinct ultraviolet-signaling pathways in bean leaves. DNA damage is associated with β-1, 3-glucanase gene induction, but not with flavonoid formation. Plant Physiol 133:1445–1452

Lamikanra O, Richard OA (2002) Effect of storage on some volatile aroma compounds in fresh-cut cantaloupe melon. J Agric Food Chem 50(14):4043–4047

Lattanzio V (2003) Bioactive polyphenols: their role in quality and storability of fruit and vegetables. J Appl Bot Food Qual 77:128–146

Lattanzio V, Di Venere D, Linsalata V, Bertolini P, Ippolito A, Salerno M (2001) Low temperature metabolism of apple phenolics and quiescence of *Phlyctaena vagabunda*. J Agric Food Chem 49:5817–5821

Lee NY, Lee MJ, Kim YK, Park J, Park HK, Choi JS, Kim JG (2010) Effect of light emitting diode radiation on antioxidant activity of barley leaf. J Korean Soc Appl Biol Chem 53:685–690

Li Q, Kubota C (2009) Effects of supplemental light quality on growth and phytochemicals of baby leaf lettuce. Environ Exp Bot 67:59–64

Li XD, Wu BH, Wang LJ, Zheng XB, Yan ST, Li SH (2009) Changes in trans-resveratrol and other phenolic compounds in grape skin and seeds under low temperature storage after post-harvest. J Hortic Sci Biotechno 84:113–118

Li X, Long Q, Gao F, Han C, Jin P, Zheng Y (2017) Effect of cutting styles on quality and antioxidant activity in fresh-cut pitaya fruit. Postharvest Biol Technol 124:1–7

Liu J, Stevens C, Khan VA, Lu JY, Wilson CL, Adeyeye O, Droby S (1993) Application of ultraviolet-C light on storage rots and ripening of tomatoes. J Food Prot 56:868–873

Liu CH, Cai LY, Lu X, Han XX, Ying TJ (2012) Effect of postharvest UV-C irradiation on phenolic compound content and antioxidant activity of tomato fruit during storage. J Integr Agric 11:159–165

López-Rubira V, Conesa A, Allende A, Artés F (2005) Shelf life and overall quality of minimally processed pomegranate arils modified atmosphere packaged and treated with UV-C. Postharvest Biol Technol 37:174–185

Ma Y, Yang M, Wang J, Jiang CZ, Wang Q (2017) Application of exogenous ethylene inhibits postharvest peel browning of 'Huangguan' pear. Front Plant Sci 7:2029

Maharaj R, Arul J, Nadeau P (1999) Effect of photochemical treatment in the preservation of fresh tomato (*Lycopersicon esculentum* cv. Capello) by delaying senescence. Postharvest Biol Technol 15:13–23

Makris DP, Rossiter JT (2001) Domestic processing of onion bulbs (*Allium cepa*) and asparagus spears (*Asparagus officinalis*): effect on flavonol content and antioxidant status. J Agric Food Chem 49:3216–3222

Marais E, Jacobs G, Holcroft DM (2001) Light and temperature affect postharvest colour development in 'Cripps pink' apple. Acta Hort 553:91–93

Massa GD, Kim HH, Wheeler RM, Mitchell CA (2008) Plant productivity in response to LED lighting. HortScience 43:1951–1956

Massolo JF, Forte LG, Concellón A, Viña SZ, Vicente AR (2019) Effects of ethylene and 1-MCP on quality maintenance of fresh cut celery. Postharv Biol Technol 148:176–183

Mishra BB, Gautam S, Sharma A (2012) Browning of fresh-cut eggplant: impact of cutting and storage. Postharv Biol Technol 67:44–51

Montero-Calderón M, Rojas-Graü MA, Martín-Belloso O (2010) Aroma profile and volatiles odor activity along gold cultivar pineapple flesh. J Food Sci 75:S506–S512

Mostafavi HA, Mirmajlessi SM, Mirjalili SM, Fathollahi H, Askari H (2012) Gamma radiation effects on physico-chemical parameters of apple fruit during commercial post-harvest preservation. Radiat Phys Chem 81:666–671

Ohashi-Kaneko K, Takase M, Kon N, Fujiwara K, Kurata K (2007) Effect of light quality on growth and vegetable quality in leaf lettuce, spinach and komatsuna. Environ Cont Biol 45(3):189–198

Olivas GI, Dávila-Aviña JE, Salas-Salazar NA, Molina FJ (2008) Use of edible coatings to preserve the quality of fruits and vegetables during storage. Stewart Posth Rev 3(6):1–10

Oufedjikh H, Mahrouz M, Amiot MJ, Lacroix M (2000) Effect of γ-irradiation on phenolic compounds and phenylalanine ammonia-lyase activity during storage in relation to peel injury from peel of *Citrus clementina* Hort. Ex Tanaka J Agric Food Chem 48:559–565

Pen LT, Jiang YM (2003) Effects of chitosan coating on shelf life and quality of fresh-cut Chinese water chestnut. LWT-Food Sci Technol 36(3):359–364

Pérez-Ilzarbe J, Hernández T, Estrella I, Vendrell M (1997) Cold storage of apples (cv. Granny Smith) and changes in phenolic compounds. Z Lebensm Unters Forsch 204:52–55

Persic M, Jakopic J, Hudina M (2019) The effect of post-harvest technologies on selected metabolites in persimmon (*Diospyros kaki Thunb.*) fruit. J Sci Food Agric 99:854–860

Pesis E (2005) The role of the anaerobic metabolites, acetaldehyde and ethanol, in fruit ripening, enhancement of fruit quality and fruit deterioration. Postharvest Biol Technol 37(1):1–19

Piljac-Žegarac J, Šamec D (2011) Antioxidant stability of small fruits in postharvest storage at room and refrigerator temperatures. Food Res Int 44:345–350

Reyes LF, Cisneros-Zevallos L (2003) Wounding stress increases the phenolic content and antioxidant capacity of purple-flesh potatoes (*Solanum tuberosum* L.). J Agric Food Chem 51:5296–5300

Rivera-Pastrana DM, Yahia E, González-Aguilar GA (2010) Phenolic and carotenoid profiles of papaya fruit (*Carica papaya* L.) and their contents under low temperature storage. J Sci Food Agric 90:2358–2365

Rivero RM, Ruiz JM, Garcıa PC, Lopez-Lefebre LR, Sánchez E, Romero L (2001) Resistance to cold and heat stress: accumulation of phenolic compounds in tomato and watermelon plants. Plant Sci 160:315–321

Rudell DR, Mattheis JP, Fan X, Fellman JK (2002) Methyl jasmonate enhances anthocyanin accumulation and modifies production of phenolics and pigments in 'Fuji' apples. J Am Soc Hortic Sci 127:435–441

Samuolienė G, Sirtautas R, Brazaitytė A, Duchovskis P (2012) LED lighting and seasonality effects antioxidant properties of baby leaf lettuce. Food Chem 134(3):1494–1499
Saniewski M, Miyamoto K, Ueda J (1998) Methyl jasmonate induces gums and stimulates anthocyanin accumulation in peach shoots. J Plant Growth Regul 17:121–124
Sari LK, Setha S, Naradisorn M (2016) Effect of UV-C irradiation on postharvest quality of 'Phulae'pineapple. Sci Hortic 213:314–320
Schotsmans W, Molan A, MacKay B (2007) Controlled atmosphere storage of rabbit eye blueberries enhances postharvest quality aspects. Postharvest Biol Technol 44:277–285
Schreiner M, Huyskens-Keil S (2006) Phytochemicals in fruit and vegetables: health promotion and postharvest elicitors. Crit Rev Plant Sci 25:267–278
Shama G (2007) Process challenges in applying low doses of ultraviolet light to fresh produce for eliciting beneficial hormetic responses. Postharvest Biol Technol 44:1–8
Shama G, Alderson P (2005) UV hormesis in fruits: a concept ripe for commercialisation. Trends Food Sci Technol 16:128–136
Siddiqui MW, Chakrabort I, Ayala-Zavala JF, Dhua RS (2011) Advances in minimal processing of fruits and vegetables: a review. J Sci Ind Res India 70:823–834
Singh SP, Pal RK (2008) Controlled atmosphere storage of guava (*Psidium guajava* L.) fruit. Postharvest Biol Technol 47:296–306
Singh SP, Singh Z (2013) Controlled and modified atmospheres influence chilling injury, fruit quality and antioxidative system of Japanese plums (*Prunus salicina* L indell). Int J Food Sci Technol 48:363–374
Steyn WJ, Holcroft DM, Wand SJE, Jacobs G (2004) Anthocyanin degradation in detached pome fruit with reference to preharvest red color loss and pigmentation patterns of blushed and fully red pears. J Am Soc Hortic Sci 129(1):13–19
Surjadinata BB, Cisneros-Zevallos L (2012) Biosynthesis of phenolic antioxidants in carrot tissue increases with wounding intensity. Food Chem 134:615–624
Talcott ST, Moore JP, Lounds-Singleton AJ, Percival SS (2005) Ripening associated phytochemical changes in mangos (*Mangifera indica*) following thermal quarantine and low-temperature storage. J Food Sci 70:C337–C341
Tamari G, Borochov A, Atzorn R, Weiss D (1995) Methyl jasmonate induces pigmentation and flavonoid gene expression in petunia corollas: a possible role in wound response. Physiol Plant 94:45–50
Tavarini S, Degl'Innocenti E, Pardossi A, Guidi L (2007) Biochemical aspects in two minimally processed lettuces upon storage. Int J Food Sci Technol 42:214–219
Tay SL, Perera CO (2004) Effect of 1-methylcyclopropene treatment and edible coatings on the quality of minimally processed lettuce. J Food Sci 69:fct131–fct135
Tomás-Barberán FA, Espín JC (2001) Phenolic compounds and related enzymes as determinants of quality in fruits and vegetables. J Sci Food Agric 81:853–876
Wang SY, Gao H (2013) Effect of chitosan-based edible coating on antioxidants, antioxidant enzyme system, and postharvest fruit quality of strawberries (*Fragaria x aranassa* Duch.). LWT-Food Sci Technol 52:71–79
Wang SY, Stretch AW (2001) Antioxidant capacity in cranberry is influenced by cultivar and storage temperature. J Agric Food Chem 49:969–974
Yahia EM (2009) The contribution of fruit and vegetable consumption to human health. In: De La Rosa LA, Alvarez-Parrilla E, González-Aguilar GA (eds) Fruit and vegetable phytochemicals, pp 3–51
Yang S, Chen Y, Feng L, Yang EN, Su X, Jiang Y (2011) Effect of methyl jasmonate on pericarp browning of postharvest lychees. J Food Process Preserv 35:417–422
Zhang Z, Nakano K, Maezawa S (2009) Comparison of the antioxidant enzymes of broccoli after cold or heat shock treatment at different storage temperatures. Postharvest Biol Technol 54:101–105

Zheng Y, Wang CY, Wang SY, Zheng W (2005) Effect of super-atmospheric oxygen on anthocyanins, phenolics and antioxidant activity of blueberries and strawberries. Acta Hort 857:475–482

Zhu F, Cai YZ, Bao J, Corke H (2010) Effect of γ-irradiation on phenolic compounds in rice grain. Food Chem 120:74–77

16 Plant Phenolics as Natural Preservatives in Food System

Pavan Kumar, Akhilesh Kumar Verma, Pramila Umaraw, Nitin Mehta, and Om Prakash Malav

Abstract

There is an increasing trend of using of natural compounds, such as plant phenolics, for their preservative effect in food due to food safety, nutritive and possible therapeutic effects. Flavonoids, small molecular weight, short carbon chain and ubiquitous compounds formed as secondary metabolites exert preservative effect due to antioxidant and antimicrobial activity. These can be added into foods as powder, gel, liquid, raw or in extract form, with latter form being more suitable to get desired antioxidant and antimicrobial effect in lower concentrations to avoid deterioration in sensory attributes of food products. There is a need to standardize efficient processing protocols for extraction of phenolics from plant material in high concentrations and active form.

Keywords

Plant phenolics · Flavonoids · Antioxidant · Antimicrobial · Phyto-extracts

16.1 Introduction

There is a continuous shift of consumer's preference for minimally processed food products and products prepared without the use of synthetic preservatives/additives due to increasing awareness about the possible health risks associated with the intake of synthetic chemicals. This preference for green label/natural food/green

P. Kumar (✉) · N. Mehta · O. P. Malav
Department of Livestock Products Technology, College of Veterinary Science Guru Angad Dev Veterinary and Animal Sciences University, Ludhiana, Punjab, India

A. K. Verma · P. Umaraw
Department of Livestock Products Technology, College of Veterinary & Animal Sciences Sardar Vallabhbhai Patel University of Agriculture & Technology, Meerut, Uttar Pradesh, India

R. Lone et al. (eds.), *Plant Phenolics in Sustainable Agriculture*,
https://doi.org/10.1007/978-981-15-4890-1_16

consumerism is likely to increase in the future. Consumer awareness and education are the key factors determining the acceptance of food products by consumers. At present, consumers have tendency to accept food products containing synthetic preservatives if these preservatives are used within prescribed limits and proper complete information regarding its uses, purpose of addition and possible ill effects have been properly displayed on packets (Hung et al. 2016). As per Food and Health Survey, 2015, conducted by International Food Information Council in 2015, consumers gave more weightage for addition of synthetic preservatives in food (36%) than foodborne illness (34%) while buying food products for consumption. It clearly indicates that consumers are nowadays more concerned about the use of synthetic preservatives in food products rather than foodborne pathogens. Similar findings and consumers concerns for food products containing synthetic additives or preservatives have also been documented by Williams et al. (2004) during a National Survey about quality and safety of foods, and they observed that more than 50% of Australians respondents showed high concern for risks associated with incorporation of synthetic additives and preservatives. Lupton (2005), in a study on discourse and beliefs to food risk in Australia during 2002–2003, also observed that most of the people were not much concerned about food poisoning but the addition of synthetic preservatives remained a critical concern, which needed to be addressed on priority basis.

The lack of proper information regarding the purpose and prescribed limits of synthetic food additives creates more fear among consumers, and if this information is properly displayed/disseminated to buyers, it encourages them to buy these products and alleviates fear for such products in consumers' minds. Wezemael et al. (2010) noted decreased perception about healthiness of beef with the addition of synthetic additives or preservatives under the EU-funded *ProSafeBeef* in eight focus groups made up of seven to nine members, selected from France, the UK, Germany and Spain. The proper approach way of addressing this issue, according to focus groups, is to promote the improved production methods and proper consumption behaviour rather than focusing on addition of synthetic preservatives in beef. Beside the absence of synthetic additives, sensory attributes such as taste and flavour of the final product also play a crucial role in determining consumer acceptance of a food, and there are instances, when consumers override the fear of synthetic chemicals in favour of a product with high sensory attributes. Thus, these aspects should not be overlooked while considering natural/green label food products.

16.2 Lipid Oxidation and Microbial Growth – Main Concerns for Food Spoilage

Deterioration of taste and nutritive quality of food caused by oxidative damage (oxidative rancidity) along with the growth of food contamination and pathogenic microbes are regarded as main culprits behind food spoilage/wastage. Lipids oxidation has an important role in determining the quality of food products even in food products containing very low amount of fat (0.5–1.0%). During oxidation of lipids, there is generation of several primary and secondary oxidation products leading to deterioration of nutritional quality and poor organoleptic attributes (especially flavour) of food products, resulting in huge economic losses and associated health hazards to consumers. This is caused due to oxidation of unsaturated fatty acids, leading to the formation of hydroperoxides which are very unstable compounds and in presence of trace elements and new free radicals, these are rapidly degraded into short carbon chain oxidation products such as ketones, aldehydes, etc. These oxidative changes have been initiated by the generation of reactive oxygen species (ROS) under favourable conditions such as the presence of air, light, metallic ions, etc. ROS comprises both oxygen-centred free radicals and non-radical derivatives. Free radical is a highly unstable, reactive and uncharged molecule which has an independent existence and contains an unpaired electron in its atomic orbit. These compounds have high affinity towards making single unpaired electron in outermost atomic orbit by accepting an electron from other molecules or by donating this extra unpaired electron, thus acting as reductants or oxidants (Cheeseman and Slater 1993). Hydroxyl radicals, hydrogen peroxides, hypochlorite, superoxide anion radical, oxygen singlet, peroxynitrite and nitric oxygen radicals are common free radicals containing oxygen. These radicals react with biologically important molecules such as nucleic acids, proteins, lipids, carbohydrates and cell membranes and damage them (Young and Woodside 2001). Besides this, these lipid oxidation products cause damage to mucous membrane of digestive tract, and thus interfere in the absorption of nutrients such as protein and folic acid. Some of these compounds are potentially toxic, such as oxysterols from cholesterol (Morel and Lin 1996).

16.3 Plant Phenolics as Natural Preservative

Phenolic compounds possess an aromatic ring and one or more hydroxyl (–OH) groups. There are about 9000 phenolic compounds reported in plant kingdom and more than 50% of these compounds are flavonoids (Wang et al. 2011). These compounds make up the most commonly distributed compounds in plants and exert wide range of bioactivity such as antioxidant, antimicrobial, anticancer, modifying gene expression, antiviral, antiulcer, protein kinase inhibition, anti-inflammatory, etc. Basically flavonoids (Fl) are made up of two benzene rings joined together by a propane unit (C6–C3–C6). Among flavonoids, flavones and flavanols are the most commonly ones present in various parts of plants. The polyphenolic nature of flavonoids enables them to possess antioxidant properties. Flavonoids are produced by plants as a natural defence against various parasites, oxidative stress and inclement climatic conditions. The human intake of flavonoids through diet varies from 20–500 mg, with red wine, apples, tomato and onion being the major source (Giuliani et al. 2014).

Flavonoids are ubiquitous, low molecular weight, indispensable polyphenolic compounds that are generated as secondary products during metabolic activity of plant. While exerting various physiological roles such as colour, pigments, stress management, disease resistance, etc, flavonoids also have several nutraceutical benefits for consumers such as antioxidant, antibacterial, antiviral, anticancer, anti-inflammatory, etc. Since ancient times, flavonoids have been integral part of our diet. Some particular diets rich in flavonoids had been linked with the sound health and high longevity of people in that specific area. Inclusion of an ample amount of red wines and red grapes by the people of southern France results in higher longevity of life and very low incidence of cardiovascular diseases despite them following some potentially unhealthy diet and lifestyle such as high fat intake, less exercise and smoking (Renaud and de Lorgeril 1992; de Lange et al. 2007; Lachman et al. 2007). Similarly the very high life longevity of Mediterranean people has been linked to high consumption of fresh fruits, vegetables and probiotic dairy products (Rice-Evans and Miller 1995). The higher longevity and very low incidence of cardiovascular diseases as observed in French and Mediterranean population despite their large intake of fat (French Paradox) could be possible due to high intake of flavonoid-rich diet (especially red grapes and red wine) (Burr 1995; Formica and Regelson 1995; de Lange et al. 2007).

Several studies have documented promising health benefits associated with consumption of flavonoids, such as inhibition of platelet activation by stimulation of inhibitor receptors of platelet endothelial cell adhesion molecules (de Lange et al. 2007), decreased mortality due to coronary heart disease (Hertog et al. 1993) and myocardial infarction in elderly population by regular intake of diet rich in flavanol (Hertog et al. 1997), etc. It has been proved that flavonoids have a possible preservative role in food due to their potential antioxidant and antibacterial activities.

Flavonoids have flavan nucleus (phenyl benzopyrone skeleton C–3C–6C–3) with chromane ring along with phenyl substituent at C2–C3 positions. Flavonoids are made up of two basic structural units (A, B) linked by heterocyclic pyran ring (C). The A, B rings are made up of 2-phenyl-benzo-γ-pyrane nucleus (Cushnie and Lamb 2005; Kumar et al. 2013; Isoda et al. 2014). Flavonoids are classified based on the oxidation status of C ring, such as anthocyanidins comprising oxidation and reduction of central ring C or the presence or absence of double bond between C2 and C3 leading to flavanones, flavanols or flavones.

16.4 Plant Phenolics as Natural Antioxidants

Flavonoids have strong antioxidant activity due to their hydrogen ion–donating ability, radical scavenging ability, inhibition of free radicals formation and by chelating/sequestering prooxidants trace metals such as Cu^{2+}, Fe^{2+}, Fe^{3+} (Ozsoy et al. 2009; Kumar et al. 2015a, b). Prior and Cao (2000) noticed the higher antioxidant capacity of flavonoids as compared to vitamin C and vitamin E. For tea flavonoids (epigallocatechin gallate) electron reduction potential (550 mV) was recorded comparable to that of alpha-tocopherol (480mV) but lower than the electron reduction potential of glutathione (920 mV) (Frei and Higdon 2003; Maeta et al. 2007). Flavonoids exert antioxidant effect by directly scavenging reactive oxygen species, chelating trace metals, increasing activity and efficacy of antioxidant enzymes, inhibition of oxidase enzyme, oxidative effect of nitric oxide, increasing uric acid levels and preventing formation of alpha-tocopherol radicals (Prochazkova et al. 2011; Nijveldt et al. 2001; Ferrali et al. 1997; Hirano et al. 2001).

16.4.1 Inactivation of Free Radicals

Flavonoids (Fl) inactivate free radicals (R°) by donating a hydrogen atom, leading to the formation of flavonoid phenoxyl radical (Fl–O°). The antioxidant activity of flavonoids depends upon its core structure comprising the configuration and total hydroxyl group counts, with maximum effect occurring due to B-ring hydroxyl group, whereas hydroxyl groups on A- and C-ring exerting comparatively less effect (Heim et al. 2002; Amic et al. 2007). The polymerization of flavonoid monomers resulted in increased antioxidant activity due to higher numbers of hydroxyl groups in these polymerized molecules such as condensed tannins. These condensed tannins, such as proanthocyanidins, possess remarkable antioxidant potential in vitro attributed to the higher hydroxyls groups in these molecules. The antioxidant potential of these molecules also depends upon the polymerization (number of oligomers present) as well as ROS against which its antioxidant potential is under consideration (Lotito et al. 2000). On the other side, glycosylation of flavonoids has been reported to lower the antioxidant potential of these molecules resulting in aglycons flavonoids, possessing higher antioxidant effect in vitro as compared to their glycan counterparts. The antioxidant potential of quercetin-3-O-rutinoside was recorded significantly lower than quercetin and rutin especially due to glycosylation at 3-OH site (Rice-Evans et al. 1996). Quercetin glycosylation at 3-OH site resulted in marked loss of its superoxide- and hypochlorite-quenching ability as well as its ferric-reducing antioxidant power (FRAP) (Sun et al. 2010a, b; Firuzi et al. 2004, 2005).

Bors et al. (1990) and Croft (2006) summarized the structural features determining radical scavenging potential of flavonoids as follows:

1. Presence of catechol (ortho-dihydroxy) structure in B-ring for efficient electron transfer.
2. Presence of 2–3 double bond with a 4-oxo function in C-ring for electron transfer from B-ring.
3. Presence of –OH group at 3 and 5 position, thus facilitating hydrogen bonding to oxo group.

16.4.2 Interaction with Antioxidant Enzymes

In addition to consuming free radicals, flavonoids interact with various antioxidant enzymes. Due to their redox properties, flavonoids increase activity of phase-II detoxifying enzymes such as NADPH-quinone oxidoreductase, UDP-glucuronosyl transferase and glutathione S-transferase. The induction of these phase II metabolic enzymes is considered major defence mechanisms against toxicants and oxidative stress. Synthesis of these phase-II metabolic enzymes is controlled by a gene sequence, Electrophile Responsive Element (EpRE), which is activated by redox properties of flavonoids (Zhu and Fahl 2001; Nerland 2007). In some situations, the intrinsic prooxidants potential of flavonoids were found beneficial in inducing expression of EpRE-mediated gene, thus exerting protective and health-promoting effect by activating detoxifying enzymes.

16.4.3 Chelation of Trace Metals

Flavonoids prevent formation of free radicals by chelating specific trace metals and prevent cells from oxidative injury. These metals are supposed to binds with catechol moiety in B-ring, 3-OH group and 4-oxo group in C-ring and 5-OH and 4-oxo groups in between C- and A-rings (Pietta 2000). The catechol moiety in B-ring is regarded as major chelation site as it binds with copper ions (Brown et al. 1998). Quercetin specifically binds with iron and known for iron-stabilizing properties. Quercetin and morin exhibited strong antioxidant potential in vitro by binding with cadmium (II), mercury (II) and lead (II) (Kopacz and Kuźniar 2003; Szeląg et al. 2003).

16.4.4 Protection of α-Tocopherol

Flavonoids protect against oxidative damage to low-density lipoprotein (LDL) and delay the lipid oxidation in cell membranes by inhibiting formation of α-tocopherol radicals through donating hydrogen (Hirano et al. 2001). This efficiency in the protection of α-tocopherol depends upon the redox potential of flavonoids such as myricetin and quercetin provide more effective protection to α-tocopherol than kaempferol and morin (Zhu et al. 2000).

16.4.5 Oxidase Inhibition

Flavonoids protect against oxidative damage by supressing oxidase enzymes which are responsible for production of superoxide radicals. Hanasaki et al. (1994) attributed the antioxidant potential of majority of flavonoids to mostly their ability to scavenge superoxide anion production through hypoxanthine–xanthine oxidase system by suppressing xanthine oxidase enzyme. Cos et al. (1998) investigated the structure–activity interrelationship and noted that –OH group at C-5 and C-7 as well as double bond between 2 and 3 carbon atoms was highly effective in suppressing xanthine oxidase activity. Flavonoids also had inhibitory effect on lipoxygenase, microsomal succinoxidase, cyclooxygenase and NADH oxidase enzymes (Korkina and Afanasev 1997) and thus exerted protective effect against oxidative damage.

16.4.6 Controlling Nitric Oxide Production

Nitric oxide in higher concentration causes oxidative stress due to formation of peroxynitrite upon reaction with superoxide anions. These are generated during oxidation of L-arginine in presence of nitric oxide synthetase (NOS) (Rubbo et al. 1994). Flavonoids regulate production of nitric oxide by indirect inhibitory effect on the expression of NOS rather than inhibiting the NOS activity as well as directly scavenging NO molecules (Olszanecki et al. 2002). Upon structure–activity studies, NOS inhibitory activity was attributed to presence of double bond at C-2 and C-3 with 4-oxo group, and –OH group at 3,5,4-trihydroxyl group; further an increased activity had been observed on methylation of this –OH group and reduced activity upon the presence of glycoside moiety and catechol or pyrogallol structure of B-ring (Matsuda et al. 2003; Kim et al. 1999).

Besides antioxidant activity, flavonoids also exhibit prooxidants activity depending upon its concentration. This is very vital for cell signalling and coordination of cell functions. This prooxidant effect helps in creating mild oxidative stress, thus leading to improved antioxidant defence and activating antioxidant enzymes that affect cell protection (Kessler et al. 2003; Halliwell 2008).

16.5 Plant Phenolics as Antimicrobial Agent

There is a growing interest in antimicrobial activity of plant-derived natural compounds rather than synthetic antimicrobials due to their associated adverse effect on consumers health. In addition to antioxidant potential, flavonoids also possess antimicrobial activity. Various ethno-medicines containing plant phenolics as active ingredients have been widely used for treatment of diseases and protection against infections in many cultures around the globe (Cushnie and Lamb 2005; Gutierrez et al. 2008). There are several published studies indicating one or more plant phenolic molecules showing antimicrobial effect by more than one mode of action/pathway. However, the sequential effects of one reaction in bacterial cells upon exposure to flavonoids have been misinterpreted with different modes of action (MAO), such as inhibition of bacterial DNA gyrase resulted in programmed cell death by preventing nucleic acid regeneration, which was previously misunderstood with the disruption of cell membranes (Gordon and Wareham 2010). Clove possesses strong antimicrobial activity among all spices due to the presence of phenolic compounds which alter the cell membrane permeability of bacteria resulting in their destruction (Bajpai et al. 2008).

The antibacterial activity of plant phenolics could be attributed to following three modes of action.

16.5.1 Damage to Cytoplasmic Membrane

Ikai et al (1993) observed strong antibacterial effect of epigallocatechin gallate (EGCs) due to binding with cytoplasmic membrane, leading to leakage of 5(6)-carboxyfluorescein from phosphatidylcholine liposomes. Catechins exhibited antibacterial activity by damaging cell membrane and this damage was observed comparatively lower in case of negatively charged lipopolysaccharides. Catechins are less effective against Gram-negative bacteria due to the presence of negatively charged lipopolysaccharides. Flavonoids decreases fluidity of inner and outer layers of cell membrane (Tsuchiya and Iinuma 2000). Flavanol, flavan-3-ol and flavolan have been reported to damage microbial cell membrane. Cushnie and Lamb (2005) noted antimicrobial activity of galangin against *Staphylococcus aureus* by disintegrating the integrity of its membrane leads to significant loss of potassium ions as well as weakening of cell wall, ultimately causing osmotic lysis (autolysis). Catechins (flavonoids of green tea and black tea) exhibited strong bactericidal activity by generating ROS and leads to production of hydrogen peroxides. The hydrogen peroxides produced were observed to exert bactericidal action, and this could be controlled by using catalase enzyme (Arakawa et al. 2004).

16.5.2 Suppression of Nucleic Acid Synthesis

Epigallocatechin exhibited antibacterial activity against *Proteus vulgaris* by inhibiting DNA synthesis and *Staphylococcus aureus* by inhibiting RNA synthesis (Mori et al. 1987). This antibacterial property of flavonoids is attributed to the free 3, 4, 5-trihydroxy B-ring and a free 3-OH group. Glycosylated flavones are known to exhibit antibacterial activity by selective inhibition of topoisomerase IV/DNA gyrase (Bernard et al. 1997). Flavonoids, especially quercetin, interact with bacterial DNA as well as with ATP-binding sites of gyrase and result in the inhibition of supercoiling of gyrase enzyme in bacteria and thus promoting DNA cleavage (Plaper et al. 2003). Various bioflavonoids also exert F1-ATP synthetase and ATP synthesis (Chinnam et al. 2010). Wang et al. (2010) observed topoisomerase I and topoisomerase II as binding sites for soybean isoflavone for inhibiting bacterial cell division. At higher concentration (12.8 mg/ml) these soybean isoflavones have been reported to cause denaturation of plasmid DNA.

16.5.3 Inhibition of Energy Metabolism

Haraguchi et al. (1998) observed the antimicrobial effects of licochalcone A and C against Gram-positive bacteria by decreasing oxygen consumption in these cells by inhibiting NADH-cytochrome C reductase enzyme. The site of action for this antibacterial inhibition by flavonoids was proposed in between CoQ and cytochrome C in electron transport chain of bacterial respiration mechanism.

Besides these aforementioned three pathways, recently another two possible pathways have been proposed for explaining antimicrobial effect of plant phenolics as inhibition of d-alanine–d-alanine ligase (Ddl), leading to the disruption of cell wall synthesis by quercetin and apigenin flavonoids, as quercetin possesses higher antibacterial activity attributed to the presence of two additional –OH groups on flavones structure (Wu et al. 2008). These flavonoids exert reversible inhibition and are competitive with substrate ATP of Ddl. Tea catechin, epigallocatechin gallate and other plant flavonoids exhibited antibacterial activity by the inhibition of FabG and FabI reductase by binding with cofactor in both enzymes (Zhang and Rock 2004) as well as inhibition of ketoacyl reductase (Li et al. 2006) of bacterial type II fatty acid synthase. Epigallocatechin gallate results in the aggregation of FabG enzyme obtained from *E. coli*, but there is a lack of studies establishing a correlation between similar findings with other enzymes or other flavonoids causing similar aggregation to this enzyme (Cushnie and Lamb 2011). Further flavonoids (flavanol, flavan-3-ol) have synergistic effect on whole microbial cell and decreased the viability of cells via potassium leakage, inhibition of nucleic acid synthesis and dihydrofolate reductase. This aggregation of bacterial cells leads to an overall decrease in surface area of bacterial cells, thus lowering oxygen consumption which was previously attributed to functioning of electron transport chain as well as lower intake of nutrients required for nucleic acid synthesis as uridine, thymidine, etc. (Ulanowska et al. 2006; Cushnie and Lamb 2011).

Rauha et al. (2000) studied the antimicrobial effect of 13 phenolic substances and 29 extracts obtained from Finnish plants against selected food pathogens and noted significant inhibition of microbial growth by flavones, quercetin and naringenin. Extracts obtained from purple loosestrife were found effective against *Candida albicans*, meadowsweet, willow herb, cloudberry and raspberry were found effective against bacteria and white birch, pine and potato were found effective against gram-positive *Staphylococcus aureus* (Fig 16.1).

16.6 Plant Phytoextracts as Source of Plant Phenolics

There is increasing interest of food industry and consumers to use natural antioxidants/preservatives in place of synthetic compounds, but this technology could only be sustainable and harnessed to its full potential by developing economical and suitable extraction technology. The traditional method of incorporating plant powders or raw form or gels in various food products could lead to alteration in sensory attributes such as colour, flavour, texture, binding and overall acceptability. Moreover, to get the desired preservative effect, these compounds have to be incorporated in food products in higher amount, leading to marked reduction of consumer acceptance and marketability of such food products due to poor sensory attributes. Also, the preparation of powder of various materials from plant sources could lead to loss of nutritive value and degradation of active compounds, but it facilitates easier transportation, storage and shelf life even at ambient temperature. In this context, extraction of various active ingredients (phenolics, essential oils) from

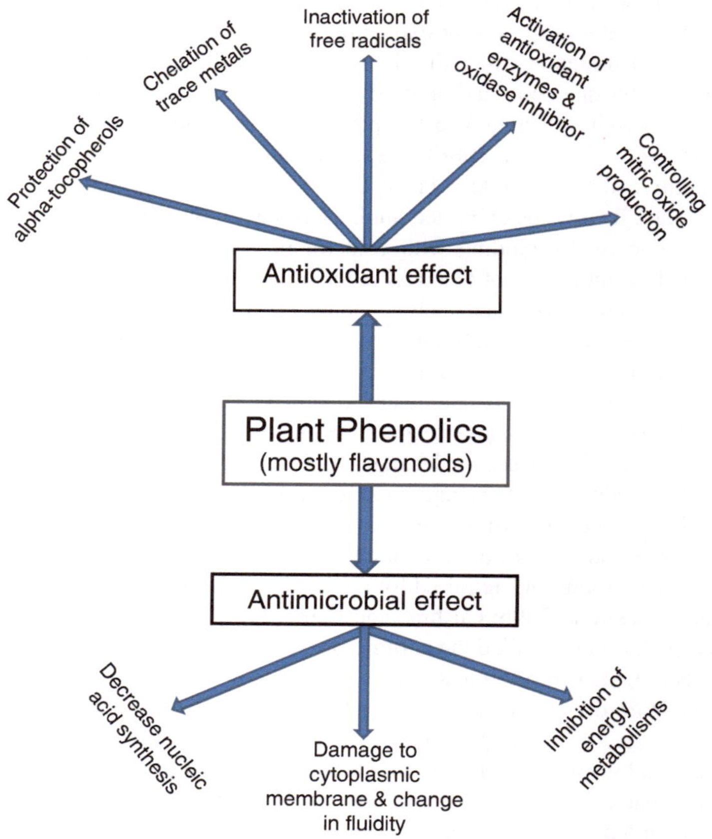

Fig. 16.1 Plant phenolics as natural preservatives in food system

plant sources such as leaves, fruits, fruit by-products, seeds, etc. by suitable extraction technology and their utilization during product preparation could be a promising approach.

At present, a lot of researches are underway to standardize the processing technologies for extraction of phytoextracts from various plant sources and characterization of these phytoextracts for their antioxidant, antimicrobial and organoleptic properties as well as optimization of their levels during development of various food products including meat and meat products. These extracts and plant powders have traditionally been used in culinary preparation as food adjunct (seasonings, flavour, colour and taste), ethno-medicines, preservatives and health supplements especially in India and China for thousands years without documenting any negative effect on consumers' health if taken within per prescribed limits. There is no documented proof of their harmful and toxicological effect on consumers' health and thus these can be utilized as natural preservatives in food products as alternatives to their synthetic counterparts. Researchers have identified various plant/vegetable parts as potential source of natural extracts exerting higher antioxidant and antimicrobial properties such as extracts from sea buckthorn, fenugreek seed, grape seed, green tea and acacia catechu (Kumar et al. 2015a, b), cinnamon bark and aloe vera gel (Rathour et al. 2017a, b), arjuna tree bark, veal meat loaves (Bishnoi et al. 2017; Birla et al. 2019), *Carica papaya* leaves (Jagtap et al. 2019), watermelon rind (Kumar et al. 2018a, b), etc.

16.6.1 Extraction Protocols

There are several phenolic compounds present in plant materials, and each has different affinity towards solvents or mixture of solvents and varied time–temperature combinations for their efficient/optimum extraction. Thus, there is need to develop suitable and rapid extraction methodology that will be suitable for the extraction of particular compounds from a wide variety of raw materials (Kumar et al. 2015a, b). As extraction protocols significantly affect the physicochemical quality and antioxidant efficacy of the extracts, it is advisable to get the maximum concentration of active ingredients in extracts with minimum denaturation or loss of activity. This facilitates the utilization of plant extracts on higher scale in food industry and improves the economics of these products (Vuong et al. 2011).

The quality of extracts depends upon several factors such as extraction protocols (nature of solvent, time–temperature combination, pressure, solvent concentration, solvent to solute–plant powder ratio), sample preparation procedure and type of samples affect the amount/concentration of phenolics and thus overall antioxidant/preservative activity. Besides this, other factors such as plant species, part of plant (stem, roots, bark, flower, fruit, leaves, etc.), developmental stage, harvesting time and stage, storage conditions, drying methods, geographical conditions, etc also play important roles in determining quality of extracts (Dumbrava et al. 2012; Negi 2012; Riahi et al. 2013; Casazza et al. 2011; Yeh et al. 2014).

16.6.2 Extraction Medium

In a single plant, there is presence of wide variety of phenolic compounds, each having different physicochemical properties, and thus it practically becomes very difficult to suggest one solvent/method which is suitable for all plant material. At present, water, ethanol, chloroform, heptane, methanol, acetone, ethyl acetate and dimethyl sulfoxide (DMSO) are widely used solvents for extracting active compounds from plant biomass. For extraction of non-polar compounds, water (aqueous) is preferred, whereas for organic/polar compounds organic solvent is preferred as solvent medium. As compared to organic solvent, water proves to be safe, economical and edible as devoid of any traces of potentially harmful organic solvents. Water can be used for extraction of wide range of plant biomass due partially hydrophilic nature of various active ingredients in plant. Khokhar and

Magnusdottir (2002) noted efficient extraction of tea catechins in aqueous medium as compared to 70% ethanol and 80% methanol. The combined use of water and organic solvent is the best approach for getting extract with higher yield and antioxidant activity. Do et al. (2013) observed highest antioxidant efficiency of extract obtained from *Limnophilia aromatica* herbs in 100% ethanol and highest yield in 50% aqueous acetone extraction medium and summarized that a combination of aqueous and organic solvent mixture facilitated the most optimal extraction of chemicals.

Organic solvents are very effective in extracting phenolic compounds but the residual effect of leftover organic solvent in extracts creates problem in utilization of these extracts in food due to harmful effect on consumers' health. These extracts are properly processed through evaporative methods to remove residue of organic solvent before incorporation in food products. Shah et al. (2014) noted organic solvent as more suitable in extraction of active compounds from plant biomass containing antioxidant properties. Turkmen et al. (2006) also reported that antioxidant activity of green and mate tea extracts obtained by extracting in different solvents were found different in various combinations and concentrations as water, acetone, N,N dimethylformamide, ethanol/methanol and highest antioxidant activity were recorded in mate tea and black tea extracts, prepared by using 50% aqueous ethanol and 50% aqueous acetone.

DMSO and methanol are very good organic solvents and widely preferred for extraction of plant biomass in various chemical reactions, but their toxic nature remains major concern for food industry. DMSO can dissolve both polar and non-polar compounds and properly mix in both organic and aqueous solutions. Methanol is very efficient in removing low molecular weight phenolic compounds from plant matrix whereas water is suitable for removal of large molecular weightage polyphenols such as flavanols (Dai and Mumper 2010). Methanol is used for extraction of polyphenols from fresh plant tissue due to inhibition of polyphenol oxidase (PPO) (Yao et al. 2004). PPO enzyme catalyse oxidation of phenolic compounds as mono-, di- and polyhydric phenols resulting in production of quinines. These quinines binds with –SH, $-NH_2$ and NH group of amino acids of proteins resulting in loss of histidine, lysine, cysteine, methionine amino acids and decreased nutritive value of proteins (Yamane et al. 2010).

16.6.3 Extraction Methods

At higher temperature for longer duration, there are chances of decreasing the activity of polyphenolic compounds which result in lower antioxidant and antimicrobial activity. Thus it is advisable to apply suitable extraction methodology at lower temperature for short duration. The bioactive compounds from plant biomass are extracted by dissolving powder or liquid form of plant material into suitable solvent generally at higher temperature for longer duration (solvent extraction/ portioning), breaking of plant matter into pieces in suitable liquid especially for retaining peculiar essence of high value herbs/spices (maceration), by using by using supercritical fluids as extracting medium at suitable temperature and pressure combinations (mostly liquefied carbon dioxide at 31°C at 74 bar pressure) (supercritical extraction), by using pressurized water at high temperature in subcritical phase (subcritical water extraction), by modifying the dielectric constant and polarity of solvent and exposing the plant material to high intensity ultrasonic waves creating tiny cavitation/bubbles around the cells followed by sudden collapse of these bubbles resulting in disruption of cell walls and releasing intracellular material (ultrasound-assisted extraction).

Under the traditional solvent extraction method, the extraction is carried out at higher temperature and involves higher solvent consumption, increased cost and degradation of activity of heat labile phenolic compounds. Subcritical water extraction method is very rapid, economical and environment-friendly. In this method, extract retains maximum activity and there is an enhanced mass transfer rate from plant biomass (Hawthorne et al. 2000; Mazaheri et al. 2010; Awaluddin et al. 2016; Zakaria and Kamal 2016). Ultrasound-assisted extraction method is very simple and relatively economical method that can be applied for a wide variety of plant materials on industrial scale (Khoddami et al. 2013).

Besides these above-mentioned extraction methods, some researchers have explored the other newer extraction methods such as microwave-assisted extraction (Tatke and Jaiswal 2011), pressurized microwave-assisted extraction (PMAE), solvent-free microwave-assisted extraction (SFMAE), squeezing and evaporation (Badr and Mahmoud 2011), extraction of polyphenolics from borage leaves by sonication (Ciriano et al. 2009), enzymatic hydrolysis (Sun et al. 2010a, b), etc. The additional microwave treatment in traditional solvent extraction resulted in significantly improving procedural and technological benefits such as reducing time of extraction, increasing yield, saving of fuel and solvent as well as lesser degradation of active compounds (Tatke and Jaiswal 2011).

16.7 Preservative Effect of Some Common Phytoextracts

16.7.1 Green Tea

Green tea extracts contain catechins showing very high antioxidant potential, and these compounds prove more effective in controlling peroxide levels in lard (rendered fat of pig) and chicken fat as compared to vitamin E and synthetic antioxidant butylated hydroxyanisole (BHA) (Chen et al. 1998). Tang et al. (2011) observed significant reduction (up to two to four times as compared to vitamin E) in lipid oxidation in terms of thiobarbituric acid reactive substances (TBARS) value in refrigerated beef, chicken, pork, duck and ostrich meat upon incorporation of tea catechins at 300 ppm levels in these meat. Tea catechins inhibited formation of putrescine and tyramine in dry fermented pork sausage without compromising pH, colour and organoleptic attributes of sausages (Bozkurt 2006). Banon et al. (2007) studied the preservative effect of green tea extract (GTE) and grape seed extract (GSE) at 300 mg/kg meat and ascorbate in low sulphite (100 mg sulphur dioxide) raw beef patties, and observed 3-day extension of shelf life of the meat patties with improved colour and oxidative stability in raw products and delay in perception of rancid flavour in cooked products. A combination of green tea catechins (GTC) along with green coffee antioxidants (GCA) in linseed oil and fish oil incorporated in fresh pork sausage at 200mg/kg were reported to prevent oxidation of lipids in both raw and cooked sausage kept under aerobic and modified atmosphere packaging at refrigeration for 7 days (Valencia et al. 2008).

16.7.2 Grape Seed

Grape seed extracts (GSE) contain higher amount of proanthocyanidins, oligomers of flavan-3-ol units (catechins and epicatechins), in addition to vitamin E, flavonoids and linoleic acid, catechins and proanthocyanidins constituting about 77.6% of total polyphenols in GSE (Silvan et al. 2013). The antioxidant potential of GSE was observed 20–50-folds more than antioxidant potential of vitamin E and vitamin C (Shi et al. 2003) and higher than gallic acid. The incorporation of GSE in raw as well as cooked products leads to inhibition of lipid oxidation and thus generation of primary (lipid hydroperoxides and hexanal) and secondary (thiobarbituric acid reactive substances TBARS) lipid oxidation products. GSE was found to improve colour, stability and freshness of beef sausage incorporated at 300 ppm GSE level stored under frozen conditions for 4 months. The treated product had significantly lower TBRAS value and better freshness as compared to control product prepared without GSE addition (Kulkarni et al. 2011). Brannan and Mah (2007) observed significant inhibition of primary and secondary oxidation products in various meat systems by incorporation of GSE at 0.1% concentration. Ozvural and Vural (2012) developed pork frankfurter with GSE addition at various concentrations and observed improvement in overall acceptability with up to 0.1% GSE incorporation, but TBARS value decreased with increased concentration of extracts. Similarly, GSE incorporation in restructured mutton slices also noted inhibited lipid oxidations (as indicated by significantly lower TBRAs and free fatty acid [FFA] values) and better retention of sensory attributes, as compared to mutton slices without extract incorporation (control) as well as mutton slices with BHA (positive control) (Reddy et al. 2013) stored under aerobic and vacuum-packaging conditions.

An addition of commercial preparation of GSE (ActiVin™) and pine bark extract (Pycnogenol®) in cooked beef was reported to inhibit lipid oxidation for 3 days under refrigeration with comparatively significantly lower values of TBARS, hexanal content and warmed-over-flavour (WOF) (Ahn et al. 2002). Similarly ActiVin™ Pycnogenol®, oleoresin rosemary (Herbalox) and synthetic antioxidant (BHA/BHT) incorporation significantly decreased TBARS value (viz. 75%, 92%, 94% and 92%, respectively) as compared to control throughout the entire storage period of 9 days (Ahn et al. 2007). Sasse et al. (2009) evaluated the comparative efficacy of various extracts (such as grape seed extract, oleoresin rosemary, oregano extract) and synthetic antioxidants (such as propyl gallate, butylated hydroxyanisole and butylated hydroxytoluene) 0.02% level of incorporation in frozen pork patties stored for 180 days and observed that patties with propyl gallate and grape seed extract better maintain colour and lower TBARS value.

16.7.3 Rosemary

Traditionally, rosemary is used as aromatic herb, food additive and culinary medicinal preparation in treating various conditions ranging from antibacterial, antidiabetic, anticancer, anti-inflammatory, maintaining cardiovascular health, reducing total salt intake, etc. Rosemary leaves and flower extract in various solvents (ethyl acetate, methanol, chloroform and ethanol) have been observed to exert functional properties such as antioxidant, antibacterial, anti-inflammatory, antinociceptive, antirheumatic, antineuralgic, antidepressant, vasodilator, etc. (Erdogrul 2002; Peng et al. 2007; Amaral et al. 2013; Ribeiro-Santos et al. 2015). This beneficial effect of rosemary extract and essential oil is contributed due to the presence of various phenolic compounds in it, such as rosmarinic acid, borneol, bornyl acetate, verbenone, caryophyllene, carnosic acid and carnosol in extract and camphor, eucalyptol and α-pinene-bornyl acetate (Arranz et al. 2015; Teixeira et al. 2013). Teixeira et al. (2013) reported monoterpenes hydrocarbons, oxygenated monoterpenes, oxygenated sesquiterpenes, sesquiterpene hydrocarbons, ketones, esters, phenol and alcohol as major compounds attributing to rosemary. Rosemary leaves contains very high amount of vitamin C as in raw material (18.51%): 0.26 mg/100 mL in aqueous extract, 0.34 mg/100 mL in alcohol extract and 0.36 mg/100 mL in acetone extract (Dumbrava et al. 2012). Peng et al. (2007) described supercritical fluids extraction as suitable technology for extraction of bioactive compounds from rosemary due to inherent benefits of rapid and fast extraction, higher yield as well as preventing the conversion of carnosic acid into carnosol in presence of oxygen. For extraction of rosemary leaves, flowers and stems, care should be taken to prevent enzymatic oxidation of phenols by phenoloxidase in the presence of water in fresh samples as it resulted in the formation of carnosol and rosamanol on enzymatic oxidation of phenol by phenoloxidase (Bellumori et al. 2015; Mulinacci et al. 2011).

16.7.4 Cinnamon Bark

Cinnamon bark has been an integral part of spices added in food preparations for taste and flavour since ancient times, and is widely used as a preservative and for medicinal purposes. It is rich source of natural phenolic compounds, and its antioxidant capacity is equivalent to synthetic antioxidants BHA/BHT/t-BHQ (Mathew and Abraham 2006; Chen et al. 2012). Cinnamon bark is dried bark of shoots of *Cinnamomum zeylanicum* F. *Lauraceae*, obtained from most of its cork and cortex, and has less than 1% of volatile oil. Cinnamon oil exerts very good antimicrobial activity against common foodborne pathogens. Cinnamon possesses methyl hydroxychalcone polymer (MHCP), which has been shown to aid in maintaining healthy blood sugar levels. Chan et al. (2007) observed the antioxidant effect of aqueous extract of cinnamon bark incorporated in chicken meat balls and noted significantly improved in redness and oxidative stability without compromising organoleptic attributes during refrigerated storage.

16.7.5 Oregano

In herbs of *Lamiaceae* family as basil (*Ocimum* spp.), rosemary (*Rosmarinus* spp.), thyme (*Thymus* spp.), mint (*Mentha* spp.) and oregano (*Origanum* spp.), rosmarinic acid, a hydroxycinnamic acid derivate, ester of caffeic acid and 3,4-dihydroxy-phenyllactic acid, is typically present that exhibits strong antioxidant properties. A study on chemical composition of 26 herbs/spice extracts and their phenolics reported high amount of rosmarinic acid (1086 to 2563 mg/100 g of dry weight) as major phenol in the *Lamiaceae* plants (Shan et al. 2005). The occurrence of two ortho-dihydroxy groups (catechol structures) in rosmarinic acid is the main structure responsible for strong antioxidant activity such as scavenging superoxide, hydroxyl radicals, and inhibiting oxidation of LDLs. Rosmarinic acid demonstrated anti-inflammatory activity through lipopolysaccharide-stimulated macrophage model, by reducing the inflammatory response with increase in secretion.

In rosemary, along with rosmarinic acid, carnosic acid and carnosol are the main antioxidant compounds present in rosemary. Carnosic acid and carnosol, together with other isoprenoids such as sterols, tocopherols or carotenoids, play a photoprotective role and are considered as bioactive constituents present in herbs. Carnosic acid originated from IPP via methylerythritol phosphate. It is located in chloroplasts and intracellular membranes, as is carnosol, formed from the oxidative degradation of carnosic acid.

16.8 Plant Phenolics in Breads and Biscuits/Cookies

Bakery products have an important place in human food basket for supply of nutrients and as an energy source. During preparation of white breads, several processing changes occur in dough, ranging from heat treatment to fermentation, overall affecting the nutrient quality of breads. Wheat bread is low in antioxidants and thus enriching bread with plant phenolics through herbs, spices, pseudocereals, fruits, vegetables, seeds, cereals and food industry by-products (such as semolina, rice bran, oat bran, barley hull extract, oat flour, ginger powder, coriander leaves, oregano leaves, green coffee, green tea, tea leaves, tomato paste, chestnut flour, yam, plum, carrot, cabbage, oregano leaves, grape pomace, etc.) could be very promising in enhancing its shelf life at ambient temperature and increasing its nutrient and sensory quality (Dziki et al. 2014).

Gawlik-Dziki et al. (2009) incorporated buckwheat plant extract containing tartary buckwheat flavonoids (TBF) during preparation of wheat bread and observed increase in phenolic contents upon extract incorporation, but at the same time a gradual concentration-dependent decrease (more in breads with 5.0% TBF than breads with 2.5% TBF incorporation) in sensory attributes of breads was recorded. The higher phenolics in enriched breads caused improved radical scavenging ability, reducing power and preventing lipid oxidation. Raba et al. (2007) also noted increased lipid stability in breads upon addition of garlic and sweet basil powder and observed increased polyphenol levels up to 0.28 mM gallic acid/100 g. Turmeric powder had also been used to increase antioxidants in bread, and addition of 4% turmeric powder did not compromise sensory attributes (Lim et al. 2011). Sikkhamondhol et al. (2009) developed functional breads by incorporation 0.10% turmeric powder, 0.10% turmeric residue and 0.10% turmeric essential oil. The developed breads with turmeric powder had highest sensory attributes in addition to higher phenolics and antioxidant potential.

Gawlik-Dziki et al. (2013) incorporated 2–3% onion skin as source of quercetins in breads with good improved bioaccessibility and bioavailability and noted decreased sensory attributes upon increasing levels of ground onion skin in bread. El-Megeid et al. (2009) studied enrichment of bread with various levels of green tea leaves and observed good acceptability of breads up to 2% level of addition as compared to breads with 4% and 6% green tea leaves. These breads were found beneficial in preventing oxidative stress and managing renal failures in rats. Peng et al. (2010) observed significant increase in antioxidant capacity of breads incorporated with grape seed extract; however this also caused marginal decline in sensory attributes, especially colour score. The addition of the powder of cherry fruit in pan breads at 3% level significantly improved DPPH radical scavenging potential of bread in a concentration-dependent manner along with acceptable organoleptic attributes (Yoon et al. 2010). Altunkaya et al. (2013) explored use of pomegranate peel powder in preparation of breads and noted a 2.5% addition of this powder in wheat bread did not affect acceptability. Pomegranate peel powder is a rich source of polyphenols. Sivam et al (2011) observed higher antioxidant potential and phenolic compounds in breads upon addition of apple pectin and extracts from various fruits.

Zilic et al. (2016) prepared cookies from anthocyanin-rich cornflour (dark red, blue and blue standard) and observed that that addition of citric acid (0.5%) in dough resulted in up to 70% higher flavonoids and anthocyanins in these biscuits. Caleja et al. (2017) studied the replacement of synthetic antioxidants (BHA) with fennel and chamomile extract in development of antioxidant-rich biscuits and observed comparable sensory and visual characteristics of both types of biscuits. Authors advocated use of natural antioxidants over synthetic antioxidants due to natural/clean image of biscuits along with environment-friendly extraction methods of obtaining plant extracts (aqueous). Mildner-Szkudlarz et al. (2009) evaluated the antioxidant effect of green tea extract upon incorporation in biscuits and noted that 1% incorporation of green tea extract in biscuits resulted significantly improving antioxidant capacity and lipid stability as well as up to 73% reduction in hydroperoxides formation. During storage, biscuits with natural antioxidants exhibited higher sensory scores as compared to biscuits with synthetic antioxidants.

16.9 Plant Phenolics in Meat and Meat Products

Meat is inherently low in antioxidants. It is a rich source of high-quality proteins, minerals, lipids and vitamins. This availability of high-quality nutrients along with water makes it highly prone for deterioration by various oxidative and microbial changes. During the processing of meat as well as development of various processed meat products, it has to undergo various processes leading to its exposure to light, metal surfaces, etc., and these condition further aggravate the risk of spoilage. Incorporation of natural additives/preservatives in place of synthetic chemicals is gaining popularity and the use of various plant phenolics has been explored for possible replacement of synthetic additives with natural compounds (Table 16.1).

16.10 Essential Oils as Source of Phenolics

Essential oils (Eos; ethereal oils/volatile oils) contain bioactive phenolics and other compounds produced as by-products of plant metabolism. EOs are obtained from vegetable raw materials either by distillation with water or steam or from the epicarp of citrus fruits by a mechanical process or by dry distillation (ISO/DIS 9235.2, 1997). In most of herbs, essential oils are the main active principal such as thymol in thyme, eugenol in clove, carvacrol in oregano and rosemary, etc. (Kumar et al. 2015a, b). Like phytoextracts, incorporation of these EOs result in better stability, easier transportation and storage, desired preservative effect in foods at very low level of incorporation, better maintenance of organoleptic attributes, easier application and standardization, better flavour and aroma, etc. (Tipsrisukond et al. 1998). These essential oils are widely used in food industry as preservatives, improving flavour and aroma, and in perfumeries and medical preparations. Essential oils retard lipid peroxidation (Mechergui et al. 2010; Viuda-Martos et al. 2010) and antimicrobial effects in meat and meat products (Ruiz-Navajas et al. 2012) due to the presence of various bioactive compounds (Table 16.2).

16.10.1 Antioxidant Effect of Essential Oils (EOs)

Essential oils are a rich source of polyphenolic compounds, and the antioxidant activity of these oils depends upon the concentration of these compounds. About more than 60 polyphenolic compounds present in various essential oils exhibit antioxidants potential but citronellal, carvacrol, isomenthone, citronellol and geraniol are the principal polyphenolic compounds exhibiting antioxidant properties. Carvacrol and thymol are active ingredients in oregano essential oil (OEO), and make up to 78–85% of OEO (Govaris et al. 2010). They do not have any toxic or ill effect on consumers' health in the present prescribed levels of incorporation in food (thymol 0.5 mg/kg and carvacrol 1 mg/kg) which is lower than their median lethal dose, viz. 565.7 mg/kg thymol and 471.2 mg/carvacrol kg. Fasseas et al. (2008) observed improved oxidative stability of raw and cooked beef and pork upon

Table 16.1 Sources of various phenolic compounds and their use in preservation of meat products

Source of Phenolic compounds	Salient findings	References
Fenugreek seeds and ginger rhizome compared to potato peel	Ginger rhizome and fenugreek seed extracts has higher antioxidant potential than potato peel extract in controlling lipid oxidation in ground beef patties	Mansour and Khalil (2000)
Aloe vera, fenugreek, ginseng, mustard, rosemary, sage and tea catechins	The optimum levels of tea catechins, rosemary and sage with 0.25%, 0.10% and 0.05% optimum levels, respectively, in pork patties product enhanced oxidative stability	McCarthy et al. (2001a)
	As compared to raw pork patties, cooked products had fourfold increase in TBARS values, and tea catechins were found to be most effective antioxidant to prevent lipid oxidation	Mccarthy et al. (2001b)
Rosemary and hyssop leaves	These extracts prevented lipid oxidation, degradation of heme pigments and helped in maintaining red colour of the cooked meat	Fernández-López et al. (2003)
Rosemary extract	Rosemary extract at 1500–2000 ppm incorporation in pork sausage was recorded more effective than synthetic antioxidants in controlling TBARS values and maintained red colour in raw frozen product	Sebranek et al. (2005)
Rosemary, green tea, coffee and grape skin extract	Rosemary extract (200 ppm) showed highest antioxidant effect followed by grape skin (200 ppm), green tea (200 ppm) and coffee extract (50 ppm) during storage of vacuum-packaged precooked pork patties.	Nissen et al. (2004)
Myrtle, rosemary nettle and lemon balm leaf	Myrtle and rosemary extracts (10%) had highest antioxidant effects than nettle and lemon balm extracts in frozen beef patties and they also better maintained the colour	Akarpat et al. (2008)
Nettle and *Hibiscus sabdariffa L.* (Roselle) flowers	Incorporation of extracts in Turkish dry fermented sausage resulted in decrease in TBARS values in treated product as compared to control during the first 4 days of storage	Karabacak and Bozkurt (2008)

(continued)

Table 16.1 (continued)

Source of Phenolic compounds	Salient findings	References
Cinnamon stick cortex, oregano leaf, clove bud, pomegranate peel and grape	Extracts in pork meat increased the oxidative stability. Clove was found very effective in retarding lipid oxidation and offered the highest antioxidant activity in raw pork	Shan et al. (2009)
Lyophilized nettle leaf extract	Ground beef treated with nettle leaves extract (500 ppm) stored under modified atmosphere packaging conditions was observed with lowest TBRAS values	Alp and Aksu (2010)
Kinnow peel, pomegranate peel and seed	Goat meat patties incorporated with the extracts exhibited higher lipid stability	Devatkal et al. (2010)
Peanut skin	Addition of extract (200 ppm total phenolics) in raw ground beef prior to cooking markedly reduced the formation of peroxides and reduced TBARS value in cooked ground beef during the refrigerated storage	Yu et al. (2010)
Red grape pomace	Addition of red grape pomace in pork nuggets improved lipid oxidative stability	Garrido et al. (2011)
Black seed, cinnamon bark, garlic aerial parts, lemon grass leaves, liquorice root and pomegranate peel	Addition of these extracts in ground beef resulted in significantly improving lipid stability	Tayel and El-Tras (2012)
Lyophilized aqueous extract of summer savoury leaf	Lipid oxidation decreased in ground beef in concentration-dependent manner	Aksu and Ozer (2013)
Date pit	Highest antioxidant was shown by the date pit extract (water–methanol–acetone–formic acid) in ground beef	Amany et al. (2012)
Broccoli powder	Broccoli powder addition at 2% level in goat meat nuggets had similar effect to BHT (100ppm).	Banerjee et al. (2012)
	Broccoli powder caused significantly reduction in TBARS value in ground beef and patties	Kim et al. (2013)
Curry and mint leaf	Ethanolic extract of curry leaf and the water extract of mint leaf had higher DPPH and ABTS activity and decreased lipid peroxidation in ground pork meat	Biswas et al. (2004)
Moringa oleifera leaf extract	Increase in TBARS value in treated samples was very low and remained lowest (0.53 mg malonaldehyde per kg sample) up to 15 days in goat meat patties, and the antioxidant	Das et al. (2012)

(continued)

Table 16.1 (continued)

Source of Phenolic compounds	Salient findings	References
	activity of extract was noted comparable to BHT at 100 mg/100 g meat	
	DPPH radical scavenging activity of both extract and BHT was recorded in raw and cooked pork patties	Muthukumar et al. (2012)
Pomegranate peel 1%	TBARS values were significantly reduced in ground goat meat and nuggets and vacuum packaging had a synergistic antioxidant effect	Devatkal et al. (2012)
Kinnow rind powder, pomegranate rind powder and pomegranate seed powder	Pomegranate rind powder extract had higher antioxidant potential in goat meat patties than kinnow rind powder and pomegranate seed powder	Devatkal et al. (2010)
Ginger, onion and garlic	Significantly reduced lipid oxidation and maintain better colour in treated stewed pork	Cao et al. (2013)
Carnosic acid	Carnosic acid (22.5 ppm and 130 ppm) extracted from dried rosemary leaves in buffalo patties resulted in significantly lower TBARS by 39–47% at lower concentration (22.5 ppm) and by 86–96% at higher concentration (130 ppm)	Naveena et al. (2013)
Cinnamon bark and aloe vera powder extracts	Incorporation of ethanolic extract of cinnamon bark (0.25%) and aloe vera powder (0.40%) in chevon rolls increased shelf life by reducing lipid oxidation in both aerobic and modified atmosphere packaging conditions during refrigerated storage	Rathour et al. (2017a, b)
Arjuna tree bark and aloe vera	Enhanced antioxidant and antimicrobial effects with better organoleptically attributes of buffalo male calf meat rolls were noted in 21 days of refrigeration storage upon incorporation of arjuna tree bark at 2% and aloe vera at 4% level	Bishnoi et al. (2017)
Leaf of butterbur, chamnamul, pak choi, Chinese chives/leek, crown daisy, fatsia, pumpkin, sesame, stonecrop *Acanthopanax* and soybean	Incorporation of extracts at 0.1% and 0.5% (w/w) and BHT in ground beef and patties resulted in concentration-dependent decreases in TBARS values and better colour stability for 12 days at 4°C	Kim et al. (2013)

(continued)

Table 16.1 (continued)

Source of Phenolic compounds	Salient findings	References
Butterbur and broccoli extracts	TBARS values of stored ground beef and patties were recorded significantly lower than the non-treated control along with better colour stability upon incorporation of butterbur (0.10% w/w) and broccoli extract (0.50% w/w)	Kim et al. (2013)
Grape seed extract	Grape seed extract (0.01% and 0.02%) had best antioxidant activity based on TBARS values and off-odours related with development of rancidity, wet cardboard off-odour in beef patties and grassy off-odour in beef and pork patties as compared to the addition of oleoresin rosemary (0.02%) and water-soluble oregano extract (0.02%)	Rojas and Brewer (2007)
	Grape seed extract provided little protection against oxidation in raw beef and pork patties	Rojas and Brewer (2008)
Red grape pomace extracts	GPI exhibited maximum colour stability, lipid oxidation inhibition and the best overall acceptability after 6 days of storage of pork burgers at 0.06 g/100 g extract incorporation	Garrido et al. (2011)
Clove essential oil and grape seed extract	Raw buffalo patties containing 0.1% clove essential oil showed TBA values 27.5–39% lower than the TBA values of samples containing 0.1% and 0.2% GSE	Tajik et al. (2014)
Green tea, rosemary and red pepper	Incorporation of extracts effectively reduced the lipid oxidation in cooked pork. Pepper extract was effective in maintaining colour and freshness in sample during chilling storage	Wojciak et al. (2011)
Lutein, sesamol, ellagic acid and olive leaf extract	Inhibition of lipid oxidation in raw and cooked patties was noted in the order: sesamol (250, 500 μg/g muscle) = ellagic acid (300, 600 μg/g muscle) > olive leaf extract (100, 200 μg/g muscle) > lutein (100, 200 μg/g muscle)	Hayes et al. (2010)
	Incorporation of sesamol, ellagic acid and olive leaf extract decreased TBARS in raw beef patties	

(continued)

Table 16.1 (continued)

Source of Phenolic compounds	Salient findings	References
Liposterine® or Exxenterol® from carob fruit with α-tocopherol	Lower TBARS value was observed in cooked pork meat systems containing Liposterine®, Exxenterol® and α-tocopherol during chilling	Bastida et al. (2009)
Radix puerariae (RP)	1% addition of RP extract in pork sausage resulted in decreased TBARS values	Jung et al. (2012)
Rosemary (260 mg/kg)	Rosemary showed comparable lipid stability and improved organoleptic attributes to SO2 (450 mg/kg) in boerewors (fresh sausage of South Africa)	Mathenjwa et al. (2012)
Mango peel	Total phenolic content was recorded significantly higher in chicken nuggets upon addition of 0.50% mango peel extract	Kadakadiyaret al. (2017)
Pediocin (10^3 AU/g) and extract of *Murraya koenigii* berries	A marked reduction in the *L. innocua* and TBARS values was observed in raw goat meat emulsion during refrigerated storage without any marked change in microstructure and colour upon addition of extract *Murraya koenigii* berries (41 ml containing 50 mg total phenolics)	Kumar et al. (2017)
Murraya koenigii Spreng berries extract	The extract (200 mg/ml) proved a good source of antioxidant (total phenolics – 9.5 ± 0.03 mg TAE/gdw and total flavonoid contents – 11.9 ± 0.66 mg CE/gdw) in chicken meat homogenates	Kumar et al. (2012)
Black currant (*Ribes nigrum L.*) extract	Extract incorporation at 5, 10 or 20 g/kg in raw pork patties kept for display significantly decreased the TBARS value and carbonyls formation and reduced the sulfhydryl loss in a dose-dependent manner	Jia et al. (2012)
Aqueous extract of lotus rhizomes knot and leaf	Lotus rhizome knot extract (3% w/w) was more effective against lipid oxidation in porcine and bovine ground meat	Huang et al. (2011)
Pomegranate peel extract	In 1.0% extract-incorporated beef ball, peroxide, malondialdehyde and carbonyl formation, loss of total protein solubility and	Turgut et al. (2017)

(continued)

Table 16.1 (continued)

Source of Phenolic compounds	Salient findings	References
	sulfhydryl groups were significantly lower even after 6 months of frozen storage	
Kitaibelia vitifolia extract	Ethanolic extract at 12.5 g/kg of meat dough resulted in moderate inhibition of *Escherichia coli* and strong antioxidant activity and in fermented dry sausage	Kurcubic et al. (2014)
Aqueous extract of litchi (*Litchi chinensis Sonn.*) seed at 50 mg/kg	Addition of aqueous extract of litchi seeds at 50 mg/kg in ground chicken meat significantly decreased cooking losses, improved water-holding capacity and pH during refrigerated storage	Yogesh et al. (2014))
Apple and olive extracts	3% olive and apple extract, onion powder and clove bud oil reduced *E. coli* O157:H7 to below detection and significantly reduced formation of heterocyclic amines, potential carcinogenic compounds in grilled ground beef patties	Rounds et al. (2013)
Citrus extract	The antioxidant active packaging with citrus extract caused significant reduction in lipid oxidation in cooked turkey meat	Contini et al. (2014)
Spray of aqueous extracts of borage and green tea	Spraying of 10% aqueous extracts of borage and 0.50% green tea extract extended shelf life of lamb leg from 8 to 11 days displayed under retail condition	Belles et al. (2017)
Heartwood of *Caesalpinia sappan*	Significant reduction in lipid oxidation, improvement in antimicrobial activity and decrease in volatile basic nitrogen formation in emulsion-type pork sausages incorporated with 0.20% *C. sappan* aqueous extract was noticed	Jin et al. (2015)
Green tea and rosemary extract	Addition of green tea (500 ppm total phenolics) and rosemary extracts (400 ppm total phenolics) in bologna-type sausage provided protection against formation of TBARS and protein carbonyls	Jongberg et al. (2013)
Green tea extract	The incorporation of 100 ppm green tea extract resulted in decreased protein oxidation and cross-linking in between myosin heavy chain without jeopardizing the textural stability	Jongberg et al. (2015)

(continued)

Table 16.1 (continued)

Source of Phenolic compounds	Salient findings	References
Propolis	Polylactic acid film containing propolis ethanolic extract, cellulose nanoparticle and *Ziziphora clinopodioides* essential oil enhanced shelf life of minced meat for 11 days	Shavisi et al. (2017)
Rosemary extract	Inhibition of lipid oxidation, whereas it had no effect on colour stability in reduced nitrite (80 mg/kg) liver pates	Doolaege et al. (2012)
Acerola fruit extract	Addition of 0.15% w/w extract resulted in 3-day extended shelf life and improved colour and lipid stability in salted beef patties	Realini et al. (2015)
Mustard ethanolic extract	Incorporation of 0.10% ethanolic extract of mustard resulted in cooked ground pork resulted in significantly lower TBARS value, peroxide values and hexanal contents	Lee et al. (2011)
Pomegranate seed extract	Total heterocyclic aromatic amine formation in beef and chicken meatball was decreased by 39% and 46% in charcoal barbecue and deep-fat frying upon addition of 0.5% (w/w) pomegranate seed extract	Keskekoglu and Uren (2014)
Grape seed and bearberry	Addition of extracts improved lipid stability in raw pork patties on days 9 and 12 of storage upon incorporation of grape seed (100–1000 μg/g muscle) and bearberry (10–1000 μg/g muscle)	Carpenter et al. (2007)
Rooibos extract	Inhibition of lipid oxidation in game meat droëwors after 14 days' storage upon incorporation of 1.0% rooibos extract	Jones et al. (2015)
Ginger extract and papain	Ginger extract (7%) caused extensive fragmentation of myofibrils whereas papain extract (0.01%) caused noticeable destructive effect on connective tissue as observed in camel meat burger	Abdel-Naeem and Mohamed (2016)
Hawthorn (*Crataegus monogyna*) leaves and flower extract	Extract obtained by supercritical fluid extraction exhibited a potential technique for obtaining antioxidants from hawthorn and its use in muscle foods	Shortle et al. (2014)

(continued)

Table 16.1 (continued)

Source of Phenolic compounds	Salient findings	References
Tea, grape, chestnut and seaweed (*Ulva lactuca* and *Ulva rigida*)	Incorporation of extract at 1000 mg/kg in pork patties resulted in inhibition of lipid oxidation as well as lower total viable counts, lactic acid bacteria, *Pseudomonas* and psychotropic aerobic bacteria	Lorenzo et al. (2014)
Plant extracts (cranberry, rosemary, lovage)	2% rosemary extract exhibited strongest activity to microbes resulting in longest shelf life of the products	Jalosinska and Wilczak (2009)
Lotus seed epicarp extract	Incorporation of extract at 6.25, 12.5, 25, 50 and 100 μg mL^{-1} in pork homogenate decreased lipid oxidation in dose-dependent manner	Qi and Zhou (2013)
Momordica grosvenori	A decreased formation of hexanal, TBARS and carbonyls and lowered sulfhydryl loss in a dose-dependent manner in dried minced pork slices was observed	Cheng et al. (2017)
Plum extract	Addition of 3% plum extract was observed to be very effective in inhibiting lipid oxidation of irradiated turkey breast meat and the production of hexanal, heptanal, octanal and nonagonal in non-irradiated meat	Lee and Ahn (2005)
Herbal extracts of marjoram, rosemary and sage	Addition of extracts at concentration of 0.04% (v/w) in irradiated ground beef lowered the TBARS values and off-odour scores and significantly increased colour and acceptability scores	Mohamed et al. (2011)
Isabel and Niagara grape seed and peel extracts	A significant reduction in lipid oxidation comparable to synthetic antioxidants at 60 mg of total phenolic compounds (PC)/kg of cooked chicken meat	Selani et al. (2011)
Green tea extract	Ginger rhizome and fenugreek seed extracts had higher antioxidant potential than potato peel extract in controlling lipid oxidation in ground beef patties	Mansour and Khalil (2000)
Propolis	The optimum levels of tea catechins, rosemary and sage with 0.25%, 0.10% and 0.05% optimum levels, respectively, in pork patties product enhanced oxidative stability	McCarthy et al. (2001a)

(continued)

Table 16.1 (continued)

Source of Phenolic compounds	Salient findings	References
Rosemary extract	As compared to raw pork patties, cooked products had fourfold increase in TBARS values and tea catechins were found to be most effective antioxidant to prevent lipid oxidation	McCarthy et al. (2001b)
Green tea extract and *Thymbra spicata* oil	These extracts prevented lipid oxidation, degradation of heme pigments and helped in maintaining red colour of the cooked meat	Fernández-López et al. (2003)
Isabel and Niagara grape seed and peel extracts	Rosemary extract at 1500–2000 ppm incorporation in pork sausage was recorded more effective than synthetic antioxidants in controlling TBARS values and maintained red colour in raw frozen product	Sebranek et al. (2005)

incorporation of 3% oregano essential oil as observed by significantly decreased production of thiobarbituric acid reacting substances (TBRAS) and 2,2-diphenyl-1-picrylhydrazyl (DPPH). Oussalah et al. (2004) noted decreased lipid oxidation in beef muscle wrapped in casein film impregnated with OEO (1% w/v). The addition of OEO in meat patties resulted in increased endogenous antioxidant glutathione content (GSH) and decreased hydrogen peroxide–induced depletion in human intestinal Caco-2 cells (Ryan et al. 2009).

Estevez et al. (2007) studied the comparative antioxidant activity of sage and rosemary essential oils with butylated hydroxyl toluene (BHT) and butylated hydroxyl anisole (BHA) and observed higher antioxidant activity of studied essential oils as compared to BHT and BHA. Incorporation of sage and rosemary essential oils at 0.1% level resulted in reduced lipid oxidation in liver pates by inhibiting degradation of polyunsaturated fatty acids (PUFAs) and thus the formation of volatile and other harmful compounds. Sage oil contains carnosic acid, which exert seven times higher antioxidant potential to BHT. Mechanically deboned meat (MDM)/mechanically recovered meat (MRM) is more prone to lipid oxidation as this meat has higher mineral content and more exposure to air. The incorporation of 200 mg/kg rosemary and marjoram essential oils in mechanically deboned beef patties and poultry meat resulted in decreased degradation of lipids and enhanced the oxidative stability of products during 3 months storage under frozen conditions (Mohamed and Mansour 2012).

Dzudie et al. (2004) observed improved oxidative stability of beef patties upon incorporation of ginger and basilica essential oils. Du and Li (2008) noted antioxidant potential of cassia essential oil even during deep fat frying of beef patties in palm oil at 150°C for 90 s and the optimum level to exert the antioxidant effect was 120 μL cassia oil in 1 l palm oil. Application of balm and thyme essential oils on chicken breast extended its shelf life by preventing lipid oxidation during

Table 16.2 Some common essential oils and their active ingredients

Sr No	Name of essential oil	Scientific name and source	Active ingredient
1.	Oregano	Leaves and flower *of Origanum vulgare* subsp. *hirtum*	Thymol carvacrol, *p*-cymene γ-terpinene, Rosamarinic acid
2.	Thyme	Leaves and flower of *Thymus vulgaris*	thymol and carvacrol, *p*-cymene and *γ*-terpinene
3.	Sage	Buds and leaves of *Salvia officinalis*	Camphor, α-thujone, viridiflorol, borneol, eucalyptol,1,8-cineole, β-thujone, bornyl acetate, carnosic acid, carnosol
4.	Rosemary	Leaf of *Rosmarinus officinalis*	Limonene, *α*-pinene, camphor and (*Z*)-linalool oxide, followed by borneol, camphene, sabinene, 1,8-cineole and *α*-terpineol
5.	Cinnamon	Bark of *Cinnamomum zeylandicum*	*trans*-Cinnamaldehyde
6.	Clove	Bud of *Syzygium aromaticum*	Eugenol
7.	Nutmeg	Ground fruit/seed of *Myristica fragrans*	d-pinene, limonene, d-borneol, l-terpineol, geraniol, safrol, and myristicin
8.	Balm	Flower, leaf of *Melissa officinalis*	Citronellal, carvacrol, *iso*-menthone citronellol, geraniol
9.	Coriander	Seeds *of Coriandrum sativum*	Linalool, camphor, geranyl acetate, *p*-Cymene
10.	Cassia	Leaves and twigs of *Cinnamomum Cassia*	(*E*)-Cinnamaldehyde, benzoic acid, cinnamaldehyde, phenylacrylic acid, Benzaldehyde
11.	Cilantro	Immature leaves *of Coriandrum sativum*	Linalool, (*E*)-2-decanal
12.	Black pepper oil	Unripe red fruit of *Piper nigrum*	Piperine and piperine isomers, monoterpene content

Source: Kumar et al. (2015a, b), Govaris et al. (2010), Jayasena and Jo (2014)

refrigerated storage for 3 weeks (Fratianni et al. 2010). Lemongrass (*Cymbopogon citratus*) essential oil exerted antioxidant activity equivalent to BHT and is widely being explored for bio-preservation.

16.10.2 Antimicrobial Activity of Essential Oils

EOs interact with lipids in cell membrane and mitochondria of microbes and alter their cell fluidity, resulting in the leakage of cellular ions in Gram-positive microbes. This ionic imbalance leads to loss of homeostasis and cell death (Cristani et al. 2007). The presence of outer layer of lipopolysaccharides around cell wall hinders the interaction of EO with cell membranes, thus resulting in deceased efficiency as compared to Gram-positive bacteria (Harpaz et al. 2003). EO inhibit synthesis of ergosterol, a principal ingredient in cell membrane of fungi, thus leading to the alteration of the integrity of cell and normal physiological characteristics (Pinto et al. 2009). Among the studied EOs, the antimicrobial efficacy of oregano, clove, coriander and cinnamon is the highest followed by thyme, mint, rosemary, mustard, cilantro and sage. Like other phenolics, EOs also have synergistic effect in combination with other EOs and hurdles, making bacterial cells more prone for destruction in conditions such as low pH, anaerobic conditions or lower oxygen concentration, high pressure, mild heat, presence of chelators and other antimicrobial agents. EOs exhibit antimicrobial effect in meat and fish products with application levels of 0.2 and 10 μl/ml against common food pathogens such as *Listeria monocytogenes*, *Salmonella typhimurium*, *Escherichia coli O157:H7*, *Shigella dysenteriae*, *Bacillus cereus* and *Staphylococcus aureus* (Burt 2004).

The overall preservative effect (antioxidant and antimicrobial activity) of essential oils depends upon fat and protein content of food products and as such in meat and fish products EOs exhibited comparatively lower efficacy than vegetable products. In general, clove, eugenol, oregano, thyme and coriander oil are more effective compared to cilantro, sage, mint and mustard oil in meat and meat products (Skandamis and Nychas 2001; Lemay et al. 2002). Hernandez-Ochoa et al. (2014) observed significant reduction in count of *Escherichia coli O157:H7*, *Salmonella*, *Listeria monocytogenes*, *Yersinia enterocolitica*, *Campylobacter jejuni*, *Clostridium perfringens*, *Staphylococcus aureus* and *Toxoplasma Gondi* in meat samples upon incorporation of essential oils (EOs) from clove, cumin and elecampane at 2250 μL level.

Cinnamon oil has cinnamaldehyde and eugenol as active ingredients and it exhibited antimicrobial activity against both Gram-positive and Gram-negative bacteria. Eugenol disintegrate bacterial cells and inactivate various metabolic enzymes. Cinnamaldehyde inhibits multiplication of *E. coli* O157:H7 and *Salmonella typhimurium*, but at 60°C it converts into benzaldehyde and binds with eugenol (Bajpai et al. 2011). Application of 2% *Rosmarinus officinalis* L. essential oil on fresh chicken breast resulted in markedly lower lactic acid bacteria and *Pseudomonas aeruginosa* counts, and extended shelf life to 16 days during retail display at 4°C (Petrova 2003). Carvacrol inhibited bacterial growth and decreased toxin production by *Bacillus cereus*. Thymol/2-isopropyl-5-methylphenol, isomers of carvacrol present in thymol oil, has been noted to possess antimicrobial effect between pH 5.5 and 6.5 against Gram-negative bacteria by binding with cell membrane lipopolysaccharide and loss of adenosine tripolyphosphates (ATP).

16.11 Conclusion

Plant phenolics are promising food additives, exerting preservative effects by inhibiting lipid oxidation and microbial growth. Further research is needed to explore newer sources of plant phenolics compounds and their incorporation in food products, their isolation and purification, and establishing the mode of action of these compounds.

References

Abdel-Naeem HH, Mohamed HM (2016) Improving the physico-chemical and sensory characteristics of camel meat burger patties using ginger extract and papain. Meat Sci 118:52–60

Ahn J, Grun IU, Fernando LN (2002) Antioxidant properties of natural plant extracts containing polyphenolic compounds in cooked ground beef. J Food Sci 67:1364–1369

Ahn J, Grun IU, Mustapha A (2007) Effects of plant extracts on microbial growth, color change, and lipid oxidation in cooked beef. Food Microbiol 24:7–14

Akarpat A, Turhan S, Ustun NS (2008) Effect of hot-water extracts from myrtle, rosemary, nettle and lemon balm leaves on lipid oxidation and color of beef patties during frozen storage. J Food Process Preserv 32:117–132

Aksu MI, Ozer H (2013) Effects of lyophilized water extract of *Satureja hortensis* on the shelf life and quality properties of ground beef. J Food Process Preserv 37:777–783

Alp E, Aksu Mİ (2010) Effects of water extract of Urtica dioica L. and modified atmosphere packaging on the shelf life of ground beef. Meat Sci 86(2):468–473

Altunkaya A, Hedegaard RV, Brimer L, Gokmen V, Skibsted LH (2013) Antioxidant capacity versus chemical safety of wheat bread enriched with pomegranate peel powder. Food Funct 4:722–727

Amany MMB, Shaker MA, Abeer AK (2012) Antioxidant activities of date pits in a model meat system. Int Food Res J 19:223–227

Amaral GP, Carvalho NR, Barcelos RP, Dobrachinski F, Portella RDL, Silva MH (2013) Protective action of ethanolic extract of *Rosmarinus officinalis* L. in gastric ulcer prevention induced by ethanol in rats. Br Ind Biol Res Assoc Food Chem Toxicol 54:48–55

Amic D, Davidovic-Amic D, Beslo D, Rastija V, Lucic B, Trinajstic N (2007) SAR and QSAR of the antioxidant activity of flavonoids. Curr Med Chem 14:827–845

Arakawa H, Maeda M, Okubo S, Shimamura T (2004) Role of hydrogen peroxide in bactericidal action of catechin. Biol Pharm Bull 27:277–281

Arranz E, Mes J, Wichers HJ, Jaime L, Mendiola JA, Reglero G (2015) Anti-inflammatory activity of the basolateral fraction of Caco-2 cells exposed to a rosemary supercritical extract. J Funct Foods 13:384–390

Awaluddin SA, Thiruvenkadam S, Izhar S, Hiroyuki Y, Danquah MK, Harun R (2016) Subcritical water technology for enhanced extraction of biochemical compounds from *Chlorella vulgaris*. BioMed Res Int. https://doi.org/10.1155/2016/5816974

Badr HM, Mahmoud KA (2011) Antioxidant activity of carrot juice in gamma-irradiated beef sausage during refrigerated and frozen storage. Food Chem 127:1119–1130

Bajpai VK, Shukla S, Kang SC (2008) Chemical composition and antifungal activity of essential oil and various extract of *Silene armeria L.* Bioresour Technol 99(18):8903–8908

Bajpai VK, Baek KH, Kang SC (2011) Control of *Salmonella* infoods by using essential oils: a review. Food Res Int 45(2):722–734

Banerjee R, Verma AK, Das AK, Rajkumar V, Shewalkar AA, Narkhede HP (2012) Antioxidant effects of broccoli powder extract in goat meat nuggets. Meat Sci 91:179–184

Banon S, Diaz P, Rodriguez M, Garrido MD, Price A (2007) Ascorbate, green tea and grape seed extracts increase the shelf life of low sulphite beef patties. Meat Sci 77:626–633

Bastida S, Sanchez-Muniz FJ, Olivero R, Perez-Olleros L, Ruiz-Roso B, Jimenez-Colmenero F (2009) Antioxidant activity of Carob fruit extracts in cooked pork meat systems during chilled and frozen storage. Food Chem 116:748–754

Belles M, Alonso V, Roncales P, Beltran JA (2017) Effect of borage and green tea aqueous extracts on the quality of lamb leg chops displayed under retail conditions. Meat Sci 129:153–160

Bellumori M, Michelozzi M, Innocenti M, Congiu F, Cencetti G, Mulinacci N (2015) An innovative approach to the recovery of phenolic compounds and volatile terpenes from the same fresh foliar sample of *Rosmarinus officinalis* L. Talanta 131:81–87

Bernard FX, Sable S, Cameron B, Provost J, Desnottes JF, Crouzet J, Blanche F (1997) Glycosylated flavones as selective inhibitors of topoisomerase IV. Antimicrob Agents Chemother 41:992–998
Birla R, Malav OP, Wagh RV, Mehta N, Chatli MK (2019) Storage stability of pork emulsion incorporated with Arjuna (Terminalia arjuna) tree bark extract. Int J Livest Res 9(2):95–109
Bishnoi S, Ahlawat SS, Bishnoi N, Kumar S, Jain S (2017) Effect of aloe vera gel and arjun tree bark extract incorporation on quality of buffalo male calf meat rolls during refrigeration storage. J Anim Res 7(1):69
Biswas AK, Keshri RC, Bisht GS (2004) Effect of enrobing and antioxidants on quality characteristics of precooked pork patties under chilled and frozen storage conditions. Meat Sci 66:733–741
Bors W, Heller W, Michel C, Saran M (1990) Flavonoids as antioxidants: determination of radical-scavenging efficiencies. Methods Enzymol 186:343–355
Bozkurt H (2006) Utilization of natural antioxidants: Green tea extract and *Thymbra spicata* oil in Turkish dry-fermented sausage. Meat Sci 73(3):442–450
Brannan R, Mah E (2007) Grape seed extract inhibits lipid oxidation in muscle from different species during refrigerated and frozen storage and oxidation catalyzed by peroxynitrite and iron/ascorbate in a pyrogallol red model system. Meat Sci 77:540–546
Brown JE, Khodr H, Hider RC, Rice-Evans C (1998) Structural dependence of flavonoid interactions with Cu2+ ions: implications for their antioxidant properties. Biochem J 330:1173–1178
Burr ML (1995) Explaining the French paradox. J R Soc Promot Heal 115:217–219
Burt S (2004) Essential oils: their antibacterial properties and potential applications in foods-a review. Int J Food Microbiol 94(3):223–253
Caleja C, Barros L, Antonio AL, Oliveira MB, Ferreira IC (2017) A comparative study between natural and synthetic antioxidants: evaluation of their performance after incorporation into biscuits. Food Chem 216:342–346
Cao Y, Gu W, Zhang J, Chu Y, Ye X, Hu Y, Chen J (2013) Effects of chitosan, aqueous extract of ginger, onion and garlic on quality and shelf life of stewed-pork during refrigerated storage. Food Chem 141:1655–1660
Carpenter R, Ogrady MN, Ocallaghan YC, Obrien NM, Kerry JP (2007) Evaluation of the antioxidant potential of grape seed and bearberry extracts in raw and cooked pork. Meat Sci 76(4):604–610
Casazza AA, Aliakbarian B, Perego P (2011) Recovery of phenolic compounds from grape seeds: effect of extraction time and solid–liquid ratio. Nat Prod Res 25:1751–1761
Chan EWC, Lim YY, Omar M (2007) Antioxidant and antibacterial activity of leaves of *Etlingera* species (*Zingiberaceae*) in Peninsular Malaysia. Food Chem 104:1586–1593
Cheeseman KH, Slater TF (1993) An introduction to free radicals chemistry. Br Med Bull 49:48–493
Chen ZY, Wang LY, Chan PT, Zhang Z, Chung HY, Liang C (1998) Antioxidative activity of green tea catechin extract compared with that of rosemary extract. J Am Oil Chem Soc 75(12):1141–1145
Chen L, Sun P, Wang T, Chen K, Jia Q, Wang H, Li Y (2012) Diverse mechanisms of antidiabetic effects of the different procyanidin oligomer types of two different cinnamon species on db/db mice. J Agric Food Chem 60(36):9144–9150
Cheng JR, Liu XM, Zhang YS, Zhang YH, Chen ZY, Tang DB, Wang JY (2017) Protective effects of *Momordica grosvenori* extract against lipid and protein oxidation-induced damage in dried minced pork slices. Meat Sci 133:26–35
Chinnam N, Dadi PK, Sabri SA, Ahmad M, Kabir MA, Ahmad Z (2010) Dietary bioflavonoids inhibit *Escherichia coli* ATP synthase in a differential manner. Int J Biol Macromol 46:478–486
Ciriano MG-I, Garcıa-Herreros C, Larequi E, Valencia I, Ansorena D, Astiasaran I (2009) Use of natural antioxidants from lyophilized water extracts of *Borago officinalis* in dry fermented sausages enriched in ω-3 PUFA. Meat Sci 83:271–277

Contini C, Álvarez R, Osullivan M, Dowling DP, Gargan SO, Monahan FJ (2014) Effect of an active packaging with citrus extract on lipid oxidation and sensory quality of cooked turkey meat. Meat Sci 96(3):1171–1176

Cos P, Ying L, Calomme M, Hu JP, Cimanga K, Van Poel B, Pieters L, Vlietinck AJ, Berghe DV (1998) Structure–activity relationship and classification of flavonoids as inhibitors of xantine oxidase and superoxide scavengers. J Nat Prod 61:71–76

Cristani M, D'Arrigo M, Mandalari G, Castelli F, Sarpietro MG, Micieli D, Venuti V, Bisignano G, Saija A, Trombetta D (2007) Interaction of four monoterpenes contained in essential oils with model membranes: implications for their antibacterial activity. J Agric Food Chem 55 (15):6300–6308

Croft KD (2006) The chemistry and biological effects of flavonoids and phenolic acids. Ann N Y Acad Sci 854:435–442

Cushnie TPT, Lamb AJ (2005) Antimicrobial activity of flavonoids. Int J Antimicrob Agents 26:343–356

Cushnie TT, Lamb AJ (2011) Recent advances in understanding the antibacterial properties of flavonoids. Int J Antimicrob Agents 38(2):99–107

Dai J, Mumper RJ (2010) Plant phenolics: extraction, analysis and their antioxidant and anticancer properties. Molecules 15:7313–7352

Das AK, Rajkumar V, Verma AK, Swarup D (2012) *Moringa oleifera* leaves extract: a natural antioxidant for retarding lipid peroxidation in cooked goat meat patties. Int J Food Sci Technol 47:585–591

de Lange DW, Verhoef S, Gorter G, Kraaijenhagen RJ, van de Wiel A, Akkerman JW (2007) Polyphenolic grape extract inhibits platelet activation through PECAM-1: an explanation for the French paradox. Alcohol Clin Exp Res 31:1308–1314

Devatkal SK, Narsaiah K, Borah A (2010) Anti-oxidant effect of extracts of kinnow rind, pomegranate rind and seed powder in cooked goat meat patties. Meat Sci 85:155–159

Devatkal SK, Thorat P, Manjunatha M (2012) Effect of vacuum packaging and pomegranate peel extract on quality aspects of ground goat meat and nuggets. J Food Sci Technol 51 (10):2685–2691

Do QD, Angkawijaya AE, Tran-Nguyen PL, Huynh LH, Soetaredjo FE, Ismadji S, Ju YH (2013) Effect of extraction solvent on total phenol content, total flavonoids content, and antioxidant activity of *Limnophila aromatic*. J Food Drug Anal 22(3):296–302

Doolaege EH, Vossen E, Raes K, de Meulenaer B, Verhe R, Paelinck H, de Smet S (2012) Effect of rosemary extract dose on lipid oxidation, colour stability and antioxidant concentrations, in reduced nitrite liver pates. Meat Sci 90(4):925–931

Du H, Li H (2008) Antioxidant effect of *Cassia* essential oil on deep-fried beef during the frying process. Meat Sci 78:461–468

Dumbrava DG, Moldovan C, Raba DN, Popa MV (2012) Vitamin C, chlorophylls, carotenoids and xanthophylls content in some basil (*Ocimum basilicum* L.) and rosemary (*Rosmarinus officinalis* L.) leaves extracts. J Agroaliment Process Technol 18(3):253–258

Dziki D, Rozylo R, Gawlik-Dziki U, Swieca M (2014) Current trends in the enhancement of antioxidant activity of wheat bread by the addition of plant materials rich in phenolic compounds. Trends Food Sci Technol 40(1):48–61

Dzudie T, Kouebou CP, Essia-Ngang JJ, Mbofung CMF (2004) Lipid sources and essential oils effects on quality and stability of beef patties. J Food Eng 65:67–72

El-Megeid AAA, AbdAllah IZA, Elsadek MF, YFA E-M (2009) The protective effect of the fortified bread with green tea against chronic renal failure induced by excessive dietary arginine in male albino rats. World J Dairy Food Sci 4:107–117

Erdogrul OT (2002) Antibacterial activities of some plant extracts used in folk medicine. Pharm Biol 40(4):269–273

Estevez M, Ramirez R, Ventanas S, Cava R (2007) Sage and rosemary essential oils versus BHTfor the inhibition of lipid oxidative reactions in liver pate. LWT-Food Sci Technol 40:58–65

Fasseas MK, Mountzouris KC, Tarantilis PA, Polissiou M, Zervas G (2008) Antioxidant activity in meat treated with oregano and sage essential oils. Food Chem 106(3):1188–1194

Fernández-López J, Sevilla L, Sayas-Barberá E, Navarro C, Marin F, Pérez-Alvarez JA (2003) Evaluation of the antioxidant potential of hyssop (*Hyssopus officinalis* L.) and rosemary (*Rosmarinus officinalis* L.) extracts in cooked pork meat. J Food Sci 68:660–664

Ferrali M, Signorini C, Caciotti B, Sugherini L, Ciccoli L, Giachetti D, Comporti M (1997) Protection against oxidative damage of erythrocyte membrane by the flavonoid quercetin and its relation to iron chelating activity. FEBS Lett 416(2):123–129

Firuzi O, Mladenka P, Petrucci R, Marrosu G, Saso L (2004) Hypochlorite scavenging activity of flavonoids. J Pharm Pharmacol 56:801–807

Firuzi O, Lacanna A, Petrucci R, Marrosu G, Saso L (2005) Evaluation of the antioxidant activity of flavonoids by "ferric reducing antioxidant power" assay and cyclic voltammetry. Biochim Biophys Acta 1721:174–184

Formica JV, Regelson W (1995) Review of the biology of quercetin and related bioflavonoids. Food Chem Toxicol 33:1061–1080

Fratianni F, Martino LD, Melone A, Feo VD, Coppola R, Nazzaro F (2010) Preservation ofchicken breast meat treated with thyme and balm essential oils. J Food Sci 75:528–535

Frei B, Higdon JV (2003) Antioxidant activity of tea polyphenols in vivo: evidence from animal studies. J Nutr 133:3275–3284

Garrido MD, Auqui M, Marti N, Linares MB (2011) Effect of two different red grape pomace extracts obtained under different extraction systems on meat quality of pork burgers. LWT Food Sci Technol 44:2238–2243

Gawlik-Dziki U, Dziki D, Baraniak B, Rufa L (2009) The effect of simulated digestion in vitro on bioactivity of wheat bread with Tartary buckwheat flavones addition. LWT Food Sci Technol 42:137–143

Gawlik-Dziki U, Swieca M, Dziki D, Baraniak B, Tomilo J, Czyz J (2013) Quality and antioxidant properties of breads enriched with dry onion (*Allium cepa* L.) skin. Food Chem 138:1621–1628

Giuliani C, Bucci I, Di Santo S, Rossi C, Grassadonia A, Piantelli M, Monaco F, Napolitano G (2014) The flavonoid quercetin inhibits thyroid-restricted genes expression and thyroid function. Food Chem Toxicol 66:23–29

Gordon NC, Wareham DW (2010) Antimicrobial activity of the green tea polyphenol (−)-epigallocatechin-3-gallate (EGCG) against clinical isolates of *Tenotrophomonas maltophilia*. Int J Antimicrob Agents 36:129–131

Govaris A, Solomakos N, Pexara A, Chatzopoulou PS (2010) The antimicrobial effectof oregano essential oil, nisin and their combination against *Salmonella enteritidis* in minced sheep meat during refrigerated storage. Int J Food Microbiol 137:175–180

Gutierrez RM, Mitchell S, Solis RV (2008) *Psidium guajava*: a review of its traditional uses, phytochemistry and pharmacology. J Ethnopharmacol 117:1–27

Halliwell B (2008) Are polyphenols antioxidants or pro-oxidants? What do we learn from cell culture and in vivo studies? Arch Biochem Biophys 476:107–112

Hanasaki Y, Ogawa S, Fukui S (1994) The correlation between active oxygens scavenging and antioxidative effects of flavonoids. Free Radic Biol Med 16:845–850

Haraguchi H, Tanimoto K, Tamura Y, Mizutani K, Kinoshita T (1998) Mode of antibacterial action of retrochalcones from Glycyrrhiza inflata. Phytochemistry 48:125–129

Harpaz S, Glatman L, Drabkin V, Gelman A (2003) Effects of herbal essential oils used to extend the shelf life of freshwater reared Asian sea bass fish (*Lates calcarifer*). J Food Prot 66 (3):410–417

Hawthorne SB, Grabanski CB, Martin E, Miller DJ (2000) Comparisons of Soxhlet extraction, pressurized liquid extraction, supercritical fluid extraction and subcritical water extraction for environmental solids: recovery, selectivity and effects on sample matrix. J Chromatogr A 892 (1):421–433

Hayes JE, Stepanyan V, Allen P, Ogrady MN, Kerry JP (2010) Evaluation of the effects of selected phytochemicals on quality indices and sensorial properties of raw and cooked pork stored in different packaging systems. Meat Sci 85:289–296

Heim KE, Tagliaferro AR, Bobilya DJ (2002) Flavonoid antioxidants: chemistry, metabolism and structure–activity relationships. J Nutr Biochem 13:572–584

Hernandez-Ochoa L, Aguirre-Prieto YB, Nevárez-Moorillon GV, Gutierrez-Mendez N, Salas-Munoz E (2014) Use of essential oils and extracts from spices in meat protection. Food Sci Technol 51(5):957–963

Hertog MG, Feskens EJ, Hollman PC, Katan MB, Kromhout D (1993) Dietary antioxidant flavonoids and risk of coronary heart disease: the Zutphen Elderly Study. Lancet 342:1007–1011

Hertog MGL, Feskens EJM, Kromhout D (1997) Antioxidant flavonols and coronary heart disease risk. Lancet 349:699

Hirano R, Sasamoto W, Matsumoto A, Itakura H, Igarashi O, Kondo K (2001) Antioxidant ability of various flavonoids against DPPH radicals and LDL oxidation. J Nutr Sci Vitaminol (Tokyo) 47:357–362

Huang B, He J, Ban X, Zeng H, Yao X, Wang Y (2011) Antioxidant activity of bovine and porcine meat treated with extracts from edible lotus (*Nelumbo nucifera*) rhizome knot and leaf. Meat Sci 87(1):46–53

Hung Y, Verbeke W, de Kok TM (2016) Stakeholder and consumer reactions towards innovative processed meat products: insights from a qualitative study about nitrite reduction and phytochemical addition. Food Control 60:690–698

Isoda H, Motojima H, Onaga S, Samet I, Villareal MO, Han J (2014) Analysis of the erythroid differentiation effect of flavonoid apigenin on K562 human chronic leukemia cells. Chem Biol Interact 220:269–277

Jagtap NS, Wagh RV, Chatli MK, Kumar P, Malav OP, Mehta N (2019) Optimisation of extraction protocol for *Carica papaya* L. to obtain phenolic rich phyto-extract with prospective application in chevon emulsion system. J Food Sci Technol 56(1):71–82

Jalosinska M, Wilczak J (2009) Influence of plant extracts on the microbiological shelf life of meat products. Polish J Food Nutr Sci 59(4)

Jayasena DD, Jo C (2014) Potential application of essential oils as natural antioxidants in meat and meat products: a review. Food Rev Int 30:71–90

Jia N, Kong B, Liu Q, Diao X, Xia X (2012) Antioxidant activity of black currant (*Ribes nigrum* L.) extract and its inhibitory effect on lipid and protein oxidation of pork patties during chilled storage. Meat Sci 91(4):533–539

Jin SK, Ha SR, Choi JS (2015) Effect of *Caesalpinia sappan* L. extract on physico-chemical properties of emulsion-type pork sausage during cold storage. Meat Sci 110:245–252

Jones M, Hoffman LC, Muller M (2015) Effect of rooibos extract (*Aspalathus linearis*) on lipid oxidation over time and the sensory analysis of blesbok (*Damaliscus pygargus phillipsi*) and springbok (*Antidorcas marsupialis*) droëwors. Meat Sci 103:54–60

Jongberg S, Torngren MA, Gunvig A, Skibsted LH, Lund MN (2013) Effect of green tea or rosemary extract on protein oxidation in Bologna type sausages prepared from oxidatively stressed pork. Meat Sci 93(3):538–546

Jongberg S, Torngren MA, Gunvig A, Skibsted LH, Lund MN (2015) Green tea extract impairs meat emulsion properties by disturbing protein disulfide cross-linking. Meat Sci 100:2–9

Jung EY, Yun IR, Go GW, Kim GD, Seo HW, Joo ST, Yang HS (2012) Effects of *Radix puerariae* extracts on physicochemical and sensory quality of precooked pork sausage during cold storage. LWT Food Sci Technol 46:556–562

Karabacak S, Bozkurt H (2008) Effects of *Urtica dioica* and *Hibiscus sabdariffa* on the quality and safety of sucuk (Turkish dry-fermented sausage). Meat Sci 78:288–296

Keskekoglu H, Uren A (2014) Inhibitory effects of pomegranate seed extract on the formation of heterocyclic aromatic amines in beef and chicken meatballs after cooking by four different methods. Meat Sci 96(4):1446–1451

Kessler M, Ubeaud G, Jung L (2003) Anti- and pro-oxidant activity of rutin and quercetin derivatives. J Pharm Pharmacol 55:131–142
Khoddami A, Wilkes MA, Roberts TH (2013) Techniques for analysis of plant phenolic compounds. Molecules 18:2328–2375
Khokhar S, Magnusdottir SGM (2002) Total phenol, catechin, and caffeine contents of teas commonly consumed in the United Kingdom. J Agric Food Chem 50:565–570
Kim HK, Cheon BS, Kim YH, Kim SY, Kim HP (1999) Effects of naturally occurring flavonoids on nitric oxide production in the macrophage cell line RAW 264.7 and their structure–activity relationships. Biochem Pharmacol 58:759–765
Kim SJ, Min SC, Shin HJ, Lee YJ, Cho AR, Kim SY, Han J (2013) Evaluation of the antioxidant activities and nutritional properties of ten edible plant extracts and their application to fresh ground beef. Meat Sci 93:715–722
Kopacz M, Kuźniar A (2003) Complexes of cadmium (II), mercury (II) and lead (II) with quercetin-5′-sulfonic acid (QSA). Pol J Chem 77:1777–1786
Korkina LG, Afanasev IB (1997) Antioxidant and chelating properties of flavonoids. Adv Pharmacol 38:151–163
Kulkarni S, Desantos FA, Kattamuri S, Rossi SJ, Brewer MS (2011) Effect of grape seed extract on oxidative, color and sensory stability of pre-cooked, frozen, re-heated beef sausage model system. Meat Sci 88:139–144
Kumar P, Kumar S, Tripathi MK, Mehta N, Ranjan R, Bhat ZF, Sing PK (2013) Flavonoids in the development of functional meat products: a review. https://doi.org/10.5455/vetworld.573-578
Kumar P, Verma AK, Mehta N, Kumar D, Malav OP (2015a) Preserving meat quality naturally. Fleishwirtschaft Int 4:46–51
Kumar Y, Yadav DN, Ahmad T, Narsaiah K (2015b) Recent trends in the use of natural antioxidants for meat and meat products. Compr Rev Food Sci Food Saf 14(6):796–812
Kumar Y, Kaur K, Shahi AK, Kairam N, Tyagi SK (2017) Antilisterial, antimicrobial and antioxidant effects of pediocin and Murraya koenigii berry extract in refrigerated goat meat emulsion. LWT-Food Sci Technol 79:135–144
Kumar P, Chatli MK, Mehta N, Malav OP, Verma AK, Kumar D, Rathour M (2018a) Antioxidant and antimicrobial efficacy of sapota powder in pork patties stored under different packaging conditions. Korean J Food Sci Anim Resour 38(3):593–603
Kumar P, Mehta N, Malav OP, Kumar Chatli M, Rathour M, Verma AK (2018b) Antioxidant and antimicrobial efficacy of watermelon rind extract (WMRE) in aerobically packaged pork patties stored under refrigeration temperature (4±1° C). J Food Process Preserv 42(10). https://doi.org/10.1111/jfpp.13757
Kurcubic VS, Maskovic PZ, Vujic JM, Vranic DV, Veskovic-Moracanin SM, Okanovic DG, Lilic SV (2014) Antioxidant and antimicrobial activity of *Kitaibelia vitifolia* extract as alternative to the added nitrite in fermented dry sausage. Meat Sci 97(4):459–467
Lachman J, Sulc M, Schilla M (2007) Comparison of the total antioxidant status of Bohemian wines during the wine-making process. Food Chem 103:802–807
Lee EJ, Ahn DU (2005) Quality characteristics of irradiated turkey breast rolls formulated with plum extract. Meat Sci 71(2):300–305
Lee MA, Choi JH, Choi YS, Kim HY, Kim HW, Hwang KE, Chung HK, Kim CJ (2011) Effects of kimchi ethanolic extracts on oxidative stability of refrigerated cooked pork. Meat Sci 89 (4):405–411
Lemay MJ, Choquette J, Delaquis PJ, Gariepy C, Rodrigue N, Saucier L (2002) Antimicrobial effect of natural preservatives in a cooked and acidified chicken meat model. Int J Food Microbiol 78:217–226
Li BH, Zhang R, Du YT, Sun YH, Tian WX (2006) Inactivation mechanism of the-ketoacyl-[acyl carrier protein] reductase of bacterial type-II fatty acid synthase by epigallocatechin gallate. Biochem Cell Biol 84:755–762
Lim HS, Park SH, Ghafoor K, Hwang SY, Park J (2011) Quality and antioxidant properties of bread containing turmeric (*Curcuma longa* L.) cultivated in South Korea. Food Chem 124:1577–1582

Lorenzo JM, Sineiro J, Amado IR, Franco D (2014) Influence of natural extracts on the shelf life of modified atmosphere-packaged pork patties. Meat Sci 96(1):526–534

Lotito SB, Actis-Goretta L, Renart ML, Caligiuri M, Rein D, Schmitz HH, Steinberg FM, Keen CL, Fraga CG (2000) Influence of oligomer chain length on the antioxidant activity of procyanidins. Biochem Biophys Res Commun 276:945–951

Lupton DA (2005) Lay discourses and beliefs related to food risks: an Australian perspective. Soc Health Illn 27(4):448–467

Maeta K, Nomur W, Takashume Y, Izawa S, Inove Y (2007) Green tea polyphenols fuction as prooxidants to active oxidative-stress-responsive transcription factors in yeasts. Appl Environ Microbiol 73:572–580

Mansour EH, Khalil AH (2000) Evaluation of antioxidant activity of some plant extracts and their application to ground beef patties. Food Chem 69:135–141

Mathenjwa SA, Hugo CJ, Bothma C, Hugo A (2012) Effect of alternative preservatives on the microbial quality, lipid stability and sensory evaluation of boerewors. Meat Sci 91(2):165–172

Mathew S, Abraham TE (2006) Studies on the antioxidant activities of cinnamon (*Cinnamomum verum*) bark extracts, through various in vitro models. Food Chem 94:520–528

Matsuda H, Morikawa T, Ando S, Toguchida I, Yoshikawa M (2003) Structural requirements of flavonoids for nitric oxide production inhibitory activity and mechanism of action. Bioorg Med Chem 11:1995–2000

Mazaheri H, Lee KT, Bhatia S, Mohamed AR (2010) Subcritical water liquefaction of oil palm fruit press fiber in the presence of sodium hydroxide: an optimisation study using response surface methodology. Bioresour Technol 101(23):9335–9341

McCarthy TL, Kerry JP, Kerry JF, Lynch PB, Buckley DJ (2001a) Assessment of the antioxidant potential of natural food and plant extracts in fresh and previously frozen pork patties. Meat Sci 57:177–184

McCarthy TL, Kerry JP, Kerry JF, Lynch PB, Buckley DJ (2001b) Evaluation of the antioxidant potential of natural food/plant extracts as compared with synthetic antioxidants and vitamin E in raw and cooked pork patties. Meat Sci 57:45–52

Mechergui K, Coelho JA, Serra MC, Lamine SB, Boukhchina S, Khouja ML (2010) Essential oils of *Origanum vulgare* L. *subsp. glandulosum* (Desf.) Ietswaart from Tunisia: chemical composition and antioxidant activity. J Sci Food Agric 90(10):1745–1749

Mildner-Szkudlarz S, Zawirska-Wojtasiak R, Obuchowski W, Gośliński M (2009) Evaluation of antioxidant activity of green tea extract and its effect on the biscuits lipid fraction oxidative stability. J Food Sci 74(8):S362–S370

Mohamed HMH, Mansour HA (2012) Incorporating essential oils of marjoram and rosemary in the formulation of beef patties manufactured with mechanically deboned poultry meat to improve the lipid stability and sensory attributes. LWT Food Sci Technol 45:79–87

Mohamed HM, Mansour HA, Farag MDEDH (2011) The use of natural herbal extracts for improving the lipid stability and sensory characteristics of irradiated ground beef. Meat Sci 87 (1):33–39

Morel D, Lin CY (1996) Cellular biochemistry of oxysterols derived from the diet or oxidation in vivo. Nutr Biochem 7:495–506

Mori A, Nishino C, Enoki N, Tawata S (1987) Antibacterial activity and mode of action of plant flavonoids against *Proteus vulgaris* and *Staphylococcus aureus*. Phytochemistry 26:2231–2234

Mulinacci N, Innocenti M, Bellumori M, Giaccherini C, Martini V, Michelozzi M (2011) Storage method, drying processes and extraction procedures strongly affect the phenolic fraction of rosemary leaves: an HPLC/DAD/ MS study. Talanta 85(1):16–176

Muthukumar M, Naveena BM, Vaithiyanathan S, Sen AR, Sureshkumar K (2012) Effect of incorporation of *Moringa oleifera* leaves extract on quality of ground pork patties. J Food Sci Technol 51(11):3172–3180

Naveena BM, Vaithiyanathan S, Muthukumar M, Sen AR, Kumar YP, Kiran M, Shaju VA, Chandran KR (2013) Relationship between the solubility, dosage and antioxidant capacity of

carnosic acid in raw and cooked ground buffalomeat patties and chicken patties. Meat Sci 95:195–202
Negi PS (2012) Plant extracts for the control of bacterial growth: efficacy, stability and safety issues for food application. Int J Food Microbiol 156(1):7–17
Nerland DE (2007) The antioxidant/electrophile response element motif. Drug Metab Rev 39:235–248
Nijveldt RJ, van Nood E, van Hoorn DEC, Boelens PG, van Norren K, van Leeuwen PAM (2001) Flavonoids: a review of probable mechanisms of action and potential applications. Am J Clin Nutr 74:418–425
Nissen LR, Byrne DV, Bertelsen G, Skibsted LH (2004) The antioxidative activity of plant extracts in cooked pork patties as evaluated by descriptive sensory profiling and chemical analysis. Meat Sci 68:485–495
Olszanecki R, Gêbska A, Kozlovski VI, Gryglewski RJ (2002) Flavonoids and nitric oxidesynthase. J Physiol Pharmacol 53:571–584
Oussalah M, Caillet S, Salmiéri S, Saucier L, Lacroix M (2004) Antimicrobial and antioxidant effects of milk protein-based film containing essential oils for the preservation of whole beef muscle. J Agric Food Chem 52(18):5598–5605
Ozsoy N, Candoken E, Akev N (2009) Implications for degenerative disorders: antioxidative activity, total phenols, flavonoids, ascorbic acid, beta-carotene and beta-tocopherol in Aloe vera. Oxidative Med Cell Longev 2(2):99–106
Ozvural EB, Vural H (2012) The effects of grape seed extract on quality characteristics of frankfurters. J Food Process Preserv 36:291–297
Peng CH, Su JD, Chyau CC, Sung TY, Ho SS, Peng CC (2007) Supercritical fluid extracts of rosemary leaves exhibit potent anti-inflammation and anti-tumor effects. Biosci Biotechnol Biochem 71(9):2223–2232
Peng X, Ma J, Cheng KW, Jiang Y, Chen F, Wang M (2010) The effects of grape seed extract fortification on the antioxidant activity and quality attributes of bread. Food Chem 119:49–53
Petrova J (2003) Microbiological quality of fresh chicken breast meat after rosemary essential oil treatment and vacuum packaging. J Anim Sci Biotechnol 46(1):140–144
Pietta PG (2000) Flavonoids as antioxidants. J Nat Prod 63:1035–1042
Pinto E, Vale-Silva L, Cavaleiro C, Salgueiro L (2009) Antifungal activity of the clove essential oil from *Syzygium aromaticum* on *Candida, Aspergillus* and dermatophyte species. J Med Microbiol 58(Pt 11):1454–1462
Plaper A, Golob M, Hafner I, Oblak M, Solmajer T, Jerala R (2003) Characterization of quercetin binding site on DNA gyrase. Biochem Biophys Res Commun 306:530–536
Prior RL, Cao G (2000) Antioxidant phytochemicals in fruits and vegetables: diet and health implications. Hortic Sci 35:588–592
Prochazkova D, Bousova I, Wilhelmova N (2011) Antioxidant and prooxidant properties of flavonoids. Fitoterapia 82(4):513–523
Qi S, Zhou D (2013) Lotus seed epicarp extract as potential antioxidant and anti-obesity additive in Chinese Cantonese Sausage. Meat Sci 93(2):257–262
Raba DN, Moigradean D, Poiana MA, Popa M, Jianu I (2007) Antioxidant capacity and polyphenols content for garlic and basil flavored bread. J Agroaliment Process Technol 13:163–168
Rathour M, Malav OP, Kumar P, Chatli MK, Mehta N (2017a) Standardization of protocols for extraction of aloe vera and cinnamon bark extracts. J Anim Res 7(1):175–182
Rathour M, Malav OP, Kumar P, Chatli MK, Mehta N (2017b) Storage stability of chevon rolls incorporated with ethanolic extracts of aloe vera and cinnamon bark at refrigeration temperature (4±1° C). J Anim Res 7(1):183–190
Rauha JP, Remes S, Heinonen M, Hopia A, Kähkönen M, Kujala T, Pihlaja K, Vuorela H, Vuorela P (2000) Antimicrobial effects of Finnish plant extracts containing flavonoids and other phenolic compounds. Int J Food Microbiol 56(1):3–12

Realini CE, Guàrdia MD, Díaz I, García-Regueiro JA, Arnau J (2015) Effects of acerola fruit extract on sensory and shelf-life of salted beef patties from grinds differing in fatty acid composition. Meat Sci 99:18–24

Reddy GB, Sen AR, Nair PN, Reddy KS, Reddy KK, Kondaiah N (2013) Effects of grape seed extract on the oxidative and microbial stability of restructured mutton slices. Meat Sci *95* (2):288–294

Renaud S, de Lorgeril M (1992) Wine, alcohol, platelets, and the French paradox for coronary heart disease. Lancet 339:1523–1526

Riahi L, Chograni H, Elferchichi M, Zaouali Y, Zoghlami N, Mliki A (2013) Variations in Tunisian wormwood essential oil profiles and phenolic contents between leaves and flowers and their effects on antioxidant activities. Ind Crop Prod 46:290–296

Ribeiro-Santos R, Carvalho-Costa D, Cavaleiro C, Costa HS, Albuquerque TG, Castilho MC, Sanches-Silva A (2015) A novel insight on an ancient aromatic plant: the rosemary (*Rosmarinus officinalis* L.). Trends Food Sci Technol 45(2):355–368

Rice-Evans CA, Miller NJ (1995) Antioxidants-the case for fruit and vegetables in the diet. Br Food J 97:35–40

Rice-Evans CA, Miller NJ, Paganga G (1996) Structure-antioxidant activity relationships of flavonoids and phenolic acids. Free Radic Biol Med 20:933–956

Rojas MC, Brewer MS (2007) Effect of natural antioxidants on oxidative stability of cooked, refrigerated beef and pork. J Food Sci 72:S282–S288

Rojas MC, Brewer MS (2008) Effect of natural antioxidants on oxidative stability of frozen, vacuum-packaged beef and pork. J Food Qual 31:173–188

Rounds L, Havens CM, Feinstein Y, Friedman M, Ravishankar S (2013) Concentration-dependent inhibition of *Escherichia coli* O157: H7 and heterocyclic amines in heated ground beef patties by apple and olive extracts, onion powder and clove bud oil. Meat Sci 94(4):461–467

Rubbo H, Radi R, Trujillo M, Telleri R, Kalyanaraman B, Barnes S, Kirk M, Freeman BA (1994) Nitric oxide regulation of superoxide and peroxynitrite-dependent lipid peroxidation. J Biol Chem 269:26066–266075

Ruiz-Navajas Y, Viuda-Martos M, Sendra E, Perez-Alvarez JA, Fernández-López J (2012) Chemical characterization and antibacterial activity of *Thymus moroderi* and *Thymus piperella* essential oils, two *Thymus* endemic species from southeast of Spain. Food Control 27(2):294–299

Ryan E, Aherne SA, O'Grady MN, McGovern L, Kerry JP, O'Brien NM (2009) Bioactivity of herb-enriched beef patties. J Med Food 12:893–901

Sasse A, Colindres P, Brewer MS (2009) Effect of natural and synthetic antioxidants on the oxidative stability of cooked, frozen pork patties. J Food Sci 74:S30–S35

Sebranek JG, Sewalt VJH, Robbins KL, Houser TA (2005) Comparison of a natural rosemary extract and BHA/BHT for relative antioxidant effectiveness in pork sausage. Meat Sci 69:289–296

Selani MM, Contreras-Castillo CJ, Shirahigue LD, Gallo CR, Plata-Oviedo M, Montes-Villanueva ND (2011) Wine industry residues extracts as natural antioxidants in raw and cooked chicken meat during frozen storage. Meat Sci 88(3):397–403

Shan B, Cai YZ, Sun M, Corke H (2005) Antioxidant capacity of 26 spice extracts and characterization of their phenolic constituents. J Agric Food Chem 53(20):7749–7759

Shah MA, Bosco SJD, Mir SA (2014) Plant extracts as natural antioxidants in meat and meat products. Meat Sci 98(1):21–33

Shan B, Cai YZ, Brooks JD, Corke H (2009) Antibacterial and antioxidant effects of five spice and herb extracts as natural preservatives of raw pork. J Sci Food Agric 89:1879–1885

Shavisi N, Khanjari A, Basti AA, Misaghi A, Shahbazi Y (2017) Effect of PLA films containing propolis ethanolic extract, cellulose nanoparticle and *Ziziphora clinopodioides* essential oil on chemical, microbial and sensory properties of minced beef. Meat Sci 124:95–104

Shi J, Yu J, Pohorly JE, Kakuda Y (2003) Polyphenolics in grape seeds-Biochemistry and functionality. J Med Food 6:291–299

Shortle E, O'grady MN, Gilroy D, Furey A, Quinn N, Kerry JP (2014) Influence of extraction technique on the anti-oxidative potential of hawthorn (*Crataegus monogyna*) extracts in bovine muscle homogenates. Meat Sci 98(4):828–834

Sikkhamondhol C, Teanpook C, Boonbumrung S, Chittrepol S (2009) Quality of bread with added turmeric (*Curcuma longa* L): powder, essential oil and extracted residues. Asian J Food Agro Ind 2:690–701

Silvan JM, Mingo E, Hidalgo M, de Pascual-Teresa S, Carrascosa AV, Martinez-Rodriguez AJ (2013) Antibacterial activity of a grape seed extract and its fractions against Campylobacter spp. Food Control 29(1):25–31

Skandamis PN, Nychas GJE (2001) Effect of oregano essential oil onmicrobiological and physico-chemical attributes of minced meat stored in airand modified atmospheres. J Appl Microbiol 91:1011–1022

Sun C, Fu J, Chen J, Jiang L, Pan Y (2010a) On-line HPLC method for screening of antioxidants against superoxide anion radical from complex mixtures. J Sep Sci 33:1018–1023

Sun W, Zhao M, Cui C, Zhao Q, Yang B (2010b) Effect of Maillard reaction products derived from the hydrolysate of mechanically deboned chicken residue on the antioxidant, textural and sensory properties of Cantonese sausages. Meat Sci 86:276–282

Szeląg A, Magdalan J, Kopacz M, Kuźniar A, Kowalski P, Pieśniewska M (2003) Assessment of efficacy of quercetin-5′-sulfonic acid sodium salt in the treatment of acute chromium poisoning: experimental studies. Pol J Pharmacol 55:1097–1103

Tajik H, Farhangfar A, Moradi M, Rohani SMR (2014) Effectiveness of clove essential oil and grape seed extract combination on microbial and lipid oxidation characteristics of raw buffalo patty during storage at abuse refrigeration temperature. J Food Process Preserv 38:31–38

Tatke P, Jaiswal Y (2011) An overview of microwave assisted extraction and its applications in herbal drug research. Res J Med Plants 5:21–31

Tayel AA, El-Tras WF (2012) Plant extracts as potent biopreservatives for *Salmonella typhimurium* control and quality enhancement in ground beef. J Food Saf 32:115–121

Teixeira B, Marques A, Ramos C, Neng NR, Nogueira JMF, Saraiva JA (2013) Chemical composition and antibacterial and antioxidant properties of commercial essential oils. Ind Crop Prod 43:587–595

Tipsrisukond N, Fernando LN, Clarke AD (1998) Antioxidant effects of essential oil and oleoresin of black pepper from supercritical carbon dioxide extractions in ground pork. J Agric Food Chem 46(10):4329–4333

Tsuchiya H, Iinuma M (2000) Reduction of membrane fluidity by antibacterial sophoraflavanone G isolated from *Sophora exigua*. Phytomedicine 7:161–165

Turgut SS, Isikci F, Soyer A (2017) Antioxidant activity of pomegranate peel extract on lipid and protein oxidation in beef meatballs during frozen storage. Meat Sci 129:111–119

Turkmen N, Sari F, Velioglu YS (2006) Effects of extraction solvents on concentration and antioxidant activity of black and black mate tea polyphenols determined by ferrous tartrate and Folin–Ciocalteu methods. Food Chem 99:835–841

Ulanowska K, Tkaczyk A, Konopa G, Węgrzyn G (2006) Differential antibacterial activity of genistein arising from global inhibition of DNA, RNA and protein synthesis in some bacterial strains. Arch Microbiol 184:271–278

Valencia I, Ogrady MN, Ansorena D, Astiasarán I, Kerry JP (2008) Enhancement of the nutritional status and quality of fresh pork sausages following the addition of linseed oil, fish oil and natural antioxidants. J Food Sci 80:1046–1054

Viuda-Martos M, Ruiz Navajas Y, Sánchez Zapata E, Fernández-López J, Perez-Alvarez JA (2010) Antioxidant activity of essential oils of five spice plants widely used in a Mediterranean diet. Flavour Fragr J 25(1):13–19

Vuong QV, Stathopoulos CE, Nguyen MH, Golding JB, Roach PD (2011) Isolation of green tea catechins and their utilization in the food industry. Food Rev Int 27:227–247

Wang Q, Wang H, Xie M (2010) Antibacterial mechanism of soybean isoflavone on *Staphylococcus aureus*. Arch Microbiol 192:893–898

Wang Y, Chen S, Yu O (2011) Metabolic engineering of flavonoids in plants and microorganisms. Appl Microbiol Biotechnol 91(4):949–956

Wezemael LV, Verbeke W, Barcellos MDD, Scholderer J, Perez-Cueto F (2010) Consumer perceptions of beef healthiness: results from a qualitative study in four European countries. BMC Public Health 10(342)

Williams P, Stirling E, Keynes N (2004) Food fears: a national survey on the attitudes of Australian adults about the safety and quality of food. Asia Pac J Clin Nutr 13(1):32–33

Wojciak KM, Dolatowski ZJ, Okon A (2011) The effect of water plant extracts addition on the oxidative stability of meat products. Acta Sci Pol Technol Aliment 10:175–188

Wu D, Kong Y, Han C, Chen J, Hu L, Jiang H, Shen X (2008) d-alanine:d-alanine ligase as a new target for the flavonoids quercetin and apigenin. Int J Antimicrob Agents 32:421–426

Yamane H, Konno K, Sabelis M, Takabayashi J, Sassa T, Oikawa H (2010) Chemical defence and toxins of plants. Comprehensive natural products II. Reference Module in Chem Molecular Sci Chem Eng 4:339–385

Yao L, Jiang Y, Datta N, Singanusong R, Liu X, Duan J, Raymont K, Lisle A, Xu Y (2004) HPLC analyses of flavonols and phenolic acids in the fresh young shoots of tea (*Camellia sinensis*) grown in Australia. Food Chem 84:253–263

Yeh HY, Chuang CH, Chen HC, Wan CJ, Chen TL, Lin LY (2014) Bioactive components analysis of two various gingers (*Zingiber officinale* Roscoe) and antioxidant effect of ginger extracts. LWT Food Sci Technol 55:329–334

Yogesh K, Ahmad T, Yadav DN, Vishwakarma RK (2014) Antioxidant effect of litchi (*Litchi chinensis* Sonn.) seed extract on raw ground chicken meat stored at $4\pm1^{\circ}$C. Int J Livestock Res 1(2):21

Yoon MH, Jo JE, Kim DM, Kim KH, Yook HS (2010) Quality characteristics of bread containing various levels of flowering cherry (*Prunus serrulata* L. var. *spontanea Max. Wils.*) fruit powder. J Korean Soc Food Sci Nutr 39:1340–1345

Young IS, Woodside JV (2001) Antioxidants in health and disease. J Clin Pathol 54:176–186

Yu J, Ahmedna M, Goktepe I (2010) Potential of peanut skin phenolic extract as antioxidative and antibacterial agent in cooked and raw ground beef. Int J Food Sci Technol 45:1337–1134

Zakaria SM, Kamal SMM (2016) Subcritical water extraction of bioactive compounds from plants and algae: applications in pharmaceutical and food ingredients. Food Eng Rev 8(1):23–34

Zhang Y-M, Rock CO (2004) Evaluation of epigallocatechin gallate and related plant polyphenols as inhibitors of the FabG and FabI reductases of bacterial type II fatty-acid synthase. J Biol Chem 279:30994–31001

Zhu M, Fahl WE (2001) Functional characterization of transcription regulators that interact with the electrophile response element. Biochem Biophys Res Commun 23:212–219

Zhu QY, Huang Y, Chen ZY (2000) Interaction between flavonoids and α- tocopherol in human low density lipoprotein. J Nutr Biochem 11:14–21

Zilic S, Kocadaglı T, Vancetovic J, Gokmen V (2016) Effects of baking conditions and dough formulations on phenolic compound stability, antioxidant capacity and color of cookies made from anthocyanin-rich corn flour. LWT-Food Sci Technol 65:597–603

Plant Phenolics for Overcoming Multidrug Resistance in Human Fungal Pathogen

17

Shweta Singh, Zeeshan Fatima, and Saif Hameed

Abstract

In spite of significant advances being made in the improvement of antifungal drugs, only limited number of drugs is currently available, and that too are not able to keep pace with the evolution of multidrug resistance (MDR). The urgent need includes the development of alternative drugs that are more efficient and tolerant than those traditionally already in use. Natural plant phenolics are among the most commonly occurring type of secondary metabolite in nature which is constantly being expanded through the discovery of new natural products. Interest in phenolics and the search for new biological activities within members of this class have intensified in recent years, as evidenced by the evaluation of their potential antifungal activities. Among most human pathogenic fungi, *Candida albicans* is of extreme importance due to their high frequency of colonization and infection in humans. Since nature has plethora of many promising natural compounds which can efficiently be exploited to improve the antifungal therapeutics, the objective of this book chapter is to describe the development of plant phenolics as antifungals for the treatment of *Candida* species and to note the most promising compounds with their diverse mechanism of actions and their uses in combination with traditional drugs.

Keywords

Candida · MDR · Natural phenolics · Efflux pumps · Cell membrane · Cell wall · Morphogenesis · Biofilm · Cell adherence

S. Singh · Z. Fatima (✉) · S. Hameed (✉)
Amity Institute of Biotechnology, Amity University Haryana, Gurugram, Haryana, India

R. Lone et al. (eds.), *Plant Phenolics in Sustainable Agriculture*,
https://doi.org/10.1007/978-981-15-4890-1_17

17.1 Introduction

Fungal infections are the fourth most leading cause of hematogenous infections worldwide, which are mostly contributed by *Candida* genus (Pierce 2005). It consists of more than 150 species including the major *Candida albicans* and non-*albicans* species like *Candida auris*, *Candida tropicalis*, and *Candida glabrata*. Generally, *Candida* is a part of human microbiota where they normally reside in the body surfaces like oral cavity, gastrointestinal tract, and vagina, in a commensal manner, but under immunocompromised conditions, they turned out to be a great menace to human population (Kim and Sudbery 2011). The magnitude of fungal infections is increasing due to increment in number of transplants and hospitalized patients. The other factors contributing to the rise in infections are the prolonged usage of antifungal drugs and long chemotherapy treatments which result in the alarming problem of drug resistance by a phenomenon known as multidrug resistance (MDR) posing additional challenge for the health industries and researchers. The available few drugs belonging to classes azoles, polyenes, echinocandins, allylamines, and flucytosine are being resisted by the *Candida* species as been reported by number of case studies in hospitals. The problem of nosocomial infections is compounded due to the biofilm formation by *Candida* species, which resides in the indwelling devices like catheters used for the treatment of patients with another disease like AIDS or transplant patients. Hence, researches are now more focused on the search for the new drugs which could be used potentially against rising *Candida* infections.

Nowadays, plant-derived products are being favored over the synthetic products, due to their minimal side effects, cost-effectiveness, and potential. Plants are a rich source of secondary metabolites which they utilize for their defense against animals or microbes. These metabolites have been shown in various studies to possess many antimicrobial properties. Therefore, these secondary metabolites can be exploited for exploring the vast potential of these compounds that may be used as antifungal drugs. Secondary metabolites consist of diverse classes like terpenoids, alkaloids, phenolic compounds, alcohols, and acids. These classes consist of many potential compounds which are being employed for various uses due to their antimicrobial, antiseptic, anticancer, antioxidant, and anti-inflammatory properties (Ksouri et al. 2012). Among these, phenolic compounds comprise a major class of plant secondary metabolites being broadly distributed over more than 8000 phenolic structures currently identified. Previously, we have reviewed the antifungal action of phenolic compounds across various pathogenic fungi (Ansari et al. 2013). The present book chapter is written with the objective to summarize at a common platform the identification of diverse phenolic compounds as promising antifungals for the treatment of *Candida* species particularly with their probable mode of actions.

17.2 Classification of Phenolic Compounds

These compounds can contain one or more aromatic rings with at least one or more hydroxyl group species. Phenolic acids, flavonoids, and tannins are the most studied groups of phenolic compounds, comprising molecules with evidenced antifungal properties. The following sections will deal with few members of each classes.

17.2.1 Hydroxycinnamic Acids

They are the most commonly occurring phenolic compounds which belong to non-flavonoid polyphenols and are also known as hydroxycinnamates (Stalmach 2014). They are hydroxy metabolites of cinnamic acids which have benzene ring in which three-carbon chain is attached because of which they are included in phenylpropanoid group. They are synthesized by mevalonate-shikimate biosynthesis pathway in plants involving the phenylalanine and tyrosine as precursors (El-Seedi et al. 2012). The derivatives are also produced such as amides (combination with amino acids or peptides) and esters (combination with hydroxyl acids or glycosides). They are most commonly produced in fruits, cereals seeds, and vegetables. Their derivatives also play a role in acting as a precursor for the formation of lignans, anthocyanins, flavonoids, etc. They have important anti-inflammatory and antioxidant properties. The antioxidant nature of hydroxycinnamic acids has broad application in diverse fields like industrial additives or as health agents. They worked effectively against the oxidation-related diseases like inflammatory injury, cardiovascular diseases, atherosclerosis, and cancer. It has radical scavenging activity and chelation of transition metals or inhibition of reactive oxygen species (ROS) enzymes by modulation of gene expression (Teixeira et al. 2013). They are widely used as health beneficiary in various metabolic syndrome like anti-inflammatory agents, cardiovascular diseases, diabetes, and dyslipidemia (Alam et al. 2016). They also have antimicrobial, anti-collagenase, anti-melanogenic, and anti-tyrosinase activities. The pharmacological properties are contributed by presence of multiple hydroxyl groups in their chemical structure. The major hydrocinnamic acids are as follows.

17.2.2 Caffeic Acid

It is the major constituent and predominant among the other forms of hydroxycinnamic acid. It is around 75% of the total hydroxycinnamic acid in citrus fruits. The synthesis of caffeic acid is due to the hydroxylation of p-coumaric acid. It has medicinal properties, such as antitumor, antioxidant, antimicrobial, anti-inflammatory, and antidiabetic activities. They have role in decreasing the inflammation by lowering the expression of inflammatory mediator TNF-α (Alam et al. 2016). It has role in inhibiting the early stage of biofilm formation in *C. albicans*. The esters of caffeic acid have found to be effective against mature biofilm and more potent in comparison with the known antifungal drug fluconazole (FLC) with lower minimum inhibitory concentration (MIC) values (De Vita et al. 2014). Caffeic acid phenethyl ester (CAPE) is the most studied derivative due to its potent antioxidant activity. It has also known to inhibit the clinical isolates of *Candida* by showing synergistic nature with known antifungal drug. The results have also shown that combination of FLC and CAPE has reduced the fungal burden and increased the longevity in nematode model, *Caenorhabditis elegans*. They have proved CAPE as promising therapeutics against resistant *C. albicans*. Another study by Sun et al. (2018) has demonstrated that CAPE shows synergism with the caspofungin and disrupts the iron homeostasis in *C. albicans*. CAPE performs this action by two ways: either by free radicals formation reactions or functional defects of mitochondrial respiratory chain CI and energy depletion in *C. albicans* [10]. Caffeic acids have been reported to inhibit the planktonic cells of *C. albicans* by targeting the glyoxylate cycle and inhibiting the activity of crucial enzyme called as isocitrate lyase (Cheah et al. 2014).

17.2.3 P-Coumaric Acid

It is one of the most dominant of all hydroxycinnamic acids. It is present in eggplants in very substantial amount, as well as broccoli and asparagus. It is synthesized by tyrosine and phenylalanine and is the major precursor in the further synthesis of other cinnamic acids. It has known antioxidant, antimicrobial, anti-inflammatory, antitumor, and antiplatelet aggregation properties. Due to their depigmenting potential and antioxidant nature, anti-collagenase, antimicrobial, and anti-inflammatory activities are being exploited to be used for cosmeceutical use. The methanolic extracts from various plants such as *Rosa rugosa*, *Ligusticum mutellina* L., *Limonium avei*, *Kitaibelia vitifolia*, and *Tamarix gallica* L. have been reported to inhibit the candidal growth and show varying MIC values (Teodoro et al. 2015a, b).

17.2.4 Ferulic Acid

It is the most common hydroxycinnamic acid which is present in plant cell walls and abundantly found in cereals as dietary source and corn bran, wheat bran, eggplant, artichokes, and beets. It is synthesized from caffeic acid by the enzyme caffeate O-methyltransferase. It has a variety of potential therapeutic effects which can be useful in the treatment of various diseases like cancer, diabetes, and lung and cardiovascular diseases. It has known antimicrobial and anti-inflammatory properties and neuroprotective and photoprotective effects. Ferulic acid has apoptotic effect on *C. albicans* and *C. glabrata*. The synergism of ferulic acid with caspofungin was found to be effective against the candidal infections. Their combinatorial action was fungicidal in nature as compared with the individual effect of ferulic acid and caspofungin. This compound has potential to be used for anticandidal treatment (Canturk 2018). The ferulic acid extracted from ethanolic and aqueous extracts from various plants has been shown to inhibit the candidal growth. In a study by Panwar et al. (2016), ferulic acid-encapsulated chitosan nanoparticles have been found to be effective against the candidal biofilm residing in dwelling devices. This approach can be a promising alternative approach for the conventional therapeutic methods.

17.2.5 Sinapic Acid

It is a 3,5-dimethoxy-4-hydroxycinnamic acid found in citrus, berry fruits, vegetables, cereals, and oilseed crops (Chen 2016). It has known antioxidant, anti-inflammatory, antiglycemic, anticancer, and antibacterial activities. It has been reported in pathological conditions such as diabetes, inflammation, oxidative stress, and neurodegeneration (Zou et al. 2002). It can exist in free form or ester form such as sinapine (sinapoylcholine), sinapoyl esters, and sinapoyl malate. The sinapic acid derivatives such as 4-vinylsyringol, sinapine, and syringaldehyde have shown antioxidant activity, acetylcholinesterase inhibition, and antimutagenicity (Kuwahara et al. 2004). The sinapic acid derivative called as syringaldehyde has been studied for its antifungal potential against *Candida guilliermondii (*Kelly et al. 2008*)*.

17.2.6 Rosmarinic Acid

It is found in *Rosmarinus officinalis* L. and synthesized from esterification of both caffeic acid and 3,4-dihydroxyphenyllactic acid. It has known antioxidant, antitumor, anti-inflammatory, and antimicrobial properties. It has high radical-scavenging activity and medicinal properties which have been exploited in pharmaceutical and cosmetic sectors. It has been reported in a study by Calixto et al. (2015) that rosmarinic acid has potential anticandidal activity against clinical isolates via its antioxidant activity which turns to be modulatory effect. It has also shown synergism with the known antifungal drug FLC providing alternative combination therapy options.

17.2.7 Hydroxybenzoic Acids

This class represents the phenolic metabolites of general structure C6 $\pm$ C. It is naturally found in *Cocos nucifera,* vanilla, wine, *Macrotyloma uniflorum* (horse gram), and *Phyllanthus acidus* (Dey et al. 2005, Tian et al. 2009). They are found in intact tissues primarily as conjugates which after hydrolysis give rise to hydroxybenzoic acids. It has very well-known antioxidant activity and low toxicity benefits used in cosmetics industry. They are considered to be effective scavengers of free radicals and are also widely used as antimicrobial food additives and are used in modern nutrition therapies (Hubková et al. 2014). They have been also tested for the antifungal activity against *Candida* species; the results have shown synergism of hydroxybenzoic acid with itraconazole (ITR) and hence can be used for improving the activity of known antifungal drugs (Faria et al. 2011). The most commonly studied hydroxybenzoic acids are as follows.

17.2.8 Gallic Acid

It is a 3,4,5-trihydroxybenzoic acid or trihydroxybenzoic acid commonly found in the oak barks, hazel nut, gallnuts, etc. They are found in free form or as tannins which can be hydrolyzed to gallic acid or gallotannins. They can form esters such as digallic and trigallic acids. It is formed from the 3-dehydroshikimate by the enzyme called as shikimate dehydrogenase which can produce 3,5-didehydroshikimate. It has known antimicrobial and anti-inflammatory activities which are being utilized in the pharmacological and cosmetics industries. The in vitro and in vivo activity of gallic acid was investigated by Li et al. (2017a, b). They have checked the antifungal activity on various fungal strains by using *Punica granatum* L. The results have shown the effect of gallic acid on the fungal cell membrane by targeting the sterol biosynthetic pathway which in turn affects the viability of *C. albicans*. The enzymatic activity of sterol 14a-demethylase P450 (CYP51), key enzyme in ergosterol biosynthesis, was reduced. Gallic acid which was extracted from *Buchenavia tomentosa* has been studied for its anticandidal activity on *C. albicans* and non-*albicans* species by Teodoro et al. (2015a, b). It has been also explored more for its mechanistic insights in *Candida* species. The extracts from *Vernonia cinerea* (L.) containing gallic acid have shown the antioxidant property against *C. albicans*. They have shown promising effect on virulence factor of *Candida* like biofilm and morphological transition (Brighenti et al. 2017). Gallic acid from acetone extract from *Buchenavia tomentosa* has shown the effect on cell adherence of *C. albicans* from oral epithelial cells and hyphal growth and disrupted biofilm (Teodoro et al. 2018). The solubility of gallic acid has been improved by using the cyclodextrins which results into the formation of HPβCD soluble inclusion complexes. The efficiency was enhanced to a great level which encourages the development of antifungal drugs (Teodoro et al. 2017). A nanoencapsulation approach has been used for the formation of chitosan-based edible films carrying the gallic acid nanoparticles for the direct delivery of gallic acid into matrix which can maintain its solubility and mechanical properties (Lamarra et al. 2017). A recent study on the extract from *Spondias tuberosa* fruit plant which is native to Brazil has shown anti-*Candida* activity probably by causing the hyperpolarization of mitochondrial and lysosomal membrane. Thus, this compound can be exploited more for its pharmacological applications (da Costa et al. 2018).

17.2.9 Vanillic Acid

It is a 4-hydroxy-3-methoxybenzoic acid which is oxidized form of vanillin, commonly used as flavoring agent. It is found in Chinese herb *Angelica sinensis* and is a natural phenol in argan oil. It is formed from ferulic acid (Lesage-Meessen et al. 1996). It has antimicrobial properties. It is used as preservatives in foods due to its antimicrobial properties (Fitzgerald et al. 2004). It has also shown antifungal properties which have been tested against yeasts and filamentous fungi (Alnuaimi et al. 2013). It has been reported to have shown anticandidal activity. The mechanisms have been studied which included loss of membrane integrity and inhibition of ergosterol biosynthesis of *C. albicans*. It has also reduced the cell adhesion by 30–40% along with the biofilm formation in a concentration-dependent manner. The morphogenetic switching from yeast to hyphae was also hampered (Raut et al. 2013). Another study on candidal biofilms has shown the inhibition of biofilm formation, and the reduction was 75–80%. It showed promising effect on oral candidiasis and can be used as a potential antifungal drug.

17.2.10 Salicylic Acid

It is a lipophilic hydroxybenzoic acid which is formed during salicin metabolism. It naturally occurs in nuts, fruits, vegetables, herbs, teas, etc. It is widely used as an important component in aspirin drug which is used as an anti-inflammatory analgesic and antipyretic and to treat pain. It is used in cosmetic industries as an anti-acne and in skincare products. It is also used as antiseptic due to its antimicrobial activity. The coating of salicylic acid prevents the adherence of candidal cells with the Silastic catheters (Farber and Wolff 1993). It was also checked to be synergistic with known antifungal drug like FLC. This suggests its use in the combination therapy against *Candida* strains (Yücesoy et al. 2000). Another drug nystatin also became more potent when worked synergistically with the salicylic acid against clinical isolates of *C. albicans* (Ibezimako et al. 2003). The derived acetylsalicylic acid has significant role in reducing the biofilm as well as planktonic growth on the abiotic surfaces like polystyrene and acrylic surfaces (Alem and Douglas 2004). A similar study was also performed which shows the inhibition of formation of candidal biofilms which needs further investigation for its use in alternative therapy (Stepanović et al. 2004). The morphogenetic ability which is switching from yeast to hyphal form was also suppressed by inhibiting the formation of 3(R)-hydroxyoxylipins which are a fatty acid (Deva et al. 2001). Another virulence factor called as phospholipases which is involved in invading the host tissue by candidal cells was also found to be reduced in a study by Trofa et al. (2009). A chemosensitizing approach can be used to enhance the antifungal potential of known antifungal drugs when used synergistically with the salicylic acid and other phenolic compounds (Faria et al. 2011). But further investigations and studies are required to look more into its mechanism of action.

17.2.11 Protocatechuic Acid

It is a 3,4-dihydroxybenzoic acid which is one of the major metabolite of polyphenols in green tea. It has potential anticancer, antioxidant, antiulcer, anti-inflammatory, antimicrobial, antidiabetic, and analgesic activities. It is extracted from barks of *Boswellia dalzielii* and leaves of *Diospyros melanoxylon*. Acai oil derived from palm tree is found to be rich in protocatechuic acid. It is also found in mushrooms and flowers which are used as beverages and in bran and grain brown rice (Kakkar and Bais 2014). It is produced during shikimate pathway and can be derived by vanillic acid and phthalic acid. It has antimicrobial activity which was tested by using the extracts from *Cirsium canum* against the *Streptococcus aureus* and *Streptococcus pneumoniae* (Kozyra et al. 2015). The extracts from barks of *Ficus ovata* containing protocatechuic acid were found to inhibit the planktonic growth of candidal cells (Kuete et al. 2009). They have been studied to possess anticandidal activity which was reviewed by Teodoro et al. (2015a, b). Further studies are required to exploit its potential to be used as potent antifungal drug.

17.2.12 Flavonoids

It is the largest family of phenolic compounds containing nearly more than 5000 hydroxylated polyphenolic compounds which have diverse important functions in plants. They are known to be involved in regulating environmental stresses, combating microbial infection, floral pigmentation, regulation of cell growth, and pollination (Kumar and Pandey 2013). They are mostly derived from fruits, vegetables, chocolate, and beverages like wine and tea. They have been shown to exhibit antidiabetic, anti-inflammatory, anticancer, antithrombogenic, antioxidant, and neuroprotective activities. They are also known to act as inhibitors for enzymes like phosphoinositide 3-kinase, xanthine oxidase (XO), cyclooxygenase (COX), and lipoxygenase. Flavonoids are considered as crucial component in a number of pharmaceutical, nutraceutical, medicinal, and cosmetic applications. The most widespread subclasses of flavonoids are flavan-3-ols (catechin), flavonols, flavanones, flavones, isoflavones, chalcones, and anthocyanins (Panche et al. 2016).

17.2.13 Epigallocatechin

It is a type of flavan-3-ol (catechin) which is highly present in green tea extract. They are also present in traces amount in fruits, nuts, etc. It has benefits for treating cancer, cardiovascular diseases, inflammatory disorders, and other ailments. It has known antibacterial, antiviral, antifungal, antitoxin, and antitumor effects too (Matsumoto et al. 2012). It is further studied for the anticandidal activity along with the other tea extracts (Chen et al. 2015).They were also known to induce apoptosis in FLC-resistant candidal strains and can also work synergistically with FLC (da Silva et al. 2014). It has been reported to inhibit the planktonic growth and can also work synergistically with the known antifungal drugs like miconazole, FLC, or amphotericin B which opens up opportunities for combination therapy for treating oral candidiasis (Ning et al. 2015). The derivative of epigallocatechin with the gallic acid, epigallocatechin gallate (EGCG), has been reported to be used as adjuvant for antifungal therapy in candidiasis. Planktonic growth of candidal cells was reduced by 75% when treated with EGCG. Interestingly, the reduction in established candidal biofilm was found to be 80%. There was also an impairment of proteasome activity which helps in degrading the host proteins by candidal cells (Evensen and Braun 2009).

17.2.14 Kaempferol

It is a 3,4′,5,7-tetrahydroxyflavone which belongs to the subclass flavonols. It occurs in fruits, vegetables, lettuce, berries, etc. It is synthesized during phenylpropanoid pathway. It has anticancer, antidiabetic, antimicrobial, and antioxidant properties (Chen and Chen 2013). The phenolic compounds extracted from *Baseonema acuminatum* leaves containing kaempferol 3-O-(6″-galloyl)-beta-D-glucopyranoside have shown anticandidal activity (De Leo et al. 2004). The ethanolic extracts from *Bryophyllum pinnatum* containing kaempferol and its rhamnoside derivatives have been tested for its antimicrobial and antioxidant properties (Tatsimo et al. 2012). A study by Yordanov M (2008) has demonstrated the inhibition of *C. albicans* extracellular enzyme activity which is crucial for the host cell penetration. The secreted aspartyl proteinase activity and rate of cell wall protein glycosylation were effectively suppressed. The extracts from the plant *Equisetum giganteum* L. which is native to America containing kaempferol and its derivatives have shown to inhibit the adherence of candidal cells to the acrylic surfaces. These findings have good implication on the inhibition of candidal biofilms in oral candidiasis and denature stomatitis (Alavarce et al. 2015). Overexpression of efflux pump is one of the major contributor for development of multidrug resistance (MDR) in *C. albicans* (Prasad and Kapoor 2005). Two exclusively studied drug transporters in *C. albicans* are ATP-binding cassette (ABC) family and major facilitator superfamily (MFS) drug transporters. The effect of kaempferol on candidal cells was further investigated on the efflux pump activity (Shao et al. 2016). It induced the suppression of kaempferol-treated FLC-resistant candidal cells. The expression of MDR1 was effectively suppressed and also showed synergism with the FLC. Hence, this could be used as an effective drug targeting the efflux pump.

17.3 Baicalein

It is a 5,6,7-trihydroxyflavone which is extracted from *Scutellaria baicalensis* and *Oroxylum indicum* plants. It can be also found in fruits, green tea, dark chocolate, and vegetables and traditionally used in Chinese herbal medicine for chronic hepatitis. It has known antifungal, anti-inflammatory, antioxidant, anti-allergic, and anti-tumorigenic actions (Donald et al. 2012). It inhibits the lipoxygenases and possess anti-inflammatory property. It was very well studied for its anticancer property by acting on the angiogenesis, metastasis, and inflammation of cancer cells (Gao et al. 2016a, b). It has known antimicrobial activity and is known to inhibit the biofilm formation and quorum-sensing signaling mechanism in *Staphylococcus aureus* (Chen et al. 2016). Baicalein inhibits the candidal growth by reducing the cell surface hydrophobicity (CSH) which is an important factor for the interaction between surface and candidal cells. It also inhibits the candidal biofilm by 70% at varying concentrations (Cao et al. 2008). Baicalein was also reported to induce programmed cell death via targeting the mitochondrial membrane potential (Dai et al. 2013). It was known to work synergistically with known antifungal drug amphotericin B which accelerates the apoptosis in *C. albicans* through the CaMCA1-mediated caspase pathway (Fu et al. 2011). Candidal growth and cell viability were inhibited by this compound which also shows synergism with FLC, but further studies are required to decipher the exact mechanism of its synergism (Serpa et al. 2012). A study was conducted on the in vitro activity of Plantago major extract in which baicalein is one of the major component and was found to be effectively reducing the candidal growth in a dose-dependent manner by various mechanisms. This includes the inhibition of biofilm formation, morphogenetic switching via inhibiting the hyphal development from yeast forms, and cell surface hydrophobicity (CSH) which is the major regulator of biofilm formation (Shirley et al. 2017).

17.4 Quercetin

It is a 2-(3,4-dihydroxyphenyl)-3,5,7-trihydroxy-4H-chromen-4-one which occurs in fruits, vegetables, leaves, and grains. It is the most abundant among flavonoids in nature. It is synthesized through phenylpropanoid pathway by phenylalanine as its precursor. It has beneficial effect on cardiovascular disease, cancers, chronic inflammatory disorders, asthma, high blood pressure, and viral infections (Kelly 2011). It has known antimicrobial activity against *Staphylococcus aureus* and *Clostridium botulinum*. It has been also found to be synergistic with the amphotericin B antifungal drug against the invasive fungus *Cryptococcus* (Oliveira et al. 2016). Quercetin extracted from edible lichen (*Usnea longissima*) has been investigated for its anticandidal activity. The results have shown that it induces FLC-mediated programmed apoptotic cell death of *Candida* cells by modulation of quorum-sensing molecule farnesol which is involved in hyphal development and biofilm formation. It also inhibits the virulence factors like phospholipase, proteinase, and esterase activity (Singh et al. 2015). The QS-mediated combined sensitizer (QC)-anticandidal agent is an effective strategy to combat candidal infections. Another study by Gao et al. (2016a, b) also demonstrated that quercetin assists FLC for inhibition of the biofilm formation in FLC-resistant candidal cells which is helpful in clinical management of the vulvovagibal candidiasis conditions. The pretreatment with the quercetin induces an anti-inflammatory effect via inhibiting the cytokine TNF-alpha production against *C. albicans* infection in macrophages (Cui et al. 2013).

17.5 Chalcones

They are open-chain flavonoids having (2E)-1,3-diphenylprop-2-en-1-one structure and occur in fruits, vegetables, and medicinal plants. They are basically characterized by the absence of "ring C" in the basic flavonoid skeleton structure. They are precursors in other flavonoid synthesis. They have antifungal, antibacterial, anti-inflammatory, and antitumor properties. They have important implications in nutraceutical industries due to nutritional and biological benefits. The hydroxylated chalcones have potential anticandidal activity (Batovska et al. 2007). A type of chalcone named 4-hydroxycordoin is found to inhibit the yeast to hyphae transition and biofilm formation of candidal cells. Thus, it can be helpful in preventing the candida from oral cavity and inhibit soft tissue invasion (Messier et al. 2011). Another chalcone called as 2′-hydroxy-4′-methoxychalcone worked synergistically with the FLC against the FLC-resistant candidal strains (Wang et al. 2016). The pharmacochemical work has been performed on the benzimidazolyl-chalcones for their antifungal activity against a clinical strain of pharmaco-chemoresistant *Candida albicans* strain and showed promising anticandidal activity (Songuigama et al. 2018). Cyclized chalcones and its derivatives are found to be effective and have synergistic activity with FLC against clinical isolates (Ahmad et al. 2017). A very recent study by Nim et al. (2018) has shown the chalcones as the potential compounds for improving the azole activity. The synthesis of chalcones with the azoles through condensation reactions triggers sensitizing of yeast strains overexpressing CaMdr1p and CaCdr1p transporters.

17.6 Lignans

They are polyphenols or diphenolic compounds which are found in flaxseeds, sesame seeds, legumes, whole grains, fruit, and vegetables. The secoisolariciresinol and matairesinol are the first precursors which have been identified in human diet. Lignans are the main source of dietary phytoestrogens. The secoisolariciresinol breaks down into enterolignans and enterolactone by human gut bacteria which are metabolites of food lignans (Peterson et al. 2010). They are strong antioxidants in nature and helpful in supporting the human health. They also act as free radical scavengers which reduce the cancer development. It is also involved in antiestrogenic effects which resists in binding of radicals to estrogen receptors and exerting the negative effects (Adlercreutz 2007). They have anti-inflammatory activity and can be categorized into eight subgroups as furofuran, furan, dibenzylbutane, dibenzylbutyrolactone, aryltetralin, arylnaphthalene, dibenzocyclooctadiene, and dibenzylbutyrolactol on the basis of cyclization pattern. They are also known for their antimicrobial properties against *C. albicans* and gram-positive bacteria. It has been reported that extract containing lignans from *Lindera erythrocarpa* has shown specific antifungal activity by inhibiting the chitin synthase 2 in *C. albicans* (Hwang et al. 2007). The lignan glycoside derived from *Styrax japonicus* plant has been demonstrated for its anticandidal mechanism via targeting candidal membranes (Park et al. 2010). Some of the extensively studied lignan compounds are as follows.

17.7 Magnolol

It is a 4-allyl-2-(5-allyl-2-hydroxy-phenyl) phenol active lignan found in the bark of *Magnolia officinalis*. It has been traditionally used in Chinese and Japanese herbal medicine. It has anti-osteoporosis activity and anti-periodontal disease activity. It has pharmacological functions and has known analgesic, antianxiety, anti-tumorigenic properties (Lee et al. 2011). It has been known for its antifungal properties (Bang et al. 2000). It is one of the most exclusively studied compounds of lignans class. Previous studies have shown inhibitory effect on *Helicobacter pylori*, *Porphyromonas gingivalis*, etc. (Chang et al. 1998). It showed inhibitory effect on planktonic growth of *C. albicans* as well as on non-*albicans* species of *Candida*. It also reduces the biofilm formation by decreasing the metabolic activity in a concentration-dependent manner (Zhou et al. 2017). It also affects the cell adhesion and yeast to hyphae transition which is a key virulence factor in *C. albicans* via Ras1-cAMP-Efg1 pathway (Sun et al. 2015a, b). It was also synergistic with azoles by inducing a higher intracellular content of antifungals. It also targets candidal cells by various mechanisms such as functioning as a Cdr1p substrate and by targeting ergosterol biosynthesis (Sun et al. 2015a, b). Magnolol was also tested against clinical isolates from patients suffering from oral candidiasis. The results demonstrated ruptures depicting the disruption of cell membrane. It also causes decrease in biofilm formation by 69.5% and release of the intracellular content, which results in the swelling of the cell wall. It is nontoxic in nature and has negligible hemolytic activity which is approx. 11.9%. The docking results have also shown magnolol interacts with ergosterol in the fungal cell membranes. All these results have indicated that magnolol is a promising antifungal agent which can be considered for using in oral candidiasis (Behbehani et al. 2017).

17.8 Honokiol

It is a 2-(4-hydroxy-3-prop-2-enyl-phenyl)- 4-prop-2-enyl-phenol which can be isolated from bark and leaves of *Magnolia grandiflora* plant. It belongs to neolignan class, and due to its small size, it can interact with cell membrane proteins through intermolecular interactions (Woodbury et al. 2013). It is a structural isomer of magnolol. It has been traditionally used as a part of Eastern traditional membrane because of its high potent nature and also for its ability to also cross the blood barrier which is helpful for using in therapies. It has known anti-inflammatory, anti-tumorigenic, antithrombotic, antiviral, and antioxidant properties. The antifungal effect of honokiol was reported in a study by (Bang et al. 2000). The synergistic activity was demonstrated with the FLC, and it was found to be effective against the FLC-resistant clinical isolates of *Candida* (Jin et al. 2010), which can be beneficial for combination therapy. Honokiol has effect on virulence factors, namely, cell adhesion, transition from yeast to hyphae, and biofilm formation through Ras1-cAMP-Efg1 pathway. It also prolonged the survival of *Candida*-infected nematodes which shows the nontoxic nature of this drug (Sun et al. 2015a, b). It also induces ROS accumulation which in turn leads to apoptosis or necrosis in candidal cells via mitochondrial dysfunction, and this will contribute to its fungicidal action (Sun et al. 2017). In a study by Sun et al. (2017), they worked on the insights into the mechanisms of its action. It has been demonstrated that honokiol induces the oxidative stress in *C. albicans*. It targets the mitochondria by mitochondrial respiratory chain C1 and induces ROS accumulation, disruption of intracellular iron homeostasis, and activation of autophagy and apoptosis signaling pathway. It can be used as a chemosensitizer and could be a cell wall perturbing agents. A recent study on the role of Hsp90-calcineurin pathway on the antifungal activity showed that Hsp90 enhances the antifungal activity of honokiol (Liao and sun 2018). Honokiol-loaded micelles were prepared by using emulsion-solvent evaporation procedure by conjugating with oligochitosan-pluronic conjugate (CS-F127) carrier. They have tested for the delivery of honokiol into the mice cells. The results have shown the increase in retention time of honokiol with low clearance rate and apparent distribution volume which enhances its pharmacokinetics property (Song et al. 2018).

17.9 Lariciresinol

It is a 4-[(2S,3R,4R)-4-[(4-hydroxy-3-methoxyphenyl)methyl]-3-(hydroxymethyl) oxolan-2-yl]-2-methoxyphenol or phenylpropanoid and found in sesame seeds and vegetables. It has known anti-inflammatory and antioxidant properties. The phenolic extract from Tylosema *esculentum* plant contains this compound that has been tested for its antimicrobial activity (Chingwaru et al. 2011). In another study, it was also tested for its antimicrobial property against the foodborne pathogen; it exerts its effect by inducing cell membrane permeabilization (Bajpai et al. 2017). The essential oil from clove containing lariciresinol has shown antifungal activity by targeting the fungal cell membrane (Pinto et al. 2009). Lariciresinol is known to be a precursor of enterolignan which is isolated from plant *Sambucus williamsii*, having medicinal property. It has anticandidal activity and thus reduces its growth and acts by disrupting the cell membrane. Its action results into depolarization of fungal membrane and lipids (Hwang et al. 2011).

17.10 Medioresinol

It is a 4-[(3S,3aR,6aR)-3-(4-hydroxy-3-methoxyphenyl)-1,3,3a,4,6,6a-hexahydrofuro [3,4-c]furan-6-yl]-2,6-dimethoxyphenol or called as a furofuran-type lignan which has been extracted from the bark of *Sambucus williamsii*. It has been used as a medicinal plant in traditional medicine due to its therapeutic properties. It has known leishmanicidal activity and also reduces the cardiovascular disease risk. It is also known for its anti-inflammatory, antivirus, and analgesic properties and also used in the treatment of edema. The antibacterial effect was evaluated on the antibiotic-resistant species, and it was found to be synergistic with the known antibiotics such as ampicillin, cefotaxime, and chloramphenicol along with the antibiofilm activity (Hwang et al. 2013). Medioresinol has a significant antifungal activity against candidal cells. In order to evaluate its antifungal effect, the metabolic, morphological, and molecular assays have been worked to find its mode of action in *C. albicans*. Its action results in cell cycle arrest and also causes apoptosis which results in cell death. Its antioxidant activity induces ROS generation, and its accumulation leads to decreased depolarization of mitochondrial membrane. This series of action results into cytochrome c release, metacaspase activation, and phosphatidylserine externalization. It also causes morphological changes in nucleic acid and cell, resulting into nuclear fragmentation and its condensation. Overall, it induces apoptosis through a mitochondria-mediated apoptosis pathway (Hwang et al. 2012).

17.11 Nyasol

It is a 4,4′-[(1Z)-1,4-pentadiene-1,3-diyl]diphenol, or it is also known as cis-hinokiresinol extracted from *Anemarrhena asphodeloides*. It is known to inhibit the production of eicosanoids and nitric oxide in vitro and has anti-inflammatory effects (Lim et al. 2009). It is also known for its antipyretic, antidiabetic, and antidepressant properties. It has anti-oocyte property against the mycelial growth of *Colletotrichum orbiculare* and *Phytophthora capsici* (Park et al. 2003). Nyasol exhibits antimicrobial activity against pathogens. Its antifungal activity has been reported in a study by Iida et al. (2000). Nyasol has been extracted from *Anemarrhena asphodeloides* and has shown antifungal activity against *C. albicans*, *Aspergillus fumigatus*, and *Trichophyton mentagrophytes*. It was also found to be synergistic with known azoles like miconazole, clotrimazole, and ketoconazole, among the other azoles. This finding can contribute for its usage as adjuvant in combination therapy.

17.12 Other Classes

17.12.1 Coumarins

They are benzopyrones which contain benzene and α-pyrone rings that can be isolated from tonka bean, vanilla grass, cinnamon, etc. The naturally occurring derivatives are umbelliferone (7-hydroxycoumarin), herniarin (7-methoxycoumarin), esculetin (6,7-dihydroxycoumarin), psoralen, and imperatorin. They have been used in the treatment of lymphedema, and they have the ability to increase the antithrombin levels in plasma. It is also used for reduction of uric acid in blood by increasing the excretion of uric acid in urine. They are also used in cosmetic, perfumery, and household products due to its biological and pharmacological properties. Warfarin is a very well-known oral anticoagulant. It also has anti-inflammatory, antiulcerogenic, antifilarial, and antibacterial properties (Kontogiorgis and Hadjipavlou-Litina 2003). Later on, its antifungal properties have been evaluated and found to be effective against *C. albicans*, *Aspergillus fumigatus*, and *Fusarium solani.* The potential coumarins and its derivatives are osthole, osthenol, xanthotoxin, oroselone, etc. (Montagner et al. 2008). Osthole is a natural coumarin which has shown strong anticandidal activity. It has also shown synergism with FLC against FLC-resistant *C. albicans* in vitro and also indicated the endogenous ROS augmentation which contributes to the synergism of FLC and osthole (Li et al. 2017a, b). Another coumarins such as robustic acid and thonningine-C which were isolated from *Millettia thonningii* have shown fungicidal activity against the *C. albicans.* The molecular modelling studies have shown the binding of these compounds to the active site in turn disrupts the sterol synthesis (Ayine-Tora et al. 2016). The phosphoramidate derivatives of coumarin have also shown promising antifungal activity by acting as chitin synthase inhibitors (Ji et al. 2016).

17.12.2 Xanthones

They are 9H-xanthen-9-ones, which are heterocyclic compounds based on the dibenzo-c-pyrone scaffold. They have pharmacological properties, including antitumor, antioxidant, anti-allergic, anti-inflammatory, and antimicrobial activities. They are categorized into simple oxygenated and prenylated polycyclic dehydroxanthones. The alpha-mangostin was known to have antifungal activity and can be considered for treating oral candidiasis due to its low toxicity (Kaomongkolgit et al. 2009). It also inhibits the biofilm formation (Kaomongkolgit and Jamdee 2017). The antifungal activity of simple oxygenated xanthones was evaluated and demonstrated to act by inhibiting the ergosterol biosynthesis (Pinto et al. 2011). The extracts from *Hypericum tetrapterum* plant containing 1,7-dihydroxyxanthone have shown strong anticandidal activity (Zubricka et al. 2015).

17.13 Conclusion

Phytophenolics have great potential to inhibit the candidal growth. Several mechanisms have been studied through which these phenolic compounds can hinder the growth of *C. albicans* or work synergistically with known antifungal drugs (Table 17.1). They are broad class of compounds which are proven to be potent in combating fungal infections. Considering the fact that synthesis of new drugs is a cumbersome process, there is a need to search more on the structural and functional mode of action of phytophenolics to exploit their phytotherapeutic potential.

Table 17.1 Summary of anticandidal phenolic compounds with their mode of actions

Name of the class	Name of the compound	Mode of action	References
Hydroxycinnamic acids	Caffeic acid	Inhibits biofilm formation	De Vita et al. (2014)
		Synergism with FLC	De Vita et al. (2014)
		Synergism with Caspofungin	Sun et al. (2018)
		Mitochondrial dysfunction	Sun et al. (2018)
		Impaired glyoxylate cycle	Cheah et al. (2014)
	p-Coumaric acid	Inhibits candidal growth	Cheah et al. (2014)
	Ferulic acid	Synergism with caspofungin	Canturk (2018)
		Induces apoptosis	Canturk (2018)
		Inhibits biofilm formation	Panwar et al. (2016)
	Sinapic acid	Inhibits oxidation	Chen (2016)
	Rosmarinic acid	Synergism with FLC	Zou et al. (2002)
		Antioxidant action	Calixto et al. (2015)
Hydroxybenzoic acids	Gallic acid	Antioxidant action	Brighenti et al. (2017)
		Inhibits biofilm formation	Brighenti et al. (2017)
		Inhibits morphogenetic switching	Teodoro et al. (2012; 2018)
		Reduces cell adherence	Brighenti et al. (2017)
		Mitochondrial dysfunction	Teodoro et al. (2012; 2018)
		Disrupts lysosomal membrane integrity	da Costa et al. (2018)
	Vanillic acid	Inhibits ergosterol biosynthesis	Raut et al. (2013)
		Disrupts cell membrane integrity	Alnuaimi et al. (2013)
		Reduces cell adherence	Raut et al. (2013)
		Inhibits morphogenetic switching	Raut et al. (2013)
		Inhibits biofilm formation	Raut et al. (2013)
	Salicylic acid	Synergism with FLC	Yücesoy et al. (2000)
		Synergism with nystatin	Alem and Douglas (2004)
		Inhibits planktonic growth	Alem and Dougles (2004)
		Inhibits biofilm formation	Alem and Dougles (2004)

(continued)

Table 17.1 (continued)

Name of the class	Name of the compound	Mode of action	References
		Inhibits morphogenetic switching	Stepanović et al. (2004)
		Reduces phospholipases activity	Deva et al. (2001) Trofa et al. (2009)
	Protocatechuic acid	Inhibits planktonic growth	Kuete et al. (2009)
Flavonoids	Epigallocatechin	Synergism with FLC Synergism with miconazole Synergism with amphotericin B Inhibits biofilm formation Reduces proteasome activity	da Silva et al. (2014) Ning et al. (2015) Ning et al. (2015) Ning et al. (2015) Evensen and Braun (2009)
	Kaempferol	Reduces aspartyl proteinase activity Reduces protein glycosylation Inhibits biofilm formation Reduces cell adherence Targets overexpression of efflux pumps	Tatsimo et al. (2012) Tatsimo et al. (2012) Alavarce et al. (2015) Alavarce et al. (2015) Shao et al. (2016)
	Baicalein	Reduces cell surface hydrophobicity Inhibits biofilm formation Mitochondrial dysfunction Induces apoptosis Synergism with amphotericin B Synergism with FLC Inhibits morphogenetic switching	Chen et al. (2016) and Shirley et al. (2017) Chen et al. (2016) and Shirley et al. (2017) Dai et al. (2013) and Fu et al. (2011) Fu et al. (2011) Fu et al. (2011) Serpa et al. (2012) Shirley et al. (2017) Shirley et al. (2017)
	Quercetin	Synergism with amphotericin B Modulation of quorum sensing Inhibits morphogenetic switching Induces apoptosis Inhibits biofilm formation Reduces phospholipase activity Reduces proteinase activity Synergism with FLC	Oliveira et al. (2016) Singh et al. (2015) Singh et al. (2015) Singh et al. (2015) Singh et al. (2015) Singh et al. (2015) Singh et al. (2015) Gao et al. (2016a, b)

(continued)

Table 17.1 (continued)

Name of the class	Name of the compound	Mode of action	References
	Chalcones	Inhibits morphogenetic switching Inhibits biofilm formation Synergism with FLC Targets efflux pumps	Messier et al. (2011) Wang et al. (2016) Songuigama et al. (2018) Nim et al. (2018)
Lignans	Magnolol	Inhibits morphogenetic switching Reduces cell adherence Inhibits biofilm formation Synergism with FLC Targets efflux pump Inhibits ergosterol biosynthesis Disrupts cell membrane integrity	Sun et al. (2015a, b) Sun et al. (2015a, b) Sun et al. (2015a, b) Behbehani et al. (2017) Sun et al. (2015a, b) Behbehani et al. (2017) Behbehani et al. (2017)
	Honokiol	Synergism with FLC Inhibits morphogenetic switching Reduces cell adherence Inhibits biofilm formation Induces apoptosis Induces oxidative stress Mitochondrial dysfunction Disrupts cell membrane	Jin et al. (2010) Sun et al. (2015a, b) Sun et al. (2015a, b) Sun et al. (2017) Sun et al. (2017) Sun et al. (2017) Sun et al. (2017) Sun et al. (2017)
	Lariciresinol	Cell membrane permeabilization Lipid depolarization	Bajpai et al. (2017) Pinto et al. (2009)
	Medioresinol	Synergism with chloramphenicol Inhibits biofilm formation Cell cycle arrest ROS generation Mitochondrial dysfunction Phosphatidylserine externalization DNA, nuclear fragmentation Induces apoptosis	Hwang et al. (2012) Hwang et al. (2012) Hwang et al. (2012) Hwang et al. (2012) Hwang et al. (2012) Hwang et al. (2012) Hwang et al. (2012)
	Nyasol	Synergism with miconazole Synergism with	Iida et al. (2000) Iida et al. (2000)

(continued)

Table 17.1 (continued)

Name of the class	Name of the compound	Mode of action	References
		clotrimazole Synergism with ketoconazole Reduces planktonic growth	Iida et al. (2000) Iida et al. (2000)
Coumarins	Osthole	Synergism with FLC ROS generation	Li et al. (2017a, b) Li et al. (2017a, b)
	Robustic acid	Inhibits sterol synthesis	Ayine-Tora et al. (2016)
	Phosphoramidate derivatives	Inhibits chitin synthesis	Ji et al. (2016)
Xanthones	Alpha-mangostin	Inhibits biofilm formation Inhibits ergosterol biosynthesis	Kaomongkolgit et al. (2009) Pinto et al. (2011)

Conflict of Interest None to declare

References

Adlercreutz H (2007) Lignans and human health. Crit Rev Clin Lab Sci 44:483–525

Ahmad A, Wani MY, Patel M, Sobral AJFN, Duse AG, Aqlan FM, Al-Bogami AS (2017) Synergistic antifungal effect of cyclized chalcone derivatives and fluconazole against *Candida albicans*. Medchemcomm 8:2195–2207

Alam MA, Subhan N, Hossain H, Hossain M, Reza HM, Rahman MM, Ullah MO (2016) Hydroxycinnamic acid derivatives: a potential class of natural compounds for the management of lipid metabolism and obesity. Nutr Metab (Lond) 11:13–27

Alavarce RA, Saldanha LL, Almeida NL, Porto VC, Dokkedal AL, Lara VS (2015) The beneficial effect of *Equisetum giganteum* L. against Candida biofilm formation: new approaches to denture stomatitis. Evid Based Complement Alternat Med 2015:939625

Alem MA, Douglas LJ (2004) Effects of aspirin and other nonsteroidal anti-inflammatory drugs on biofilms and planktonic cells of *Candida albicans*. Antimicrob Agents Chemother 48:41–47

Alnuaimi AD, O'Brien-Simpson NM, Reynolds EC, McCullough MJ (2013) Clinical isolates and laboratory reference Candida species and strains have varying abilities to form biofilms. FEMS Yeast Res 13:689–699

Ansari MA, Anurag A, Fatima Z, Hameed S (2013) Natural phenolic compound: A potential antifungal target. In: Mendez-Vilas A (ed) Microbial pathogens and strategies for combating them: science, technology and education. Formatex Research Center, Badajoz, pp 1189–1195

Ayine-Tora DM, Kingsford-Adaboh R, Asomaning WA, Harrison JJ, Mills-Robertson FC, Bukari Y, Sakyi PO, Kaminta S, Reynisson J (2016) Coumarin antifungal Lead compounds from *Millettia thonningii* and their predicted mechanism of action. Molecules 21:E1369

Bajpai VK, Shukla S, Paek WK, Lim J, Kumar P, Kumar P, Na M (2017) Efficacy of (+)-Lariciresinol to control bacterial growth of *Staphylococcus aureus* and *Escherichia coli* O157: H. Front Microbiol 8:804

Bang KH, Kim YK, Min BS, Na MK, Rhee YH, Lee JP, Bae KH (2000) Antifungal activity of magnolol and honokiol. Arch Pharm Res 23:46–49

Batovska D, Parushev S, Slavova A, Bankova V, Tsvetkova I, Ninova M, Najdenski H (2007) Study on the substituents' effects of a series of synthetic chalcones against the yeast *Candida albicans*. Eur J Med Chem 42:87–92

Behbehani J, Shreaz S, Irshad M, Karched M (2017) The natural compound magnolol affects growth, biofilm formation, and ultrastructure of oral Candida isolates. Microb Pathog 113:209–217

Brighenti FL, Salvador MJ, Gontijo AVL, Delbem ACB, Delbem ÁCB, Soares CP, de Oliveira MAC, Girondi CM, Koga-Ito CY (2017) Plant extracts: initial screening, identification of bioactive compounds and effect against *Candida albicans* biofilms. Future Microbiol 12:15–27

Calixto Júnior JT, Morais SM, Martins CG, Vieira LG, Morais-Braga MF, Carneiro JN, Machado AJ, Menezes IR, Tintino SR, Coutinho HD (2015) Phytochemical analysis and modulation of antibiotic activity by Luehea paniculata Mart. & Zucc. (Malvaceae) in multiresistant clinical isolates of Candida spp. Biomed Res Int 2015:807670

Canturk Z (2018) Evaluation of synergistic anticandidal and apoptotic effects of ferulic acid and caspofungin against *Candida albicans*. J Food Drug Anal 1:439–443

Cao Y, Dai B, Wang Y, Huang S, Xu Y, Cao Y, Gao P, Zhu Z, Jiang Y (2008) In vitro activity of baicalein against *Candida albicans* biofilms. Int J Antimicrob Agents 32:73–77

Chang B, Lee Y, Ku Y, Bae K, Chung C (1998) Antimicrobial activity of magnolol and honokiol against periodontopathic microorganisms. Planta Med 64:367–369

Cheah HL, Lim V, Sandai D (2014) Inhibitors of the glyoxylate cycle enzyme ICL1 in *Candida albicans* for potential use as antifungal agents. PLoS One 9:e95951

Chen C (2016) Sinapic acid and its derivatives as medicine in oxidative stress-induced diseases and aging. Oxidative Med Cell Longev 2016:3571614

Chen AY, Chen YC (2013) A review of the dietary flavonoid, kaempferol on human health and cancer chemoprevention. Food Chem 138:2099–2107

Chen M, Zhai L, Arendrup MC (2015) In vitro activity of 23 tea extractions and epigallocatechin gallate against Candida species. Med Mycol 53:194–198

Chen Y, Liu T, Wang K, Hou C, Cai S, Huang Y, Du Z, Huang H, Kong J, Chen Y (2016) Baicalein inhibits *Staphylococcus aureus* biofilm formation and the quorum sensing system in vitro. PLoS One 11:e0153468

Chingwaru W, Duodu G, Zyl YV, Schoeman CJ, Majinda RT, Yeboah SO, Jackson J, Kapewangolo PT, Kandawa-Schulz M, Minnaar A, Cencic A (2011) Antibacterial and anticandidal activity of *Tylosema esculentum* (marama) extracts. S Afr J Sci 107:1–11

Cui S, Qian J, Bo P (2013) Inhibitive effect on phagocytosis of *Candida albicans* induced by pretreatment with quercetin via actin cytoskeleton interference. J Tradit Chin Med 33:804–809

da Costa Cordeiro BMP, de Lima Santos ND, Ferreira MRA, de Araújo LCC, Junior ARC, da Conceição Santos AD, de Oliveira AP, da Silva AG, da Silva Falcão EP, Dos Santos Correia MT, da Silva Almeida JRG, da Silva LCN, Soares LAL, Napoleão TH, da Silva MV, Paiva PMG (2018) Hexane extract from Spondias tuberosa (Anacardiaceae) leaves has antioxidant activity and is an anti-Candida agent by causing mitochondrial and lysosomal damages. BMC Complement Altern Med 18:284

da Silva CR, de Andrade Neto JB, de Sousa CR, Figueiredo NS, Sampaio LS, Magalhães HI, Cavalcanti BC, Gaspar DM, de Andrade GM, Lima IS, de Barros Viana GS, de Moraes MO, Lobo MD, Grangeiro TB, Nobre Júnior HV (2014) Synergistic effect of the flavonoid catechin, quercetin, or epigallocatechin gallate with fluconazole induces apoptosis in *Candida tropicalis* resistant to fluconazole. Antimicrob Agents Chemother 58:1468–1478

Dai BD, Wang Y, Zhao LX, Li DD, Li MB, Cao YB, Jiang YY (2013) Cap1p attenuates the apoptosis of *Candida albicans*. FEBS J 280:2633–2643

De Leo M, Braca A, De Tommasi N, Norscia I, Morelli I, Battinelli L, Mazzanti G (2004) Phenolic compounds from *Baseonema acuminatum* leaves: isolation and antimicrobial activity. Planta Med 70:841–846

De Vita D, Friggeri L, D'Auria FD, Pandolfi F, Piccoli F, Panella S, Palamara AT, Simonetti G, Scipione L, Di Santo R, Costi R, Tortorella S (2014) Activity of caffeic acid derivatives against *Candida albicans* biofilm. Bioorg Med Chem Lett 24:1502–1505

Deva R, Ciccoli R, Kock L, Nigam S (2001) Involvement of aspirin-sensitive oxylipins in vulvovaginal candidiasis. FEMS Microbiol Lett 198:37–43

Dey G, Chakraborty M, Mitra A (2005) Profiling C6-C3 and C6-C1 phenolic metabolites in *Cocos nucifera*. J Plant Physiol 162:375–381

Donald G, Hertzer K, Eibl G (2012) Baicalein--an intriguing therapeutic phytochemical in pancreatic cancer. Curr Drug Targets 13:1772–1776

El-Seedi HR, El-Said AM, Khalifa SA, Göransson U, Bohlin L, Borg-Karlson AK, Verpoorte R (2012) Biosynthesis, natural sources, dietary intake, pharmacokinetic properties, and biological activities of hydroxycinnamic acids. J Agric Food Chem 60:10877–10895

Evensen NA, Braun PC (2009) The effects of tea polyphenols on *Candida albicans*: inhibition of biofilm formation and proteasome inactivation. Can J Microbiol 55:1033–1039

Farber BF, Wolff AG (1993) Salicylic acid prevents the adherence of bacteria and yeast to silastic catheters. J Biomed Mater Res 27:599–602

Faria NC, Kim JH, Gonçalves LA, Martins Mde L, Chan KL, Campbell BC (2011) Enhanced activity of antifungal drugs using natural phenolics against yeast strains of Candida and Cryptococcus. Lett Appl Microbiol 52:506–513

Fitzgerald DJ, Stratford M, Gasson MJ, Ueckert J, Bos A, Narbad A (2004) Mode of antimicrobial action of vanillin against *Escherichia coli, Lactobacillus plantarum* and *Listeria innocua*. J Appl Microbiol 97:104–113

Fu Z, Lu H, Zhu Z, Yan L, Jiang Y, Cao Y (2011) Combination of baicalein and Amphotericin B accelerates *Candida albicans* apoptosis. Biol Pharm Bull 34:214–218

Gao Y, Snyder SA, Smith JN, Chen YC (2016a) Anticancer properties of baicalein: a review. Med Chem Res 25:1515–1523

Gao M, Wang H, Zhu L (2016b) Quercetin assists fluconazole to inhibit biofilm formations of fluconazole-resistant *Candida albicans* in in vitro and in vivo antifungal managements of Vulvovaginal candidiasis. Cell Physiol Biochem 40:727–742

Hubková B, Veliká B, Birková A, Guzy J, Mareková M (2014) Hydroxybenzoic acids and their derivatives as peroxynitrite scavengers. Free Radic Biol Med 75:S33–S34

Hwang EI, Lee YM, Lee SM, Yeo WH, Moon JS, Kang TH, Park KD, Kim SU (2007) Inhibition of chitin synthase 2 and antifungal activity of lignans from the stem bark of Lindera erythrocarpa. Planta Med 73:679–682

Hwang B, Cho J, Hwang IS, Jin HG, Woo ER, Lee DG (2011) Antifungal activity of lariciresinol derived from *Sambucus williamsii* and their membrane-active mechanisms in *Candida albicans*. Biochem Biophys Res Commun 410:489–493

Hwang JH, Hwang IS, Liu QH, Woo ER, Lee DG (2012) (+)-Medioresinol leads to intracellular ROS accumulation and mitochondria-mediated apoptotic cell death in *Candida albicans*. Biochimie 94:1784–1793

Hwang JH, Choi H, Hwang IS, Kim AR, Woo ER, Lee DG (2013) Synergistic antibacterial and antibiofilm effect between (+)-medioresinol and antibiotics in vitro. Appl Biochem Biotechnol 170:1934–1941

Ibezimako EC, Esimone CO, Ofokansi KC, Oragui CI (2003) Evaluation of the antifungal properties of nystatin-salicylic acid combinations against clinical isolates of *Candida albicans*. J Biomed Invest 2003:10–17

Iida Y, Oh KB, Saito M, Matsuoka H, Kurata H (2000) In vitro synergism between nyasol, an active compound isolated from *Anemarrhena asphodeloides*, and azole agents against *Candida albicans*. Planta Med 66:435–438

Ji Q, Ge Z, Ge Z, Chen K, Wu H, Liu X, Huang Y, Yuan L, Yang X, Liao F (2016) Synthesis and biological evaluation of novel phosphoramidate derivatives of coumarin as chitin synthase inhibitors and antifungal agents. Eur J Med Chem 108:166–176

Jin J, Guo N, Zhang J, Ding Y, Tang X, Liang J, Li L, Deng X, Yu L (2010) The synergy of honokiol and fluconazole against clinical isolates of azole-resistant *Candida albicans*. Lett Appl Microbiol 51:351–357

Kakkar S, Bais S (2014) A review on protocatechuic acid and its pharmacological potential. ISRN Pharmacol 2014:952943

Kaomongkolgit R, Jamdee K (2017) Inhibitory effect of alpha-mangostin on Candida biofilms. Odontology 105:248–253

Kaomongkolgit R, Jamdee K, Chaisomboon N (2009) Antifungal activity of alpha-mangostin against *Candida albicans*. J Oral Sci 51:401–406

Kelly GS (2011) Quercetin. Monograph Altern Med Rev 16:172–194

Kelly C, Jones O, Barnhart C, Lajoie C (2008) Effect of furfural, vanillin and syringaldehyde on *Candida guilliermondii* growth and xylitol biosynthesis. Appl Biochem Biotechnol 148:97–108

Kim J, Sudbery P (2011) *Candida albicans*, a major human fungal pathogen. J Microbiol 49:171–177

Kontogiorgis C, Hadjipavlou-Litina D (2003) Biological evaluation of several coumarin derivatives designed as possible anti-inflammatory/antioxidant agents. J Enzyme Inhib Med Chem 18:63–69

Kozyra M, Biernasiuk A, Malm A, Chowaniec M (2015) Chemical compositions and antibacterial activity of extracts obtained from the inflorescences of Cirsium canum (L.) all. Nat Prod Res 29:2059–2063

Ksouri R, Ksouri WM, Jallali I, Debez A, Magné C, Hiroko I, Abdelly C (2012) Medicinal halophytes: potent source of health promoting biomolecules with medical, nutraceutical and food applications. Crit Rev Biotechnol 32:289–326

Kuete V, Nana F, Ngameni B, Mbaveng AT, Keumedjio F, Ngadjui BT (2009) Antimicrobial activity of the crude extract, fractions and compounds from stem bark of Ficus ovata (Moraceae). J Ethnopharmacol 124:556–561

Kumar S, Pandey AK (2013) Chemistry and biological activities of flavonoids: an overview. Sci World J 2013:162750

Kuwahara H, Kanazawa A, Wakamatu D, Morimura S, Kida K, Akaike T, Maeda H (2004) Antioxidative and antimutagenic activities of 4-vinyl-2,6-dimethoxyphenol (canolol) isolated from canola oil. J Agric Food Chem 52:4380–4387

Lamarra J, Giannuzzi L, Rivero S, Pinotti A (2017) Assembly of chitosan support matrix with gallic acid-functionalized nanoparticles. Mater Sci Eng C Mater Biol Appl 79:848–859

Lee YJ, Lee YM, Lee CK, Jung JK, Han SB, Hong JT (2011) Therapeutic applications of compounds in the Magnolia family. Pharmacol Ther 130:157–176

Lesage-Meessen L, Delattre M, Haon M, Thibault JF, Ceccaldi BC, Brunerie P, Asther M (1996) A two-step bioconversion process for vanillin production from ferulic acid combining *Aspergillus niger* and *Pycnoporus cinnabarinus*. J Biotechnol 50:107–113

Li ZJ, Liu M, Dawuti G, Dou Q, Ma Y, Liu HG, Aibai S (2017a) Antifungal activity of Gallic acid in vitro and in vivo. Phytother Res 31:1039–1045

Li DD, Chai D, Huang XW, Guan SX, Du J, Zhang HY, Sun Y, Jiang YY (2017b) Potent in vitro synergism of fluconazole and Osthole against fluconazole-resistant *Candida albicans*. Antimicrob Agents Chemother 61:e00436–e00417

Liao K, Sun L (2018) Roles of the Hsp90-Calcineurin pathway in the antifungal activity of Honokiol. J Microbiol Biotechnol 28:1086–1093

Lim H, Nam JW, Seo EK, Kim YS, Kim HP (2009) (−)-Nyasol (cis-hinokiresinol), a norneolignan from the rhizomes of *Anemarrhena asphodeloides*, is a broad spectrum inhibitor of eicosanoid and nitric oxide production. Arch Pharm Res 32:1509–1514

Matsumoto Y, Kaihatsu K, Nishino K, Ogawa M, Kato N, Yamaguchi A (2012) Antibacterial and antifungal activities of new acylated derivatives of epigallocatechin gallate. Front Microbiol 3:53

Messier C, Epifano F, Genovese S, Grenier D (2011) Inhibition of *Candida albicans* biofilm formation and yeast-hyphal transition by 4-hydroxycordoin. Phytomedicine 18:380–383

Montagner C, de Souza SM, Groposoa C, Delle Monache F, Smânia EF, Smânia A Jr (2008) Antifungal activity of coumarins. Z Naturforsch C 63:21–28

Nim S, Baghel P, Tran-Nguyen VK, Peres B, Nguyen KA, Pietro AD, Falson P, Prasad R, Boumendjel A (2018) Make azoles active again: chalcones as potent reversal agents of transporters-mediated resistance in *Candida albicans*. Future Med Chem 10:2177–2186

Ning Y, Ling J, Wu CD (2015) Synergistic effects of tea catechin epigallocatechin gallate and antimycotics against oral Candida. Arch Oral Biol 60:1565–1570

Oliveira VM, Carraro E, Auler ME, Khalil NM (2016) Quercetin and rutin as potential agents antifungal against Cryptococcus spp. Braz J Biol 76:1029–1034

Panche AN, Diwan AD, Chandra SR (2016) Flavonoids: an overview. J Nutr Sci 5:e47

Panwar R, Pemmaraju SC, Sharma AK, Pruthi V (2016) Efficacy of ferulic acid encapsulated chitosan nanoparticles against *Candida albicans* biofilm. Microb Pathog 95:21–31

Park HJ, Lee JY, Moon SS, Hwang BK (2003) Isolation and anti-oomycete activity of nyasol from Anemarrhena asphodeloides rhizomes. Phytochemistry 64:997–1001

Park C, Woo ER, Lee DG (2010) Anti-Candida property of a lignan glycoside derived from *Styrax japonica S.* et Z. via membrane-active mechanisms. Mol Cells 29:581–584

Peterson J, Dwyer J, Adlercreutz H, Scalbert A, Jacques P, McCullough ML (2010) Dietary lignans: physiology and potential for cardiovascular disease risk reduction. Nutr Rev 68:571–603

Pierce GE (2005) Pseudomonas aeruginosa, Candida albicans, and device-related nosocomial infections: implications, trends, and potential approaches for control. J Ind Microbiol Biotechnol 32:309–318

Pinto E, Vale-Silva L, Cavaleiro C, Salgueiro L (2009) Antifungal activity of the clove essential oil from *Syzygium aromaticum* on Candida, Aspergillus and dermatophyte species. J Med Microbiol 58:1454–1462

Pinto E, Afonso C, Duarte S, Vale-Silva L, Costa E, Sousa E, Pinto M (2011) Antifungal activity of xanthones: evaluation of their effect on ergosterol biosynthesis by high-performance liquid chromatography. Chem Biol Drug Des 77:212–222

Prasad R, Kapoor K (2005) Multidrug resistance in yeast Candida. Int Rev Cytol 242:215–248

Raut JS, Rajput SB, Shinde RB, Surwase BS, Karuppayil SM (2013) Vanillin inhibits growth, morphogenesis and biofilm formation by *Candida albicans*. T.B.P.A 3:130–138

Serpa R, França EJ, Furlaneto-Maia L, Andrade CG, Diniz A, Furlaneto MC (2012) In vitro antifungal activity of the flavonoid baicalein against Candida species. J Med Microbiol 61:1704–1708

Shao J, Zhang M, Wang T, Li Y, Wang C (2016) The roles of CDR1, CDR2, and MDR1 in kaempferol-induced suppression with fluconazole-resistant *Candida albicans*. Pharm Biol 54:984–992

Shirley KP, Windsor LJ, Eckert GJ, Gregory RL (2017) In vitro effects of Plantago Major extract, Aucubin, and Baicalein on *Candida albicans* biofilm formation, metabolic activity, and cell surface hydrophobicity. J Prosthodont 26:508–515

Singh BN, Upreti DK, Singh BR, Pandey G, Verma S, Roy S, Naqvi AH, Rawat AK (2015) Quercetin sensitizes fluconazole-resistant Candida albicans to induce apoptotic cell death by modulating quorum sensing. Antimicrob Agents Chemother 59:2153–2168

Song Z, Sun J, Deng P, Zhou F, Xu H, Wen Y, Teng F, Ge D, Feng R (2018) Oligochitosan-pluronic 127 conjugate for delivery of honokiol. Artif Cells Nanomed Biotechnol 5:1–11

Songuigama C, Jean-Paul ND, Mamidou KW, Drissa S, Mahama Q (2018) Anticandidosic activities of new Chalcones vectorised by Benzimidazole against strain of *Candida Albicans* Pharmacoresistance to azoles. IOSR-JPBS 13:11–16

Stalmach A (2014) Bioavailability of dietary anthocyanins and hydroxycinnamic acids trans. In: Watson RR, Preedy VR, Zibadi S (eds) . Polyphenols in Human Health and Disease Academic, San Diego, pp 561–576

Stepanović S, Vuković D, Jesić M, Ranin L (2004) Influence of acetylsalicylic acid (aspirin) on biofilm production by Candida species. J Chemother 16:134–138

Sun L, Liao K, Wang D (2015a) Effects of magnolol and honokiol on adhesion, yeast-hyphal transition, and formation of biofilm by *Candida albicans*. PLoS One 10:e0117695

Sun LM, Liao K, Liang S, Yu PH, Wang DY (2015b) Synergistic activity of magnolol with azoles and its possible antifungal mechanism against *Candida albicans*. J Appl Microbiol 118:826–838

Sun L, Liao K, Wang D (2017) Honokiol induces superoxide production by targeting mitochondrial respiratory chain complex I in *Candida albicans*. PLoS One 12:e0184003

Sun L, Liao K, Hang C (2018) Caffeic acid phenethyl ester synergistically enhances the antifungal activity of fluconazole against resistant *Candida albicans*. Phytomedicine 40:55–58

Tatsimo SJ, Tamokou Jde D, Havyarimana L, Csupor D, Forgo P, Hohmann J, Kuiate JR, Tane P (2012) Antimicrobial and antioxidant activity of kaempferol rhamnoside derivatives from *Bryophyllum pinnatum*. BMC Res Notes 5:158

Teixeira J, Gaspar A, Garrido EM, Garrido J, Borges F (2013) Hydroxycinnamic acid antioxidants: an electrochemical overview. Biomed Res Int 2013:251754

Teodoro GR, Ellepola K, Seneviratne CJ, Koga-Ito CY (2015a) Potential use of phenolic acids as anti-Candida agents: A review. Front Microbiol 6:1420

Teodoro GR, Brighenti FL, Delbem AC, Delbem ÁC, Khouri S, Gontijo AV, Pascoal AC, Salvador MJ, Koga-Ito CY (2015b) Antifungal activity of extracts and isolated compounds from *Buchenavia tomentosa* on *Candida albicans* and non-albicans. Future Microbiol 10:917–927

Teodoro GR, Gontijo AVL, Borges AC, Tanaka MH, Lima GMG, Salvador MJ, Koga-Ito CY (2017) Gallic acid/hydroxypropyl-β-cyclodextrin complex: improving solubility for application on in vitro/ in vivo *Candida albicans* biofilms. PLoS One 12:e0181199

Teodoro GR, Gontijo AVL, Salvador MJ, Tanaka MH, Brighenti FL, Delbem ACB, Delbem ÁCB, Koga-Ito CY (2018) Effects of acetone fraction from Buchenavia tomentosa aqueous extract and Gallic acid on *Candida albicans* biofilms and virulence factors. Front Microbiol 9:647

Tian RR, Pan QH, Zhan JC, Li JM, Wan SB, Zhang QH, Huang WD (2009) Comparison of phenolic acids and Flavan-3-ols during wine fermentation of grapes with different harvest times. Molecules 14:827–838

Trofa D, Agovino M, Stehr F, Schäfer W, Rykunov D, Fiser A, Hamari Z, Nosanchuk JD, Gácser A (2009) Acetylsalicylic acid (aspirin) reduces damage to reconstituted human tissues infected with Candida species by inhibiting extracellular fungal lipases. Microbes Infect 11:1131–1139

Wang YH, Dong HH, Zhao F, Wang J, Yan F, Jiang YY, Jin YS (2016) The synthesis and synergistic antifungal effects of chalcones against drug resistant *Candida albicans*. Bioorg Med Chem Lett 26:3098–3102

Woodbury A, Yu SP, Wei L, García P (2013) Neuro-modulating effects of honokiol: a review. Front Neurol 4:130

Yücesoy M, Oktem IM, Gülay Z (2000) In-vitro synergistic effect of fluconazole with nonsteroidal anti-inflammatory agents against *Candida albicans* strains. J Chemother 12:385–389

Zhou P, Fu J, Hs H, Liu X (2017) In vitro inhibitory activities of magnolol against Candida spp. Drug Des Devel Ther 11:2653–2661

Zou Y, Kim AR, Kim JE, Choi JS, Chung HY (2002) Peroxynitrite scavenging activity of sinapic acid (3,5-dimethoxy-4-hydroxycinnamic acid) isolated from *Brassica juncea*. J Agric Food Chem 50:5884–5890

Zubricka D, Mišianikova A, Henzelyova J, Valletta A, De Angelis G, D'Auria FD, Simonetti G, Pasqua G, Čellarova E (2015) Xanthones from roots, hairy roots and cell suspension cultures of selected Hypericum species and their antifungal activity against *Candida albicans*. Plant Cell Rep 34:1953–1962

Plant Phenolics: Their Biosynthesis, Regulation, Evolutionary Significance, and Role in Senescence

18

Kamal Tyagi, Pawan Shukla, Gulab Khan Rohela, Aftab Ahmad Shabnam, and Ranjana Gautam

Abstract

Phenolics are one of the most abundant groups of secondary metabolites found throughout the plant kingdom. Phenolic compounds are widely distributed in plant tissues, particularly contributing to color, flavor, taste, and astringency to fruits. Phenolics are aromatic benzene ring compounds with one or more hydroxyl groups attached to it. Phenolics constitute most likely the largest group of plant secondary metabolites, varying in size from a simple structure with an aromatic ring to complex ones such as lignins. Most phenolic compounds are thought to be by-products of the metabolism of the aromatic amino acid phenylalanine. Phenolics are considered to play a vital role in the evolution of plants by providing an adaptive advantage to cope up with the changing environments. Furthermore, they are believed to be involved in leaf and petal senescence. Change in the leaf or petal color during senescence is also due to anthocyanin pigments, a kind of plant phenolics.

Keywords

Mulberry · Plant phenolics · Senescence · Anthocyanin

K. Tyagi
Department of Viticulture and Enology, University of California, Davis, CA, USA

P. Shukla (✉) · G. K. Rohela · A. A. Shabnam
Biotechnology Section, Moriculture Division, Central Sericultural Research and Training Institute, Central Silk Board, Ministry of Textiles, Government of India, Pampore, India

R. Gautam
Department of Plant Sciences, University of Hyderabad, Hyderabad, Telangana, India

R. Lone et al. (eds.), *Plant Phenolics in Sustainable Agriculture*,
https://doi.org/10.1007/978-981-15-4890-1_18

18.1 Introduction

The plant phenolics play an important role in the survival of plants by providing specific adaptation to the changing environmental conditions. Phenolics are defined as the molecule that is not directly involved in the internal economy of the plant that produces it. They are essential for plant survival and defense mechanisms and act like an immune system of the plants. Phenolics are diverse compounds that differ from species to species and from tissue to tissue. They are produced by plants and compartmented in plant vacuole. They occur ubiquitously in plants in a variety of tissues. The characteristic feature of all phenolics is the presence of one or more benzene rings with one or more hydroxyl groups that may be variously elaborated with methyl, methoxyl, amino, or glycosyl groups. Phenolics size is ranging from the simple phenol (C_6H_6O) of 94.1 g/mol to large complex molecules such as tannins having more than 1000 g/mol (Hagerman et al. 1998). Based on their chemical composition, their function in plants is important in plant development, growth, and defense mechanisms (Coley et al. 1985). They provide an adaptive advantage to the plants itself that produce it, e.g., wade off competition, antigerminative or toxic for other plants (allelopathic), better adapted to harsh environments, and able to combat feeding animals such as insects or cattle (antifeedant) (Delgoda and Murray 2017). Several conceivable and clearly functional roles for some of these compounds have been proposed and documented (Beckman 2000). They include the following:

(1) Protection from injurious UV radiation
(2) Coloration of flowers that attract pollinating animals
(3) Deterrence to grazing animals and feeding insects because of their astringent, toxic nature
(4) Providing a tanglefoot mechanism for trapping feeding insects
(5) Resistance to pathogens
(6) Volatiles that affect the growth of other neighboring plants

Thus, the primary established roles are clearly ecological in nature, some having dual or even multiple functions (Beckman 2000). In addition to their role in defense, plant growth, and development, plant phenolics also have taxonomic and evolutionary importance. The metabolic pathway that produces phenolics and their link with taxonomic distribution displayed that phenolics play a crucial role in gradual diversification of plants to suit varying lifestyles during plant evolution (Delgoda and Murray 2017). Indeed, when the first plants moved from aquatic life to terrestrial life form, they faced many different stressful conditions, such as ultraviolet (UV) radiation which is involved in damaging of the biomolecules (Caretto et al. 2015). To cope up with this condition, land plants started synthesizing the phenolic compounds such as phenylpropanoids include flavonoids, monolignols, phenolic acids, stilbenes, and caumarins which act as sunscreens for the plants (Caretto et al. 2015).

The sequential action of different biosynthetic pathways in plants leads to the production of phenolics. The glycolysis and pentose phosphate pathways provide precursors (phosphoenolpyruvate and erythrose 4-phosphate, respectively) to the shikimate pathway (Caretto et al. 2015). Phenylalanine, produced by the shikimate pathway, is the precursor of phenylpropanoid metabolism which, in turn, feeds the diverse specific flavonoid pathways (Caretto et al. 2015). The biosynthesis of phenolics is an energetically expensive process even though plants produce it due to their advantageous nature (Delgoda and Murray 2017). To make this process more favorable to plants, the genes involved in the production of these molecules reside next to each other including the resistance genes that demonstrate the natural selection of it (Delgoda and Murray 2017). They are not merely artifacts of isolation procedures and do in fact exist in nature (Delgoda and Murray 2017).

18.2 Biosynthesis of Phenolics

Plants have developed the ability to produce large number of secondary metabolites that are not required in the primary growth and development but are of vital importance for their interaction with the environment (Cheynier et al. 2013). This "chemodiversity" is mainly developed in land plants, which have to encounter numerous environmental challenges, and in particular in vascular plants that also need to maintain efficient transport of the metabolites, structural rigidity, and a fine regulation of homeostasis (Caputi et al. 2012). Several classes of phenolics have been categorized on the basis of their basic skeletons like C6 (phenol), C6-C1 (phenolic acid and aldehyde), C6-C2 (phenylacetic acid), C6-C3 (hydroxycinnamic acids), C6-C1-C6 (xanthones), C6-C2-C6 (stilbenes), and C6-C3-C6 (flavonoids) (Cheynier et al. 2013). Phenolics are formed by three different biosynthetic pathways: (i) the shikimate/chorismate or succinylbenzoate pathway, which produces the phenylpropanoid derivatives (C6-C3); (ii) the acetate/malonate or polyketide pathway, which produces the side-chain-elongated phenylpropanoids, including the large group of flavonoids (C6-C3-C6) and some quinones; and (iii) the acetate/mevalonate pathway, which produces the aromatic terpenoids, mostly monoterpenes, by dehydrogenation reactions (Bhattacharya et al. 2010).

Phenylpropanoid-derived metabolites are a huge class of secondary metabolites produced from primary metabolites, phenylalanine or tyrosine, through a series of enzymatic reactions. Based on chemical structures, phenylpropanoids can be divided into five groups, including flavonoids, monolignols, phenolic acids, stilbenes, and coumarins (Noel et al. 2005; Vogt 2010; Liu et al. 2015). Among them, flavonoids, monolignols, and phenolic acids are three most common groups which can be found in almost all land plants. In the past decades, phenylpropanoids have stirred a wide concern in the fields of plant biology and medicines for their essential roles in plant survival and broad biological activities in the human body. Many studies in the recent years have shown that some chronic diseases such as cardiovascular, diabetic, and some cancer can be reduced by taking phenylpropanoid-pathway-derived metabolites (Roupe et al. 2006; Ojala et al. 2000; Kumar and Pandey 2013; Venugopala et al. 2013;Yang et al. 2016; Xia et al. 2017).

The initial steps of the phenylpropanoid pathway are collectively referred to as the general phenylpropanoid pathway (GPP) (Winkel 2004; Ferrer et al. 2008; Vogt 2010). Derivation of phenylpropanoids from phenylalanine undergoes a series of enzymatic reactions to be converted to other metabolites. In the first enzymatic step of phenylpropanoid pathway, phenylalanine is deaminated by phenylalanine ammonia lyase (PAL) to yield cinnamic acid, which is then hydroxylated and transformed to p-coumaric acid by cinnamate-4-hydroxylase (C4H) in the second enzymatic step. On the contrary, in some monocots, fungi and bacterial species, tyrosine ammonia lyase (TAL), or bifunctional ammonia lyase (PTAL) can directly convert tyrosine into p-coumaric acid, which bypasses the C4H intermediate (Watts et al. 2004; Jendresen et al. 2015; Barros et al. 2016). Following this, 4-coumaroyl-CoA ligase (4CL) catalyzes the conversion of p-coumaric acid into p-coumaroyl-CoA, which is an important branch point leading to generate various phenylpropanoid compounds (Vogt 2010; Liu et al. 2015). One of the well-understood downstream branches following the GPP is the flavonoid pathway. In *Arabidopsis thaliana* and *Vitis vinifera*, the majority of enzymatic steps have been described and well understood (Appelhagen et al. 2014; Ichino et al. 2014; Petrussa et al. 2013).

Biosynthesis of flavonoids starts with one p-coumaroyl-CoA and three malonyl-CoA molecules by committed enzymatic step and chalcone synthase (CHS) resulting in the production of naringenin chalcone (Falcone Ferreyra et al. 2012). In the second enzymatic step, chalcone isomerase (CHI) catalyzes the stereospecific cyclization of naringenin chalcone or isoliquiritigenin chalcone to flavanone (naringenin flavanone or isoliquiritigenin flavanone) (Ngaki et al. 2012). Naringenin flavanone is a common intermediate of numerous flavonoid subgroups, including flavones, flavanonols, flavonols, anthocyanins, proanthocyanidins (PA), and isoflavones. However, isoliquiritigenin flavanone can be only converted to legume-specific isoflavonoids (Ngaki et al. 2012).

Dihydroflavonols, also called flavanonols or 3-hydroxy-flavanone, possess a hydroxyl group at C3 position, but there is no C2-C3 double bond. Dihydrokaempferol possesses a hydroxyl group at 4′ position of the B-ring. Dihydroquercetin has two hydroxyl groups, which are located at 3′ and 4′ positions, respectively. Dihydromyricetin has three hydroxyl groups located at 3′, 4′, and 5′ positions, respectively. Dihydrokaempferol is synthesized by hydroxylation at C3 position of naringenin flavanone under the catalysis of a 2-ODD superfamily member, flavanone 3-hydroxylase (F3H) (Turnbull et al. 2004). Dihydrokaempferol can be subsequently hydroxylated into dihydroquercetin or dihydromyricetin by P450 superfamily members flavonoid 3′-hydroxylase (F3'H) or flavonoid 3′5'--hydroxylase (F3'5'H), respectively. F3'H catalyzes the hydroxylation at C3' position of B-ring, whereas F3'5'H catalyzes the hydroxylation at both C3' and C5' positions. Except for dihydrokaempferol, naringenin flavanone can also be direct substrate of F3'H and F3'5'H, which convert it into eriodictyol and pentahydroxyflavone, respectively. The products are subsequently hydroxylated to dihydroquercetin or dihydromyricetin by F3H (Petrussa et al. 2013). Interestingly, dihydroflavonols are main branch-point intermediates of the flavonoid pathway. One branch leads to the production of flavonols, and another one flows into the biosynthesis of anthocyanins and proanthocyanidins.

Flavonols are the most abundant flavonoids in plants. The 3-hydroxyflavone backbone of flavonols is characterized by a hydroxyl group at C3 position. Flavonol synthase (FLS) is a key enzyme in the flavonol branch, and under the catalysis of FLS, different dihydroflavonols (dihydrokaempferol, dihydroquercetin, and dihydromyricetin) can be oxidized into the corresponding flavonols (kaempferol, quercetin, and myricetin) (Petrussa, et al. 2013; Saito et al. 2013).

Anthocyanins are one of the classes of flavonoids accountable for the color of grape berries. Anthocyanins are the glycosylated products of aglycone, known as anthocyanidins. Six different types of anthocyanidins are found in nature, and they include cyanidin (Cy), pelargonidin (Pg), delphinidin (Dp), malvidin (Mv), peonidin (Pn), and petunidin (Pt). The first committed step directed into the biosynthesis of Cy, Pg, and Dp is catalyzed by dihydroflavonol reductase (DFR) with its cofactor, NAPDH (Petrussa et al. 2013; Shi and Xie 2014). DFR competes with FLS on the substrate (dihydroflavonols) and catalyzes the reduction of the 4-keto group of dihydroflavonol to leucoanthocyanidins (flavan-3,4-diols), such as leucocyanidin, leucopelargonidin, and leucodelphinidin. The second step in this branch is catalyzed by anthocyanidin synthase (ANS), which is also known as leucoanthocyanidin dioxygenase, LDOX. ANS/LDOX is one of the four 2-ODD enzymes in the flavonoid pathway (Turnbull et al. 2004; Cheng et al. 2014). The other three 2-ODDs are F3H, FLS, and FNSI. Using 2-oxoglutarate and oxygen as co-substrates, ANS/LDOX oxidizes leucoanthocyanidins into the corresponding anthocyanidins (Cy, Pg, and Dp). UDP glucose/flavonoid-3-O-glucosyltransferase (UFGT) is critical for conversion of anthocyanidin to anthocyanin biosynthesis (Zheng et al. 2013).

Proanthocyanidins (PA), also known as condensed tannins (CTs), are named because they release anthocyanidins on acid hydrolysis. They are usually found in seed coat but also exist in leaves, flowers, and stems (Dixon et al. 2013). Based on the cis- or trans-stereochemistry at C2 and C3 positions of the C-ring, flavan-3-ols can be divided into 2,3-trans-flavanols and 2,3-cis-flavanols, which are represented by (+)-catechin and (−)-epicatechin, respectively. 2,3-trans-flavanols and 2,3-cis-flavanols share the same upstream biosynthetic pathway with anthocyanins but have different downstream branches catalyzed by distinct enzymes. NADPH-dependent leucoanthocyanidin reductase (LAR) competes with ANS/LDOX on the leucoanthocyanidin (flavan-3,4-diols) substrate and catalyzes the reduction reaction to produce 2,3-trans-flavanols, such as (+)-catechin, (+)-afzelechin, and (+)-gallocatechin. In contrast, 2,3-cis-flavan-3-ols are synthesized from a distinct branch with different substrates (Tanner et al. 2003; Dixon et al. 2013; Zhou et al. 2015). NADPH-dependent anthocyanidin reductase (ANR) converts anthocyanidins (pelargonidin, cyanidin, or delphinidin) to the corresponding 2,3-cis-flavan-3-ols (e.g., (−)-epiafzelechin, (−)-epicatechin, or (−)-epigallocatechin) (Xie et al. 2003; Zhou et al. 2015). Subsequently, flavan-3-ol units are condensed into PAs. Polymerization of PA is probably catalyzed by laccase-like polyphenol oxidases, but the real enzyme and the specific clustering mechanism remain in study.

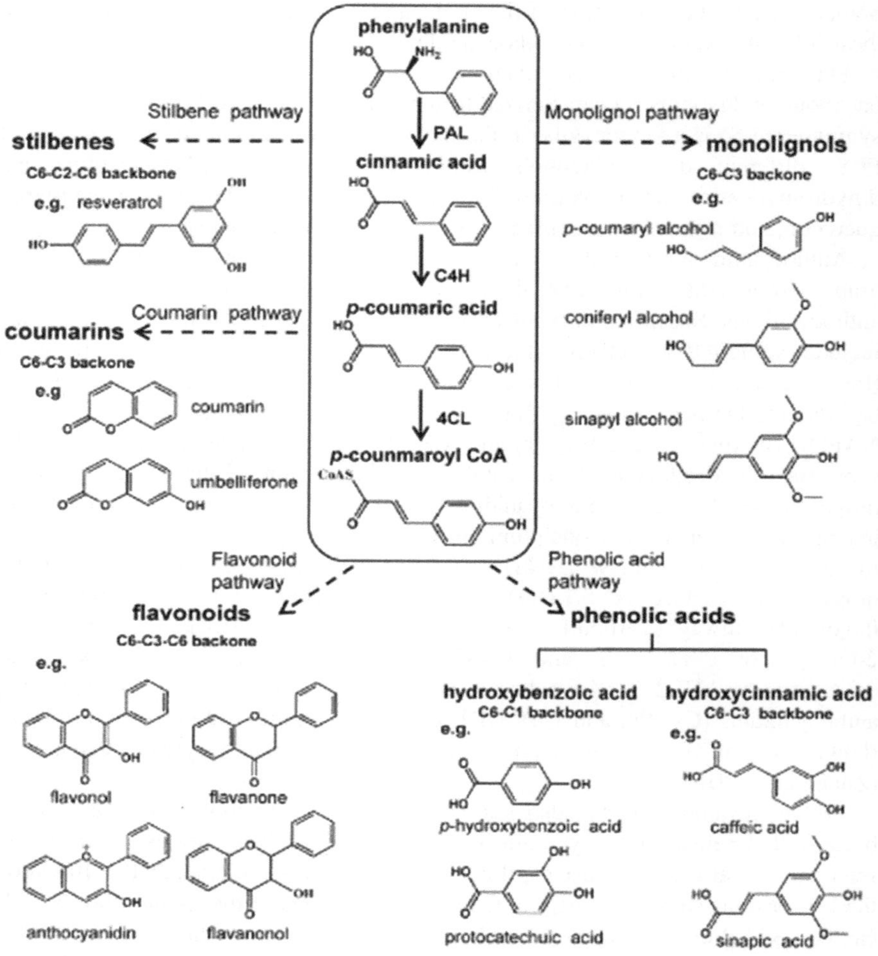

Fig. 18.1 Schematic representation of biosynthetic pathways of phenolic acids (Deng and Lu 2017). *PAL*, phenylalanine ammonia lyase; *C4H*, cinnamate 4-hydroxylase; *4CL*, 4-coumaroyl-CoA ligase

Furthermore, methylation catalyzed by O-methyltransferases and acylation and glycosylation by acyltransferases (AT) and glycosyltransferases (GTs) respectively of secondary metabolites, including phenylpropanoids and various other derived phenolic compounds, are the fundamental chemical modifications which generate enormous diversity of secondary metabolites (Tanaka et al. 2008). The complete biosynthetic pathway of phenolic acids is depicted in Fig. 18.1.

18.3 Regulation of Phenolics

The control of the synthesis of polyphenols involves a cascade of overlapping regulatory signals. These include developmental signals, such as during lignification of new growth or the production of anthocyanins during fruit and flower development and ripening, and environmental signals for protection against abiotic and biotic stresses. For some polyphenols, such as the flavonoids and lignin, there is now an excellent understanding on the regulation of biosynthetic genes.

Studies across a range of systems, including berry ripening, flower development, and stress-induced vegetative pigmentation, have found that the regulation of flavonoid production occurs principally through changes in the level of transcription of the biosynthetic genes. This is accomplished by the transcription factors (TFs). MYBs are the largest transcription factor (TF) families involved in activation or inhibition of gene transcription in the plant kingdom. Among them, R2R3MYBs are the largest MYB subfamily characterized in plants.

R2R3MYBs not only contribute in cell fate and identity, plant developmental processes, and responses to biotic and abiotic stresses but also play regulatory roles in primary and secondary metabolism, particularly in the biosynthesis of phenylpropanoids, such as flavonoids and monolignols (Jin and Martin 1999; Dubos et al. 2010). MYB-involved regulation of anthocyanin, proanthocyanidin, flavonol, and other flavonoid biosynthesis is largely conserved in plants (Hichri et al. 2011; Liu et al. 2015; Xu et al. 2015; Zhou et al. 2015). In general, MYBs involved in flavonoid biosynthesis can be divided into two types, independent MYB TFs and transcriptional complexes (MYB-bHLH-WDR complex (MBW) or MYB-bHLH complex). AtFLS, AtCHS, AtCHI, AtF3H, and AtF3'H are activated by independent MYB TFs, such as AtMYB12/PFG1 (production of flavonol glycoside 1), AtMYB11/PFG2, and AtMYB111/PFG3 that belong to subgroup 7 (Stracke et al. 2007, 2010). AtCHS and AtF3H are also activated by MYB-bHLH heterodimers consisting of AtMYB12 or AtMYB111 and a plant-specific bHLH TF, AtTCP3 (Li and Zachgo 2013). It has been reported that regulation of late biosynthetic genes such as production of anthocyanins and PAs required MBW complex (Dubos et al. 2010) which is composed of specific members of the R2R3MYB and basic helix-loop-helix (bHLH) TF families in conjunction with a WD-repeat (WDR; tryptophan-aspartic acid (W-D) dipeptide repeat) protein. Variant MBW complexes can form from different MYB and bHLH components, and these can have different target genes and vary in their activation or repression actions. The complex can activate the expression of late anthocyanin biosynthetic genes, such as AtDFR, AtANS/LDOX, and UF3GT (Gonzalez et al. 2008; Shi and Xie 2014; Xu et al. 2015, Kuhn et al. 2013). In grapes, there are positive transcriptional regulators involved in the general flavonoid pathway (VvMYB5a and VvMYB5b; Deluc et al. 2006, 2008) and flavonols (VvMYBF1; Czemmel et al. 2009), anthocyanins (VlMYBA1 and VlMYBA2; Kobayashi et al. 2004), and synthesis of tannins (VvMYBPA1; Bogs et al. 2007). In tomato fruit, anthocyanin synthesis is absent because of the bottleneck of *CHI* expression. Recently, Butelli et al. (2008) have selected two different snapdragon TF *Delia* (*Del*) which is bHLH and *Rossea 1* (*Ros1*) R2R2MYB-type TF and overexpressed in tomato (cv. Micro-Tom) using fruit-specific E8 promoter. Whereas transgenic plants show normal vegetative growth, transgenic fruit started to synthesize anthocyanin at the end of green stage and continued to accumulate these pigment during subsequent ripening ultimately reaching an intense, uniform purple coloration both in peel and the skin (Butelli et al. 2008).

The PA-related R2R3MYBs were first described for *A. thaliana*, with the R2R3MYB TT2 interacting with TT8 (bHLH) and TTG1 (WDR) to form the MBW complex (Feller et al. 2011; Hichri et al. 2011). Subsequently, the genes involved in regulation of PA biosynthesis have been characterized for other species such as grape, in which several PA-related R2R3MYBs have been identified (Bogs et al. 2007; Deluc et al. 2008; Terrier et al. 2009) and various legumes (Dixon et al. 2013). A significant recent breakthrough is the identification of an R2R3MYB from rabbit's foot clover (*Trifolium arvense* L.) that when overexpressed in *Medicago sativa* L. or *Trifolium repens* L. induces foliar PA accumulation at up to 1.8% dry weight (Hancock et al. 2012). In grape, PAs are major influences on the sensory qualities of the resultant wine. Persimmon is a very interesting study system, as in persimmon PAs can accumulate to such levels that they render the fruit highly astringent (Akagi et al. 2012).

Except anthocyanins, PA, and flavonols, there are only a few polyphenol biosynthetic pathways for which TFs have been characterized. Most notably, the regulation of lignin biosynthesis has been extensively studied in *A. thaliana* and to a lesser extent the grasses (Gray et al. 2012; Zhao and Dixon 2011). It includes AtMYB4, AtMYB26, AtMYB32, AtMYB43, AtMYB46, AtMYB52, AtMYB54, AtMYB58, AtMYB61, AtMYB63, AtMYB75/PAP1, AtMYB83, AtMYB85, and AtMYB103 (Newman et al. 2004; Zhong et al. 2008; McCarthy et al. 2009; Kim et al. 2012; Romano et al. 2012; Zhong and Ye 2014; Ko et al. 2014; Yoon et al. 2015; Nakano et al. 2015). These MYBs work together with other TFs, such as the secondary wall NAC (SWN). They form a complicated transcriptional network to regulate the transcription of lignin biosynthetic genes (Zhao and Dixon 2011; Ko et al. 2014; Yoon et al. 2015). Overexpression of AtMYB85 results in specific lignin ectopic deposition in the epidermal and cortical cells of stems (Zhong et al. 2008).

The knowledge of MYBs regulating other phenylpropanoids is limited and fragmented, and they may act as activators or repressors in the biosynthetic pathway of phenolic acids. One such example is AtMYB4 which acts as an active repressor controlling sinapate ester biosynthesis in a UV-dependent manner (Jin et al. 2000) and *Petunia hybrida* (Hook.) Vilm. MYB4 represses formation of phenylpropanoid scent compounds in the petals (Colquhoun et al. 2011). Moreover in *Salvia miltiorrhiza*, SmMYB39 represses phenolic acid biosynthesis through negatively regulating the expression of SmC4H and tyrosine aminotransferase gene (SmTAT) (Zhang et al. 2013). There has been recent progress on understanding the regulation of another branch of the phenylpropanoid pathway that is producing the isoflavonoids. Changes in TF gene expression associated with isoflavonoid production were characterized for the *Lotus japonicus* L. and a group of R2R3MYBs (subgroup 2) identified as candidate activators for the necessary biosynthetic genes (Shelton et al. 2012).

Several TFs with a repressive effect on flavonoid biosynthesis are now known. The first repressor TF to be identified from plants was the R2R3MYB mentioned earlier, AtMYB4. AtMYB4 production is downregulated by exposure to UV-B irradiation, which in turn releases the cinnamate-4-hydroxylase gene from AtMYB4-based suppression, thus stimulating the production of sinapate esters for UV-B protection (Jin et al. 2000). A potential repressor for anthocyanin production was identified soon after AtMYB4, FaMYB1 from strawberry. Expression of FaMYB1 in tobacco (Aharoni et al., 2001) or *Lotus corniculatus* L. (Paolocci et al. 2011) represses anthocyanin or PA biosynthesis, respectively. Both AtMYB4 and FaMYB1 possess residues required for R2R3MYB interaction with the bHLH partners and also have a domain in the C-terminal that mediates active transcriptional repression, called the ethylene-responsive transcription factor (ERF)-associated amphiphilic repression (EAR) domain. This suggests that they can form MBW complexes that repress, rather than activate, target genes.

While many of the recent advances have come from the model species, such as *A. thaliana*, *Petunia*, and *Antirrhinum majus* L., there have also been significant advances in the depth of understanding of regulation of phenylpropanoid biosynthesis in important horticultural species, in particular grape and apple (Hichri et al. 2011; Allan et al. 2008). The completion of the genome sequences of these two species has enabled more extensive analysis of the phenylpropanoid-related TF families. The R2R3MYBs regulating anthocyanin production in grape have been extensively studied over the last decade (Hichri et al. 2011), but more recently, those for the PA (Bogs et al. 2007; Deluc et al. 2008; Terrier et al. 2009) and flavonol (Czemmel et al. 2009) pathway branches have been identified, as well as the phenylpropanoid-related bHLHs and WDR (Hichri et al. 2011; Matus et al. 2010). Lin-Wang et al. (2010) characterized a large number of anthocyanin-related R2R3MYB activators from Rosaceae species, with a focus on apple, and also identified a range of apple R2R3MYBs containing the EAR domain that were able to inhibit anthocyanin gene activation in a transient assay system (Lin-Wang et al. 2011).

The use of TFs as transgenes for modification of the biosynthesis of polyphenolics is now well proven, with many different genes shown to be effective and a large number of species successfully targeted (Hichri et al. 2011, Dixon et al. 2013). However, there are still issues to be resolved in using TFs for pathway regulation. In particular, great variation is observed in transgenic phenotype depending on the specific transgene/host species combination. For example, in a given species, it may vary whether the R2R3MYB or bHLH of the flavonoid-related MBW complex is more effective, or two R2R3MYBs from the same gene family may generate dramatically different phenotypes. These variations likely reflect differences in the activation strength and specific target DNA motifs of the individual TF proteins and/or hierarchies of regulation among the TFs (Feller et al. 2011). Some TFs may activate a wide range of target genes when overexpressed, including in pathways for which they are not normally key regulatory factors. For example, AtMYB12 can induce production of both flavonols and caffeoylquinic acids when overexpressed in tomato (Luo et al. 2008). For some target pathways, the correct combination of activator and repressor TFs may be required if the desired level of phenolic production is to be achieved (Shelton et al. 2012).

The understanding of the regulation of polyphenol production is well advanced compared to the state of knowledge for other plant secondary metabolite pathways. However, there are still many areas in which progress is limited. In particular, there is little depth of knowledge outside of the phenylpropanoid pathway, with information lacking for groups of compounds as significant as the coumarins, anthraquinones, naphthoquinones, xanthones, and hydrolyzable tannins.

18.4 Evolutionary Significance of Plant Phenolics

Phenolic compounds have been produced in plants to provide adaptive advantage and metabolic plasticity to survive under different environmental challenges throughout the course of evolution. Plant phenolics played a significant role in the colonization of land plant from aquatic life. This transition faced three major challenges: large and rapid temperature changes, desiccation, and direct harsh ultraviolet light (Delgoda and Murray 2017). Additional challenges included the need for anchorage on soil and rock, the combat of new forms of microorganic occupancies in the new soil environment, and eventually competition by other plants for space and resources as well as grazing predators (Delgoda and Murray 2017). It appears that the production of phenolics has played a crucial role in adaption to new environment and overcoming these abiotic and biotic hurdles.

The biosynthesis of the phenolics, phenylpropanoids, flavonoids, and sporopollenins, present in land plants, provided protection against UV radiation, which enable them to survive in the harsh terrestrial landscape with direct exposure to sunlight (Delgoda and Murray 2017). These metabolites have played a critical role in helping the early algal forms to transition into the mosses and liverworts found in terrestrial environments (Delgoda and Murray 2017). Cuticle development in bryophyte also helps in the avoidance of some level of desiccation, although their approach was a rather rudimentary on and off method for metabolism, for wet and dry times, respectively (Delgoda and Murray 2017).

Nearly 40 million years ago, the production of lignin was first observed in tracheophytes to overcome the problems of desiccation (Delgoda and Murray 2017). In addition to protection against desiccation, lignin, a phenolic, complex polymer, provided physical strength to the plant and enables them to adopt erect and upright form against soil and also rigidity required for vasculature. It also made the cell wall rigid and tough required for vasculature and internal irrigation through xylem and phloem cell formations (Delgoda and Murray 2017).

The formation of true leaves having trichomes (in addition to mechanical defenses such as thorns, hairs, and crystals) occurs in late Paleozoic Era (Levin 1973). Trichome provided defense against insect feeder. Different types of trichomes were also storage place for exuding terpenes, phenolics, alkaloids, and others with repellent properties (Levin 1973).

By the late Devonian Era, a well-develop root system in plants was observed which helps the plant for its establishment in the soil and taking water from the soil. Phenolics like isoflavonoids used to be produced by many plant roots against soilborne microorganism during its establishment in the soil (Delgoda and Murray 2017).

During the Mesozoic Era, tannin biosynthesis in the seed coat was observed with the appearance of seed plants in gymnosperms (Delgoda and Murray 2017). Tannin biosynthesis acts as antifeedant and protects the seed. With the evolution of the angiosperms around 145 million years ago, many compounds like anthocyanins and carotenoids (terpenoids) appeared to provide color and fragrance to the flower which help them for pollination and reproductive success of the flowering plants (Delgoda and Murray 2017).

18.5 Role of Plant Phenolics in Senescence

Plant senescence is a degenerative and highly regulated process which occurs at last stage of development and required for fitness and survival of the plant. It involves the recycling and reallocation of nutrients to actively growing parts of plant. The senescence is governed by external as well as internal stimuli in a coordinated manner. The expression of many different genes and transcription factors (TF) was also observed during the courses of senescence at molecular level. The total process of senescence can be divided into three phases: initiation, reorganization, and termination (Bresson et al. 2017). The initiation phase involves the integration of different external as well as internal signals at cellular, tissue, and organ levels. During the reorganization phase, breakdown of macromolecules and remobilization of nutrients occur in the cell (Bresson et al. 2017). Many different proteases and their regulator get activated and are involved in breakdown of proteins. During this process, large macromolecules were converted into transportable and useful by-products. However, macromolecule degradation also resulted in the generation of many toxic intermediates and by-products (Bresson et al. 2017). The removal of these toxic intermediates is essential to maintain nucleus and mitochondria remain functional in order to make a transcription control and provide sufficient energy till the end of senescence. Therefore, antioxidative enzymes get activated to detoxify these toxic intermediate compounds (Bresson et al. 2017). Anthocyanin, a natural antioxidant, was also used to protect against oxidative stress-induced damage and increases during senescence (He and Giusti 2010). During the termination phase, disruption of vacuole resulted in gradual degradation of the cytoplasm due to release of nuclease and protease (Bresson et al. 2017). Finally, Cell death occurs (Bresson et al. 2017).

Terpene compounds also play an important role in plant senescence. They are synthesized by two pathways in the cytosol, endoplasmic reticulum, peroxisomes, and plastids and stored in glandular cells of leaves and resin ducts of needles (Korankye et al. 2017). Terpene synthesis and plant senescence have been tied to photosynthesis of a plant since photosynthesis is reported to serve as a carbon source in initiating terpene biosynthesis (Korankye et al. 2017). However, there have been speculations of alternative carbon sources such as xylem and chloroplast in studies where plants showed reduced rates of photosynthesis but increased in terpene biosynthesis (Korankye et al. 2017). Role of terpenes in plant senescence was reported in *Arabidopsis thaliana* exposed to citral, peppermint, β-pinene, α-pinene, and camphene (Korankye et al. 2017). Postharvest studies have also shown that after trees such as balsam fir are cut, excessive synthesis and/or emission of terpene compounds such as β-pinene, β-terpinene, camphene, and 3-carene is induced prior to needle abscission (Korankye et al. 2017).

18.5.1 Leaf Senescence

In plant leaves, the common physical indicators of senescence most often occur in a much later stage than the actual onset but are usually characterized when the mesophyll tissue begins to lose its greenness and turn to yellow or red (Korankye et al. 2017). The color change is due to both preferential degradation of chlorophyll compared to carotenoids and synthesis of new compounds, such as anthocyanins and phenolics (Matile 1980), and then the ultimate consequence of senescence, which may or may not result in organ abscission (Korankye et al. 2017). Plants, being a sessile in nature, have to respond rapidly to the adverse environmental conditions. In general, they respond by shedding off the affected part of plants and saving the rest of the plant. For instance, a leaf affected with disease undergoes senescence and finally drops off the plant, thereby preventing the spread of disease and allowing the rest of the plant to continue in its development. Leaf senescence occurs in response to the many factors like disease, drought, and nutrient deficiency. A study on green leaves and red leaves of *Prunus* indicated that the anthocyanins help the red leaves to perform better during early stage of senescence and also in photoprotection (Lo Piccolo et al. 2018). Metabolomics study of natural early tobacco leaf senescence indicates decreases in membrane lipids, free sterols, and acylated sterol glucosides along with the accumulation of sterol esters and alkaloids. The amino acid levels were significantly decreased, particularly those of N-rich amino acids (glutamine and asparagine), thus reflecting N translocation. Sugar alcohols and polyphenols accumulated when the lower leaves turned yellow (Li et al. 2016).

18.5.2 Autumn Senescence

Deciduous tree commonly shed their leaves during autumn, and this process is known as autumn senescence. This process is important for winter storage of plants in which remobilization of nutrient occurs from leaves. Autumn senescence is also a kind of leaf senescence which is characterized by chlorophyll degradation and flavonoid synthesis (Mattila et al. 2018). The spectacular autumn colors are partly caused by exposure of carotenoids due to faster degradation of chlorophyll (Mattila et al. 2018). In addition, some species also synthesize specific flavonoids, e.g., anthocyanins (Falcone Ferreyra et al. 2012) which are largely responsible for the red colors in senescing leaves (Mattila et al. 2018). Anthocyanins are suggested to act as antioxidants or metal ion chelators, in blocking ultraviolet (UV) or visible radiation or playing roles in defense and signalling (Archetti 2009; Landi et al. 2015). The amounts of other flavonoid compounds, flavonols or flavonol glycosides, have been reported to increase during age-related senescence (Torras-Claveria et al. 2012), but their roles are poorly understood. In contrast to anthocyanins, flavonols do not absorb visible light. They are found in the cuticle and sometimes inside the cells, and they block UV radiation (Solovchenko and Merzlyak 2008) and may act as ROS scavengers or signal molecules (Pollastri and Tattini 2011). Flavonol glycosides have also been suggested to be involved in supercooling of xylem parenchyma cells of *Cercidiphyllum japonicum* (Kasuga et al. 2008).

Keskitalo et al. (2005) described the cellular time table of autumn senescence in *Populus tremula*. Similarly, Mattila et al. (2018) studied the autumn senescence in *Sorbus aucuparia*, *Acer platanoides*, *Betula pendula*, and *Prunus padus* and measured the chlorophyll and flavonol contents every morning and evening during the whole autumn with a nondestructive method from individual leaves. Increase in flavonols is commonly observed during senescence which is accompanied with the rapid degradation of chlorophyll. Moy et al. (2015) recorded increase in anthocyanin content during late autumn in maple tree. Autumn senescence is considered to be coordinated and synchronized at the organism level; neither all leaves of a tree nor all cells of a single leaf senesce simultaneously (Keskitalo et al. 2005).

18.5.3 Petal Senescence

Petal senescence is a rapid and synchronous process that represents the final stage of petal development. Prior to senescence, petals play an important role in pollination of the plants by producing scent and color that attract pollinators. Anthocyanins, betalains, and carotenoids are the major flower pigment and provide color to the petals and other organs of the plants. Color change, rolling, wilting, etc. are the visible signs of petal senescence, while collapse of mesophyll cells occurs before the appearance of visible morphological changes (Ma et al. 2018). Petals are evolutionarily derived from leaves; therefore, many events are common in both senescence; these includes disintegration of intracellular structures, degradation of macromolecules and membrane systems, and recycling of substances (Ma et al. 2018). However, petal and leaf senescence are also distinct from each other in a number of ways (Ma et al. 2018).

First, leaf senescence can be reversible in some species, and re-greening is also possible, before a point of "no return," while the process of petal senescence can only be delayed, but not arrested or reversed (Ma et al. 2018). Petal senescence is greatly activated or accelerated by pollination and is thus tightly regulated by developmental signals, rather than environmental signals, which contrasts with leaf senescence (Ma et al. 2018). Secondly, petals are decorative organs that contain relatively lower levels of nutrients and are considered to be "sink" organs, while leaves are "source" organs and the primary site of photosynthesis, so nutrient remobilization is a critical event during leaf senescence (Ma et al. 2018). Nutrient remobilization from petals is therefore less critical during senescence, although nutrients can be transferred from petals to the ovary during seed development (Ma et al. 2018). Thirdly, petal senescence is relatively rapid, and petal abscission can occur even while the petal is still turgid in some plant species (Ma et al. 2018). In contrast, leaf senescence involves gradual yellowing and wilting, and the leaf abscission occurs after nutrient recycling has been completed (Ma et al. 2018).

The genus *Oenothera*, evening primrose, is known to undergo a flower color change during senescence. The flowers of this genus bloom in the evening and fade in the morning (Teppabut et al. 2018). When fully opened, the petals of *O. tetraptera* are white, and then, they become pink in the morning (Teppabut et al. 2018). Those of *O. laciniata* as well as *O. stricta* are yellow, and then, they turn orange as they fade. These phenomena strongly indicate that an anthocyanin is biosynthesized during senescence (Teppabut et al. 2018).

18.6 Conclusions

Phenolics are ubiquitously present throughout the plant kingdom and play a significant role in plant development and defense by acting as an immune system of plant. In recent past, many researchers have focus on identification and extraction of different plant phenolics due to their antioxidant properties and remarkable effect in preventing many diseases including cancer. Plant phenolics also played a critical role in diversification of plant during evolution. In addition to this, their role in different types of plant senescence indicates their importance for plant life cycle. However, more findings are required to understand their involvement during senescence.

References

Aharoni A, De Vos CR, Wein M, Sun Z, Greco R, Kroon A, Mol JN, O'connell AP (2001) The strawberry FaMYB1 transcription factor suppresses anthocyanin and flavonol accumulation in transgenic tobacco. Plant J 28(3):319–332

Akagi T, Katayama-Ikegami A, Kobayashi S, Sato A, Kono A, Yonemori K (2012) Seasonal abscisic acid signal and a basic leucine zipper transcription factor, DkbZIP5, regulate proanthocyanidin biosynthesis in persimmon fruit. Plant Physiol 158(2):1089–1102

Allan AC, Hellens RP, Laing WA (2008) MYB transcription factors that colour our fruit. Trends Plant Sci 13(3):99–102

Appelhagen I, Thiedig K, Nordholt N, Schmidt N, Huep G, Sagasser M, Weisshaar B (2014) Update on transparent testa mutants from *Arabidopsis thaliana*: characterisation of new alleles from an isogenic collection. Planta 240(5):955–970

Archetti M (2009) Classification of hypotheses on the evolution of autumn colours. Oikos 118:328–333

Barros J, Serrani-Yarce JC, Chen F, Baxter D, Venables BJ, Dixon RA (2016) Role of bifunctional ammonia-lyase in grass cell wall biosynthesis. Nature Plants 2(6):16050

Beckman CH (2000) Phenolic-storing cells: keys to programmed cell death and periderm formation in wilt disease resistance and in general defence responses in plants? Physiol Mol Plant Pathol 57:101–110

Bhattacharya A, Sood P, Citovsky V (2010) The roles of plant phenolics in defence and communication during agrobacterium and rhizobium infection. Mol Plant Pathol 11(5):705–719

Bogs J, Jaffé FW, Takos AM, Walker AR, Robinson SP (2007) The grapevine transcription factor VvMYBPA1 regulates proanthocyanidin synthesis during fruit development. Plant Physiol 143 (3):1347–1361

Bresson J, Bieker S, Riester L, Doll J, Zentgraf U (2017) A guideline for leaf senescence analyses: from quantification to physiological and molecular investigations. J Exp Bot 69:769–786

Butelli E, Titta L, Giorgio M, Mock HP, Matros A, Peterek S, Schijlen EG, Hall RD, Bovy AG, Luo J, Martin C (2008) Enrichment of tomato fruit with health-promoting anthocyanins by expression of select transcription factors. Nat Biotechnol 26(11):1301

Caputi L, Malnoy M, Goremykin V, Nikiforova S, Martens S (2012) A genome-wide phylogenetic reconstruction of family 1 UDP-glycosyltransferases revealed the expansion of the family during the adaptation of plants to life on land. Plant J 69(6):1030–1042

Caretto S, Linsalata V, Colella G, Mita G, Lattanzio V (2015) Carbon fluxes between primary metabolism and phenolic pathway in plant tissues under stress. Int J Mol Sci 16:26378–26394

Cheng A-X, Han X-J, Wu Y-F, Lou H-X (2014) The function and catalysis of 2-oxoglutarate-dependent oxygenases involved in plant flavonoid biosynthesis. Int J Mol Sci 15(1):1080–1095

Cheynier V, Comte G, Davies KM, Lattanzio V, Martens S (2013) Plant phenolics: recent advances on their biosynthesis, genetics, and ecophysiology. Plant Physiol Biochem 72:1–20

Coley PD, Bryant JP, Chapin FS (1985) Resource availability and plant antiherbivore defense. Science 230:895–899

Colquhoun TA, Kim JY, Wedde AE, Levin LA, Schmitt KC, Schuurink RC, Clark DG (2011) PhMYB4 fine-tunes the floral volatile signature of Petunia× hybrida through PhC4H. J Exp Bot 62(3):1133–1143

Czemmel S, Stracke R, Weisshaar B, Cordon N, Harris NN, Walker AR, Robinson SP, Bogs J (2009) The grapevine R2R3-MYB transcription factor VvMYBF1 regulates flavonol synthesis in developing grape berries. Plant Physiol 151(3):1513–1530

Delgoda R, Murray J (2017) Evolutionary perspectives on the role of plant secondary metabolites, Pharmacognosy. Elsevier, New York, pp 93–100

Deluc L, Barrieu F, Marchive C, Lauvergeat V, Decendit A, Richard T, Carde JP, Mérillon JM, Hamdi S (2006) Characterization of a grapevine R2R3-MYB transcription factor that regulates the phenylpropanoid pathway. Plant Physiol 140(2):499–511

Deluc L, Bogs J, Walker AR, Ferrier T, Decendit A, Merillon JM, Robinson SP, Barrieu F (2008) The transcription factor VvMYB5b contributes to the regulation of anthocyanin and proanthocyanidin biosynthesis in developing grape berries. Plant Physiol 147(4):2041–2053

Deng Y, Lu S (2017) Biosynthesis and regulation of phenylpropanoids in plants. Crit Rev Plant Sci 36:257–290

Dixon RA, Liu C, Jun JH (2013) Metabolic engineering of anthocyanins and condensed tannins in plants. Curr Opin Biotechnol 24(2):329–335

Dubos C, Stracke R, Grotewold E, Weisshaar B, Martin C, Lepiniec L (2010) MYB transcription factors in Arabidopsis. Trends Plant Sci 15(10):573–581

Falcone Ferreyra ML, Rius S, Casati P (2012) Flavonoids: biosynthesis, biological functions, and biotechnological applications. Front Plant Sci 3:222

Feller A, Machemer K, Braun EL, Grotewold E (2011) Evolutionary and comparative analysis of MYB and bHLH plant transcription factors. Plant J 66(1):94–116

Ferrer JL, Austin MB, Stewart C Jr, Noel JP (2008) Structure and function of enzymes involved in the biosynthesis of phenylpropanoids. Plant Physiol Biochem 46(3):356–370

Gonzalez A, Zhao M, Leavitt JM, Lloyd AM (2008) Regulation of the anthocyanin biosynthetic pathway by the TTG1/bHLH/Myb transcriptional complex in Arabidopsis seedlings. Plant J 53 (5):814–827

Gray J, Caparrós-Ruiz D, Grotewold E (2012) Grass phenylpropanoids: regulate before using! Plant Sci 184:112–120

Hagerman AE, Riedl KM, Jones GA, Sovik KN, Ritchard NT, Hartzfeld PW, Riechel TL (1998) High molecular weight plant polyphenolics (tannins) as biological antioxidants. J Agric Food Chem 46:1887–1892

Hancock KR, Collette V, Fraser K, Greig M, Xue H, Richardson K, Jones C, Rasmussen S (2012) Expression of the R2R3-MYB transcription factor TaMYB14 from *Trifolium arvense* activates proanthocyanidin biosynthesis in the legumes *Trifolium repens* and *Medicago sativa*. Plant Physiol 159(3):1204–1220

He J, Giusti MM (2010) Anthocyanins: natural colorants with health-promoting properties. Annu Rev Food Sci Technol 1:163–187

Hichri I, Barrieu F, Bogs J, Kappel C, Delrot S, Lauvergeat V (2011) Recent advances in the transcriptional regulation of the flavonoid biosynthetic pathway. J Exp Bot 62(8):2465–2483

Ichino T, Fuji K, Ueda H, Takahashi H, Koumoto Y, Takagi J, Tamura K, Sasaki R, Aoki K, Shimada T, Hara-Nishimura I (2014) GFS9/TT9 contributes to intracellular membrane trafficking and flavonoid accumulation in Arabidopsis thaliana. Plant J 80(3):410–423

Jendresen CB, Stahlhut SG, Li M, Gaspar P, Siedler S, Förster J, Maury J, Borodina I, Nielsen AT (2015) Highly active and specific tyrosine ammonia-lyases from diverse origins enable enhanced production of aromatic compounds in bacteria and Saccharomyces cerevisiae. Appl Environ Microbiol 81(13):4458–4476

Jin H, Martin C (1999) Multifunctionality and diversity within the plant MYB-gene family. Plant Mol Biol 41(5):577–585

Jin H, Cominelli E, Bailey P, Parr A, Mehrtens F, Jones J, Tonelli C, Weisshaar B, Martin C (2000) Transcriptional repression by AtMYB4 controls production of UV-protecting sunscreens in Arabidopsis. EMBO J 19(22):6150–6161

Kasuga J, Hashidoko Y, Nishioka A, Yoshiba M, Arakawa K, Fujikawa S (2008) Deep supercooling xylem parenchyma cells of katsura tree (*Cercidiphyllum japonicum*) contain flavonol glycosides exhibiting high anti-ice nucleation activity. Plant Cell Environ 31:1335–1348

Keskitalo J, Bergquist G, Gardeström P, Jansson S (2005) A cellular timetable of autumn senescence. Plant Physiol 139:1635–1648

Kim WC, Ko JH, Han KH (2012) Identification of a cis-acting regulatory motif recognized by MYB46, a master transcriptional regulator of secondary wall biosynthesis. Plant Mol Biol 78 (4–5):489–501

Ko JH, Jeon HW, Kim WC, Kim JY, Han KH (2014) The MYB46/MYB83-mediated transcriptional regulatory programme is a gatekeeper of secondary wall biosynthesis. Ann Bot 114 (6):1099–1107

Kobayashi S, Goto-Yamamoto N, Hirochika H (2004) Retrotransposon-induced mutations in grape skin color. Science 304(5673):982–982

Korankye EA, Lada R, Asiedu S, Caldwell C (2017) Plant senescence: the role of volatile terpene compounds (VTCs). Am J Plant Sci 8:3120–3139

Kuhn N, Guan L, Dai ZW, Wu BH, Lauvergeat V, Gomès E, Li SH, Godoy F, Arce-Johnson P, Delrot S (2013) Berry ripening: recently heard through the grapevine. J Exp Bot 65 (16):4543–4559

Kumar S, Pandey AK (2013) Chemistry and biological activities of flavonoids: an overview. The Scientific World Journal 2013:1–16

Landi M, Tattini M, Gould KS (2015) Multiple functional roles of anthocyanins in plant-environment interactions. Environ Exp Bot 119:4–17

Levin DA (1973) The role of trichomes in plant defense. Q Rev Biol 48:3–15

Li S, Zachgo S (2013) TCP3 interacts with R2R3-MYB proteins, promotes flavonoid biosynthesis and negatively regulates the auxin response in *Arabidopsis thaliana*. Plant J 76(6):901–913

Li L, Zhao J, Zhao Y, Lu X, Zhou Z, Zhao C, Xu G (2016) Comprehensive investigation of tobacco leaves during natural early senescence via multi-platform metabolomics analyses. Sci Rep 6:37976

Lin-Wang K, Bolitho K, Grafton K, Kortstee A, Karunairetnam S, McGhie TK, Espley RV, Hellens RP, Allan AC (2010) An R2R3 MYB transcription factor associated with regulation of the anthocyanin biosynthetic pathway in Rosaceae. BMC Plant Biology 10(1):50

Lin-Wang KUI, Micheletti D, Palmer J, Volz R, Lozano L, Espley R, Hellens RP, Chagne D, Rowan DD, Troggio M, Iglesias I (2011) High temperature reduces apple fruit colour via modulation of the anthocyanin regulatory complex. Plant Cell Environ 34(7):1176–1190

Liu J, Osbourn A, Ma P (2015) MYB transcription factors as regulators of phenylpropanoid metabolism in plants. Mol Plant 8(5):689–708

Lo PE, Landi M, Pellegrini E, Agati G, Giordano C, Giordani T, Lorenzini G, Malorgio F, Massai R, Nali C (2018) Multiple consequences induced by epidermally-located anthocyanins in young, mature and senescent leaves of Prunus. Front Plant Sci 9:917

Luo J, Butelli E, Hill L, Parr A, Niggeweg R, Bailey P, Weisshaar B, Martin C (2008) AtMYB12 regulates caffeoylquinic acid and flavonol synthesis in tomato: expression in fruit results in very high levels of both types of polyphenol. Plant J 56(2):316–326

Ma N, Ma C, Liu Y, Shahid MO, Wang C, Gao J (2018) Petal senescence: a hormone view. J Exp Bot 69:719–732

Matile P (1980) Catabolism of chlorophyll: involvement of peroxidase? Z Pflanzenphysiol 99:475–478

Mattila H, Valev D, Havurinne V, Khorobrykh S, Virtanen O, Antinluoma M, Mishra KB, Tyystjärvi E (2018) Degradation of chlorophyll and synthesis of flavonols during autumn senescence—the story told by individual leaves. AoB Plants 10:ply028

Matus JT, Poupin MJ, Cañón P, Bordeu E, Alcalde JA, Arce-Johnson P (2010) Isolation of WDR and bHLH genes related to flavonoid synthesis in grapevine (Vitis vinifera L.). Plant Mol Biol 72(6):607–620

McCarthy RL, Zhong R, Ye ZH (2009) MYB83 is a direct target of SND1 and acts redundantly with MYB46 in the regulation of secondary cell wall biosynthesis in Arabidopsis. Plant Cell Physiol 50(11):1950–1964

Moy A, Le S, Verhoeven A (2015) Different strategies for photoprotection during autumn senescence in maple and oak. Physiol Plant 155:205–216

Nakano Y, Yamaguchi M, Endo H, Rejab NA, Ohtani M (2015) NAC-MYB-based transcriptional regulation of secondary cell wall biosynthesis in land plants. Front Plant Sci 6:288

Newman LJ, Perazza DE, Juda L, Campbell MM (2004) Involvement of the R2R3-MYB, AtMYB61, in the ectopic lignification and dark-photomorphogenic components of the det3 mutant phenotype. Plant J 37(2):239–250

Ngaki MN, Louie GV, Philippe RN, Manning G, Pojer F, Bowman ME, Li L, Larsen E, Wurtele ES, Noel JP (2012) Evolution of the chalcone-isomerase fold from fatty-acid binding to stereospecific catalysis. Nature 485(7399):530

Noel JP, Austin MB, Bomati EK (2005) Structure–function relationships in plant phenylpropanoid biosynthesis. Curr Opin Plant Biol 8(3):249–253

Ojala T, Remes S, Haansuu P, Vuorela H, Hiltunen R, Haahtela K, Vuorela P (2000) Antimicrobial activity of some coumarin containing herbal plants growing in Finland. J Ethnopharmacol 73 (1-2):299–305

Paolocci F, Robbins MP, Passeri V, Hauck B, Morris P, Rubini A, Arcioni S, Damiani F (2011) The strawberry transcription factor FaMYB1 inhibits the biosynthesis of proanthocyanidins in *Lotus corniculatus* leaves. J Exp Bot 62(3):1189–1200

Petrussa E, Braidot E, Zancani M, Peresson C, Bertolini A, Patui S, Vianello A (2013) Plant flavonoids—biosynthesis, transport and involvement in stress responses. Int J Mol Sci 14 (7):14950–14973

Pollastri S, Tattini M (2011) Flavonols: old compounds for old roles. Ann Bot 108:1225–1233

Romano JM, Dubos C, Prouse MB, Wilkins O, Hong H, Poole M, Kang KY, Li E, Douglas CJ, Western TL, Mansfield SD (2012) AtMYB61, an R2R3-MYB transcription factor, functions as a pleiotropic regulator via a small gene network. New Phytol 195(4):774–786

Roupe KA, Remsberg CM, Yáñez JA, Davies NM (2006) Pharmacometrics of stilbenes: seguing towards the clinic. Curr Clin Pharmacol 1(1):81–101

Saito K, Yonekura-Sakakibara K, Nakabayashi R, Higashi Y, Yamazaki M, Tohge T, Fernie AR (2013) The flavonoid biosynthetic pathway in Arabidopsis: structural and genetic diversity. Plant Physiol Biochem 72:21–34

Shelton D, Stranne M, Mikkelsen L, Pakseresht N, Welham T, Hiraka H, Tabata S, Sato S, Paquette S, Wang TL, Martin C (2012) Transcription factors of Lotus: regulation of isoflavonoid biosynthesis requires coordinated changes in transcription factor activity. Plant Physiol 159(2):531–547

Shi M-Z, Xie D-Y (2014) Biosynthesis and metabolic engineering of anthocyanins in Arabidopsis thaliana. Recent Pat Biotechnol 8(1):47–60

Solovchenko A, Merzlyak M (2008) Screening of visible and UV radiation as a photoprotective mechanism in plants. Russ J Plant Physiol 55:719

Stracke R, Ishihara H, Huep G, Barsch A, Mehrtens F, Niehaus K, Weisshaar B (2007) Differential regulation of closely related R2R3-MYB transcription factors controls flavonol accumulation in different parts of the *Arabidopsis thaliana* seedling. Plant J 50(4):660–677

Stracke R, Jahns O, Keck M, Tohge T, Niehaus K, Fernie AR, Weisshaar B (2010) Analysis of PRODUCTION OF FLAVONOL GLYCOSIDES-dependent flavonol glycoside accumulation in *Arabidopsis thaliana* plants reveals MYB11-, MYB12-and MYB111-independent flavonol glycoside accumulation. New Phytol 188(4):985–1000

Tanaka Y, Sasaki N, Ohmiya A (2008) Biosynthesis of plant pigments: anthocyanins, betalains and carotenoids. Plant J 54(4):733–749

Tanner GJ, Francki KT, Abrahams S, Watson JM, Larkin PJ, Ashton AR (2003) Proanthocyanidin biosynthesis in plants purification of legume leucoanthocyanidin reductase and molecular cloning of its cDNA. J Biol Chem 278(34):31647–31656

Teppabut Y, Oyama K-i, Kondo T, Yoshida K (2018) Change of Petals′ Color and Chemical Components in Oenothera Flowers during Senescence. Molecules 23:1698.

Terrier N, Torregrosa L, Ageorges A, Vialet S, Verries C, Cheynier V, Romieu C (2009) Ectopic expression of VvMybPA2 promotes proanthocyanidin biosynthesis in grapevine and suggests additional targets in the pathway. Plant Physiol 149(2):1028–1041

Torras-Claveria L, Jáuregui O, Codina C, Tiburcio AF, Bastida J, Viladomat F (2012) Analysis of phenolic compounds by high-performance liquid chromatography coupled to electrospray ionization tandem mass spectrometry in senescent and water-stressed tobacco. Plant Sci 182:71–78

Turnbull JJ, Nakajima J-i, Welford RWD, Yamazaki M, Saito K, Schofield CJ (2004) Mechanistic studies on three 2-oxoglutarate-dependent oxygenases of flavonoid biosynthesis anthocyanidin synthase, flavonol synthase, and flavanone 3β-hydroxylase. J Biol Chem 279(2):1206–1216

Venugopala KN, Rashmi V, Odhav B (2013) Review on natural coumarin lead compounds for their pharmacological activity. BioMed Res Int 1:1–15

Vogt T (2010) Phenylpropanoid biosynthesis. Mol Plant 3:2–20

Watts KT, Lee PC, Schmidt-Dannert C (2004) Exploring recombinant flavonoid biosynthesis in metabolically engineered *Escherichia coli*. Chembiochem 5(4):500–507

Winkel BS (2004) Metabolic channeling in plants. Annu Rev Plant Biol 55:85–107

Xia N, Daiber A, Förstermann U, Li H (2017) Antioxidant effects of resveratrol in the cardiovascular system. Br J Pharmacol 174(12):1633–1646

Xie D-Y, Sharma SB, Paiva NL, Ferreira D, Dixon RA (2003) Role of anthocyanidin reductase, encoded by BANYULS in plant flavonoid biosynthesis. Science 299(5605):396–399

Xu W, Dubos C, Lepiniec L (2015) Transcriptional control of flavonoid biosynthesis by MYB–bHLH–WDR complexes. Trends Plant Sci 20(3):176–185

Yang L, Yang C, Li C, Zhao Q, Liu L, Fang X, Chen XY (2016) Recent advances in biosynthesis of bioactive compounds in traditional Chinese medicinal plants. Sci Bull 61(1):3–17

Yoon J, Choi H, An G (2015) Roles of lignin biosynthesis and regulatory genes in plant development. J Integr Plant Biol 57(11):902–912

Zhang S, Ma P, Yang D, Li W, Liang Z, Liu Y, Liu F (2013) Cloning and characterization of a putative R2R3 MYB transcriptional repressor of the rosmarinic acid biosynthetic pathway from Salvia miltiorrhiza. PloS One 8(9):e73259

Zhao Q, Dixon RA (2011) Transcriptional networks for lignin biosynthesis: more complex than we thought? Trends Plant Sci 16(4):227–233

Zheng Y, Li JH, Xin HP, Wang N, Guan L, Wu B-H, Li SH (2013) Anthocyanin profile and gene expression in berry skin of two red *Vitis vinifera* grape cultivars that are sunlight dependent versus sunlight independent. Aust J Grape Wine Res 19(2):238–248

Zhong R, Ye ZH (2014) Complexity of the transcriptional network controlling secondary wall biosynthesis. Plant Sci 229:193–207

Zhong R, Lee C, Zhou J, McCarthy RL, Ye ZH (2008) A battery of transcription factors involved in the regulation of secondary cell wall biosynthesis in Arabidopsis. Plant Cell 20(10):2763–2782

Zhou M, Wei L, Sun Z, Gao L, Yu M, Tang Y, Wu Y (2015) Production and transcriptional regulation of proanthocyanidin biosynthesis in forage legumes. Appl Microbiol Biotechnol 99 (9):3797–3806

Plant Phenolics Under Water-Deficit Conditions: Biosynthesis, Accumulation, and Physiological Roles in Water Stress Alleviation

19

Satish Kumar, Bharat Bhushan, G. C. Wakchaure, Kamlesh K. Meena, Mahesh Kumar, Nand Lal Meena, and Jagadish Rane

Abstract

Phenolic compounds play important role as a structural component of secondary cell wall and non-enzymatic antioxidant system of the plants. The encounter to the drought stress conditions bring about the imbalance in the rate of ROS (reactive oxygen species) production and their quenching by antioxidant machinery of plants which puts plant under oxidative stress. The plant phenolics specially flavonoids play important role in neutralizing these harmful ROS and protect the plant from oxidative damage of ROS. Further, the controlled imposition of drought stress to plants is successfully used as a strategy to enhance the content of the bioactive compounds and phenylpropanoids in economically important food and medicinal plants. In this chapter we present the overview of phenylpropanoids biosynthesis and also present a brief account of the protective roles of phenolics specially flavonoids in drought stress alleviation. A concise description of the molecular interventions attempted so far, to regulate the synthesis of phenylpropanoids in different plants, is also presented.

Keywords

Phenylpropanoids · ROS (reactive oxygen species) · Flavonoids · PAL (phenylalanine ammonia-lyase) · Chalcone synthase · Drought stress

S. Kumar · G. C. Wakchaure · K. K. Meena · M. Kumar · J. Rane
ICAR-National Institute of Abiotic Stress Management, Pune, India

B. Bhushan (✉)
ICAR-Indian Institute of Maize Research, Ludhiana, Punjab, India
e-mail: bharat.bhushan@icar.gov.in

N. L. Meena
ICAR-National Bureau of Plant Genetic Resources, New Delhi, India

R. Lone et al. (eds.), *Plant Phenolics in Sustainable Agriculture*,
https://doi.org/10.1007/978-981-15-4890-1_19

19.1 Introduction

Plants synthesize several thousand phenol derivatives collectively referred to as plant phenolics. The common structural feature of the diverse plant phenolics is the presence of hydroxy-substituted benzene ring within their chemical structure. In fact, the expression "plant phenolics" encompasses a highly diverse group with enormous structural diversity that includes multitudes of plant secondary metabolites like flavonoids, stilbenes, tannins, and many cell wall components like lignin, lignans, suberin, and cutins. Due to their vast structural diversity and their very broad range of physiological functions as cell wall structural component and as secondary metabolites, the straightforward definition of plant phenolics becomes evasive. However, Quideau et al. have suggested the use of the term "plant phenolics" strictly for secondary natural metabolites synthesized via shikimate/phenylpropanoid pathway, forming phenylpropanoids or the "polyketide" acetate/malonate pathway, which can produce simple phenols, or both of them (Quideau et al. 2011).

The polyphenols are known to play diverse roles in all the plant forms including primitive forms as bryophytes to most advanced angiosperms. In fact, phenolic compounds do not play a direct role in primary photosynthetic or respiratory metabolism of plants but still have an important role in plant survival. The diverse physiological roles played by the plant phenolic compounds include their role as fungicides (stilbenes), natural pesticides (tannins), signaling molecules for establishing symbiotic associations, attracting the pollinators, cell wall constituent (lignin), and forming impermeable layers of the cell wall (suberin and cutin). Moreover, the flavonoids like chalcones, flavones, and flavonols are known to absorb UV light and have been reported to act as UV photo screens (Hertweck 2009). In addition to their important role for the survival of plants, the plant phenolics act as bioactive compounds having great human health benefits and better nutraceutical properties. The occurrence of the low-molecular-weight plant phenolics is universal in higher plants; however, some of them are species-specific phenolics and thus are of great taxonomic relevance and also exploited as plant species/family-specific biomarkers (Ishimaru et al. 1987; Veit et al. 1995; Almaraz-Abarca et al. 2006; Kharazian 2014; Ávila-Reyes et al. 2018).

The production of the phenolic compounds by plants is often affected by the external environmental conditions like exposure to UV light, an encounter with a pathogen, wounding, the growth conditions like moisture status of the soil, soil pH and salinity, and many other biotic and abiotic factors. A great deal of information has been generated about the effect of the water stress on primary metabolism (photosynthesis and respiration), but the information regarding flavonoid metabolism in response to drought stress remains fragmented. In this chapter, we discuss various aspects of phenolics biosynthesis under water stress condition. We here also present the description of the physiological role played by plant phenolics in drought stress alleviation.

19.2 Biosynthesis of Plant Phenolics: An Overview

Biosynthesis of the plant phenolics takes place through the extension of the shikimate pathway. The synthesis of plant phenolic compounds (phenylpropanoids) does take place either through the shikimate/chorismate pathway or malonate pathway or involving both of them (Knaggs 2001). The aromatic terpenoids are also synthesized by acetate/mevalonate pathway (Bhattacharya et al. 2010), but it will not be dealt with in this description. The phenylalanine, an aromatic amino acid, synthesized in the shikimic acid pathway, is the common precursor of all phenylpropanoid compounds. The aromatic amino acid phenylalanine synthesized through the shikimic acid pathway is deaminated to cinnamic acid by the action of the phenylalanine ammonia-lyase (PAL). This step catalyzed by the phenylalanine ammonia-lyase (PAL) is the committed step in the phenylpropanoid biosynthesis pathway. The PAL is the most intensively studied enzyme of the plant secondary metabolism. Subsequently, the hydroxylation (introduction of the hydroxyl group at carbon 4) of the phenyl ring of the cinnamic acid is catalyzed by P_{450}-monooxygenase to generate p-coumaric acid. The p-coumaric acid further undergoes hydroxylation and methylation of the newly added hydroxyl group at positions 3 and 5 by the sequential enzymatic action of the P_{450}-monooxygenase (hydroxylation) and O-methyl transferases (methylation) to generate the ferulic acid and sinapic acid.

The carboxyl groups of the p-coumaric acid, ferulic acid, and sinapic acid are reduced to their corresponding alcohol forms in a multistep conversion to generate the alcohol, namely, p-coumarin alcohol, coniferyl alcohol, and the sinapyl alcohol, respectively. These alcohol forms are collectively known as monolignols. This conversion basically involves the activation of the COOH group of p-coumaric acid, ferulic acid, and sinapic acid by the corresponding *hydroxycinnamate*: *CoA ligase* at the expense of CoASH and ATP resulting in the formation of thioester to produce p-coumaroyl-CoA, coniferyl-CoA acid, and sinapyl-CoA, respectively.

The thioester p-coumaroyl-CoA serves as a branch point from which other metabolic pathways in the phenylpropanoid, network diverge (Vogt 2010). At this branch point, the generated thioester can either be used by a lignin-specific branch of the phenylpropanoid pathway forming p-coumaryl, coniferyl, and sinapyl alcohol or else may be diverted to the flavonoid biosynthesis through malonate pathway (Fig. 19.1).

For lignin biosynthesis, the generated thioesters are further reduced to the corresponding alcohol by the sequential action of NADP oxidoreductase and the corresponding dehydrogenase where two molecules of $NADPH^{+}$+ H^{+} are used to produce corresponding alcohols. The coumaryl alcohol, coniferyl alcohol, and sinapyl alcohol are commonly referred to as "monolignols." The monolignols are essential building block for the subsequent synthesis of the structural phenylpropanoids like lignin, lignan, suberin, and cutin.

The lignans are the dimerization product of monolignols. The monolignols can undergo dimerization supposed to be mediated through the formation of free radicals resulting in the formation of lignans. The lignans mainly involve the formation of the linkage either through their side chains (e.g., as in pinoresinol) or the linkage

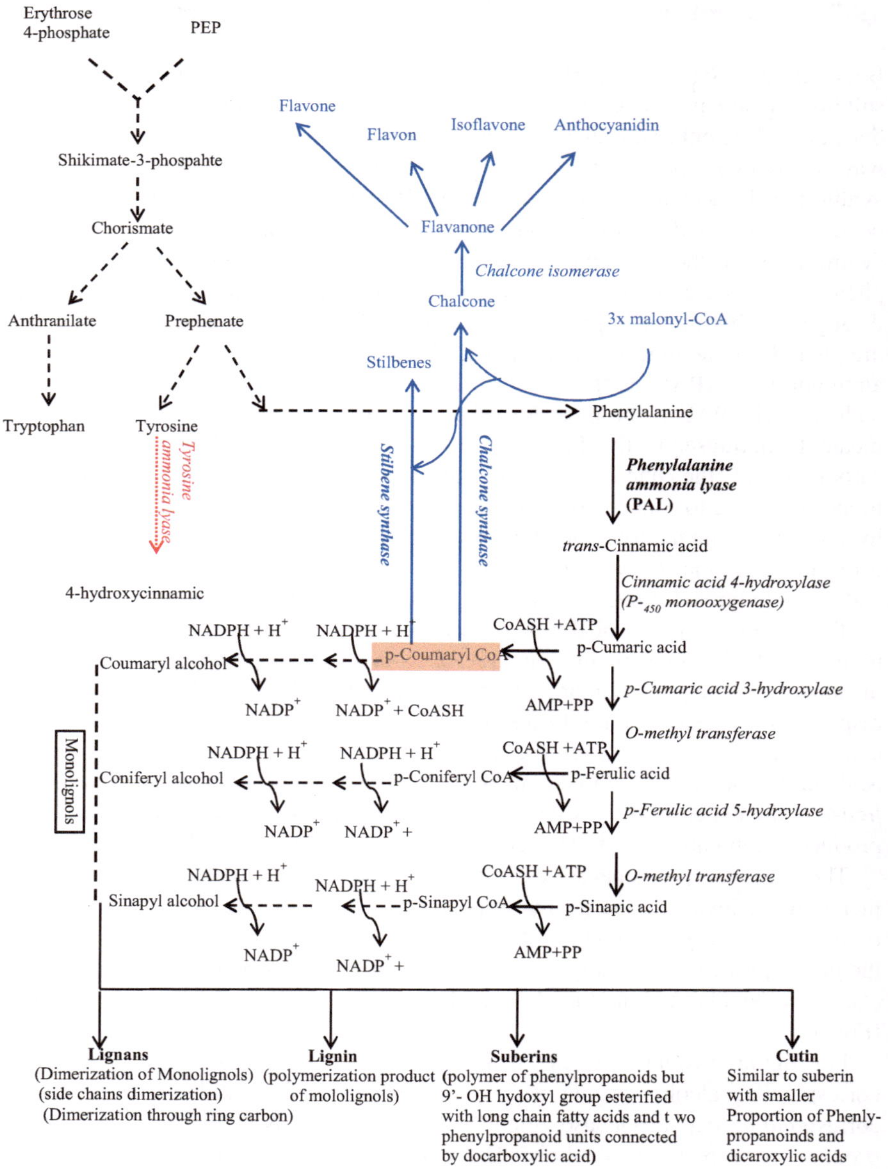

Fig. 19.1 An overview of the biosynthesis of phenylpropanoids in plants. The lignin-specific branch of phenyl propanoids diverge at p-coumaryl-CoA (highlighted in enclosed red color box), while the flavonoid branch is shown with blue color arrows and labels. The dotted arrow indicates that the actual biosynthetic steps are summed up, and only key steps are shown for establishing the context. The details of these steps are omitted for sake of presenting precise overview:

involving phenyl ring carbons (e.g., as in malognol). Unlike lignans which are the product of dimerization of monolignol units, the lignin is the polymerization product of monolignols where polymerization is supposedly mediated by laccases and peroxidases, resulting in the formation of lignin. Lignin is the second most abundant plant polymer after cellulose. Hence, 40% of the total flux of organic compounds circulating in the biosphere is being channeled through phenylpropanoid biosynthesis (Zhang and Liu 2015). In suberins, monolignols are connected in a manner similar to lignin, but the 9′- OH group is esterified with long-chain fatty acids, and two phenylpropanoid units are also connected by dicarboxylic acids forming ester linkage.

Flavonoids, which include a vast array of low-weight phenolics, are synthesized through the malonate pathway. The p-Coumaroyl-CoA, (a thioester of coumaric acid) and three molecules of malonyl-CoA are used to form a molecule of chalcone (contain two phenyl rings) by the enzymatic action of the chalcone synthase (Ververidis et al. 2007). Chalcone synthase has low catalytic activity and hence plant cells produce a higher quantity of the chalcone synthase. Chalcone is acted upon by chalcone isomerase to form flavanone which serves as the precursor of so many flavonoids (refer Fig. 19.1). Some plants like peanut and grapevine have another enzyme stilbene synthase which also uses p-Coumaroyl-CoA and three molecules of malonyl-CoA to form stilbenes. Some stilbenes like resveratrol and vinifera are known to have very strong antifungal activity and thus play an important role in plant protection (Lee and Lee 2015; Gabaston et al. 2017).

19.3 Phenylpropanoids and Their Role in Water Stress Alleviation

Water stress is one of the most significant abiotic stresses that affect the many physiological and biochemical processes of plants and phenylpropanoid metabolism is no exception to this. Under water stress conditions, there is increased in the production of reactive oxygen species (ROS) like superoxide anion O^{2-}, hydrogen peroxide H_2O_2, singlet oxygen Ȯ, and hydroxyl radical OH (Zhang and Kirkham 1994). These reactive oxygen species may cause lipid peroxidation, membrane damage, protein degradation, breakage of DNA, and cell mortality, if not taken care by the antioxidant system of plants (Shao et al. 2008). The detoxification of ROS in plants is mediated by enzymatic and non-enzymatic antioxidant system. The enzymatic antioxidant system includes the action of many enzymes like *superoxide dismutase* (SOD), *ascorbate peroxidase* (APX), *catalase* (CAT), and *peroxidase* (POD) to neutralize the deleterious effect of the ROS. The non-enzymatic antioxidant includes the vitamin C, tocopherols, carotenoids, glutathione, and phenol derivatives (Ashraf et al. 2019). The enhanced production of phenolics (a prominent component of non-enzymatic antioxidant system of plants) under water-deficit condition is obvious physiological demand of the plant under water stress exposure.

The overproduction of the ROS (reactive oxygen species) under water stress is counterbalanced by the induction of phenylpropanoid pathway resulting in increased flavonoid biosynthesis and other phenolics. For instance, the increased accumulation of flavonoids under drought stress conditions has been reported in willow leaves (Akula and Ravishankar 2011) and Arabidopsis (Nakabayashi et al. 2014). The Nakabayashi and coworkers measured the anthocyanin and flavonol levels in Arabidopsis plants at intermittent intervals of 2 days under imposed drought stress conditions and noticed the increase in the content of anthocyanin and flavonol in response to drought. In particular, anthocyanins A5, A8, A9, A10, and A11 and flavonols F6 and F8 were first to accumulate under water-deficit conditions, while flavonols F1, F2, and F3 accumulated later stage of drought (Nakabayashi et al. 2014). Plant tissues containing a higher accumulation of phenolics like anthocyanins exhibit better desiccation tolerance.

The changes in the accumulation of a higher amount of phenolics under water-deficit conditions are often brought about by the altered activity of enzymes of the phenylpropanoid biosynthesis pathway or increased de novo synthesis of these enzymes. The concentration of flavonoids in plant cell often exceeds 1 mM and may even reach up to 10 mM under water stress in some specialized cells (Larson 1988). Phenylalanine ammonia-lyase (PAL) is the gateway enzyme of phenylpropanoid biosynthesis which diverts the central flux of carbon from the primary metabolism to the synthesis of a vast array of phenolics like lignin, anthocyanin and flavonoids, pigments, and phytoalexins. PAL is also used as a biochemical marker whose overexpression indicates the onset of plant antioxidant defense mechanism in response to biotic and abiotic stresses. PAL activity does exist in all the higher plants, and in some fungi and a few bacteria, but have not been reported in animals (Xiang and Moore 2005). PAL activity may be regulated by feedback inhibition by its own product, cinnamic acid, which may also modify the expression of the PAL gene (Boudet 2007). Hence, any change in the enzymatic activity of PAL or the change in expression of genes encoding PAL (de novo synthesis of PAL) leads to the changes in phenolic content of the plant. In fact, there is an ample number of reports describing the multilevel regulation of PAL including change in enzymatic activity of PAL, altered gene expression, de novo synthesis, and post-translational modifications in response to certain external environmental variable. The increased activity of PAL under water stress conditions and consequent increased production of phenolic compounds like ferulic acid has been shown in the leaves of maize (Hura et al. 2008) and fruit of the capsicum (Phimchan et al. 2014). However, there are contrary reports also which have shown the marked decrease in the activity of the PAL in response to the water-deficit condition in maize (Bardzik et al. 1971).

The increased expression of another important enzyme of flavonoid biosynthesis pathway, chalcone synthase, has also been reported in Arabidopsis under water stress conditions (Nakabayashi et al. 2014). In fact, chalcone synthase (CHS), chalcone isomerase (CHI), and flavanone 3-hydroxylase (F3H) are the three key enzymes of flavonoids biosynthesis. Chalcone synthase acts on the CoA-ester of cinnamic acid and uses three molecules of malonyl-CoA to form chalcone. Chalcone is further isomerized to flavanone by the enzyme chalcone flavanone isomerase (CHI). This flavonnone is the precursor for the synthesis of a myriad of flavonoids compounds. The increased expression of the chalcone synthase under water-deficit conditions confirms its role in water stress alleviation. The proteomic studies have also shown the changes in the flavonoid biosynthesis enzymes, namely, chalcone isomerase (CHI) and dihydroflavonol-4-reductase in response to drought stress conditions. The synthesis of dihydroflavonol-4-reductase, an important enzyme of flavonoid biosynthesis pathway, has been reported to decrease in drought-sensitive *Z. mays* cultivar but was higher in the tolerant genotype of the *Z. mays* (Benesova et al. 2012). Pandey and coworker reported a decrease in levels of CHI (chalcone isomerase), a key enzyme of flavonoid biosynthesis in *O. sativa* under drought stress conditions (Pandey et al. 2010). The changes in the enzyme activity/gene expression/de novo synthesis of key enzymes of phenylpropanoid biosynthesis in drought stress response have been summarized in Table 19.1.

19.4 Controlled Drought Stress for Enhancing Antioxidant Potential of Food and Medicinal Plants

The food substances rich in flavonoids and phenolic content are considered as nutraceuticals. Such antioxidants-rich foods impart the protection against many serious human ailments like arthritis, emphysema, retinopathy, neurodegenerative cardiovascular diseases, atherosclerosis, cataracts, and even cancer (Repo-Carrasco-Valencia et al. 2010). The food items with a high content of flavonoids are regarded as functional foods. The deliberate controlled imposition of the water stress has been shown to enhance the overall content of the bioactive compounds in medicinal plants (Kleinwachter and Selmar 2014). The accumulation of the higher amount of plant phenolics under water stress conditions in fruits, vegetable, and even cereals crops is widely reported (Gharibi et al. 2016; Siracusa et al. 2017). The natural drought stress-tolerant plants accumulate a good amount of antioxidants including flavonoids and other phenolics. For instance, the leafy vegetables like *Amaranthus tricolor* are well acclimated to drought stress and hence accumulate good amount of flavonoids (Sarker and Oba 2018). The higher metabolic plasticity and increased accumulation of flavonoids in the leaves has been attributed to the capability of plants like *Moringa olefera* to establish in xeric and water-scarce environments (Brunetti et al. 2018b). Even the drought-sensitive plants grown under the arid or semiarid conditions often accumulate a higher amount of the flavonoids and other phenols compared to the same plants grown under moderate conditions. The phenolic compounds play an important role as an antioxidant system for neutralizing the deleterious effect of the ROS produced under water stress conditions.

As a primary physiological response to the drought stress conditions, plants tend to partially close their stomata to reduce the water losses through transpiration. As a consequence of this, the CO_2 uptake and fixation through Calvin cycle also decreases in the leaves. The $NADPH^{+}$+H^{+} produced during the light reaction of the photosynthesis starts accumulating making the internal environment of the plant

Table 19.1 The effect of the drought stress on the enzymatic activity/gene expression/de novo synthesis of the key enzymes of phenylpropanoid biosynthesis pathway in different crops

S. No.	Type of function studied	Crop/plant	Plant part assayed	Salient finding	Reference
1	PAL enzymatic activity	*Olea europaea* L (olive)	Fruit	PAL activity increased with water deficit	Tovar et al. (2002)
2	PAL enzymatic activity	*Zea mays* (maize)	Leaf	PAL activity increased with deficit irrigation and correlated well with ferulic acid content	Hura et al. (2008)
3	PAL enzymatic activity	*Trifolium repens* (white clover)	Leaf	PAL activity increased with drought	Lee et al. (2007)
4	PAL enzymatic activity	*Triticum aestivum* (wheat)	Seedling	PAL activity increased with water deficit	Tian and Lei (2006)
5	PAL enzymatic activity	*Labisia pumila* (kacip fatimah)	Leaf	PAL activity increased with water deficit	Jaafar et al. (2012)
6	PAL enzymatic activity	*Capsicum annum* (Capsicum)	Fruit	PAL activity increased with water deficit	Phimchan et al. (2014)
7	[a]PAL enzymatic activity	*Zea mays* (maize)	Leaf	10% to 20% water deficit resulted in decreased activity of PAL	Bardzik et al. (1971)
8	PAL enzymatic activity	*Zea mays* (maize)	Leaf and root	The PAL activity sharply increased in leaves during drought stress but remained constant in root	Gholizadeh (2011),
9	PAL gene expression (*pal* gene)	*Betula pendula* (birch)	Leaf	Drought stress induced PAL gene expression	Paakkonen et al. (1998)
10	Chalcone synthase (*chs*) gene expression	*Achillea pachycephala*	Leaf	The CHS gene expression increased with drought stress	Gharibi et al. (2019)
11	Chalcone synthase gene expression (*chs* gene)	*Triticum aestivum* (wheat)	Leaf	The *chs* gene expression increased under water stress	Ma et al. (2014)

(continued)

Table 19.1 (continued)

S. No.	Type of function studied	Crop/plant	Plant part assayed	Salient finding	Reference
12	Chalcone isomerase (*chi*) expression	*Chrysanthemum morifolium*	Leaf	The *chi* gene expression increased under water stress	Hodaei et al. (2018)
13	*Flavanone 3-hydroxylase f3 h gene expression*	*Chrysanthemum morifolium*	Leaf	The *f3h* gene expression increased under water stress	Hodaei et al. (2018)
14	*Flavanone 3-hydroxylase f3 h gene expression*	*Vitis vinifera*	Berry skin	The *f3h* gene expression increased under water stress	Castellarin et al. (2007)
15	*Dihydroflavonol 4-reductase (DFR) gene expression*	*Vitis vinifera*	Berry skin	DFR gene was upregulated under early season water deficit	Castellarin et al. (2007)
16	*Flavonol synthase (FLS) gene expression*	*Triticum aestivum* (wheat)	Leaf	*fls* gene expression increased under drought stress	Ma et al. (2014)
17	*Flavone synthase (FNS) gene expression*	*Triticum aestivum* (wheat)	Leaf	*fns* gene upregulated post drought stress	Ma et al. (2014)
18	*Anthocyanidin synthase (ANS) gene expression*	*Triticum aestivum* (wheat)	Leaf	Higher expression level of *ans* gene under drought stress conditions	Ma et al. (2014)

[a]Indicates unexpected trend

cell highly reducing. As a consequence of reduced uptake of CO_2 and elevated concentrations of $NADPH+H^+$ in leaves, the metabolic processes are pushed toward the synthesis of highly reduced compounds, like isoprenoids, phenols, or alkaloids (Selmar and Kleinwächter 2013). However, there exist few contradictory reports also which suggest the reduced level of phenolic compounds like caffeic acid, *p*-coumaric acid, and ferulic acid during water stress in grapevine (Król et al. 2014). Nonetheless, the increased accumulation of antioxidant has been reported in many crops including wheat (Keleş and Öncel 2002), *Lavender* (Munné-Bosch et al. 2001), *Ligustrum vulgare* (Tattini et al. 2004).

However, the application of drought stress for enhancing the antioxidants and phytochemicals is not without risk. The application of the deliberate water stress, its intensity, frequency, and intervals is crucial to attaining the desired level of accumulation of bioactive compounds. In lettuce (*Lactuca sativa*), the exposure of plants to multiple water stress resulted in a significant reduction in shoot growth, whereas mild stress imposed before harvest resulted in an enhanced concentration of phytochemical (Oh et al. 2010).

In real field conditions, drought stress is often accompanied by higher temperatures. The increased accumulation of the bioactive compounds and plant phenolics under heat conditions has also been reported in major cereal grains. For instance, Shewry et al. (2010) evaluated 26 wheat cultivars by growing them at 6 different locations spread across Hungary, France, Poland, and the UK for 2 years and found that the phytochemicals like stanols, alkylresorcinols, and bound phenolic acids had strong positive correlations with the mean temperature between heading and harvest. The whole-wheat grain does contain significant amounts of antioxidants like carotenoids and polyphenols that can effectively scavenge many free radicals formed during various metabolic reactions. Particularly the polyphenols of the wheat have hydroxyl groups linked to the aromatic rings that can react and stabilize free radicals. The effect of environmental conditions is generally more pronounced on the soluble fraction of phenolic compounds compared to bound forms of polyphenols in wheat grain (Di Silvestro et al. 2017). In addition to the wheat, barley is also considered a good source of bioactive phenols. Martinez et al. evaluated 27 barley genotypes (*Hordeum vulgare* L.) under two different environmental regimes in the Czech Republic and Spain and reported the good amount of tocopherols (ranging between 39.9 and 81.6 μg/g) in barley (Martinez et al. 2017). Phytochemicals present in barley are categorized into several major classes, like polyphenols, flavonoids, phytosterols, lignans, tocols, and folates. Authors identified 64 bioactive compounds in the barley that included 19 phenolic acids and aldehydes, 9 flavan 3-ols, 9 flavone glycosides, and 27 anthocyanins indicating barley as the good source of antioxidants.

19.5 Trickling Phenylpropanoids Biosynthesis Pathway for Enhancing Drought Stress Tolerance

In the past decade, there is a growing demand for the food rich in flavonoids and other antioxidants. This has attracted the attention of the researcher to increase the flavonoid content in the food item. Further, the increased production of the non-enzymatic antioxidants particularly flavonoids is also seen as the pragmatic strategy to combat the deleterious effects of water stress on plant growth and development. Earlier, structural and regulatory genes of the maize were transferred into rice, which resulted in the increased expression of the genes of the anthocyanin pathway (Gandikota et al. 2001).

Due to the fine understanding of the biochemical pathways of phenylpropanoid metabolism, the efforts are on to trickle this pathway at the molecular level for enhanced production of flavonoids. The transgenic modifications targeted at phenylpropanoid biosynthesis are likely to be the most direct way of enhancing flavonoid biosynthesis. The scientific study across diverse systems reflects the role of transcription factors in the regulation of flavonoid biosynthesis. The transcription factors often regulate the gene expression of multiple genes and hence targeted for transgenic research. For instance, Naing and coworkers recently reported the enhanced drought stress tolerance in transgenic tobacco by overexpressing snapdragon-derived *Ros1* (Naing et al. 2018). The overexpression of *Ros1* resulted in enhanced anthocyanin accumulation and elevated expression of stress-responsive genes in tobacco plants. The overexpression of *SbMYB8* gene derived from *Scutellaria baicalensis* (a traditional Chinese medicinal plant) in tobacco has been shown to regulate the chalcone synthase activity for increased production of flavonoid like caffeoylquinic acid. The *SbMYB8* transgenic tobacco plants synthesized higher amount of flavonoids, displayed higher gene expression of flavonoid synthesis genes, and displayed better drought tolerance compared to wild-type plants (Yuan et al. 2015). Very recently, the transgenic Arabidopsis for SsMAX2 *gene (a key component of strigolactones signaling)* has been shown to accumulate higher anthocyanin content under drought stress conditions leading to significant improvement in drought stress tolerance compared to wild-type plants (Wang et al. 2019).

The genetic manipulations of the phenylpropanoid biosynthesis pathway have been a prime target of researchers engaged in biofuel research for engineering plants for low lignin content. The thioester p-coumaroyl-CoA serves as a branch point in the phenylpropanoid pathway from which lignin-specific branch and flavonoid-specific branch of the phenylpropanoid network diverge (Vogt 2010). At this branch point, the generated thioester can either be used by a lignin-specific branch of the phenylpropanoid pathway forming p-coumaryl, coniferyl, and sinapyl alcohol monolignols or else may be diverted to the flavonoid biosynthesis through malonate pathway. The downregulation of genes of lignin-specific branch can alter the carbon flux through the phenylpropanoid pathway possibly toward flavonoid biosynthesis and can also modulate the synthesis of other secondary metabolites. For producing the plant material with better pulp properties, the downregulation/knockdown of the genes of lignin biosynthesis is attempted. The lignin modification has been shown to be associated with activation of the genes involved in oxidative stress in tobacco, poplar, and Arabidopsis (Baxter and Stewart Jr 2013). However, the lignin is important component of plant secondary cell wall. The reduction in the lignin content of the plant cell wall may not only negatively affect the drought stress tolerance of the plant but also increase the possibility of pathogen attack.

19.6 Conclusion and Future Perspectives

The unavailability of the required amount of water is the major constraint in achieving the ideal threshold yields in any agricultural or horticultural cropping system. Plants are often confronted with the limited supply of the water at some stage of their life cycle. However, some plants are evolutionary well adopted to water scarce conditions and have the natural ability to tackle the drought-induced dysregulation of metabolism often indicated by the abrupt generation of a large number of reactive oxygen species (ROS). Such tolerant plants are equipped with higher levels of enzymatic (peroxidases, superoxide dismutase, and catalase) and non-enzymatic antioxidant (phytochemicals) system as a physiological weapon for quenching of the ROS produced during the drought stress. The plant phenolics have been a subject of scientific investigation for the last many decades. The plant phenolics represent the most abundant and the most versatile natural products of plants. The understanding of the plant phenolics biosynthesis, accumulation, and its regulation is of utmost importance for devising the strategy to minimize the severe losses caused due to the various abiotic stresses. Extensive research has unraveled the molecular mechanisms of drought and desiccation tolerance with more emphasis on the role of plant flavonoids and other antioxidants. In response to water stress, plants typically accumulate a wide range of antioxidants, including enzymatic antioxidants and non-enzymatic. By now, a great deal of information has been generated concerning the biosynthesis, regulation, and genetic manipulation of plant phenolics. Therefore, there are many opportunities to exploit accumulated information to overcome the great agricultural loses by enhancing the phenylpropanoid contents ensuring the minimal trade-off for primary metabolites.

Does the role of flavonoids are simply limited to the chemical quenching of ROS, or they play a more versatile role in plants? This is the question which is pertinent mainly because of the exhaustive chemical diversity of the flavonoids in nature. It is biologically irrational that plants would channel 40% of the total carbon flux through phenylpropanoid biosynthesis pathways for the mere quenching of the ROS species if the other enzymatic antioxidant system is also operational under stress conditions. The vast chemical diversity of flavonoids (more than 7000) that plant has evolved also suggest about their multifaceted role, else such great chemical diversity was not required for ROS detoxification. The role of the flavonoids as signaling molecules is also being discussed in plants. Recently, the flavonoids like quercetin derivatives and their role in signaling cascades that regulate cell growth have been discussed (Hou and Kumamoto 2010). It is also proposed that the flavonoids modulate the phytohormone signaling by inhibiting the activity of a wide range of protein kinases, including mitogen-activated protein kinases, that operate downstream of ROS in the regulation of cell growth and differentiation (Brunetti et al. 2018a). With the continuously expanding knowledge about the chemical diversity of flavonoids and their proposed wider role in modulating the phytohormones signaling and cell growth, the plant flavonoids research would continue to assume central focal theme for plant abiotic stress tolerance, even in remaining part of twenty-first century.

References

Akula R, Ravishankar GA (2011) Influence of abiotic stress signals on secondary metabolites in plants. Plant Signal Behav 6(11):1720–1731

Almaraz-Abarca N, González-Elizondo MS, Tena-Flores JA, Ávila-Reyes JA, Herrera-Corral J, Naranjo-Jiménez N (2006) Foliar flavonoids distinguish Pinus leiophylla and Pinus chihuanuana (Coniferales: Pinaceae). Proc Biol Soc Wash 119(3):426–437

Ashraf MA, Riaz M, Arif MS, Rasheed R, Iqbal M, Hussain I, Salman M (2019) 9 the role of non-enzymatic antioxidants in improving abiotic stress tolerance in plants. In: Plant tolerance to environmental stress: role of phytoprotectants

Ávila-Reyes JA, Almaraz-Abarca N, Chaidez-Ayala AI, Ramírez-Noya D, Delgado-Alvarado EA, Torres-Ricario R et al (2018) Foliar phenolic compounds of ten wild species of Verbenacea as antioxidants and specific chemomarkers. Braz J Biol 78(1):98–107

Bardzik JM, Marsh HV, Havis JR (1971) Effects of water stress on the activities of three enzymes in maize seedlings. Plant Physiol 47(6):828–831

Baxter HL, Stewart CN Jr (2013) Effects of altered lignin biosynthesis on phenylpropanoid metabolism and plant stress. Biofuels 4(6):635–650

Benesova M, Holá D, Fischer L, Jedelský PL, Hnilička F, Wilhelmová N et al (2012) The physiology and proteomics of drought tolerance in maize: early stomatal closure as a cause of lower tolerance to short-term dehydration? PLoS One 7(6):e38017

Bhattacharya A, Sood P, Citovsky V (2010) The roles of plant phenolics in defence and communication during agrobacterium and rhizobium infection. Mol Plant Pathol 11(5):705–719

Boudet AM (2007) Evolution and current status of research in phenolic compounds. Phytochemistry 68(22–24):2722–2735

Brunetti C, Fini A, Sebastiani F, Gori A, Tattini M (2018a) Modulation of Phytohormone signaling: a primary function of flavonoids in plant–environment interactions. Front Plant Sci 9

Brunetti C, Loreto F, Ferrini F, Gori A, Guidi L, Remorini D et al (2018b) Metabolic plasticity in the hygrophyte Moringa oleifera exposed to water stress. Tree Physiol 38(11):1640–1165

Castellarin SD, Matthews MA, Di Gaspero G, Gambetta GA (2007) Water deficits accelerate ripening and induce changes in gene expression regulating flavonoid biosynthesis in grape berries. Planta 227(1):101–112

Di Silvestro R, Di Loreto A, Bosi S, Bregola V, Marotti I, Benedettelli S et al (2017) Environment and genotype effects on antioxidant properties of organically grown wheat varieties: a 3-year study. J Sci Food Agr 97(2):641–649

Gabaston J, Cantos-Villar E, Biais B, Waffo-Teguo P, Renouf E, Corio-Costet MF et al (2017) Stilbenes from Vitis vinifera L. waste: a sustainable tool for controlling Plasmopara viticola. J Agric Food Chem 65(13):2711–2718

Gandikota M, de Kochko A, Chen L, Ithal N, Fauquet C, Reddy AR (2001) Development of transgenic rice plants expressing maize anthocyanin genes and increased blast resistance. Mol Breed 7(1):73–83

Gharibi S, Tabatabaei BES, Saeidi G, Goli SAH (2016) Effect of drought stress on total phenolic, lipid peroxidation, and antioxidant activity of Achillea species. Appl Biochem Biotechnol 178 (4):796–809

Gharibi S, Tabatabaei BES, Saeidi G, Talebi M, Matkowski A (2019) The effect of drought stress on polyphenolic compounds and expression of flavonoid biosynthesis related genes in Achillea pachycephala Rech. f. Phytochemistry 162:90–98

Gholizadeh A (2011) Effects of drought on the activity of phenylalanine ammonia lyase in the leaves and roots of maize inbreds. Aust J Basic Appl Sci 5:952–956

Hertweck C (2009) The biosynthetic logic of polyketide diversity. Angew Chem Int Ed 48 (26):4688–4716

Hodaei M, Rahimmalek M, Arzani A, Talebi M (2018) The effect of water stress on phytochemical accumulation, bioactive compounds and expression of key genes involved in flavonoid biosynthesis in Chrysanthemum morifolium L. Ind Crop Prod 120:295–304

Hou DX, Kumamoto T (2010) Flavonoids as protein kinase inhibitors for cancer chemoprevention: direct binding and molecular modeling. Antioxid Redox Signal 13(5):691–719

Hura T, Hura K, Grzesiak S (2008) Contents of total phenolics and ferulic acid, and PAL activity during water potential changes in leaves of maize single-cross hybrids of different drought tolerance. J Agron Crop Sci 194(2):104–112

Ishimaru K, Nonaka GI, Nishioka I (1987) Tannins and related compounds. LV. Isolation and characterization of acutissimins a and B, novel tannins from Quercus and Castanea species. Chem Pharm Bull 35(2):602–610

Jaafar HZ, Ibrahim MH, Fakri M, Farhana N (2012) Impact of soil field water capacity on secondary metabolites, phenylalanine ammonia-lyase (PAL), maliondialdehyde (MDA) and photosynthetic responses of Malaysian Kacip Fatimah (Labisia pumila Benth). Molecules 17 (6):7305–7322

Keleş Y, Öncel I (2002) Response of antioxidative defence system to temperature and water stress combinations in wheat seedlings. Plant Sci 163(4):783–790

Kharazian N (2014) Chemotaxonomy and flavonoid diversity of Salvia L.(Lamiaceae) in Iran. Acta Bot Bras 28(2):281–292

Kleinwachter M, Selmar D (2014) Influencing the product quality by applying drought stress during the cultivation of medicinal plants. In: Physiological mechanisms and adaptation strategies in plants under changing environment. Springer, New York, pp 57–73

Knaggs AR (2001) The biosynthesis of shikimate metabolites. Nat Prod Rep 18(3):334–355

Król A, Amarowicz R, Weidner S (2014) Changes in the composition of phenolic compounds and antioxidant properties of grapevine roots and leaves (Vitis vinifera L.) under continuous of long-term drought stress. Acta Physiol Plant 36(6):1491–1499

Larson RA (1988) The antioxidants of higher plants. Phytochemistry 27(4):969–978

Lee J, Lee DG (2015) Novel antifungal mechanism of resveratrol: apoptosis inducer in Candida albicans. Curr Microbiol 70(3):383–389

Lee BR, Kim KY, Jung WJ, Avice JC, Ourry A, Kim TH (2007) Peroxidases and lignification in relation to the intensity of water-deficit stress in white clover (Trifolium repens L.). J Exp Bot 58 (6):1271–1279

Ma D, Sun D, Wang C, Li Y, Guo T (2014) Expression of flavonoid biosynthesis genes and accumulation of flavonoid in wheat leaves in response to drought stress. Plant Physiol Biochem 80:60–66

Martinez M, Motilva MJ, de las Hazas MCL, Romero MP, Vaculova K, Ludwig IA (2017) Phytochemical composition and β-glucan content of barley genotypes from two different geographic origins for human health food production. Food Chem 245:61–170

Munné-Bosch S, Schwarz K, Alegre L (2001) Water deficit in combination with high solar radiation leads to midday depression of a-tocopherol in field-grown lavender (Lavandula stoechas) plants. Funct Plant Biol 28(4):315–321

Naing AH, Ai TN, Lim KB, Lee IJ, Kim CK (2018) Overexpression of Rosea1 from snapdragon enhances anthocyanin accumulation and abiotic stress tolerance in transgenic tobacco. Front Plant Sci 9

Nakabayashi R, Yonekura-Sakakibara K, Urano K, Suzuki M, Yamada Y, Nishizawa T et al (2014) Enhancement of oxidative and drought tolerance in Arabidopsis by over accumulation of antioxidant flavonoids. Plant J 77(3):367–379

Oh MM, Carey EE, Rajashekar CB (2010) Regulated water deficits improve phytochemical concentration in lettuce. J Am Soc Hortic Sci 135(3):223–229

Paakkonen E, Seppänen S, Holopainen T, Kokko H, Kärenlampi S, Kärenlampi L, Kangasjärvi J (1998) Induction of genes for the stress proteins PR-10 and PAL in relation to growth, visible injuries and stomatal conductance in birch (Betula pendula) clones exposed to ozone and/or drought. New Phytol 138(2):295–305

Pandey A, Rajamani U, Verma J, Subba P, Chakraborty N, Datta A et al (2010) Identification of extracellular matrix proteins of rice (Oryza sativa L.) involved in dehydration-responsive network: a proteomic approach. J Proteome Res 9(7):3443–3464

Phimchan P, Chanthai S, Bosland PW, Techawongstien S (2014) Enzymatic changes in phenylalanine ammonia-lyase, cinnamic-4-hydroxylase, capsaicin synthase, and peroxidase activities in Capsicum under drought stress. J Agric Food Chem 62(29):7057–7062

Quideau S, Deffieux D, Douat-Casassus C, Pouysegu L (2011) Plant polyphenols: chemical properties, biological activities, and synthesis. Angew Chem Int Ed 50(3):586–621

Repo-Carrasco-Valencia R, Hellström JK, Pihlava JM, Mattila PH (2010) Flavonoids and other phenolic compounds in Andean indigenous grains: quinoa (Chenopodium quinoa), kañiwa (Chenopodium pallidicaule) and kiwicha (Amaranthus caudatus). Food Chem 120(1):128–133
Sarker U, Oba S (2018) Drought stress enhances nutritional and bioactive compounds, phenolic acids and antioxidant capacity of Amaranthus leafy vegetable. BMC Plant Biol 18(1):258
Selmar D, Kleinwächter M (2013) Influencing the product quality by deliberately applying drought stress during the cultivation of medicinal plants. Ind Crop Prod 42:558–566
Shao HB, Chu LY, Lu ZH, Kang CM (2008) Primary antioxidant free radical scavenging and redox signaling pathways in higher plant cells. Int J Biol Sci 4(1):8
Shewry PR, Piironen V, Lampi AM, Edelmann M, Kariluoto S, Nurmi T et al (2010) The HEALTHGRAIN wheat diversity screen: effects of genotype and environment on phytochemicals and dietary fiber components. J Agric Food Chem 58:9291–9298
Siracusa L, Gresta F, Sperlinga E, Ruberto G (2017) Effect of sowing time and soil water content on grain yield and phenolic profile of four buckwheat (Fagopyrum esculentum Moench.) varieties in a Mediterranean environment. J Food Compos Anal 62:1–7
Tattini M, Galardi C, Pinelli P, Massai R, Remorini D, Agati G (2004) Differential accumulation of flavonoids and hydroxycinnamates in leaves of Ligustrum vulgare under excess light and drought stress. New Phytol 163(3):547–561
Tian X, Lei Y (2006) Nitric oxide treatment alleviates drought stress in wheat seedlings. Biol Plant 50(4):775–778
Tovar MJ, Romero MP, Girona J, Motilva MJ (2002) L-Phenylalanine ammonia-lyase activity and concentration of phenolics in developing olive (Olea europaea L cv Arbequina) fruit grown under different irrigation regimes. J Sci Food Agric 82(8):892–898
Veit M, Beckert C, Höhne C, Bauer K, Geiger H (1995) Interspecific and intraspecific variation of phenolics in the genus Equisetum subgen us Equisetum. Phytochemistry 38(4):881–891
Ververidis F, Trantas E, Douglas C, Vollmer G, Kretzschmar G, Panopoulos N (2007) Biotechnology of flavonoids and other phenylpropanoid-derived natural products. Part I: chemical diversity, impacts on plant biology and human health. Biotechnol J: Healthcare Nutr Technol 2 (10):1214–1234
Vogt T (2010) Phenylpropanoid biosynthesis. Mol Plant 3(1):2–20
Wang Q, Ni J, Shah F, Liu W, Wang D, Yao Y et al (2019) Overexpression of the stress-inducible SsMAX2 promotes drought and salt resistance via the regulation of redox homeostasis in Arabidopsis. Int J Mol Sci 20(4):837
Xiang L, Moore BS (2005) Biochemical characterization of a prokaryotic phenylalanine ammonia lyase. J Bacteriol 187(12):4286–4289
Yuan Y, Qi L, Yang J, Wu C, Liu Y, Huang L (2015) A Scutellaria baicalensis R2R3-MYB gene, SbMYB8, regulates flavonoid biosynthesis and improves drought stress tolerance in transgenic tobacco. Plant Cell Tiss Org Cult (PCTOC) 120(3):961–972
Zhang J, Kirkham MB (1994) Drought-stress-induced changes in activities of superoxide dismutase, catalase, and peroxidase in wheat species. Plant Cell Physiol 35(5):785–791
Zhang X, Liu CJ (2015) Multifaceted regulations of gateway enzyme phenylalanine ammonia-lyase in the biosynthesis of phenylpropanoids. Mol Plant 8(1):17–27

Plants as Biofactories for Phenolic Compounds

20

Sara Aviles-Gaxiola, Gabriela Olivo-Vázquez, Luis A. Cabanillas-Bojórquez, Erick P. Gutiérrez-Grijalva, and J. Basilio Heredia

Abstract

For a long time, plants have been considered as healing agents for treating and preventing different diseases. These beneficial effects are mainly attributed to phenolic compounds. Phenolic compounds have as a basic structure one or more hydroxyl groups (–OH) attached to an aromatic ring. They are classified in two groups: flavonoids and phenolic acids. The regular intake of plant-based foods rich in phenolic compounds has been associated to the prevention and treatment of diseases such as Alzheimer's, Parkinson's, cardiovascular problems, atherosclerosis, metabolic syndrome, diabetes, and many types of cancer. Therefore, their demand has increased, so plants are now being used as biofactories for their production. By exposing plants to abiotic and biotic stress, phenolic compounds are synthesized in response to ROS overproduction. Their production has also been raised by increasing the expression of genes involved in phenolic compounds metabolic pathways. Present chapters address the use of biotic and abiotic stress as well as genetic engineering tools for increasing phenolic content in different plants.

Keywords

Plant · Phenolics · Biofactory · Stress · Biotic · Abiotic · ROS

S. Aviles-Gaxiola · G. Olivo-Vázquez · L. A. Cabanillas-Bojórquez · J. B. Heredia (✉)
Department of Food Science and Technology, Centro de Investigación en Alimentación y Desarrollo, A.C, Culiacán, Sinaloa, Mexico
e-mail: jbheredia@ciad.mx

E. P. Gutiérrez-Grijalva
Cátedras CONACYT – Centro de Investigación en Alimentación y Desarrollo, AC, Culiacán, Sinaloa, México

R. Lone et al. (eds.), *Plant Phenolics in Sustainable Agriculture*,
https://doi.org/10.1007/978-981-15-4890-1_20

20.1 Introduction

Around 80% of people living in developing countries consider their native plants as the main healing agent for treating and preventing different diseases. This healing power has been attributed to their antioxidant activity and associated to the presence of phenolic compounds (Abdul Qadir et al. 2017).

Phenolic compounds are divided in two groups: flavonoids and phenolic acids. They are among the most investigated phytochemicals in both medicinal and nutritional areas (Del Rio et al. 2010).

Phenols were first isolated in 1834 by Friedlieb Ferdinand Runge and used as a wound dressing, antiseptic, and disinfectant in Germany. One of the first registers was in 1865 where the surgeon Joseph Lister employed them in a tibia fracture surgery (Hugo 1978). Currently, over 8000 different naturally occurring phenolic compounds are known, and the list grows (Tsao 2010).

The main sources of flavonoids are parsley (*Petroselinum crispum*), celery (*Apium graveolens*), chamomile (*Matricaria chamomilla*), *Ginkgo biloba*, mint (*Mentha arvensis*), and red pepper (*Capsicum annum*) (Panche et al. 2016). While the main sources of phenolic acids are green tea (*Camellia sinensis*), coffee, berries, red wine, and whole cereals such as wheat (*Triticum* spp.), rice (*Oryza sativa*), corn (*Zea mays*), and oats (*Avena sativa*) (Weichselbaum and Buttriss 2010).

20.2 Chemical Characteristics

Phenolics are comprised in a wide array of compounds derived from the secondary metabolism of plants (Kutchan et al. 2015). Phenolic compounds have as a basic structure one or more hydroxyl groups (–OH) attached to an aromatic ring (Fig. 20.1). The presence of the hydrogen in the –OH groups makes phenolic compounds weak labile acids. Phenolic compounds can be ubiquitously found throughout the plant kingdom where they are usually present as esters or glycosides.

Nowadays, scientific literature uses the terms "phenolic compounds" and "polyphenols" to refer to the widespread subclasses of phenolic compounds, which are grouped according to their chemical characteristics related to their structure and type of compound. However, phenolic compounds are metabolically speaking, only metabolites derived from the shikimate phenylpropanoid and/or the polyketide pathways, with one or more phenolic rings and lacking of nitrogen-based functional groups (Quideau et al. 2011).

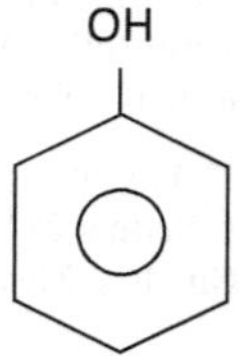

Fig. 20.1 Graphical representation of a phenol conformed of an aromatic ring with an hydroxyl group (–OH) attached

Phenolic compounds are the subject of many studies since they have been implicated in a myriad of health-promoting properties. The chronic intake of plant-based foods rich in phenolic compounds has been associated to an onset/prevention of noncommunicable diseases such as Alzheimer's, Parkinson's, cardiovascular problems, atherosclerosis, metabolic syndrome, diabetes, and many types of cancer (Del Rio et al. 2010; Fraga et al. 2010; Ozcan et al. 2014). These bioactive effects are attributed to the physicochemical properties related within the phenol functional group, which allows the –OH group to act as hydrogen-bond donor or as an acceptor (Quideau et al. 2011).

20.3 Classification

Phenolic compounds are a diverse group of organic compounds; at least 8000 phenolic structures have been elucidated. Phenolics have an important role in the protection of the plants against biotic or abiotic stress. These molecules are present in most plant tissues, for example, leaves, roots, and edible parts of the plant such as fruits (de la Rosa et al. 2019). Furthermore, phenolics can be found in nature conjugated to sugar moieties and organic acids.

There are several ways to classify phenolic compounds, and nowadays there is not a general agreement in how to categorize them. Besides, their classification can be dependent on the molecule's different properties. For example, (1) classifying phenolic compounds by their solubility is convenient for explaining the metabolic fate of phenolic compounds since it depends greatly on this property (Santana-Gálvez and Jacobo-Velázquez 2018); (2) based on the chemical structure, phenolic compounds can be categorized by the number of aromatic rings (Fig. 20.2), and this classification has been previously detailed (Santana-Gálvez and Jacobo-Velázquez 2018).

Phenolic compounds can be also classified in flavonoids and non-flavonoids. Flavonoids are the most abundant phenolics in plant kingdom. Its presence in nature is more commonly found in leaves and the skin of fruits (Crozier et al. 2009). The basic structure of flavonoids consists of 15 carbon atoms arranged in 2 phenyl rings named as A and B, linked through a heterocyclic pyran ring named as ring C (Fig. 20.3).

Flavonoids are subdivided into six groups of families, differing in their degree of oxidation and pattern of substitution of the C ring, whereas individual compounds in each family are differentiated by the pattern of hydroxylation and methylation of the rings A and B. The six groups of flavonoids are flavones, flavonols, flavonones, isoflavones, anthocyanidins, and flavan-3-ols (de la Rosa et al. 2019). From these subgroups, flavonols are the most abundant flavonoids; the most representative flavonols are kaempferol, myricetin, quercetin, and isorhamentin (Crozier et al. 2009).

On the other hand, non-flavonoids consist of phenolic acids, stilbenes, lignans, and tannins (Oliveira et al. 2014). The most representative phenolic acid is gallic

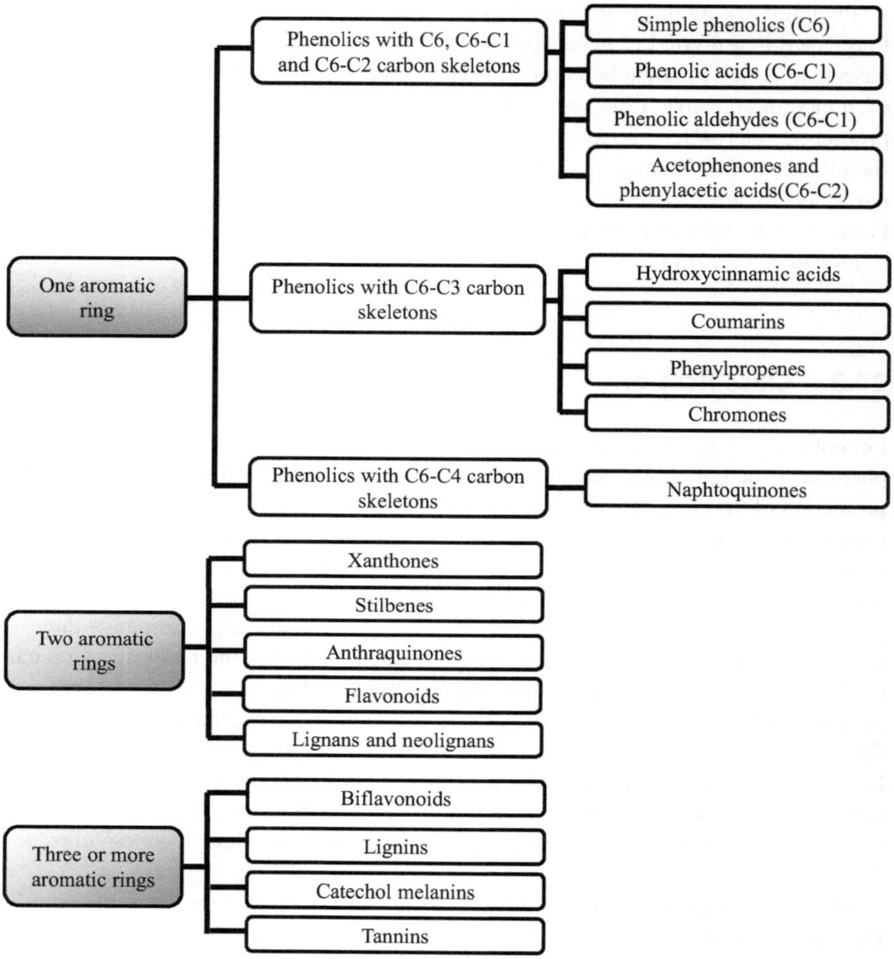

Fig. 20.2 Classification of phenolic compounds based on their basic chemical structure. Source: (Santana-Gálvez and Jacobo-Velázquez 2018)

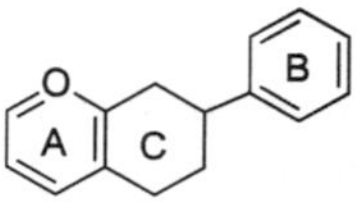

Fig. 20.3 Basic flavonoid skeleton

acid and acts as a precursor for the synthesis of other non-flavonoids such as hydrolysable tannins and stilbenes (Crozier et al. 2009).

20.4 Health-Promoting Properties

20.4.1 Neurodegenerative Diseases

The overrun of the endogenous antioxidant system due to free radicals is one of the main factors involved in the onset of neurodegenerative diseases (Poprac et al. 2017). The adoption of a Mediterranean diet has been associated with a lower incidence of neurodegenerative diseases due to the inclusion of foods such as wine (Valls-Pedret et al. 2015).

The total phenolic compounds in red wine reaches up to 1000 mg/kg. These components exert their antioxidant activity by directly scavenging the reactive oxygen species (ROS) and nitrogen oxygen species (NOS), and by inducing phase II detoxifying enzymes via nuclear factor E2-related factor 2 (Nrf2) activation over in vitro models (Rodriguez-Morato et al. 2015). For the particular case of Parkinson's disease, (−)-epigallocatechin-3-gallate, the most abundant polyphenol found in tea, and quercetin also found in red wine, protects against oxidative hippocampal neurodegeneration at the time that inhibits L-DOPA (dopamine precursor) methylation on in vivo models (Kang et al. 2013). As for Alzheimer's, phenolic acids and polyphenols modified by microbiota inhibit amyloid β aggregation, which is associated to cognitive decile in Alzheimer's disease (Pistollato et al. 2016).

20.4.2 Inflammation

Inflammation is a natural defense process that recognizes harmful environmental stimuli (Neagu et al. 2018). Phenolic compounds such as epicatechin and caffeic acid, found in red Cannonau wine, diminish the intestinal inflammation induced by dietary oxysterols, which are cholesterol auto-oxidation products that may lead to colorectal cancer by preventing the induction of NF-κB-dependent inflammatory events (Guina et al. 2015).

Oleocanthal is a phenolic compound isolated from virgin olive oil with an anti-inflammatory property similar to ibuprofen. It represents 10% of the olive oil phenolic fraction, and it inhibits cyclooxygenase pathway over in vitro monocytes and monocyte-derived macrophages (Lucas et al. 2011). Kaempferol is another phenolic compound with autoinflammatory activity and highly found in golden root (*Rhodiola sachalinensis*). It reduced nitric oxide production by inhibiting nitric oxide synthase over in vitro LPS-stimulated RAW 264.7 cells (Choe et al. 2012).

20.4.3 Cancer

Cancer is characterized by a loss of cell death capacity and an uncontrolled cell division (Wong 2011). One of the main responsible for cancer onset is ROS, known to regulate signaling molecules required for cell cycle progression and to control the expression of various tumor suppressor genes. Phenolic compounds such as the anthocyanin delphinidin induce apoptosis and cell cycle arrest in several types of cancer by the suppression of the NF-κB pathway. Another target for cancer treatment is matrix metalloproteinase, an enzyme capable of degrade extracellular matrix. It is regulated by peonidin-3-glucoside (Zhou et al. 2016). Many other phenolic compounds have been studied, mainly over in vitro cancer models where cinnamic acid, caffeic acid, coumaric acid, and ferulic acid outstand for their antiproliferative effect over melanoma, prostate, and breast cancer, respectively (Anantharaju et al. 2016).

20.4.4 Cardiovascular Health

The term "cardiovascular diseases" encompasses a set of conditions among which peripheral arterial disease, coronary heart disease, venous thromboembolism, and congenital heart diseases stand out. Foods such as garlic, onion, ginger, and dark chocolate were consumed for a long time for their cardioprotective effect. Know it is known that these effects are due to the high antioxidant contents mainly due to phenolic acid, flavonoids, lignans, and stilbenes (Olas 2017). A high source of stilbenes is grape pomace. Consuming it for 7 days has cardioprotective effects against isoprenaline-induced infarct-like lesion. The effect is associated to the stilbenes trans- and cis-resveratrol, glycosylated derivatives of resveratrol, trans- and cis-piceid, piceatannol, and viniferins (resveratrol dimmers), which reduce oxidative stress (Balea et al. 2018).

20.4.5 Diabetes

The fungi *Phellinus igniarius* is a rich source of 7,8-dihydroxycoumarin, 3,4-dihydroxybenzalacetone, 7,3′-dihydroxy-5′-methoxyisoflavone, and inoscavin C. These phenolics are considered as potent antidiabetic compounds. All of them increase glucose uptake over in vitro models by up to 2.34-fold. They activate GLUT4 translocation via AMPK pathway modulation (Zheng et al. 2018). Another source of antidiabetic phenolic compounds is *Senecio biafrae* leaves rich in gallic, chlorogenic, and caffeic acid as well as rutin, quercetin, and kaempferol. Together, these compounds inhibit the activity of alpha-glucosidase preventing the conversion of complex carbohydrates in to simple glucose (Ajiboye et al. 2018).

20.4.6 Obesity

Obesity is a preventable condition characterized by overweight (Hruby and Hu 2015). Hydroxycinnamic acid is considered as a potential compound for the management of obesity health complications. It inhibits macrophage infiltration and NF-kB activation in obese animals. Also, it inhibits the expression of adipokines

TNF-α at the time that prevents adipocyte differentiation (Alam et al. 2016). Also, the polyphenols flavan-3-ols, epigallocatechin gallate, genistein, daidzein, curcumin, resveratrol, and quercetin glycosides found mainly in black tea, green tea, soybean, turmeric, grapes, and onion, respectively, have been largely studied and recognized for their control in body weight and fat accumulation. This compound promotes energy dissipation by activating brown adipose tissue, increasing energy expenditure (Mele et al. 2017).

20.4.7 Hyperlipidemia

Hyperlipidemia is caused by an excess of cholesterol, triglycerides, or both, in the bloodstream (Karr 2017). Cinnamon polyphenols are able to inhibit hyperlipidemia in high-fat diet-fed rats by enhancing the expression of the transcription factors SREBP-1c, LXRs, NF-κB, and Nrf2 as well as enzymes such as ACLY and FAS involved in lipid metabolism (Tuzcu et al. 2017).

The most abundant phenolics identified in cinnamon are catechin, (−)-epigallocatechin gallate, syringic acid, gallic acid, vanillic acid, and p-coumaric acid (Lv et al. 2012). Recent studies have shown that *Clinacanthus nutans* leaves, a medicinal plant endemic to Asia, given to rats for 7 weeks, attenuate the oxidative stress through increasing serum antioxidant activity and upregulating the expression of hepatic antioxidant genes. The effects are attributed to protocatechuic acid (Sarega et al. 2016).

20.4.8 Osteoprotective

Osteoporosis is characterized by bone mass decrease, which can lead to fractures. For a long time, ferns such as *Drynaria* species have been used to prevent bone loss, specially its rhizome. The active compounds were unknown until recently, where chlorogenic acid, syringic acid, trans-ferulic acid, (−)-epigallocatechin, epigallocatechin gallate, quercetin dehydrate, and luteolin were identified in *Drynaria* extracts. These extracts increase the cell viability of osteoblastic cells by around 40%, promoting a high cellular density (Kang et al. 2014).

Also, genistein aglycone, a soybean isoflavone, has been known for a long time as a phytochemical that induces bone formation and resorption through the osteoprotegerin-sRANKL system (Bitto et al. 2010). The medicinal effect of some phenolic compounds is described in Table 20.1.

20.5 Strategies to Enhance Phenolic Compounds in Plants

Nowadays phenolic compounds are of greater interest. Their demand has increased by several factors such as a lot of studies attributing them potential health benefits, consumer health concerns, as well as an increased rate of diseases related to aging.

Table 20.1 Medicinal effect of phenolic compounds

Disease	Phenolic compounds	Source	Mechanism	References
Neurodegeneration	Hydroxytyrosol	Red wine	Scavenging of ROS and phase II detoxifying enzymes induction via Nrf2 activation	Rodriguez-Morato et al. (2015)
Parkinson	(−)Epigallocatechin-3-gallate and querceti	Green tea and red wine, respectively	Inhibition of L-DOPA methylation	Kang et al. (2013)
Inflammation	Epicatechin and caffeic acid Oleocanthal Kaempferol	Red Cannonau wine Virgin olive oil Golden rice (*Rhodiola sachalinensis*)	Prevention over the induction of NF-κB-dependent inflammatory events via ROS scavenging Inhibition of cyclooxygenase pathway Inhibition of nitric oxide synthase	Guina et al. (2015), Lucas et al. (2011), Choe et al. (2012)
Cancer	Delphinidin Peonidin-3-glucoside		Suppression of the NF-κB pathway Inhibition of matrix metalloproteinase	Zhou et al. (2016)
Cardiovascular health	Stilbenes	Grape pomace	Isoprenaline-induced infarct-like lesion reduction via oxidative stress reduction	Balea et al. (2018)
Diabetes	7,8-dihydroxycoumarin, 3,4-dihydroxybenzalacetone, 7,3′-dihydroxy-5′-methoxyisoflavone and inoscavin C Rutin, quercetin and kaempferol	*Phellinus igniarius* *Senecio biafrae*	Glucose uptake increment by activation of GLUT4 translocation Inhibition of alpha-glucosidase activity	Zheng et al. (2018) Ajiboye et al. (2018)
Obesity	Hydroxycinnamic acid, epigallocatechin gallate, genistein, daidzein, curcumin, and resveratrol	Black tea, green tea, soybean, turmeric, grapes, and onion, respectively	Inhibition of macrophage infiltration and NF-kB activation in obese animals, inhibition of adipokines *TNF*-α expression, prevention of adipocyte differentiation, and energy dissipation promotion by brown adipose tissue activation, promoting energy expenditure	Alam et al. (2016), Mele et al. (2017)

Hyperlipidemia	Catechin, (−)-epigallocatechin gallate, syringic acid, gallic acid, vanillic acid, and p-coumaric acid Protocatechuic acid	Cinnamon *Clinacanthus nutans*	Enhancement in the expression of the transcription factors SREBP-1c, LXRs, NF-κB, and Nrf2 Increased serum antioxidant activity and upregulation of the expression of hepatic antioxidant genes	Tuzcu et al. (2017)
Osteoporosis	Chlorogenic acid, syringic acid, trans-ferulic acid, (−)-epigallocatechin, epigallocatechin gallate, quercetin dehydrate, and luteolin	*Drynaria* species	Induction of bone formation and resorption through the osteoprotegerin-sRANKL system	Bitto et al. (2010)

Plants are the main source for obtaining phenolic compounds, so techniques have been introduced to increase the plant phenolic content (Ameer et al. 2017). Among them, the use of abiotic and biotic conditions has been found to increase the bioactive compounds such as polyphenols and flavonoids in response to ROS overproduction. By exposing different types of plants to these types of stresses, vegetable tissues with higher antioxidant capacity can be obtained (Abdi et al. 2019; Khan et al. 2019). Another strategy for phenolic increase in plants is genetic manipulation, which has recently emerged since the biosynthetic pathways for phenolic production have been unveiled. Therefore, the expression of genes involved in all of them, especially transcription factors that allow the control of specific enzymes, has been engineered as a promising strategy for medicinal plant production.

These three topics will be discussed in detail throughout this chapter.

20.5.1 Abiotic Stress and Phenolic Compounds in Plants

Plants are continuously subjected to different types of stresses, affecting their normal growth, for example, adverse abiotic conditions such as cold, heat, drought, excess of salt or toxic metals, as well as nutrient imbalance (Zhu 2016).

Abiotic stresses are major limiting factors and responsible for enormous crop yield losses worldwide (Jorge et al. 2016), so omics and biology tools have been used to elucidate the pathways responding to different abiotic stresses conditions.

Over the time, it has been found that these types of stresses are mainly mediated by the overproduction of signaling molecules such as ROS and phytohormones like methyl jasmonate and ethylene. These trigger the overproduction of some phytochemical compounds, for example phenolics, promoting its accumulation (Jacobo-Velázquez and Cisneros-Zevallos 2012).

Phenolic compounds are highly valued biomolecules, so different strategies have been applied to increase its production, including the induction of adverse abiotic conditions. This strategy has been suggested as a replacement of genetic tools; technologies are still considered hazardous.

In Fig. 20.4, the abiotic stress strategies employed for phenolic overproduction in plants are classified.

20.6 Radiation

20.6.1 Light

Visible light is the electromagnetic spectrum visible to the human eye, from 380 to 740 nm. Visible light overexposure triggers strategies of photoprotection to counteract the damage, so it induces photosynthesis to occur, therefore promoting other metabolic pathways to increase (Younis et al. 2010).

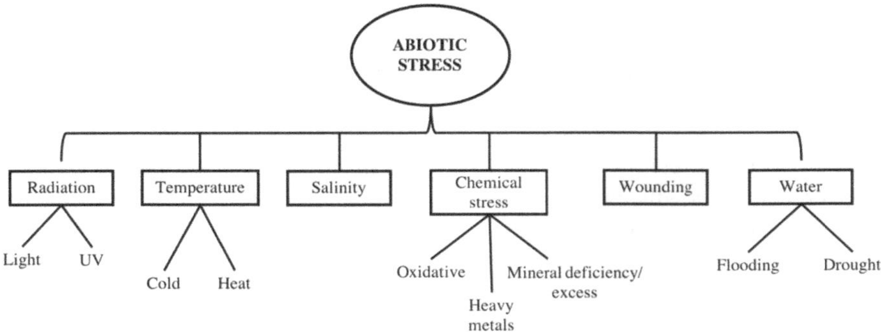

Fig. 20.4 The abiotic stress strategies employed for phenolic production in plants

Given the above, light-emitting diodes (LEDs) have been used lately to produce crops with special features, including improved nutritional profiles. Red and blue light going from 625–740 to 400–500 nm, respectively, is the most used spectra.

Monochromatic blue LED stimulates the biosynthesis of total phenolics and flavonoids in 2-week-old cherry tomato seedlings after 4 weeks exposure. Total phenolic and flavonoid content increased up to 33.33 and 75%, respectively (Kim et al. 2014). The same happens when applying blue LED over red leaf lettuce plants (Johkan et al. 2010).

In addition of being used for the increase of total phenolic content, LEDs have also been tested for raising particular compounds concentration. Blue and red LEDs tested separately promoted the accumulation of particular compounds in peas. After 4 days exposure, red LED increased catechin content by 14.3%, while chlorogenic acid and rutin content were increased by 26.3 and 18%, respectively, when applying blue LED (Liu et al. 2016).

Besides blue and red LEDs, the white one has been tested. Kale sprouts grown under white or blue LEDs showed an increase of total phenolic compounds of 34.55% and 69.09%, respectively. On the other side, the particular contents of anthocyanin in white, red, and blue LED-exposed grown sprouts were 8.69-, 5.94-, and 18.16-fold higher, respectively, as compared to nonexposed sprouts (Qian et al. 2016).

The combined effect of LEDs has also been evaluated. Blue and red LED increases total phenolic concentration in Sunmang and Grand Rapid lettuces by 55% and twofold, respectively (Son and Oh 2013). Also, chlorogenic acid, vitexin, rutin, quercetin, cyanidin 3-*O*-glucoside, and cyanidin 3-*O*-rutinoside concentrations are increased in Tartary buckwheat submitted to blue- and red-combined LED treatment (Seo et al. 2015).

The mechanism by which light radiation exposure works is by overexpressing *FtDFR* and *FtANS. These genes are overexpressed up to 7.1-fold.*

20.6.2 UV

Ultraviolet (UV) radiation is classified as UV-A, UV-B, and UV-C, in the range of 320–400, 280–320, and 100–280 nm, respectively (Edward et al. 2014). The application of UV rays affects plants by changing the normal metabolic pathways of photosynthesis, cell division, and any other process related to plant growth and development (Surjadinata et al. 2017). These effects are mainly promoted by reactive species accumulation which induces the expression of antioxidant genes already linked to the UV resistance 8 locus pathway. The above results in the activation of the plant defense system, including the triggering of the phenylpropanoid metabolism which results in the accumulation of secondary metabolites such as phenolic antioxidants. Consequently, UV radiation has been proposed, alone or in combination, as a cheap tool to enhance phenolic compound content in horticultural crops during postharvest life (Formica-Oliveira et al. 2017).

Lettuce, a very important crop worldwide, has been used as a plant factory to accumulate phenolic compounds. Total phenolic content (especially anthocyanins) and total antioxidant capacity were increased by 30 and 40%, respectively, without growth inhibition after 4 d UV-A exposure. The above was attributed to an increase in phenylalanine ammonia lyase (PAL) gene expression (a key gateway gene for the biosynthesis of phenolic compounds), and therefore PAL activity was increased (Lee et al. 2013). PAL is the first enzyme in the phenylpropanoid pathway.

On the other hand, UV-B applied to plants such as *Vitis vinifera* grape skin after defoliation and until harvest, alters the expression of 121 phenylpropanoid biosynthesis-related genes, including VvFLS*1*, *VvGT5*, *VvGT6*, *VvHY5–1*, *VvHY5–2*, and *VvRU*. Also, PAL and chalcone synthase gene expression was increased. These gene products are involved in the flavonol and monoterpenoid biosynthetic pathways. Among the compounds which concentration was highly increased by UV-B light application, there was syringic acid, caffeoyl tartaric acid, and resveratrol, which increased by 13%, 40%, and twofold, respectively, as compared to nonexposed plants (Rodríguez-Calzada et al. 2019).

The exposure to UV-C also increases resveratrol production in grapes, especially in *Vitis amurensis* and *Vitis labrusca* grapes. The mechanism is by PAL and stilbene synthase increased expression, where PAL is the first enzyme of the phenylpropanoid pathway, and stilbene synthase is responsible for catalyzing the reaction of resveratrol formation. Also, chalcone synthase gen had higher expressions. As for epigenetic changes, UC-V decreases methylation levels of *VaSTS2*, *VaSTS6*, and *VaSTS10* genes (Tyunin and Kiselev 2016). The above has also been documented for other plants such as shell ginger (*Alpinia zerumbet*) and in lemon pomace (Xuan et al. 2016; Papoutsis et al. 2016).

The effect of different types of UV radiation has been compared. For example, UV-A and UV-B over broccoli sprouts both particularly change their phenolic profiles. Applying UV-A during 120 min increased gallic acid hexoside I (~14%), 4-*O*-caffeoylquinic acid (~42%), gallic acid derivative (~48%), and 1-sinapoyl-2,2-diferulolyl-gentiobiose (~61%), and by applying UV-B, sinapoyl malate (~12%), gallotannic acid (~48%), and 5-sinapoyl-quinic acid (~121%) were increased (Moreira-Rodríguez et al. 2017).

UV irradiation must be exploited for the production of functional foods rich in bioactive phytochemicals. This has been highly recognized and utilized for sunscreens and cosmetics production (Takshak and Agrawal 2019).

The results obtained in terms of phenolic accumulation in different plant materials by applying radiation are summarized in Table 20.2.

20.7 Temperature

20.7.1 Cold

Cold stress has been applied to different plant materials. It is classified as either chilling or freezing, above and below 0 °C, respectively (Pareek et al. 2017). This condition promotes stress, affecting physiological, biochemical, molecular, and developmental processes of plants. Different results have been found when it comes to cold stress and phenolics plant production. In some cases, the content increases, but in others it is reduced or maintained. These will depend mainly on the intensity and duration of the stress, as well as on the plant material (Hajihashemi et al. 2018).

For example, when exposing *Vitis vinifera* leaves at 7 °C during a week, a reduction in total phenolic content of up to 22.05% has been observed, especially in caffeic acid, p-coumaric acid, ferulic acid, and caffeic acid. Also, tannin content of *Vitis vinifera* leaves exposed below 0 °C was reduced up to 15.9% (Król et al. 2015).

On the contrary, low temperatures increase phenolic content in other plants such as kale, cabbage, papaya, and olive tree leaves. In Kale and cabbage stored at 12 °C, phenolic content was increased by 65.71 and 60%, respectively, as compared to normal temperature conditions (Soengas et al. 2018). Regarding to Maradol papaya, storing it at chilling conditions (1 °C) promotes phenolic accumulation compared with those stored at normal 25 °C conditions. The compounds increased after 8 days of storage were ferulic acid and caffeic acid by 40 and by twofold, respectively. As for freezing, olive tree leaves submitted below −7 °C for 15 days showed an increase in PAL expression by up to 4.8-fold, promoting a total phenolic increase of 12% (Ortega-García et al. 2008).

It was found that low-temperature exposure changed gene expression after 24 h treatment. Genes codifying for enzymes such as PAL and chalcone synthase were increased by 8- and 50-fold, respectively, and 4-coumarate, CoA ligase and chalcone isomerase, increased by threefold. The expression of these genes is back to normal after returning the plants to 25 °C during 2 days (Christie et al. 1994).

Also, in other plants such as chickpea, submitting the seeds to cold stress promoted an increased expression of PAL and cinnamyl alcohol dehydrogenase genes. This indicates the crucial role of phenylpropanoid pathway in creating cold tolerance (Rakei et al. 2016).

Table 20.2 Phenolic accumulation in plant materials by radiation exposure

Radiation type	Plant material	Effects	Mode of action	References
Visible light				
Blue	Tomato seedlings and red leaf Lettuce Peas Kale sprouts	↑ Phenolic and flavonoid content by 33.33 and 75%, respectively ↑ Chlorogenic acid and rutin content by 26.3 and 18%, respectively ↑ Phenolic and anthocyanin content by 69.9% and 18.16-fold, respectively	Overexpression of *FtDFR* and *FtANS genes by 7.1-fold*	Kim et al. (2014) Liu et al. (2016), Johkan et al. (2010) Qian et al. (2016)
Red	Peas Kale sprouts	↑ Chlorogenic acid by 14.3% ↑ Anthocyanin content by 5.94-fold		Liu et al. (2016), Qian et al. (2016)
White	Kale sprouts	↑ Phenolic and anthocyanin content by 34.55% and 8.69-fold, respectively		Qian et al. (2016)
Blue + red	Sunmang lettuces Grand Rapid Tartary buckwheat	↑ Phenolic content by 55% ↑ Phenolic content by twofold ↑ Chlorogenic acid, vitexin, rutin, quercetin, cyanidin 3-*O*-glucoside, and cyanidin 3-*O*-rutinoside content		Seo et al. (2015)
UV				
UV-A	Lettuce Broccoli sprouts	↑ Phenolic content by 30% ↑ Gallic acid hexoside, 4-*O*-caffeoylquinic acid, gallic acid derivative, and 1-sinapoyl-2,2-diferulolyl-gentiobiose by 14, 42, 48, and 81%, respectively	Overexpression of *PAL* and *chalcone synthase* genes	Lee et al. (2013)
UV-B	*Vitis vinifera* grapes skin Broccoli sprouts	↑ Syringic acid, caffeoyl tartaric acid and resveratrol, by 13%, 40%, and twofold ↑ Sinapoyl malate, gallotannic acid, and 5-sinapoyl-quinic acid by 12, 48, and 121%, respectively		Rodríguez-Calzada et al. (2019), Moreira-Rodríguez et al. (2017)

20.7.2 Heat

The main use of heat in food industry is for reducing microbial levels. Nevertheless, it has been found that a lot of biochemical pathways are overexpressed in plants subjected to high temperature which in turn increases phytochemicals synthesis (Zhu 2016).

Heat shock at 100 °C was applied over fresh-cut carrots, increasing their phenolic content by 73.98% after 7 days of the heat shock application. Nevertheless, after 10 days the content starts to decrease. Phenolic compound accumulation was due to phenylpropanoid pathway activation. The most increased compound in wounded carrot is caffeoylquinic acid (Alegria et al. 2012). Six wheat genotype seedlings submitted to 45 °C during 20 h showed an increase of total phenolic content by about 40% in all genotypes (Ahmad et al. 2014).

The production and accumulation of phenolics according due to high temperatures will depend on the vegetable tissue. For example, fresh-cut carrots and onions accumulate more phenolics at 20 °C, while celery does it at 10 °C (Xiaoan et al. 2017).

Heat effect, as for the other stresses, is mainly the same. During heat, cell disruption may occur, liberating cytosolic ATP into the extracellular matrix. This ATP is recognized by receptors in the far cell's membrane increasing cytosolic $Ca2+$ concentrations, which activates NADPH oxidase and thus ROS and NOS production. ROS and NOS act as signals to activate PAL, and the plants answers back by producing antioxidant molecules as a defense system (Jacobo-Velázquez et al. 2011). The results obtained in terms of phenolic accumulation in some plant materials by applying different temperatures are summarized in Table 20.3.

20.8 Salinity

Around 20% of the world's irrigated lands have an excess of salt content, known as salinity. Salinity affects plants in multiple ways, for example, blocking respiration, inhibiting photosynthesis, and disordering metabolism and growth processes (Zhao et al. 2017). Nutritional quality of buckwheat sprouts under salinity was estimated. Here, the exposure time (from 1 to 7 days) and the salt concentrations (10, 50, and 100 mM) were evaluated. It was concluded that the most extreme conditions (100 mM for 7 days) resulted in the most effective condition to increase the phenolics content by 153%. The main increased phenolics were isoorientin, orientin, rutin, and vitexin. Previous studies have shown most of the phenolic compounds are generated by the phenylpropanoid pathway, which is stimulated by biotic and abiotic stresses (Lim et al. 2012). This effect is also associated to the increased in plant hormones production, including jasmonic acid and its methylated derivate (methyl jasmonic acid). These hormones induce enzyme production involved in the phenylpropanoid pathway, including PAL, thereby resulting in the accumulation of phenolic compounds (Lim et al. 2012).

Table 20.3 Phenolic accumulation in plant materials by low- or high-temperature exposure

Temperature type	Plant material	Effect	Mode of action	References
Cold				
Chilling	*Vitis vinifera* leaves Kale Cabbage Papaya	↓ Phenolic content by 22.05% ↑ Phenolic content by 65.71% ↑ Phenolic content by 60% ↑ Ferulic acid and caffeic acid by 40- and twofold, respectively	Overexpression of *PAL* and *chalcone synthase* genes by up to 8- and 50-fold, respectively	Król et al. (2015)
Freezing	*Vitis vinifera* leaves Olive tree leaves	↓ Phenolic content by 15.9% ↑ Phenolic content by 12%		Król et al. (2015) Ortega-García et al. (2008)
Heat				
100 °C	Fresh-cut carrots	↑ Phenolic content by 73.98%	Increasing cytosolic Ca^{2+} concentrations, which activates NADPH oxidase and thus ROS and NOS production overexpressing *PAL* gene	Alegria et al. (2012)
45 °C	Wheat	↑ Phenolic content by 40%		Ahmad et al. (2014)

Biosynthesis of phenolics was also stimulated in other plants such as olive cultivar leaves by irrigating them every 3 days during 1 month with 125 mM NaCl. Salinity stimulated the biosynthesis of phenolics by increasing its total content up to 129%. From them, the content of oleuropein, the most abundant phenolic from olive tree and oil, was increased by 3.8-fold (Petridis et al. 2012a). Similarly, a high salinity concentration of 300 mM Nacl was evaluated over *Salvia coccinea* leaf, increasing the total phenolic content by 17% (Grzeszczuk et al. 2018).

Salinity stress (5 dS/m) was applied to both susceptible and tolerant rice varieties. Total phenolics and flavonoids, as well as vanillin, ferulic acid, p-coumaric acid, and protocatechuic acid, were increased in tolerant varieties. Phenolic compounds were increased around 48% while flavonoids only 17.15%. In contrast they were markedly reduced in the susceptible cultivars (Jamalian et al. 2013). The enhancement in the synthesis of flavonoids and phenolics of strong tolerant varieties might be due to the adaptive mechanism of rice under salt stress. The presence of ferulic acid under osmotic stress may be related to the strengthening of the plant cell wall and the overall cell elongation. Also, they are possibly found in bounded forms with sugar or glycoside (Minh et al. 2016). It is believed that the salinity stress also enhances the phenolic compound production by the abscisic acid increased content in plants. The results obtained in terms of the phenolic accumulation in different plant materials by applying different salinity stresses are summarized in Table 20.4.

Table 20.4 Phenolic accumulation in plant materials by high salinity exposure

Salinity (NaCl)	Plant material	Effects	Mode of action	References
100 mM	Buckwheat sprouts	↑ Phenolic content by 153%, mainly isoorientin, orientin, rutin and vitexin	Increased plant hormones production, including jasmonic acid and methyl jasmonic acid which overexpresses *PAL* gene	Lim et al. (2012)
125 mM	Olive leaves	↑ Phenolic and oleuropein content by 129% and 3.8-fold, respectively		Petridis et al. (2012a)
300 mM	*Salvia coccinea* leaves	↑ Phenolic content by 17%.		Grzeszczuk et al. (2018)
5 dS/m	·Susceptible rice varieties Tolerant rice varieties	↓ Phenolic content ↑ Phenolic and flavonoid content by 48 and 17.15%, respectively ↑ Flavonoids, vanillin, ferulic acid, p-coumaric acid, and protocatechuic acid		Minh et al. (2016)

20.9 Chemical Stress

20.9.1 Oxidative

Oxidative stress is the result of unfavorable biotic and abiotic conditions. It also can be artificially induced by ROS such as superoxide (O_2^-) and hydrogen peroxide (H_2O_2) addition.

Common daisy (*Bellis perennis L.*) callus was submitted to 10 mM H_2O_2 for 10 h. The treatment with H_2O_2 allowed the detection of kaempferol, myricetin, quercetin, and isorhamnetin, compounds not detected in the control (0 mM H_2O_2). Also, total phenolic content was increased by 29.25% after H_2O_2 exposure. Furthermore, as a result of oxidative stress, SOD and CAT were increased.

It was found that CAT, SOD, total phenolic, total flavonoid, and proline activity had a significant positive correlation with the phenolic accumulation under H_2O_2 pretreatment. The in vitro culture methods under controlled laboratory conditions, such as the one studied in this investigation, have been of great interest for the production of bioactive molecules (Karakas et al. 2015).

20.9.2 Heavy Metals

Heavy metals are considered as one of the most dangerous pollutants (Morkunas et al. 2018). When plants are in the presence of these compounds, phytochemicals such as phenolic compounds are overproduced and accumulated to protect the plant against toxicity. When plants are exposed to heavy metals, phenolic compounds are mostly increased, since they function as metal chelators and participate in ROS scavenging (Malčovská et al. 2014).

Cadmium (Cd^{2+}) and lead (Pb^{2+}) exposure (50 and 10 ppm, respectively) during 2 weeks increased the total phenolics in corn (*Zea mays*) leaves by around 41.17 and 47%, respectively. The main phenolic compounds increased were chlorogenic acid and rutin; on the other hand, the level of other compounds such as caffeic acid and ferulic acid was decreased (Kısa et al. 2016). Another heavy metal evaluated over plants is nickel (Ni). When exposing chamomile (*Matricaria chamomilla)* to 120 μM Ni during 10 days, polyphenol oxidase activity was decreased, but an increase in the total phenolic compounds (by 18%) and in PAL activity as well as in shikimate dehydrogenase was observed. Another enzyme production, such as cinnamyl alcohol dehydrogenase, was not affected. Also, in leaf rosettes it was detected an increase in chlorogenic acid accumulation (9.02%), as well as protocatechuic acid (7.47%) (a compound with chelating strength) and caffeic acid (153%) (Kováčik et al. 2007).

Transcriptional and post-transcriptional analysis of soybean and lupine roots exposed to Cd^{2+} (25 mg/l) or Pb^{2+} (350 mg/l) was carried out. Here, it was found that under heavy metal stress, there was a transcriptional and post-transcriptional control of PAL expression. For soybean, both metals increased PAL expression, while for lupin only Pb^{2+}. Due to the above, it is concluded that heavy metal stress imposed by Cd^{2+} and Pb^{2+} can cause an induction in the phenylpropanoid pathways of soybean and lupin (Pawlak-Sprada et al. 2011).

Cd^{2+} effect (5 μg/g soil) was also evaluated over *Erica andevalensis*. This heavy metal increased cinnamic acid derivatives (by threefold), epigallocatechin (81.78%), and rutin (32.32%). Nevertheless, when cadmium concentration was increased up to 50 ug/g, the synthesis and release of phenolics were reduced. So, the excess of cadmium may reduce the synthesis of phenolics to avoid deleterious effect caused by the phenoxyl radicals' production. The overall results showed that the phenolic compounds play an important role in the cadmium defense of *E. andevalensis* (Márquez-García et al. 2012).

Maize plants (*Zea mays L.*) have also been exposed to a Cd^{2+} stress (50 μM). Hydroponically grown maize superoxide radicals were increased by up to 1.6-fold. Nevertheless, total phenolic content in maize leaf was barely increased by this Cd^{2+} concentration. It is thought that the presence of high amounts of Cd^{2+} is related to the phenolics involvement in lignin biosynthesis, affecting the pool of free phenols in the plant tissue, which has also been reported in cucumber plants (Malčovská et al. 2014).

Besides Cd^{2+}, Pb^{2+}, and Ni, other heavy metals such as cobalt (Co2+) and silver (Ag+) have been tested. Solutions of 5 μM of Co2+, Ag + , and Cd2+ have been

Table 20.5 Phenolic accumulation in plant materials by heavy metal exposure

Heavy metals	Plant material	Effects	Mode of action	References
Cadmium	Corn leaves *Erica andevalensis*	↑ Phenolic content by 41.17% ↑ Cinnamic acid derivatives, epigallocatechin, and rutin by threefold, 81.78%, and 32.32%, respectively	Overexpression of *PAL* and *shikimate dehydrogenase* genes	Kısa et al. (2016) Márquez-García et al. (2012)
Lead	Corn leaves	↑ Phenolic content by 47%.		Kısa et al. (2016)
Nickel	Chamomile leaf rosettes	↑ Phenolic content by 18% and ↑chlorogenic acid, protocatechuic acid, and caffeic acid by 9.02, 7.47, and 153%, respectively		Kováčik et al. (2007)
Cobalt and silver	Vitis *vinifera*	↑ 3-*O*-glucosyl-resveratrol by 1.6-fold		Cai et al. (2013)

added to *Vitis vinifera* cell suspension cultures. An increase of 3-O-glucosyl-resveratrol up to 1.6-fold after 4 h exposure was observed, not affecting cell viability. Cell viability was only affected when heavy metal concentrations increased up to 25 μM (Cai et al. 2013). Other studies in terms of phenolic accumulation in different plant materials by applying different heavy metals are summarized in Table 20.5.

20.10 Mineral Deficiency and Excess

Plants need at least 14 minerals to grow optimally. If there is a deficit or an excess of one of them, crop yield may be affected. There are some minerals required in large amounts (nitrogen, phosphorus, potassium, calcium, magnesium, and sulfur), while others are needed to a lesser extent (chlorine, boron, iron, manganese, copper, zinc, nickel, and molybdenum). It has been observed that the concentration of nutrient elements, such as minerals, is related with the accumulation of plant defense metabolites.

The excess of minerals such as copper (Cu) has a pronounced effect increasing PAL, shikimate dehydrogenase, cinnamyl alcohol dehydrogenase, polyphenol oxidase, and ascorbate peroxidase activity in chamomile plants exposed to 10 uM Cu during 7 days. Due to this, protocatechuic acid, p-OH benzoic, vanillic acid, syringic acid, p-coumaric acid, ferulic acid, and chlorogenic acid were increased by 63.47, 137.5, 44, 33.55, 76.38, 58.33, and 107%, respectively (Kováčik et al. 2009).

Cu excess effect has also been evaluated over other plants. It has been reported that 1.0 uM of Cu increases total phenolic content by 19% in rosemary leaves (*Rosmarinus officinalis* L.), compared to normal levels. Cu excess in sorghum also increases total phenolic content (Hejazi et al. 2012). Cu levels of 0.5 mM in red cabbage increased anthocyanin concentrations by 126.7% due to oxidative stress

Table 20.6 Phenolic accumulation in plant materials submitted to mineral excess or deficiency

Mineral	Plant material	Effects	Mode of action	References
Excess				
Copper	Chamomile Rosemary leaves Red cabbage	↑ Vanillic acid, syringic acid, p-coumaric acid, ferulic acid, and chlorogenic acid by 63.47, 137.5, 44, 33.55, 76.38, 58.33, and 107%, respectively ↑ Phenolic content by 19% ↑ Anthocyanin by 126.7%	Overexpression of *PAL*, *shikimate dehydrogenase*, *cinnamyl alcohol dehydrogenase*, *polyphenol oxidase*, and *ascorbate peroxidase* genes expression	Kováčik et al. (2009) HEJAZI et al. (2012)
Deficiency				
Nitrogen	Tomato leaf	↑ Chlorogenic acid and rutin by 136 and 117%, respectively	Nothing reported yet	Bénard et al. (2011)
Nitrogen and potassium	American ginsengs	↑ Vanillic acid, p-coumaric acid, and trans-cinnamic acid		Du et al. (2011)
Phosphorus	American ginsengs	↓ Phenolic content		Du et al. (2011)

induction and enhancement of thiobarbituric acid reactive substances in all the plant, suggesting that these flavonoids are synthesized to neutralize ROS (Posmyk et al. 2009). The excess of some minerals induces other many alterations in plants, for example, it affects the uptake of other essential nutrients.

Mineral deficiency also alters the content of phenolic compounds. Nitrogen and potassium deficiency in American ginseng increases the concentration of vanillic acid, p-coumaric acid, and trans-cinnamic acid. Nevertheless, the opposite effect was observed after phosphorus deficiency, where total phenolic content decreased (Du et al. 2011).

Nitrogen deprivation during 19 days on tomato (*Solanum lycopersicum*) leaf increases chlorogenic acid and rutin by 136 and 117%, respectively. It was reported that this increase in phenolic compounds was maintained even when the plant was again exposed to normal nitrogen levels (Bénard et al. 2011). When nitrogen deprivation is combined with an exposure to 12 °C for 8 days, flavonoids levels are also enhanced. The responsible are structural genes in the phenylpropanoid and flavonoid pathway enhancement. From this, eight of the ten structural genes involved in flavonol metabolism showed enhanced expression, including phenylalanine ammonia lyase, chalcone synthase, flavanone 3-hydroxylase, and flavonol synthase (Løvdal and Lillo 2009).

Data in terms of phenolic accumulation in different plant materials by submitting them to mineral excess or deficiency is summarized in Table 20.6.

20.11 Water

20.11.1 Flooding

Flooding, also known as submergence, is a scenario where the plant is totally or partially immersed in water. Due to limited oxygen uptake, anaerobic metabolism may result in the accumulation of phytochemical end-products generated for the plant to survive. Different varieties of sweet potatoes, Taoyuan 2, Sushu 18, and Simon 1, have been tested after flooding treatments where flavonoid content was increased by 11, 4, and 4%, respectively (Lin et al. 2006). More studies are required in this and other plant species to investigate the underlying mechanism.

20.11.2 Drought

Water deficiency promotes a change in biological processes such as photosynthesis, respiration, nutrient metabolism, and secondary metabolism which results in reduced plant growth and also in phytochemicals accumulation (Jaleel et al. 2009; Marchese et al. 2010).

Agave salvia, a plant used in the central and north regions of Mexico for the elaboration of alcoholic beverages, has been exposed to drought conditions not showing differences in its phenolic profile, nor in the concentration as compared to normal conditions. This is because *Agave salmiana* is a stress-tolerant plant, so its phenylpropanoid biosynthesis is not affected (Puente-Garza et al. 2017).

In the case of plants such as *Vitis vinifera*, a change in their phenolic content was seen after drought stress. *Vitis vinifera* roots and leaves under drought stress showed a reduction in total phenolic compounds by 9 and 30%, respectively. Caffeic acid, p-coumaric acid, and ferulic acid were the main phenolic compounds affected. The above promote a lowering on the antioxidant potential (Król et al. 2014). This reduction may be explained by the phenylpropanoid and shikimate pathway enzymes reducing their activity under drought stress, which has been observed in other plants.

On the contrary, other plants increase its phenolic compound production to avoid the oxidative damage caused by the lack of water, mainly those not used to drought conditions.

Plants such as *Aristotelia chilensis*, a berry native to Chile, increased anthocyanin concentration by upregulation of key anthocyanin pathway genes such as *dihydroflavonol 4-reductase*, *UDP-glucose: flavonoid 3-O-glucosyl transferase (UFGT)* and MYB R2R3-type transcription factors such as Myeloblastosis A1 (MybA1) and Myeloblastosis 5A (González-Villagra et al. 2018). This mechanism of enhanced phenolic compounds due to drought stress has also been observed in *Achillea* species and *Amaranthus tricolor*, where a 75% of water supply reduction increased the total phenolic and total flavonoid content by up and twofold and 38%, respectively (Sarker and Oba 2018). Also, a reduction of 67% water supply in olive

Table 20.7 Phenolic accumulation in plant materials submitted to water deficiency

Plant material	Effects	Mode of action	References
Agave salmiana	No changes in phenolic content or profile		Puente-Garza et al. (2017)
Vitis vinifera roots	↓ Phenolic content by 9%	Under expression of phenylpropanoid and shikimate pathway genes	Król et al. (2014)
Vitis vinifera leaves	↓ Phenolic content by 30%		Król et al. (2014)
Aristotelia chilensis	↑ Anthocyanin	Upregulation of key anthocyanin pathway genes such as *dihydroflavonol 4-reductase, UDP-glucose: flavonoid 3-O-glucosyl transferase* (*UFGT*) and MYB R2R3-type transcription factors such as Myeloblastosis A1 (MybA1) and Myeloblastosis 5A	González-Villagra et al. (2018)
Achillea species	↑ Phenolic content by twofold		Sarker and Oba (2018)
Amaranthus tricolor	↑ Phenolic content by 38%		Sarker and Oba (2018)
Olive tree	↑ Phenolic content by 35%		Petridis et al. (2012b)

tree (*Olea europaea*) increased the phenolic compounds by around 35% (Petridis et al. 2012a).

After a long drought stress exposure of 3 weeks, *Codonopsis lanceolata* leaf and roots showed a total phenolic content increase up to 150%, being the predominant compounds catechin, benzoic acid, chlorogenic acid, ferulic acid, gallic acid, rutin, and vanillic acid. The responsible mechanism was an overexpression of 5-enolpyruvylshikimate-3-phosphate (EPSP) synthase, an enzyme in the shikimate pathway, involved in the synthesis of aromatic amino acids such as phenylalanine and tyrosine, the precursors in the phenylpropanoid biosynthesis (Ghimire et al. 2017).

Gene expression is also very affected after drought stress in plants such as potato, where PAL, HCT, HQT, C3H, CHS, CHI, F3H, DFR, and AN1expression was induced, which is related to an altered sucrose flux. The overexpression of HQT has also been documented in Solanaceae species exposed to drought, like in the case of tomatoes, where chlorogenic acid is accumulated in response that overexpression of HQT (André et al. 2009). The results obtained in terms of phenolic accumulation in different plant materials by submitting them to water deficiency are summarized in Table 20.7.

20.12 Wounding

Wounding methods have been applied as postharvest strategies for increasing phenolic content in different plants. Potato tubers are one of them. They were sliced, pie-cutted, or shredded followed by 144 h at 10 °C storage. The highest phenolic compound content was found in sliced potato tubers, where the content of these compounds increased by twofold. The most abundant phenolic compounds were chlorogenic acid (CGA), neo-chlorogenic acid (neo-CGA), and crypto-chlorogenic acid (crypto-CGA). Wounding stressed potato could be used as a starting material for the extraction of high-value antioxidant phenolic compounds with potential applications in the pharmaceutical and dietary supplement industries (Torres-Contreras et al. 2014).

One of the vegetables more studied in terms of wounding is carrots, where it has been found that the higher the wounding intensity, the higher the phenolic accumulation. It is possible to increase the total phenolic content by 2.5-fold and the antioxidant capacity to 12.4-fold when wounding at 23.5 cm^2/g. By this it is induced the synthesis of chlorogenic acid and 3,5-dicaffeoylquinic acid (Surjadinata and Cisneros-Zevallos 2012).

Other vegetables such as lettuce, cilantro, cabbage, green beans, apples, plums, peaches, strawberries, bell peppers, asparagus, celery, carrots, radishes, potatoes, and jicama have been submitted to wounding stress and exogenous ethylene and methyl jasmonate. The combination of wound stress and phytohormones resulted in increased PAL activity in some vegetables. Nevertheless, the response was tissue-dependent since the phenolic content of asparagus, cabbage, apples, tomatoes, and pears was not affected (Heredia and Cisneros-Zevallos 2009). Wounded vegetables that are included in most people regular diet may be considered as an inexpensive source to obtain phenolic compounds and thereby to promote the increase in antioxidants ingest.

20.13 Biotic Stress and Phenolic Compounds in Plants (Gabriela)

Phenolic compounds are synthesized by the secondary metabolism of plants, and it is used as a mechanism of defense against different biotic and abiotic stress. The metabolic pathways for the synthesis of phenolic compounds and other secondary metabolites can be manipulated, specifically promote their synthesis. Therefore, the study of the methods that enhance the production of secondary metabolites is currently of scientific interest. Biotic elicitors are molecules produced by living organisms, such fungus or formed by the plant itself. Plant cell cultures have been used to study the potential of some elicitors in the production of phenolic compounds. Chitosan has been the most reported biotic elicitor in the recent years. Chitosan is a polysaccharide composed of β-(1–4)-linked d-glucosamine and N-acetyl-d-glucosamine randomly distributed within the polymer. This polysaccharide is a structural component of the cell wall of fungi which are pathogens to plants. In plants, chitosan acts as elicitor of plant defensive mechanisms (Cheung et al. 2015; Ferri and Tassoni 2011).

The application of chitosan and irrigation frequencies has been evaluated for the enhancement of essential oils and phenolic content of *Salvia officinalis* L. The foliar application of chitosan (0.5 g/L) and reduced irrigation showed the maximum concentration of phenolic compounds in *Salvia officinalis* L. (Vosoughi et al.

2018). Chitosan has been applied as elicitor in many plants. For instance, in spinach, chitosan (0.1 mg/mL) increased the content of phenolic and flavonoids, with an increase in the antioxidant activity (Singh 2016). Other biotic elicitors are yeasts, and they have been evaluated and compared to the action of chitosan. Yeast and chitosan were evaluated as the biotic stressors on *Rumex cyprius* and showed to enhance the phenolic accumulation with an increase in the antioxidant activity. The maximum antioxidant activity was obtained with yeasts at 200 and 400 mg/mL, whereas the level of chitosan was not a determinant of the antioxidant activity since the low and high levels showed significant results. Furthermore, yeast showed to promote the formation of new phenolics such as gallic and chlorogenic acid (Al Khateeb et al. 2016). Furthermore, low concentration of yeast (50 mg/L) has been reported to enhance the production of phenolic compounds in adventitious root cultures of *Polygonum multiflorum* (Ho et al. 2018b).

The mechanism of action of fungal chitosan is the induction of the activities of defensive enzymes, as showed in *Zanthoxylum bungeanum* stems, where this effect triggers the total phenolic content production (Li et al. 2016). Therefore, fungal elicitors can stimulate the secondary metabolite production. These elicitors can enhance the enzymatic activity of defensive enzymes such as PAL and chalcone isomerase (Simic et al. 2015).

20.14 Genetic Enhancement of Phenolic Compounds in Plants

As it has been previously mentioned, several strategies have been studied to enhance the content of phenolic compounds in plants, most of them involve physiological approaches such as manipulation of biotic and abiotic stresses in plants to trigger the metabolic biosynthetic pathways leading to the overproduction of phenolic compounds. These works have combined high-throughput massive DNA sequencing and metabolic profiling technologies, leading to the development of a new area of study called phytochemical genomics. Thus, the physiological manipulation of phenolic compounds enhancement has led to the discovery of the metabolic pathway of phenolics biosynthesis through the identification of key genes involved in the phenolic pathway (Saito 2013). On this subject, there are different genomic techniques from which plants can be genetically modified. Eckerstorfer et al. (2019) listed some of the most widely used GM techniques such as "genome editing with site-directed nucleases, genome editing directed by synthetic oligonucleotides, RNA directed DNA methylation, cisgenesis and intragenesis, transgrafting, agro-infiltration, haploid induction and accelerated breeding." For a more detailed research on this matter, we recommend the work by Eckerstorfer et al. (2019). A summary of recent studies regarding functional genomics on enhancement of phytochemical content of plants is found on Table 20.8. However, even though genomic manipulation is a promising technique to enhance phenolic compounds in crops and medicinal plants, the regulatory framework in the world limits its widespread distribution. In this regard, some authors have published strategies and recommendations for countries that desire to regulate GMO applications;

Table 20.8 Summary of some works of functional genomics applied to enhance phenolic content

Plant	Genomic modification	Result	References
Glycine max L.	Not specified	Non-GM soybean had highest polyphenol and sterol content of than GM soybean; non-GM methanol extract was more potent against colon carcinoma cells	Marrelli et al. (2013)
Brassica rapa ssp. *Rapa*	Increased expression of the genes BrMYB28, BrMYB29, BrMYB34, BrMYB51, BrMYB122, CYP79, and CYP83 in turnip hairy root	Higher flavonol, hydroxybenzoic acid, and hydroxycinnamic acid content in GM turnip hairy root; GM turnip also showed higher antioxidant, antimicrobial, and antiproliferative activity	Chung et al. (2016)
Codonopsis lanceolata	Transgenic plants with overexpression of γ-tocopherol methyl transferase gene	Transgenic plants had higher chlorogenic acid, luteolin, apigenin, protocatechuic acid, rutin, quercetin, salicylic acid, and caffeic acid. Also transgenic plants showed increased antimicrobial activity against *Staphylococcus aureus*, *Salmonella typhimurium*, *Klebsiella pneumoniae*, and *Escherichia coli*	Ghimire et al. (2017)
Strawberry	Transformed strawberry with a construct containing an anthocyanidin synthase gene	GM strawberry line ANS L18 had the highest total phenolic, total flavonoid content. Line ANS L15 had highest anthocyanin concentration	Giampieri et al. (2018)
Artemisia dubia WALL	Transgenic lines of *Artemisia dubia* WALL transformed with *Agrobacterium tumefaciens* harboring *rol ABC* genes	Increased production of flavonoids up to 71.1% and up to 110.8% higher total phenolic content in transgenic plants. GM plants showed higher antibacterial, antifungal, cytotoxicity, antitumor, and antioxidant activities	Kiani et al. (2015)
Nitraria schoberi L.	Transformed primary leaves with a wild strain of *Agrobacterium rhizogenes* 15,834 SWISS	Increased biosynthesis of flavonoids and hydroxycinnamic acids. 3.8-fold higher content of catechin in ethanol extracts of GM *N. schoberi*; higher antiviral activity in GM plants against influenza [A(H5N1), A (H3N2)]	Zheleznichenko et al. (2018)

(continued)

Table 20.8 (continued)

Plant	Genomic modification	Result	References
Polygonum multiflorum	Transgenic transformation of hairy root lines of *P. multiflorum* with *A. rhizogenes* strain KCCM 11879, followed by exposure to methyl jasmonate	GM HR-01 line showed increased total phenolic content and higher quercetin, myricetin, kaempferol, cinnamic acid, chlorogenic acid, caffeic acid, ferulic acid, coumaric acid, gallic acid, protocatechuic acid, biochanin, hesperidin, naringenin	Ho et al. (2018a)
Taraxacum antungense	Induced overexpression of chlorogenic acid biosynthetic gene (HQT)	Chlorogenic acid levels increased up to 82.49% in transgenic lines	Liu et al. (2018)
Mitracarpus hirtus	Hairy root of *M. hirtus* was modified utilizing a root-inducing plasmid using *A. rhizogenes* A13-mediated transformation and also treated with 2-chloro-4-pyridyl-N-phenylurea	Increased secondary metabolite production; higher levels of chrysophanol (2.23-fold) and 2-methoxy-4-vinylphenol (1.95-fold)	Pansuksan et al. (2014)
Nicotiana tabacum cv. Petit Havana	Agrobacterium-mediated transformation of tobacco plants with MYB transcription factor AtMYB11	Flavonols such as quercetin, rutin, kaempferol and kaempferol-3-rutinoside, and chlorogenic acid biosynthesis was enhanced in GM tobacco plants	Pandey et al. (2015)

nonetheless, regulation frameworks vary among countries; thus there is currently no international harmonization (Eckerstorfer et al. 2019). Release of GMOs is regulated by government agencies; for instance, the US Department of Agriculture, the Food and Drug Administration, and the Environmental Protection Agency are the organizations that regulate release of GMO in the USA; Europe is regulated by the European Food Safety Authority; England is regulated by the Department for Environment Food and Rural Affairs; Mexico is regulated by the Federal Commission for the Protection Against Sanitary Risk (Eckerstorfer et al. 2019; Estados Unidos Mexicanos 2005). On this subject, Ichim (2019) stated that Romania was one of the first European countries adopting GM crops for commercial cultivation; however after 17 years, cultivation stopped, but it's excepted to continue with this activity in spite of the European Union block on authorization regarding the cultivation of GM crops. Up to date, France, Germany, Poland, Italy, Austria, Hungary, Lithuania, Latvia, Bulgaria, and Greece are European countries that still restrict the cultivation of GM crops (Ichim 2019). Restrictions arise not due to concerns regarding safety to human and animal health but to concerns regarding the release in the environment of authorized or unauthorized GM crops (Rostoks et al. 2019).

20.15 Conclusions

Phenolic compounds are in great demand, mainly for their potential in treating and preventing noncommunicable diseases. Harvest and postharvest exposure to unfavorable conditions such as abiotic stress can be used as an alternative to accumulate total or specific phenolic compounds in different plant materials. By abiotic stress exposure, high commercially valued plants with high antioxidant activity can be produced. These adverse conditions mediate the overproduction of signaling molecules, such as reactive oxygen species and phytohormones like methyl jasmonate and ethylene, triggering the overproduction of phenolics, mainly by the overexpression of phenylalanine ammonia lyase, the enzyme involved in the first step of the phenylpropanoid metabolism. The production and accumulation of phenolic compounds by abiotic stress induction depend on the stimulus intensity as well as on the vegetable tissue. Generally, the more the plant is removed from its comfort zone, the greater the production of these phytochemicals. One of the most promising strategies of abiotic stress induction is by wounding plant matrices. This strategy that can be even carried out household to increase the consumption of antioxidant compounds.

Another tool for phenolic compound overproduction is biotic stress, where elicitors are mainly used for increasing the production of phenolic compounds in plants cell culture. This is a promising strategy for the production and subsequent recovery and purification of compounds of interest, aiming to generate food supplements or drugs.

Because of the metabolic pathway for phenolic compounds synthesis is now known, genetic engineering has emerged as another option to increase their production. However, since the regulatory framework has not been developed enough, the use of this particular tool is still limited. Numerous in vivo and in vitro studies have shown the bioactivity of polyphenols, so increasing the consumption of these natural compounds emerges as a promising alternative for treating and preventing health disorders related to oxidative stress. Here, we discussed three alternatives to achieve it.

References

Abdi G, Shokrpour M, Karami L, Salami SA (2019) Prolonged water deficit stress and methyl Jasmonate-mediated changes in metabolite profile, flavonoid concentrations and antioxidant activity in peppermint (Mentha x piperita L.). Not Bot Horti Agrobot Cluj Napoca 47(1):70–80

Abdul Qadir M, Shahzadi SK, Bashir A, Munir A, Shahzad S (2017) Evaluation of phenolic compounds and antioxidant and antimicrobial activities of some common herbs. Int J Anal Chem 2017:1–6

Ahmad HS, Abbasi KY, Hameed HA, Hussain J (2014) Survey and documentation of medicinal plants traditionally used for different ailments in district Lodhran, Punjab. Pak J Med Plants Res 3(4):142

Ajiboye BO, Ojo OA, Okesola MA, Akinyemi AJ, Talabi JY, Idowu OT, Fadaka AO, Boligon AA, Anraku de Campos MMJFS, Nutrition (2018) In vitro antioxidant activities and inhibitory effects of phenolic extract of Senecio biafrae (Oliv and Hiern) against key enzymes linked with type II diabetes mellitus and Alzheimer's disease. Food Sci Nutr 6(7):1803–1810

Al Khateeb W, Alu'datt M, Al Zghoul H, Kanaan R, El-Oqlah A, Lahham J (2016) Enhancement of phenolic compounds production in in vitro grown Rumex cyprius Murb. Acta Physiol Plant 39 (1):14

Alam MA, Subhan N, Hossain H, Hossain M, Reza HM, Rahman MM, Ullah MOJN (2016) Hydroxycinnamic acid derivatives: a potential class of natural compounds for the management of lipid metabolism and obesity. Nutr Metab 13(1):27

Alegria C, Pinheiro J, Duthoit M, Gonçalves EM, Moldão-Martins M, Abreu M (2012) Fresh-cut carrot (cv. Nantes) quality as affected by abiotic stress (heat shock and UV-C irradiation) pre-treatments. LWT Food Sci Technol 48(2):197–203

Ameer K, Shahbaz HM, Kwon J-H (2017) Green extraction methods for polyphenols from plant matrices and their byproducts: a review. Compr Rev Food Sci Food Saf 16(2):295–315

Anantharaju PG, Gowda PC, Vimalambike MG, Madhunapantula SV (2016) An overview on the role of dietary phenolics for the treatment of cancers. Nutr J 1(1):99

André CM, Schafleitner R, Legay S, Lefèvre I, Aliaga CA, Nomberto G, Hoffmann L, Hausman J-F, Larondelle Y, Evers D (2009) Gene expression changes related to the production of phenolic compounds in potato tubers grown under drought stress. Phytochemistry 70 (9):1107–1116

Balea ŞS, Pârvu AE, Pop N, Marín FZ, Pârvu M (2018) Polyphenolic compounds, antioxidant, and cardioprotective effects of pomace extracts from Fetească Neagră cultivar. Oxidative Med Cell Longev 2018:1–11

Bénard C, Bourgaud F, Gautier H (2011) Impact of temporary nitrogen deprivation on tomato leaf phenolics. Int J Mol Sci 12(11):7971–7981

Bitto A, Polito F, Squadrito F, Marini H, D'Anna R, Irrera N, Minutoli L, Granese R, Altavilla D (2010) Genistein aglycone: a dual mode of action anti-osteoporotic soy isoflavone rebalancing bone turnover towards bone formation. Curr Med Chem 17(27):3007–3018

Cai Z, Kastell A, Speiser C, Smetanska I (2013) Enhanced resveratrol production in Vitis vinifera cell suspension cultures by heavy metals without loss of cell viability. Appl Biochem Biotechnol 171(2):330–340

Cheung RC, Ng TB, Wong JH, Chan WY (2015) Chitosan: an update on potential biomedical and pharmaceutical applications. Mar Drugs 13(8):5156–5186

Choe KI, Kwon JH, Park KH, Oh MH, Kim MH, Kim HH, Cho SH, Chung EK, Ha SY, Lee MW (2012) The antioxidant and anti-inflammatory effects of phenolic compounds isolated from the root of Rhodiola sachalinensis A. BOR Mol 17(10):11484–11494

Christie PJ, Alfenito MR, Walbot V (1994) Impact of low-temperature stress on general phenylpropanoid and anthocyanin pathways: enhancement of transcript abundance and anthocyanin pigmentation in maize seedlings. Planta 194(4):541–549

Chung IM, Rekha K, Rajakumar G, Thiruvengadam M (2016) Production of glucosinolates, phenolic compounds and associated gene expression profiles of hairy root cultures in turnip (Brassica rapa ssp. rapa). 3Biotech 6(2):175

Crozier A, Jaganath IB, Clifford MN (2009) Dietary phenolics: chemistry, bioavailability and effects on health. Nat Prod Rep 26(8):1001–1043

de la Rosa LA, Moreno-Escamilla JO, Rodrigo-García J, Alvarez-Parrilla E (2019) Phenolic compounds. In: Yahia EM, Carrillo-Lopez A (eds) Postharvest physiology and biochemistry of fruits and vegetables. Woodhead Publishing, Duxford, pp 253–271

Del Rio D, Costa LG, Lean MEJ, Crozier A (2010) Polyphenols and health: what compounds are involved? Nutr Metab Cardiovasc Dis 20(1):1–6

Du J, Yang J, Jiao X, Gao W (2011) Effect of nitrogen, phosphorus and potassium deficiency on content of phenolic compounds in exudation of American ginseng. Zhongguo Zhong Yao Za Zhi 36(3):326–329

Eckerstorfer MF, Engelhard M, Heissenberger A, Simon S, Teichmann H (2019) Plants developed by new genetic modification techniques – comparison of existing regulatory frameworks in the EU and non-EU countries. Front Bioeng Biotechnol 7(26):1–16

Edward TL, Kirui M, Omolo JO, Ngumbu RG, Odhiambo PM (2014) Effect of ultraviolet-A (UV-A) and ultraviolet-C (UV-C) light on mechanical properties of oyster mushrooms during growth. Biophys J 2014:1–6

Ferri M, Tassoni A (2011) Chitosan as elicitor of health beneficial secondary metabolites in in vitro plant cell cultures. In: Richard GM, Jennifer MT (eds) Handbook of chitosan research and application, pp 1–25

Formica-Oliveira AC, Martínez-Hernández GB, Díaz-López V, Artés F, Artés-Hernández F (2017) Effects of UV-B and UV-C combination on phenolic compounds biosynthesis in fresh-cut carrots. Postharvest Biol Technol 127:99–104

Fraga CG, Galleano M, Verstraeten SV, Oteiza PI (2010) Basic biochemical mechanisms behind the health benefits of polyphenols. Mol Asp Med 31(6):435–445

Ghimire BK, Seong ES, Yu CY, Kim SH, Chung IM (2017) Evaluation of phenolic compounds and antimicrobial activities in transgenic Codonopsis lanceolata plants via overexpression of the γ-tocopherol methyltransferase (γ-tmt) gene. S Afr J Bot 109:25–33

Giampieri F, Gasparrini M, Forbes-Hernandez TY, Mazzoni L, Capocasa F, Sabbadini S, Alvarez-Suarez JM, Afrin S, Rosati C, Pandolfini T, Molesini B, Sánchez-Sevilla JF, Amaya I, Mezzetti B, Battino M (2018) Overexpression of the Anthocyanidin synthase gene in strawberry enhances antioxidant capacity and cytotoxic effects on human hepatic cancer cells. J Agric Food Chem 66(3):581–592

González-Villagra J, Cohen JD, Reyes-Díaz MM (2018) Abscisic acid is involved in phenolic compounds biosynthesis, mainly anthocyanins, in leaves of Aristotelia chilensis plants (Mol.) subjected to drought stress. Physiol Plant 165(4):855–866

Grzeszczuk M, Salachna P, Meller E (2018) Changes in photosynthetic pigments, total phenolic content, and antioxidant activity of Salvia coccinea Buc'hoz ex Etl. induced by exogenous salicylic acid and soil salinity. Molecules 23(6):1296–1307

Guina T, Deiana M, Calfapietra S, Cabboi B, Maina M, Tuberoso CI, Leonarduzzi G, Gamba P, Gargiulo S, Testa GJF, Poli G, Biasi F (2015) The role of p38 MAPK in the induction of intestinal inflammation by dietary oxysterols: modulation by wine phenolics. Food Funct 6 (4):1218–1228

Hajihashemi S, Noedoost F, Geuns JM, Djalovic I, Siddique KH (2018) Effect of cold stress on photosynthetic traits, carbohydrates, morphology, and anatomy in nine cultivars of Stevia rebaudiana. Front Plant Sci 9:1–12

Hejazi MM, Shariatmadari H, Khoshgoftarmanesh A, Dehghani F (2012) Copper effects on growth, lipid peroxidation, and total phenolic content of rosemary leaves under salinity stress. J Agric Sci Technol 14:205–212

Heredia JB, Cisneros-Zevallos L (2009) The effects of exogenous ethylene and methyl jasmonate on the accumulation of phenolic antioxidants in selected whole and wounded fresh produce. Food Chem 115(4):1500–1508

Ho TT, Lee JD, Ahn MS, Kim SW, Park SY (2018a) Enhanced production of phenolic compounds in hairy root cultures of Polygonum multiflorum and its metabolite discrimination using HPLC and FT-IR methods. Appl Microbiol Biotechnol 102(22):9563–9575

Ho TT, Lee JD, Jeong CS, Paek KY, Park SY (2018b) Improvement of biosynthesis and accumulation of bioactive compounds by elicitation in adventitious root cultures of Polygonum multiflorum. Appl Microbiol Biotechnol 102(1):199–209

Hruby A, Hu FB (2015) The epidemiology of obesity: a big picture. PharmacoEconomics 33 (7):673–689

Hugo W (1978) Phenols: a review of their history and development as antimicrobial agents. Microbios 23(92):83–85

Ichim MC (2019) The Romanian experience and perspective on the commercial cultivation of genetically modified crops in Europe. Transgenic Res 28(1):1–7

Jacobo-Velázquez DA, Cisneros-Zevallos L (2012) An alternative use of horticultural crops: stressed plants as biofactories of bioactive phenolic compounds. Agriculture 2(3):259–271

Jacobo-Velázquez DA, Martínez-Hernández GB, del C. Rodríguez S, Cao CM, Cisneros-Zevallos L (2011) Plants as biofactories: physiological role of reactive oxygen species on the accumulation of phenolic antioxidants in carrot tissue under wounding and hyperoxia stress. J Agric Food Chem 59(12):6583–6593

Jaleel CA, Manivannan P, Wahid A, Farooq M, Al-Juburi HJ, Somasundaram R, Panneerselvam R (2009) Drought stress in plants: a review on morphological characteristics and pigments composition. Int J Agric Biol 11(1):100–105

Jamalian S, Gholami M, Esna-Ashari M (2013) Abscisic acid-mediated leaf phenolic compounds, plant growth and yield is strawberry under different salt stress regimes. Theor Exp Plant Phys 25 (4):291–299

Johkan M, Shoji K, Goto F, Hashida S-N, Yoshihara T (2010) Blue light-emitting diode light irradiation of seedlings improves seedling quality and growth after transplanting in red leaf lettuce. HortScience 45(12):1809–1814

Jorge TF, Rodrigues JA, Caldana C, Schmidt R, van Dongen JT, Thomas-Oates J, António C (2016) Mass spectrometry-based plant metabolomics: metabolite responses to abiotic stress. Mass Spectrom Rev 35(5):620–649

Marchese JA, Ferreira JFS, Rehder VLG, Rodrigues O (2010) Water deficit effect on the accumulation of biomass and artemisinin in annual wormwood(Artemisia annua L., Asteraceae). Braz J Plant Physiol 22(1):1–9

Kang KS, Yamabe N, Wen Y, Fukui M, Zhu BT (2013) Beneficial effects of natural phenolics on levodopa methylation and oxidative neurodegeneration. Brain Res 1497:1–14

Kang SN, Lee J, Park JH, Cho JH, Park JH, Cho KK, Lee OH, Kim IS (2014) In vitro anti-osteoporosis properties of diverse Korean Drynariae rhizoma phenolic extracts. Nutrients 6 (4):1737–1751

Karakas FP, Yidirim AB, Bayram R, Yavuz MZ, Gepdiremen A, Turker AU (2015) Antiproliferative activity of some medicinal plants on human breast and hepatocellular carcinoma cell lines and their phenolic contents. Trop J Pharm Res 14(10):1787–1795

Karr S (2017) Epidemiology and management of hyperlipidemia. Am J Manag Care 23(9):139–148

Khan T, Khan T, Hano C, Abbasi BH (2019) Effects of chitosan and salicylic acid on the production of pharmacologically attractive secondary metabolites in callus cultures of Fagonia indica. Ind Cop Prod 129:525–535

Kiani BH, Ullah N, Haq I, Mirza B (2015) Transgenic Artemisia dubia WALL showed altered phytochemistry and pharmacology. Arab J Chem 2:1–11

Kim EY, Park SA, Park BJ, Lee Y, Oh MM (2014) Growth and antioxidant phenolic compounds in cherry tomato seedlings grown under monochromatic light-emitting diodes. Hortic Environ Biotechnol 55(6):506–513

Kısa D, Elmastaş M, Öztürk L, Kayır Ö (2016) Responses of the phenolic compounds of Zea mays under heavy metal stress. Appl Biol Chem 59(6):813–820

Kováčik J, Klejdus B, Bačkor M, Repčák M (2007) Phenylalanine ammonia-lyase activity and phenolic compounds accumulation in nitrogen-deficient Matricaria chamomilla leaf rosettes. Plant Sci 172(2):393–399

Kováčik J, Klejdus B, Hedbavny J, Štork F, Bačkor M (2009) Comparison of cadmium and copper effect on phenolic metabolism, mineral nutrients and stress-related parameters in Matricaria chamomilla plants. Plant Soil 320(1–2):231

Król A, Amarowicz R, Weidner S (2014) Changes in the composition of phenolic compounds and antioxidant properties of grapevine roots and leaves (Vitis vinifera L.) under continuous of long-term drought stress. Acta Physiol Plant 36(6):1491–1499

Król A, Amarowicz R, Weidner S (2015) The effects of cold stress on the phenolic compounds and antioxidant capacity of grapevine (Vitis vinifera L.) leaves. J Plant Physiol 189:97–104

Kutchan TM, Gershenzon J, Møller BL, Gang DR (2015) Natural products. In: Buchanan BB, Gruissem W, Jones RL (eds) Biochemistry & molecular biology of plants. American Society of Plants, Rockville, pp 1250–1318

Lee MJ, Son JE, Oh M-M (2013) Growth and phenolic content of sowthistle grown in a closed-type plant production system with a UV-A or UV-B lamp. Hortic Environ Biotechnol 54(6):492–500

Li P, Linhardt RJ, Cao Z (2016) Structural characterization of Oligochitosan elicitor from Fusarium sambucinum and its elicitation of defensive responses in Zanthoxylum bungeanum. Int J Mol Sci 17(12):2076

Li X, Li M, Han C, Jin P, Zheng Y (2017) Increased temperature elicits higher phenolic accumulation in fresh-cut pitaya fruit. Postharvest Biol Technol 129:90–96

Lim JH, Park KJ, Kim BK, Jeong JW, Kim HJ (2012) Effect of salinity stress on phenolic compounds and carotenoids in buckwheat (Fagopyrum esculentum M.) sprout. Food Chem 135(3):1065–1070

Lin KH, Chao PY, Yang CM, Cheng WC, Lo HF, Chang TR (2006) The effects of flooding and drought stresses on the antioxidant constituents in sweet potato leaves. Bot Stud 47(4):417–426

Liu H, Chen Y, Hu T, Zhang S, Zhang Y, Zhao T, Yu H, Kang Y (2016) The influence of light-emitting diodes on the phenolic compounds and antioxidant activities in pea sprouts. J Funct Foods 25:459–465

Liu Q, Liu Y, Xu Y, Yao L, Liu Z, Cheng H, Ma M, Wu J, Wang W, Ning W (2018) Overexpression of and RNA interference with hydroxycinnamoyl-CoA quinate hydroxycinnamoyl transferase affect the chlorogenic acid metabolic pathway and enhance salt tolerance in Taraxacum antungense Kitag. Phytochem Lett 28:116–123

Løvdal T, Lillo C (2009) Reference gene selection for quantitative real-time PCR normalization in tomato subjected to nitrogen, cold, and light stress. Anal Biochem 387(2):238–242

Lucas L, Russell A, Keast R (2011) Molecular mechanisms of inflammation. Anti-inflammatory benefits of virgin olive oil and the phenolic compound oleocanthal. Curr Pharm Des 17 (8):754–768

Lv J, Huang H, Yu L, Whent M, Niu Y, Shi H, Wang TT, Luthria D, Charles D, Yu LLJFC (2012) Phenolic composition and nutraceutical properties of organic and conventional cinnamon and peppermint. Food Chem 132(3):1442–1450

Malčovská SM, Dučaiová Z, Maslaňáková I, Bačkor M (2014) Effect of silicon on growth, photosynthesis, oxidative status and phenolic compounds of maize (Zea mays L.) grown in cadmium excess. Water Air Soil Pollut 225(8):2056

Márquez-García B, Fernández-Recamales M, Córdoba F (2012) Effects of cadmium on phenolic composition and antioxidant activities of Erica andevalensis. J Bot 2012:1–6

Marrelli M, Tudisco R, Mastellone V, Conforti F (2013) A comparative study of phytochemical composition of genetically and non-genetically modified soybean (Glycine max L.) and evaluation of antitumor activity. Nat Prod Res 27(6):574–578

Mele L, Bidault G, Mena P, Crozier A, Brighenti F, Vidal-Puig A, Del Rio D (2017) Dietary (poly) phenols, brown adipose tissue activation, and energy expenditure: a narrative review. Adv Nutr 8(5):694–704

Estados Unidos Mexicanos (2005) Ley de bioseguridad de organismos genéticamente modificados. Diario Oficial de la Federción

Minh L, Khang D, Ha PTT, Tuyen PT, Minh TN, Quan NV, Xuan TD (2016) Effects of salinity stress on growth and phenolics of rice (Oryza sativa L.). Int Lett Nat Sci 57:1–10

Moreira-Rodríguez M, Nair V, Benavides J, Cisneros-Zevallos L, Jacobo-Velázquez D (2017) UVA, UVB light doses and harvesting time differentially tailor glucosinolate and phenolic profiles in broccoli sprouts. Molecules 22(7):1065

Morkunas I, Woźniak A, Mai V, Rucińska-Sobkowiak R, Jeandet P (2018) The role of heavy metals in plant response to biotic stress. Molecules 23(9):2320

Neagu M, Constantin C, Caruntu C, Dumitru C, Surcel M, Zurac S (2018) Inflammation: a key process in skin tumorigenesis. Oncol Lett 17:4068–4084

Olas B (2017) The multifunctionality of berries toward blood platelets and the role of berry phenolics in cardiovascular disorders. Platelets 28(6):540–549
Oliveira LL, Carvalho MV, Melo L (2014) Health promoting and sensory properties of phenolic compounds in food. Revista Ceres 61:764–779
Ortega-García F, Blanco S, Peinado MÁ, Peragón J (2008) Polyphenol oxidase and its relationship with oleuropein concentration in fruits and leaves of olive (Olea europaea) cv.'Picual' trees during fruit ripening. Tree Physiol 28(1):45–54
Ozcan T, Akpinar-Bayizit A, Yilmaz-Ersan L, Delikanli B (2014) Phenolics in human health. Int J Chem Eng Appl 5(5):393–396
Panche A, Diwan A, Chandra S (2016) Flavonoids: an overview. J Nutr Sci 5:1–15
Pandey A, Misra P, Trivedi PK (2015) Constitutive expression of Arabidopsis MYB transcription factor, AtMYB11, in tobacco modulates flavonoid biosynthesis in favor of flavonol accumulation. Plant Cell Rep 34(9):1515–1528
Pansuksan K, Mii M, Supaibulwatana K (2014) Phytochemical alteration and new occurring compounds in hairy root cultures of Mitracarpus hirtus L. induced by phenylurea cytokinin (CPPU). Plant Cell Tissue Organ Cult 119(3):523–532
Papoutsis K, Vuong Q, Pristijono P, Golding J, Bowyer M, Scarlett C, Stathopoulos C (2016) Enhancing the total phenolic content and antioxidants of lemon pomace aqueous extracts by applying UV-C irradiation to the dried powder. Foods 5(3):55
Pareek A, Khurana A, Sharma AK, Kumar R (2017) An overview of signaling regulons during cold stress tolerance in plants. Curr Genomics 18(6):498–511
Pawlak-Sprada S, Arasimowicz-Jelonek M, Podgórska M, Deckert J (2011) Activation of phenylpropanoid pathway in legume plants exposed to heavy metals. Part I. effects of cadmium and lead on phenylalanine ammonia-lyase gene expression, enzyme activity and lignin content. Acta Biochim Pol 58(2):211–216
Petridis A, Therios I, Samouris G (2012a) Genotypic variation of total phenol and Oleuropein concentration and antioxidant activity of 11 Greek olive cultivars (Olea europaea L.). HortScience 47(3):339–342
Petridis A, Therios I, Samouris G, Koundouras S, Giannakoula A (2012b) Effect of water deficit on leaf phenolic composition, gas exchange, oxidative damage and antioxidant activity of four Greek olive (Olea europaea L.) cultivars. Plant Physiol Biochem 60:1–11
Pistollato F, Sumalla Cano S, Elio I, Masias Vergara M, Giampieri F, Battino M (2016) Role of gut microbiota and nutrients in amyloid formation and pathogenesis of Alzheimer disease. Nutr Rev 74(10):624–634
Poprac P, Jomova K, Simunkova M, Kollar V, Rhodes CJ, Valko M (2017) Targeting free radicals in oxidative stress-related human diseases. Trends Pharmacol Sci 38(7):592–607
Posmyk M, Kontek R, Janas K (2009) Antioxidant enzymes activity and phenolic compounds content in red cabbage seedlings exposed to copper stress. Ecotoxicol Environ Saf 72 (2):596–602
Puente-Garza CA, Meza-Miranda C, Ochoa-Martínez D, García-Lara S (2017) Effect of in vitro drought stress on phenolic acids, flavonols, saponins, and antioxidant activity in Agave salmiana. Plant Physiol Biochem 115:400–407
Qian H, Liu T, Deng M, Miao H, Cai C, Shen W, Wang Q (2016) Effects of light quality on main health-promoting compounds and antioxidant capacity of Chinese kale sprouts. Food Chem 196:1232–1238
Quideau S, Deffieux D, Douat-Casassus C, Pouységu L (2011) Plant polyphenols: chemical properties, biological activities, and synthesis. Angew Chem 50(3):586–621
Rakei A, Maali-Amiri R, Zeinali H, Ranjbar M (2016) DNA methylation and physio-biochemical analysis of chickpea in response to cold stress. Protoplasma 253(1):61–76
Rodríguez-Calzada T, Qian M, Strid Å, Neugart S, Schreiner M, Torres-Pacheco I, Guevara-González RG (2019) Effect of UV-B radiation on morphology, phenolic compound production, gene expression, and subsequent drought stress responses in chili pepper (Capsicum annuum L.). Plant Physiol Biochem 134:94–102

Rodriguez-Morato J, Xicota L, Fito M, Farre M, Dierssen M, de la Torre R (2015) Potential role of olive oil phenolic compounds in the prevention of neurodegenerative diseases. Molecules 20 (3):4655–4680

Rostoks N, Grantiņa-Ieviņa L, Ieviņa B, Evelone V, Valciņa O, Aleksejeva I (2019) Genetically modified seeds and plant propagating material in Europe: potential routes of entrance and current status. Heliyon 5(2):1–21

Saito K (2013) Phytochemical genomics – a new trend. Curr Opin Plant Biol 16(3):373–380

Santana-Gálvez J, Jacobo-Velázquez DA (2018) Classification of phenolic compounds. In: Nollet LM, Gutierrez-Uribe JA (eds) Phenolic compounds in food: characterization and analysis, vol 1, 1st edn. CRC Press, Boca Raton, pp 3–20

Sarega N, Imam MU, Ooi D-J, Chan KW, Md Esa N, Zawawi N, Ismail M (2016) Phenolic rich extract from Clinacanthus nutans attenuates hyperlipidemia-associated oxidative stress in rats. Oxid Med Cell Longev 2016:1–17

Sarker U, Oba S (2018) Response of nutrients, minerals, antioxidant leaf pigments, vitamins, polyphenol, flavonoid and antioxidant activity in selected vegetable amaranth under four soil water content. Food Chem 252:72–83

Seo JM, Arasu MV, Kim YB, Park SU, Kim SJ (2015) Phenylalanine and LED lights enhance phenolic compound production in Tartary buckwheat sprouts. Food Chem 177:204–213

Simic SG, Tusevski O, Maury S, Hano C, Delaunay A, Chabbert B, Lamblin F, Lainé E, Joseph C, Hagège D (2015) Fungal elicitor-mediated enhancement in phenylpropanoid and naphtodianthrone contents of Hypericum perforatum L. cell cultures. Plant Cell Tissue Organ Cult 122(1):213–226

Singh S (2016) Enhancing phytochemical levels, enzymatic and antioxidant activity of spinach leaves by chitosan treatment and an insight into the metabolic pathway using DART-MS technique. Food Chem 199:176–184

Soengas P, Rodríguez VM, Velasco P, Cartea ME (2018) Effect of temperature stress on antioxidant defenses in Brassica oleracea. ACS Omega 3(5):5237–5243

Son K-H, Oh M-M (2013) Leaf shape, growth, and antioxidant phenolic compounds of two lettuce cultivars grown under various combinations of blue and red light-emitting diodes. HortScience 48(8):988–995

Surjadinata BB, Cisneros-Zevallos L (2012) Biosynthesis of phenolic antioxidants in carrot tissue increases with wounding intensity. Food Chem 134(2):615–624

Surjadinata B, Jacobo-Velázquez D, Cisneros-Zevallos L (2017) UVA, UVB and UVC light enhances the biosynthesis of phenolic antioxidants in fresh-cut carrot through a synergistic effect with wounding. Molecules 22(4):668

Takshak S, Agrawal S (2019) Defense potential of secondary metabolites in medicinal plants under UV-B stress. J Photochem Photobiol B 193:51–88

Torres-Contreras AM, Nair V, Cisneros-Zevallos L, Jacobo-Velázquez DA (2014) Plants as biofactories: stress-induced production of chlorogenic acid isomers in potato tubers as affected by wounding intensity and storage time. Ind Crop Prod 62:61–66

Tsao R (2010) Chemistry and biochemistry of dietary polyphenols. Nutrients 2(12):1231–1246

Tuzcu Z, Orhan C, Sahin N, Juturu V, Sahin K (2017) Cinnamon polyphenol extract inhibits hyperlipidemia and inflammation by modulation of transcription factors in high-fat diet-fed rats. Oxid Med Cell Longev 2017

Tyunin AP, Kiselev KV (2016) Alternations in VaSTS gene cytosine methylation and t-resveratrol production in response to UV-C irradiation in Vitis amurensis Rupr. Cells. Plant Cell Tissue Organ 124(1):33–45

Valls-Pedret C, Sala-Vila A, Serra-Mir M, Corella D, De la Torre R, Martínez-González MÁ, Martínez-Lapiscina EH, Fitó M, Pérez-Heras A, Salas-Salvadó J (2015) Mediterranean diet and age-related cognitive decline: a randomized clinical trial. JAMA Intern Med 175(7):1094–1103

Vosoughi N, Gomarian M, Ghasemi Pirbalouti A, Khaghani S, Malekpoor F (2018) Essential oil composition and total phenolic, flavonoid contents, and antioxidant activity of sage (Salvia

officinalis L.) extract under chitosan application and irrigation frequencies. Ind Crop Prod 117:366–374
Weichselbaum E, Buttriss J (2010) Polyphenols in the diet. Nutr Bull 35(2):157–164
Wong RS (2011) Apoptosis in cancer: from pathogenesis to treatment. J Exp Clin Cancer Res 30 (1):87
Xuan TD, Khanh TD, Khang DT, Quan NT, Elzaawely AA (2016) Changes in chemical composition, total phenolics and antioxidant activity of Alpinia (Alpinia zerumbet) leaves exposed to UV. Int Lett Nat Sci 55:25–34
Younis MB, Hasaneen MNA-G, Abdel-Aziz HMM (2010) An enhancing effect of visible light and UV radiation on phenolic compounds and various antioxidants in broad bean seedlings. Plant Signal Behav 5(10):1197–1203
Zhao CY, Si JH, Feng Q, Deo RC, Yu TF, Du Li P (2017) Physiological response to salinity stress and tolerance mechanics of Populus euphratica. Environ Monit Assess 189(11):533
Zheleznichenko T, Banaev E, Asbaganov S, Voronkova M, Kukushkina T, Filippova E, Mazurkova N, Shishkina L, Novikova T (2018) Nitraria schoberi L. hairy root culture as a source of compounds with antiviral activity against influenza virus subtypes A(H5N1) and A(H3N2). 3Biotech 8(6):260
Zheng S, Deng S, Huang Y, Huang M, Zhao P, Ma X, Wen Y, Wang Q, Yang XJF (2018) Antidiabetic activity of a polyphenol-rich extract from Phellinus igniarius in KK-Ay mice with spontaneous type 2 diabetes mellitus. Food Funct 9(1):614–623
Zhou Y, Zheng J, Li Y, Xu D-P, Li S, Chen Y-M, Li H-B (2016) Natural polyphenols for prevention and treatment of cancer. Nutrients 8(8):515
Zhu J-K (2016) Abiotic stress signaling and responses in plants. Cell 167(2):313–324

Quantitative Genetics and the Genetic Basis for Polyphenolics Trait in Plants

21

Gaurav Sharma, Sonia, and Atul Kumar Upadhyay

Abstract

Plant phenolics have been found to influence both plant biological processes and human health as a dietary component. Plants produce an enormous number of phenolic secondary metabolites which are involved in interaction with the environment, for their reproductive strategy and for their defence mechanisms. These include developmental signals, such as during lignification of new growth or the production of anthocyanins/carotene during fruit and flower development, and environmental signals for protection against abiotic and biotic stresses. Due to this importance, many QTLs and genes are been found responsible for synthesis of plant phenolics. Not only QTLs but dietary polyphenols are been reported to bring certain kind of epigenetic changes and are either eliciting array of gene expression in the body to influence many changes in metabolism. Studies on other dietary plant phenolics to cause male sterility and improving fertility by preventing DNA damage have been discussed in this book chapter.

Keywords

Genetics · Traits · Plants · Secondary metabolites

G. Sharma (✉)
ICAR-Accredited School of Agriculture, Lovely Professional University, Phagwara, India

Sonia
School of Pharmaceutical and Healthcare Sciences, CT University, Ludhiana, India

A. K. Upadhyay
Department of Biotechnology, Thapar Institute of Engineering and technology, Patiala, Punjab, India

R. Lone et al. (eds.), *Plant Phenolics in Sustainable Agriculture*,
https://doi.org/10.1007/978-981-15-4890-1_21

21.1 Introduction

Plants have the ability to produce different phenolic compounds for the growth and development and reproductive processes and for their defence mechanisms that are very important for the plants. These processes involve different biochemical transformations. Methylation, methyltransferases, acylation and glycosylation are involved at their respective stages in the process of production of various phenolic compounds that are considered basic for the fundamental chemical modifications. These modified metabolites show different polarity, volatility and chemical stability and interact with various compounds (co-pigmentation) and biological activity. In the process of production of plant phenolics, a matrix of potentially overlapping regulatory signals are involved as in case of lignification of new growth or the production of anthocyanins during the development of fruit and different environmental signals as defence mechanism against abiotic and biotic stresses. There is now a good perception of the nature of environmental signals for the flavonoids, the signal transduction pathway interaction by the activation of the phenolic biosynthetic genes. Different microorganisms can coexist within the plant environment that can establish various interactions with the host plant and that are often the basis for the synthesis of targeted phenolic metabolites in response to these interactions. The increasing evidence suggests that root-specific chemicals (exudates) might start and influence biological and physical interactions within the roots and soil organisms. These associations include signal traffic between the roots of competing plants, roots and soil microbes and one-way signals that transmit the nature of chemical and physical soil characteristics to the roots that are very necessary to operate such type of processess. Transcriptional regulation and signal transduction can also be modulated by plant phenolics for essential physiological processes. Plant phenolics are also linked with the growth hormone auxin in certain cases. The development of pollens has been observed by certain phenolic compounds. Flavonoids help in imparting the orange, red and blue/purple colours to many plant tissues. The chapter will provide deeper insights into the quantitative genetics and the genetic basis for polyphenolics trait in plants.

21.2 Synthesis of Polyphenolic Compounds, Genetic Regulation and Engineering

21.2.1 Genetic Basis for Polyphenolics Synthesis in Plants

The ability to produce various secondary metabolites by plants is not necessary in the primary processes of growth and development but is important for their interaction with the environment. Chemodiversity is well-developed in terrestrial plants having different environmental challenges that need to maintain well-organized transport of the metabolites, structural rigidity and a fine regulation of homeostasis (Caputi et al. 2012). The change in natural compounds with glycosyl and acyl moieties is catalyzed by acyltransferases (AT) and glycosyltransferases (GTs) and produces thousands of molecular variants (Tanaka et al. 2008). These steps occur after the completion of the biosynthesis of the respective aglycone known as regiospecific reactions (Ono et al. 2010). The wide array of cDNAs encoding members from different plant species and tissues of both enzyme groups has been cloned. The catalytical function is only explained for a minor part of gene products by using recombinant enzymes in vitro (Caputi et al. 2012).

Catalytic versatility is characterized by both classes of enzymes that normally make functional prophecy from primary sequence alone unfeasible.

The accessibility of more and more plant genomes and protein crystal structures presents potential to acquire a wider view of the families of these two important protein classes in terms of functional evolution and the key structural elements recognizing the specificities, especially regiospecificities and catalytical properties. Metabolite profiling, transcriptomics, sequence assessment, modelling, biochemical characterization and the use of relative genomics plus genetic engineering can be effectively useful to recognize substrate requirements and movement of the large number of candidate genes predicted in plant genomes (Caputi et al. 2012; D'Auria 2006; Tuominen et al. 2011).

21.3 Role of Acyltransferases (AT) and Glycosyltransferases (GT) in Polyphenol Synthesis

Enzymatic acylation determines aliphatic and/or aromatic acyl moieties onto the nucleophile (OHe or NHe) of acceptor molecules having ester and amide bonds. The groups involved in aromatic acylation are coumaroyl or sinapoyl groups that induce stabilization of the molecule and the color intensity of anthocyanins. Malonylation is important for stability of the molecules and for the r solubility of water and for the protection upon degradation by enzymes (Luo et al. 2007). Dissimilar acylations are universal in the secondary metabolism of plants. N- or O-acylation is implicated in forming and modifying phenolics and polyphenolics primarily and is also involved in the ecophysiological roles of alkaloids and terpenoids (D'Auria 2006; Tuominen et al. 2011; Yu et al. 2000).

The BAHD-ATs are a large family of plant-specific monomeric acyl-CoA-utilizing and functionally diverse enzymes, which depicts 10–30% resemblance at the amino acid level. The five angiosperm taxa supported a refined grouping of eight major clades for this family based on genome analysis. Recently, revealed clade-specific motifs ought to assist functional studies of substrate and donor specificities among diverse BAHD enzymes.

The development of BAHD family in altered lineages is associated to taxon-specific metabolic diversity (Tuominen et al. 2011). The second well-recognized functional subfamily within BAHD-ATs is fashioned by anthocyanin/flavonoid ATs, such as hydroxycinnamoyltransferases and malonytransferases involved in modification of various polyphenols after alcohol acetyltransferases (Yu et al. 2000). It is assumed that these proteins may be located to the cytosol as transit peptide for translocation to other subcellular sites is yet to be recognized. Products produced include modified anthocyanins, lignin, small green-leaf volatile esters, suberin and defence compounds and phytoalexins (D'Auria 2006). Some members have a more limited and others a wider preference for in vitro experiments. The flexibility in their substrate specificities, the prediction of function based only on structural information, is a major hurdle, and, hence, the great majority of BAHD-ATs are uncharacterized till date on the basis of their substrates and products (Luo et al. 2007).

Recently a novel class of ATs, serine carboxypeptidase-like (SCPL) acyltransferases, was characterized. Proteins belonging to ATs, serine carboxypeptidase-like (SCPL) acyltransferases, assist transacylation reactions of various secondary metabolites using energy-rich 1-O-b-glucose esters in the synthesis (Stehle et al. 2009). These vacuolar proteins follow a substitute cellular route of transacylation spatially isolated from the cytoplasmic enzymes of the BAHD-AT family. Efforts in cloning and classification led to the recognition of diagnostic peptides for SCPL acyltransferases, which enables the detection of candidate genes in some plant genomes. Biochemical analysis of SCPL acyl transferases is dependent on complete heterologous expression systems, efficient protein purification protocols, and the supply of appropriate substrates. Clearly, SCPL ATs have evolved from a hydrolytic ancestor by adapting functional elements of the proteases such as the catalytic triad, oxyanion hole, and substrate recognition H-bond network to their new function (Mugford and Milkowski 2012).

A ubiquitous group of enzymes that have the ability to catalyze the transfer of a sugar moiety to various acceptor molecules from an activated donor molecule are known as GTs. Sugars, lipids, proteins, nucleic acids, antibiotics and other small molecules may also act as substrates for GT proteins other than secondary metabolites that result in the formation of polyglycosides, disaccharides by glycosylation and various glycosides of non-carbohydrate moieties. The synthesis of such molecules is biologically important as it may carry different functions within the organism and involves the action of hundreds of different such chemicals that are more or less selective,

GTs (Caputi et al. 2012; Coutinho et al. 2003). 94 GT families are also included in family 1 of a classification scheme that have the ability to transfer sugars onto small molecules. UDP glycosyl-transferase (UGTs) is the largest enzyme family that has different enzymes from plants, animals, fungi, bacteria, and also viruses. UGTs utilize UDP-activated sugars (glucose, galactose, arabinose) as the major donor molecule that contain the well-preserved UDP-glycosyltransferases-defining motif as a unique character that is one of the few regions of significant sequence similarity. The so-called plant secondary product glycosyltransferase motif is a carboxyl-terminal consensus sequence of 44 amino acid residues believed to represent the nucleotide-diphosphate-sugar-binding site of the enzymes (Caputi et al. 2012). UGTs that are concerned in plant secondary metabolism showed broad substrate specificity, during in vitro experiments, with recombinant proteins that are documented as acceptor molecules. This promiscuity could also donate to the enormous skeletal variations of small molecules concerning their glycosylation prototype. Nevertheless, UGTs can also be discerning, and the substrate specificities of some enzymes are defined by regiospecific or regioselective characteristics of the aglycones (Lim et al. 2003). A vast number of genes responsible for encoding of UGTs have been found in plant genomes that are usually consistent with structural diversity of secondary metabolites of plants.

The cloned UGTs are functionally characterized by in vitro and/or in vivo studies, but the number of proteins is relatively very less, and it is argued that in vivo activity of the UGTs is complicated for in vitro activity studies can sometimes be ambiguous.

Researchers advocated that the enzyme is also able to transfer different flavonoid aglycones, like quercetin, apigenin kaempferol and genistein to glycosylated products. On the other hand, flavones (apigenin) and isoflavones (genistein) have not been isolated from naturally occurring metabolites in *A. thaliana*.

The selective approach to highlight the problems of different enzymes can display regio- and stereospecificity, and in several cases, it is very wanton; identifying a range of substrates and producing multiple products have been observed in some plants. A pool of information is now available about plant genomes that provide bulk of understanding in multigene families for secondary metabolites. The knowledge can be used that has helped us to explore molecular adaptations occurred during plant evolution, the elucidation of new or not well-known metabolic pathways and the identification of genes that could be exploited for biotechnological applications.

It is revealed that UGT79 B1 encodes for an anthocyanin 3-O-glucoside:2 00 -O-xylosyltransferase has drastically reduced the anthocyanin content and enzyme assay with the respective recombinant proteins by knockout mutants. The recombinant F3GGT1 is a substrate for transfer of UDP-xylose to the galactose moiety. It appears that the first glycosylation step in which F3GT1 is responsible for the formation of cyanidin 3-O-galactoside is an important step for anthocyanin accumulation in the red-fleshed fruit of *A. chinensis* (Montefiori et al. 2011). 3-Deoxyanthocyanins are recognized in few plant species but also help in flower pigments in Gesneriaceae and Bignoniaceae, which gives bright red or orange-red colour. 5-O-Glucosyltransferase was cloned from *Sinningia cardinalis* (Lehm.) that concentrates 3-deoxyanthocyanins in these flowers. 3-Hydroxyanthocyanidins or flavones, flavonols and flavanones are not accepted by recombinant protein but only transferred the glucosyl moiety from UDP-glucose to the 3-deoxyanthocyanidins apigeninidin and luteolinidin (Nakatsuka and Nishihara 2010). The pattern on flower colour which is affected by glycosylation is indicated in blue-flowered *Veronica persica Poiret.* The glycosylated anthocyanin is accompanied by apigenin 7-O-(2-O-glucuronosyl)-glucuronide that gives it a bathochromic shift towards blue. With the help of reverse genetics, two UGTs encoded an anthocyanin 3-O-glucoside-2 00 -O- glucosyltransferase and flavonoid 7-O-glucuronosyltransferase. This leads to an understanding that bluish coloration in flowers is due to expression of these two genes and bathochromic effect (Ono et al. 2010).

Similarly, in grapevine (*V. vinifera*), 181 putative UGTs were recognized (Caputi et al. 2012), but only some of these are functionally characterized. Two of them, VvGT5 and VvGT6, were described as UDP-glucuronic acid:flavonol-3-O-glucuronosyltransferase (GAT) and bifunctional UDP-glucose/UDP-galactose:flavonol-3-O-glucosyltransferase/galactosyltransferase, respectively.

It was revealed that the genes have adequate sequence similarity due to gene duplication and neofunctionalization with slight amino acid modifications that in turn contribute to structural variation of flavonols in the grape fruit (Ono et al. 2010). *Ectopic overexpression of two R2R3MYB (myeloblastosis) transcription factors involved proanthocyanidin (PA) biosynthesis in grape fruits indicated that there are three expressed cytoplasmatic UGTs when compared to control groups. The synthesis of 1-O-acyl-glucose esters of various hydroxybenzoic and hydroxycinnamic acids is due to the catalyzation by respective recombinant proteins. Flavonoids or stilbenes as substrate were not accepted by the proteins, leading to the arguments that the enzymes could be involved in PA galloylation using glucose esters for the formation of hydroxycinnamic esters in vivo as transcripts were only detectable in early stages of berry development in skins and seeds (Khater et al. 2012).

The bitterness in certain citrus fruits is a result of flavanone glycosides. The biosynthetic step is catalyzed by two rhamnosyltransferases that utilize, beside others, flavanone 7-O-glucose as the main substrate. It is well-known fact that flavanone 7-O-neohesperidosides (rhamnose-2-glucose; e.g. neohesperidin and naringin) play a major role in imparting a bitter taste, while flavanone 7-O-rutinosides (rhamnose-6-glucose; e.g. hesperidin and narirutin) do not have any taste. O-glucosylation at position 7, which is catalyzed by a 7-O-glycosyl-transferase, is the initial precursor. Subsequently, a rhamnosyltransferase is involved in the second step. Accordingly, a 1,2-rhamnosyltransferase (1,2RhaT) imparted bitter-taste pummelo (*Citrus maxima Burm.*) when cloned, and the encoded enzyme was found to specifically catalyzer hamnosylation of flavanone and flavone 7-O-glucosides at position 2 of the glucose moiety. The counterpart from non-bitter oranges (Citrus sinensis (L.) Osbeck) was recognized as 1,6-rhamnosyltransferase (1,6RhaT) that triggers the biosynthesis of the tasteless flavanones and is a promiscuous enzyme for substrate specificity, since rhamnosylation was found with 3- or 7-O-glycosylated flavanones, flavones, flavonols and anthocyanins at position 6 of the glucose. The encoding genes are related as they originated independently before specification of the citrus genus (Frydman et al. 2013).

Two novel acyl-glucose-dependent anthocyanin glycosyltransferases (AAGTs) from *Dianthus caryophyllus* L. (Carnation) and *Delphinium grandiflorum* L. (Delphinium) were described as new reaction mechanism for the sugar transfer. These enzymes were shown to catalyze the 5- and 7-O-glucosylation of anthocyanidins, respectively, which had been generally been thought to be catalyzed by UGTs. The petal color variation of carnations was found to be due to the occurrence or nonexistence of the glucosylation at the position 5 (Nishizaki et al. 2011) as both of the enzymes use aromatic acyl-glucose instead of UDP-glucose as sugar donor to impart color. In addition, a strict acceptor preference for anthocyanins was observed with no activity found with various other flavonoids and phenolic compounds. Both enzymes were classified as members of glycoside hydrolase family 1 (GH1), which act as beta-glycosidases based on the obtained amino acid sequences. Residual GH activity was obvious with the recombinant protein but was negligible compared to GT activity. GH1s are involved in several primary processes like plant defence, lignification, hydrolysis of oligosaccharides and phytohormone regulation (Matsuba et al. 2010).

However, GH1 proteins have putative transit peptide sequences that are very important for localization in the endoplasmic reticulum, peroxisomes, mitochondrion and plasma membrane and for the sequestration into vacuoles. The activity of an AAGT was detected also in flowers of *Agapanthus africanus* (L.) Hoffmanns. The isolated cDNA was named as an AA7GT and the recombinant protein shows akin biochemical properties as the Dianthus and Delphinium proteins (Miyahara et al. 2011, 2012). Separately, phylogenetic analysis helped in recognition of up to 11 GH1 candidate genes from the genetic makeup of *A. thaliana* with unknown function based on bioinformatic approach. These approaches postulated that AAGTs were derived from glucosidases early during the evolution of angiosperm (Miyahara et al. 2011).

21.4 Gene Regulation of Polyphenol Production in Plants

The release of polyphenols involves a variety of potentially overlapping regulatory signals like developmental signals as in lignification of new growth or the release of anthocyanins during fruit and flower growth that help to defend against abiotic and biotic stresses. There is an outstanding perception of the nature of the regulatory signals for some flavonoids and the transduction pathway that connects to the activation of the biosynthetic genes.

It has been observed that processess like flower development, berry ripening, and vegetative pigmentation has primarily through the action of transcription factors (TFs). Several TF have been recognized to control flavonoid biosynthesis, monocots and eudicots.

The fundamental to the direct directive of anthocyanin and proanthocyanidin (PA) biosynthetic genes is core "MBW" regulation complexes, comprising of detailed members of the R2R3MYB and basic helix-loop-helix (bHLH) TF families in combination with a WD repeat (WDR; tryptophan-aspartic acid (W-D) dipeptide repeat) protein. Different MBW complexes can be made from various MYB and bHLH components, and these can have diverse target genes and diverge in their activation or repression actions. The genes responsible for control of PA biosynthesis have been characterized for some species like grape, in which several PA-related R2R3-MYBs were recognized (Bogs et al. 2007; Terrier et al. 2009), other species like legumes (Dixon et al. 2013), and persimmon (*Diospyros kaki* Thunb.) as well. PAs are important for the development of legumes and the regulation of proteins in ruminants. R2R3MYBs-regulating PA production was recognized from the legumes of Medicago and Lotus and used in genetic alteration to enhance PA levels in forage crops (Dixon et al. 2013). An important advancement is the recognition of an R2R3MYB from rabbit's foot clover (*Trifolium arvense* L.) that when overexpressed in *Medicago sativa* L. or *Trifolium repens* L. brings foliar PA build-up at up to 1.8% dry weight (Hancock et al. 2012). In grape fruits, PAs manipulated the sensory qualities of the end product as wine. In Persimmon PAs may accumulate to concentrations that make fruit highly mordant (Akagi et al. 2012).

In PAs and flavonols, there are only a few polyphenol biosynthetic pathways for which TFs have been characterized outside of the anthocyanin. The control of lignin biosynthesis is extensively researched in *A. thaliana* and to a lesser extent the grasses (Gray et al. 2012; Zhao and Dixon 2011). In *A. thaliana*, MYB85 and members of R2R3MYB subgroup 3 directly enhance lignin biosynthesis under the influences of upstream MYB and NAC. Further, members of R2R3MYB subgroup 4 inhibit the synthesis of the antecedents for the branch of the phenylpropanoid pathway leading to lignin and a range of other compounds. Changes in TF gene is linked with isoflavonoid release well-described for the *Lotus japonicus* L. and a group of R2R3MYBs (subgroup 2) recognized as a candidate activators for the essential biosynthetic genes (Shelton et al. 2012). In combination with the induction of the subgroup 2 genes, there was downregulation of Tfs that control branches of the phenylpropanoid passageway that might compete with isoflavonoid production, especially for the production of flavonols and PAs and the likely bHLH component of the anthocyanin MBW complex. Several additional mechanisms and TFs are recognized that control flavonoid biosynthesis, most of which are thought to act by influencing the MBW complex activity by the mechanisms that provoke MBW activator complex. The different mechanisms that help in the degradation of TF protein are due to the response pathway for the induction of anthocyanins in response to light in *A. thaliana* leaves and apple fruit.

The bZIP TF protein LONG HYPOCOTYL5 (HY5) is a key coordinator of the light response in *A. thaliana* and HY5 will enhance flavonoid biosynthesis through direct binding at promoters of anthocyanin biosynthetic genes and the induction of MYB12, the direct activator of flavonol biosynthesis. Over the years, the first repressor was the R2R3MYB identified from plants. The exposure to UV-B irradiation downregulates AtMYB4, and that AtMYB4 production is downregulated by exposure to UV-B irradiation, which in turn relieves the cinnamate-4-hydroxylase gene from AtMYB4-based suppression and promotes production of sinapate esters for UV-B protection (Jin et al. 2000). A potential repressor for anthocyanin production was recognized soon after AtMYB4, FaMYB1 from strawberry (Aharoni et al. 2001).

The plant-specific EAR domain which has the core sequence of LxLxL or DLNxxP occurs in various TF families in addition to the R2R3MYBs (Kagale and Rozwadowski 2011). Directions of confirmed and candidate R2R3MYB-EAR repressors for flavonoid biosynthesis from a range of species indicated various conserved areas, mostly (P/L)DLNL(E/D)L chain that is EAR domain. The EAR domain makes it simple to recognize the repressors of flavonoid biosynthesis from genomic sequence data. R2R3MYB-EAR repressors also include AtMYBL2 of *A. thaliana* (Dubos et al. 2008; Matsui et al. 2008) and MYB27 of petunia (Albert et al. 2011; Kroon 2004.). AtMYBL2 and PhMYB27 have unique roles to control vegetative pigmentation that is expressed at various levels in leaves of shade-grown plants, and in contrary plants are exposed to light stress.

Apart from AtMYBL2 that is a truncated protein MYBs of the R3MYB type containing the R3-region of the MYB DNA-binding domain, six R3MYBs have been identified as being involved in this process in *A. Thaliana*, having TRIPTYCHON (TRY) and CAPRICE (CPC) with best characterization process (Feller et al. 2011; Balkunde et al. 2010). The regulation of epidermal cell fates CPC has a role in anthocyanin regulation as revealed by transgenic overexpression (Zhang et al. 2009,Zhu et al. 2009). A homolog from petunia, MYBX, suggests its primary role is in negatively regulating anthocyanin biosynthesis (Kroon 2004; Albert et al. 2012), and the candidate R3MYBs for flavonoid regulation is also present in the sequence databases for other species.

This clearly indicates that the regulation of polyphenol production is well-established contrary to the state of knowledge of secondary metabolite pathways. However, there are still many areas that need proper exploration.

21.5 Plant Phenolics, Their Effect on Human Health and Its Genetic Basis

Several exceptional reviews are focused on the health-related effects and bioavailability of polyphenols in human nutrition (5–9). Originally, the main health claims for polyphenols were based on their properties as scavengers of free radicals and reactive oxygen species (ROS). It is well-established fact that plant phenolics play a vital role in human health by way of human nutrition and the mechanism by which free radicals which are reactive oxygen species are influenced.

The increased utilization of fruits and vegetables for fibre, micronutrients and antioxidant compounds that are beneficial to human health is recommended by nutritionists (Willett 2002). Apple (Malus x domestica Borkh.) is a good source of dietary fibre (2 g/100 g fresh fruit) (McGhie et al. 2005; Boyer and Liu 2004) and is enriched with high concentrations of quercetin, epicatechin and procyanidin polymers but is lower in vitamin C than other fruits (5–25 mg/100 g depending on the cultivar (Davey et al. 2006), (McGhie et al. 2005; Hertog et al. 1993; Vinson et al. 2001). The presence of polyphenolic compounds help in the prevention of epithelial cancers and cardiovascular diseases, thrombotic stroke, obesity, type 2 diabetes and neurodegenerative diseases associated with aging and infections (Arts and Hollman 2005). Polyphenolic compounds like flavonoids, phenolic acids, lignans and stilbenes are secondary metabolites involved in defence against aggression by pathogens or ultraviolet radiation.

Similarly, Sophorae Radix (*Sophora flavescens* Aiton) belongs to the Fabaceae family which have been used as a functional food ingredient and traditional herbal medicine as an antipyretic, diuretic and anthelmintic, in addition to these, it is also used for the treatments of diarrhea, gastrointestinal haemorrhage and eczema (Bae 2000).

About 200 compounds have been recognized from Sophorae Radix, including alkaloids, flavonoids, terpenoids and other compounds (He et al. 2015; Piao et al. 2006; Ling et al. 2007) with alkaloids and flavonoids as the main active compounds

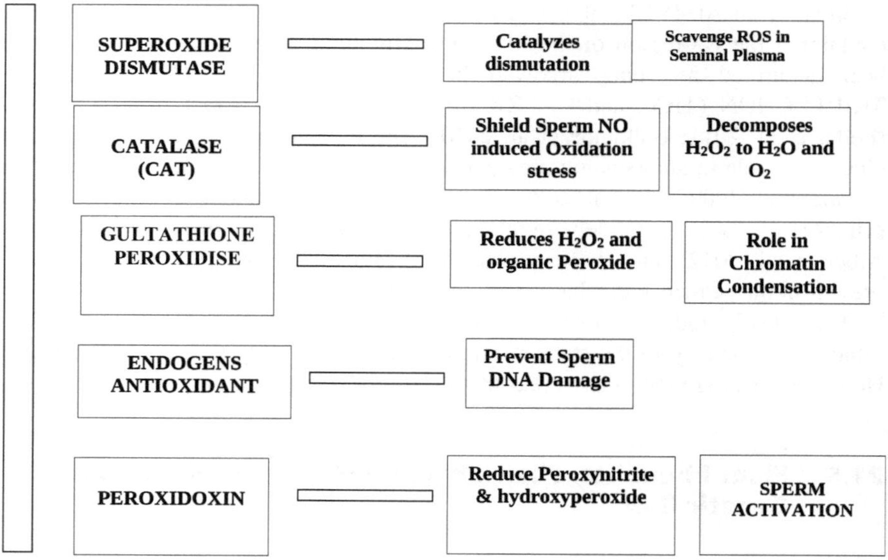

Fig. 21.1 Critical role of antioxidant enzymes in spermatogenesis. *ROS* reactive oxygen species, *NO* nitric oxide. (Source: https://www.ncbi.nlm.nih.gov/pmc/articles/PMC5456340/)

found in roots, stems, leaves, flowers and seeds of Sophorae Radix (Kuroyanagi et al. 1999). The rare alkaloids like sophocarpine, oxymatrine, matrine, sophoridine, kuraridin, kurarinone, isokurarinine, norkurarinone, pterocarpin, formononetin, trifolirhizin, daidzein, umbelliferone, maackiain, kuraridinol, kurarinol, neo-kurinol and norkurarinol of the plant kingdom are mainly found in Sophora species (Kim et al. 2004).

The prevention of cancer by green tea polyphenols is an induced antioxidative or pro-oxidative effects, and its necessity is dependent on the stage of carcinogenesis (Lambert and Elias 2010). The enhanced endogenous antioxidant is an important factor before carcinogen exposure, whereas pro-oxidant cell killing are necessary in clearing transformed cells from the body and thus restrictive tumour development (Fig. 21.1).

Epigenetic processes is well-regulated by several dietary polyphenols like caffeic acid, catechin, chlorogenic acid, etc. that play important roles in gene expression regulation such as DNA methylation processes (Link et al. 2010), histone modification and microRNA expression (Pan et al. 2013). Polyphenol-induced epigenetic changes may explain the chemopreventive roles to avert and other diseases in humans. The non-glycosylated form of phlorizin, phloretin, influences epigenetic processes, and heritable changes that are not encoded in the DNA sequence play vital role in gene expression regulation in breast cancer cells (Paluszczak et al. 2010). Quercetin are efficient inhibitors of sulfotransferases (Otake et al. 2000; Marchetti et al. 2001), which effect the activity of thyroid hormones, steroids and catecholamines (Coughtrie et al. 1998).

Oligomeric procyanidins present in apples affect multiple and important mechanisms of cancer. In vitro antimutagenic activity, antioxidant activity, anti-inflammatory mechanisms, anti-proliferative, modulation of carcinogen metabolism, modulation of signal transduction pathways and apoptosis-inducing activity, as well as novel mechanisms on epigenetic events and innate immunity, are well embraced (Gerhauser 2008). Alzheimer's disease is well-controlled by apple polyphenols, and apple juice helps in cognitive decline of normal aging suppressing overexpression of presenilin-1, which is concurrent to the production of amyloid-b peptide (Chan and Shea 2009).

Sperm cells are highly sensitive to reactive oxygen species (ROS), which are produced during cellular oxidation. In normal cell biology, ROS levels increase with a decreasing antioxidant response, resulting in oxidative stress which threatens sperm biology. The Chinese plant *Tripterygium wilfordii* and its botanical cousins contains several orally active compounds that provide male contraception (Qian 1986). Pills made from *Tripterygium* have been used in traditional Chinese medicine for over 1000 years. The extracts from the seeds of papaya fruits (*Carica papaya*) administrated to monkeys revealed that monkeys did not ejaculated the sperms and, hence, can be used as a contraceptive in humans as well (Verma and Chinoy 2001). Similarly, leaf extract of bael (*Aegle marmelos*) diminished the activity of germ cells as well (Das et al. 2006).

It is revealed that increased oxidative damage to sperm membranes is linked with changes in signal transductions that will affect the fertility. The male and female fertility cells possess an inherent limited capacity to release ROS that aids in fertilization process. Hence, it is very much argued that antioxidant enzymes, (like SOD catalase, glutathione peroxidase and reductases), vitamins (E, C, carotenoids also called tetraterpenoids) and few molecules involved in certain important biological systems are released from plants, mostly from fruits and vegetables. Hence, this is a way forwards that the phenolics compounds may treat different infertility problems in humans (Adewoyin et al. 2017). There is a relationship between reactive oxygen species, oxidative stress and infertility as shown in Fig. 21.2.

21.6 Conclusion

Till date many QTLs and candidate genes such as acyltransferases and glycosyltransferases have been reported to known to regulate polyphenol synthesis in plants. There is a requirement for in-depth knowledge on the molecular mechanism specific to their mode of action. Phenolics compounds are been well-established in fruits, while study about their shelf life in fruits should be improved in order to preserve and conserve these compounds. High variation has been reported in case of polyphenolics localization in fruit peels, and flesh among cultivar can be useful for breeding programme. However, progress in genomics and biotechnology application will boost genetic improvement of elite genotype for highly specific polyphenolic traits. Apart from above polyphenol has been part of our diet and plays

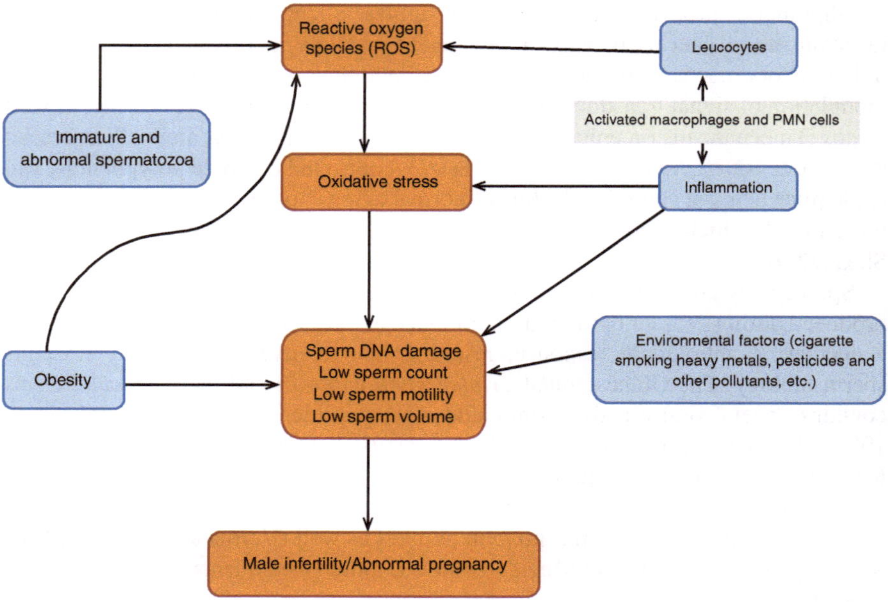

Fig 21.2 Adopted from Adewoyin et al. (2017)

a crucial role against many health alignments. Not only QTLs but also dietary polyphenols are been reported to bring certain kind of epigenetic changes and are either eliciting array of gene expression in the body to influence many changes in metabolism. Studies are required on other dietary plant phenolics to cause male sterility and improving fertility by preventing DNA damage.

References

Adewoyin M, Ibrahim M, Roszaman R, Isa M, Alewi N, Rafa A, Anuar M (2017) Male infertility: the effect of natural antioxidants and phytocompounds on seminal oxidative stress. Diseases 5 (1):9. https://doi.org/10.3390/diseases5010009

Aharoni A, De Vos CHR, Wein M, Sun Z, Greco R, Kroon A, Mol JNM, O'Connell AP (2001) The strawberry FaMYB1 transcription factor suppresses anthocyanin and flavonol accumulation in transgenic tobacco. Plant J 28:319–332

Akagi T, Katayama-Ikegami A, Kobayashi S, Sato A, Kono A, Yonemori K (2012) Seasonal abscisic acid signal and a basic leucine zipper transcription factor, DkbZIP5, regulate proanthocyanidin biosynthesis in persimmon fruit. Plant Physiol 158:1089–1102

Albert NW, Lewis DH, Zhang H, Schwinn KE, Jameson PE, Davies KM (2011) Members of an R2R3-MYB transcription factor family in Petunia are developmentally and environmentally regulated to control complex floral and vegetative pigmentation patterning. Plant J 65:771e784

Albert NW, Davies KM, Schwinn KE (2012) Repression e the dark side of anthocyanin regulation? Acta Hortic. in press

Arts ICW, Hollman PCH (2005) Polyphenols and disease risk in epidemiologic studies. Am J Clin Nutr 81(1):317S–325S. https://doi.org/10.1093/ajcn/81.1.317S

Bae K (2000) In: Bae K (ed) Medicinal plants of Korea. Kyo-Hak Pub, Seoul
Balkunde R, Pesch M, Hülskamp M (2010) Trichome patterning in *Arabidopsis thaliana* from genetic to molecular models. Curr Top Dev Biol 91:299–321
Bogs J, Jaffé FW, Takos AM, Walker AR, Robinson SP (2007) The grapevine transcription factor VvMybPA1 regulates proanthocyanidin synthesis during fruit development. Plant Physiol 143:1347–1361
Boyer J, Liu RH (2004) Apple phytochemicals and their health benefits. Nutr J 2004:3(5)
Caputi L, Malnoy M, Goremykin V, Nikiforova S, Martens S (2012) A genome-wide phylogenetic reconstruction of family 1 UDP-glycosyltransferases revealed the expansion of the family during the adaptation of plants to life on land. J Plant 69:1030–1042
Chan A, Shea T (2009) Dietary supplementation with apple juice decreases endogenous amyloid-beta-β levels in murine brain. J Alzheimers Dis 16(1):167–171
Coughtrie MW, Sharp S, Maxwell K, Innes NP (1998) Biology and function of the reversible sulfation pathway catalysed by human sulfotransferases and sulfatases. Chem Biol Interact 109:3–27
Coutinho PM, Stam M, Blanc E, Henrissat B (2003) Why are there so many carbohydrate-active enzyme-related genes in plants? Trends Plant Sci 8:563–565
D'Auria J (2006) Acyltransferases in plants: a good time to be BAHD. Curr Opin Plant Biol 9:331–340
Das UK, Maiti R, Jana D, Ghosh D (2006) Effect of aqueous extract of leaf of Aegle marmelos on testicular activities in rats. Int J Clin Pharmacol Ther 5:21–25
Davey MW, Kenis K, Keulemans J (2006) Genetic control of fruit vitamin C contents. Plant Physiol 142(1):343–351
Dixon RA, Liu C, Jun JH (2013) Metabolic engineering of anthocyanins and condensed tannins in plants. Curr Opin Biotechnol 24:329–335
Dubos C, Le Gourrierec J, Baudry A, Huep G, Lanet E, Debeaujon I, Routaboul JM, Alboresi A, Weisshaar B, Lepiniec L (2008) MYBL2 is a new regulator of flavonoid biosynthesis in *Arabidopsis thaliana*. Plant J 55:940–953
Feller A, Machemer K, Braun EL, Grotewold E (2011) Evolutionary and comparative analysis of MYB and bHLH plant transcription factors. Plant J 66:94e116
Frydman A, Liberman R, Huhman DV, Carmeli-Weissberg M, Sapir-Mir M, Ophir R, Sumner LW, Eyal Y (2013) The molecular and enzymatic basis of bitter/non-bitter flavor of citrus fruit: evolution of branch-forming rhamnosyltransferases under domestication. Plant J 73:166–178
Gerhauser C (2008) Cancer chemopreventive potential of apples, apple juice, and apple components. Planta Med 74:1608–1624
Gray J, Caparros-Ruiz D, Grotewold E (2012) Grass phenylpropanoids: regulate before using! Plant Sci 184:112–120
Hancock KR, Collette V, Fraser K, Greig M, Xue H, Richardson KCJ, Rasmussen S (2012) Expression of the R2R3-MYB transcription factor TaMYB14 from Trifolium arvense activates proanthocyanidin biosynthesis in the legumes Trifolium repens and Medicago sativa. Plant Physiol 159:1204–1220
He X, Fang J, Huang L, Wang J, Huang X (2015) Sophora flavescens Ait.: traditional usage, phytochemistry and pharmacology of an important traditional Chinese medicine. J Ethnopharmacol 172:10–29
Hertog MGL, Hollman PCH, Katan MB, Kromhout D (1993) Intake of potentially anticarcinogenic flavonoids and their determinants in adults in the Netherlands. Nutr Cancer 20(1):21–29
Jin H, Cominelli E, Bailey P, Parr A, Mehrtens F, Jones J, Tonelli C, Weisshaar B, Martin C (2000) Transcriptional repression by AtMYB4 controls production of UV-protecting sunscreens in Arabidopsis. EMBO J 19:6150–6161
Kagale S, Rozwadowski K (2011) EAR motif-mediated transcriptional repression in plants: an underlying mechanism for epigenetic regulation of gene expression. Epigenetics 6:141–146
Khater F, Fournand D, Vialet S, Meudec E, Cheynier V, Terrier N (2012) Identification and functional characterization of cDNAs coding for hydroxybenzoate/hydroxycinnamate

glucosyltransferases co-expressed with genes related to proanthocyanidin biosynthesis. J Exp Bot 63:1201–1214

Kim YC, Kim HS, Wataya Y (2004) Antimalarial activity of lavandulyl flavanones isolated from the roots of Sophora flavescens. Biol Pharm Bull 27(5):748–750

Kroon AR (2004) Transcription regulation of the anthocyanin pathway in Petunia hybrida. PhD thesis, Vrije Universiteit, Amsterdam, NL

Kuroyanagi M, Arakawa T, Hirayama Y, Hayashi T (1999) Antibacterial and antiandrogen flavonoids from Sophora flavescens. J Nat Prod 62(12):1595–1599

Lambert JD, Elias (2010) RJ antioxidant and pro-oxidant activities of green tea polyphenols: a role in cancer prevention. Arch Biochem Biophys 501:65–72

Lim EK, Baldauf S, Li Y, Elias L, Worrall D, Spencer SP, Jackson RG, Taguchi G, Ross J, Bowles DJ (2003) Evolution of substrate recognition across a multigene family of glycosyltransferases in Arabidopsis. Glycobiology 13:39–145

Ling JY, Zhang GY, Cui ZJ, Zhang CK (2007) Supercritical fluid extraction of quinolizidine alkaloids from Sophora flavescens Ait. and purification by high-speed counter-current chromatography. J Chromatogr A 1145(1–2):123–127

Link A, Balaguer F, Goel A (2010) Cancer chemoprevention by dietary polyphenols: promising role for epigenetics. Biochem Pharmacol 80:1771–1792

Luo J, Nishiyama Y, Fuell C, Taguchi G, Elliott K, Hill L, Tanaka Y, Kitayama M, Yamazaki Y, Bailey P, Parr A, Michael AJ, Saito K, Martin C (2007) Convergent evolution in the BAHD family of acyl transferases: identification and characterization of anthocyanin acyl transferases from *Arabidopsis thaliana*. J Plant 50:678–695

Marchetti F, de Santi C, Vietri M, Pietrabissa A, Spisni R, Mosca F, Pacifici GM (2001) Differential inhibition of human liver and duodenum sulphotransferase activities by quercetin, a flavonoid present in vegetables, fruit and wine. Xenobiotica 31:841–847

Matsuba Y, Sasaki N, Tera M, Okamura M, Abe Y, Okamoto E, Nakamura H, Funabashi H, Takatsu M, Saito M, Matsuoka H (2010) A novel glucosylation reaction on anthocyanins catalyzed by acyl-glucose–dependent glucosyltransferase in the petals of carnation and delphinium. Plant Cell 22(10):3374–3389

Matsui K, Umemura Y, Ohme-Takagi M (2008) AtMYBL2, a protein with a single MYB domain, acts as a negative regulator of anthocyanin biosynthesis in Arabidopsis. Plant J 55:954–967

McGhie TK, Hunt M, Barnet LE (2005) Cultivar and growing region determine the antioxidantpolyphenolic concentration and composition of apples grown in New Zealand. J Agric Food Chem 53:3065–3070

Miyahara T, Matsuba Y, Ozeki Y, Sasaki N (2011) Identification of genes in *Arabidopsis thaliana* with homology to a novel acyl-glucose dependent gluco-syltransferase of carnations. Plant Biotechnol 28:311–315

Miyahara T, Nishizaki Y, Takahashi V, Ozeki Y, Sasaki N (2012) Isolation of an acyl-glucose-dependent anthocyanin glucosyltransferase from the monocot africans and expression analysis of its homologus genes in Arabidopsis thaliana. In: Lattanzio V, Mulinacci N, Pinelli P, Romani A (eds) Polyphenols communications 2012. Stabilimento Grafico Rindi, Prato, pp 351–352

Montefiori M, Espley RV, Stevenson D, Cooney J, Datson PM, Saiz A, Atkinson RG, Hellens RP, Allan AC (2011) Identification and characterisation of F3GT1 and F3GGT1, two glycosyltransferases responsible for anthocyanin biosynthesis in red-fleshed kiwifruit (Actinidia chinensis). J Plant 65:106–118

Mugford ST, Milkowski C (2012) Serine carboxypeptidase-like acyltransferases from plants. Methods Enzymol 516:279–297

Nakatsuka T, Nishihara M (2010) UDP-glucose:3-deoxyanthocyanidin 5-O-gluco-syltransferase from Sinningia cardinalis. Planta 232:383–392

Nishizaki Y, Matsuba Y, Okamoto E, Okamura M, Ozeki Y, Sasaki N (2011) Structure of the acyl-glucose-dependent anthocyanin 5-O-glucosyltransferase gene in carnations and its disruption by transposable elements in some varieties. Mol Gen Genomics 286:383–394

Ono E, Homma Y, Horikawa M, Kunikane-Doi S, Imai H, Takahashi S, Kawai Y, Ishiguro M, Fukui Y, Nakayama T (2010) Functional differentiation of the glycosyltransferases that contribute to the chemical diversity of bioactive flavonol glycosides in grapevines (*Vitis vinifera*). Plant Cell 22:2856–2871

Otake Y, Nolan AL, Walle UK, Walle T (2000) Quercetin and resveratrol potently reduce estrogen sulfotransferase activity in normal human mammary epithelial cells. J Steroid Biochem Mol Biol 73(5):265–270

Paluszczak J, Krajka-Kuzniak V, Baer-Dubowska W (2010) The effect of dietary polyphenols on the epigenetic regulation of gene expression in MCF7 breast cancer cells. Toxicol Lett 192:119–125

Pan MH, Lai CS, Wu JC, Ho CT (2013) Epigenetic and diseases targets by polyphenols. Curr Pharm Des. in press 19:6156

Piao XL, Piao XS, Kim SW, Park JH, Kim HY, Cai SQ (2006) Identification and characterization of antioxidants from Sophora flavescens. Biol Pharm Bull 29(9):1911–1915

Qian SZ (1986) Antifertility effect of Tripterygium wilfordii in men. Contraception 23:22–24

Shelton D, Stranne M, Mikkelsen L, Pakseresht N, Welham T, Hiraka H, Tabata S, Sato S, Paquette S, Wang TL, Martin C, Bailey P (2012) Transcription factors of Lotus: regulation of isoflavonoid biosynthesis requires coordinated changes in transcription factor activity. Plant Physiol 159:531–547

Stehle F, Brandt W, Stubbs MT, Milkowski C, Strack D (2009) Sinapoyl- transferases in the light of molecular evolution. Phytochemistry 70:1652–1662

Tanaka Y, Sasaki N, Ohmiya A (2008) Biosynthesis of plant pigments: anthocyanins, betalains and carotenoids. Plant J 54:733–749

Terrier N, Torregrosa L, Ageorges A, Vialet S, Verriès C, Cheynier V, Romieu C (2009) Ectopic expression of VvMybPA2 promotes proanthocyanidin biosynthesis in Vitis vinifera L. and suggests additional targets in the pathway. Plant Physiol 149:1028–1041

Tuominen LK, Johnson VE, Tsai CJ (2011) Differential phylogenetic expansions in BAHD acyltransferases across five angiosperm taxa and evidence of divergent expression among Populus paralogues. BMC Genomics 12:236

Verma RJ, Chinoy NJ (2001) Effect of papaya seed extract on microenvironment of cauda epididymis. Asian J Androl 3:143–146

Vinson JA, Su XH, Zubik L, Bose P (2001) Phenol antioxidant quantity and quality in foods: fruits. J Agric Food Chem 49(11):5315–5321

Willett WC (2002) Balancing life-style and genomics research for disease prevention. Science 296 (5568):695–698

Yu XH, Gou JY, Liu CY (2000) BAHD superfamily of acyl-CoA dependent acyltransferases in Populus and Arabidopsis: bioinformatics and gene expression. Plant Mol Biol 70:421–442

Zhang W, Ning G, Lv H, Liao L, Bao M (2009) Single MYB-type transcription factor AtCAPRICE: a new efficient tool to engineer the production of anthocyanin in tobacco. Biochem Biophys Res Commun 388:742–747

Zhao Q, Dixon RA (2011) Transcriptional networks for lignin biosynthesis: more complex than we thought? Trends Plant Sci 4:227–233

Zhu HF, Fitzsimmons K, Khandelwal A (2009) CPC, a single-repeat R3 MYB, is a negative regulator of anthocyanin biosynthesis in Arabidopsis. Mol Plant 2:790–802

Role of Phenolic Compounds in Plant-Defensive Mechanisms

22

Santosh Kumar, Md. Minhajul Abedin, Ashish Kumar Singh, and Saurav Das

Abstract

Phenolics are ubiquitous secondary metabolites found in plant. They are aromatic compounds synthesized by phenylpropanoid pathway. Phenolics have been in the focus of many findings on plant-defenses mechanisms to pathogens, including bacteria, fungi, and viruses, and major abiotic stresses like drought, salinity, and UV. Phenolic compound exhibits antimicrobial and antioxidant properties which helps plant to evade pathogenic infections as well as protect the major tissues from toxic effect of reactive oxygen species. Rapid upregulation of genes in the phenylpropanoid pathway and the accumulation of phenolics can be observed in response to environmental stress. Phenolic compounds also play an important role in protecting the plant from insect herbivory. Phenolic compounds are diverse and classified based on the number of carbons. Structural diversity of the phenolic compound defines its functional properties and distribution in different plant species. Beside defensive mechanisms, phenolic compounds are also important in cross-talk or plant-microbe interaction and communications. Structural diversity of flavonoids from the leguminous plant is important in species-specific symbiotic relationships.

Keywords

Phenolics · Ubiquitous · Drought · Salinity · Antioxidant · Stress

S. Kumar · M. M. Abedin · A. K. Singh
Department of Microbiology, School of Life Sciences, Sikkim University, Gangtok, Sikkim, India

S. Das (✉)
Department of Agronomy and Horticulture, University of Nebraska – Lincoln, Lincoln, NE, USA
e-mail: sdas4@unl.edu

R. Lone et al. (eds.), *Plant Phenolics in Sustainable Agriculture*,
https://doi.org/10.1007/978-981-15-4890-1_22

22.1 Introduction

Phenols are compounds with one or more hydroxyl groups attached directly to an aromatic ring. They are similar to alcohols of aliphatic structure where hydroxyl group is attached to a chain of carbon. However, hydrogen of the phenolic hydroxyl is influenced by the presence of an aromatic ring, and they are labile, which makes phenols as weak acids. The term phenolic covers a diverse group of chemical compounds. On the basis of structure, phenolic compounds comprise an aromatic ring, bearing one or more hydroxyl substituent's which range from simple phenolic molecules to highly polymerized compounds (Balasundram et al. 2006). Harborne and Simmonds (1964) classified phenolic compounds into different groups based on the number of carbons in the molecule (Table 22.1) (Harborne and Simmonds 1964).

Table 22.1 Classification of phenolic compounds

No. of carbon atoms	Structure	No. of phenolic cycles	Class	Examples
6	C_6	1	Simple phenols, benzoquinones	Catechol, hydroquinone, 2,6-dimethoxybenzoquinone
7	C_6–C_1	1	Phenolic acids, phenolic aldehydes	Gallic, salicylic acids
8	C_6–C_2	1	Acetophenones, tyrosine derivatives, phenylacetic acids	3-Acetyl-6-methoxybenzaldehyde, tyrosol, p-hydroxyphenylacetic acid, homogentisic acid
9	C_6–C_3	1	Hydroxycinnamic acids, phenylpropenes, coumarins, isocoumarins, chromones	Caffeic, ferulic acids, myristicin, eugenol, umbelliferone, aesculetin, bergenin, eugenin
10	C_6–C_4	1	Naphthoquinones	Juglone, plumbagin
13	C_6–C_1–C_6	2	Xanthonoids	Mangiferin
14	C_6–C_2–C_6	2	Stilbenoids, anthraquinones	Resveratrol, emodin
15	C_6–C_3–C_6	2	Chalconoids, flavonoids, isoflavonoids, neoflavonoids	Quercetin, cyanidin, genistein
16	C_6–C_4–C_6	2	Halogenated algal phenolic compounds	Kaviol A, colpol
18	$(C_6–C_3)_2$	2	Lignans, neolignans	Pinoresinol, eusiderin
30	$(C_6–C_3–C_6)_2$	4	Biflavonoids	Amentoflavone
Many	$(C_6–C_3)_n$, $(C_6)_n$, $(C_6–C_3–C_6)_n$	$n > 12$	Lignins, catechol melanins, flavolans, polyphenolic proteins, polyphenols	Raspberry ellagitannin, tannic acid

Phenolic compounds are secondary metabolites, universally present in plants. They are the derivatives of pentose phosphate, shikimate, and phenylpropanoid pathways (Balasundram et al. 2006; Cheynier 2012). They are present in plant seeds, leaves, bark, and flowers. More than 8000 phenolic structures are currently known. They can range from simple molecules of low-molecular-weight (phenolic acids, flavonoids, phenylpropanoids) to highly polymerized compounds (lignins, melanins, lignans, tannins). With at least 4000 identified molecules, flavonoids represent the most common and widely distributed subgroups comprising flavones, isoflavones, and 2,3-dihydro derivatives of flavone (Cheynier 2012). Besides flavonoids, lignans are one of the most distinctive groups of phenylpropanoids. They are known to possess a variety of biological activities including antibacterial, antifungal, antiviral, anti-inflammatory, anti-cancerous, and antioxidant effects (Pereira et al. 2016). Phenolic compounds also relate to the human health benefits which derived from consuming high levels of fruits and vegetables. The beneficial effects derived from phenolics have been attributed to their antioxidant activity. Phenolic compounds act as a natural source of antioxidants (Parr and Bolwell 2000).

The term polyphenol is often used as an alternative of phenolic compounds; however, it should be restricted to molecules bearing at least two or more phenolic rings in their structure (Quideau et al. 2011). Polyphenols or polyhydroxyphenols are organic chemicals with multiple phenols as structural units. It was first defined by Bate-Smith (1962) as "water soluble phenolic compounds having molecular weights between 500–3,000 Da." They have special properties such as the ability to precipitate alkaloids, gelatin, and other proteins from solution (Cheynier 2012). These compounds are one of the most widely occurring groups of extracted phytochemicals and are of considerable morphological and physiological importance in plants. Phenolics play a major role in growth and reproduction, and also it provides protection against pathogens and predators (Balasundram et al. 2006). It also contributes toward the color and sensory characteristics of vegetables and fruits (Balasundram et al. 2006). Some phenolic compounds are widely distributed, while others are specific to certain families of plant or found only in certain plant organs at certain developmental stages. Structural diversity of phenolic compound defines its functional attributes and its specific distribution. For example, anthocyanin is the pigment of most red and blue plant organs. They are found in flowers and in mature fruits and play an important role in attracting pollinators and helping in pollination. Presence of anthocyanin in young leaves protect them from photodamage and help in normal growth. A red anthocyanin coating of leaves also protects the young leaves from insect herbivory (Karageorgou and Manetas 2006). Phenolic compounds like proanthocyanidins (flavan-3-ol oligomers and polymers) and hydrolyzable tannins (i.e., gallotannins and ellagitannins, based on multiple esters of gallic) participate in plant defense against herbivores, fungi, and viruses. They are mostly found in early developmental stages of plant based on their functional properties (Cheynier 2012). Flavonoids have been identified to guard the tissue against UV radiation. The depletion of ozone layer over the last three decades has created UV radiation stress

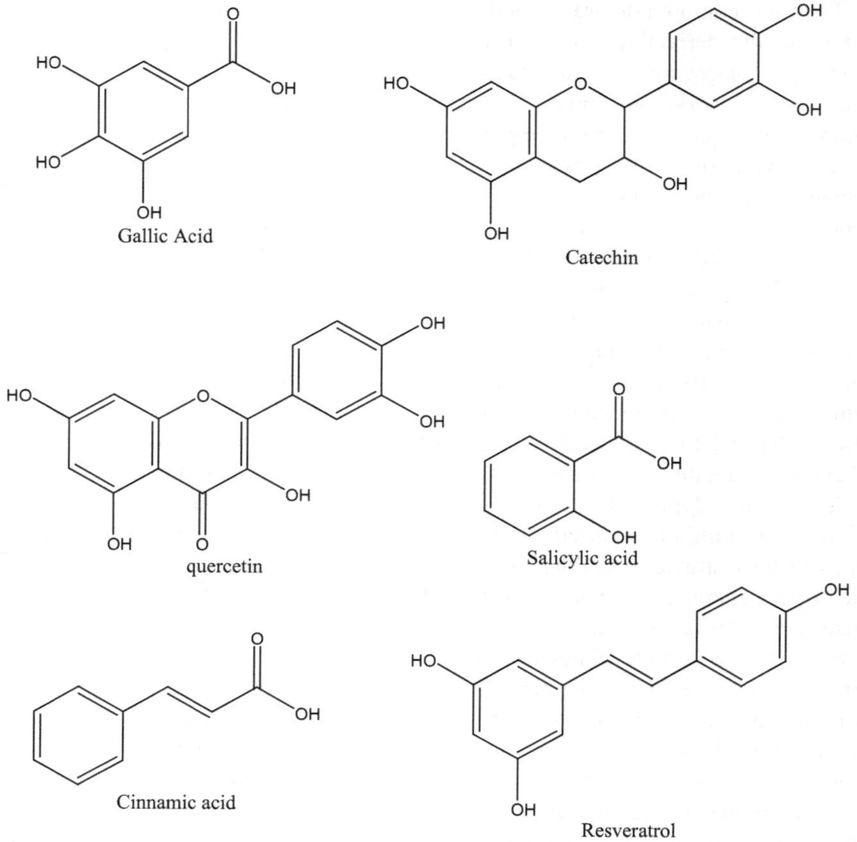

Fig. 22.1 Structure of some of the common plant phenolic compounds

on plant. Flavonoids were identified to reduce the reactive oxidative species generated over the UV-B stress (Agati and Tattini 2010) (Fig. 22.1).

22.2 Functional Properties of Phenolic Compounds

Phenolic compounds are mainly synthesized from cinnamic acid, which is formed from phenylalanine by the action of phenylalanine ammonia-lyase (PAL) through phenylpropanoid pathway (Heleno et al. 2015). The importance of this pathway can be supported by the fact that, in normal growth conditions, 20% of carbon fixed by plants flows through this pathway (Michalak 2006a, b). Phenolic compounds have various functions in plants. An enhancement of phenylpropanoid metabolism and accumulation of phenolic compounds can be observed under different environmental stress conditions (Bergmann et al. 1994; Król et al. 2014; Caliskan et al. 2017). The combination of isoflavones and some other flavonoids is induced when plants are infected, wounded, under low temperatures, low nutrient conditions, and UV, metal stress condition (Mierziak et al. 2014; Schulz et al. 2016). Plants accumulate UV-absorbing flavonoids and additional phenolic compounds mainly in vacuoles of epidermal cells, to avoid the penetration of UV-B into the deeper tissues of the plant (Rodríguez-Calzada et al. 2019). The initiation of phenolic compound biosynthesis was observed in wheat in response to nickel toxicity and in maize in response to aluminum. *Phaseolus vulgaris* exposed to cadmium accumulate soluble and insoluble phenolics. *Phyllanthus tenellus* leaves contain more phenolics than control plants after being sprayed with copper sulfate (Michalak 2006a, b). Phenolic compounds have also an important role in pathogenic resistance as bioactive or antimicrobial compounds. In response to microbial attack, induce defense mechanism synthesizes broad-spectrum antibacterial compounds which initiate site-specific hypersensitive response to protect the spread of infection and future attack (Mandal et al. 2010).

Besides the stress response metabolites, phenolic acid has also an important role in plant-microbe interaction. Legume plants release phenolic acids as root exudates during germination and seedling growth stage. *Rhizobium* community of rhizosphere respond to the phenolic acid (flavonoids) and undergo metabolic changes, which initiate the nodulation process in legumes. The structural and functional diversity of phenolics secreted by legumes provides a competitive advantage for nodulation by selective rhizobial strains (Blum et al. 2000; Mandal et al. 2010). A current study also suggested that root exudates contribute to plant pathogenic resistance via secretion of antimicrobial compounds. These findings point to the importance of plant root exudates which are mainly phenolic compounds as belowground signaling molecules, particularly in defense responses (Lanoue et al. 2010). The study showed that under *Fusarium* attack, the barley root system secrets t-cinnamic acid which has antimicrobial activity. The secretion of de novo biosynthesized t-cinnamic acid induced within 2 days and shows dynamic plant-defense mechanisms at root level (Lanoue et al. 2010).

Plants grow and live in very complex and varying ecosystems. Because plants lack the mobility to escape from pathogens or herbivores attack, they have developed constitutive and inducible defenses mechanisms that are triggered by spatio-temporally dynamic signaling mechanisms. These defenses counteract pathogens directly via the production of toxins or defense plant structures. Root exudates play an important role in induced plant-defense mechanisms. The roots of *Ocimum basilicum* secrete rosmarinic acid when challenged by the pathogenic fungus *Pythium ultimum* (Bais et al. 2002). Stimulation of iridoid glycosides in root exudates of *Plantago lanceolata* in the presence of nematodes acts as nematicides (Wurst et al. 2010) (Fig. 22.2 and Table 22.2).

Several simple and complex phenolic compounds also act as allelochemicals, phytoalexin, phytoanticipins, and nematicides. Phenolic compounds like terpenoids, hydroquinones, hydroxybenzoates, hydroxycinnamates, and hydroxynaphthoquinones are known for allelochemical activity for competitive plants and weeds (Weir et al. 2004; Jilani et al. 2008). Compounds like cajanin, medicarpin, coumestrol, and

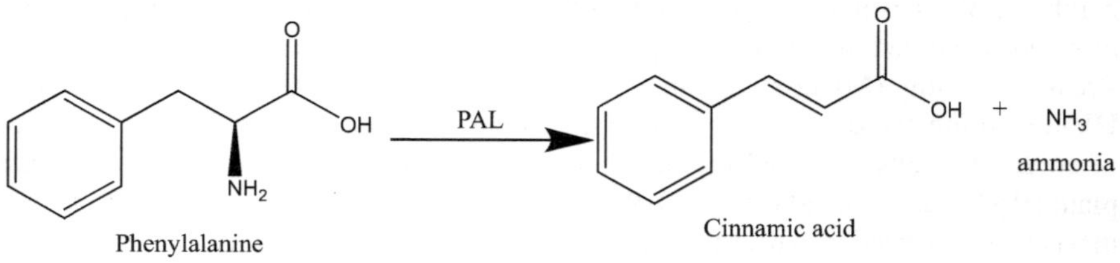

Fig. 22.2 Enzymatic conversion of phenylalanine to cinnamic acid. [Phenylalanine ammonia-lyase (PAL)]

limonoids can act as nematicides in plant-defense mechanisms. Plants respond to pathogen attack by accumulating phytoalexins, such as hydroxycoumarins and hydroxycinnamate conjugates (Karou et al. 2005). The synthesis, release, and accumulation of phenolics in particular salicylic acid (SA) are central to many defense strategies against pathogenic invaders (Boller and He 2009; Bhattacharya et al. 2010). De novo synthesis of SA occurs followed by infection or stress response. Accumulation of SA is a key molecule of systemic-acquired resistance. The volatile derivative, viz., salicylic acid methyl ester, has the ability to induce protection in other parts of the plant and to neighboring plants (Horvath and Chua 1994; Heil 1999; Bhattacharya et al. 2010).

22.3 Phenolics and Derivatives

22.3.1 Flavonoids

Flavonoids are the largest group of secondary metabolites, which are derivatives of simple phenols. Structurally they have a 15-carbon skeleton, consisting of two aromatic rings connected by three carbon bridge (C6–C3–C6) (Kulbat 2016). Structural and functional differentiation of flavonoids depends on its substituents (Martens et al. 2010). Flavonoids can be basically classified into three classes based on structural skeleton, viz., *flavonoids* or *bioflavonoids*, *isoflavonoids* derived from 3-phenylchromen-4-one (3-phenyl-1,4-benzopyrone), and *neoflavonoids*, derived from 4-phenylcoumarine (4-phenyl-1,2-benzopyrone) (Fig. 22.3). Flavonoids can be further subgrouped into anthocyanidins, anthoxanthins, chalcones, flavones, flavonols, flavandiols, flavans, proanthocyanidins, and tannins (Falcone Ferreyra et al. 2012).

In higher plants, flavonoids are involved in floral pigmentation, symbiotic relationships with prokaryotes, protection from pathogenic infection, insect herbivory, UV filtration, and antioxidant activity. Characteristics of unique color of plants like yellow are due to quercetin, or shades of blue color are due to malvidin (Mol et al. 1998). The flavonoids secreted by leguminous plant roots act as a messenger for the initiation of rhizobial infection stage toward the symbiotic relationships. Plants accumulate flavonoids in vacuoles of epithelial cells to protect tissues against

Table 22.2 Function of some of the naturally occurring phenolic compounds. (Bhattacharya et al. 2010)

Phenolic compounds	Function
Coniferyl alcohol, sinapinic acid, cinnamic acid	*vir* gene inducers, determinants of scent and attractants of pollinators and symbiotic microbes in plants, etc.
Hydroxybenzoate, hydroxycinnamates, 5-hydroxyanthraquinones	Allelochemicals for plant competition
Umbelliferone, *p*-hydroxybenzoic acid, vanillyl alcohol, isoflavones	Chemoattractant in *Rhizobium*
Sinapinic acid, syringic acid, ethylsyringamide, propylsyringamide, carbethoxyethylensyringamide, parahydroxybenzoate, ferulic acid	*vir* gene inducers in *Agrobacterium*
Vanillyl alcohol, bromo acetosyringone	Inhibitors of *vir* gene induction in *Agrobacterium*
Acetosyringone, α-hydroxyacetosyringone, *p*-hydroxybenzoate	Chemoattractant in *Agrobacterium* and *Rhizobium* and *vir* gene inducers in *Agrobacterium*
Salicylic acid	Quorum quencher in *Agrobacterium*
Hydroquinone	Allelochemical for plant competition
Coumarins, xanthones, anthocyanidins	Determinants of color and attractants of pollinators in plants
Caffeic acid	*vir* gene inducer in *Agrobacterium*
3,4-Dihydroxybenzoic acid	Chemoattractant in *Agrobacterium* and *Rhizobium*
Protocatechuic acid, β-resorcylic acid, protocatechuate, *p*-resorcylate, catechol	*vir* gene inducer in *Agrobacterium*
Chlorogenic acid	Precursor for lignin and suberin synthesis in plants
Lignin, tannins, and suberins	Structural components of plant cells
Catechins	Plant defense
Flavonoids, flavonols, flavones, genistein, daidzein, *O*-acetyldaidzein, 6-*O*-malonylgenistin, 6-*O*-malonyl daidzin, glycitin, 6-*O*-malonylglycitin	*nod* gene inducers in *Rhizobium*
Apigenin, naringenin, luteolin	Chemoattractant in *Agrobacterium* and *Rhizobium* and *nod* gene inducers in *Rhizobium*
Gallate, gallic acid, pyrogallic acid, syringic acid, kaempferol	*vir* gene inducers in *Agrobacterium*
Flavanones, quercetin	*nod* gene inducers in *Rhizobium*
Isoflavonoids	Chemoattractant and *nod* gene inducers in *Rhizobium*
Cajanin, medicarpin, glyceoline, rotenone, coumestrol, phaseolin, phaseolin, limonoids, tannins, flavonoids	Phytoalexins, phytoanticipins, and nematicides in plant defense

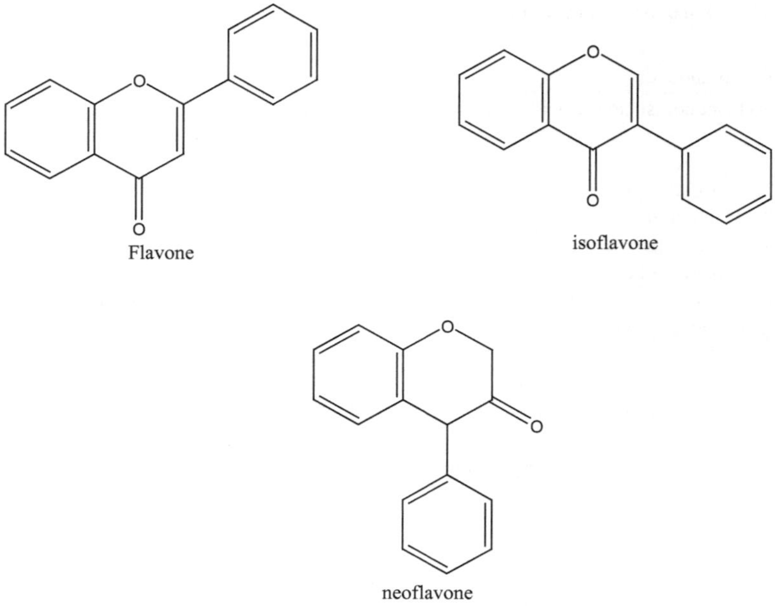

Fig. 22.3 Flavonoids classes

UV stress. Flavonoids can absorb radiation of high energy (maximum absorbance at 250–270 nm and 335–360 nm) (Winkel-Shirley 2002; Liang et al. 2006). Flavonoids also protect the plants from metal toxicity by metal – chelation. Metal – flavonoids chelates displays superoxide dismutase activity (A. 2006).

22.3.2 Coumarins

Coumarin acid is another important product of shikimic acid pathway, which is produced by hydroxylation of cinnamic acid or deamination of tyrosine by tyrosine ammonia – lyase (TAL) enzyme. It has a basic structure of $C_9H_6O_2$. Coumarins exhibit toxicity against herbivores. Bitter-taste coumarin also discourages animals from eating plants containing large amount of these compounds (Kulbat 2016) (Fig. 22.4).

22.3.3 Tannins

Tannins are polyphenolic compound that protects plants from herbivory. There are three classes of tannins: hydrolyzable polymer synthesized from gallic acid, phlorotannins, and condensed tannins and phlobatannins (Haslam 1966; Mueller-Harvey 2001). Tannins are generally found in high concentration in the bark and

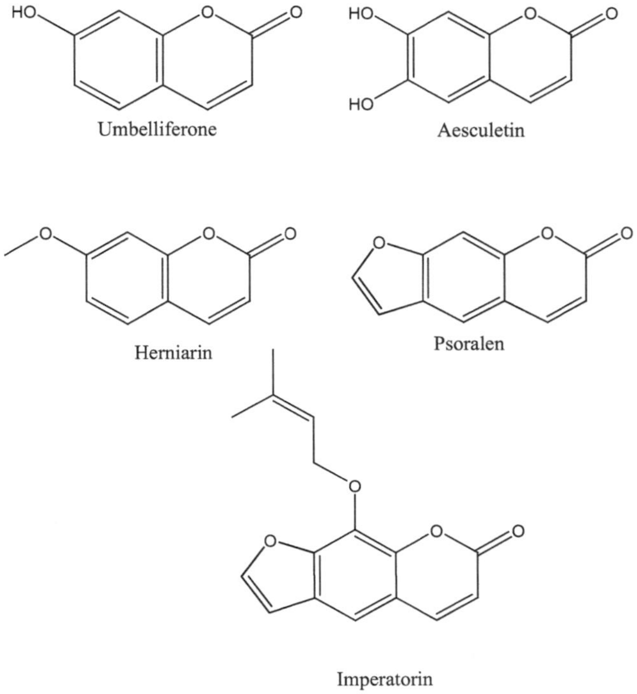

Fig. 22.4 Some naturally occurring coumarin derivatives

leaves of trees. Unpleasant, bitter taste and ability to denature protein makes tannin an exceptional compound to provide plant protection against insects (Kulbat 2016).

22.3.4 Salicylic Acid

Pathogenic infection initiates the accumulation of pathogenesis-related (PR) protein at a distant location from the infection site to enhance the state of resistance. Along with the PR protein, accumulation of salicylic acid (SA) and hydrogen peroxides also occurs at the site of infection as a systemic-acquired resistance, which is mediated by different signaling molecules. SA acid is one of the major signaling molecule for initiation of systemic-acquired resistance (SAR) (Fig. 22.5). It has been found that the exogenous supply of SA can induce SAR and provide resistance to infected plants.

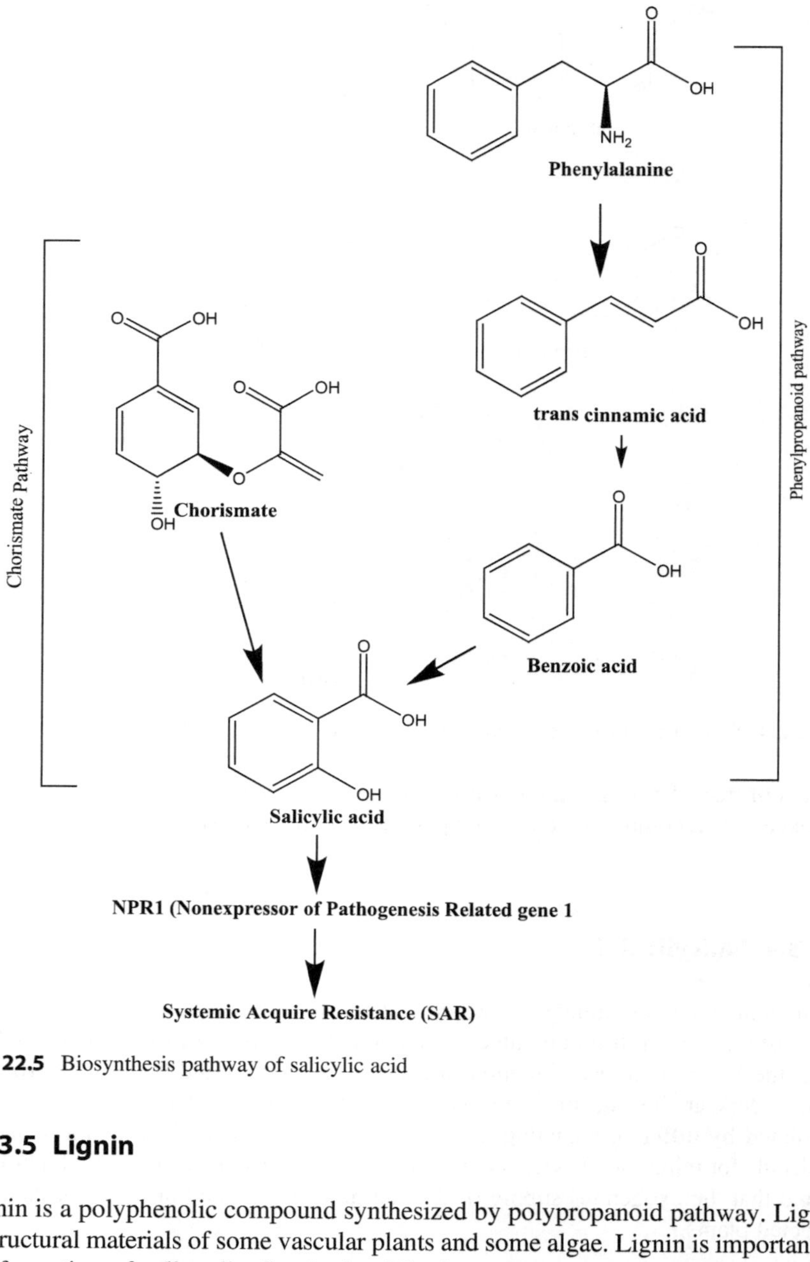

Fig. 22.5 Biosynthesis pathway of salicylic acid

22.3.5 Lignin

Lignin is a polyphenolic compound synthesized by polypropanoid pathway. Lignin is structural materials of some vascular plants and some algae. Lignin is important in the formation of cell walls. Synthesis of lignin can be de novo and specifically in response to pathogenic attack. In cultivar variety of wheat which are resistant to *Puccinia recondite* f. sp. *tritici*, fungal pathogen-causing leaf rust shows more accumulation of lignin compared to susceptible varieties (Southerton and Deverall

1990). A similar reaction can be observed in wheat in response to *Fusarium graminearum* infections which cause *Fusarium* head blight. Several researchers showed the importance of lignin biosynthesis in response to pathogenic attack. Inhibition of the enzyme PAL and cinnamyl alcohol dehydrogenase which are important in lignin biosynthesis showed reduced resistance to stem rust (*Puccinia graminis* Pers f. sp. *triti* Erics. & E. Henn.) (Moldenhauer et al. 2006). Coordinated expression of lignin biosynthesis is required against pathogenic infections.

22.3.6 Phytoalexin

Stress response initiates formation of two different types of compounds. In one response, the plant can form compounds which are off-target and work from considerable distance from the infection site. In other response, plant forms compounds which specifically act at the site of infection. In general, the stress metabolites or the compounds formed under stress and infections are referred to as phytoalexin. Phytoalexin shows specific toxicity against pathogens. Most of the phytoalexins belong to flavonoids and isoflavonoids group and exhibit antimicrobial and antioxidant activity (Vermerris and Nicholson 2006). Phytoalexins can breakdown cell wall, disrupt metabolism, and prevent growth of the pathogenic microorganisms. The importance of phytoalexin can be gleaned from the fact mutants for phytoalexin exhibit more pathogenic colonization compared to wild types, and susceptibility of plant tissues increases when phytoalexin biosynthesis is inhibited (Glazebrook and Ausubel 1994). Synthesis of phytoalexins is important for defensive mechanisms against fungal and other microorganisms.

- Trans-resveratrol is a phytoalexin which can inhibit the growth of *Botrytis cinerea* (Timperio et al. 2012).
- Grapevine secretes delta viniferin against the infection of *Plasmopara viticola* (Favaron et al. 2017).
- Danielone, a phytoalexin found in papaya fruit, shows antifungal activity against *Colletotrichum gloeosporioides* (Echeverri et al. 1997).
- Chlorogenic acid present in the periderm of potato tubers is toxic to the *Streptomyces scabies*, which causes potato scab (Villegas and Kojima 1985) (Fig. 22.6).

22.4 Phenolic Compounds in Abiotic Stress

Environmental condition governing the agricultural lands is often inadequate for crop production. Abiotic stress including drought, salinity, and water scarcity causes huge economic and yield losses every year around the world. Climate change and improper land use have also aggravated the land degradation. Drought and salinity are still major abiotic stress for many agricultural lands. Research showed that

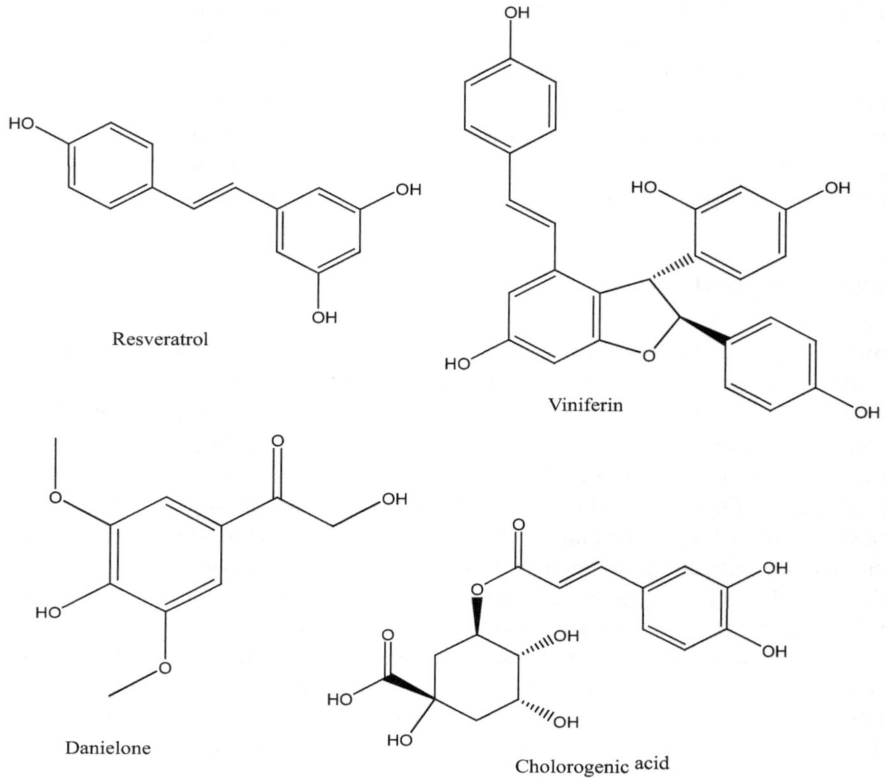

Fig. 22.6 Structure of few of the phytoalexin

increased abiotic stresses like drought, salinity, low/high temperature, and nutrient deprivation are associated with oxidative stress and production of reactive oxygen species (ROS) (Schulz et al. 2016; Caliskan et al. 2017). Reactive oxygen species are superoxide anion (O_2^-), hydrogen peroxide (H_2O_2), and hydroxyl radical (HO) which are formed in cells during partial reduction of oxygen species. Plant has evolved different antioxidant mechanisms to mitigate the cytotoxic effect of ROS. Plant can use enzymatic and non-enzymatic components for antioxidant activity. In enzymatic components, superoxide dismutase, ascorbate peroxidase, and glutathione reductase are most important. In non-enzymatic antioxidant mechanisms, a polyphenolic compound like phenolic acids, flavonoids, proanthocyanidins, and anthocyanins play an important role to balance the ROS effect. Evidence suggest the accumulation of antioxidant metabolites in major cellular tissues like mesophyll cells, chloroplast, and mitochondria which can suffer major damage from ROS (Martinez et al. 2016).

22.5 Phenolic Compounds in Insect Herbivory

A complex interaction can be observed between the insect pest and phenolics. A positive correlation has been recorded for the insect herbivory of spotted spider mites (*Tetrancychus urticae*) and constitutive concentration of catechol phenolics in strawberry leaves (Luczynski et al. 1990). Gossypol, a phenolic pigment of cotton, has a deterrent effect on numerous insect pest and is toxic to *Heliothis virescens* (tobacco bollworm) and *Heliothis zea* (bollworm) (Maxwell et al. 1965). Catechol can control the mite's population. It binds to mite's digestive system and inactivates them (Rehman et al. 2012). Another example is tannin; Feeny reported the tannin can control the larvae of the oak month (*Opheropthera brumata*) (Feeny 1970). Wheat cultivars containing phenolics are much less attractive to *Rhopalosiphum padi* (cereal aphid) (Fürstenberg-Hägg et al. 2013).

22.6 Conclusions

Phenolic compounds play an important role in plant growth and development, particularly in defense mechanisms. Most of the phenolic compounds have potent antioxidant properties, neutralizing the effects of oxidative stress. Some of them exhibit ability to chelate heavy metal ions. Importantly, phenolic phytoalexins exhibit antibiotic and antifungal activity. Coumarins and tannins repel herbivores, whereas phenylpropanoids are starting molecules for the synthesis of lignin and suberin, in order to strengthen cell walls. Flavonoids act as a chemoattractant and initiate the symbiotic relationship between Rhizobium and leguminous plant.

References

Agati G, Tattini M (2010) Multiple functional roles of flavonoids in photoprotection. New Phytol 186:786–793. https://doi.org/10.1111/j.1469-8137.2010.03269.x

Bais HP, Walker TS, Schweizer HP, Vivanco JM (2002) Root specific elicitation and antimicrobial activity of rosmarinic acid in hairy root cultures of Ocimum basilicum. Plant Physiol Biochem 40:983–995. https://doi.org/10.1016/S0981-9428(02)01460-2

Balasundram N, Sundram K, Samman S (2006) Phenolic compounds in plants and Agri-industrial by-products: antioxidant activity, occurrence, and potential uses. Food Chem 99:191–203. https://doi.org/10.1016/j.foodchem.2005.07.042

Bate-Smith EC (1962) The phenolic constituents of plants and their taxonomic significance. I. Dicotyledons. J Linn Soc Lond Bot 58:95–173. https://doi.org/10.1111/j.1095-8339.1962.tb00890.x

Bergmann H, Leinhos V, Machelett B (1994) Increase of stress resistance in crop plants by using phenolic compounds. Acta Hortic:390–397. https://doi.org/10.17660/ActaHortic.1994.381.50

Bhattacharya A, Sood P, Citovsky V (2010) The roles of plant phenolics in defence and communication during agrobacterium and rhizobium infection. Mol Plant Pathol 11:705–719

Blum U, Staman KL, Flint LJ, Shafer SR (2000) Induction and/or selection of phenolic acid-utilizing bulk-soil and rhizosphere Bacteria and their influence on phenolic acid Phytotoxicity. J Chem Ecol 26:2059–2078. https://doi.org/10.1023/A:1005560214222

Boller T, He SY (2009) Innate immunity in plants: an arms race between pattern recognition receptors in plants and effectors in microbial pathogens. Science (80–) 324:742–744. https://doi.org/10.1126/SCIENCE.1171647

Caliskan O, Radusiene J, Temizel KE, Staunis Z, Cirak C, Kurt D, Odabas MS (2017) The effects of salt and drought stress on phenolic accumulation in greenhouse-grown Hypericum pruinatum. Ital J Agron 12. https://doi.org/10.4081/ija.2017.918

Cheynier V (2012) Phenolic compounds: from plants to foods. Phytochem Rev 11:153–177. https://doi.org/10.1007/s11101-012-9242-8

Echeverri F, Torres F, Quiñones W, Cardona G, Archbold R, Roldan J, Brito I, Luis JG, Lahlou EH (1997) Danielone, a phytoalexin from papaya fruit. Phytochemistry 44:255–256

Falcone Ferreyra ML, Rius SP, Casati P (2012) Flavonoids: biosynthesis, biological functions, and biotechnological applications. Front Plant Sci 3. https://doi.org/10.3389/fpls.2012.00222

Favaron F, Lucchetta M, Odorizzi S, Pais da Cunha A, Sella L (2017) The role of grape polyphenols on trans-resveratrol activity against Botrytis cinerea and of fungal laccase on the solubility of putative grape PR protein. J Plant Pathol 91:579–588

Feeny P (1970) Seasonal changes in oak leaf tannins and nutrients as a cause of spring feeding by winter moth caterpillars. Ecology 51:565–581. https://doi.org/10.2307/1934037

Fürstenberg-Hägg J, Zagrobelny M, Bak S (2013) Plant defense against insect herbivores. Int J Mol Sci 14:10242–10297. https://doi.org/10.3390/ijms140510242

Glazebrook J, Ausubel FM (1994) Isolation of phytoalexin-deficient mutants of Arabidopsis thaliana and characterization of their interactions with bacterial pathogens. Proc Natl Acad Sci USA 91:8955–8959. https://doi.org/10.1073/pnas.91.19.8955

Harborne JB, Simmonds NW (1964) Biochemistry of phenolic compounds. Academic Press, London

Haslam E (1966) Chemistry of vegetable tannins. Chem Veg Tann

Heil M (1999) Systemic acquired resistance: available information and open ecological questions. J Ecol 87:341–346. https://doi.org/10.1046/j.1365-2745.1999.00359.x

Heleno SA, Martins A, Queiroz MJRP, Ferreira ICFR (2015) Bioactivity of phenolic acids: metabolites versus parent compounds: a review. Food Chem 173:501–513. https://doi.org/10.1016/J.FOODCHEM.2014.10.057

Horvath DM, Chua N-H (1994) The role of salicylic acid in systemic acquired resistance. Curr Opin Biotechnol 5:131–136. https://doi.org/10.1016/S0958-1669(05)80025-4

Jilani G, Mahmood S, Chaudhry AN, Hassan I, Akram M (2008) Allelochemicals: sources, toxicity and microbial transformation in soil-a review. Ann Microbiol 58:351–357. https://doi.org/10.1007/BF03175528

Karageorgou P, Manetas Y (2006) The importance of being red when young: Anthocyanins and the protection of young leaves of Quercus coccifera from insect herbivory and excess light. Tree Physiol 26:613–621. https://doi.org/10.1093/treephys/26.5.613

Karou D, Dicko MH, Simpore J, Traore AS (2005) Antioxidant and antibacterial activities of polyphenols from ethnomedicinal plants of Burkina Faso. Afr J Biotechnol 4:823–828. https://doi.org/10.5897/AJB09.1302

Król A, Amarowicz R, Weidner S (2014) Changes in the composition of phenolic compounds and antioxidant properties of grapevine roots and leaves (Vitis vinifera L.) under continuous of long-term drought stress. Acta Physiol Plant 36:1491–1499. https://doi.org/10.1007/s11738-014-1526-8

Kulbat K (2016) The role of phenolic compounds in plant resistance. Biotechnol Food Sci 80:97–108

Lanoue A, Burlat V, Schurr U, Röse USR (2010) Induced root-secreted phenolic compounds as a belowground plant defense. Plant Signal Behav 5:1037–1038. https://doi.org/10.4161/psb.5.8.12337

Liang B, Huang X, Zhang G, Zhang F, Zhou Q (2006) Effect of lanthanum on plants under supplementary ultraviolet-B radiation: effect of lanthanum on flavonoid contents in soybean seedlings exposed to supplementary ultraviolet-B radiation. J Rare Earths 24:613–616. https://doi.org/10.1016/S1002-0721(06)60174-9

Luczynski A, Isman MB, Raworth DA (1990) Strawberry foliar Phenolics and their relationship to development of the two spotted spider mite. J Econ Entomol 83:557–563. https://doi.org/10.1093/jee/83.2.557

Mandal SM, Chakraborty D, Dey S (2010) Phenolic acids act as signaling molecules in plant-microbe symbioses. Plant Signal Behav 5:359–368

Martens S, Preuß A, Matern U (2010) Multifunctional flavonoid dioxygenases: Flavonol and anthocyanin biosynthesis in Arabidopsis thaliana L. Phytochemistry 71:1040–1049. https://doi.org/10.1016/j.phytochem.2010.04.016

Martinez V, Mestre TC, Rubio F, Girones-Vilaplana A, Moreno DA, Mittler R, Rivero RM (2016) Accumulation of Flavonols over Hydroxycinnamic acids favors oxidative damage protection under abiotic stress. Front Plant Sci 7:838. https://doi.org/10.3389/fpls.2016.00838

Maxwell FG, Lafever HN, Jenkins JN (1965) Blister beetles on glandless Cotton1. J Econ Entomol 58:792–793. https://doi.org/10.1093/jee/58.4.792

Michalak A (2006a) Phenolic compounds and their antioxidant activity in plants growing under heavy metal stress. Pol J Environ Stud 15:523–530

Michalak A (2006b) Phenolic compounds and their antioxidant activity in plants growing under heavy metal stress. Pol J Environ Stud 15:523–530. https://doi.org/10.1016/j.fitote.2011.01.018

Mierziak J, Kostyn K, Kulma A (2014) Flavonoids as important molecules of plant interactions with the environment. Molecules 19:16240–16265

Mol J, Grotewold E, Koes R (1998) How genes paint flowers and seeds. Trends Plant Sci 3:212–217. https://doi.org/10.1016/S1360-1385(98)01242-4

Moldenhauer J, Moerschbacher BM, van der Westhuizen AJ (2006) Histological investigation of stripe rust (Puccinia striiformis f.sp. tritici) development in resistant and susceptible wheat cultivars. Plant Pathol 55:469–474. https://doi.org/10.1111/j.1365-3059.2006.01385.x

Mueller-Harvey I (2001) Analysis of hydrolysable tannins. Anim Feed Sci Technol 91:3–20. https://doi.org/10.1016/S0377-8401(01)00227-9

Parr AJ, Bolwell GP (2000) Phenols in the plant and in man. The potential for possible nutritional enhancement of the diet by modifying the phenols content or profile. J Sci Food Agric 80:985–1012. https://doi.org/10.1002/(SICI)1097-0010(20000515)80:7<985::AID-JSFA572>3.0.CO;2-7

Pereira C, Barros L, Ferreira IC (2016) Extraction, identification, fractionation and isolation of phenolic compounds in plants with hepatoprotective effects. J Sci Food Agric 96:1068–1084. https://doi.org/10.1002/jsfa.7446

Quideau S, Deffieux D, Douat-Casassus C, Pouységu L (2011) Plant polyphenols: chemical properties, biological activities, and synthesis. Angew Chemie Int Ed 50:586–621. https://doi.org/10.1002/anie.201000044

Rehman F, Khan FA, Badruddin SMA (2012) Role of Phenolics in plant defense against insect herbivory. In: Chemistry of Phytopotentials: health, energy and environmental perspectives. Springer, Berlin/Heidelberg, pp 309–313

Rodríguez-Calzada T, Qian M, Strid Å, Neugart S, Schreiner M, Torres-Pacheco I, Guevara-González RG (2019) Effect of UV-B radiation on morphology, phenolic compound production, gene expression, and subsequent drought stress responses in chili pepper (Capsicum annuum L.). Plant Physiol Biochem 134:94–102. https://doi.org/10.1016/j.plaphy.2018.06.025

Schulz E, Tohge T, Zuther E, Fernie AR, Hincha DK (2016) Flavonoids are determinants of freezing tolerance and cold acclimation in Arabidopsis thaliana. Sci Rep 6:34027. https://doi.org/10.1038/srep34027

Southerton SG, Deverall BJ (1990) Changes in phenylalanine ammonia-lyase and peroxidase activities in wheat cultivars expressing resistance to the leaf-rust fungus. Plant Pathol 39:223–230. https://doi.org/10.1111/j.1365-3059.1990.tb02496.x

Timperio AM, D'Alessandro A, Fagioni M, Magro P, Zolla L (2012) Production of the phytoalexins trans-resveratrol and delta-viniferin in two economy-relevant grape cultivars

upon infection with Botrytis cinerea in field conditions. Plant Physiol Biochem 50:65–71. https://doi.org/10.1016/J.PLAPHY.2011.07.008
Vermerris W, Nicholson R (2006) Phenolic compound biochemistry. Springer, Dordrecht
Villegas RJA, Kojima M (1985) Sweet potato root enzyme which catalyzes the formation of Chlorogenic acid from1 - *O* -Caffeoyl- d –glucose and d -Quinic acid. Agric Biol Chem 49:263–265. https://doi.org/10.1080/00021369.1985.10866714
Weir TL, Park S-W, Vivanco JM (2004) Biochemical and physiological mechanisms mediated by allelochemicals. Curr Opin Plant Biol 7:472–479. https://doi.org/10.1016/J.PBI.2004.05.007
Winkel-Shirley B (2002) Biosynthesis of flavonoids and effects of stress. Curr Opin Plant Biol 5:218–223. https://doi.org/10.1016/S1369-5266(02)00256-X
Wurst S, Wagenaar R, Biere A, van der Putten WH (2010) Microorganisms and nematodes increase levels of secondary metabolites in roots and root exudates of Plantago lanceolata. Plant Soil 329:117–126. https://doi.org/10.1007/s11104-009-0139-2

Role of Salicylic Acid in Biotic and Abiotic Stress Tolerance in Plants

23

Heba I. Mohamed, Hanaa H. El-Shazly, and Abdelfattah Badr

Abstract

Salicylic acid (SA) is an endogenous growth regulator of phenolic nature and also a signaling molecule, which participates in the regulation of physiological processes in plants such as growth, photosynthesis, and other metabolic processes. Several studies support a major role of SA in modulating the plant response to various biotic and abiotic stresses. Its role in plant disease resistance is well documented for dicotyledonous plants, where it is required for basal resistance against pathogens as well as for the inducible defense mechanism and systemic acquired resistance (SAR); this confers resistance against a broad spectrum of pathogens. The activation of SAR is associated with the heightened level of expression of the pathogenesis-related proteins, some of which possess antimicrobial activity. Also, SA potentially generates a wide array of metabolic responses in plants and also affects plant-water relations. This molecule also found to be very active in mitigating oxidative stress under adverse environmental conditions. Hence, understanding the physiological role of SA would help in developing biotic and abiotic stress tolerance in plants.

Keywords

Salicylic acid · Cold · Heat · Salinity · Drought · Pathogen · Herbivores · Genes

H. I. Mohamed (✉) · H. H. El-Shazly
Biological and Geological sciences Department, Faculty of Education, Ain Shams University, Cairo, Egypt

A. Badr
Botany and Microbiology Department, Faculty of Science, Helwan University, Cairo, Egypt

R. Lone et al. (eds.), *Plant Phenolics in Sustainable Agriculture*,
https://doi.org/10.1007/978-981-15-4890-1_23

23.1 Introduction

Abiotic stresses determine plant productivity and plant distribution. In particular, drought and salinity affect more than 10% of arable land, and desertification and salinization are rapidly increasing on a global scale declining average yields for most major crop plants by more than 50% (Bartels and Sunkar 2005; Mantri et al. 2012; Mohamed et al. 2016; Akladious and Mohamed 2018; Mohamed et al. 2018; El-Beltagi et al. 2020). Traditional plant breeding approaches to improve abiotic stress tolerance of crops had limited success due to multigenic nature of stress tolerance (Mantri et al. 2012). In the last decade, molecular techniques have been used to understand the mechanisms by which plants perceive environmental signals and further their transmission to cellular machinery to activate adaptive responses (Mantri et al. 2012).

Plants have evolved complex defense system to overcome the biotic stresses as a natural system poses plenty of opposing forces on plants. Some secondary metabolites help plant to communicate with other organisms, and some protect plants from abiotic stress (Schafer and Wink 2009). These secondary metabolites are significantly important for growth and development. Three major types of secondary metabolites are produced in plant's body (phenolics, terpenes, and nitrogen/sulfur containing compounds). Thus, to improve plant performance and to reduce the loss of productivity, this can be implemented through various approaches, and one of those is the application of secondary metabolites (Ballhorn et al. 2009; Bastam et al. 2013; Liu et al. 2014; Mohamed and Akladious 2017; Mohamed et al. 2018).

Salicylic acid (SA), chemically known as 2-hydroxybenzoic acid, is one of a diverse groups of phenolic compounds, consisting of an aromatic ring bearing a hydroxyl group or its functional derivative, which is synthesized by plants. Phenolic compound has role in activating plant defenses especially systemic acquired resistance (SAR) (Shah 2003; Vlot et al. 2009). SA and its derivative (aspirin: acetyl SA) have been widely used for years as an anti-inflammatory drug.

23.2 Salicylic Acid Biosynthesis in Plants

Biosynthesis of SA is suggested to occur via two alternate pathways, viz., phenylalanine pathway (Ogawa et al. 2006; Sawada et al. 2006) and isochorismate pathway (Wildermuth et al. 2001; Garcion et al. 2008). In phenylalanine pathway, phenylalanine ammonia lyase (PAL) is a chief regulator of phenylpropanoid pathway and plays a key role in SA biosynthesis. Phenylalanine acts as a substrate of PAL and gets converted to *trans*-cinnamic acid. The decarboxylation of side chain of *trans*-cinnamic acid leads to the formation of benzoic acid through two probable routes (Fig. 23.1). In the first route of β-oxidation pathway, benzoic acid is formed from *trans*-cinnamic acid through cinnamoyl-CoA and benzoyl-CoA intermediates (Lee et al. 1995). The second route for the formation of benzoic acid involves non-β-oxidation pathway in which *para*-hydroxybenzaldehyde is the intermediate (Ribnicky et al. 1998; Chong et al. 2001). Benzoic acid formed from either of the

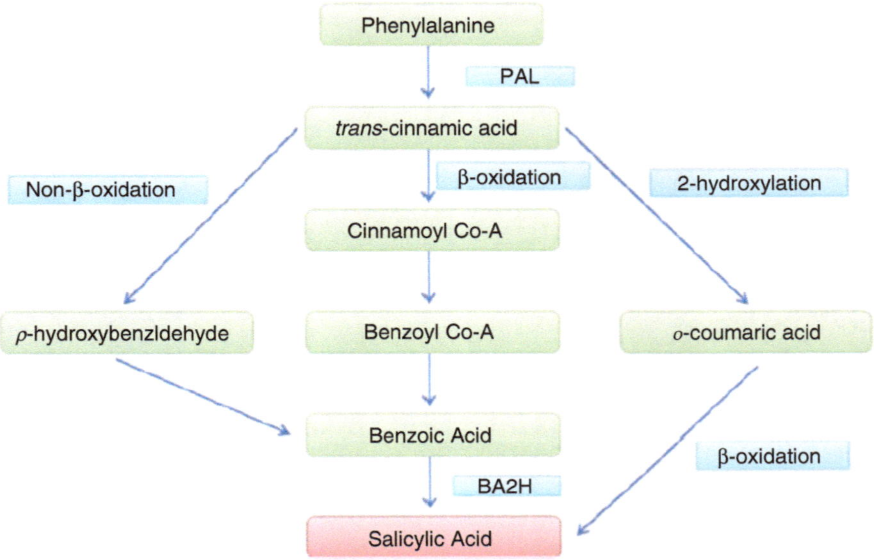

Fig. 23.1 Schematic outline of phenylalanine pathway in SA biosynthesis (*PAL* phenylalanine ammonia lyase, *BA2H* benzoic acid 2-hydroxylase) (Lee et al. 1995)

two mechanisms is converted to SA in the presence of the enzyme benzoic acid 2-hydroxylase (BA2H) (Leon et al. 1993, 1995). In the same phenylalanine pathway, the formation of SA can also take place from *trans*-cinnamic acid by its 2-hydroxylation to form *ortho*-coumaric acid, followed by decarboxylation mechanism by β-oxidation to produce SA (Lee et al. 1995). The second pathway of SA biosynthesis has been extensively studied in bacteria (Fig. 23.2) and starting with chorismate from shikimic acid pathway. It is converted to isochorismate, and the reaction is catalyzed by the enzyme isochorismate synthase (ICS) (Gaille et al. 2002, 2003). The formation of SA from isochorismate is thought to be catalyzed by isochorismate pyruvate lyase (IPL). This enzyme or similar proteins are bacterial IPL, however, has not been detected in plants (Strawn et al. 2007).

23.3 Physiological Role of Salicylic Acid

Salicylic acid has also been studied for its effects on various physiological processes related to growth and development of plants under normal conditions. Among these effects are the induction of flowering in herbaceous species (Hegazi and El-Shrayi 2007); stimulation of root development, stomatal closure, and reduced transpiration (Singh and Usha 2003); reversal of the effects of abscisic acid (Davies 2004); regulation of gravitropism (Hussein et al. 2007); and fruit ripening inhibition (Srivastava and Dwivedi 2000). Also, Khan et al. (2003) reported increases in leaf

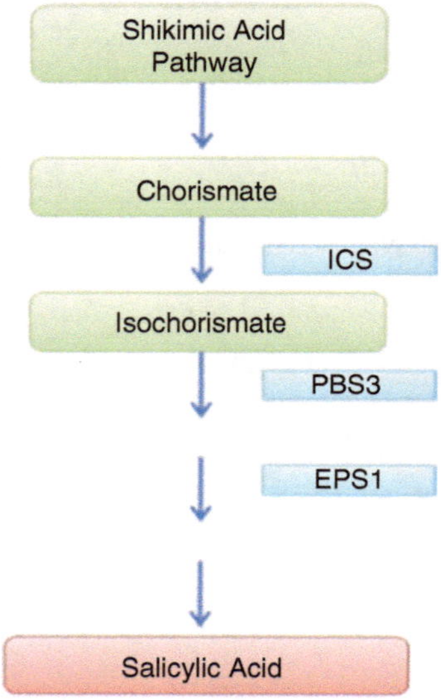

Fig. 23.2 Schematic outline of isochorismate pathway in SA biosynthesis (Wildermuth et al. 2001; Gaille et al. 2003; Chen et al. 2009)

area and shoot dry weight of soybean and corn plants treated with 10-5 M of SA. The increase in bioproductivity of plants is mainly due to the positive effect of SA on root length and its density (Larqué-Saavedra and Martin-Méx 2007). There is evidence of a cross talk between the SA and auxin signaling pathways during plant vegetative growth (Rivas-San and Plasencia 2011). The promotion of wheat growth in response to the application of SA was attributed to the increase of photosynthetic tissues, since there was a positive correlation between photosynthesis and leaf area on wheat plants treated with SA, both under stress and normal growth conditions (Arfan et al. 2007). The observed increase in photosynthesis rate in plants sprayed with SA can be assigned to metabolic changes at the chloroplasts level (efficiency of photosystem II, Rubisco enzyme activity, and supply of ATP and NADPH for the carbon reduction cycle) (Rivas-San and Plasencia 2011). However, the stimulatory effect of SA on gas exchange parameters and plant development is dependent on several factors such as application mode, exposure time, and ontogenetic stage of the plant (Vanacker et al. 2001; Horváth et al. 2007), and the effective concentrations of SA differ among species and their domestication stage (Arfan et al. 2007).

The flowering-promoting effect after SA application can also be indirect as SA alters the synthesis and/or signaling pathways of other plant hormones including jasmonic acid, ethylene, and auxin (Vlot et al. 2009). In addition, the exogenous application of SA can rise the content of endogenous bioactive GA in response, changing the hormonal status of the plant (Mukherjee and Kumar 2007; Kim et al. 2009). Increased content of endogenous SA as a result of its exogenous application was correlated to the positive influence on the plant growth and flowering (Kim et al. 2009). In addition, flavonoid levels in plants are significantly affected by plant growth regulators (PGRs) (Klessig and Malamy 1994). Increased endogenous levels of SA can trigger cell signaling pathways which regulate the expression of genes encoding enzymes related to the phenylpropanoid pathway production (among them the flavonoids), increasing the amount or the activity of these enzymes. For example, the chalcone synthase activity (the first enzyme to branch off from phenylpropanoid metabolism to flavonoid metabolism) was increased in plants treated with SA (Ghasemzadeh et al. 2012).

23.4 Role of SA in Biotic Stress

23.4.1 Role of SA Against Pathogens

Early last century, several studies showed that when a plant was infected by a pathogen, some systemic defense mechanisms were activated involving an increased resistance against subsequent pathogen attacks (Gozzo and Faoro 2013). Hypersensitive response (HR) is one early response associated with necrotic lesions at the site of pathogen entry, ROS accumulation, and activation of defense-related genes that (among others) encode several families of PR proteins (Fu and Dong 2013). Increased levels of pathogenesis-related protein (PR) gene expression are observed in non-inoculated tissues and the development of SAR, as a broad resistance to different pathogens (Conrath et al. 1995; Vlot et al. 2009). So far, 17 families of PR proteins have been identified (Hoffmann-Sommergruber 2000). Among PR genes, PR1, PR2, and PR5 are strongly induced upon infection by biotrophic and semi-biotrophic pathogens. The expression of PR1, PR2, and PR5 is dependent on SA (Selitrennikoff 2001; Zhang et al. 2010), and these genes are often used as markers of the SA pathway. PR1 R1 has sterol-binding activity, which inhibits pathogen growth by sequestrating sterol from pathogens (Gamir et al. 2017).

SA was described as endogenous signal in the resistance response at first time in 1979 in tobacco when White (1979) observed that acetylsalicylic acid (aspirin) induced resistance to *tobacco mosaic virus* (TMV), increasing PR protein accumulation and reducing lesion numbers. Subsequently, Malamy et al. (1990) observed that the endogenous salicylic acid levels in resistant but not susceptible cultivars increased in infected and uninfected leaves after TMV inoculation. Moreover, SA levels increase in both inoculated and non-inoculated systemic tissues (Kessmann et al. 1994; Sticher et al. 1997). SA at low concentrations also promotes the faster and stronger activation of callose deposition and gene expression in response to pathogen or microbial elicitors, a process called "priming," which contributes to induced defense mechanisms (Kohler et al. 2002).

Salicylic acid affects the activity of mitochondrial alternative oxidase. Alternative oxidase is an enzyme that directly links the oxidation of the ubiquinol/ubiquinone pool in mitochondria to the reduction of oxygen to water, without the synthesis of ATP (Millar et al. 2011). This ability of SA to affect alternative oxidase has been associated with SA-mediated resistance to some viruses (Singh et al. 2004). It has been suggested that SA action on the alternative oxidase capacity could affect ROS levels in the mitochondria, which in turn could activate antiviral defenses (Singh et al. 2004). SA also binds enzymes like catalase, ascorbate peroxidase, aconitase, and a carbonic anhydrase from tobacco (Durner and Klessig 1995; Slaymaker et al. 2002). Some of these enzymes are involved in the metabolism of reactive oxygen species (ROS) and in redox homeostasis. Redox changes are associated with plant defense responses (Torres et al. 2002; Durrant and Dong 2004). SA binding affects the biochemical activity of catalase and ascorbate peroxidase (Durner and Klessig 1995) and may have a role in the vicinity of the HR, where SA concentrations are high. In addition, SA also affects lipid peroxidation, which may have a role in plant defense (Anderson et al. 1998). Indeed, genetic studies in *Arabidopsis* have demonstrated that genes associated with lipid metabolism and putative lipid transfer proteins are required for SAR (Maldonado et al. 2002; Shah 2005). Salicylic acid activates a group of events resulting in the inhibition of viral replication and their cell-to-cell and long-distance transmission in plants (Singh et al. 2011).

Salicylic acid is an endogenous compound that operates in the signaling pathway for plant defense. After TMV infection, salicylic acid accumulates to high levels at the site of infection, with a subsequent, but much smaller rise, in the uninfected systemic tissues (Malamy et al. 1990). In tobacco, this increase paralleled the transcriptional activation *of PR* genes in both the inoculated and uninoculated leaves. Strikingly, exogenously supplied SA induced the same set of nine genes that are activated systemically upon TMV infection. An increase in salicylic acid levels in the phloem of cucumber plants infected with either tobacco necrosis virus or *Colletotrichum lagenarium* was also shown to precede the development of systemic acquired resistance (Ryals et al. 1996; Wobbe and Klessig 1996). The infection of resistant plants by pathogens generally results in hypersensitive response – the formation of necrotic lesions and restricted pathogen growth and spread. A variety of defense responses is induced locally around the sites of infection. An oxidative burst precedes the formation of necrotic lesions (Chen et al. 1993). Salicylic acid binds and inhibits tobacco catalase activity both in vitro and in vivo. Thus, one possible function of salicylic acid is to inhibit the hydrogen peroxide (H_2O_2) degrading the activity of catalase, thereby leading to an increase in the endogenous level of H_2O_2, which is generated by photorespiration, photosynthesis, oxidative phosphorylation, and the hypersensitive response-associated oxidative burst. The H_2O_2 and other ROS derived from it could then serve as second messengers to activate the expression of plant defense-related genes, such as *PR-1* (Chen et al. 1993; Conrath et al. 1995).

Salicylic acid has been shown to induce the expression of many defense-related genes and to be significant in the production of H_2O_2, the induction of cell death, and the activation of several genes induced by fungal elicitors and wounding (Mur et al. 1996). It was also demonstrated that salicylic acid treatment of tobacco leaves increases an *as-1*-binding activity and that phosphatase treatment of nuclear extracts decreases it (Jupin and Chua 1996). From these results, it was proposed that the *as-1*-binding activity is sequestered by an inhibitory protein that is released after salicylic acid treatment, probably via a phosphorylation event(s). This in turn leads to activation of promoters containing the *as-l-like* element. A MAP kinase that can be activated by salicylic acid and TMV has been identified and purified from tobacco extracts (Zhang and Klessig 1997).

Additional defense responses in surrounding cells include the induction of genes for pathogenesis-related (PR) proteins, peroxidases, and enzymes involved in cell wall strengthening and the biosynthesis of phytoalexins. Some of these genes are also activated systemically and are believed to play a role in the development of systemic acquired resistance (Bowles 1990). The synthesis and accumulation of salicylic acid appear to be necessary for the activation of several of these defense responses, both locally and systemically. A substantial amount of salicylic acid is converted to salicylic acid β-glucoside, a probable storage form (Ryals et al. 1996; Wobbe and Klessig 1996).

After a pathogen attack, changes in the concentration of SA have an effect on the maintenance of the redox state of the cell, probably by regulating the expression of genes encoding antioxidants (Rao and Davis 1999; Vanacker et al. 2000). It has been described that biotic stress situations increase the ROS production which could act as second messengers mediating SA pathways for expression of defense genes (Yoshioka et al. 2008; Torres 2010). In this sense, it has been also reported that high concentrations of SA can act uncoupling oxidative phosphorylation and hence the respiration chain, stimulating ROS generation in mitochondria and also inducing the alternative respiratory pathway (Moore et al. 2002). However, GSH levels increased significantly in soy cells after incubation for 2 days with SA or its analogues (Knorzer et al. 1999). In addition, it has been described that GSH could regulate the expression of SA-dependent genes via NPR1, after exposure to the pathogen (Urbanek Krajnc and Müller 2006). NPR1 protein is a transcriptional factor whose location or activity was influenced by the redox state of the cell (Mou et al. 2003). However, overexpression of the NPR1 protein leads to constitutive expression of PR genes in the absence of inducers, suggesting that NPR1 is a positive regulator of SAR required for the translation of the signal accumulation of SA and expression of resistance genes (Cao et al. 1994; Mou et al. 2003). SA accumulation induces redox changes leading to the monomerization of NPR1, probably by intermolecular disulfide bond reduction, allowing it to be transported to the nucleus. Then NPR1 in the nucleus promotes the binding of transcription factors to SA-responsive promoters, regulating the expression of *PR* genes (Mou et al. 2003; Després et al. 2003). The inhibition of the reduction of NPR1 and therefore its monomerization lead to a decrease in the expression of PR genes.

23.4.2 Role of SA Against Herbivores

Salicylates are also responsible for protecting plants from herbivores by stimulating affected plants (plants experiencing insect damage) to increase the production of their volatile secretions which are responsible for attracting anti-herbivores (Thaler 1999). Responses to SA depend on a regulatory protein called Nonexpressor of Pathogenesis-Related Genes1 (*NPR1*) (Pieterse and Van Loon 2004). The *NPR1* gene is activated through redox pathways by SA accumulation and is translocated to the nucleus; however, it does not bind to DNA directly, but acts through transcription factors (Pieterse and Van Loon 2004). SA induces greater defense against piercing and sucking type of insect pests than the chewing ones (Zhao et al. 2009). SA signaling molecule is involved in local defense as well as induction of systemic resistance. Production of ROS by SA pathway has been proposed to induce resistance in plants against insect pests, e.g., in tomato plants against *Helicoverpa armigera* (Peng et al. 2004). H_2O_2 induced by SA in plants defends them against various insect pests since H_2O_2 actively damages the digestive system of insects leading to reduced growth and development (Peng et al. 2004). Furthermore, SA signals the release of plant volatiles that attract the natural enemies of insect pests, e.g., lima bean and tomato plants infested by spider mite attract the natural enemies of spider mite (de Boer et al. 2004). However SA and jasmonic acid (JA) act antagonistically, where SA inhibits the activity of JA and vice versa (Maffei et al. 2007). SA serves as a volatile signal to trigger induced defenses in plants, including HIPV emission, and a number of predaceous arthropods are attracted to SA under field conditions (de Boer et al. 2004; Maffei et al. 2007).

23.5 Role of SA in Abiotic Stress

The role of SA in mitigating abiotic stress has widely been studied in the last few decades. A large volume of research reports indicate that both endogenous SA synthesis and exogenous application enhance plant tolerance to salinity (Hasanuzzaman et al. 2014), drought (Kang et al. 2012; Alam et al. 2013), extreme temperature (Kumar et al. 2015; Khanna et al. 2016), and toxic metal and metalloids (Singh et al. 2015; Islam et al. 2016). Exogenous SA showed enhanced plant growth, photosynthesis, and decreased ROS production under various abiotic stresses (Fig. 23.3).

23.5.1 Salt Stress

The effect of exogenous SA treatments, in the response to NaCl stress, seems to be dependent of the SA concentrations used the plant species, the application mode of the treatment, and the physiological state of the plant during the application as well as the level of salinity and the exposure time to NaCl. Exogenous SA treatments improved plant growth under saline stress (Szepesi 2006; Bastam et al. 2013; Liu et al. 2014) as well as the seed germination process in the presence of NaCl (Rajjou et al. 2006; Lee et al. 2010). The pre-treatment of tomato plants with low SA concentrations (10^{-4} M) improved the acclimation of tomato plants to 100 mM NaCl. Also, SA pre-treatment improved the photosynthetic efficiency, enhanced ascorbate peroxidase and guaiacol peroxidase activity in roots, and induced an accumulation of polyamines (Szepesi 2006). SA pre-treatment also provided protection against salinity in tomato plants, probably due to the increased activation of aldose reductase and APX enzymes and to the accumulation of osmolytes, such as sugars, sugar alcohol, or proline (Tari et al. 2002, 2004; Szepesi et al. 2005).

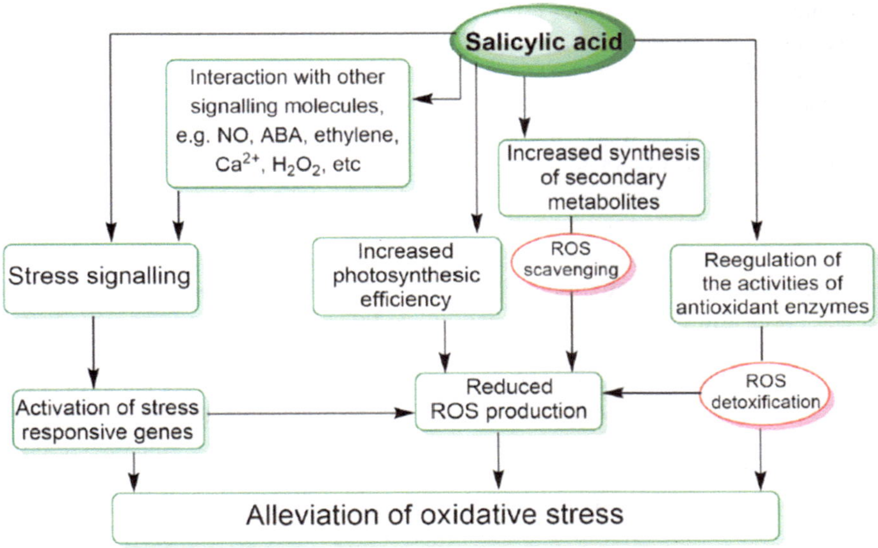

Fig. 23.3 Some possible ways of SA-induced oxidative stress tolerance to plants (Hasanuzzaman et al. 2017)

The soaking of wheat seeds in 0.05 mM SA also reduced the salinity damage on seedling growth and accelerated the growth processes (Shakirova et al. 2003). The treatment of wheat plants with SA increased the level of cell division within the apical meristem of seedling roots, causing an increase in plant growth and elevated wheat productivity (Shakirova et al. 2003). A high ABA level was also maintained in wheat seedlings, treated with SA. The SA-induced increase in ABA might contribute to the preadaptation of plants to stress, since ABA is known to have a key role in inducting the synthesis of a range of stress proteins ensuring the development of anti-stress reactions, for example, the maintenance of proline accumulation (Bandurska and Stroinski 2005). The stress-induced accumulation of active oxygen species and, therefore, the level of SOD and peroxidase activity in the roots of young wheat seedlings, pre-treated with SA, were significantly lower than in untreated plants, indicating that these enzymes contribute to the protective effect of SA on plants under conditions of salination (Sakhabutdinova et al. 2004). It was also shown that a salt-induced protein (SALT) was present in the xylem parenchyma cells of vascular bundles in the major and minor leaf veins. The expression of the gene encoding SALT was upregulated following the treatment with a fungal elicitor, JA, ABA, or NaCl. However, SA alone or in combination with one of the other elicitors not only strongly inhibited SALT gene expression but also exhibited an antagonistic effect in suspension cells and leaves (Kim et al. 2004).

The foliar application 1 mM SA alleviated the NaCl (100 mM) damage in tomato plants and plant growth was strongly reduced in salt-stressed plants, but this reduction was less pronounced in SA-treated plants (He and Zhu 2008). SA alleviates NaCl stress by lower levels of lipid peroxidation and H_2O_2 accumulation as well as the enhancement of the antioxidant capacity of plants by increases in CAT, APX, DHAR, and ASC and GSH contents (Lee et al. 2010). SA promotes germination under saline conditions by reducing the NaCl-induced oxidative damage (Rajjou et al. 2006; Lee et al. 2010). In addition, high SA concentrations can have a toxic effect on plant growth and development possibly due to its reported effect inducing ROS accumulation (Rao et al. 1997). Borsani et al. (2001) found that SA enhanced the deleterious effect of NaCl or drought stress in *Arabidopsis* seedlings by increasing the rate of ROS generation in photosynthetic tissues. NaCl-adapted tomato cells contained a lower concentration of SA than an adapted cells. The adaptation process to NaCl was also related with a higher antioxidative capacity because salt-adapted cells also contained higher basal levels of APX and GR activities (Molina et al. 2002). In a more recent work, Bastam et al. (2013) reported that the exogenous application of SA improved the tolerance of pistachio seedlings to NaCl stress (up to 90 mM NaCl). The SA-treated plants showed lower NaCl-induced injury symptoms, a better growth rate, and higher chlorophyll contents and photosynthetic capacity than the non-treated plants. In addition, the foliar application of 0.1 mM SA improved the growth of cotton seedlings in the presence of 100 mM NaCl. The SA-treated plants displayed better growth and photosynthetic rates and showed low ROS accumulation ($O_2^{\cdot-}$ and H_2O_2) and lipid peroxidation that correlated with a significant enhancement of CAT activity (Liu et al. 2014).

Salicylic acid mitigated salinity stress-injury in *Solanum lycopersicum* by causing characteristic changes in the expression pattern of GST gene family members such as *SlGSTT2*, *SlGSTT3*, and *SlGSTF4* (Csiszár et al. 2014). Exogenously sourced SA (0.5 mM) was reported to improve salt tolerance in *Triticum aestivum* due to an enhanced transcript level of antioxidant genes; *GPX1*, *GPX2*, *DHAR*, *GR*, *GST1*, *GST2*, *MDHAR*, and *GS*, and an increased activity of ascorbate (AsA)-GSH pathway enzymes (Li et al. 2013). In another instance, SA-mediated restoration of membrane potential and prevention of salt-induced K^{+-}loss via GORK channel and eventually improved salinity-tolerance were evinced in *A. thaliana* (Jayakannan et al. 2013).

23.5.2 Drought Stress

The effect of SA on water stress is more homogeneous than its effect on salt stress, and some early reports showed that the SA treatment could improve the response to drought stress (Munne-Bosch and Penuelas 2003; Bechtold et al. 2010; Khokon et al. 2011; Ying et al. 2013). In general, the improved drought response induced by SA is associated with an increase or maintenance of plant growth, Rubisco activity, and the antioxidative capacity. SA-treated bayberry plants displayed better RWC (relative water content), photosynthetic rates, as well as higher CAT and SOD activity and proline contents than non-treated plants (Ying et al. 2013). In addition, SA attenuated the drought-induced oxidative stress as recorded by a decrease in some oxidative stress parameters such as lipid peroxidation and electrolyte leakage, suggesting that SA can partially protect the membrane integrity. SA increased

Rubisco and SOD activities as well as chlorophyll contents in drought-stressed wheat seedlings (Singh and Usha 2003). The improvement of drought tolerance of barley plants by SA application was associated with an increase in the antioxidative defenses and the maintenance of photosynthesis under water stress conditions (Habibi 2012). Miura et al. (2013) also observed that drought stress induced the expression of *PR-1* and *PR-2*, two typical SA-inducible genes, suggesting that SA may be required for drought tolerance. In addition, the treatment of wheat seedlings with 0.5 mM SA alleviated the growth inhibition induced by drought. This response was linked to an increase in ASC and GSH as well as an increase in the transcription of *GST1*, *GST2*, *GR*, and *MDHAR* genes (Kang et al. 2013).

It has been suggested that SA may act as an ROS scavenger (Kang et al. 2013). SA treatment increased the ASC-GSH cycle enzymes along with SOD and CAT in two maize cultivars, showing different sensibility to water stress, after 10 days of withholding water, suggesting that ASC-GSH cycle can act to remove the H_2O_2 generated during the early phase of water stress (Saruhan et al. 2012). In fact, SA reduced the stomatal conductance in a dose-dependent manner in different plant species, including *Vicia faba* (Mori et al. 2001), *Commelina communis* (Lee 1998), and *Arabidopsis* (Khokon et al. 2011). The SA-induced stomatal closure is dependent on ROS generation, because the application of antioxidant enzymes such as catalase and SOD suppressed the stomatal closure. In addition, the stomatal closure induced by SA was completely suppressed by the action of salicylhydroxamic acid (SHAM), a cell wall peroxidase inhibitor. These results suggested that SA induced stomatal closure by means of the ROS generated by cell wall peroxidases (Khokon et al. 2011; Miura et al. 2013). However, the treatment of two maize cultivars with 1 μM SA by foliar spraying reversed the drought-induced stomatal closure (Saruhan et al. 2012). Therefore, it can be suggested that the induction of drought tolerance by exogenous SA application may have a significant practical application in agriculture, horticulture, and forestry.

23.5.3 Heat Stress

Phytohormones help plant in adjusting to heat stress by modifying the plants' ability to adapt under changing environments either through induction of osmolyte synthesis, increase in antioxidants, or utilization of nutrients (Ahammed et al. 2015). Endogenous levels of SA and its glucoside have been shown to increase in plants subjected to heat stress. Khan et al. (2013) reported that SA treatment under heat stress increased proline production which increased osmotic potential enabling the plants for higher water intake resulting in positive influence on stomatal aperture and photosynthetic machinery leading to higher efficiency of PS II and increased Rubisco activity that cumulatively resulted in increased photosynthesis. Foliar spraying of cucumber plants with 1 mM SA induced heat tolerance, as shown by the lower electrolyte leakage parameter, lower H_2O_2, higher catalase activity, and lipid peroxide levels (Shi et al. 2006). SA increases the photosynthetic rate in grape leaves under heat stress (Wang et al. 2010) and can alleviate the heat-induced damage in plants by upregulating the antioxidant system (Wang and Li 2006). SA is also known to stabilize the trimers of heat shock transcription factors and to aid them in binding to the heat shock element in the promoter of HSP genes (Jurivich et al. 1992). Wang et al. (2010) found that the HSP21 immune signal increased in both SA-treated and control leaves during heat stress. Exogenous application of SA has been shown to enhance thermotolerance in tobacco and *Arabidopsis* (Lopez-Delgado et al. 1998; Clarke et al. 2004). Wang and Li (2006) reported that spraying grape plant with 0.1 mM SA causes reduction in thiobarbituric acid reactive substances and relative electrolyte leakage in young grape leaves and increased antioxidative enzymes.

Liu et al. (2008) proposed another strategy by which SA induces heat tolerance through the changes in the activities of plasma membrane H^+ and Ca^{2+} ATPase. Liu and Huang (2005) reported that the thermotolerance induced by exogenous SA was related to the Ca^{2+}-CaM system in grape plants. SA increased the activity of H^+-ATPase in grape plant leaves and kept the stability of the plasma membrane H^+-ATPase following heat stress. SA induced an increase of the Ca^{2+} level in cells and activated the Ca^{2+}-CaM system, which further stimulated the activity of Ca^{2+}-ATPase. Finally, the Ca^{2+} level in the cytoplasm was regulated and maintained at a normal level by Ca^{2+}-ATPase to protect cells and to avoid injury resulting from heat shock. Khan et al. (2015) reported that SA proves beneficial for plants both under optimal and stress environments. It modulates the production of osmolytes and other metabolites and plant nutrient status to protect plants under abiotic stress conditions. Galani et al. (2016) suggested that the exogenous application of SA mitigated the deleterious effects of heat stress on *Gossypium hirsutum* seedlings by reducing membrane damage through reduced electrolyte leakage, MDA content and H_2O_2 content. It reduced dehydration losses through triggering antioxidative defense system and increase proline accumulation. Thus SA enhances heat tolerance in cotton in efficient and simple means.

23.5.4 Cold Stress

Plants are said to be cold-sensitive if they die or suffer severe damage at temperatures between 0 and 15 °C. Cold-tolerant plants, on the other hand, are still able to grow near freezing point and are capable of surviving temperatures as low as 10–15 °C below zero (Sanghera et al. 2011). Apart from genetic factors, cold sensitivity also depends on the stage of development and the level of metabolic activity. Results achieved so far suggest that differences in membrane composition make a substantial contribution to sensitivity to low-temperature stress. The quantity of unsaturated fatty acids has been shown to be much lower in cold-sensitive plants than in those tolerant of low temperatures (Steponkus et al. 1993). The most important way of avoiding freezing is the accumulation of osmolytes induced by low temperature (sugars, polyalcohols, amino acids, polyamines, quaternary ammonium compounds, etc.), which cause the freezing point within the cell to drop. This process prevents the dehydration of the cytoplasm as the result of frost. SA and other phenol derivatives are known to improve the cold tolerance of plants (Luo et al. 2012).

It was shown that the addition of 0.5 mM SA to the hydroponic growth solution of young maize plants under normal growth conditions provided protection against subsequent low-temperature stress. Besides the obvious visual symptoms, this observation was confirmed by chlorophyll fluorescence parameters and electrolyte leakage measurements from the leaves (Janda et al. 1997, 1999). Pre-treatment of the leaves of chilling-sensitive banana seedlings with 0.5 mM SA solution by spraying the foliage or irrigating the roots for 1 day induced an increase in chilling tolerance during subsequent 5 °C chilling stress (Kang et al. 2003). However, SA pre-treatment caused an activation of SOD, catalase, and APX activities during a period of 5 °C chilling stress, while it did not change the peroxidase activity. These effects of SA on activities of various protective enzymes could be associated with H_2O_2 metabolism. Presoaking seeds before sowing could also be an effective way of improving cold tolerance. In tomato and bean plants, 0.1 mM and 0.5 mM concentrations of both SA proved effective not only against heat and drought stress but also against low-temperature stress (Senaratna et al. 2000).

Priming pepper (*Capsicum annuum* L.) seeds imbibed in KNO_3 solution containing aspirin at a concentration range of 0.05–0.5 mM resulted in an increase in the germination percentage, and the germination at low temperature became faster and better (Korkmaz 2005). The positive effect of SA on cold tolerance was shown not only under chilling but also under freezing conditions in winter wheat (Tasgin et al. 2003). Exogenous SA used at 0.01, 0.1, and 1 mM concentrations decreased freezing injury in the winter wheat leaves of plants grown under cold and control conditions. Cold conditions caused an increase in ice nucleation activity by apoplastic proteins, which were isolated from the leaves. Exogenous SA caused an increase in ice nucleation activity under cold and control conditions. These results show that SA can increase freezing tolerance in winter wheat by affecting apoplastic proteins. However, this effect cannot be generalized as experiments with winter rye showed that the apoplastic proteins accumulated after SA treatment had no antifreeze activity (Yu et al. 2001).

23.5.5 Heavy Metal

The prime response of plants to heavy metals exposure is overproduction of ROS. Some metals generate ROS directly, by participating in Haber-Weiss reactions or indirectly by disturbing antioxidative defense system, electron transport chain, or metabolism of essential elements (Yadav 2010). Overproduction of ROS, such as superoxide radical, hydrogen peroxide, hydroxyl radical, and singlet oxygen, is highly toxic and causes oxidative damage to nucleic acids, proteins, carbohydrates, and lipid peroxidation (Gill and Tuteja 2010). To combat uncontrolled oxidation, plants possess antioxidative defense system that includes enzymatic (superoxide dismutase, SOD; guaiacol peroxidase, POD; catalase, CAT; dehydroascorbate reductase, DHAR; monodehydroascorbate reductase, MDHAR; glutathione reductase, GR; ascorbate peroxidase, APEX; glutathione peroxidase, GPX; and glutathione S-transferase, GST) and nonenzymatic antioxidants (glutathione, GSH; ascorbic acid, ASA; and phenolic compounds, α-tocopherols) for scavenging and detoxification of ROS (Gill and Tuteja 2010; Sharma et al. 2012). SA application at a concentration of 0.1 or 0.2 mM reduced the inhibitory effect of Pb^{2+} and Hg^{2+} at seed germination and early seedling growth of two rice cultivars (Mishra and Choudhuri 1997).

Pre-treatment of SA at 0.5 mM concentration protected the barely seedlings from cadmium (Cd) toxicity during the following growth period (Metwally et al. 2003). However, SA treatments strongly or completely suppressed the Cd-induced upregulation of the antioxidant enzyme activities. It can be concluded that SA alleviates Cd toxicity not at the level of antioxidant defense but by affecting other mechanisms of Cd detoxification. One mechanism may include binding of Cd resulting in a lowered level of plasmatic free Cd. Alternatively, SA stimulates the expression of certain ABC transporters (Eichhorn et al. 2006). Such transporters have been implicated in the vacuolar sequestering of the products of Cd action therefore releasing the Cd stress (Rea et al. 1998).

However, SA induces activities of antioxidative enzymes under metal stress (Chen et al. 2007; Guo et al. 2007; Wang et al. 2009; Parashar et al. 2014; Zhang et al. 2015), some have reported decline in CAT activity under metal stress (Pandey et al. 2013). Exogenous application of SA slightly induced NADH oxidase activity (which stimulates H_2O_2 generation) in mercury-exposed roots of alfalfa. H_2O_2 acts as secondary messenger and triggers signaling cascades for activating defense mechanisms under stress when produced at moderate levels (Zhou et al. 2009). Few studies indicate that SA being an iron-chelating molecule acts as an antioxidant as it can directly scavenge hydroxyl radicals which may result in lowering of metal-induced increase in antioxidative enzyme activity (Ahmad et al. 2011).

The enhanced growth characteristics of Cd-fed plants sprayed with 10^{-5} M of SA might also be due to the fact that SA acts at the level of transcription and/or translation, thereby increasing the activity of various other enzymes necessary for growth of plants (Hayat et al. 2010). Treatment with SA has been reported to increase crop yield. The plausible reason for increasing the crop yield might be due to delayed senescence of leaves and flowers in response to exogenous SA (Imran et al. 2007) that will automatically help in extending the duration of photosynthetically active sites and also prevent the premature loss of flowers and fruits.

23.6 Conclusion

Salicylic acid induces resistance against a particular pathogenic attack by increasing defense enzymes/PR proteins and increases the growth and development by increasing proteins, carbohydrates, chlorophyll contents, etc. Exogenous salicylic acid augments the internal glutathione cycle and thus boosts the antioxidants and metal detoxification systems. Moreover, exogenous salicylic acid alleviates the stresses in a dose-dependent manner, depending on the type of stress as well as the plant species. Salicylic acid being a scavenger of hydroxyl radical and an iron-chelating compound inhibits the direct impact of hydroxyl radicals as well as their generation on plant growth.

References

Ahammed GJ, Li X, Yu J, Shi K (2015) NPR1-dependent salicylic acid signaling is not involved in elevated CO_2 induced heat stress tolerance in Arabidopsis thaliana. Plant Signal Behav 10(6): e101194

Ahmad P, Nabi G, Ashraf M (2011) Cadmium-induced oxidative damage in mustard [*Brassica juncea* (L.) Czern. & Coss.] plants can be alleviated by salicylic acid. S Afr J Bot 77:36–44

Akladious SA, Mohamed HI (2018) Ameliorative effects of calcium nitrate and humic acid on the growth, yield component and biochemical attribute of pepper (*Capsicum annuum*) plants grown under salt stress. Sci Hortic 236:244–250

Alam MM, Hasanuzzaman M, Nahar K, Fujita M (2013) Exogenous salicylic acid ameliorates short-term drought stress in mustard (*Brassica juncea* L.) seedlings by upregulating the antioxidant defense and glyoxalase system. Aust J Crop Sci 7:1053–1063

Anderson MD, Chen Z, Klessig DF (1998) Possible involvement of lipid peroxidation in salicylic acid-mediated induction of PR1 gene expression. Phytochemistry 47:555–566

Arfan M, Athar H, Ashraf M (2007) Does exogenous application of salicylic acid through the rooting medium modulate growth and photosynthetic capacity in two differently adapted spring wheat cultivars under salt stress? J Plant Physiol 164:685–694

Ballhorn DJ, Kautz S, Heil M, Hegeman AD (2009) Cyanogenesis of wild lima bean (*Phaseolus lunatus* L.) is an efficient direct defence in nature. PLoS One 4(5):e5450

Bandurska H, Stroinski A (2005) The effect of salicylic acid on barley response to water deficit. Acta Physiol Plant 27:379–386

Bartels D, Sunkar R (2005) Drought and salt tolerance in plants. Crit Rev Plant Sci 24:23–58

Bastam N, Baninasab B, Ghobadi C (2013) Improving salt tolerance by exogenous application of salicylic acid in seedlings of pistachio. Plant Growth Regul 69:275–284

Bechtold U, Lawson T, Mejia-Carranza J, Meyer RC, Brown IR, Altmann T, Ton J, Mullineaux PM (2010) Constitutive salicylic acid defences do not compromise seed yield, drought tolerance and water productivity in the Arabidopsis accession C24. Plant Cell Environ 33:1959–1973

Borsani O, Valpuesta V, Botella MA (2001) Evidence for a role of salicylic acid in the oxidative damage generated by NaCl and osmotic stress in Arabidopsis seedlings. Plant Physiol 126:1024–1030

Bowles DJ (1990) Defense-related proteins in higher plants. Annu Rev Biochem 59:873–907

Cao H, Bowling SA, Gordon AS, Dong X (1994) Characterization of an Arabidopsis mutant that is nonresponsive to inducers of systemic acquired resistance. Plant Cell 8:1583–1592

Chen Z, Silva H, Klessig DF (1993) Active oxygen species in the induction of plant systemic acquired resistance by salicylic acid. Science 262:1883–1886

Chen J, Zhu C, Li L-P, Sun Z-Y, Pan X-B (2007) Effects of exogenous salicylic acid on growth and H_2O_2 metabolizing enzymes in rice seedlings under lead stress. J Environ Sci 19(1):44–49

Chen Z, Zheng Z, Huang J, Lai Z, Fan B (2009) Biosynthesis of salicylic acid in plants. Plant Signal Behav 4:493–496

Chong J, Pierrel M-A, Atanassova R, Werck-Reichhart D, Fritig B, Saindrenan PS (2001) Free and conjugated benzoic acid in tobacco plants and cell cultures. Induced accumulation upon elicitation of defense responses and role as salicylic acid precursors. Plant Physiol 125:318–328

Clarke SM, Mur LA, Wood JE, Scott IM (2004) Salicylic acid dependent signaling promotes basal thermotolerance but is not essential for acquired thermotolerance in *Arabidopsis thaliana*. Plant J 38:432–437

Conrath U, Chen Z, Ricigliano JR, Klessig DF (1995) Two inducers of plant defense responses, 2,6-dichloroisonicotinic acid and salicylic acid, inhibit catalase activity in tobacco. Proc Natl Acad Sci U S A 92:7143–7147

Csiszár J, Horváth E, Váry Z, Gallé Á, Bela K, Brunner S, Tari I (2014) Glutathione transferases uperupper gene family in tomato: salt stress- regulated expression of representative genes from distinct GST classes in plants primed with salicylic acid. Plant Physiol Biochem 78:15–26

de Boer JG, Posthumus MA, Dicke M (2004) Identification of volatiles that are used in discrimination between plants infested with prey or nonprey herbivores by a predatory mite. J Chem Ecol 30:2215–2230

Davies PJ (2004) Plant hormones: their nature, occurrence and functions. In: Davies PJ (ed) Plant hormones: biosynthesis, signal transduction, action. Kluwer Academic Publishers, London, pp 1–15

Després C, Chubak C, Rochon A, Clark R, Bethune T, Desveaux D, Fobert PR (2003) The Arabidopsis NPR1 disease resistance protein is a novel cofactor that confers redox regulation of DNA binding activity to the basic domain/leucine zipper transcription factor TGA1. Plant Cell 15:2181–2191

Durner J, Klessig DF (1995) Inhibition of ascorbate peroxidase by salicylic acid and 2, 6-dichloroisonicotinic acid, two inducers of plant defense responses. Proc Natl Acad Sci U S A 92:11312–11316

Durrant WE, Dong X (2004) Systemic acquired resistance. Annu Rev Phytopathol 42:185–209

Eichhorn H, Klinghammer M, Becht P, Tenhaken R (2006) Isolation of a novel ABC-transporter gene from soybean induced by salicylic acid. J Exp Bot 57:2193–2201

El-Beltagi HS, Mohamed HI, Sofy MR (2020) Role of Ascorbic acid, glutathione and proline applied as singly or in sequence combination in improving chickpea plant through physiological change and antioxidant defense under different levels of irrigation intervals. Molecules 25(7): E1702. https://doi.org/10.3390/molecules25071702

Fu ZQ, Dong X (2013) Systemic acquired resistance: turning local infection into global defense. Annu Rev Plant Biol 64:839–863

Gaille C, Kast P, Hass D (2002) Salicylate biosynthesis in *Pseudomonas aeruginosa*. Purification and characterization of PchB, a novel bifunctional enzyme displaying isochorismate pyruvatelyase and chorismate mutase activities. J Biol Chem 277:21768–21775

Gaille C, Reimmann C, Haas D (2003) Isochorismate synthase (PchA), the first and rate-limiting enzyme in salicylate biosynthesis of *Pseudomonas aeruginosa*. J Biol Chem 278:16893–16898

Galani S, Hameed S, Ali MK (2016) Exogenous application of salicylic acid: inducing thermotolerance in cotton (*Gossypium Hirsutum* L.) seedlings. Int J Agric Food Res 5:9–18

Gamir J, Darwiche R, van't Hof P, Choudhary V, Stumpe M, Schneiter R, Mauch F (2017) The sterol-binding activity of pathogenesis-related protein 1 reveals the mode of action of an antimicrobial protein. Plant J 89:502–509

Garcion C, Lohmann A, Lamodiere E, Catinot J, Buchala A, Doermann P, Métraux J-P (2008) Characterization and biological function of the isochorismate synthase2 gene of the Arabidopsis. Plant Physiol 147:1279–1287

Ghasemzadeh A, Jaafar HZE, Karimi E, Ibrahim MH (2012) Combined effect of CO_2 enrichment and foliar application of salicylic acid on the production and antioxidant activities of anthocyanin, flavonoids and isoflavonoids from ginger. BMC Complement Altern Med 12:229–23911

Gill SS, Tuteja N (2010) Reactive oxygen species and antioxidant machinery in abiotic stress tolerance in crop plants. Plant Physiol Biochem 48(12):909–930

Gozzo F, Faoro F (2013) Systemic acquired resistance (50 years after discovery): moving from the lab to the field. J Agric Food Chem 61(51):12473–12491

Guo B, Liang YC, Zhu YG, Zhao FJ (2007) Role of salicylic acid in alleviating oxidative damage in rice roots (*Oryza sativa*) subjected to cadmium stress. Environ Pollut 147:743–749

Habibi G (2012) Exogenous salicylic acid alleviates oxidative damage of barley plants under drought stress. Acta Biol Szeged 56:57–63

Hasanuzzaman M, Alam MM, Nahar K, Mahmud JA, Ahamed KU, Fujita M (2014) Exogenous salicylic acid alleviates salt stress-induced oxidative damage in *Brassica napus* by enhancing the antioxidant defense and glyoxalase systems. Aust J Crop Sci 8:631–639

Hasanuzzaman M, Nahar K, Bhuiyan TF, Anee TI, Inafuku M, Oku H, Fujita M (2017) Salicylic acid: an all-rounder in regulating abiotic stress responses in plants. In: Phytohormones – signaling mechanisms and crosstalk in plant development and stress responses. INTECH, Rijeka, pp 31–57

Hayat Q, Hayat S, Irfan M, Ahmad A (2010) Effect of exogenous salicylic acid under changing environment: a review. Environ Exp Bot 68(1):14–25

He Y, Zhu ZJ (2008) Exogenous salicylic acid alleviates NaCl toxicity and increases antioxidative enzyme activity in Lycopersicun esculentum. Biol Plant 52:792–795

Hegazi AM, El-Shrayi AM (2007) Impact of salicylic acid and paclobutrazol exogenous application on the growth, yield and nodule formation of common bean. Aust J Basic Appl Sci 1:834–840

Hoffmann-Sommergruber K (2000) Plant allergens and pathogenesis related proteins. What do they have in common? Int Arch Allergy Immunol 122:155–166

Horváth E, Szalai G, Janda T (2007) Induction of abiotic stress tolerance by salicylic acid signaling. J Plant Growth Regul 26:290–300

Hussein MM, Balbaa LK, Gaballah MS (2007) Salicylic acid and salinity effects on growth of maize plants. Res J Agric Biol Sci 3:321–328

Imran H, Zhang Y, Du G, Wang G, Zhang J (2007) Effect of salicylic acid (SA) on delaying fruit senescence of Huang Kum pear. Front Agric China 1(4):456–459

Islam F, Yasmeen T, Arif MS, Riaz M, Shahzad SM, Imran Q, Ali I (2016) Combined ability of chromium (Cr) tolerant plant growth promoting bacteria (PGPB) and salicylic acid (SA) in attenuation of chromium stress in maize plants. Plant Physiol Biochem 108:456–467

Janda T, Szalai G, Tari I, Páldi E (1997) Exogenous salicylic acid has an effect on chilling symptoms in maize (*Zea mays* L.) plants. In: Sowinski P, Zagdanska B, Aniol A, Klaus P (eds) Crop development for cool and wet European climate. ECSP-EEC-EAEC, Brussels, pp 179–187

Janda T, Szalai G, Tari I, Páldi E (1999) Hydroponic treatment with salicylic acid decreases the effect of chilling injury in maize (*Zea mays* L.) plants. Planta 208:175–180

Jayakannan M, Bose J, Babourina O, Rengel Z, Shabala S (2013) Salicylic acid improves salinity tolerance in *Arabidopsis* by restoring membrane potential and preventing salt-induced K+ loss via a GORK channel. J Exp Bot 64:2255–2268

Jupin I, Chua N-H (1996) Activation of the CaMV as-1 ciselement by salicylic acid: differential DNA-binding of a factor related to TGAla. EMBO J 15:5679–5689

Jurivich DA, Sistonen L, Kroes RA, Morimoto RI (1992) Effect of sodium salicylate on the human HS response. Science 255:1243–1245

Kang GZ, Wang CH, Sun GC, Wang ZX (2003) Salicylic acid changes activities of H_2O_2-metabolizing enzymes and increases the chilling tolerance of banana seedlings. Environ Exp Bot 50:9–15

Kang G, Li G, Xu W, Peng X, Han Q, Zhu Y (2012) Proteomics reveals the effects of salicylic acid on growth and tolerance to subsequent drought stress in wheat. J Proteome Res 11:6066–6079

Kang GZ, Li GZ, Liu GQ, Xu W, Peng XQ, Wang CY, Zhu YJ, Guo TC (2013) Exogenous salicylic acid enhances wheat drought tolerance by influence on the expression of genes related to ascorbate-glutathione cycle. Biol Plant 57:718–724

Kessmann H, Staub T, Hofmann C, Maetzke T, Herzog J, Ward E, Uknes S, Ryals J (1994) Induction of systemic acquired disease resistance in plants by chemicals. Annu Rev Phytopathol 32:439–459

Khan W, Prithiviraj B, Smith DL (2003) Photosynthetic responses of corn and soybean to foliar application of salicylates. J Plant Physiol 160:485–492

Khan MIR, Iqbal N, Masood A, Per TS, Khan NA (2013) Salicylic acid alleviates adverse effects of heat stress on photosynthesis through changes in proline production and ethylene formation. Plant Signal Behav 8:e26374

Khan MIR, Fatma M, Per TS, Anjum NA, Khan NA (2015) Salicylic acid-induced abiotic stress tolerance and underlying mechanisms in plants. Front Plant Sci 6:462

Khanna P, Kaur K, Gupta AK (2016) Salicylic acid induces differential antioxidant response in spring maize under high temperature stress. Indian J Exp Biol 54:386–393

Khokon MDAR, Okuma E, Hossain MA, Munemasa S, Uraji M, Nakamura Y, Mori IC, Murata Y (2011) Involvement of extracellular oxidative burst in salicylic acid-induced stomatal closure in Arabidopsis. Plant Cell Environ 34:434–443

Kim ST, Kim SG, Hwang DH, Kang SY, Koo SC, Cho MJ, Kang KY (2004) Expression of a salt-induced protein (SALT) in suspension-cultured cells and leaves of rice following exposure to fungal elicitor and phytohormones. Plant Cell Rep 23:256–262

Kim YH, Hamayun M, Khan AL, Na CI, Kang SM, Han HH, Lee IJ (2009) Exogenous application of plant growth regulators increased the total flavonoid content in *Taraxacum officinale* (Wigg). Afr J Biotechnol 8:5727–5732

Klessig DF, Malamy J (1994) The salicylic acid signal in plants. Plant Mol Biol 26:1439–1458

Knorzer OC, Lederer B, Durner J, Boger P (1999) Antioxidative defense activation in soybean cells. Physiol Plant 107:294–302

Kohler A, Schwindling S, Conrath U (2002) Benzothiadiazole-induced priming for potentiated responses to pathogen infection, wounding, and infiltration of water into leaves require the NPR1/NIM1 gene in Arabidopsis. Plant Physiol 128:1046–1056

Korkmaz A (2005) Inclusion of acetyl salicylic acid and methyl jasmonate into the priming solution improves low temperature germination and emergence of sweet pepper. HortScience 40:197–200

Kumar RR, Sharma SK, Goswami S, Verma P, Singh K, Dixit N, Pathak H, Viswanathan C, Rai RD (2015) Salicylic acid alleviates the heat stress-induced oxidative damage of starch biosynthesis pathway by modulating the expression of heat-stable genes and proteins in wheat (*Triticum aestivum*). Acta Physiol Plant 37:1–12

Larqué-Saavedra A, Martin-Méx R (2007) Effects of salicylic acid on bioproductivity of plants. In: Hayat S, Ahmad A (eds) Salicylic acid: a plant hormone. Springer, Dordrecht, pp 15–23

Lee JS (1998) The mechanism of stomatal closing by salicylic acid in *Commelina communis* L. J Plant Biol 41:97–102

Lee H-I, Leon J, Raskin I (1995) Biosynthesis and mechanism of salicylic acid. Proc Natl Acad Sci U S A 92:4076–4079

Lee S, Kim SG, Park CM (2010) Salicylic acid promotes seed germination under high salinity by modulating antioxidant activity in *Arabidopsis*. New Phytol 188:627–637

Leon J, Yalpani N, Raskin I, Lawton MA (1993) Induction of benzoic acid 2-hydroxylase in virusinoculated tobacco. Plant Physiol 103:323–328

Leon J, Lawton MA, Raskin I (1995) Hydrogen peroxide stimulates salicylic acid biosynthesis in tobacco. Plant Physiol 108:1673–1678

Li G, Peng X, Wei L, Kang G (2013) Salicylic acid increases the contents of glutathione and ascorbate and temporally regulates the related gene expression in salt-stressed wheat seedlings. Gene 529:321–325

Liu X, Huang B (2005) Root physiological factors involved in cool-season grass response to high soil temperature. Environ Exp Bot 53:233–245

Liu Y, Zhang J, Liu H, Huang W (2008) Salicylic acid or heat acclimation pre-treatment enhances the plasma membrane associated ATPase activities in young grape plants under heat shock. Sci Hortic 119:21–27

Liu S, Dong Y, Kong XJ (2014) Effects of foliar application of nitric oxide and salicylic acid on salt-induced changes in photosynthesis and antioxidative metabolism of cotton seedlings. Plant Growth Regul 73:67–78

Lopez-Delgado H, Dat JF, Foyer CH, Scot IM (1998) Induction of thermotolerance in potato microplants by acetylsalicylic acid and H_2O_2. J Exp Bot 49:713–720

Luo Z, Wu X, Xie Y, Chen C (2012) Alleviation of chilling injury and browning of postharvest bamboo shoot by salicylic acid treatment. Food Chem 131:456–461

Maffei ME, Mithofer A, Boland W (2007) Insects feeding on plants: rapid signals and responses preceding the induction of phytochemical release. Phytochemistry 68:2946–2959

Malamy J, Carr JP, Klessig DF, Raskin I (1990) Salicylic acid: a likely endogenous signal in the resistance response of tobacco to viral infection. Science 250:1002–1004

Maldonado AM, Doerner P, Dixon RA, Lamb CJ, Cameron RK (2002) A putative lipid transfer protein involved in systemic resistance signaling in *Arabidopsis*. Nature 419:399–403

Mantri N, Patade V, Penna S, Ford R, Pang E (2012) Abiotic stress responses in plants: present and future. In: Ahmad P, Prasad MNV (eds) Abiotic stress responses in plants: metabolism to productivity. Springer, New York, pp 1–20

Metwally A, Finkemeier I, Georgi M, Dietz KJ (2003) Salicylic acid alleviates the cadmium toxicity in barley seedlings. Plant Physiol 132:272–281

Millar AH, Whelan J, Soole KL, Day DA (2011) Organization and regulation of mitochondrial respiration in plants. Annu Rev Plant Biol 62:79–104

Mishra A, Choudhuri M (1997) Ameliorating effects of salicylic acid on lead and mercury: induced inhibition of germination and early seedling growth of two rice cultivars. Seed Sci Technol 25 (2):263–270

Miura K, Okamoto H, Okuma E, Shiba H, Kamada H, Hasegawa PM, Murata Y (2013) SIZ1 deficiency causes reduced stomatal aperture and enhanced drought tolerance via controlling salicylic acid-induced accumulation of reactive oxygen species in *Arabidopsis*. Plant J 73:91–104

Mohamed HI, Latif HH, Hanafy RS (2016) Influence of nitric oxide application application on some biochemical aspects, endogenous hormones, minerals and phenolic compounds of *Vicia faba* plant grown under arsenic stress. Gesunde Pflanzen 68:99–107

Molina A, Bueno P, Mar�n MC, Rodriguez-Rosales MP, Belver A, Venema K, Donaire JP (2002) Involvement of endogenous salicylic acid content, lipoxygenase and antioxidant enzyme activities in the response of tomato cell suspension cultures to NaCl. New Phytol 156:409–415

Moore AL, Albury MS, Crichton PG, Affourtit C (2002) Function of the alternative oxidase: is it still a scavenger? Trends Plant Sci 7:478–481

Mou Z, Fan W, Dong X (2003) Inducers of plant systemic acquired resistance regulate NPR1 function through redox changes. Cell 113:935–944

Mukherjee D, Kumar R (2007) Kinetin regulates plant growth and biochemical changes during maturation and senescence of leaves, flowers, and pods of *Cajanus cajan* (L.). Biol Plant 50:80–85

Munne-Bosch S, Penuelas J (2003) Photo-and antioxidative protection, and a role for salicylic acid during drought and recovery in field-grown *Phillyrea angustifolia* plants. Planta 217:758–766

Mur LAJ, Naylor G, Warner SAJ, Sugars JM, White RF, Draper J (1996) Salicylic acid potentiates defense gene expression in tissue exhibiting acquired resistance to pathogen attack. Plant J 9:559–571

Mohamed HI, Latif HH (2017) Improvement of drought tolerance of soybean plants by using methyl jasmonate. Physiol Mol Biol Plants 23:545–556

Mohamed HI, Akladious SA, Ashry NA (2018) Evaluation of water stress tolerance of soybean using physiological parameters and retrotransposon-basedmarkers. Gesunde Pflanzen 70:205–215

Mohamed HI, Akladious SA (2017) Changes in antioxidants potential, econdary metabolites and plant hormones induced by different fungicides treatment in cotton plants. Pestic Biochem Physiol 142:117–122

Mohamed HI, El-Beltagi HS, Aly AA, Latif HH (2018) The role of systemic and non systemic fungicides on the physiological and biochemical parameters in Gossypium hirsutum plant, implications for defense responses. Fresenius Environ Bull 27(12):8585–8593

Ogawa D, Nakajima N, Seo S, Mitsuhara I, Kamada H, Ohashi Y (2006) The phenylalanine pathway is the main route of salicylic acid biosynthesis in tobacco tobacco mosaic virus-infected tobacco leaves. Plant Biotechnol 23:395–398

Pandey P, Srivastava RK, Dubey RS (2013) Salicylic acid alleviates aluminum toxicity in rice seedlings better than magnesium and calcium by reducing aluminum uptake, suppressing oxidative damage and increasing antioxidative defense. Ecotoxicology 22(4):656–670

Parashar A, Yusuf M, Fariduddin Q, Ahmad A (2014) Salicylic acid enhances antioxidant system in *Brassica juncea* grown under different levels of manganese. Int J Biol Macromol 70:551–558

Peng J, Deng X, Huang J, Jia S, Miao X, Huang Y (2004) Role of salicylic acid in tomato defense against cotton bollworm, *Helicoverpa armigera* Hubner. Z Naturforsch C 59:856–862

Pieterse CMJ, Van Loon LC (2004) NPR1: the spider in the web of induced resistance signaling pathways. Curr Opin Plant Biol 7:456–464

Rajjou L, Belghazu M, Huget R, Robin C, Moreau A, Job C, Job D (2006) Proteomic investigation of the effect of salicylic acid on *Arabidopsis* seed germination and establishment of early defense mechanisms. Plant Physiol 141:910–923

Rao MV, Davis KR (1999) Ozone-induced cell death occurs via two distinct mechanisms in *Arabidopsis*: the role of salicylic acid. Plant J 17:603–614

Rao MV, Paliyat G, Ormrod D, Murr DP, Watkin CB (1997) Influence of salicylic acid on H_2O_2 production, oxidative stress and H_2O_2 metabolizing enzymes. Salicylic acid-mediated oxidative damage requires H_2O_2. Plant Physiol 115:137–149

Rea PA, Li ZS, Lu YP, Drozdowicz YM, Martinoia E (1998) From vacuolar GS-X pumps to multispecific ABC transporters. Annu Rev Plant Physiol Plant Mol Biol 49:727–760

Ribnicky DM, Shulaev V, Raskin I (1998) Intermediates of salicylic acid biosynthesis in tobacco. Plant Physiol 118:565–572

Rivas-San VM, Plasencia J (2011) Salicylic acid beyond defence: its role in plant growth and development. J Exp Bot 62:3321–3338

Ryals JA, Neuenschwander UH, Willits MG, Molina A, Steiner H-Y, Hunt MD (1996) Systemic acquired resistance. Plant Cell 8:1809–1819

Sakhabutdinova AR, Fatkhutdinova DR, Shakirova FM (2004) Effect of salicylic acid on the activity of antioxidant enzymes in wheat under conditions of salination. Appl Biochem Microbiol 40:501–505

Sanghera GS, Wani SH, Hussain W, Singh NB (2011) Engineering cold stress tolerance in crop plants. Curr Genomics 12:30–43

Saruhan N, Saglam A, Kadioglu A (2012) Salicylic acid pretreatment induces drought tolerance and delays leaf rolling by inducing antioxidant systems in maize genotypes. Acta Physiol Plant 34:97–106

Sawada H, Shim IS, Usui K (2006) Induction of benzoic acid 2-hydroxylase and salicylic acid biosynthesis-modulation by salt stress in rice seedlings. Plant Sci 171:263–270

Schafer H, Wink M (2009) Medicinally important secondary metabolites in recombinant microorganism or plants: progress in alkaloid biosynthesis. Biotechnol J 4(12):1684–1703

Selitrennikoff CP (2001) Antifungal proteins. Appl Environ Microbiol 67:2883–2894

Senaratna T, Touchell D, Bunn E, Dixon K (2000) Acetyl salicylic acid (Aspirin) and salicylic acid induce multiple stress tolerance in bean and tomato plants. Plant Growth Regul 30:157–161

Shah J (2003) The salicylic acid loop in plant defense. Curr Opin Plant Biol 6:365–371

Shah J (2005) Lipids, lipases and lipid modifying enzymes in plant disease resistance. Annu Rev Phytopathol 43:229–260

Shakirova FM, Sakhabutdinova AR, Bezrukova MV, Fatkhutdinova RA, Fatkhutdinova DR (2003) Changes in the hormonal status of wheat seedlings induced by salicylic acid and salinity. Plant Sci 164:317–322

Sharma P, Jha AB, Dubey RS, Pessarakli M (2012) Reactive oxygen species, oxidative damage, and antioxidative defense mechanism in plants under stressful conditions. J Bot 2012:1–26

Shi Q, Bao Z, Zhu Z, Ying Q, Qian Q (2006) Effects of different treatments of salicylic acid on heat tolerance, chlorophyll fluorescence, and antioxidant enzyme activity in seedlings of *Cucumis sativa* L. Plant Growth Regul 48:127–135

Singh B, Usha K (2003) Salicylic acid induced physiological and biochemical changes in wheat seedlings under water stress. Plant Growth Regul 39:137–141

Singh DP, Moore CA, Gilliland A, Carr JP (2004) Activation of multiple antiviral defence mechanisms by salicylic acid. Mol Plant Pathol 5:57–63

Singh A, Srivastava AK, Singh AK (2011) Exogenous application of salicylic acid to alleviate the toxic effects of insecticides in *Vicia faba* L. Environ Toxicol 28(12):666–672

Singh AP, Dixit G, Mishra S, Dwivedi S, Tiwari M, Mallick S, Pandey V, Trivedi PK, Chakrabarty D, Tripathi RD (2015) Salicylic acid modulates arsenic toxicity by reducing its root to shoot translocation in rice (*Oryza sativa* L.). Frontiers. Plant Sci 6:340

Slaymaker DH, Navarre DA, Clark D, Del Pozo O, Martin GB, Klessig DF (2002) The tobacco salicylic acid-binding protein 3 (SABP3) is the chloroplast carbonic anhydrase, which exhibits antioxidant activity and plays a role in the hypersensitive defense response. Proc Natl Acad Sci U S A 99:11640–11645

Srivastava MK, Dwivedi UN (2000) Delaying ripening of banana fruits by salicylic acid. Plant Sci 158:87–96

Steponkus PL, Uemura M, Webb MS (1993) A contrast of the cryostability of the plasma membrane of winter rye and spring oat-two species that widely differ in their freezing tolerance and plasma membrane lipid composition. In: Steponkus PL (ed) Advances in low-temperature biology, vol 2. JAI Press, London, pp 211–312

Sticher L, Mauch-Mani B, Métraux JP (1997) Systemic acquired resistance. Annu Rev Phytopathol 35:353–270

Strawn MA, Marr SK, Inoue K, Inada N, Zubieta C, Wildermuth MC (2007) *Arabidopsis* isochorismate synthase functional in pathogen-induced salicylate biosynthesis exhibits properties consistent with a role in diverse stress responses. J Biol Chem 282:5919–5933

Szepesi A (2006) Salicylic acid improves the acclimation of *Lycopersicon esculentum* mill. L. to high salinity by approximating its salt stress response to that of the wild species *L. pennellii*. Acta Biol Szeged 50:177

Szepesi Á, Csiszár J, Sz B, Gémes K, Horváth F, Erdei L, Deér A, Simon LM, Tari I (2005) Role of salicylic acid pre-treatment on the acclimation of tomato plants to salt- and osmotic stress. Acta Biol Szeged 49:123–125

Tari I, Csiszár J, Szalai G, Horváth F, Pécsváradi A, Kiss G, Szepesi Á, Szabó M, Erdei L (2002) Acclimation of tomato plants to salinity stress after a salicylic acid pre-treatment. Acta Biol Szeged 46:55–56

Tari I, Simon LM, Deér KA, Csiszár J, Sz B, Gy K, Szepesi Á (2004) Influence of salicylic acid on salt stress acclimation of tomato plants: oxidative stress responses and osmotic adaptation. Acta Physiol Plant 26S:237

Tasgin E, Atici O, Nalbantoglu B (2003) Effects of salicylic acid and cold on freezing tolerance in winter wheat leaves. Plant Growth Regul 41:231–236

Thaler JS (1999) Jasmonate-inducible plant defenses cause increased parasitism of herbivores. Nature 399:686–688

Torres MA (2010) ROS in biotic interactions. Physiol Plant 138:414–429

Torres MA, Dangl JL, Jones JDG (2002) *Arabidopsis* gp91phox homologues AtrbohD and AtrbohF are required for accumulation of reactive oxygen intermediates in the plant defense response. Proc Natl Acad Sci U S A 99:517–522

Urbanek Krajnc A, Müller M (2006) An insight into the defense mechanisms and the role of glutathione during advance ZYMV infection in Styrian oil pumpkin. Agricultura 4:27–35

Vanacker H, Carver TLW, Foyer CH (2000) Early H_2O_2 accumulation in mesophyll cells leads to induction of glutathione during the hypersensitive response in the barley-powdery mildew interaction. Plant Physiol 123:1289–1300

Vanacker H, Lu H, Rate DN, Greenberg JT (2001) A role for salicylic acid and NPR1 in regulating cell growth in *Arabidopsis*. Plant J 28:209–216

Vlot AC, Dempsey MA, Klessig DF (2009) Salicylic acid, a multifaceted hormone to combat disease. Annu Rev Phytopathol 47:177–206

Wang LJ, Li SH (2006) Salicylic acid-induced heat or cold tolerance in relation to Ca^{2+} homeostasis and antioxidant systems in young grape plants. Plant Sci 170:685–694

Wang H, Feng T, Peng X, Yan M, Tang X (2009) Up-regulation of chloroplastic antioxidant capacity is involved in alleviation of nickel toxicity of Zea mays L. by exogenous salicylic acid. Ecotoxicol Environ Saf 72:1354–1362

Wang LJ, Fan L, Loescher W, Duan W, Liu GJ, Cheng JS, Luo HB, Li SH (2010) Salicylic acid alleviates decreases in photosynthesis under heat stress and accelerates recovery in grapevine leaves. BMC Plant Biol 10:34

White RF (1979) Acetylsalicylic acid (aspirin) induces resistance to tobacco mosaic virus in tobacco. Virology 99:410–412

Wildermuth MC, Dewdney J, Wu G, Ausubel FM (2001) Isochorismate synthesis is required to synthesize salicylic acid for plant defense. Nature 414:562–565

Wobbe KK, Klessig DF (1996) Salicylic acid – an important signal in plants. In: Dennis ES et al (eds) Plant gene research. Springer Verlag, Wien/New York, pp 167–196

Yadav SK (2010) Heavy metals toxicity in plants: an overview on the role of glutathione and phytochelatins in heavy metal stress tolerance of plants. S Afr J Bot 76(2):167–179

Ying Y, Yue Y, Huang X, Wang H, Mei L, Yu W, Zheng B, Wu J (2013) Salicylic acid induces physiological and biochemical changes in three Red bayberry (*Myric rubra*) genotypes under water stress. Plant Growth Regul 71:181–189

Yoshioka H, Bouteau F, Kawano T (2008) Discovery of oxidative burst in the field of plant immunity. Plant Signal Behav 3:143–155

Yu X-M, Griffith M, Wiseman SB (2001) Ethylene induces antifreeze activity in winter rye leaves. Plant Physiol 126:1232–1240

Zhang S, Klessig DF (1997) Salicylic acid activates a 48 kilodalton MAP kinase in tobacco. Plant Cell 9:809–824

Zhang J, Du X, Wang Q, Chen X, Lv D, Xu K, Qu S, Zhang Z (2010) Expression of pathogenesis related genes in response to salicylic acid, methyl jasmonate and 1-aminocyclopropane-1-carboxylic acid in *Malus hupehensis* (Pamp.) Rehd. BMC Res Notes 3:208

Zhang Y, Xu S, Yang S, Chen Y (2015) Salicylic acid alleviates cadmium-induced inhibition of growth and photosynthesis through upregulating antioxidant defense system in two melon cultivars (*Cucumis melo* L.). Protoplasma 2(3):911–924

Zhao LY, Chen JL, Cheng DF, Sun JR, Liu Y, Tian Z (2009) Biochemical and molecular characterizations of *Sitobion avenae* induced wheat defense responses. Crop Prot 28:435–442

Zhou ZS, Guo K, Elbas AA, Yang ZM (2009) Salicylic acid alleviates mercury toxicity by preventing oxidative stress in roots of Medicago sativa. Environ Exp Bot 65:27–34

Root Phenolics Profile Modulates Microbial Ecology of Rhizosphere 24

Anil Kumar Singh and Poonam Singla

Abstract

Rhizosphere is a narrow zone of soil surrounding plant roots. Plants release an array of organic and inorganic compounds in rhizosphere to influence microbial community, manage soil herbivores, attract beneficial symbionts, alter chemical and physical properties of the soil, and hinder growth of competitor plants. Phenolic compounds are important plant secondary metabolites released in the rhizosphere. Chemically, plant phenolics are organic compounds of plant origin having one or more functional hydroxyl group bonded directly to aromatic hydrocarbon. In last few decades our understanding of phenolic compounds and their role in aboveground plant parts have increased tremendously. Relatively, our knowledge about phenolic compounds and their role in biology and ecology of rhizosphere is not very clear. Plants can alter amount as well as composition of phenolic compounds in rhizodeposits to bring about desirable changes in the micro-environment of rhizosphere. Profiles of phenolic compounds in plant are not only regulated by inherited genetic information but also influenced by environmental cues. The rhizosphere has significant influence on vigor and ecological fitness of their host plant. Understanding the role of phenolic compounds in rhizosphere will allow us to design and develop sustainable agriculture practices. The current chapter describes the role of root exudates component, particularly the phenolic compounds, on plant–microbe interaction and functioning of rhizosphere.

A. K. Singh (✉)
Department of Botany, Sant Baba Bhag Singh University, Khiala, Punjab, India

P. Singla
Department of Chemistry and Centre for Advanced Studies in Chemistry, Panjab University, Chandigarh, India

R. Lone et al. (eds.), *Plant Phenolics in Sustainable Agriculture*,
https://doi.org/10.1007/978-981-15-4890-1_24

Keywords

Phenolics · Root · Rhizosphere · Ecology · Plant–microbe interaction

24.1 Introduction

In natural environment, plants interact with several living organisms daily. These interactions are more frequent in the underground plant parts since soil harbors extremely diverse community of macro- and microorganisms. The narrow zone of soil lying in the vicinity of plant root is directly influenced by the physiological state of plant. Plants release an array of compounds through roots to constitute plant-influenced habitat for soil-dwelling organisms (Baetz and Martinoia 2014). Lorenz Hiltner used the term "rhizosphere" to describe this narrow region of soil present around plant roots (Hiltner 1904). Rhizosphere is inhabited by unique population of microorganisms gathered under the influence of chemicals released by plants roots (Baetz and Martinoia 2014). Since its inception in 1904, the definition of rhizosphere has become narrower and more definitive.

In the recent past, the importance of studying rhizosphere has extended from basic biological sciences to applied sciences as it has potential to serve agriculture (Haldar and Sengupta 2015; Ahkami et al. 2017). The rhizosphere is a highly dynamic system shaped by complex interaction of biotic and abiotic components (McNear 2013). Rhizosphere-residing biotic components have significant influence on the vigor and ecological fitness of their host (Haldar and Sengupta 2015). The study of interaction among the biotic and abiotic components within the narrow region of soil surrounding plant root is called "Rhizosphere Ecology." Exploring ecology of rhizosphere is more challenging than aboveground study. This has resulted in relative lesser information about rhizosphere ecology as compared to the aboveground ecology. The major difficulty in exploring ecology of rhizosphere is the technical challenges associated with complex soil matrix (Broeckling et al. 2008). Microorganisms are important biotic components of rhizosphere and very small fraction of microbial species is amenable to laboratory culture (Philippot et al. 2013). Irrespective of the challenges, efforts are being made to unleash the hidden knowledge of rhizosphere ecology. Ecological knowhow of rhizosphere can be harnessed in agriculture, environmental remediation, water filtration, medical discoveries, and industrial applications (De-la-Pena and Loyola-Vargas 2014; Haldar and Sengupta 2015; Ahkami et al. 2017).

24.2 The Rhizosphere Effect

The narrow zone of soil surrounding plant roots has comparatively higher microbial population density as compared to the bulk soil (Haichar et al. 2014). This phenomenon is often referred as "Rhizosphere Effect." The microorganisms residing in the bulk soil experience carbon and energy starvation–like condition. However, root

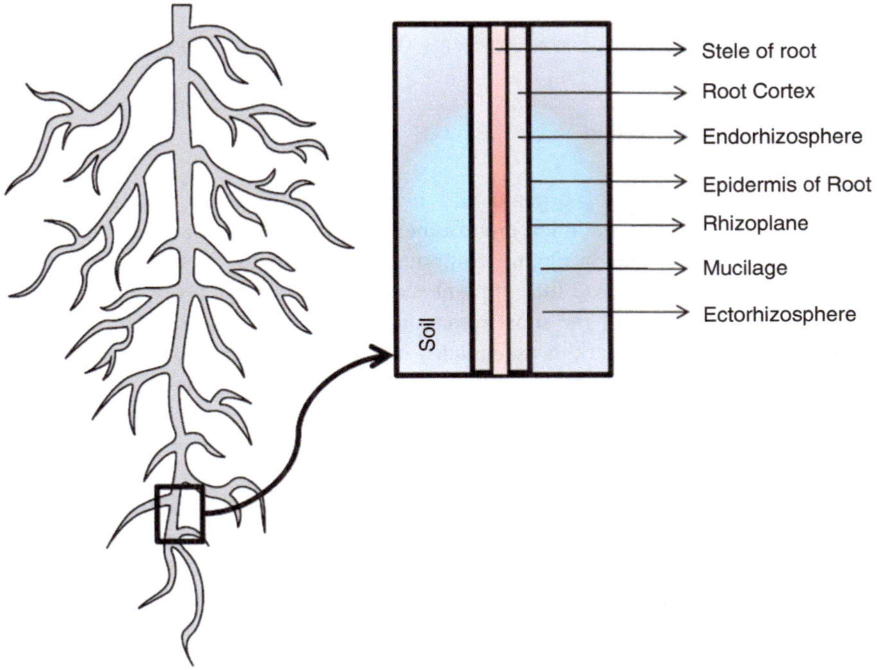

Fig. 24.1 Schematic representation of a plant root and rhizosphere showing different zones, namely, endorhizosphere, rhizoplane and ectorhizosphere

secretion in rhizosphere makes it a nutrient-rich zone which attracts microorganism from bulk soil. Hinsinger et al. (2009) reported the presence of 10^4 protozoa per gram of rhizosphere soil. The number of bacteria, nematodes, and fungi were found to be in the range of 10^6–10^9, 10^1-10^2, and 10^5–10^6 per gram of rhizosphere soil (Watt et al. 2006; Mendes et al. 2013). It appears that plants recruit useful microorganisms from the available microbial diversity present in bulk soil. In view of this, rhizosphere can be best explained as a narrow zone of soil surrounding plant root where physical, biological, and chemical parameters of soil are influenced by root secretions and associated soil microorganisms (Baetz and Martinoia 2014). Rhizosphere can be distinguished into three zones namely: endorhizosphere, rhizoplane, and ectorhizosphere (Fig. 24.1). Endorhizosphere is the innermost region of rhizosphere consisting of root cortex and endodermis. This zone is marked by the presence of microbes and mineral ions in the apoplastic spaces available between the cells (McNear 2013). Rhizoplane is the middle zone of rhizosphere consisting of the epidermal cells and mucilaginous layer present on the epidermal layer of root (McNear 2013). Ectorhizosphere is the outer most part of rhizosphere starting from rhizoplane and gradually extending outward into the bulk soil. Ectorhizosphere does not have sharp limiting boundary but rather exists as a gradient of chemical, biological, and physical properties (McNear 2013). Generally, ectorhizosphere region is discernible by biological indicators such as microbial density, enzymatic activity, or mapping root-derived gradients (Broeckling et al. 2008).

24.3 Rhizodeposits

Plants secrete an array of organic and inorganic compounds through roots to organize as well as alter the microenvironment of rhizosphere (Baetz and Martinoia 2014). Both organic and inorganic compounds released by roots are collectively called "Root Exudes." The ability of plant root to secrete a vast array of compounds into rhizosphere is one of the most remarkable metabolic features. Through these secretions, plant tries to mark its ascendancy in underground habitat. Secretions from root are known to alter chemical, physical, and biological processes in the rhizosphere (Jones et al. 2009). Extent of change induced by root exudes is highly dependent on the quality and quantity of root secretions as well as intrinsic characteristic of the soil (Jones et al. 2004).

Some research groups have extended the concept of root exudes to rhizodeposits (Baetz and Martinoia 2014; Haichar et al. 2014). Rhizodeposits are root secretions that get deposited around the roots and influence microenvironment of rhizosphere. Rhizodeposits are mainly composed of root exudes, mucilage, border cells, and gas metabolites (Haichar et al. 2014). The quality and quantity of rhizodeposits influence the soil microbial community present in rhizosphere, manage soil herbivores, attract beneficial symbionts, alter chemical and physical properties of the soil, and hinder the growth of rival plant species (Haichar et al. 2014).

The importance of root secretions can be highlighted by the fact that plants release significant portion of carbon fixed during photosynthesis in form of rhizodeposits. Quantifying amount of carbon released by root in the soil-less culture is easy but has no ecological significance (Jones et al. 2009). However, development of tracer technique has provided numerous options for quantifying carbon released by root. Nardi et al. (2000) reported that nearly 5–21% of all photosynthetically fixed carbon is released in rhizosphere through root exudes. According to Newman (1985), root of plant species can release 10–250 mg carbon per gram of root through root exudes. Kuzyakov and Domanski (2000) reported that cereal crops, like wheat and barley, can transfer 20–30% of total assimilated carbon into the rhizosphere. Similarly, pasture plants can transfer 30–50% of total assimilated carbon into the rhizosphere (Kuzyakov and Domanski 2000). The carbon fixed during photosynthesis is generally transferred to rhizosphere within few hours (grasses) or few days (trees). Using $13CO_2$ labeling method, Derrien et al. (2004) reported that nearly 16% of the carbon fixed by wheat plant is released into the soil within 5 h.

Root exudates are a major portion of rhizodeposits and major contributor of organic carbon in rhizosphere. Exudates are water-soluble low-molecular weight components of rhizodeposits (Ahkami et al. 2017). They are released from root by passive diffusion. Plants have very little control on the release of root exudes (Bais et al. 2006). A major portion of root exudes are released along the steep concentration gradient prevailing between the cytoplasm of undamaged root cells and the external soil solution (Philippot et al. 2013). Usually the concentration of molecules in cytoplasm is in millimolar range and concentration of molecules in external soil solution is in micromolar range (Ahkami et al. 2017). Movement of root exudes from cytoplasm of root cell to surrounding soil solution through the phospholipid bilayer of the plasma membrane is determined by membrane permeability and polarity of compounds. The physiological state of a plant particularly of root cells has significant influence on membrane permeability (Rudrappan et al. 2007). Thus, factors affecting cell membrane integrity such as extreme temperature, nutrient deficiency, or pathogen attack can influence root exudes (Jones et al. 2009). Root exudates serve as energy source as well as communicating molecules for plant and microorganisms. Exudates attract several microorganisms as well as repel some other microorganisms, and thus influence the assemblage of microorganisms in the rhizosphere (Haichar et al. 2014). Communicating molecules initiate physiological interaction between the soil microorganisms and the plant roots by changing the environment of rhizosphere (De-la-Pena and Loyola-Vargas 2014).

Root mucilage is a gelatinous layer covering the root tips (Jones et al. 2009). Mucilage is primarily composed of polysaccharides, proteins, and some phospholipids (Read et al. 2003). Epidermal cells and root cap along with sloughed-off root surface cells are major producer of mucilage. While growing, roots apply enormous amount of pressure to push the root tip through the soil. Growth of root tip through sturdy soil with such a high pressure causes sloughing of root cap. The cells that are sloughed off may remain functional for several days and secrete mucilage to attract beneficial microorganisms (Jones et al. 2009). The main purpose of mucilage is to reduce the friction between root tip and adjacent soil, thus reducing the wear and tear of root. Root mucilage also serves to protect root from desiccation and assists in nutrient acquisition (Jones et al. 2009). Under certain conditions, root mucilage is known to protect young root meristem from heavy metal toxicity (Mendes et al. 2013). Mucilage secreted by root helps in holding the soil particles together (Fig. 24.2). Mucilage secretion also helps in improving the soil quality by enhancing aeration and water infiltration (Read et al. 2003).

Root border cells are metabolically active cell programmed to separate from main plant root into the rhizosphere (Hawes et al. 2000). The amount of border cells produced and released in rhizosphere depends on plant type and environmental condition. In case of water stress, border cells generally remain adhered to the root tips (Hawes et al. 2000). According to Bais et al. (2006), roots exudes are also released from root border cells and root border–like cells. Later, they gradually separate from surface of root as they grow and deposits their content into the rhizosphere (Bais et al. 2006).

Through their metabolic activity, plant roots and microorganisms release carbon dioxide in rhizosphere. Due to release and accumulation in rhizosphere, the concentration of carbon dioxide can go high up to 17.5% in the root zone (Dakora and Phillips 2002). High concentration of carbon dioxide in root zone increases calcium carbonate dissolution, to release Ca^{2+} ions (Golding et al. 2012). Plants can easily uptake Ca^{2+} ions through roots. Hydrogen gas is another gas metabolite released as a major byproduct in rhizosphere during N_2 fixation in leguminous plants (Dong and

Fig. 24.2 Root system of Congress grass (*Parthenium hysterophorus*) with adhering rhizosphere soil. Root exudes, particularly mucilage, help in holding soil particles together and adhered to the root

Layzell 2001). Hydrogen-oxidizing bacterial community present in rhizosphere support root elongation by decreasing ethylene level in host plant (Maimaiti et al. 2007).

24.4 Composition of Root Exudes

Identifying the chemical composition and concentration of root exudes in soil is very difficult due to technical constrains. Several researchers have extrapolated the finding of sterile hydroponic study to natural conditions so as to have more insight into the natural role of root exudes (Feng et al. 2014; Kawasaki et al. 2016; Kawasaki et al. 2018). Root exudes are a cocktail of organic chemicals like sugars, amino acids, organic acids, fatty acids, sterols, growth factors, phenolics compounds, and enzymes (Table 24.1).

The organic compounds present in root exudes can be grouped into high and low molecular weight compounds (Badri and Vivanco 2009). The high molecular weight compounds are made up of complex molecules like cellulose and constitute the major form of carbon released from root. Usually, the soil microorganisms are not so efficient in utilizing high molecular weight compounds as carbon and energy source (McNear 2013). The low molecular weight compounds are more diverse fraction of root exudes. Owing to molecular diversity, these molecules perform or influence wide array of functions in rhizosphere (Baetz and Martinoia 2014). The list of low molecular weight compounds present in root exudes is lengthy; however in general sense it can be characterized into water-soluble sugars, organic acids, amino acids, hormones, vitamins, phenolics compounds, and other secondary metabolites (Uren

Table 24.1 Some organic compounds released by plant roots in rhizosphere

Organic compounds	Examples
Sugars	Arabinose, deoxyribose, fructose, galactose, glucose, maltose, mannose, mucilages of various compositions, oligosaccharides, raffinose, rhamnose, ribose, sucrose, xylose
Amino acids	α-alanine, β-alanine, α-amino adipic, γ -amino butyric, arginine, asparagine, aspartic, citrulline, cystathionine, cysteine, cystine, deoxymugineic, 3-epihydroxymugineic, glutamine, glutamic, glycine, homoserine, histidine, isoleucine, leucine, lysine, methionine, mugineic, ornithine, phenylalanine, proline, serine, threonine, tryptophan, tyrosine, valine
Organic acids	Acetic, aconitic, aldonic, ascorbic, benzoic, butyric, caffeic, citric, *p*-coumaric, erythonic, ferulic, formic, fumaric, glutaric, glycolic, glyoxylic, lactic, malic, malonic, oxaloacetic, oxalic, *p*-hydroxy benzoic, piscidic, propionic, pyruvic, succinic, syringic, tartaric, tetronic, valeric, vanillic
Fatty acids	Linoleic, linolenic, oleic, palmitic, stearic
Sterols	Campesterol, cholesterol, sitosterol, stigmasterol
Growth factors	*p*-amino benzoic acid, biotin, choline, *n*-methyl nicotinic acid, niacin, pantothenic, vitamin B_1 (thiamine), B_2 (riboflavin), and B_6 (pyridoxine)
Phenolic compounds	Phenolic acid, hydroxycinnamic acid, flavonoids, isoflavonoids, neoflavonoids
Enzymes	Amylase, invertase, peroxidase, phenolase, phosphatases, polygalacturonase, protease
Nucleotides	Adenine, guanine, uridine or cytidine
Miscellaneous	Auxins, *p*-benzoquinone, scopoletin, hydrocyanic acid, glucosides, hydroxamic acids, luteolin, unidentified ninhydrin-positive compounds, unidentified soluble proteins, reducing compounds, ethanol, glycinebetaine, inositol and myo-inositol-like compounds, dihydroquinone

Modified from Uren (2007), Lattanzio (2013)

2001). Most of the low molecular weight compounds present in root exude are easily utilized by microorganisms (McNear 2013). The natural role of low molecular weight compounds present in root exudes is not fully deciphered (Ahkami et al. 2017), though available literature confers that low molecular weight compounds of root exudes are most important in constituting the microbial community in rhizosphere (Hinsinger et al. 2005; Ahkami et al. 2017). The low molecular weight compounds of root exudes have been reported as sources of nutrients for microorganisms, agents of invasion, and mediators signaling molecules to attract symbiotic partners or encourage root surface colonization by plant growth–promoting microorganisms (Berg and Smalla 2009; Lynch et al. 2012; Haichar et al. 2014).

The qualitative and quantitative compositions of root exudates are influenced by several biotic and abiotic components (Hinsinger et al. 2005). Several environmental factors like soil pH, soil type, oxygen concentration, light intensity, soil temperature, and nutrient accessibility for microorganisms are known to influence composition of root exudates (Hinsinger et al. 2005). Soil physical and chemical properties as well as microbial diversity and density have more pronounced impact on qualitative and quantitative compositions of root exudates (Haichar et al. 2014). The composition of root exudes varies with plant species as well as with the age of plant. Generally, root exudes of a young plant has high carbon percentage. The carbon-rich exudes attract large number of microorganisms. However, with maturation the percentage of carbon in root exudes decreases and the percentage of nitrogen-containing compound increases (Lareen et al. 2016).

Availability of nutrients also has profound effect on root exudes. Nutrient elements are required for normal growth of plant. Plants respond to nutrient deficiency by altering root exudes, which in turn change the microenvironment of the rhizosphere (Badri and Vivanco 2009; Lareen et al. 2016). Root exudates facilitate plants in acquiring nutrients by acidifying or changing the redox conditions of nutrients present in the rhizosphere (Badri and Vivanco 2009). For acquiring certain elemental nutrients, plant releases chelating agent in root exudates to directly capture and acquire. Exudates aid in the release of nutrients either by dissolution of insoluble mineral phases or desorption nutrients from soil particles or organic matter (Badri and Vivanco 2009). After the release, nutrients enter the soil solution and then are taken up by the plant.

Nitrogen and phosphorus are the most important plants nutrients. Plants alter root exudes to ensure the availability of nitrogen and phosphorus. They can take nitrogen in form of NO_3^- and NH_4^+ ions (McNear 2013), and alter root exudes depending upon the form in which N is available in the rhizosphere (Liu et al. 2016). For acquiring NH_4^+ ion, plants release H^+ in the rhizosphere resulting in decrease in the soil pH (McNear 2013). The NO_3^- ions are acquired from rhizosphere by release of bicarbonate ions in the root exudes. Accumulation of bicarbonate ions in the soil increase pH of soil. Green leaves and decomposing litter can influence rhizosphere nitrogen through phenolics such as chlorogenic acid (McNear 2013).

Low availability of P is known to stimulate releases of organic acid anions in root exudes. Plants take phosphorus in form of PO_4^{3-} ions. For overcoming deficiency of phosphorus, plants have evolved mechanisms that depend upon the plant type, species, and genotype (McNear 2013). Plants release organic acid into the rhizosphere for decreasing the pH of soil, which in turn facilitates the solubilization of soil-bounded phosphorus (Wang et al. 2015; Vengavasi and Pandey 2018). Some plant species may release piscidic acid in root exudes in response to phosphorus deficiency (Dakora and Phillips 2002). Piscidic acid chelates to $FePO_4$ and releases PO_4^{3-} for making the ions available for plants.

Iron deficiency can also influence composition of root exudes (Lareen et al. 2016). Dicotyledonous and non-graminaceous plants respond to iron deficiency by releasing protons into the microenvironment of rhizosphere and increase the reducing ability of the epidermal cells of root (Zheng 2010; Tripathi et al. 2018). They may also secrete organic acids or phenolic compounds into rhizosphere for acquiring iron from soil. Change in soil pH not only influences the availability of other micronutrients but also impacts the microbial community structure of rhizosphere (Lareen et al. 2016). Graminaceous plants overcome iron deficiency by increasing the secretion of phytosiderophores in rhizosphere (Tripathi et al. 2018). Siderophores chelate strongly with iron and is then brought back into cells of roots through iron-chelated shuttle transporters present in the plasma membrane.

Microorganisms are also known to produce siderophores. Roots of plant attract such microbes through root exudes to enhance the iron availability (Carvalhais et al. 2013). Plants improve phosphate supply by releasing strigolactones in rhizosphere, which attract mycorrhiza in rhizosphere (Akiyama et al. 2002, 2005). Strigolactones are known to stimulate spore germination and cell proliferation of arbuscular mycorrhizal fungi like *Gigaspora rosea*, *Glomus intraradices* and *Gl. claroideum* (Besserer et al. 2006). To enhance nitrogen supply, plant roots release flavonoids to attract nitrogen-fixing symbionts in rhizosphere (Hassan and Mathesius 2011). Flavonoids are important signaling molecules in developing symbiosis between legumes and nitrogen-fixing microorganisms (Cooper 2004).

24.5 Phenolic Compounds Present in Root Exudes

Phenolic acids are low molecular mass compounds, consisting of a hydroxyl group bonded directly to an aromatic hydrocarbon(s). Plants produce diverse class of phenolic compounds as secondary metabolites to fulfill a very broad range of physiological activities (Lattanzio 2013). Aforementioned description of plant phenols does not encompass all known plant phenolic compounds. Quideau et al. (2011) proposed that plant phenolics can be best described as secondary natural metabolites produced by plants either through shikimate–phenylpropanoid pathway or the polyketide acetate–/malonate pathway for performing diverse physiological activities. Phenolics compounds produced by plants can be categorized on the basis of their basic skeleton. As evident from Fig. 24.3, the simplest phenolic compound produced by plant is phenol and most complex phenolic compound is condensed tannin.

Phenolics compounds are ubiquitous secondary metabolites of plants. Generally it is believed that about 2% of all carbon photosynthesized by plants is converted into flavonoids (phenolic compound) or closely related compounds (Robards and Antolovich 1997). Phenolic compounds account for nearly 40% of all circulating organic carbons present in biosphere (Lattanzio 2013). Naturally phenolic compounds are derived from either the shikimic acid pathway or the malonate acetate pathway or both. Majority of phenolic compounds in plants, fungi, and bacteria are synthesized by shikimic acid pathway. The malonic acid pathway forms significant amount of phenolic acids for fungi and bacteria as compared to higher plants. Phenolic acids are produced in plants via shikimic acid through the phenylpropanoid pathway, as by-products of the monolignol pathway and as breakdown products of lignin and cell wall polymers in vascular plant (Chapman and Regan 1980; Croteau et al. 2000). Some of the phenolics present in rhizosphere may have microbial origin (Lareen et al. 2016).

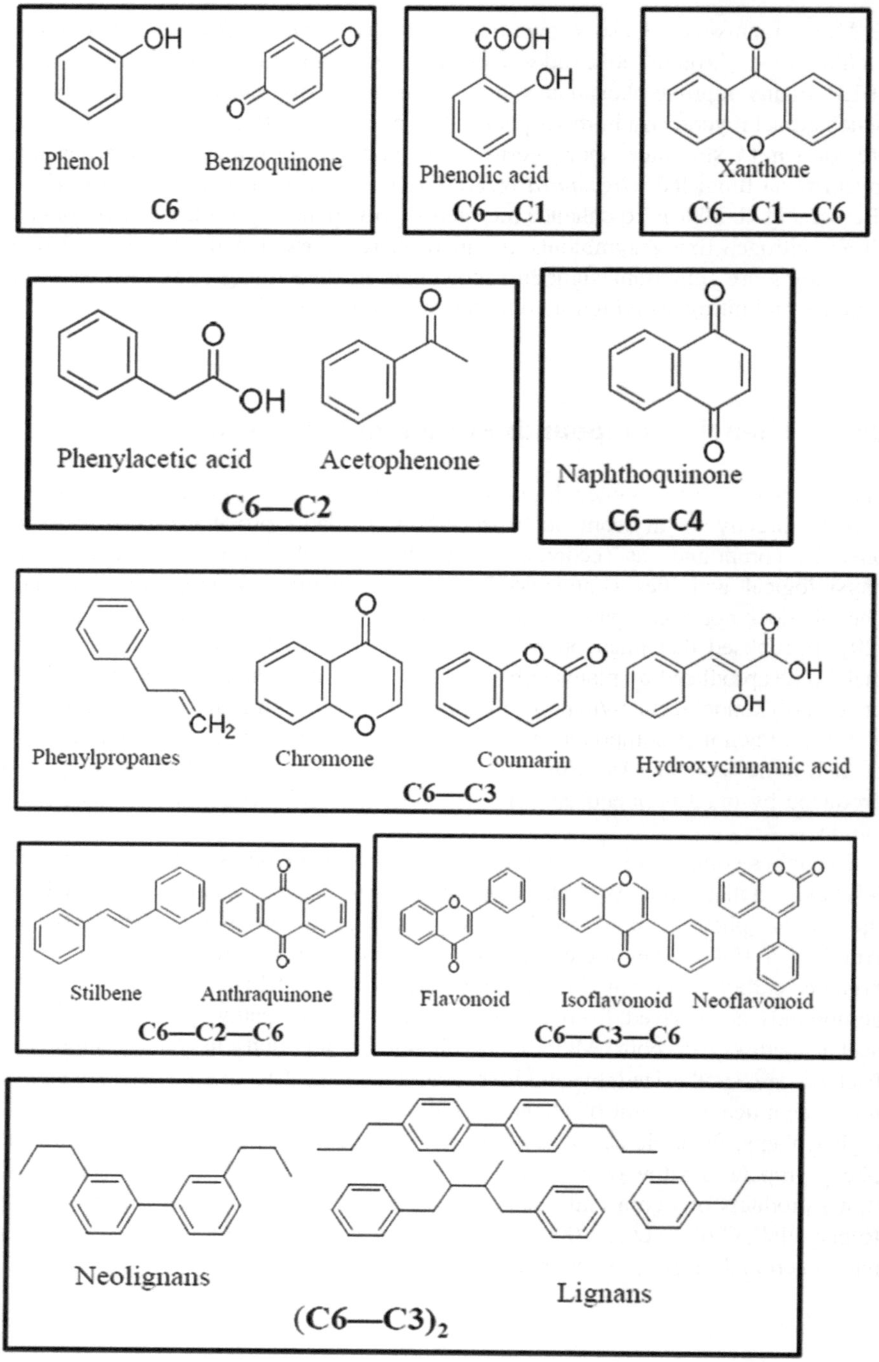

Fig. 24.3 Classification of plant phenolic compounds on the basis of basic skeleton structure (Lattanzio 2013)

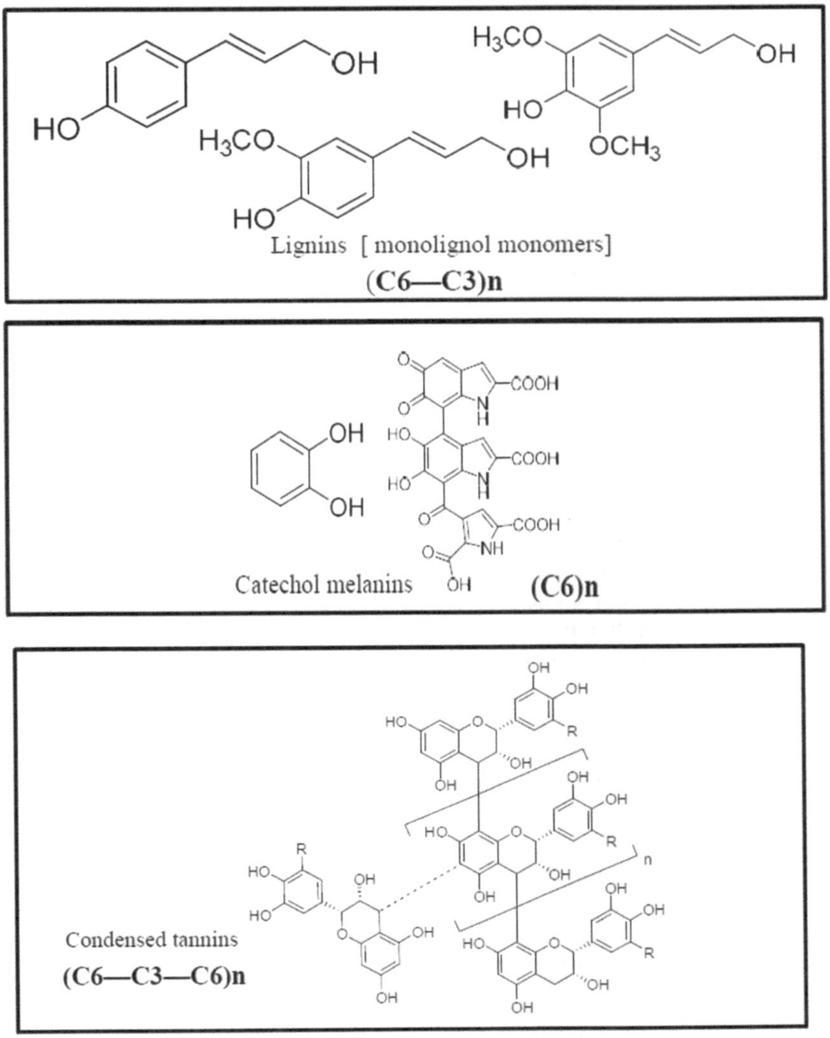

Fig. 24.3 (continued)

24.6 Ecological Perspective of Phenolic Compounds Present in Rhizosphere

The carbon-rich zone present around root supports the assemblage and growth of diverse microbial flora and fauna. Interaction between flora and fauna within rhizosphere is largely influenced by the plant root exudes (Fig. 24.4). Phenolic compounds released in rhizosphere bestow metabolic plasticity essential for

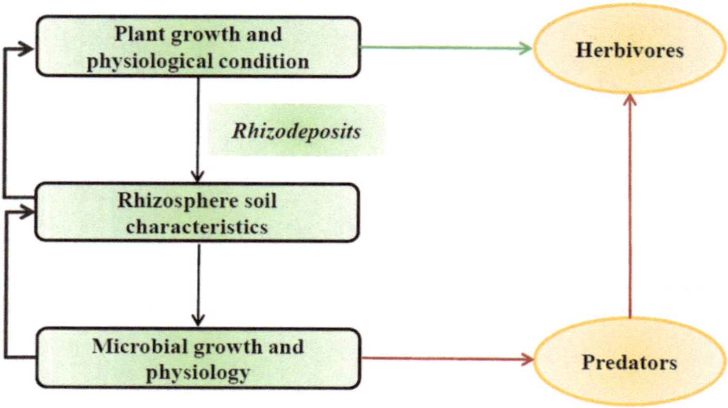

Fig. 24.4 Simple model representing the biotic interactions that influence the rhizosphere

perceiving and responding to biotic and abiotic stress (Baetz and Martinoia 2014). They are known to play multifunctional roles in rhizosphere particularly in the plant–microbe interactions. Plant phenolic compounds released by roots can execute both positive and negative interactions in the rhizosphere (Bais et al. 2006; Philippot et al. 2013). The positive interactions mainly revolve around assemblage of beneficial microbes in the vicinity of root. Phenolic compound mediated negative interactions, include interaction of host plant with parasitic plants, pathogenic microbes, and invertebrate herbivores. Figure 24.5 shows some phenolic compounds well known to act as signaling molecules in rhizosphere. Phenolics compounds (like chrysoeriol and luteolin) act as positive signaling molecules by inducing transcription of nodulation (*nod*) gene in *Rhizobium meliloti* (Hartwig et al. 1990). Juglone (5-hydroxy-1,4-naphthalenedione) is phenolic compound well known to have negative impact on growth of other plants. Aliskan and Terzi (2001) have demonstrated allelopathic effects of juglone on seed germination and seedling growth of tomato, cucumber, garden cress, and alfalfa. Phenolic compounds from plant parts like root and seed exudates, leaf leachates, and decaying plant matters play crucial roles in determining the physical, chemical, and biological features of soil (Philippot et al. 2013). The process of soil formation is also influenced by phenolic compounds (Inderjit and Mallik 1997).

24.7 Mutualism

Plants develop beneficial relations with microorganisms to acquire extra armor for luxurious growth. Some bacteria (like rhizobia) and fungi (like vesicular–arbuscular mycorrhiza) have the ability to develop mutual symbiotic relationship with roots of plants. There are also some rhizosphere-dwelling microorganisms that benefit plants without developing direct contact with the plants. Within rhizosphere there can be

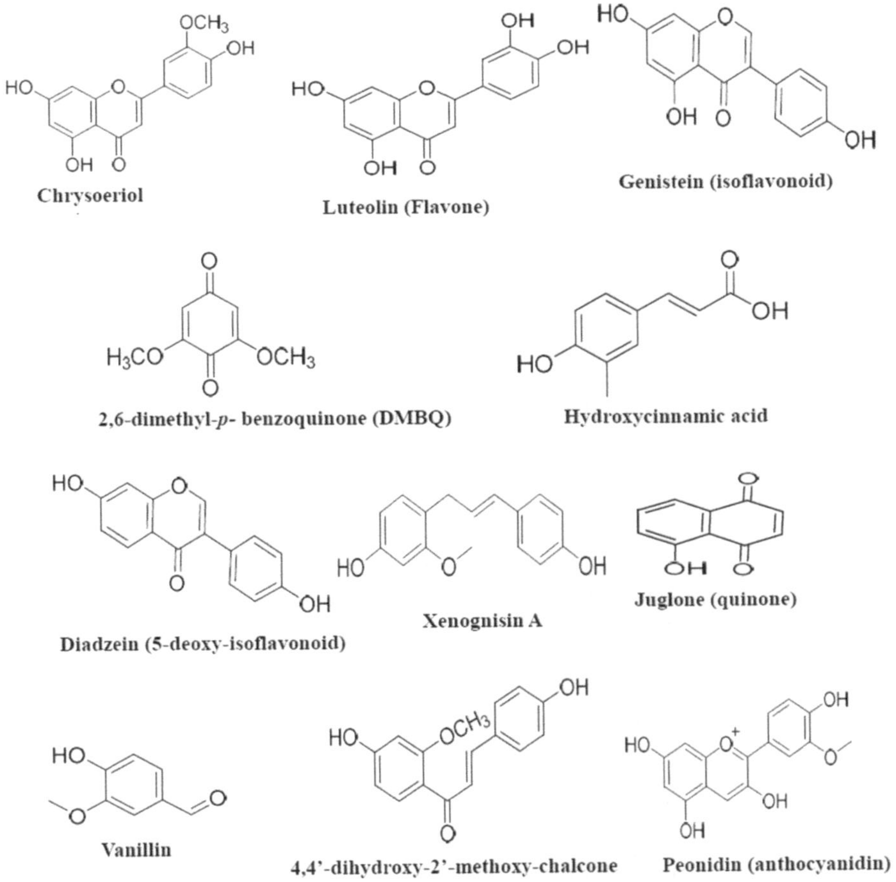

Fig. 24.5 Some phenolic compounds that act as signaling molecule between plant roots and other organisms present in the rhizosphere

three categories of microorganism. First category of microorganisms benefit plant by enhancing the nutrient supply to plants. There may or may not be a direct contact between microorganisms and plants. These microorganisms alter biotic and abiotic characteristics of rhizospheric soil by their activities, which further enhances nutrient availability for plants. Second category of microorganisms prevents growth or activity of pathogenic organisms in the rhizosphere. They act as biocontrol agents in rhizosphere and support plant growth. Such microorganisms can also induce systematic resistance in plants and suppress disease occurrence or progression. Third category of microorganisms directly helps in plant growth by producing growth hormones in the vicinity of roots or within the plant. Rhizosphere also harbors microorganisms that are neutral in interaction with plants. Such microorganisms neither harm nor directly benefit plants. Saprophytes present in rhizosphere are important in mobilization of nutrient in the soil. The absence of saprophytes in the soil are known to retard plant growth and crop productivity (Ahkami et al. 2017).

In recent past the importance of plant growth–promoting bacteria residing in the rhizosphere has been widely studied. These bacteria belong to diverse taxa. Some of the commonly encountered plant growth–promoting rhizosphere bacteria are *Azospirillum*, *Bacillus*, *Pseudomonas*, *Rhizobium*, *Serratia*, *Stenotrophomonas*, and *Streptomyces* (Haldar and Sengupta 2015). Usually, rhizobacteria produce secondary metabolites that mimic plant hormones such as auxin and gibberellins, thus directly aiding in host plant growth (McNear 2013). These exogenously produced plant hormones can have significant influence on root growth, dynamics, and evolution of rhizosphere as a microenvironment. Some fungi belonging to the genera *Ampelomyces*, *Coniothyrium*, and *Trichoderma* have also been reported to benefit host plant by nutrient mobilization and protection from pathogens (Lareen et al. 2016).

Association between *Rhizobia* (nitrogen-fixing bacteria) and leguminous plants is one of the most well-known mutual symbiotic associations between plants and bacteria. Symbiotic bacteria reside in root nodules and fix atmospheric nitrogen for plants. In return plants provide the bacteria with nutrients. Nonleguminous plants can develop symbiotic association with nitrogen-fixing actinomycetes, and such plants are called "actinorhizal" plants. Symbiotic association is the outcome of a series of molecular dialogue between bacteria and plants. The first event in developing symbiotic association is attracting bacteria toward rhizosphere by releasing attractant in the rhizosphere. Once the bacterium is in rhizosphere zone, the molecular signal activates genes responsible for nodule formation. Phenolic compounds, particularly flavonoids, are important molecules of signaling system. Flavonoids act as chemo-attractants for nitrogen-fixing bacteria and provide host specificity (Lareen et al. 2016). Flavonoids present in root exudes are known to regulate genes involve in root nodule formation. Most of the flavonoids are active as nod gene inducers at nanomolar to low micromolar concentrations. The first flavonoids reported to induce nod gene were luteolin and 7,4′ dihydroxyflavone (DHF) isolated from *M. sativa* and *Trifolium repens* (white clover) respectively (Peters et al. 1986; Redmond et al. 1986). Cooper (2004) reported isolation of 30 nod gene–inducing flavonoids from nine leguminous plants belonging to different genera.

Soybean (*Glycine max*) produces two isoflavonoids, namely daidzein and genistein, for inducing nod gene of *Bradyrhizobium japonicum*. However, daidzein and genistein inhibit nod gene expression of *Sinorhizobium meliloti*. Nod gene of *S. meliloti* can be induced by luteolin released by *Medicago sativa*. Flavonoids can also elevate intracellular calcium level in rhizobia and consequently stimulate NodD proteins for Nod factor expression (Moscatiello et al. 2010).

Roots of leguminous plants continuously release flavonoids into the surrounding soil. Concentration of flavonoids increases significantly if the compatible rhizobium species are present in rhizosphere. Flavonoids induce expression of nod gene in rhizobia. Product of nod gene activates synthesis of Nod factor, which in turn triggers a sequence of events required for nodule formation. Initially, the root hair curls around the specific rhizobia, followed by entry of rhizobia into root via infection thread and development of nodules within the root. It has been suggested that a mixture of flavonoids is more effective in inducing nod genes as opposed to a single compound (Bolanos-Vasquez and Warner 1997; Begum et al. 2001). The specific exudation of flavonoid (mixtures) from legume hosts together with the specific perception of flavonoids by NodD proteins of different rhizobia is partially responsible for the host specificity of the symbiosis. Apart from flavonoids, simple phenolic compounds like vanillin and isovanillin produced by roots of wheat have been reported to induce nod genes in *Rhizobium* sp. NGR234 (Kobayashi et al. 2004).

Specificity of association between rhizobium and leguminous plants has also been assigned to the phenolic compounds produced by host plant. Nodulation has been reported to be influenced by host-derived phenolics namely cinnamic, benzoic, and hydroxybenzoic acids. In Alnus seed, flavonoid-like compounds (flavanone and isoflavanone) have been reported to induce nodule formation (Benoit and Berry 1997).

Frankia, an actinobacterium, can develop symbiotic association with eight different dicotyledonous families. Flavonoids produced by plants mediate host recognition and association. The specificity in Myricaceae–Frankia symbiosis has been reported to be linked to plant root phenolics (Popovici et al. 2010). Phenolic compounds, namely phenols, flavonoids, and hydroxycinnamic acids, were found to be the major compounds differentially affected by Frankia inoculation in Myricaceae plants (Popovici et al. 2010).

Mycorrhiza is a symbiotic association between fungus and plants. Generally, phosphorus deficiency in the soil stimulates mycorrhizal symbiosis. The associated fungus may colonize internal tissues of host plant roots and exist as arbuscular mycorrhizal fungi. Some fungi may also exist extracellular on root surface as ectomycorrhizal fungi. The successful establishment of symbiotic association between fungus and host plants involve four principal steps, namely germination of spore, growth of hyphae, recognition of host, and finally formation of appressoria. Attraction of fungi by virtue of host plant roots exudes is an important event for symbiotic association between fungi and host plant. Once the fungi are in the vicinity of roots, exudes elicit hyphal growth and branching. There are evidences showing that root exudes from host plant can stimulate better growth of fungi as compared to nonhost exudates. Plant phenolics play a key role as signaling molecules during mycorrhizal formation. Phenolic compounds present in root exudes have been shown to support fungal spore germination, enhance hyphal growth, hyphal branching, and secondary spore formation. The coumestrol has been reported to initiate hyphal growth (Morandi et al. 1984). Morandi et al. (2009) have reported that coumestrol-hyperaccumulating mutant of *M. truncatula* can exhibit hyperinfection through its mycorrhizal symbionts.

Plant phenolic compound profiles of root exudes also change upon the establishment of successful association. Flavonoids gradually accumulate before the start of infection and after initiation concentration of phenolics compound vary with different stages of infection. The expression of genes involved in phenylpropanoid, flavonoid, and isoflavonoid metabolism changes upon arbuscular microbial association. Scervino et al. (2005) demonstrated that *Trifolium repens* infection by *Glomus intraradices* changes the composition of flavonoids present in roots. Generally, infection by arbuscular mycorrhizal fungus increases phenolic compound content in root exudes. The entry of mycorrhizal fungus into root is liable to induce activation of flavonoid phytoalexins. However, the intensity of induction of phytoalexins is relatively low in mycorrhizal fungus infection as compared to the infection by pathogens. Some flavonoids like pyranoisoflavones produced by white lupin have been reported to inhibit hyphal branching in non-mycorrhizal fungi, thus accounting for the host specificity during symbiotic association between fungus and plant.

Giovannetti (2000) proposed that the perception of the right molecular signals from the host roots promotes differential morphogenesis of arbuscular mycorrhizal hyphae in the rhizosphere. However, fungal penetration into the root cells depends on the host genome. Therefore, the initial stages of arbuscular mycorrhizal establishment is regulated by phenolic compounds, and root penetration as well as arbuscular mycorrhizal development are likely to be regulated by the interactions of host plant and the fungal partner. Phenolic compounds like *p*-coumaric acid, *p*-hydroxybenzoic acid, and quercetin have been reported to stimulate growth and colonization of clover and sorghum root by the arbuscular mycorrhizal fungus.

Root colonization by rhizosphere-dwelling beneficial microorganisms can stimulate and induce systematic resistance in plants. Induced systematic resistance (ISR) is the state of plant that protects nonexposed plant parts against the future attack by microbial pathogens (Corne et al. 2014). Usually plants develop ISR upon infection by pathogens, root colonization by specific beneficial microorganisms, or after treatment with specific chemicals. Induced resistance activates latent defense mechanism of plants. Induced state is not only expressed at site of attack or exposure but also systematically in plant parts separated from site of induction. ISR provides broad spectrum protection.

ISR-inducing beneficial microorganisms must colonize roots of host plants. For establishing prolong mutualistic interaction with host, microorganism must evade plant immunity response. Mycorrhizal fungus *Rhizophagus intraradices* employs symbiosis effectors MiSSP7 to suppress effector-triggered mediated immunity in host plant before inducing systematic resistance. Similarly, ectomycorrhizal fungus *Laccaria bicolor* suppresses salicylic acid–dependent defenses for establishing itself on host plant. The plant growth–promoting fungi like *Piriformospora indica* initiates jasmonic acid–mediated signaling pathway in Arabidopsis roots to suppress both early and late defense responses (Corne et al. 2014).

24.8 Predation

Predation is a biological interaction between organisms, where an organism known as predator kills and eats another organism known as prey. Seed herbivores can also be included in predation. There are soil fauna that feed on pathogenic microbes and act as biological control system for plants (Ingham et al. 1985). Protists and protozoans present in the soil feed on rhizosphere-dwelling fungi and bacteria, thus influencing the assemblage of microbiome near root. Consumption of bacteria or fungi by protists reduces microbial population of rhizosphere, thus increasing the nutrient availability for leftover microorganisms (Ingham et al. 1985). Protists have higher C:N ratio than fungi or microorganisms, and to maintain this protists release excess of nitrogen in the surroundings, thus making it available to host plant and other microorganisms (Ingham et al. 1985). Predation of bacteria by nematodes has exhibited to influence nitrogen availability and plant growth. Soil fauna also help in releasing locked-up nutrient in dormant cell by consuming them (Ingham et al. 1985).

Excessive predation of beneficial microorganisms by soil fauna in rhizosphere may be harmful to plant (Sparling et al. 1981). To avoid such conditions, plants release nematicides and phytoalexins (Elhady et al. 2018). Root leachates of several plants have been reported to exhibit nematicidal activity (Elhady et al. 2018). Active components of these root leachates were found to be phenolic compounds like *p*-hydroxybenzoic acid, vanillic acid, caffeic acid, and ferulic acid (Iannucci et al. 2013). Apart from nematicidal activity, phenolic compounds can also act as repellents or motility inhibitors (Iannucci et al. 2013).

24.9 Competition

Competition is an interaction between organisms or species, in which both interacting partners are harmed. Rhizosphere-inhabiting organisms compete for food, water, and space. Competition is one of the important factors that determine the structure of rhizospheric community. Allelopathy and autotoxicity are two of the important forms of competition prevalent in rhizosphere. Phenolic compounds possessing allelopathic properties inhibit growth of adjacent plants (Fig. 24.6).

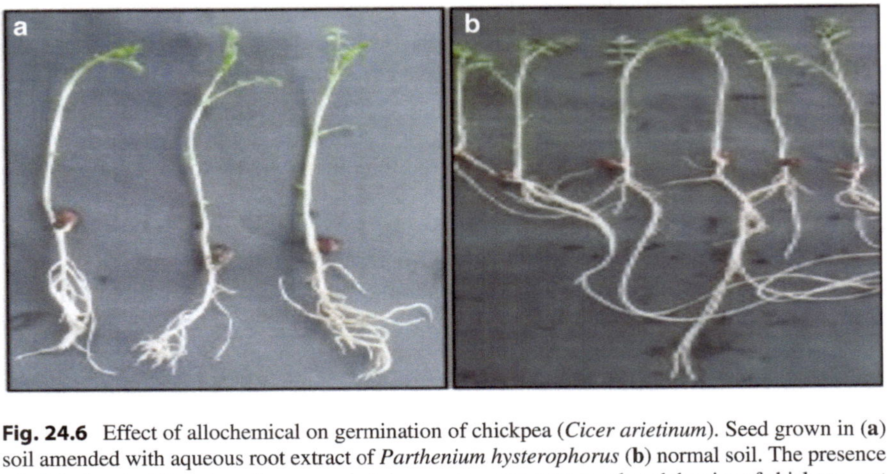

Fig. 24.6 Effect of allochemical on germination of chickpea (*Cicer arietinum*). Seed grown in (**a**) soil amended with aqueous root extract of *Parthenium hysterophorus* (**b**) normal soil. The presence of allelochemical like phenolic compounds in aqueous root extract reduced the size of chickpea root as compared to the control

Phenolic compounds can be produced actively by plant cells in rhizosphere or decaying plant parts can also added phenolic compounds to microenvironment of rhizosphere (Bertin et al. 2003). The presence of allelopathic phenolic compounds like benzoic acid, cinnamic acid, coumarins, flavonoids, isoflavinoids, and tannins are reported to inhibit seed growth as well as harm normal growth of plant post germination (Bertin et al. 2003). Forage legume *Desmodium uncinatum* inhibit post-germination and attachment of *Striga* by producing several (iso)flavonoids in its root exudates (Hooper et al. 2010; Khan et al. 2010). The success of several invasive weeds has been attributed to flavonoids with allelochemicals. Spotted knapweed (invasive weed of North America), produces catechin in its root exudes (Bais et al. 2002). Catechins induce generation of reactive oxygen species in susceptible plants, leading locals to cell death and subsequently the death of entire the root system (Bais et al. 2003).

Phenolic compounds present in root exudes have been reported to act as phytoalexins and nematicides (Hassan and Mathesius 2011). Phytoalexins exhibit both antifungal and antibacterial activities. They are generally synthesized in response to attack by pathogens. Nematicides have lethal activity against nematodes. Under certain conditions, phytoalexins and nematicides can be stored in plant root in an inactive form. Upon attack, they release as broad-spectrum phytoanticipins to escalate a quick defense against pathogen. Phenolic compounds such as cajanin, medicarpin, glyceollin, rotenone, coumestrol, phaseolin, and phaseolinin are known to inhibit growth and proliferation of soilborne pathogens (Hassan and Mathesius 2011). Isoflavonoids such as medicarpin, pisatin, and maackiain have been reported to exhibit antimicrobial activity against soil-dwelling microorganisms (Nihorimbere et al. 2010). Roots of alfalfa and pea produce medicarpin to protect plants from the pathogenic soil fungus *Rhizoctonia solani* (Kapulnik et al. 2001). Medicarpin prevents fungal infection by inhibiting fungal germ tube and mycelial growth (Blount et al. 1992). Quercetin, another well-known plant phenolic compound executes antimicrobial activity by binding to GyrB subunit of *E. coli* DNA gyrase and inhibits the ATPase activity. Quercetin has also been reported to inhibit the growth of the fungus *Neurospora crassa* (Parvez et al. 2004).

Some plants release pterocarpans around the roots when they come in contact with pathogenic microorganisms. Armero et al. (2001) reported that exposing chickpea seedlings to elevated level of glutathione increases pterocarpan biosynthesis. These synthesized pterocarpans are released by the roots into the surroundings, to prevent the damage caused by the pathogens.

Extracts from aboveground plant parts also demonstrated insecticidal activity by the virtue of phenolic compound like phenols, tannins, and several flavonoids. However, there is a lack of direct evidence exhibiting phenolic compounds from underground plant parts demonstrating insecticidal activity. Singh et al. (2014) demonstrated that tomato root hair extracts have phenolic compounds such as rutin, quercetin, kaempferol, gallic acid, protocatechuic acid, ferulic acid, colorogenic acid, and caffeic acid. These compounds were able to inhibit growth of larvae of *Helicoverpa armigera and Spodoptera litura.*

24.10 Effect of Phenolic Compounds on Rhizosphere Food Web

The rhizosphere is a nutrient-rich region as compared to bulk soil. It supports a dense and diverse population of primary consumers depended upon root, which acts as primary carbon source (Moore et al. 2003). The length of trophic chain depends upon the amount of carbon release into the rhizosphere. Theoretically, the more the amount of carbon in root exudes, the longer would be the tropic length. Form of carbon can also have significant influence on the food web. In general the rhizosphere food web is believed to support three to eight trophic nodes (Fig. 24.7).

The primary consumers of plant root exudes are microorganisms (bacteria and fungi). The composition of root exudes directly determine the microbial community. Microorganisms exhibit specificity in their ability to utilize various constituents of root exudes. Root exudes from two different varieties of same plant can select different genotypes of same bacterial species (Walker et al. 2003). Larger secondary metabolites such as polysaccharides and polypeptides are preferred carbon source for microbial community as compared to phenolic compounds. The presence of phenolic rings makes phenolic compounds less susceptible to general microbial degradation. Phenolic compounds act as microbial attractant or replant and thus influence the microbial community structure. Usually root exudes repel pathogenic microorganisms. However, under certain conditions pathogenic microorganisms can

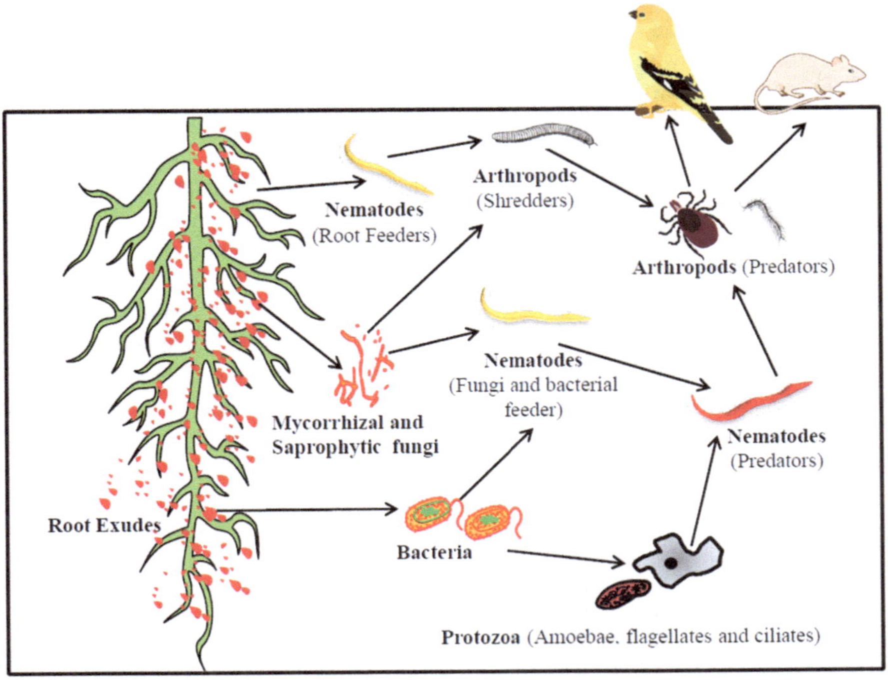

Fig. 24.7 Simple food web involving rhizosphere, soil and above ground dwelling organisms

infect root and cause the death of entire root system. Dead roots support detritus nutrient pool that is utilized by saprophytic microorganisms (Moore et al. 2003).

Invertebrates like insects, arthropods, and nematodes can also feed on live, dead, or decaying plant material. Nematodes feeding on root can induce its death. Arthropods known as shredders feed on dead or decaying vegetation or rhizosphere microorganisms. Protozoans like amoeba feed on soil bacteria and are further consumed by nematodes. Nematodes can feed on bacteria and fungi present in rhizosphere. Predatory nematodes can feed on nonpredatory nematodes. Predatory nematodes may also feed on soil-dwelling protozoan. Predatory arthropods feed on nematodes. Arthropods are further consumed by aboveground organisms like mice and birds, extending the food chain.

24.11 Conclusion and Future Perspectives

Undoubtedly, phenolic compounds are most extensively studied plant secondary metabolites. During last few decades several studies have demonstrated the multiple roles of plant phenols. Phenolic compound profiles of a plant are not only determined by genetic composition but also influenced by the environmental factors. It appears that phenolic content of plant is shaped by the interaction between plants and their surrounding environment in such a way that plants can have luxurious growth in the available external conditions. Most of the studies have concentrated on aboveground parts with very little information about the role of phenolic compounds in roots and their surrounding. Available literature suggest that phenolic compounds released in rhizosphere play important role in plant–microbe interaction, and thus they shape the microenvironment lying in close vicinity of roots. Interestingly, most of the research has focused on the regulation by flavonoid signaling while ignoring other phenolic compounds, although they are the major component of the rhizodeposits. Phenolic compounds released by roots influence the food web of soil by controlling the rhizosphere microbial community.

The idea of engineering rhizospheres for promoting plant health and growth necessitates the elaborate study of phenolic compounds in rhizosphere. Phenolic compounds that attract beneficial microorganisms and repel harmful organisms may help in designing the rhizosphere in such a way that can bestow good growth and better yield. The use of metabolomics approach for identifying root exudes' molecules at different stages of plant development and under different environmental conditions may open up new avenues in rhizosphere engineering.

Acknowledgments The authors thank the Chancellor and Vice-Chancellor of Sant Baba Bhag Singh University, Jalandhar, for their continuous support and encouragement. This work is based upon the research supported by Apollo BioLife Fund (Project Number: TBB/4/2017).

References

Ahkami AH, Allen White Iii R, Handakumbura PP, Jansson C (2017) Rhizosphere engineering: enhancing sustainable plant ecosystem productivity. Rhizosphere 3(2):233–243

Akiyama K, Matsuoka H, Hayashi H (2002) Isolation and identification of a phosphate deficiency-induced C-glycosyl flavonoid that stimulates Arbuscular mycorrhiza formation in melon roots. Mol Plant-Microbe Interact 15:334–340

Akiyama K, Matsuzaki K, Hayashi H (2005) Plant sequiterpines induce hyphal branching in arbuscular mycorrhizal fungi. Nature 435:824–827

Aliskan IK, Terzi I (2001) Allelopathic effects of walnut leaf extracts and juglone on seed germination and seedling growth. J Hortic Sci Biotechnol 76(4):436–440

Armero J, Requejo R, Jorrin J, Lopez-Valbuena R, Tena M (2001) Release of phytoalexins and related isoflavonoids from intact chickpea seedlings elicited with reduced glutathione at root level. Plant Physiol Biochem 39:785–795

Badri DV, Vivanco JM (2009) Regulation and function of root exudates. Plant Cell Environ 32:666–681

Baetz U, Martinoia E (2014) Root exudates: the hidden part of plant defense. Trends Plant Sci 19 (2):90–98

Bais HP, Walker TS, Stermitz FR, Hufbauer RA, Vivanco JM (2002) Enantiomeric dependent phytotoxic and antimicrobial activity of catechin; a rhizosecreted racemic mixture from *Centaurea maculosa* (spotted knapweed). Plant Physiol 128:1173–1179

Bais HP, Vepachedu R, Gilroy S, Callaway RM, Vivanco JM (2003) Allelopathy and exotic plant invasion: from molecules and genes to species interactions. Science 301:1377–1380

Bais HP, Weir TL, Perry LG, Gilroy S, Vivanco JM (2006) The role of root exudates in rhizosphere interactions with plants and other organisms. Annu Rev Plant Biol 57:233–266

Begum AA, Leibovitch S, Migner P, Zhang F (2001) Specific flavonoids induced nod gene expression and pre-activated nod genes of *Rhizobium leguminosarum* increased pea (*Pisum sativum* L.) and lentil (*Lens culinaris* L.) nodulation in controlled growth chamber environments. J Exp Bot 52:1537–1543

Benoit LF, Berry AM (1997) Flavonoid-like compounds from seeds of red alder (*Alnus rubra*) influence host nodulation by *Frankia* (*Actinomycetales*). Physiol Plant 99:588–593

Berg G, Smalla K (2009) Plant species and soil type cooperatively shape the structure and function of microbial communities in the rhizosphere. FEMS Microbiol Ecol 68(1):1–13

Bertin C, Yang X, Weston LA (2003) The role of root exudates and allelochemicals in the rhizosphere. Plant Soil 256:67–83

Besserer A, Puech-Pagès V, Kiefer P et al (2006) Strigolactones stimulate arbuscular mycorrhizal fungi by activating mitochondria. PLoS Biol 4(7):e226

Blount JW, Dixon RA, Paiva NL (1992) Stress responses in alfalfa (*Medicago sativa* L). Antifungal activity of medicarpin and its biosynthetic precursors—implications for the genetic manipulation of stress metabolites. Physiol Mol Plant Pathol 41:333–349

Bolanos-Vasquez MC, Warner D (1997) Effects of *Rhizobium tropici*, *R. etli*, and *R. leguminosarum* bv phaseoli on nod gene-inducing flavonoids in root exudates of *Phaseolus vulgaris*. Mol Plant-Microbe Interact 10:339–346

Broeckling CD, Manter KD, Paschke M, Vivanco JM (2008) Rhizosphere ecology. In: Faith BD (ed) Encyclopedia of ecology, 2nd edn. Elsevier Science, Amsterdam/Boston, pp 574–578

Carvalhais LC, Dennis PG, Fan B, Fedoseyenko D, Kierul K, Becker A, von Wiren N, Borriss R (2013) Linking plant nutritional status to plant-microbe interactions. PLoS One 8(7):68555

Chapman DJ, Regan MA (1980) Evolution of a biochemical pathway: evidence from comparative biochemistry. Annu Rev Plant Physiol 31:639–645

Cooper JE (2004) Multiple responses of rhizobia to flavonoids during legume root infection. Adv Bot Res 41:1–62

Corne MJP, Christos Z, Roeland LB, David MW, Saskia CMVW, Peter AHMB (2014) Induced systemic resistance by beneficial microbes. Annu Rev Phytopathol 52(1):347–375

Croteau R, Kutchan TM, Lewis NG (2000) Natural products (secondary metabolites). In: Buchanan B, Gruissem W, Jones R (eds) Biochemistry and molecular biology of plants. American Society of Plant Physiologist, Rockville, pp 1250–1318

Dakora FD, Phillips DA (2002) Root exudates as mediators of mineral acquisition in low-nutrient environments. Plant Soil 245:35–47

De-la-Pena C, Loyola-Vargas VM (2014) Biotic interactions in the rhizosphere: a diverse cooperative enterprise for plant productivity. Plant Physiol 166(2):701–719

Derrien D, Marol C, Balesdent J (2004) The dynamics of neutral sugars in the rhizosphere of wheat: an approach by 13C pulse-labelling and GC/C/IRMS. Plant Soil 267:243–253

Dong Z, Layzell DB (2001) H_2 oxidation, O_2 uptake and CO_2 fixation in hydrogen treated soils. Plant Soil 229:1–12

Elhady A, Adss S, Hallmann J, Heuer H (2018) Rhizosphere microbiomes modulated by pre-crops assisted plants in defense against plant-parasitic nematodes. Front Microbiol 9:1133

Feng H, Qiu M, Zhang N, Shen Q, Liu Y, Zhang R, Vivanco JM (2014) Enhanced rhizosphere colonization of beneficial *Bacillus amyloliquefaciens* SQR9 by pathogen infection. FEMS Microbiol Lett 353(1):49–56

Giovannetti M (2000) Spore germination and pre-symbiotic Mycelial growth. In: Kapulnik Y, Douds DD (eds) Arbuscular mycorrhizas: physiology and function. Springer, Dordrecht

Golding A, Zou Y, Yang X, Flynn B, Dong Z (2012) Plant growth promoting H_2- oxidizing bacteria as seed inoculants for cereal crops. Agric Sci 3:510–516

Haichar FZ, Santaella C, Heulin T, Achouak W (2014) Root exudates mediated interactions belowground. Soil Biol Biochem 77:69–80

Haldar S, Sengupta S (2015) Plant-microbe cross-talk in the Rhizosphere: insight and biotechnological potential. Open Microbiol J 9:1–7

Hartwig UA, Maxwell CA, Joseph CM, Phillips DA (1990) Chrysoeriol and Luteolin released from alfalfa seeds induce nod genes in *Rhizobium meliloti*. Plant Physiol 92(1):116–122

Hassan S, Mathesius U (2011) The role of flavonoids in root-rhizosphere signalling: opportunities and challenges for improving plant-microbe interactions. J Exp Bot 63(9):3429–3444

Hawes MC, Gunawardena U, Miyasaka S, Zhao X (2000) The role of root border cells in plant defense. Trends Plant Sci 5(3):128–133

Hiltner L (1904) Über neuere Erfahrungen und Probleme auf dem Gebiete der Bodenbakteriologie unter besonderer Berücksichtigung der Gründüngung und Brache. Arb Dtsch Landwirtsch Ges 98:59–78

Hinsinger P, Gobran GR, Gregory PJ, Wenzel WW (2005) Rhizosphere geometry and heterogeneity arising from root-mediated physical and chemical processes. New Phytol 168(2):293–303

Hinsinger P, Bengough A, Vetterlein D, Young I (2009) Rhizosphere: biophysics, biogeochemistry and ecological relevance. Plant Soil 321:117–152

Hooper AM, Tsanuo MK, Chamberlain K, Tittcomb K, Scholes J, Hassanali A, Khan ZR, Pickett JA (2010) Isoschaftoside, a C-glycosylflavonoid from *Desmodium uncinatum* root exudate, is an allelochemical against the development of Striga. Phytochemistry 71:904–908

Iannucci A, Fragasso M, Platani C, Papa R (2013) Plant growth and phenolic compounds in the rhizosphere soil of wild oat (*Avena fatua L.*). Front Plant Sci 4:509

Inderjit, Mallik AU (1997) Effect of phenolic compounds on selected soil properties. Forest Ecol Manag 92(1):11–18

Ingham RE, Trofymow JA, Ingham ER, Coleman DC (1985) Interactions of bacteria, fungi, and their nematode grazers: effects on nutrient cycling and plant growth. Ecol Monogr 55 (1):119–140

Jones DL, Hodge A, Kuzyakov Y (2004) Plant and mycorrhizal regulation of rhizodeposition. New Phytol 163:459–480

Jones D, Nguyen C, Finlay DR (2009) Carbon flow in the rhizosphere: carbon trading at the soil-root interface. Plant Soil 321:5–33

Kapulnik Y, Guenoune D, Galili S, Phillips DA, Volpin H, Chet I, Okon Y (2001) The defense response elicited by the pathogen *Rhizoctonia solani* is suppressed by colonization of the AM-fungus *Glomus intraradices*. Plant Sci 160:925–932

Kawasaki A, Donn S, Ryan PR, Mathesius U, Devilla R, Jones A et al (2016) Microbiome and exudates of the root and Rhizosphere of *Brachypodium distachyon*, a model for wheat. PLoS One 11(10):0164533

Kawasaki A, Okada S, Zhang C, Delhaize E, Mathesius U, Richardson AE, Watt M, Gilliham M, Ryan PR (2018) A sterile hydroponic system for characterizing root exudates from specific root types and whole-root systems of large crop plants. Plant Methods 14(1):114

Khan ZR, Midega CAO, Bruce TJA, Hooper AM, Pickett JA (2010) Exploiting phytochemicals for developing a 'push–pull' crop protection strategy for cereal farmers in Africa. J Exp Bot 61:4185–4196

Kobayashi H, Graven YN, Broughton WJ, Perret X (2004) Flavonoids induce temporal shifts in gene-expression of nod-box controlled loci in *Rhizobium* sp. NGR234. Mol Microbiol 51 (2):335–347

Kuzyakov Y, Domanski G (2000) Carbon input by plants into the soil. J Plant Nutr Soil Sci 163:421–431

Lareen A, Burton F, Schäfer P (2016) Plant root-microbe communication in shaping root microbiomes. Plant Mol Biol 90(6):575–587

Lattanzio V (2013) Phenolic compounds: introduction. In: Ramawat KG, Mérillon JM (eds) Natural products. Springer-Verlag, Berlin/Heidelberg, pp 1544–1580

Liu H, Tang C, Li C (2016) The effects of nitrogen form on root morphological and physiological adaptations of maize, white lupin and faba bean under phosphorus deficiency. AoB Plants 8: plw058

Lynch, James M, de Leij F (2012) Rhizosphere. In: eLS. Wiley, Chichester. https://doi.org/10.1002/9780470015902.a0000403.pub2

Maimaiti J, Zhang Y, Yang J, Cen YP, Layzell DB, Peoples M, Dong Z (2007) Isolation and characterization of hydrogen-oxidizing bacteria induced following exposure of soil to hydrogen gas and their impact on plant growth. Environ Microbiol 9:435–444

McNear DH Jr (2013) The Rhizosphere – roots, soil and everything in between. Nat Educ Knowl 4 (3):1

Mendes R, Garbeva P, Raaijmakers JM (2013) The rhizosphere microbiome: significance of plant beneficial, plant pathogenic, and human pathogenic microorganisms. FEMS Microbiol Rev 37:634–663

Moore JC, McCann K, Setälä H, De Ruiter PC (2003) Top-down is bottom-up: does predation in the rhizosphere regulate aboveground dynamics?. Ecology 84 (4):846-857

Morandi D, Bailey JA, Gianinazzipearson V (1984) Isoflavonoid accumulation in soybean roots infected with vesicular arbuscular mycorrhizal fungi. Physiol Plant Pathol 24:357–364

Morandi D, le Signor C, Gianinazzi-Pearson V, Duc G (2009) A *Medicago truncatula* mutant hyper-responsive to mycorrhiza and defective for nodulation. Mycorrhiza 19:435–441

Moscatiello R, Squartini A, Mariani P, Navazio L (2010) Flavonoid-induced calcium signalling in *Rhizobium leguminosarum* bv. Viciae. New Phytol 188:814–823

Nardi S, Concheri G, Pizzeghello D, Sturaro A, Rella R, Parvoli G (2000) Soil organic matter mobilization by root exudates. Chemosphere 5:653–658

Newman EI (1985) The rhizosphere: carbon sources and microbial populations. In: Fitter AH (ed) Ecological interactions in soil. Blackwell Scientific Publications, Oxford, p 107

Nihorimbere V, Ongena M, Smargiassi M, Thonart P (2010) Beneficial effect of the rhizosphere microbial community for plant growth and health. Biotechnol Agron Soc Environ 15:327–337

Parvez MM, Tomita-Yokotani K, Fujii Y, Konishi T, Iwashina T (2004) Effects of quercetin and its seven derivatives on the growth of *Arabidopsis thaliana* and *Neurospora crassa*. Biochem Syst Ecol 32:631–635

Peters NK, Frost JW, Long SR (1986) A plant flavone, luteolin, induces expression of *Rhizobium meliloti* nodulation genes. Science 233:977–980

Philippot L, Raaijmakers JM, Lemanceau P, van der Putten WH (2013) Going back to the roots: the microbial ecology of the rhizosphere. Nat Rev Microbiol 11:789

Popovici J, Comte G, Bagnarol E, Alloisio N, Fournier P, Bellvert F, Bertrand C, Fernandez MP (2010) Differential effects of rare specific flavonoids on compatible and incompatible strains in the *Myrica gale–Frankia actinorhizal* symbiosis. Appl Environ Microbiol 76:2451–2460

Quideau S, Deffieux D, Douat-Casassus C, Pouységu L (2011) Plant polyphenols: chemical properties, biological activities, and synthesis. Angew Chem Int Ed 50:586–621

Read DB, Bengough AG, Gregory PJ, Crawford JW, Robinson D, Scrimgeour CM, Young IM, Zhang K, Zhang X (2003) Plant roots release phospholipid surfactants that modify the physical and chemical properties of soil. New Phytol 157(2):315–326

Redmond JW, Batley M, Djordjevic MA, Innes RW, Kuempel PL, Rolfe BG (1986) Flavones induce expression of nodulation genes in *Rhizobium*. Nature 323:632–635

Robards R, Antolovich M (1997) Analytical chemistry of fruit bioflavonoids. A review. Analyst 122:11R–34R

Rudrappan T, Quinn WJ, Stanley-Wall NR, Bais HP (2007) A degradation product of the salicylic acid pathway triggers oxidative stress resulting in down-regulation of *Bacillus subtilis* biofilm formation on *Arabidopsis thaliana* roots. Planta 226:283–297

Scervino JM, Ponce MA, Erra-Bassells R, Vierheilig H, Ocampo JA, Godeas A (2005) Flavonoids exclusively present in mycorrhizal roots of white clover exhibit a different effect on arbuscular mycorrhizal fungi than flavonoids exclusively present in non-mycorrhizal roots of white clover. J Plant Interact 1(1):15–22

Singh H, Dixit S, Verma PC, Singh PK (2014) Evaluation of Total phenolic compounds and insecticidal and antioxidant activities of tomato hairy root extract. J Agric Food Chem 62 (12):2588–2594

Sparling GP, Ord BG, Vaughan D (1981) Changes in microbial biomass and activity in soils amended with phenolic acids. Soil Biol Biochem 13(6):455–460

Tripathi DK, Singh S, Gaur S, Singh S, Yadav V, Liu S, Singh VP, Sharma S, Srivastava P, Prasad SM, Dubey NK, Chauhan DK, Sahi S (2018) Acquisition and homeostasis of Iron in higher plants and their probable role in abiotic stress tolerance. Front Environ Sci 5:86

Uren NC (2001) Types, amounts and possible functions of compounds released into the rhizosphere by soil-grown plants. In: Pinton R, Varanini Z, Nannipieri P (eds) The rhizosphere. Biochemistry and organic substances at the soil-plant interface. Marcel Dekker, New York, pp 19–40

Uren NC (2007) Types, amounts, and possible functions of compounds released into the Rhizosphere by soil-grown plants. In: Pinton R, Varanini Z, Nannipieri P (eds) The Rhizosphere: biochemistry and organic substances at the soil-plant interface. CRC Press, Hoboken, pp 1–22

Vengavasi K, Pandey R (2018) Root exudation potential in contrasting soybean genotypes in response to low soil phosphorus availability is determined by photo-biochemical processes. Plant Physiol Biochem 124:1–9

Walker TS, Bais HP, Grotewold E, Vivanco JM (2003) Root exudation and rhizosphere biology. Plant Physiol 132(1):44–51

Wang YL, Almvik M, Clarke N et al (2015) Contrasting responses of root morphology and root-exuded organic acids to low phosphorus availability in three important food crops with divergent root traits. AoB Plants 7:plv097

Watt M, Wendy KS, Passioura JB (2006) Rates of root and organism growth, soil conditions, and temporal and spatial development of the rhizosphere. Ann Bot 97:839–855

Zheng SJ (2010) Iron homeostasis and iron acquisition in plants: maintenance, functions and consequences. Ann Bot 105(5):799–800

Defensive Role of Plant Phenolics Against Pathogenic Microbes for Sustainable Agriculture

25

Ajay Kumar Gautam, Pramod Kumar Singh, and M. Aravind

Abstract

Plant phenolics are secondary metabolites of plants which have different functions in plantlike giving color and odor to the fruit and flower, defense against biotic and abiotic stresses. Studies in the field of utilization of plant phenolics are scanty. The plant microbe pathogens include bacteria, fungi, and viruses caused a lot of economical loss in agricultural field. The reduced output, rising prices and side effects of chemical pesticides are pushing the farmers to find an alternative. Plant phenols which act as an innate defensive mechanism in many plants against plant micro pathogens can be standardized and used as pesticides. Biological pesticides made of plant phenolics can effectively replace the chemical pesticides. This review says about the general defensive properties of plant phenolics, studies on the defensive role of plant phenolics in both intact plants, and effect of extracted phenolics in crop plants. The review also suggests more studies in the field of defensive utility of plant phenolics.

Keywords

Plant phenolics · Flavonoids · Pathogens · Antioxidant · Fungal pathogens

A. K. Gautam (✉)
Department of Plant Sciences, Central University of Punjab, Bathinda, Punjab, India

P. K. Singh
Department of Biosciences, Christian Eminent College, Indore, India

M. Aravind
Department of Plant Sciences, Central University of Kerala, Kasaragod, India

R. Lone et al. (eds.), *Plant Phenolics in Sustainable Agriculture*,
https://doi.org/10.1007/978-981-15-4890-1_25

25.1 Introduction

Plant secondary metabolites are groups of the chemical compounds that synthesized inside the plant in biochemical pathways. Phenolic is one of the widespread groups of compounds which possess an aromatic ring. In addition, phenols include methyl ethers, esters, and glycosides groups known as polyphenols (Harborne 1989). Most of the phenolic compounds are soluble in the polar organic solvent (chloroform, ether, ethanol, ethyl acetate, and methanol), but some phenolic glycosides are soluble in water. Almost all the phenolic compounds have unique absorption characteristics, most of the phenolic compounds show in the UV spectrum, while colored compound absorbs in the visible spectrum as well (Van Sumere 1989). Phenolic compounds released by the plant part such as leaves and roots which enter into the soil as leachates and influencing the activity of harmful microbes nutrient cycling. Some of the natural compounds like gallic acid, flavonoids, and ferulic acid are responsible for the inhibition of growth of saprophytic fungi and spore germination (Hättenschwiler and Vitousek 2000). Naturally, different types of secreted plant phytochemicals (juglone, quercetin, 2,4-dihydroxy-1,4(2H) benzoxazine-3-one, etc.) in the form of volatile emissions, root exudation, and bark breakdown interact with other plants that may be beneficial or harmful to them (Inderjit and Gross 2002; Weir et al. 2004).

The use of plant and plant products as medicines may be traced as far back as the beginning of human civilization (Kelly 2009). The earliest mention of the medicinal use of the plant in India is found in "'Rigveda," which said to have been written between 4500 and 1600 B.C. and is supposed to be the oldest repository of human knowledge (Tucakov 1964). "Ayurveda," the base of Hindu culture medicinal science in India, in the eighth division deals with the drug's basic properties and different aspects of life science and the healing art. The inhibition produced by the plant extracts against a particular organism depends upon various extrinsic and intrinsic parameters (Rastogi and Malhotra 2002).

Production of modern drugs from natural products is widely done now by mainly exploiting traditional medicinal property of the material, which has been using since thousands of years ago (Prakash and Gupta 2005). From ancient times the use of medicinal plants to treat disease was done all over the world (Richard and Forrest 1982). The holly texts like Vedas and Bible tells about the usage of plant products as herbal remedies, which shows its medicinal properties was recognized in ancient times also (Parekh and Chanda 2006). In fact, plants produce a diverse range of bioactive molecules, making them a rich source of different types of medicines (Mukherjee 2003). Since ancient times, higher plants play role as a sources of medicinal compounds that have continued for the maintenance of human health (Mallikhrjuna et al. 2007). Over 50% of all modern clinical drugs are of natural product origin (Lapornik et al. 2005), and these natural products play an important role in drug development programs in the pharmaceutical industries (Igbinosa et al. 2009).

Plants are capable of synthesizing diverse types of organic molecules, which may be divided into two major groups: primary and secondary metabolites (Handa et al. 2008). Primary metabolites are synthesized by the plants for their growth and

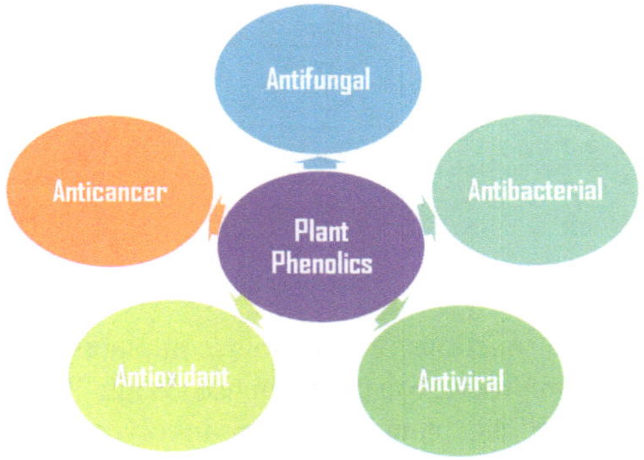

Fig. 25.1 Diagram represents the various biological potential of plant phenolics

sustained for life. These metabolites are used in the plants for respiration and photosynthesis (Goyal et al. 2007). Secondary metabolites are metabolic products which are not essential for growth and life sustains, but it is used by plants as defensive molecules against predation by microorganisms, insects, and herbivores (Gopi and Vatsala 2006). However, some of which may involve in plant odor (terpenoides), pigmentation (tannins and quinones), and flavor (capsaicin) (Ahmad and Beg 2001). However, these defensive molecules give plants their medicinal value which is appreciated by human beings because of their great importance in the health care of individuals and communities (Uma Devi et al. 2007).

Approximately, 12,000 secondary metabolites are known from the different plant which is estimated to be less than 10% of the total phytochemicals found in plants. These secondary metabolites are synthesized in the plants in response to stress condition (Venkatesan and Kannabiran 2008). In recent times, phytochemicals have drawn attention due to their potent antioxidant properties (see Fig. 25.1) and roles in the protection of human health, when their dietary intake is significant (Vidyadhar et al. 2010; Wurochekke et al. 2008).

Plant secondary metabolites can be divided into four major groups: terpenes, phenolics, glycosides, and alkaloids (Wurochekke et al. 2008).

25.2 Terpenes

Terpenes are defined as secondary metabolites built up from isoprene units. Terpenes contain only hydrocarbon chain, while terpenoides are oxygenated hydrocarbons (Wang 2010). The formula of terpenes is $(C_5H_8)n$.

25.2.1 Classification of Terpenes

The classification is based on the number of isoprene units (Ashour et al. 2010).

Number of isoprene units	Name	Carbon atoms
2 unit	Monoterpenes	C_{10}
3 unit	Sesquiterpenes	C_{15}
4 unit	Diterpenes	C_{20}
5 unit	Triterpenes	C_{30}
6 unit	Tetraterpenes	C_{40}
More than 8 unit	Polyterpenes	–

The functions of terpenoides in plants have been recognized as growth, development, reproduction, and defense (Wink 2008). A best known example of diterpene is gibberellins, plant hormones involved in the control of seed germination, stem elongation, and flower induction (Thomas et al. 2005). Another example of terpenoides hormone is abscisic acid (ABA), is not properly considered a lower terpenoid, since it is formed from the oxidative cleavage of a C40 carotenoid precursor (Schwartz et al. 1997).

25.3 Polyphenols

Phenols comprise a class of aromatic organic compounds having at least one hydroxyl group attached directly to the benzene ring. The presence of hydroxyl groups in the molecules of phenols means that phenols are like alcohols is being able to form strong intermolecular hydrogen bonds (Wang et al. 2010). This hydrogen bonding causes phenols to be associated and therefore to have higher boiling points than hydrocarbons of the same molecular weight. The ability to form strong hydrogen bonds to molecules of water confers on phenols a modest solubility in water (Vollhardt and Schore 1998).

With over 8000 variations have been identified, polyphenols are common substances throughout the plant kingdom. They provide an important role in plant metabolism, provide some defense against predators (by their astringency) from the brilliant colors in many fruits and vegetables, and prevent premature seed germination. In general, plant phenolics are involved in defense against ultraviolet radiation and act as protective compounds against herbivores and mechanical supporting pathogens. Phenolics are widespread constituents of plant foods (fruits, vegetables, cereals, olive, legumes, etc.) and beverages (tea, coffee, etc.). It is also helpful in attracting pollinators and fruit dispersers or in reducing the growth of nearby competing plants. Being so ubiquitous, polyphenols naturally form an integral role in human and animal diets. Until recently, their role had been classified by animal nutritionists as “antinutrients” because of their ability to reduce the digestibility of proteins and subsequent increase in fecal nitrogen excretion (Longstaff and McNabb 1991). Certain polyphenols may form complexes with metal cations, thereby interfering with the intestinal absorption of minerals such as iron and copper (Brune et al. 1989). This effect has been attributed to the galloyl and catechyl groups of polyphenolic compounds, phenolic acids in coffee (Brune et al. 1989), and

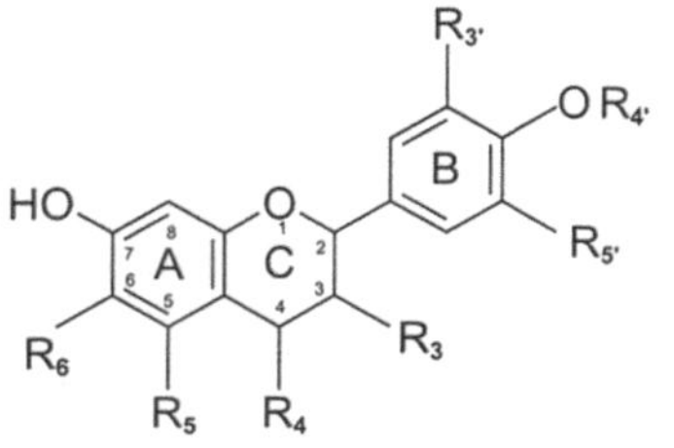

Fig. 25.2 Basic structure of flavonoids

polymerized products in black tea and cocoa (Hurrell et al. 2002). Based upon their structure, phenolics have been divided into two main categories, i.e., flavonoids and non-flavonoids like tannins (Chandrasekara and Shahidi 2010).

25.4 Flavonoids

Flavonoids are polyphenolic compounds containing 15 carbon atoms with two aromatic rings connected through a 3-carbon bridge shown in Fig. 25.2 (Graf et al. 2005).

It is the most important families of natural pigments in the plant kingdom. They are responsible for the blue, purple, red, and orange colors of many fruits and vegetables (Kondratyuk and Pezzuto 2004). These pigments provide color and promote health benefits to consumers due to their antioxidant capacity (De Groot and Rauen 1998). Mostly used and well-known secondary metabolites (flavonoids) include quercetin, catechins, anthocyanidins, and kaempferol.

25.5 Synergistic Effects of Polyphenols

Phenolic compounds are primarily of interest with regard to human health because of their antioxidant activity. Because of their structure, they are very efficient scavengers of free radicals and are also metal chelators (Shahidi and Naczk 1995). In addition to the antioxidant characteristics of flavonoids, other potential health-promoting bioactivities include anti-allergic, anti-inflammatory, antimicrobial, and anticancer properties (Cody et al. 1986). There are many ways in which flavonoids may act to prevent cancer, including inducing detoxification enzymes, inhibiting cancer cell proliferation, and promoting cell differentiation (Kalt 2001). Some flavonoids are also beneficial against heart disease because they inhibit blood platelet aggregation and provide antioxidant protection to LDL (Frankel et al. 1993). Studies on the health benefits of the phenolic acids to date have largely focused on their antioxidant activity.

25.6 Pharmacologic Effects of Plant Phenolics

25.6.1 Polyphenols as Antioxidants

The human body is exposed to a large number of foreign chemicals every day (Santhakumari et al. 2003). These chemicals can generate free radicals inside the human body and show negative impact on human health by affecting the metabolism (Carmen and Florin 2009). The oxygen consumption inherent in cells growth leads to the generation of a series of free radicals. Highly active free radicals and their uncontrolled production are responsible for numerous pathological manifestations such as cell tumor (prostate and colon cancers) and coronary heart diseases (Karadenz et al. 2005). Antioxidants can significantly delay or prevent the oxidation of easily oxidizable substances (Atrooz 2009). Natural antioxidants are classified according to their mechanism of action as chain-breaking antioxidants which scavenge free radicals or inhibit the initiation step or interrupt the propagation step of oxidation of lipid and as preventive antioxidants which somehow slow the rate of oxidation but do not convert free radicals (Semalty et al. 2009).

Dietary polyphenols are the most abundant antioxidants included in human diets. With over 8000 structural variants, these are secondary metabolites of plants with aromatic ring(s) bearing one or more hydroxyl moieties. They are subdivided depending upon the number of phenolic rings and the structural elements that link these rings (Butterfield and Lauderback 2002). The phenolic acids with the subclasses derived from hydroxybenzoic acids such as gallic acid and from hydroxycinnamic acid, containing caffeic, coumaric acid, and ferulic. Plant flavonoids are divided into six subclasses which comprises flavonols, isoflavones, anthocyanidins, flavanones, and flavones, flavanols.

The antioxidant properties of polyphenols exhibit a wide array of biological potential. They inhibit LDL oxidation in vitro (Frankel et al. 1993). LDL isolated from volunteers supplemented with red wine or red wine polyphenols show reduced susceptibility to oxidation (Nigdikar et al. 1998). Therefore, polyphenols possibly protect LDL oxidation in vivo and reduce the problems of atherosclerosis significantly and also prevent the development of some cancers (Halliwell 1999). In addition, flavonoids have antithrombotic and anti-inflammatory effects (Gerritsen et al. 1995). The antimicrobial property of polyphenolic compounds has been well documented (Chung et al. 1998).

Several types of polyphenols (phenolic acids, hydrolyzable tannins, and flavonoids) show anticarcinogenic and antimutagenic effects. Most of the polyphenols might inhibit the development of malignant cancer, expression of mutant genes, and the activation of procarcinogens (Bravo 1998). Nevertheless, some polyphenols have been reported to be mutagenic in microbial assays and cocarcinogens or promoters in inducing skin carcinogenesis in the presence of other carcinogens (Chung et al. 1998). Recently the anti-hemorrhoidal effect of phenolic compounds from the bark of *Acacia ferruginea* DC is reported (Faujdar et al. 2019). The *Pereskia bleo* methanol extract, a cactus, contains several phenolic compounds with abundant antioxidant and antibacterial activity (Johari and Khong 2019).

25.6.2 Polyphenols as Anticarcinogenic

Compound generated in the body after metabolism called oxidant molecules and cause damage to cells, particularly DNA, proteins, lipids, and the degeneration of somatic cells (Barbacanne et al. 2000). Due to this effect, different diseases arise such as cancer, cardiovascular disease, cataracts, and suppression of the immune system. Antioxidants molecules present in the plants can neutralize these damages, by scavenging reactive oxygen species or by inducing endogenous defense systems (Bortoletto et al. 2015). Polyphenol compounds are recognized as the most powerful antioxidant molecules in plants (Barbacanne et al. 2000). Antioxidant activities of polyphenols have been attributed to their reducing capacities or through their possible influences on oxidant molecules (Saito and Hayashi 2015).

25.7 Phenolic Compounds and Plant Defense

Plants trip up quite a lot of pests and pathogens which can lead to the resistance against them and protect them by increasing the amounts of the chlorophyll contents, size of leaves, new branches, and increasing uptake of nutrients (Paul et al. 2000; Dietrich et al. 2005). In spite of this, most of the phenolics can induce plant defense against environmental stress and protect plants (Koricheva et al. 2004).

25.7.1 Phenolic Compounds Against Fungal Pathogens

Fungi are unable to photosynthesize and rely on plants for their nutrition, there are around 1.5 million species of fungi in existence, six times more than the current 250,000 plant species. The plants are continuously exposed to the various types of stress exerted by pathogens to generate resistance against those pathogens (Lattanzio and Cardinali 2006).

25.7.2 Phenolics Compounds Against Fungal Attack

A large number of plant compounds have been reported to have antifungal activity. In that especially, phenols and phenolic glycosides have high resistance against fungal pathogens. Mainly simple phenols, phenolic acids, flavonols, and dihydrochalcones come underactive antifungal phenols. Typically, plants respond to fungal infection by producing more antifungal phenols at the infection site if the plant site contains antifungal phenols that are not sufficient to stop the fungal infection at that time (Slatnar et al. 2016). Usually, plants store the antifungal phenolic compounds in the vacuoles or organelles. So the concentration of the phenols that affect the invading fungus will be proportional to the tissue damage done by the fungus. Usually, a group of fungi called biotrophs may avoid the release of phenolic compounds by decreasing the damage to the host, whereas necrotrophs invade more into the cell and cause the substantial release of these compounds. The factors like environmental conditions, age, and host's genotype will cause variation in the level and nature of affection of the fungal pathogen (Price et al. 1987; Davis 1991).

A lot of previous studies are there which shows the phenolic compounds in several crops are effective against several fungal diseases affecting that particular crop. Two species of fungi *Fusarium oxysporum* and *Rhizoctonia solani* cause root rot-wilt complex disease in cultivated flower crop geranium (or rose geranium *Pelargonium graveolens* L. Hert); phenolic acids such as tannic, gallic, caffeic, ferulic, and benzoic are seen to be produced at the time of infection of the pathogen and are very much effective against them (Prasad et al. 2008). *Venturia inaequalis* (Cooke) G. Winter, which is generally called scab fungus which affects the apple leaves, is affected by some phenolic compounds like flavanols (epicatechin, procyanidin B1, catechin), dihydrochalcones, and hydroxycinnamic acids (Mikulič-Petkovšek et al. 2009; Slatnar et al. 2012). Hydroxycinnamic acid (HCA) derivatives inhibit the growth and sporulation of fungi in some crops (Sammi and Masud 2009). The role of catechin in the plant's pathogen defense might be their interaction with proteins and inhibition of the enzymes secreted by pathogenic fungi which attacks the crops (Bors and Michel 2002). Vitaceae family of plants which includes grapes shows increased metabolism and accumulation of phytoalexins and synthesis of PR proteins against pathogenic fungi as a defense metabolism (Jeandet et al. 2002). The flavonoid sakuranetin which is accumulated highly in the rice leaf on infection of *Pyricularia oryzae*, a pathogen which causes blast disease shows they are important in defense against that fungus (Kodama et al. 1992). *Sorghum bicolor* shows change in metabolism at the time of attack of the fungi *Colletotrichum sublineolum*, the plant produces an array of defensive phenolic compounds like apigeninidin, luteolinidin, 3-deoxyanthocynidin phytoalexins, etc. (Tugizimana et al. 2019).

Some phenolic compounds extracted from the plants can be applied to the crops to decrease the level of fungal infection; this kind of studies will be pathbreaking in the field of pesticides particularly in the scenario where the artificial pesticides are very expensive and less effective. *Alternaria alternata* which causes soft rot disease on ripe cherry tomatoes showed reduced infection when phenolic compounds like ferulic acid, gallic acid, catechin, chlorogenic acid, caffeic acid, and p-coumaric acid from rude foliar extracts of a wild *Capsicum annuum* was applied on artificially infected ripe cherry tomatoes (Pane et al. 2016). Same like that the extracts from the leaf extracts of the giant knotweed (*Reynoutria sachalinensis*, Polygonaceae) showed disease resistance against common pathogens of cucumber like *Pythium ultimum*, *P. aphanidermatum*, and *Botrytis cinerea* further analysis shown that the phenolic compounds (caffeic and ferulic acids, p-coumaric, and p-coumaric acid methyl ester) are the reason behind it (Daayf et al. 2000).

25.7.3 Phenolic Compounds Against Bacterial Pathogen

A large number of bacterial pathogens are attacking our crop plants which cause a rapid loss in the field of agriculture. Plants have some innate disease resistance against the pathogenic bacteria also. One of the important disease resistances is by the production of phenolic compounds and oxidative products of phenols. These compounds show differential metabolism and accumulation in different periods of bacterial infection in the host plant. So the study of this kind of changes in the phenolic content in the plants by bacterial infection is important to study their role in disease resistance and also to produce value-added natural products derived from the plant phenolics.

Xylella fastidiosa is a bacterium that causes severe diseases in crops such as variegated chlorosis in citrus, almond leaf scorch, and grape pierce disease (Maddox et al. 2010). The phenolic compound sakuranetin shows high accumulation and defense against the important rice bacterial pathogens *Burkholderia glumae* and *Xanthomonas oryzae*, the former causing bacterial grain rot and the latter causing blight and leaf streak (Cho and Lee 2015).The green walnut husk is affected by blight disease by *Xanthomonas arboricola* pv. juglandis (Xaj) bacteria. The affected plant showed fivefold more hydroxycinnamic acids, up to threefold more gallic acid, up to 4.3-fold more quercetins, and up to 23-fold more catechin than the normal plant. These statistics show their defensive role against the bacteria (Mikulic-Petkovsek et al. 2011).

Polyphenol oxidases are copper-containing enzymes which are catalyzing the oxygen-dependent oxidation of phenols to quinones seen commonly in angiosperms and thereby regularly involving in plant defense against pests and pathogens. Artificially increasing the production of the PPO by genetic engineering can increase the resistance against pest in crop plants by increasing the rate of oxidation of phenols to quinones. The development of transgenic tomato plants (*Lycopersicon esculentum* Mill. cv. Moneymaker) with PPO cDNA by using the cauliflower mosaic virus 35S promoter highly increased the resistance against the bacterial pathogen *Pseudomonas syringae* (John and Steffens 2002). *Xanthomonas citri* subsp. *citri* (Xcc) causes citrus canker, affects sweet orange-producing areas around the world, plant phenol-derived alkyl dihydroxybenzoates when added to citrus plants showed resistance to bacteria by preventing their cell division in the fruit (Nazaré et al. 2018). Designed products made up of plant phenolics which have antibacterial property can effectively act as an alternative to the artificial antibacterial products in the near future.

25.7.4 Phenolic Compound Against Viral Pathogens

One of the most affected microbe pathogens to crop plants is viruses. Plant viral diseases are very harmful to the crop plants, and they are easily spreadable in the field. Several plant phenolic compounds show defense against the viral pathogen;

salicylic acid is one of the important compounds with the high antiviral property. The discovery of the disease resistance function of salicylates was founded in Xanthi-nc tobacco (*Nicotiana tabacum* which contains the "N" gene, which originates from *Nicotiana glutinosa*) in 1979. The plant shows hypersensitive reaction (HR reaction) response by restricting the viral attack to a small area of initial attack (keep the other parts safe) in response to tobacco mosaic virus (TMV) attack (Holmcs 1938). There was a considerable reduction in the number of lesions formed when the tobacco plant leaves are injected salicylic acid (0.01% solution) and aspirin (0.02% solution) in prior to the infection of the virus; the same result was there when aspirin (acetylsalicylic acid) is sprayed to the leaf also (White 1979). In addition to the decrease in the lesion number, the salicylic acid reduced the lesion size in Xanthi-nc tobacco (Wieringa-Brants and Schets 1988). Tobacco necrosis virus (TNV) another pathogen in tobacco plant also showed reduced lesion size when treated with salicylic acid (Pennazio et al. 1985).

Salicylate also reduced the symptoms of infection with the tobacco necrosis virus (TNV) in asparagus beans (Pennazio et al. 1987), and salicylates also induced PR proteins and thereby giving resistance to alfalfa mosaic virus in bean and cowpea (Hooft van Huijsduijnen et al. 1986). In a study was the total phenol content of the tomato plant checked before and after cauliflower mosaic virus (CMV)-16 affection showed an increase of antioxidant phenolic compounds after the attack (Nayef et al. 2018). Application of the phenolics by different methods into crops can decrease the viral infection to a larger margin.

25.8 Phenolics in Plant-Insect Interactions

Phenolic compounds play a physical and chemical interaction between plants and pathogens, owing to the increase in hypersensitive reaction and plant resistance to pathogens. The basic mechanism behind the resistance against pathogens may be altering the genetic factors (Painter 1941). Phenolics that have to be used in plant protection have historically relied on the synthesis of the plant's total phenolic content that is active against a specific organism. The mechanism has been developed to recognize the plant-insect interactions by increasing the level of particular phenolic toxins with respect to the host. Owing to the increasing the level of phenolics such as salicylates in *Salix* leaves which protects against larvae of *Operophtera brumata* by developing the barriers (Castellanos and Espinosa-García 1997; Harborne 2001).

25.8.1 Discussion and Future Perspective

The increased resistance of the pathogens to chemical pesticides caused the decreased effect of these pesticides and great financial loss in the field of agriculture. This scenario is important for rediscovering the local plant-based pesticides and also to find some biologically active alternatives for the chemical pesticides. Plant phenolics which are of great effect in the field of pesticide manufacturer are only less utilized or studied. Value-added products of phenolics will be a perfect replacement for the chemical pesticides. Even though the plant extracts are used against pathogenic microbes, less interest is seen to extract the biologically active material from the plant and to market it.

Some crop plants have inborn phenolic compounds in it to prevent the fungal microbe attack like catechol 1 and protocatechol 2 against onion smudge disease caused by *Colletotrichum circinans*. These two phenolics reduce the spore germination of the fungi to 2% in the onion tuber (Links et al. 1929; Angell et al. 1930; Walker and Stahmann 1955). Chlorogenic acid 3 (Johnson and Schaal 1952; Lee and LeTourneau 1958) and benzaldehyde 4 (Wilson and Wisniewski 1989) in potato; the flavanol naringin and polymethoxyflavone tangeretin in bitter orange (*Citrus aurantium*) (Arcas et al. 2000); tyrosol, catechin, and oleuropein from olive tree (Del Río et al. 2003); salicylic acid in woody plants; etc. are such phenolic compounds. The extraction of this kind of phenolic compounds and their utilization in other crops which is affected by the same fungal pathogen is very important in the future. Only few studies are done on the plant extract and identify these compounds for their mechanism of action against fungal pathogens. Some phenolic compound ursolic acid extracted from the leaf of *Breonadia salicina* for antifungal activity against plant pathogenic fungi: *Aspergillus parasiticus*, *Aspergillus niger*, *Colletotrichum gloeosporioides*, *Penicillium janthinellum*, *Penicillium expansum*, *Fusarium oxysporum*, and *Trichoderma harzianum* (Mahlo et al. 2013).

Researchers should be encouraged to do research on the production of the biological pesticides from plant phenolics especially in the present scenario where a large percentage of the population is facing life-threatening diseases like cancer from chemical pesticides. The government should give proper funding to these kinds of studies. Repeated testing, field experiments, biochemical tests, and quality checks should be done to evaluate the possible side effect of the compound to the particular crop in interest. Sometimes some compounds can act antagonistically in case of other pathogen and can support its growth in the crop; the quantity of application of the tested phenolics or its compounds can also produce similar effects. So the government should implement proper testing and clearance methods to manufacture the phenolic products.

Fungal, viral, and bacterial pathogens are the main cause of diseases in the crop plants, of which the fungal pathogen causes 85% of the diseases followed by virus and bacteria. The plants have their own defensive mechanism to cope up with these micro pathogens up to some level, like production of secondary metabolites including phenolic compounds. When the biotic stress of the pathogens are far more than the plant can resist, it can cause very fatal diseases which can decrease the production of the plant and sometimes can lead to its death. So many artificial pesticides are used to cure these diseases, but nowadays it has mainly two disadvantages, one is that it is not that much effective than earlier because the micro pathogens got resistance against these pesticides, especially viral pathogens. The second thing is the increased price of these pesticides which keeps the common farmers out of the plan. The hazardous poisonous effect of pesticides in eatable crops is another aspect.

The effect of chemical pesticide in the land and water is also very important. So an alternative to the chemical pesticides should be there in this scenario. Phenolic compounds can be used as an alternative; these compounds are natural, less expensive, and nonpoisonous. The studies of antimicrobial activity of the phenols should be done, and keen interest should be given to produce and manufacture natural pesticides based on phenolics.

References

Ahmad AJ, Beg AZ (2001) Antimicrobial and phytochemical studies on 45 Indian medicinal plants against multi drug resistance human pathogens. J Ethnopharmacol 74:113–123

Angell HR, Walker JC, Link KP (1930) The relation of protocatechuic acid to disease resistance in the onion. Phytopathology 20:431–438

Arcas MC, Botia JM, Ortuño AM, Del Rio JA (2000) UV irradiation alters the levels of flavonoids involved in the defense mechanism of *Citrus aurantium* fruits against *Penicillium digitatum*. Eur J Plant Pathol 106(7):617–622

Ashour M, Wink M, Gershenzon JG (2010) Biochemistry of terpenoids: monoterpenes, sesquiterpenes and diterpenes. In: Wink M (ed) , vol 40, 2nd edn. Wiley, New Jersey

Atrooz OM (2009) The antioxidant activity and polyphenolic contents of different plant seeds extracts. Pak J Biol Sci 12:1063–1068

Barbacanne MA, Souchard JP, Darblade B, Iliou JP, Nepveu F, Pipy B, Bayard F, Arnal JF (2000) Detection of superoxide anion released extracellularly by endothelial cells using cytochrome c reduction, ESR, fluorescence and lucigenin-enhanced chemiluminescence techniques. Free Radic Biol Med 29:388–396

Bors W, Michel C (2002) Chemistry of the antioxidant effect of polyphenols. Ann N Y Acad Sci 957:57–69

Bortoletto P, Lyman K, Camacho A, Fricchione M, Khanolkar A, Katz BZ (2015) Chronic granulomatous disease. Pediatr Infect Dis J 34:1110–1114

Bravo L (1998) Polyphenols: chemistry, dietary sources, metabolism, and nutritional significance. Nutr Rev 56:317–333

Brune M, Rossander L, Hallberg L (1989) Iron absorption and phenolic compounds: importance of different phenolic structures. Eur J Clin Nutr 43(8):547–557

Butterfield DA, Lauderback CM (2002) Lipid peroxidation and protein oxidation in Alzheimer's disease brain: potential causes and consequences involving amyloid-peptide-associated free radical oxidative stress. Free Radic Biol Med 32:1050–1060

Carmen MM, Florin A (2009) Total antioxidant capacity of some fruit juices: electrochemical and spectrophotometrically approaches. Molecules 14:480–493

Castellanos I, Espinosa-García FJ (1997) Plant secondary metabolite diversity as a resistance trait against insects: a test with *Sitophilus granarius* (Coleoptera: Curculionidae) and seed secondary metabolites. Biochem Syst Ecol 25(7):591–602

Chandrasekara A, Shahidi F (2010) Content of insoluble bound phenolics in millets and their contribution to antioxidant capacity. J Agric Food Chem 58:6706–6714

Cho M, Lee S (2015) Phenolic phytoalexins in rice: biological functions and biosynthesis. Int J Mol Sci 16(12):29120–29133

Chung KT, Wong TY, Wei CI, Huang YW, Lin NY (1998) Tannins and human health: a review. Crit Rev Food Sci Nutr 38:421–464

Cody V, Middleton E Jr, Harbome JB (1986) Plant flavonoids in biology and medicine, vol 1. R Liss New York, Alan

Daayf F, Ongena M, Boulanger R, Hadrami IE, Bélanger RR (2000) Induction of phenolic compounds in two cultivars of cucumber by treatment of healthy and powdery mildew-infected plants with extracts of *Reynoutria sachalinensis*. J Chem Ecol 26(7):1579–1593

Davis RH (1991) Glucosinolates. In: DMello JP, Duffus CM, Duffus JH (eds) Toxic substances in crop plants. Cambridge, Royal Society of Chemistry, pp 202–225
De Groot H, Rauen U (1998) Tissue injury by reactive oxygen species and the protective effects of flavonoids. Fundam Clin Pharmacol 12:249–255
Del Río JA, Báidez AG, Botía JM, Ortuño A (2003) Enhancement of phenolic compounds in olive plants (*Olea europaea* L.) and their influence on resistance against Phytophthora sp. Food Chem 83(1):75–78
Dietrich R, Ploss K, Heil M (2005) Growth responses and fitness costs after induction of pathogen resistance depend on environmental conditions. Plant Cell Environ 28:211–222
Faujdar S, Sati B, Sharma S, Pathak AK, Paliwal SK (2019) Phytochemical evaluation and anti-hemorrhoidal activity of bark of *Acacia ferruginea* DC. J Tradit Complement Med 9(2:85–89
Frankel EN, Kanner J, German JB, Parks E, Kinsella JE (1993) Inhibition of oxidation of human low-density lipoprotein by phenolic substances in red wine. Lancet 341:454–457
Gerritsen ME, Carley WW, Ranges GE, Shen C-P, Phan SA, Ligon GF, Perry CA (1995) Flavonoids inhibit cytokine-induced endothelial cell adhesion protein gene expression. Am J Pathol 147:278–292
Gopi C, Vatsala TM (2006) In vitro studies on effects of plant growth regulators on callus and cell suspension culture biomass yield from *Gymnema sylvestre* R Br. Afr J Biotechnol 5 (12):1215–1219
Goyal BR, Goyal RK, Mehta AA (2007) Phyto-pharmacology of *Achyranthes aspera*: a review. Pharmacol Rev 1:143–153
Graf BA, Milbury PE, Blumberg JB (2005) Flavonols, flavonones, flavanones and human health: Epidemiological evidence. J Med Food 8:281–290
Halliwell B (1999) Establishing the significance and optimal intake of dietary antioxidants: the biomarker concept. Nutr Rev 57:104–113
Handa SS, Khanuja SPS, Longo G, Rakesh DD (2008) Extraction technologies for medicinal and aromatic plants. International Centre for Science and High Technology, Trieste, pp 21–25
Harborne JB (1989) In: Dey PM, Harborne JB (eds) Plethora of polyphenols plant phenolics. Volume 1 in methods in plant biochemistry. Academic, London, p 552
Harborne JB (2001) Twenty-five years of chemical ecology. Nat Prod Rep 18:361–379
Hättenschwiler S, Vitousek PM (2000) The role of polyphenols in terrestrial ecosystem nutrient cycling. Trends Ecol Evol 15(6):238–243
Holmcs FO (1938) Inheritance of resistance to tobacco mosaic disease in tobacco. Phytopathology 28:553–561
Hooft van Huijsduijnen RAM, Alblas SW, de Rijk RH, Bol JF (1986) Induction by salicylic acid of pathogenesis-related proteins and resistance to alfalfa mosaic virus infection in various plant species. J Gen Virol 67(2):135–143
Hurrell RF, Hurrell RF, Reddy MB, Burri J, Cook JD (2002) Phytate degradation determines the effect of industrial processing and home cooking on iron absorption from cereal-based foods. Br J Nutr 88:117–123
Igbinosa OO, Igbinosa EO, Aiyegoro OA (2009) Antimicrobial activity and phytochemical screening of stem bark extracts from *Jatropha curcas* (Linn). Afr J Pharm Pharmacol 3(2):058–062
Inderjit, Gross E (2002) Plant phenolics: potential role in aquatic and terrestrial ecosystems. In: Martens S, Treutter D, Forkmann G (eds) Polyphenols 2000. Technische Universität Müchen, Technische Universität München, Freising, pp 206–234
Jeandet P, Douillet-Breuil AC, Bessis R, Debord S, Sbaghi M, Adrian M (2002) Phytoalexins from the Vitaceae: biosynthesis, phytoalexin gene expression in transgenic plants, antifungal activity, and metabolism. J Agric Food Chem 50(10):2731–2741
Johari MA, Khong HY (2019) Total phenolic content and antioxidant and antibacterial activities of *Pereskia bleo*. Adv Pharmacol Sci 2019:Article ID 7428593, 4 pages
John LL, Steffens C (2002) Overexpression of polyphenol oxidase in transgenic tomato plants results in enhanced bacterial disease resistance. Planta 215:239–247

Johnson G, Schaal LA (1952) Relationship of chlorogenic acid to scab resistance in potatoes. Science 115:627–629

Kalt W (2001) Health functional phytochemicals of fruit. Hortic Rev 27:269–315

Karadenz F, Burdurlu HS, Koca N (2005) Antioxidant activity of selected fruits and vegetables grown in Turkey. Turk J Agric For 29:297–303

Kelly K (2009) History of medicine. Facts on file, New York, pp 29–50

Kodama O, Miyakawa J, Akatsuka T, Kiyosawa S (1992) Sakuranetin, a flavanone phytoalexin from ultraviolet-irradiated rice leaves. Phytochemistry 31(11):3807–3809

Kondratyuk TP, Pezzuto JM (2004) Natural product polyphenols of relevance to human health. Pharm Biol 42:46–63

Koricheva J, Nykanen H, Gianoli E (2004) Meta-analysis of trade-offs among plant antiherbivore defenses: are plants jacks-of-all-trades, masters of all? Am Nat 163(4):E64–E75

Lapornik B, Prosek M, Wondra AG (2005) Comparison of extracts prepared from plant by-products using different solvents and extraction time. J Food Eng 71:214–222

Lattanzio V, Cardinali A (2006) Role of phenolics in the resistance mechanisms of plants against fungal pathogens and insects. Phytochem Adv Res 661:23–67

Lee SF, LeTourneau DJ (1958) Chlorogenic acid content and verticillium wilt resistance of potatoes. Phytopathology 48:268–274

Links KP, Dickson AD, Walker JC (1929) Further observations on the occurrence of protocatechuic acid in pigmented onion scales and its relation to disease resistance in onions. J Biol Chem 84:719–725

Longstaff MA, McNabb JM (1991) The inhibitory effects of hull polysaccharides and tannins of field beans (*Vicia faba* L.) on the digestion of amino acids, starch and lipid, and on digestive activities in young chicks. Br J Nutr 65:199–216

Maddox CE, Laur LM, Tian L (2010) Antibacterial activity of phenolic compounds against the Phytopathogen *Xylella fastidiosa*. Curr Microbiol 60(1):53–58

Mahlo SM, McGaw LJ, Eloff JN (2013) Antifungal activity and cytotoxicity of isolated compounds from leaves of *Breonadia salicina*. J Ethnopharmacol 148(3):909–913

Mallikhrjuna PB, Rajanna LN, Seetharam YN, Sharanabassapa GK (2007) Phytochemical studies of *Strychnos potatorum* L F- A medicinal plant. J Chem 4(4):510–518

Mikulič-Petkovšek M, Štampar F, Veberič R (2009) Accumulation of phenolic compounds in apple in response to infection by the scab pathogen, *Venturia inaequalis*. Physiol Mol Plant Pathol 74:60–67

Mikulic-Petkovsek M, Slatnar A, Veberic R, Stampar F, Solar A (2011) Phenolic response in green walnut husk after the infection with bacteria *Xanthomonas arboricola* pv. juglandis. Physiol Mol Plant Pathol 76(3–4):159–165

Mukherjee PK (2003) GMP for Indian Systems of Medicine. Business Horizons, New Delhi, pp 99–112

Nayef N, Montasser MS, Afzal M (2018) A comparative study of the influence of cucumber mosaic virus on free radical scavengers of tomato and squash plants. J Plant Chem Ecophysiol 3 (1):1020

Nazaré AC, Polaquini CR, Cavalca LB, Anselmo DB, Saiki MFC, Monteiro DA, Zielinska A, Rahal P, Gomes E, Scheffers DJ, Ferreira H, Regasini LO (2018) Design of antimicrobial agents: alkyl dihydroxybezoatesa gainst *Xanthomonas citri* subsp *citri*. Int J Mol Sci 19(10)

Nigdikar SV, Williams NR, Griffin BA, Howard AN (1998) Consumption of red wine polyphenols reduces the susceptibility of low-density lipoproteins to oxidation in vivo. Am J Clin Nutr 68:258

Painter RH (1941) The Economic Value and Biologic Significance of Insect Resistance in Plants. J Econ Entomol 34(3):358–367

Pane C, Fratianni F, Parisi M, Nazzaro F, Zaccardelli M (2016) Control of alternaria post-harvest infections on cherry tomato fruits by wild pepper phenolic-rich extracts. Crop Prot 84:81–87

Parekh J, Chanda S (2006) Invert antimicrobial activity of extracts of *Leunaea procumbens* Roxb. (Labiateae), *Vitis vinifera* (Vitaceae) and *Cyperus rotundus* L. (Cyperaceae). Afr J Biomed Res 9:89–93

Paul ND, Hatcher PE, Taylor JE (2000) Coping with multiple enemies: an integration of molecular and ecological perspectives. Trends Plant Sci 5(5):220–225

Pennazio S, Roggero P, Gentile IA (1985) Effects of salicylate on virus infected tobacco plants. J Phytopathol 1(14):203–213

Pennazio S, Colariccio D, Roggero P, Lenzi R (1987) Effect of salicylate stress on the hypersensitive reaction of asparagus bean to tobacco necrosis virus. Physiol Mol Plant Palhol 30:347–357

Prakash P, Gupta N (2005) Therapeutic uses of *Ocimum sanctum* (Tulsi) with a note on eugenol and its pharmacological actions: a short review. Indian J Physiol Pharmacol 49(2):125–131

Prasad D, Singh A, Singh KP, Bist S, Tewari A, Singh UP (2008) The role of phenolic compounds in disease resistance in geranium. Arch Phytopathol Plant Protect 43(7):615–623

Price KR, Johnson IT, Fenwick GR (1987) The chemistry and biological significance of saponins in food and feedingstuffs. Crit Rev Food Sci Nutr 26:27–133

Rastogi RP, Malhotra BN (2002) Glossary of Indian medicinal plants. National Institute of Science Communication, New Delhi

Richard D, Forrest MB (1982) Early history of wound treatment. J R Soc Med 75:198–205

Saito H, Hayashi H (2015) Transformation rate between ferritin and hemosiderin assayed by serum ferritin kinetics in patients with normal iron stores and iron overload. Nagoya J Med Sci 77:571–583

Sammi S, Masud T (2009) Effect of different packaging systems on the quality of tomato (*Lycopersicon esculentum* var. Rio Grande) fruits during storage. Int J Food Sci Technol 44:918–926

Santhakumari P, Prakasam A, Pugalendi KV (2003) Modulation of oxidative stress parameters by treatment with piper leaf in streptozotocin induced diabetic rats. Indian J Pharm 35:373–378

Schwartz SH, Tan BC, Gage DA, Zeevaart JAD, McCarty DR (1997) Specific oxidative cleavage of carotenoids by VP14 of maize. Science 276:1872–1874

Semalty M, Semalty A, Joshi GP, Rawat MSM (2009) Comparison of in vitro antioxidant activity of *Trigonella foenum-graecum* and *T. corniculata* seed. Res J Phytochem 3(3):63–67

Shahidi F, Naczk M (1995) Food phenolics, sources, chemistry, effects, applications. Technomic Publishing Co Inc, Lancaster

Slatnar A, Mikulič-Petkovšek M, Halbwirth H, Štampar F, Stich K, Veberič R (2012) Polyphenol metabolism of developing apple skin of a scab resistant and a susceptible apple cultivar. Trees Struct Funct 26:109–119

Slatnar A, Mikulič-Petkovšek M, Veberič R, Štampar F (2016) Research on the involvement of phenolics in the defense of horticultural plants. Acta Agriculturae Slovenica 107(1):183–189

Thomas SG, Rieu I, Steber CM (2005) Gibberellin metabolism and signaling. Vitam Horm 72:289–338

Tucakov J (1964) Pharmacognosy. Institute for text book issuing in SR Srbije, Beograd, pp 11–30

Tugizimana F, Djami-Tchatchou AT, Steenkamp PA, Piater LA, Dubery IA (2019) Metabolomic analysis of defense-related reprogramming in *Sorghum bicolor* in response to *Colletotrichum sublineolum* infection reveals a functional metabolic web of phenylpropanoid and flavanoid pathways. Front Plant Sci 4(9):1840

Uma Devi P, Murugan S, Puja S, Selvi S, Chinnaswamy P, Vijayanand E (2007) Antibacterial, in vitro lipid per oxidation and phytochemical observation on *Achyranthes bidentata* Blume. Pak J Nutr 6:447–451

Van Sumere CF (1989) Methods in plant biochemistry. In: Dey PM, Harborne JB (eds) Plant phenolics, vol 1. Academic, London, p 29

Venkatesan GK, Kannabiran K (2008) Antimicrobial activity of saponin fractions of the leaves of *Gymnema sylvestre* and *Eclipta prostrate*. World J Microbiol Biotechnol 24:2737–2740

Vidyadhar S, Saidulu M, Gopal TK, Chamundeeswari D, Rao U, Banji D (2010) In vitro anthelmintic activity of the whole plant of *Enicostemma littorale* by using various extracts. Int J Appl Biol Pharm Technol 1(3):1119–1125

Vollhardt KPC, Schore NE (1998) Organic chemistry: structure and function. W.H. Freeman, New York

Walker JC, Stahmann MA (1955) Chemical nature of disease resistance in plants. Annu Rev Plant Physiol 6:351–366

Wang GX (2010) In vivo anthelmintic activity of five alkaloids from *Macleaya microcarpa* (Maxim) Fedde against *Dactylogyrus intermedius* in *Carassius auratus*. Vet Parasitol 171:305–313

Wang GS, Han J, Zhao LW, Jiang DX, Liu YT, Liu XL (2010) Anthelmintic activity of steroidal saponins from *Paris polyphylla*. Phytomedicine 17:1102–1105

Weir TL, Park SW, Vivanco JM (2004) Biochemical and physiological mechanisms mediated by allelochemicals. Curr Opin Plant Biol 7(4):472–479

White RF (1979) Short communications: acetylsalicylic acid (aspirin) induces resistance to tobacco mosaic virus in tobacco. Virology 99:410–412

Wieringa-Brants DH, Schets FM (1988) Effect of salicylic acid and phenylserine on the hypersensitive reaction of tobacco to tobacco mosaic virus. J Phytopathol 1(23):333–343

Wilson CL, Wisniewski ME (1989) Biological control of postharvest diseases of fruits and vegetables: an emerging technology. Annu Rev Phytopathol 27(1):425–441

Wink M (2008) Plant secondary metabolism: diversity, function and its evolution. Nat Prod Commun 3(8):1205–1216

Wurochekke AU, Anthony EA, Obadiah W (2008) Biochemical effects on the liver and kidney of rats administered aqueous stem bark extract of *Ximenia Americana*. Afr J Biotechnol 7 (16):2777–2780